Paleozoic Sequence Stratigraphy, Biostratigraphy, and Biogeography: Studies in Honor of J. Granville ("Jess") Johnson

Edited by

Gilbert Klapper
Department of Geology
University of Iowa
Iowa City, Iowa 52242

Michael A. Murphy
Geology Department
University of California
Davis, California 95616

and

John A. Talent
Centre for Ecostratigraphy and Palaeobiology
School of Earth Sciences
Macquarie University
North Ryde, N.S.W. 2109, Australia

THE GEOLOGICAL SOCIETY OF AMERICA 1888

SPECIAL PAPER

321

1997

Published by The Geological Society of America, Inc.
3300 Penrose Place, P.O. Box 9140, Boulder, Colorado 80301

Printed in U.S.A.

GSA Books Science Editor Abhijit Basu

Library of Congress Cataloging-in-Publication Data

Paleozoic sequence stratigraphy, biostratigraphy, and biogeography : studies in honor of J. Granville ("Jess") Johnson / edited by Gilbert Klapper, Michael A. Murphy, and John A. Talent.
p. cm. -- (Special paper ; 321)
Includes bibliographical references and index.
ISBN 0-8137-2321-3
1. Geology, Stratigraphic--Paleozoic. 2. Paleontology--Paleozoic.
I. Johnson, J. G. II. Klapper, Gilbert. III. Murphy, Michael A.
IV. Talent, John A. V. Series: Special papers (Geological Society of America) ; 321.
QE654.P2415 1997
560'.172--dc21 97-34722
CIP

Cover: Upper Devonian (Frasnian) depositional sequences of the Guilmette Formation of eastern Nevada. Represented are Sequences 4 and 5, and the adjacent parts of Sequences 3 and 6, at sections in the Egan Range (Gap Mountain, GM, and southern Egan Range, ER) and the Schell Creek Range (southern part of the range, SCH, and Sidehill Pass, SP). SBZ = sequence boundary zone, TST = transgressive systems tract, MFZ = maximum flooding zone, and HST = highstand systems tract. *See* Sequence stratigraphy of the Middle to Upper Devonian Guilmette Formation, southern Egan and Schell Creek ranges, Nevada, by Todd A. LaMaskin and Maya Elrick.

10 9 8 7 6 5 4 3 2 1

Contents

BIOGEOGRAPHY AND PALEOGEOGRAPHY

PALEOBIOLOGY

Preface

Jess Johnson, scholar and theorist, was a person with great integrity, an unusual capacity for observation, and a profound sensitivity toward others. His work ranged from Middle Paleozoic biostratigraphy, taxonomy, and sedimentary tectonics to sequence stratigraphy. He often probed fundamental questions of philosophy and procedure, illuminating whatever issues he addressed in meticulously crafted prose, with each word weighed for its impact.

Jess interacted with and drew inspiration from a wide circle of colleagues from around the globe. To express our gratitude for the illumination he brought to so many areas of the earth sciences, and to honor his remarkable life's journey, we invited contributions on topics within his sphere of interest. We present these papers in memory of Jess Johnson, in tribute to the spirit of questioning that he so embodied.

Geological Society of America
Special Paper 321
1997

The inimitable Jess Johnson

John A. Talent
Centre for Ecostratigraphy and Palaeobiology, School of Earth Sciences, Macquarie University, N.S.W. 2109, Australia

At about 1 P.M. on July 5, 1994, with wife, Miriam, and daughter, Jeanette, at his side, John Granville ("Jess") Johnson, hero to brachiopod and conodont workers and to lovers of things Devonian, passed away painlessly from pneumonia at his home in Corvallis, Oregon. Jess's broad spectrum of elegant contributions in the fields of paleontology, stratigraphy, paleogeography, paleobiogeography, and sedimentary tectonics—and especially his holistic approach to such matters—marks him as having been one of the more seminal minds in contemporary sedimentary geology.

Jess produced almost 200 publications that included some 25 incisive contributions on eustatic events, extinctions, speciation, tectonics, stratigraphic concepts, and faunal provinciality. These were backed by numerous monographic and other major contributions on brachiopods, conodonts, and, to a lesser degree, gastropods. Above all, he authored or coauthored three of the most outstanding works of grand-scale synthesis on things mid-Paleozoic of this generation: his global analysis (with Gilbert Klapper, 1980) of the Devonian conodont record as regards endemism and dispersal, his much cited elucidation (with Charlie Sandberg and Gil Klapper, 1985) of the Devonian eustatic pattern, and more recently his major synthesis (1990, the *Journal of Paleontology* Outstanding Paper Award for the year) on Early and Middle Devonian brachiopod paleocommunities of Nevada in relation to a biofacies-realm-province model. In his later years Jess had become involved with event stratigraphy, especially in carbonate sequences, and undoubtedly would have gone on to have written perceptively in that area, too. Jess had a remarkable capacity to shed light where there didn't appear to be any. He was a fount of wisdom and unpretentious charm, totally lacking in chutzpah. He uniquely stimulated all who came in contact with him and was always generous in giving prominence to his coworkers. He was the powerhouse in most of the enterprises in which he became involved.

Jess (Fig. 1) was born in the midcontinent, in Jefferson City, Missouri, on 13 January 1932. His father was of Swedish descent, though born in the United States. Jess's wife, Miriam Anderson Johnson, a unique person whom he had the good fortune to marry in 1954, was also of Swedish parentage; her parents had migrated to the United States from northern Finland in the early 1920s. Miriam was to devote a large slice of her life to looking after Jess; together they survived the raising of five children. There were already three children—Marlene, Jeanette, and Caroline Eve ("Carrie"), the last only six weeks old—when, in early October 1959, Jess contracted the bulbar-spinal poliomyelitis that was to leave him permanently debilitated, paralyzed from neck to waist and with virtually no diaphragm function. The discovery of the first effective polio vaccine by Jonas Salk in 1954 had resulted in a grand-scale program of vaccination, but after a number of children in California and Idaho had contracted the disease from a defective batch of vaccine, there had been reluctance by many to undergo vaccination. Jess had been among those disinclined to be vaccinated; he gambled and lost.

It was 11 months before Jess emerged from hospital. For the rest of his life, except for brief intervals, he had to be attached to a respirator, night and day. He required regular attention from Miriam and other family members or, at work, from a medical attendant. A few of these, especially Claudia (née Dubois) Regier, his medical attendant for his last 21 years, were also skilled research assistants. It was no easy task. As Miriam says, "Some lasted only a few days or weeks. The good ones we could count on our 10 fingers. Claudia was the best! And the most reliable." Through it all there was always Miriam, a bastion of patience and good cheer, tirelessly looking after him; it is hard to imagine the creativity of his research, his phenomenal productivity, or, for that matter, his survival without her. There were to be two more children: Valerie (deceased) and Ian.

Jess outsurvived nearly all those who had been afflicted with bulbar-spinal poliomyelitis during the 1950s–early 1960s polio era. The 1954 discovery of the Salk vaccine, followed by discovery of the first oral vaccine by Albert B. Sabin between 1957 and 1958 (but not licensed until 1961, after Jess had contracted the disease), eliminated the need for production of respirators. As a consequence, Jess's last years were punctuated by increasingly frequent servicing of respirators, construction of replacement parts, and a quest for respirators (for spare parts) from the families of polio victims who had passed away. At the time of his

Talent, J. A., 1997, The inimitable Jess Johnson, *in* Klapper, G., Murphy, M. A., and Talent, J. A., eds., Paleozoic Sequence Stratigraphy, Biostratigraphy, and Biogeography: Studies in Honor of J. Granville ("Jess") Johnson: Boulder, Colorado, Geological Society of America Special Paper 321.

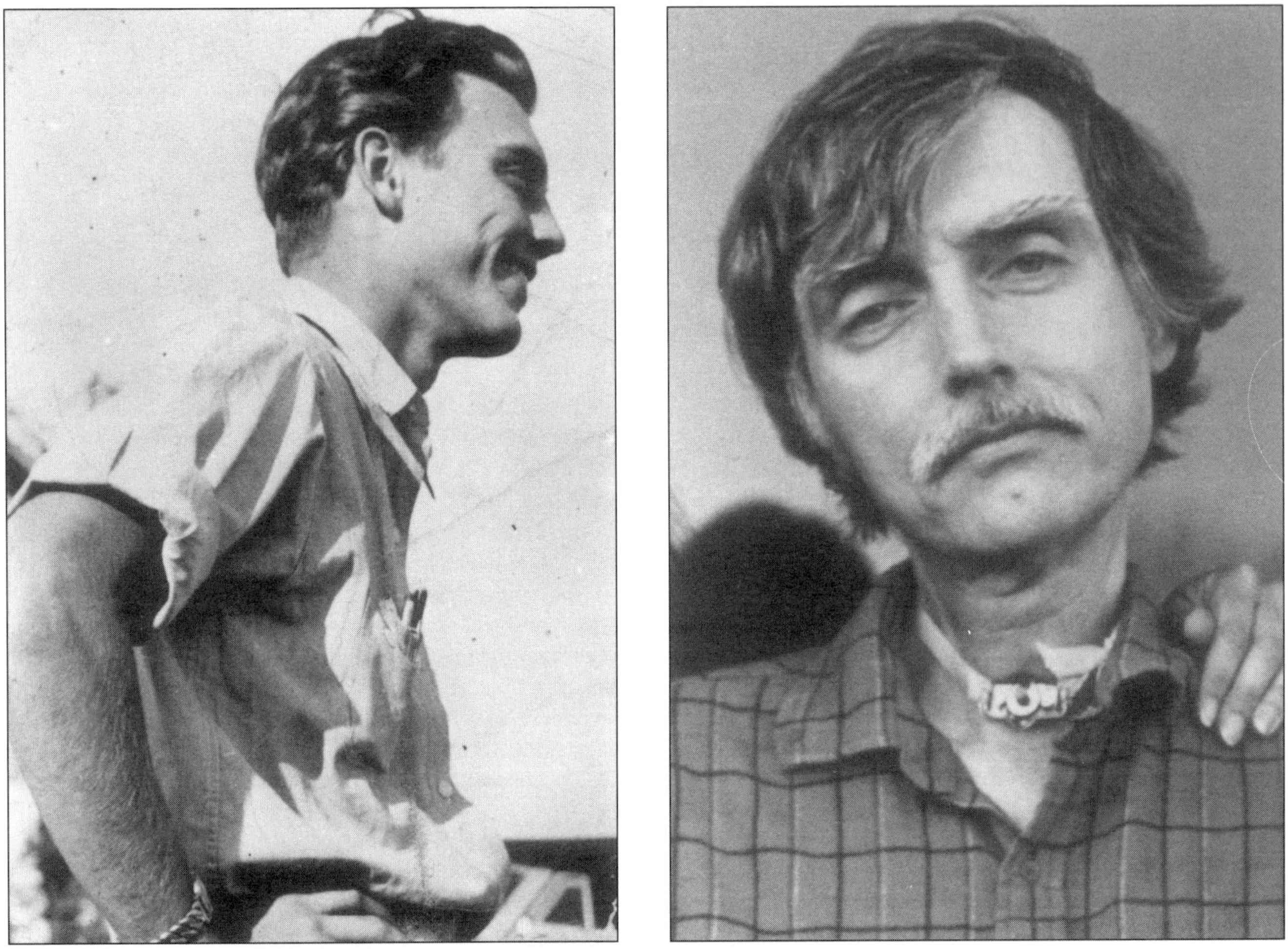

Figure 1. J. Granville ("Jess") Johnson (1932–1994). Picture on the left was taken in the mid-1950s, that on the right in the early 1990s. Hand on Jess's shoulder is that of his oldest daughter, Marlene.

death a respirator was being custom built for him by Peter Meyer, an Oregon State University machinist. Pete, who took care of Jess's respirators after the Johnsons moved to Corvallis, designed and built several rock-splitters and other devices for Jess.

Jess's tertiary education was undertaken entirely at the University of California, Los Angeles, where he commenced the study of engineering in September 1949 but switched to a B.A. in geology. The latter work was interrupted by military service (February 1954 to January 1956) as an air traffic controller. For a time he thought his career might be in that direction, but he abruptly abandoned this dream after attempting to land two U.S. Air Force planes from opposite directions, simultaneously! We who have been involved with Devonian and Silurian brachiopods, conodonts, and associated stratigraphies are grateful for this near tragedy; it brought Jess into our mid-Paleozoic bailiwicks.

Returning to UCLA, Jess completed his B.A. and then did an M.S. in geology (1957–1959). Nevada figured prominently in Jess's geologic career. He had been camp manager for a UCLA summer field camp in central Nevada in 1957 run by Jerry Winterer. Jerry was to supervise Jess's M.S. dissertation; it was on Nevada geology, based mainly on mapping in the Coal Canyon area. Jerry recalls:

> My recollections of Jess are strongly imbued with my impressions of his enormous physical vitality and his early keenness for a career in structural geology and tectonics. He loved geology, and he loved being in the mountains. In the field, Jess was a natural: tireless both physically and mentally, driving himself to do all the laborious and painstaking tasks of gathering good field data; Jess seemed to have legs that started up near his collarbone, moving him effortlessly up the mountains. In the field and at his desk, his creative engine was ever pumping to invent and test hypotheses about local structural problems and about the wider problems of the paleogeography and tectonic evolution of the Great Basin.

Jess's Ph.D. program (1959 to 1964), focused on the Lower Devonian brachiopods of central Nevada, was under the supervision of Jerry Winterer until Jerry left UCLA to go to Scripps; it was completed with crinoid worker Gary Lane as UCLA supervisor, although effectively it was supervised by Art Boucot.

Jess had had the good fortune to correspond with Art Boucot while the latter was still at MIT. In July 1961 Art had relocated to the Division of Geological Sciences at Cal Tech and swiftly set up a factory operation that included an underground acid-leaching facility built to Art's specifications in the corner of a Cal Tech car park. Art soon made Jess's acquaintance and, with his enterprise expanding impressively with NSF and other grants, was able to offer Jess a position as his research assistant. In July 1962 Jess joined the Boucot "stable." As Jess was no longer able to go to the field, Art collected copiously from silicified horizons in the Lower Devonian of central Nevada and dissolved a considerable tonnage of such materials for Jess's thesis in the subterranean acid-leaching facility. The resulting cornucopia of taxonomically diverse material enabled Jess to complete an exceptionally large and elegant dissertation for which he was awarded a Ph.D. by UCLA in October 1964. The degree-conferring ceremony and Jess's triumph over adversity were featured on the CBS evening news.

Jess had already become the second pivot in the Boucot group. Now, with his Ph.D. behind him and with the title of research associate, his productivity seemed to accelerate. Commencing in 1963, a remarkable series of 54 papers, coauthored with Boucot, poured out in journals around the globe; 11 of them were of monographic proportions. The rate of production was prodigious. Jess proved to be hyperproductive in his own right. Monograph after monograph came from his pen, commencing in 1970 with his Geological Society of America (GSA) memoir, *Great Basin Lower Devonian Brachiopoda*. The latter was supplemented in 1973 and 1976 by two University of California volumes, coauthored with Art Boucot and Mike Murphy, dealing with silicified Wenlock to early Lochkovian brachiopods from the Roberts Mountains Formation, also in the Great Basin region of Nevada.

By the early 1960s, Mike Murphy, who had come to know Jess as an undergraduate at UCLA, was the only remaining member of the UCLA faculty still taking students to Nevada. He had noticed brachiopods in the Roberts Mountain Formation, until then thought to be entirely Silurian, but had not collected them because no one seemed interested. It was a query from Art Boucot regarding the possibility of brachiopods in the Roberts Mountains Formation and a positive response from Mike that led to a joint field trip from which one limestone block produced 80 species of brachiopods by Jess's first count. A formidable team came into being: Jess and Art on brachiopods, Bill Berry on graptolites, and Mike, who until then had focused much of his research on Cretaceous ammonoids but soon took up conodonts. As Mike says,

> Before I knew it Art had me working for Jess most of the time and I finally abandoned the Cretaceous because I couldn't do both. From the time Jess became an assistant professor at Corvallis, we collaborated and worked in tandem. He took most of the inner shelf areas with his students and I took the basin with my students. I tried to visit all of his students in the field at least one time since he couldn't and we worked that way to the end. Of course, all the brachs any of us found went to Jess.

Jess supervised four graduate students for the Ph.D.: Richard Aerts Flory, Roy Edward Smith, Thomas Lee DeKeyser, and Constance Jefferson Sansome. He supervised 27 for the M.S. and influenced a substantial number of undergraduates as well; most worked on aspects of the geology of Nevada. Notable among the latter on whom he left his stamp and who went on to make their mark elsewhere have been Maya Elrick, Dave Johnson, and Jed Day. A dozen of his graduate students coauthored substantial papers with him. Jess was an inspiration not only to his students but to countless fellow scientists around the globe: by letter, by E-mail, and, by repute, to many who never had the chance to pass through Corvallis. Many never knew that a pilgrimage to Corvallis was necessary to meet him.

Jess's many reports on brachiopods (including five of monographic scale) contributed greatly to refining understanding of the mid-Paleozoic geology of western North America, especially as regards its chronologic and bathymetric framework. His impact became even greater when, in the mid-1970s, he branched into conodonts. His first paper on conodonts was a seminal revision (with Willi Ziegler and Gil Klapper, 1976) of the *varcus* Zone. From then on he continued to work with brachiopods and conodonts in tandem.

From October to December 1967 I had the pleasure of spending nearly three months with Art Boucot and Jess at Cal Tech. Jess and I had coauthored four papers and were in the process of cleaning up a manuscript with Art on Early Devonian brachiopod zoogeography for a GSA special paper. Although wasted in appearance, confined to a wheelchair, and attached to a respirator, as though to a trailer, Jess had exceptional presence. Most who met him for the first time were struck, as I was, by his large, pale green eyes and the twinkle in them. He had a dry sense of humor and relished incongruity. His son, Ian, has said that

> It was a most astounding experience, even for us in the family, to walk into his office, and have him turn and smile. He didn't have contact with a lot of people, and appreciated visitors very much, I think. He was, as Marlene [his eldest daughter] observed, more focused on the experience of human interaction than anyone we have met. I think part of that was shown in his eyes.

Jess did not have the strength to lift anything heavier than a coffee cup nor the dexterity to handle one well, but he would dictate manuscripts and letters (great numbers of them) into a dictaphone to be typed by the ever-cheerful and positive Miriam. Later he regained sufficient coordination to be able to type.

As a consequence of a general shrinkage in NSF funding and the unwillingness of Cal Tech to pick up the shortfall, Boucot sought a professorship elsewhere, with a condition that Jess also be given an academic appointment. Jess was the difficult item to sell because there was no way he could pass any sort of medical fitness examination; he had already, independently, contacted several institutions but without success. Art negotiated a deal with the University of Pennsylvania in which Art would take up a professorship and Jess would be appointed assistant professor of geology. The head of the Geology Department at the University of Pennsylvania, the late Henry Faul, a pioneer geochronologist of some distinction, had been instrumental in engineering this offer.

In April 1968 Art left for six months in the USSR as a National Academy of Sciences exchange fellow. In July of the same year, with high hopes, Jess and Peggy Losee, Art's aficionado of isopach compilations, and a trainload of Art's collections moved from Pasadena to Philadelphia. By the time Art returned in October the disastrous nature of the move had become apparent. The energies that Henry Faul had put into persuading the university to hire Art and Jess were now focused, as Art says, "on putting them in their place." Henry announced that they were to take his orders and that Art's NSF grant money would be utilized by Henry. It was thus vital that alternative employment be found, quickly. Again Art sought a university that would offer appointments to both him and Jess. Oregon State University (OSU) made such an offer.

I visited Philadelphia in June 1969 to find that Henry was away on fieldwork in New Mexico or elsewhere in the Southwest, seemingly happy that he had triumphed over his unruly underlings. However, not only Boucot and Johnson but also sedimentologist George de Vries Klein and structural geologist Reg Shagam, having lined up employment elsewhere, were in the midst of a packing frenzy, determined that when Henry returned he would find the department virtually empty except for a few bits of paper, sawdust, and splinters from crates. Which is exactly what happened!

So, after not quite 12 months, Art and Jess and the inevitable trainload of fossils headed back westward, this time to Corvallis, but without Peggy Losee, for whom the Philadelphia experience had been too traumatic. The transcontinental transfers, moreover, had been highly disturbing for Jess. He continued to despair of making a career in geology, deprived of contact with rocks in the field, and believed he was fated to remain without tenure for the rest of his days. In Art Boucot's view, however, the battle to get Jess established in academia had been won from the time OSU hired him, and especially when he received his first NSF grant in 1970 and was no longer dependent on the grant-winning capabilities of others—until then, specifically Art. Among the people who buoyed Jess up in the early years at Corvallis when his spirits were down were geochemist E. Julius Dasch, then on the geology staff at OSU; Paul Ritcher, a colleague from another department at OSU who, though crippled from having contracted poliomyelitis from live Sabin vaccine, had been able to establish a research profile that earned him a full professorship; Arlene Demaris, one of Jess's research assistant–cum–medical aides (May 1971 to February 1973); and Sallie Hee, who worked in the Department of Oceanography at OSU. Sallie was killed by an avalanche on Mt. Hood; Jess had thought a lot of her. All were supportive; all were insistent that there was indeed a career for him in science.

Jess, trying his best not to be involved in departmental politics, poured his energies into research and teaching, punctuated by generation of his autobiography. He continued to teach and do research at OSU for almost 25 years, until death took him. His physical condition notwithstanding, Jess was promoted to associate professor in 1973 and then full professor in September 1978. During a highly successful interlude, from January 1986 to August 1988, he was an efficient and more than usually compassionate head of department. Something of the latter role continued on. Right to the end, his office at OSU was a bastion of sanity and humanity for students and staff beset by personal problems. This I noted when I had the good fortune to spend a few weeks with Jess and Miriam in September and October 1993. Serendipitously, in March 1973, Jess had obtained the services of Claudia Regier as research assistant–cum–medical aide. Claudia, who was quietly unflappable, acid-leached samples, picked conodonts, chased literature, and in myriad other ways meticulously helped his research go forward.

Despite being confined to a wheelchair, Jess was an uncommonly good teacher, always well prepared, always quick with illuminating insights and droll repartee. Typical was one occasion when, while he was lecturing, one of the students noticed that Jess's respirator had come free from the power point and swiftly plugged it in again. "That student gets an A!" dryly observed Jess. In 1977 Jess received the OSU College of Science Carter Teaching Award for "inspirational teaching"; he was a finalist for this award on many occasions.

It was necessary for Jess to follow a precise timetable at home and at work. The family car was fitted with a respirator. Morning and evening, five days a week, he would be driven by Miriam to a side door of Wilkinson Hall, the geosciences building at OSU. There he would be met by Claudia Regier, who would usher him to his room; he always walked in, breathing from a hand pump. Jess would pass the day in his office in a wheelchair, linked to a respirator. In the late afternoon Miriam would call for him and drive him home, where he would be linked to another respirator, by which time too much of his energies had been drained to contemplate further work. After dinner he would relax by watching the news on TV; watching videos, generally of recent movies; or listening to music. He liked a wide range of rock music, including Annie Lennox, Van Morrison, and Pink Floyd, especially their album *The Wall*. He had earlier taken a liking to folk music, especially the Weavers, and early Bob Dylan, but never developed empathy for "heavy metal." Other favorites included Big Bill Broonzy, Dire Straits, early Fleetwood Mac, Joe Cocker, Nina Simone's jazz-soul-folk meld, and the pianistics of Keith Jarrett, especially his *Köln Concert*.

Jess was a lover of quiet satire, a connoisseur of comedy. He loved the irreverent Jay Leno and, perhaps even more so, Dave Letterman and his theatrical gestures but insisted that no comedian was the best. It was a ritual when I stayed with the Johnsons in 1993 to record the Dave Letterman shows each night and watch them the following evening, zapping the commercials by remote control.

Jess enjoyed company, but it was impossible for him to

participate in social gatherings other than those organized at the Johnson home. At work, he enjoyed having lunch with friends. Some of them lunched with him regularly for many years, among them Anita Grunder and John Dilles of the OSU Geology Department, Bob Frank of the OSU English Department, Elizabeth Mannarino of the Corvallis Public Library, and a neighbor, Linda Barstow. Discussions would range widely, rarely focusing on geologic matters.

Amazingly, during the 1970s, Jess recovered limited diaphragm function. On weekends, in sunny weather, he would sometimes sit outside for an hour or so, simply feeling glad to be alive, but always with a respirator at his side, pumping. On such occasions, completely free of stress and not directly hooked up to a respirator, he could go for several minutes without taking a machine-reinforced breath.

Jess was an efficient worker, an innovator, ready to probe unexplored pathways. There were times when he felt he was no longer a creative force, but there were soon new challenges. He received continuous NSF support, commencing in 1970; dollar-for-dollar he must have been one of their most economic producers. Project after project was always swiftly and neatly carried through. His office was always laid out for maximum efficiency. The heavier specimens and tomes had, however, to be brought to his desk, overlooking which were two large inspirational portraits: one of Los Angeles poet Charles Bukowski, the other of actress Faye Dunaway as she had appeared in the film *Bonnie and Clyde*. A large-screen state-of-the art computer, an Apple Macintosh Quadra 700, housed in his office at OSU, became a great love. He delighted in having an almost endless choice of fonts and the complete *Random House* dictionary at his fingertips. The instantaneity of E-mail communication with a wide spectrum of colleagues around the world especially delighted him. At one point during my 1993 visit to Corvallis, Jess opined that he was giving some thought to retirement. There were symptoms, he said, that his system was starting to slow down, and he was not sure how much longer the NSF might continue funding his research. In October 1993 he bought an Apple Notebook that enabled him to give more attention at home to his poetry and letters to newspapers.

Jess was a lover of words, both prose and poetry. The region around Corvallis, Benton County, is one of the few relatively liberal counties in Oregon; much of the state is a land of lumberjacks, farmers, and pickup trucks. It has therefore more than the usual share of conservatives who see little value in paying taxes to be used on water, power, national parks, and education, including tertiary education. Some of them have a remarkable penchant for expressing their antiprogressive views in letters to the editor. In his later years Jess stepped into the fray with gusto. It seemed that every week or so he would pump out an elegantly worded letter to the *Corvallis Gazette-Times*, taking the conservatives to task, demolishing their logic, and, inter alia, defending the core values of a liberal society. In this, as in everything else, he was fearless, like a latter-day Ambrose Bierce. Jess developed excellent rapport with the editor of the *Corvallis Gazette-Times*, Norm Lewis. Like Bierce, Jess was always seeking new ways to phrase double entendres and turn cliches inside out, unleashing them with devastating effect. To some of his targets he must have been as popular as a hedgehog in a long-jump pit. None of them bested him. Miriam would deliver his letters to the *Corvallis Gazette-Times* by hand to save a day in the mails, but by late 1993 he had started firing them off by E-mail. Having his letters appear in print the following day delighted him. "I love instant publication!" he would say.

There is a legacy of unpublished J. G. Johnson material, much of it not in the domain of scientific writing. The major scientific item left for publication is a voluminous manuscript on the Silurian and Devonian spiriferidines for the revised volumes on Brachiopoda of the *Treatise on Invertebrate Paleontology*. For this he had been organizer and chief contributor. Typically, the text had been completed well ahead of schedule and, apart from a few figures, was ready to go to the editors. An extension of the *Treatise* enterprise, a planned monograph covering nearly 600 nominal genera of spiriferidines, sadly will not now come to fruition.

An as yet unpublished autobiography of considerable length, *The Second Life is Harder*, details Jess's relentless fight for physical survival and professional satisfaction; he began the manuscript in December 1971. He wrote several short stories and had completed his part of a novel in which he and his son, Ian, were contributing alternate chapters. As Ian says, "He was able to produce five times as much material as I was, despite his one-fingered approach to typing." Jess was a great admirer of the prose works of Anaïs Nin; he had a copy of almost everything written by her.

Over many years, much of Jess's innermost responses to the world around him had been distilled into poetry, vast amounts of it: spare, unrhymed, often passionate, singularly lacking in self-pity. It, too, is a legacy that must be put into print. Jess had a special penchant for the writings of antiestablishment Californian poet Charles Bukowski; he had a large collection of Bukowski's uninhibited writings and taped interviews. "Buk" (or Hank, from Hank Chinaski, his fictional alter ego) predeceased Jess, on 9 March.

And now that Jess is no longer with us, who will take his place? I, for one, can't think of anyone! We have lost not only a great mind but a powerful catalyst as well. He has established a lofty benchmark for future workers concerned with sedimentary geology in general and with things mid-Paleozoic in particular. I suspect it will be a long time before anyone in the earth sciences with such a physically fragile frame will have a comparable commitment to productivity and excellence. The last words are best left to long-standing friend and confidant, Mike Murphy, who accompanied him on his final voyage to Nevada.

> We (Gil and I) met Miriam in Eureka [Nevada] on the 29th of July . . . and the following morning we [the family and students] went with them to the Roberts Mountains. One of the students was doing fieldwork in the area and had a Jeep. He ferried us to a high saddle below Roberts Creek Mountain (10,200 feet) and we then climbed to the

peak, which is the highest Devonian outcrop in the state. In the saddle we scattered most of Jess's ashes. . . . [We] scattered the remaining ashes from the peak. It was a beautiful morning and storm clouds were gathering. As we scattered the ashes, cliff swallows swirled and dived around us just as though they were celebrating the fact that his spirit was now free—no more of the indignities of respirators, tubes, loss of privacy, dependency—the constant bonds of his life. It was releasing for all of us. Miriam and the children read some of his poetry and some other pieces or said something about their father and it was very wrenching, but all in all it was more like a wake in which there were moments of sadness, but everyone was glad to have come.

ACKNOWLEDGMENTS

Miriam Johnson and Gil Klapper provided photos, including those in Figure 1, for this celebration of a unique life. Miriam, Gil, Ian Johnson, Art Boucot, Mike Murphy, and Jerry Winterer, all of whom knew Jess's world better than I, provided corrections and indispensable comments. Stephen Shank retouched the photos.

PUBLICATIONS BY J. GRANVILLE JOHNSON

Johnson, J. G., 1962, Brachiopod faunas of the Nevada Formation (Devonian) in central Nevada: Journal of Paleontology, v. 36, p. 165–169.

Johnson, J. G., 1962, Lower Devonian–Middle Devonian boundary in central Nevada: American Association of Petroleum Geologists Bulletin, v. 46, p. 542–546.

Johnson, J. G., and Boucot, A. J., 1963, *Cloudella nom.nov.* for *Pleurothyris* Cloud (Terebratuloidea): Bulletins of American Paleontology, v. 46, p. 123.

Boucot, A. J., and Johnson, J. G., 1964, Brachiopods of the Ede Quartzite (lower Llandovery) of Norderön, Jämtland: Uppsala University Geological Institutions Bulletin, v. 42, no. 9, 11 p., 7 plates.

Boucot, A. J., and Johnson, J. G., 1964, Additional occurrences of eospiriferid brachiopods: Journal of Paleontology, v. 38, p. 605–606.

Boucot, A. J., Harper, C., Johnson, J. G., and Walmsley, V. G., 1964, Revision of the family Rhipidomellidae: Geological Society of America Special Paper 76, p. 18A.

Boucot, A. J., and Johnson, J. G., 1964. Devonian brachiopods from the Mina Plomosas–Placer de Guadalupe area, Chihuahua, Mexico and their paleogeographic significance, *in* Geology of Mina Plomosas–Placer de Guadalupe area, Chihuahua, Mexico—Field trip guidebook, 1964: West Texas Geological Society Publication 64-50, p. 104–109.

Boucot, A. J., Johnson, J. G., and Staton, R. D., 1964, On some atrypoid, retzioid and athyridoid Brachiopoda: Journal of Paleontology, v. 38, p. 805–822, plates 125–128.

Johnson, J. G., and Reso, A., 1964, Probable Ludlovian brachiopods from the Sevy Dolomite of Nevada: Journal of Paleontology, v. 38, p. 74–84, plates 19, 20.

Boucot, A. J., Johnson, J. G., and Walmsley, V. G., 1965, Revision of the Rhipidomellidae and the affinities of *Mendacella* and *Dalejina*: Journal of Paleontology, v. 39, p. 331–340, plates 45, 46.

Johnson, J. G., 1965, Lower Devonian stratigraphy and correlation, northern Simpson Park Range, Nevada: Bulletins of Canadian Petroleum Geology, v. 13, p. 365–381.

Boucot, A. J., Johnson, J. G., and Staton, R. D., 1965, Atrypidina, Retziidina, Athyrididina, *in* Moore, R. C., ed., Treatise on invertebrate paleontology, Part H, Brachiopoda, Volume 2: New York, Geological Society of America (and University of Kansas Press), p. H632–H667.

Johnson, J. G., 1965, Articulate brachiopods, *in* Doumani, G. A., Boardman, R. S., Fisher, D. W., Johnson, J. G., McAlester, A. L., Miles, R. S., Rowell, A. J., Saul, J., and Tasch, P., Lower Devonian fauna of the Horlick Formation, Ohio Range, Antarctica: Antarctic Research Series (American Geophysical Union, National Academy of Science, National Research Council, Publication 1299), v. 6, p. 255–261, plates 3–8.

Johnson, J. G., and Reso, A., 1966, Brachiopods from the Pilot Shale (Devonian) in southeastern Nevada: Journal of Paleontology, v. 40, p. 125–129, plate 14.

Johnson, J. G., 1966, Middle Devonian brachiopods from the Roberts Mountains, central Nevada: Palaeontology, v. 9, p. 152–181, plates 23–27.

Boucot, A. J., and Johnson, J. G., 1966, *Ehlersella*, a new Silurian stricklandiid brachiopod genus: Journal of Paleontology, v. 40, p. 569–572, plates 67, 68.

Gill, E. D., Boucot, A. J., and Johnson, A. J., 1966. The brachiopod genus *Maoristrophia* Allan (Lower Devonian, Strophomenacea) redescribed: Proceedings of the Royal Society of Victoria, v. 79, p. 355–361, plate 39.

Boucot, A. J., Gill, E. D., Johnson, J. G., Lenz, A. C., and Talent, J. A., 1966, *Skenidioides* and *Leptaenisca* in the Lower Devonian of Australia (Victoria, Tasmania) and New Zealand, with notes on other Devonian occurrences of *Skenidioides*: Proceedings of the Royal Society of Victoria, v. 79, p. 363–369, plate 40.

Anderson, M. M., Boucot, A. J., and Johnson, J. G., 1966, Devonian terebratulid brachiopods from the Accraian series of Ghana: Journal of Paleontology, v. 40, p. 1365–1367, plate 172.

Boucot, A. J., Johnson, J. G., and Jones, C. R., 1966, Silurian brachiopods from Malaya: Journal of Paleontology, v. 40, p. 1027–1031, plate 120.

Boucot, A. J., Johnson, J. G., and Talent, J. A., 1966, Early Devonian brachiopod zoogeography: Geological Society of America Special Paper 87, p. 15–16.

Boucot, A. J., Gauri, K. L., and Johnson, J. G., 1966, New subfamily Proschizophoriinae of dalmanellid brachiopods: Paläontologische Zeitschrift, v. 40, p. 155–172, plates 12–15.

Boucot, A. J., Johnson, J. G., Harper, C., and Walmsley, V. G., 1966, Silurian brachiopods and gastropods of southern New Brunswick: Geological Survey of Canada Bulletin 140, 45 p., plates I–XVIII.

Boucot, A. J., Johnson, J. G., and Struve, W., 1966, *Stringocephalus* ontogeny and distribution: Journal of Paleontology, v. 40, p. 1349–1364, plates 169–171.

Johnson, J. G., 1966, Two new spiriferid brachiopod genera from the Lower Devonian of Nevada: Journal of Paleontology, v. 40, p. 1043–1050, plates 127–129.

Boucot, A. J., and Johnson, J. G., 1966, *Lissocoelina* and its homeomorph *Plicocoelina* n. gen. (Silurian, Pentameracea): Journal of Paleontology, v. 40, p. 1037–1042, plates 123–126.

Johnson, J. G., 1966, *Parachonetes*, a new Lower and Middle Devonian brachiopod genus: Palaeontology, v. 9, p. 365–370, plates 62, 63.

Harper, C. W., Jr., Johnson, J. G., and Boucot, A. J., 1967, The Pholidostrophiinae (Brachiopoda; Ordovician, Silurian, Devonian): Senckenbergiana Lethaea, v. 48, p. 403–441, 10 plates.

Boucot, A. J., and Johnson, J. G., 1967, *Allanetes*, a new Lower Devonian chonetid brachiopod genus: Proceedings of the Royal Society of Victoria, v. 80, p. 141–145, plate 22.

Johnson, J. G., and Talent, J. A., 1967, *Muriferella*, a new genus of Lower Devonian septate dalmanellid: Proceedings of the Royal Society of Victoria, v. 80, p. 43–49, plates 9, 10.

Johnson, J. G., and Talent, J. A., 1967, Cortezorthinae, a new subfamily of Siluro-Devonian dalmanellids: Palaeontology, v. 10, p. 142–170, plates 19–22.

Boucot, A. J., and Johnson, J. G., 1967, Paleogeography and correlation of Appalachian Province Lower Devonian sedimentary rocks: Tulsa Geological Society Digest, v. 35, p. 35–87, 2 plates.

Boucot, A. J., and Johnson, J. G., 1967, Species and distribution of *Coelospira* (Brachiopoda): Journal of Paleontology, v. 41, p. 1226–1241, plates 163–166.

Johnson, J. G., 1967, Systematic position of the genus *Irboskites* (Brachiopoda): Journal of Paleontology, v. 41, p. 1003–1004.

Johnson, J. G., 1967, *Toquimaella*, a new genus of karpinskiinid brachiopod:

Journal of Paleontology, v. 41, p. 874–880, plate 111.

Johnson, J. G., and Boucot, A. J., 1967, *Gracianella*, a new Late Silurian genus of atrypoid brachiopods: Journal of Paleontology, v. 41, p. 868–873, plates 109, 110.

Amsden, T. W., Boucot, A. J., and Johnson, J. G., 1967, *Conchidium* and its separation from the subfamily Pentamerinae: Journal of Paleontology, v. 41, p. 861–867, plates 106–108.

Boucot, A. J., and Johnson, J. G., 1967, Silurian and Upper Ordovician atrypids of the genera *Plectatrypa* and *Spirigerina*: Norsk Geologisk Tidsskrift, v. 47, p. 79–101.

Boucot, A. J., Johnson, J. G., Doumani, G. A., and Webers, G. F., 1967, Devonian of Antarctica, *in* Oswald, D. H., ed., International Symposium on the Devonian System, Volume 1: Calgary, Alberta Society of Petroleum Geologists, p. 639–648.

Johnson, J. G., Boucot, A. J., and Murphy, M. A., 1967, Lower Devonian faunal succession in central Nevada, *in* Oswald, D. H., ed., International Symposium on the Devonian System, Volume 2: Calgary, Alberta Society of Petroleum Geologists, p. 679–691.

Boucot, A. J., Johnson, J. G., and Talent, J. A., 1967, Lower and Middle Devonian faunal provinces based on Brachiopoda, *in* Oswald, D. H., ed., International Symposium on the Devonian system, Volume 2: Calgary, Alberta Society of Petroleum Geologists, p. 1239–1254.

Boucot, A. J., and Johnson, J. G., 1967, Appalachian Province Early Devonian palaeogeography and brachiopod zonation, *in* Oswald, D. H., ed., International Symposium on the Devonian system, Volume 2: Calgary, Alberta Society of Petroleum Geologists, p. 1255–1267.

Boucot, A. J., Johnson, J. G., and Murphy, M. A., 1968, Great Basin Silurian to Gedinnian shelly faunas: Geological Society of America Special Paper 115, p. 19–20.

Boucot, A. J., Berry, W. B. N., and Johnson, J. G., 1968, The crust of the Earth from a Lower Paleozoic point of view, *in* Phinney, R. A., ed., The history of the Earth's crust, a symposium: Princeton, Princeton University Press, p. 208–225.

Johnson, J. G., 1968, A new species of *Vagrania* (Devonian, Brachiopoda) from Nevada: Journal of Paleontology, v. 42, p. 1200–1204, plate 139.

Boucot, A. J., and Johnson, J. G., 1968, Brachiopods of the Bois Blanc Formation in New York: U.S. Geological Survey Professional Paper 548B, p. B1–B27, 8 plates.

Johnson, J. G., Boucot, A. J., and Gronberg, E. C., 1968, A new genus of stringocephaloid brachiopod from the Middle Devonian of Nevada: Journal of Paleontology, v. 42, p. 406–414, plates 59, 60.

Johnson, J. G., and Boucot, A. J., 1968, External morphology of *Anatrypa* (Devonian, Brachiopoda): Journal of Paleontology, v. 42, p. 1205–1207, plate 160 (part).

Boucot, A. J., and Johnson, J. G., 1968, Evidence of color banding in a Lower Devonian rhynchonellid brachiopod: Journal of Paleontology, v. 42, p. 1208–1209, plate 160 (part).

Johnson, J. G., and Murphy, M. A., 1968, Age and position of Lower Devonian graptolite zones relative to the Appalachian standard succession: Geological Society of America Special Paper 121, p. 151.

Johnson, J. G., and Murphy, M. A., 1969, Llandovery to Givetian brachiopod zonal sequence in the Silurian and Devonian of central Nevada: Geological Society of America Special Paper 121, p. 518.

Boucot, A. J., Hazzard, J. C., and Johnson, J. G., 1969, Reworked Lower Devonian fossils, Nopah Range, Inyo County, California: American Association of Petroleum Geologists Bulletin, v. 53, p. 163–167.

Anderson, M. M., Boucot, A. J., and Johnson, J. G., 1969, Eifelian brachiopods from Padaukpin, Northern Shan States, Burma: British Museum (Natural History) Bulletin, Geology, v. 18, p. 105–163, 10 plates.

Boucot, A. J., Johnson, J. G., and Link, A. G., 1969, Redescription of *Aliconchidium* (Silurian, Pentameridae) and its variation: Palaeontographica, Abteilung A, v. 133, p. 177–183, plates 24–26.

Johnson, J. G., and Boucot, A. J., 1969, Givetian brachiopod zonal sequence in the Silurian and Devonian of central Nevada: Geological Society of America Special Paper 121, 518 p.

Boucot, A. J., Johnson, J. G., and Talent, J. A., 1969, Early Devonian brachiopod zoogeography: Geological Society of America Special Paper 119, 113 p., 20 plates.

Johnson, J. G., Reso, A., and Stephens, M., 1969, Late Upper Devonian brachiopods from the West Range Limestone of Nevada: Journal of Paleontology, v. 43, p. 1351–1368, plates 157–160.

Johnson, J. G., and Murphy, M. A., 1969, Age and position of Lower Devonian graptolite zones relative to the Appalachian standard succession: Geological Society of America Bulletin, v. 80, p. 1275–1282.

Johnson, J. G., 1969, Origin of *Gruenewaldtia* (Devonian, Brachiopoda): Journal of Paleontology, v. 43, p. 947–951.

Johnson, J. G., 1969, Some North American rensselandiid brachiopods: Journal of Paleontology, v. 43, p. 829–837, plates 105, 106.

Johnson, J. G., and Lane, N. G., 1969, Two new Devonian crinoids from central Nevada: Journal of Paleontology, v. 43, p. 69–73, plate 14.

Boucot, A. J., and Johnson, J. G., 1969, Position of the Siluro-Devonian boundary in the Appalachian province: Geological Society of America Abstracts with Programs 1969, pt. 1, Northeastern Section, p. 4.

Berdan, J. M., Berry, W. B. N., Boucot, A. J., Cooper, G. A., Jackson, D. E., Johnson, J. G., Klapper, G., Lenz, A. C., Martinsson, A., Oliver, W. A., Jr, Rickard, L. V., and Thorsteinsson, R., 1969, Siluro-Devonian boundary in North America: Geological Society of America Bulletin, v. 80, p. 2165–2174, plate 1.

Johnson, J. G., Reso, A., and Stephens, M., 1970, *Crurithyris subfissa*, new name for *Crurithyris fissa* Johnson, Reso, and Stephens, 1969: Journal of Paleontology, v. 44, p. 1146.

Johnson, J. G., 1970, Great Basin Lower Devonian Brachiopoda: Geological Society of America Memoir 121, 421 p., 74 plates.

Boucot, A. J., and Johnson, J. G., 1970, Redescription of *Clintonella* Hall (Silurian, Brachiopoda): Journal of Paleontology, v. 44, p. 893–897, plate 128.

Johnson, J. G., Taghanic onlap and the end of North American Devonian provinciality: Geological Society of America Bulletin, v. 81, p. 2077–2105, 4 plates.

Johnson, J. G., and Boucot, A. J., 1970, Brachiopods and age of the Tor Limestone of central Nevada: Journal of Paleontology, v. 44, p. 265–269, plate 54.

Johnson, J. G., 1970, Early Middle Devonian brachiopods from central Nevada: Journal of Paleontology, v. 44, p. 252–264, plates 51–53.

Boucot, A. J., Johnson, J. G., and Shagam, R., 1971, Braquiopodos del silurico de Los Andes de Merida, Venezuela (Silurian brachiopods from the Andes of Merida, Venezuela): Memoirs, Geological Congress, Venezuela, 4th, Volume 1, Bulletin of Geology Special Publication 5, p. 101–102.

Johnson, J. G., and Westbroek, P., 1971, Cardinalia terminology of rhynchonellid brachiopods: Geological Society of America Bulletin, v. 82, p. 1699–1701.

Johnson, J. G., 1971, A quantitative approach to faunal province analysis: American Journal of Science, v. 270, p. 257–280.

Johnson, J. G., 1971, Lower Givetian brachiopods from central Nevada: Journal of Paleontology, v. 45, p. 301–326, plates 38–46.

Johnson, J. G., 1971, Some new and problematical brachiopods from the Lower Devonian of Nevada: Journal of Paleontology, v. 45, p. 95–99, plates 21, 22.

Boucot, A. J., Johnson, J. G., and Rubel, M., 1971, Descriptions of brachiopod genera of subfamily Virgianinae Boucot et Amsden, 1963: Eesti NSV Teaduste Akadeemia Toimetised, Keemia-Geoloogia, v. 20, p. 271–280, 6 plates.

Johnson, J. G., 1971, *Living and fossil brachiopods* by M. J. S. Rudwick: Review: American Journal of Science, v. 271, p. 551–553.

Johnson, J. G., 1971, Timing and coordination of orogenic, epeirogenic, and eustatic events: Geological Society of America Bulletin, v. 82, p. 3263–3298.

Berdan, J. M., Berry, W. B. N., Boucot, A. J., Cooper, G. A., Jackson, D. E., Johnson, J. G., Klapper, G., Lenz, A. C., Martinsson, A., Oliver, W. A., Jr., Rickard, L. V., and Thorsteinsson, R., 1971, Granitsa silura i devona v Severnoy Amerike (Silurian-Devonian boundary in North America), *in* Granitsa silura i devona i biostratigrafiya silura (Silurian-Devonian boundary and Silurian biostratigraphy), Trudy III Mezhdunarodnogo Simpoziuma, Proceedings, Third International Symposium, Volume 1. Leningrad, Nauka, p. 49–63.

Johnson, J. G., and Boucot, A. J., 1972, Origin and composition of the Carinatininae (Devonian Brachiopoda): Journal of Paleontology, v. 46, p. 31–38, 3 plates.

Johnson, J. G., 1972, The *Antistrix* brachiopod faunule from the Middle Devonian of central Nevada: Journal of Paleontology, v. 46, p. 120–124, 3 plates.

Johnson, J. G., and Ludvigsen, R., 1972, *Carinagypa*, a new genus of pentameracean brachiopod from the Devonian of western North America: Journal of Paleontology, v. 46, p. 125–129.

Boucot, A. J., and Johnson, J. G., 1972, *Callicalyptella*, a new genus of notanopliid brachiopod from the Devonian of Nevada: Journal of Paleontology, v. 46, p. 299–302, 2 plates.

Johnson, J. G., and Norris, A. W., 1972, *Tecnocyrtina*, a new genus of Devonian brachiopods: Journal of Paleontology, v. 46, p. 565–572, 2 plates.

Johnson, J. G., 1972, *Teichertina*, the last dicaelosiid brachiopod: Journal of Paleontology, v. 46, p. 830–835, 2 plates.

Johnson, J. G., and Florey, R. A., 1972, A Rasenriff fauna from the Middle Devonian of Nevada: Journal of Paleontology, v. 46, p. 892–899, 4 plates.

Boucot, A. J., Johnson, J. G., and Shagam, R., 1972, Braquiopodos siluricos de Los Andes meridenos de Venezuela (Silurian brachiopods from the Merida Andes of Venezuela): Memoirs, Geological Congress, Venezuela, 4th, Volume 1, Bulletin of Geology Special Publication 5, p. 585–659, 34 plates.

Johnson, J. G., 1972, Antler Effect equals Haug Effect: Geological Society of America Bulletin, v. 83, p. 2497–2498.

Johnson, J. G., and Dasch, E. J., 1972, Origin of the Appalachian Faunal Province of the Devonian: Nature (Physical Sciences), v. 236, p. 125–126.

Johnson, J. G., 1972, *Faunal provinces in space and time*, edited by F. A. Middlemiss, P. F. Rawson, and G. Newall: Essay review: American Journal of Science, v. 272, p. 894–898.

Johnson, J. G., 1973, Review of *Early Devonian brachiopods from the Lesser Khingan district of northeast China*, by Takashi Hamada: Journal of Paleontology, v. 47, p. 140–141.

Johnson, J. G., 1973, Some North America rensselandiid brachiopods, Part 2: Journal of Paleontology, v. 47, p. 1102–1107, 2 plates.

Boucot, A. J., and Johnson, J. G., 1973, Silurian brachiopods, *in* Hallam, A., ed., Atlas of palaeobiogeography: Amsterdam, Elsevier, p. 59–65.

Johnson, J. G., and Boucot, A. J., 1973, Devonian brachiopods, *in* Hallam, A., ed., Atlas of palaeobiogeography: Amsterdam, Elsevier, p. 89–96.

Johnson, J. G., 1973, *Tecnocyrtina beckmanni*, a new species of brachiopod from the lower Upper Devonian of Germany: Senckenbergiana Lethaea, v. 53, p. 517–520, 1 plate.

Johnson, J. G., 1973, Late Early Devonian rhynchonellid genera from Arctic and western North America: Journal of Paleontology, v. 47, p. 465–472, 3 plates.

Johnson, J. G., Boucot, A. J., and Murphy, M. A., 1973, Pridolian and early Gedinnian age brachiopods from the Roberts Mountains Formation of central Nevada: University of California Publications in Geological Sciences, v. 100, 75 p., 31 plates.

Johnson, J. G., 1973, Mid-Lochkovian brachiopods from the Windmill Limestone of central Nevada: Journal of Paleontology v. 47, p. 1013–1030, 5 plates.

Johnson, J. G., 1974, Affinity of dayiacean brachiopods: Palaeontology, v. 17, p. 437–439.

Johnson, J. G., 1974, *Oriskania* (terebratulid brachiopod) in the Lower Devonian of central Nevada: Journal of Paleontology, v. 48, p. 1207–1212, 1 plate.

Johnson, J. G., 1974, Extinction of perched faunas: Geology, v. 2, p. 479–482.

Johnson, J. G., 1974, Early Devonian brachiopod biofacies of western and Arctic North America: Journal of Paleontology, v. 48, p. 809–819.

Johnson, J. G., 1974, Shorelines of epeiric seas: American Journal of Science, v. 274, p. 465–470.

Johnson, J. G., 1974, Sea-floor spreading and orogeny; correlation or anticorrelation?: Geology , v. 2, p. 199–201.

Johnson, J. G., 1974, Great Basin Silurian to Lower Devonian—a biostratigraphic case history: American Association of Petroleum Geologists Bulletin, v. 58, p. 139–141.

Johnson, J. G., 1974, Steiner's galacto-geologic correlations: Discussion: Geology, v. 2, p. 25.

Johnson, J. G., 1974, Middle Devonian Givetian brachiopods from the *Leiorhynchus castanea* Zone of Nevada: Geologica et Palaeontologica, v. 8, p. 49–96, 12 plates.

Johnson, J. G., 1975, Late Early Devonian brachiopods from the Disappointment Bay Formation, Lowther Island, Arctic Canada: Journal of Paleontology, v. 49, p. 947–978, 10 plates.

Johnson, J. G., 1975, Asian stringocephaloid brachiopod genera *Omolonia* and *Paracrothyris* in the Middle Devonian of western North America: Journal of Paleontology, v. 49, p. 979–983, 1 plate.

Johnson, J. G., 1975, Allopatric speciation in fossil brachiopods: Journal of Paleontology, v. 49, p. 646–661, 3 plates.

Johnson, J. G., 1975, Devonian brachiopods from the *Quadrithyris* Zone (upper Lochkovian), Canadian Arctic Archipelago: Geological Survey of Canada Bulletin 235, p. 5–57, plates 2–11.

Johnson, J. G., 1975, Roberts Mountains Thrust: gravity slide or underthrust?: Discussion: Geology, v. 3, p. 219–220.

Johnson, J. G., and Potter, E. C., 1975, Silurian (Llandovery) downdropping of the western margin of North America: Geology, v. 3, p. 331–334.

Johnson, J. G., 1975, Extinction of perched faunas: Reply to C. W. Pitrat: Geology, v. 3, p. 229–230.

Johnson, J. G., and Neibuhr, W. W., II, 1976, Anatomy of an assemblage zone: Geological Society of America Bulletin, v. 86, p. 1693–1703.

Johnson, J. G., 1976, *Antistrix* (Middle Devonian brachiopod) is a terebratulid: Journal of Paleontology, v. 50, p. 355–356.

Johnson, J. G., and Perry, D. G., 1976, Middle Devonian brachiopods from the Bird Fiord Formation of Bathurst Island, Arctic Canada: Canadian Journal of Earth Sciences, v. 13, p. 615–635, 14 plates.

Johnson, J. G., and Potter, E. C., 1976, The depths inhabited by Silurian brachiopod communities: Reply to S. V. Shabica and A. J. Boucot: Geology, v. 4, p. 132, 189–191.

Johnson, J. G., and Potter, E. C., 1976, Late Ordovician–Early Silurian glaciation and the Ordovician-Silurian boundary in the northern Canadian Cordillera: Comment: Geology, v. 4, p. 795.

Johnson, J. G., 1976, *Chattertonia* n. gen. (Brachiopoda), the chonetid-productid link: Journal of Paleontology, v. 50, p. 789–791.

Johnson, J. G., and Kendall, G. W., 1976, Late Early Devonian brachiopods and biofacies from central Nevada: Journal of Paleontology, v. 50, p. 1113–1128, 3 plates.

Johnson, J. G., Boucot, A. J., and Murphy, M. A., 1976, Wenlockian and Ludlovian age brachiopods from the Roberts Mountains Formation of central Nevada: University of California Publications in Geological Sciences, v. 115, 102 p., 55 plates.

Ziegler, W., Klapper, G., and Johnson, J. G., 1976, Redefinition and subdivision of the *varcus* Zone (conodonts, Middle-?Upper Devonian) in Europe and North America: Geologica et Palaeontologica, v. 10, p. 109–140, 4 plates.

Smith, R. E., and Johnson, J. G., 1977, *Atrypella scheii* (Holtedahl) and *Atrypella phoca* (Salter), (Silurian Brachiopoda): Journal of Paleontology, v. 51, p. 350–356, 1 plate.

Johnson, J. G., and Sandberg, C. A., 1977, Lower and Middle Devonian continental-shelf rocks of the western United States: Geological Society of America Abstracts with Programs, v. 9, p. 1037–1038.

Johnson, J. G., 1977, Lower and Middle Devonian faunal intervals in central Nevada, based on brachiopods: Geological Society of America Abstracts with Programs, v. 9, p. 1038.

Johnson, J. G., 1977, Status of Devonian studies in western and Arctic North America: Geological Society of America Abstracts with Programs, v. 9, p. 1038.

Johnson, J. G., 1977, Status of Devonian studies in western and Arctic North

America, *in* Murphy, M. A., Berry, W. B. N., and Sandberg, C. A., Western North America: Devonian: University of California, Riverside Campus Museum Contribution 4, p. 1–15.

Johnson, J. G., 1977, Lower and Middle Devonian faunal intervals in central Nevada based on brachiopods, *in* Murphy, M. A., Berry, W. B. N., and Sandberg, C. A., Western North America: Devonian: University of California, Riverside Campus Museum Contribution 4, p. 16–32.

Oliver, W. A., Jr., and Johnson, J. G., 1977, Notes on Upper Silurian and Devonian coral zones in the Great Basin, *in* Murphy, M. A., Berry, W. B. N., and Sandberg, C. A., Western North America: Devonian: University of California, Riverside Campus Museum Contributions 4, p. 107–111.

Johnson, J. G., and Sandberg, C. A., 1977, Lower and Middle Devonian continental shelf rocks of the western United States, *in* Murphy, M. A., Berry, W. B. N., and Sandberg, C. A., Western North America: Devonian: University of California, Riverside Campus Museum Contributions 4, p. 121–143.

Johnson, J. G., and Oliver, W. A., Jr., 1977, Silurian and Devonian coral zones in the Great Basin, Nevada and California: Geological Society of America Bulletin, v. 88, p. 1462–1468.

Johnson, J. G., 1978, Devonian, Givetian age brachiopods and biostratigraphy, central Nevada: Geologica et Palaeontologica, v. 12, p. 117–150, 8 plates.

Johnson, J. G., and Klapper, G., 1978, Devonian brachiopods and conodonts from central Oregon: Journal of Paleontology, v. 52, p. 295–299, 1 plate.

Johnson, J. G., Penrose, N. L., and Wise, M. T., 1978, Biostratigraphy, biotopes and biogeography in the Lower Devonian (upper Lochkovian, lower Pragian) of Nevada: Journal of Paleontology, v. 52, p. 793–806, 1 plate.

Johnson, J. G., 1979, The method of multiple working hypotheses in science?: Geological Society of America News and Information, v. 1, p. 22.

Johnson, J. G., 1979, Intent and reality in biostratigraphic zonation: Journal of Paleontology, v. 53, p. 931–942.

Johnson, J. G., 1979, Devonian brachiopod biostratigraphy: Special Papers in Palaeontology 23, p. 291–306.

Boucot, A. J., and Johnson, J. G., 1979, The Pentamerinae (Silurian Brachiopoda): Palaeontographica, Abteilung A, v. 163, p. 87–129, 15 plates.

Johnson, J. G., 1979, Devonian, *in* McGraw-Hill yearbook of science and technology, 1979: New York, McGraw-Hill, p. 159–161.

Klapper, G., and Johnson, J. G., 1979, *Late Silurian and Early Devonian graptolite, brachiopod and coral faunas from northwestern and Arctic Canada*, by D. E. Jackson, A. C. Lenz, and A. E. H. Pedder: Book review: Geoscience Canada, v. 6, p. 224.

Johnson, J. G., Klapper, G., and Trojan, W. R., 1980, Brachiopod and conodont successions in the Devonian of the northern Antelope Range, central Nevada: Geologica et Palaeontologica, v. 14, p. 77–116, 4 plates.

Johnson, J. G., Klapper, G., and Trojan, W. R., 1980, Upper range of *Stringocephalus* (Devonian Brachiopoda): Newsletters on Stratigraphy, v. 8, p. 232–235.

Klapper, G., and Johnson, J. G., 1980, Endemism and dispersal of Devonian conodonts: Journal of Paleontology, v. 54, p. 400–455, 4 plates.

Johnson, J. G., and Pendergast, A., 1981, Timing and mode of emplacement of the Roberts Mountains Allochthon, Antler Orogeny: Geological Society of America Bulletin, v. 92, p. 648–658.

Johnson, J. G., Hildreth, G. D., Kendall, G. W., and Trojan, W. R., 1981, Biostratigraphy and biotopes of the lower Middle Devonian *Leptathyris circula* Zone, central Nevada: Journal of Paleontology, v. 55, p. 814–825, 1 plate.

Boucot, A. J., and Johnson, J. G., 1981, Emendations to the Pentamerinae (Silurian Brachiopoda): Journal of Paleontology, v. 55, p. 681.

Johnson, J. G., 1982, Chronozones and other misapplications of chronostratigraphic concept: Lethaia, v. 14, p. 285–286. [imprint 1981]

Johnson, J. G., 1982, *Davidsonia* and Davidsoniacea, Devonian strophomenid brachiopods: Journal of Paleontology, v. 56, p. 973–976, 1 plate.

Johnson, J. G., 1982, Occurrence of phyletic gradualism and punctuated equilibria through geologic time: Journal of Paleontology, v. 56, p. 1329–1331.

Murphy, M. A., Power, J. D., and Johnson, J. G., 1982, A suggested Late Devonian overlap of the Roberts Mountains Allochthon, Roberts Mountains, Eureka County, Nevada: Geological Society of America Abstracts with Programs, v. 14, p. 219.

Johnson, J. G., and Trojan, W. R., 1982, The *Tecnocyrtina* brachiopod fauna (?Upper Devonian) of central Nevada: Geologica et Palaeontologica, v. 16. p. 119–150, 8 plates.

Johnson, J. G., and Heckel, P. H., 1982, Devonian, *in* McGraw-Hill encyclopaedia of science and technology (fifth edition), Volume 4, p. 147–151.

Johnson, J. G., 1983, Early Paleozoic Cordilleran biofacies, provinces, and realms, *in* Stevens, C. H., ed., Pre-Jurassic rocks in western North American suspect terranes: Los Angeles, Society of Economic Paleontologists and Mineralogists, Pacific Section, p. 1–5.

Johnson, J. G., and Pendergast, A., 1983, Antler Orogeny and foreland basin, a model: Discussion: Geological Society of America Bulletin, v. 94, p. 684.

Johnson, J. G., 1983, Mid-Paleozoic age of the Roberts thrust unsettled by new data from northern Nevada: Comment: Geology, v. 11, p. 60–62.

Johnson, J. G., 1983, Eustatic control of turbidites and winnowed turbidites: Comment: Geology, v. 11, p. 58.

Kendall, G. W., Johnson, J. G., Brown, J. O., and Klapper, G., 1983, Stratigraphy and facies across Lower Devonian–Middle Devonian boundary, central Nevada: American Association of Petroleum Geologists Bulletin, v. 67, p. 2199–2207.

Sandberg, C. A., Gutschick, R. C., Poole, F. G., Sando, W. J., and Johnson, J. G., 1983, Middle Paleozoic history interpretation as an aid to Western Overthrust Belt petroleum exploration, *in* Geological Survey Research 1982: U.S. Geological Survey Professional Paper 1375, p. 26–27.

Sandberg, C. A., Gutschick, R. C., Johnson, J. G., Poole, F. G., and Sando, W. J., 1983, Middle Devonian to Late Mississippian geologic history of the Overthrust Belt region, western United States, *in* Powers, R. B., ed., Geologic studies of the Cordilleran thrust belt: Denver, Rocky Mountains Association of Geologists, p. 691–719. [imprint 1982]

Murphy, J. D., Power, J. D., and Johnson, J. G., 1984, Evidence for Late Devonian movement within the Roberts Mountains Allochthon, Roberts Mountains, Nevada: Geology, v. 12, p. 20–23.

Johnson, J. G., and Murphy, M. A., 1984, Time-rock model for Siluro-Devonian continental shelf, western United States: Geological Society of America Bulletin, v. 95, p. 1349–1359.

Johnson, J. G., 1984, Temperature and biotic crises in the marine realm: Discussion: Geology, v. 12, p. 741–743.

Murphy, M. A., Power, J. D., and Johnson, J. G., 1984, Evidence for Late Devonian movement within the Roberts Mountains Allochthon, Roberts Mountains, Nevada: Comment: Geology, v. 12, p. 445–446.

Johnson, J. G., 1985, Silurian carbonate shelf and slope evolution in Nevada; a history of faulting, drowning, and progradation: Discussion: Geology, v. 13, p. 744–746.

Johnson, J. G., Klapper, G., and Sandberg, C. A., 1985, Devonian eustatic fluctuations in Euramerica: Geological Society of America Bulletin, v. 96, p. 567–587.

Johnson, J. G., Klapper, G., Murphy, M. A., and Trojan, W. R., 1986, Devonian series boundaries in central Nevada and neighboring regions, western North America: Courier Forschungsinstitut Senckenberg, v. 75. p. 177–196. [imprint 1985]

Johnson, J. G., 1986, *Aspects of a stratigraphic system: The Devonian*, by D. L. Dineley: Review: American Scientist, v. 74, p. 84.

Johnson, J. G., Klapper, G., and Sandberg, C. A., 1986, Late Devonian eustatic cycles around margin of Old Red Continent: Annales de la Société Géologique de Belgique, v. 109, p. 141–147.

Sandberg, C. A., Gutschick, R. C., Johnson, J. G., Poole, F. G., and Sando, W. J., 1986, Middle Devonian to Late Mississippian event stratigraphy of Overthrust Belt region, western United States: Annales de la Société Géologique de Belgique, v. 109, p. 205–207.

Johnson, J. G., 1986, Revision of Lower Devonian (Emsian) brachiopod biostratigraphy and biogeography, central Nevada: Journal of Paleontology, v. 60, p. 825–844.

Johnson, J. G., 1987, Unconformity-bounded stratigraphic units: Discussion: Geological Society of America Bulletin, v. 99, p. 443.

Johnson, J. G., and Cowell, P. F., 1987, Elimination of anomalous Devonian stratigraphy south of Eureka, Nevada: Palaios, v. 2, p. 85–86.

Johnson, J. G., and Heckel, P. H., 1987, Devonian, *in* McGraw-Hill encyclopaedia of science and technology (sixth edition), Volume 5, p. 163–167.

Johnson, J. G., 1987, On selecting biostratigraphic boundaries: Subcommission on Devonian Stratigraphy Newsletter, no. 3, p. 3–4.

Boucot, A. J., Johnson, J. G., and Ning Zhang, 1988, Silurian (Wenlockian) brachiopods from southeastern California: Palaeontographica, Abteilung A, v. 201, p. 103–127.

Johnson, J. G., 1988, Volcanism, eustasy, and extinctions: Geology, v. 16, p. 573–574.

Johnson, J. G., 1988, *Global bio-events, a critical approach*, edited by O. H. Walliser: Review: Earth Science Reviews, v. 25, p. 247–248.

Johnson, J. G., Sandberg, C. A., and Poole, F. G., 1989, Early and Middle Devonian paleogeography of western United States, *in* McMillan, N. J., Embry, A. F., and Glass, D. J., eds., Devonian of the world: Calgary, Canadian Society of Petroleum Geologists Memoir 14, v. 1, p. 161–182. [imprint 1988]

Sandberg, C. A., Poole, F. G., and Johnson, J. G., 1989, Upper Devonian of western United States, *in* McMillan, N. J., Embry, A. F., and Glass, D. J., eds., Devonian of the world: Calgary, Canadian Society of Petroleum Geologists Memoir 14, v. 1, p. 183–220. [imprint 1988]

Johnson, J. G., and Sandberg, C. A., 1989, Devonian eustatic events in the Western United States and their biostratigraphic responses, *in* McMillan, N. J., Embry, A. F., and Glass, D. J., eds., Devonian of the world: Calgary, Canadian Society of Petroleum Geologists Memoir 14, v. 3, p. 171–178. [imprint 1988]

Johnson, J. G., and Murphy, M. A., 1989, Roberts Mountains Allochthon; stratigraphic comparison with Lower Paleozoic outer continental margin strata of the northern Canadian Cordillera: Comment: Geology, v. 17, p. 1063.

Johnson, J. G., 1989, Base of the Upper Devonian in the conodont zonation: Newsletters on Stratigraphy, v. 21, p. 11–14.

Johnson, J. G., 1990, Lower and Middle Devonian brachiopod-dominated communities of Nevada, and their position in a biofacies-province-realm model [with a section on revision of Middle Devonian conodont zones, by G. Klapper and J. G. Johnson]: Journal of Paleontology, v. 64, p. 902–941.

Johnson, J. G., 1990, Method of multiple working hypotheses; a chimera: Geology, v. 18, p. 44–45.

Johnson, J. G., 1990, *Talentella*, a new Siluro-Devonian genus of enteletacean brachiopods: Journal of Paleontology, v. 64, p. 489–490.

Johnson, J. G., 1990, Method of multiple working hypotheses; a chimera: Reply: Geology, v. 18, p. 918.

Johnson, J. G., Sandberg, C. A., and Poole, F. G., 1991, Devonian lithofacies of western United States, *in* Cooper, J. D., and Stevens, C. H., eds, Paleozoic paleogeography of the western United States, II: Society of Economic Paleontologists and Mineralogists, Pacific Section, v. 67, p. 83–105.

Johnson, J. G., and Bird, J. M., 1991, History of Lower Devonian basin to platform transects in Nevada, *in* Cooper, J. D., and Stevens, C. H., eds, Paleozoic paleogeography of the western United States, II: Society of Economic Paleontologists and Mineralogists, Pacific Section, v. 67, p. 311–315.

Johnson, J. G., 1992, Belief and reality in biostratigraphic zonation: Newsletters on Stratigraphy, v. 26, p. 41–48.

Johnson, J. G., and Visconti, R., 1992, Roberts Mountains Thrust relationships in a critical area, northern Sulphur Spring Range, Nevada: Geological Society of America Bulletin, v. 104, p. 1208–1220.

Johnson, J. G., and Lenz, A. C., 1992, *Eoplicoplasia*, a new genus of Silurian-Lower Devonian ambocoeliid brachiopods: Journal of Paleontology, v. 66, p. 530–531.

Johnson, J. G., and Klapper, G., 1992, North American Midcontinent Devonian T-R cycles, *in* Chaplin, J. R., and Barrick, J. E., eds., Special papers in paleontology and stratigraphy: A tribute to Thomas W. Amsden: Oklahoma Geological Survey Bulletin, v. 145, p. 127–135.

Johnson, J. G., and Heckel, P. H., 1992, Devonian, *in* McGraw-Hill encyclopaedia of science and technology (seventh edition), Volume 5, p. 165–169.

Blodgett, R. B., and Johnson, J. G., 1993, Early Middle Devonian (Eifelian) gastropods of central Nevada: Palaeontographica, Abteilung A, v. 222, p. 83–139.

Johnson, J. G., 1993, Gigantic Ordovician volcanic ash fall in North America and Europe; biological, tectonomagmatic, and event-stratigraphic significance: Comment: Geology, v. 21, p. 383.

Johnson, J. G., and Blodgett, R. B., 1993, Russian Devonian brachiopod genera *Cyrtinoides* and *Komiella* in North America: Journal of Paleontology, v. 67, p. 952–958.

Blodgett, R. B., and Johnson, J. G., 1994, First recognition of the genus *Verneuilia* Hall and Clarke (Brachiopoda, Spiriferida) from North America (west-central Alaska): Journal of Paleontology, 68, p. 1240–1242.

Elrick, M., and Johnson, J. G., 1994, Sequence stratigraphy of Middle Devonian ramp-to-basin carbonates of the eastern Great Basin: Geological Society of America Abstracts with Programs, v. 26, p. 50.

Johnson, J. G., Klapper, G., and Elrick, M., 1996, Devonian transgressive-regressive cycles and biostratigraphy, northern Antelope Range, Nevada: Establishment of reference horizons for global cycles: Palaios, v. 11, p. 3–14.

STUDENTS AT ORGEON STATE UNIVERSITY FOR WHOM JESS JOHNSON SERVED AS MAJOR ADVISER (LIST PROVIDED BY CLAUDIA REGIER)

PH.D.

Richard A. Flory, 1974
Roy E. Smith, 1976
Thomas L. DeKeyser, 1978
Constance J. Sansome, 1986

M.S.

Walter W. Niebuhr II, 1973
George W. Kendall, 1975
Eric C. Potter, 1975
Diane C. Wise, 1976
Robert A. Schalla, 1978
Edwin A. Drake, 1978
Thomas R. Peargin, 1978
William R. Trojan, 1978
Linda J. Jarvis, 1981
M. Anne Pendergast, 1981
Thurman L. Heironimus, 1981
James O. Brown, 1982
John P. Graham, 1982
Robert V. Visconti, 1982
Roger S. Sans, 1985
John H. Whitaker, 1985
Maya Elrick, 1986
Peter F. Cowell, 1986
Joseph T. Lipka II, 1987
F. Michael DeDen, 1987
Jill M. Bird, 1989
Sherri L. Lee, 1989
Heidi Hoffmann, 1989
Jeff L. Waldman, 1990
James M. Helgeson, 1993
Storr Nelson, 1993
Paul Mayer, 1994

UNDERGRADUATE HONOR'S THESIS

Gail D. Hildreth, 1978

Manuscript Accepted by the Society February 27, 1997

Printed in U.S.A.

Geological Society of America
Special Paper 321
1997

Use of K-bentonite beds as time-planes for high-resolution lithofacies analysis and assessment of net rock accumulation rate: An example from the upper Middle Ordovician of eastern North America

Stephen A. Leslie
Department of Earth Science, University of Arkansas at Little Rock, Little Rock, Arkansas 72204-1099
Stig M. Bergström
Department of Geological Sciences, The Ohio State University, Columbus, Ohio 43210

ABSTRACT

Regional correlation of the vertically closely spaced Mohawkian Deicke and Millbrig K-bentonite (altered volcanic ash) beds in eastern North America provides a unique stratigraphic framework in which lithofacies distributions, changes of lithofacies distributions through time, and net rock accumulation rates can be examined. Four distinct lithofacies are present in the interval just below the Deicke, namely the Wackestone Facies, Mudstone Facies, Nodular Wackestone/Calcareous Shale Facies, and Sandstone/Siltstone/Claystone Facies. Shortly after Millbrig time the Wackestone Facies was replaced by the Green Shale/Packstone Facies, and the Mudstone Facies was replaced by the Wackestone/Packstone Facies. This lithofacies change was associated with an increase in water depth in the epicontinental sea that covered much of eastern North America at that time. The distribution of the other lithofacies changed little in the approximately 0.2 to 1.4 m.y. between the deposition of the Deicke and the Millbrig K-bentonite beds.

The net rock accumulation between the Deicke and Millbrig K-bentonite beds in the sections studied varies from as little as 0.37 m in the Upper Mississippi Valley to as much as 14.62 m in southwestern Virginia. The area of greatest net rock accumulation is within the Sandstone/Siltstone/Claystone Facies and represents molasse-filling of the Sevier Basin. Net rock accumulation in the deeper-water setting of Nodular Wackestone/Calcareous Shale Facies was relatively small because of low sedimentation rate caused in part by trapping of clastic sediments in the Sevier Basin. Small net rock accumulation in central Kentucky is possibly due to subaerial exposure and erosion of sediment that accumulated after deposition of the Deicke K-bentonite Bed and before deposition of the Millbrig K-bentonite Bed. The smallest net accumulation, which is in the Upper Mississippi Valley, is due to compaction of the green shale that is the dominant lithology between the Deicke and Millbrig K-bentonite beds, extremely slow sedimentation rate, and generally lower subsidence rates in this more cratonic setting.

Leslie, S. A., and Bergström, S. M., 1997, Use of K-bentonite beds as time-planes for high-resolution lithofacies analysis and assessment of net rock accumulation rate: An example from the upper Middle Ordovician of eastern North America, *in* Klapper, G., Murphy, M. A., and Talent, J. A., eds., Paleozoic Sequence Stratigraphy, Biostratigraphy, and Biogeography: Studies in Honor of J. Granville ("Jess") Johnson: Boulder, Colorado, Geological Society of America Special Paper 321.

INTRODUCTION

Modeling of regional lithofacies distribution and net rock accumulation rates based on conventional lithologic and biostratigraphic correlation is limited by the resolution of correlation. Even between sections correlated by the best constrained biostratigraphic schemes, stratigraphic resolution is often blurred by difficulties in correlating across biofacies and lithofacies boundaries. In an effort to construct a high-resolution chronostratigraphic framework, Sweet (1979, 1984, 1995) proposed a conodont graphic correlation scheme for Ordovician rocks in North America that synthesizes conodont range data from many sections representing different biofacies and lithofacies. Although this method of correlation provides relatively good resolution, it is still no better than 6 m (Sweet, 1984). In order to examine facies relations at a finer scale over long distances, a more precise means of stratigraphic correlation is needed.

The purpose of this chapter is to document the lithofacies distributions and net rock accumulation rates in the upper Middle Ordovician of eastern North America, using a novel stratigraphic approach. The deposition of a volcanic ash bed is geologically instantaneous, on the scale of days to weeks. Therefore, if such an ash bed can be correlated regionally with certainty, it is essentially a geologic time-plane. Two beds of altered volcanic ash (K-bentonite) in latest Turinian and earliest Chatfieldian strata in eastern North America (Fig. 1) are particularly well suited to serve as regional time-planes. These beds, the Deicke and the Millbrig, have been correlated successfully throughout much of eastern North America (Fig. 2) using traditional stratigraphy and biostratigraphy (Templeton and Willman, 1963) in combination with chemical fingerprinting (Huff, 1983; Kolata et al., 1986; McVey, 1993), wire-line logs (Huff and Kolata, 1990; Kolata et al., 1996), and petrography (Haynes, 1992, 1994). The vast regional distribution of two K-bentonite beds that are close in age provides a unique opportunity to examine broad lithofacies changes and net rock accumulation with great precision in the interval between, and adjacent to, these ash beds over a region covering some 1.5 million km^2.

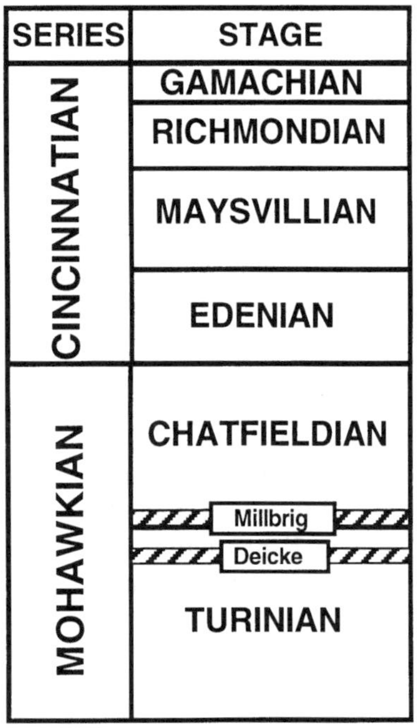

Figure 1. Mohawkian and Cincinnatian stage-level nomenclature (Leslie and Bergström, 1995). The base of the Chatfieldian is the Millbrig K-bentonite Bed.

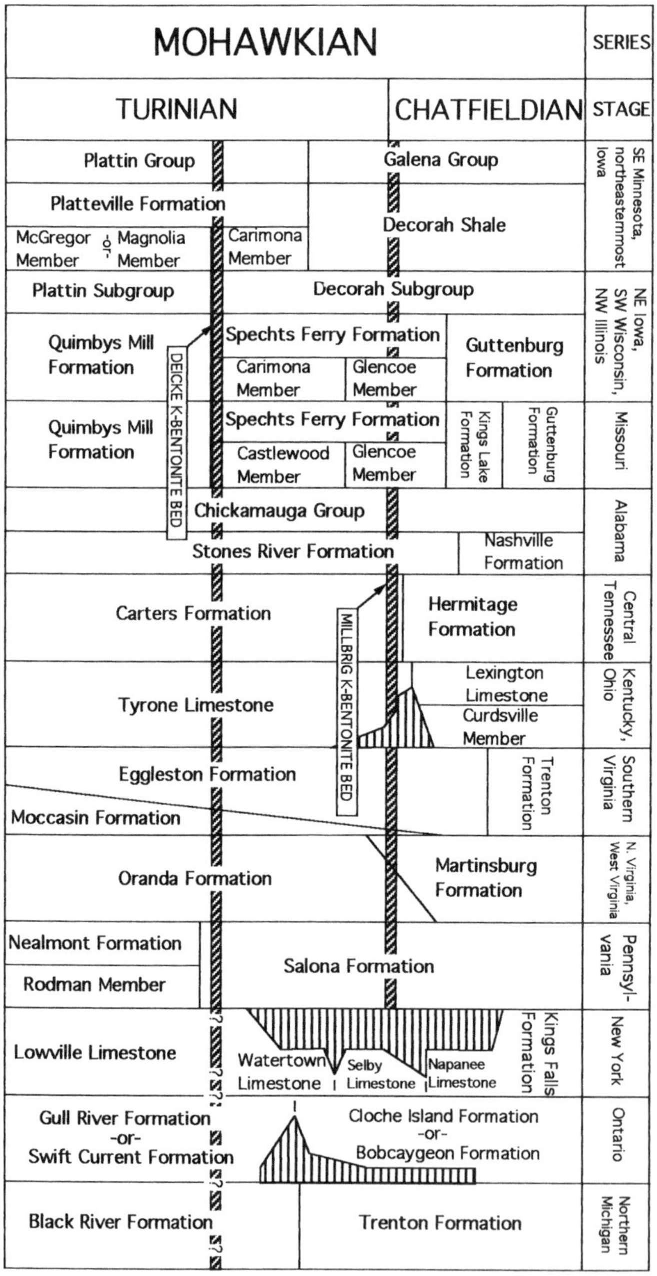

Figure 2. Regional correlation of the Deicke and Millbrig K-bentonite beds and late Middle Ordovician stratigraphic nomenclature in eastern North America.

This chapter is dedicated to the late J. G. Johnson in recognition of his important work on Paleozoic rocks and faunas. He also took interest in the stratigraphic use of the Middle Ordovician K-bentonite beds dealt with herein (Johnson, 1993).

METHODS

Data from 45 investigated localities were used to construct lithofacies and net rock accumulation maps that represent regional "snapshot" images of the upper Middle Ordovician of eastern North America. The lithologic classification follows Dunham (1962). All sections contain either the Deicke K-bentonite Bed, the Millbrig K-bentonite Bed, or both. K-bentonite bed correlations are based on Kolata et al. (1986), Huff and Kolata (1990), Haynes (1992, 1994), McVey (1993), Leslie (1995), and Kolata et al. (1996).

The appendix contains locality information for all sections that are not described in these references.

LITHOFACIES

North America occupied a low-latitude position during the late Turinian and early Chatfieldian. Most of the midcontinent area was covered by a shallow epicontinental sea that stretched from at least central Alabama to southern Ontario (Keith, 1988, figs. 1-3). To the east, water depth increased into a series of northeast-trending basins, representing the foreland basins (Read, 1980) that resulted from the arc collision of the Taconic Orogeny. East of this volcanic arc was the Iapetus Ocean. One piece of convincing evidence for this arc-continent collision event is the numerous K-bentonite beds preserved in Ordovician strata throughout eastern North America that have chemical signatures consistent with silicic arc volcanism (Samson et al., 1989; Delano et al., 1990; Huff et al., 1992). The Deicke and Millbrig K-bentonite beds are the most widespread and prominent ash beds produced by this episode of volcanism (Huff et al., 1992; Kolata et al., 1996).

In the following section "pre-Deicke" refers to the time represented by rocks immediately below the Deicke K-bentonite Bed, and "post-Millbrig" refers to the time represented by the rocks immediately above the Millbrig K-bentonite Bed. The corresponding terms expressing stratigraphic position are "sub-Deicke" and "supra-Millbrig," respectively.

Sub-Deicke lithofacies

Four distinct lithofacies are present in the beds immediately below the Deicke K-bentonite Bed in eastern North America. The regional distribution of these lithofacies is illustrated in Figure 3. The Upper Mississippi Valley is dominated by the Wackestone Facies. This facies is characterized by medium-bedded brachiopod and gastropod wackestone that commonly contains thin brown shaly partings. The eastern midcontinent area is dominated by the Mudstone Facies. This facies is characterized by massive, sparsely fossiliferous, fenestral mudstone. The Mudstone Facies encompasses much of what is commonly referred to as "Black River" lithology. Immediately south of the Mudstone Facies belt in Alabama and east of the Mudstone Facies belt in the north-central Appalachians are areas characterized as the Nodular Wackestone/Calcareous Shale Facies. This facies consists of dark gray, calcareous shale interbedded with rubbly, nodular, dark gray wackestone. In some areas, as in central Alabama and north-central Virginia, the nodular wackestone is the dominant lithology, whereas in central Pennsylvania calcareous shale is dominant. A Sandstone/Siltstone/Claystone Facies is present in the south-central Appalachians.

Supra-Millbrig lithofacies

Five distinct lithofacies are present in beds immediately above the Millbrig K-bentonite Bed (Fig. 4). The Upper Mississippi Valley is dominated by the Green Shale/Packstone Facies. This facies is characterized by green shale interbedded with variably discontinuous brachiopod and bryozoan packstone beds. The eastern midcontinent area is dominated by the Wackestone/Packstone Facies. The rocks of this lithofacies are medium-gray brachiopod wackestone and packstone with intermittent grainstone beds that are classified as the Trenton and Lexington limestones. The Millbrig K-bentonite Bed is missing throughout much of the Jessemine Dome in Kentucky, where it was either eroded or never preserved (Kolata et al., 1996). Although the facies map (Fig. 4) illustrates the Wackestone/Packstone facies as present in the northern midcontinent area, this is speculative. The Millbrig has not yet been identified in this area (northern Michigan, southern Ontario, New York) and there may be a stratigraphic gap at this level in part of this region, especially in north-central New York. However, based on the lack of clear evidence for a stratigraphic gap in the Lake Simcoe and Manitoulin Island areas of southern Ontario, this facies is tentatively projected into the north and north-central limit of the outcrop belt. The Nodular Wackestone/Calcareous Shale Facies is little changed from its sub-Deicke distribution in central Alabama and in the north-central Appalachian region. A narrow belt of Mudstone Facies stretches from north-central Alabama to the West Virginia–Virginia southern border that is lithologically the same as the pre-Deicke Mudstone Facies. The siliciclastic Sandstone/Siltstone/Claystone Facies in the south-central Appalachians is in about the same geographic region as it occupied in pre-Deicke time.

Several lithofacies changes are apparent when the lithofacies distributions immediately below the Deicke K-bentonite Bed (Fig. 3) are compared with those immediately above the Millbrig K-bentonite Bed (Fig. 4). Figure 5 shows two schematic cross sections that illustrate these lithofacies changes. The Wackestone Facies in the Upper Mississippi Valley in pre-Deicke time is replaced by the Green Shale/Packstone Facies in post-Millbrig time. The Mudstone Facies that in pre-Deicke time dominated the central midcontinent area is present only in a narrow band in the southern Appalachians and is replaced over much of its pre-

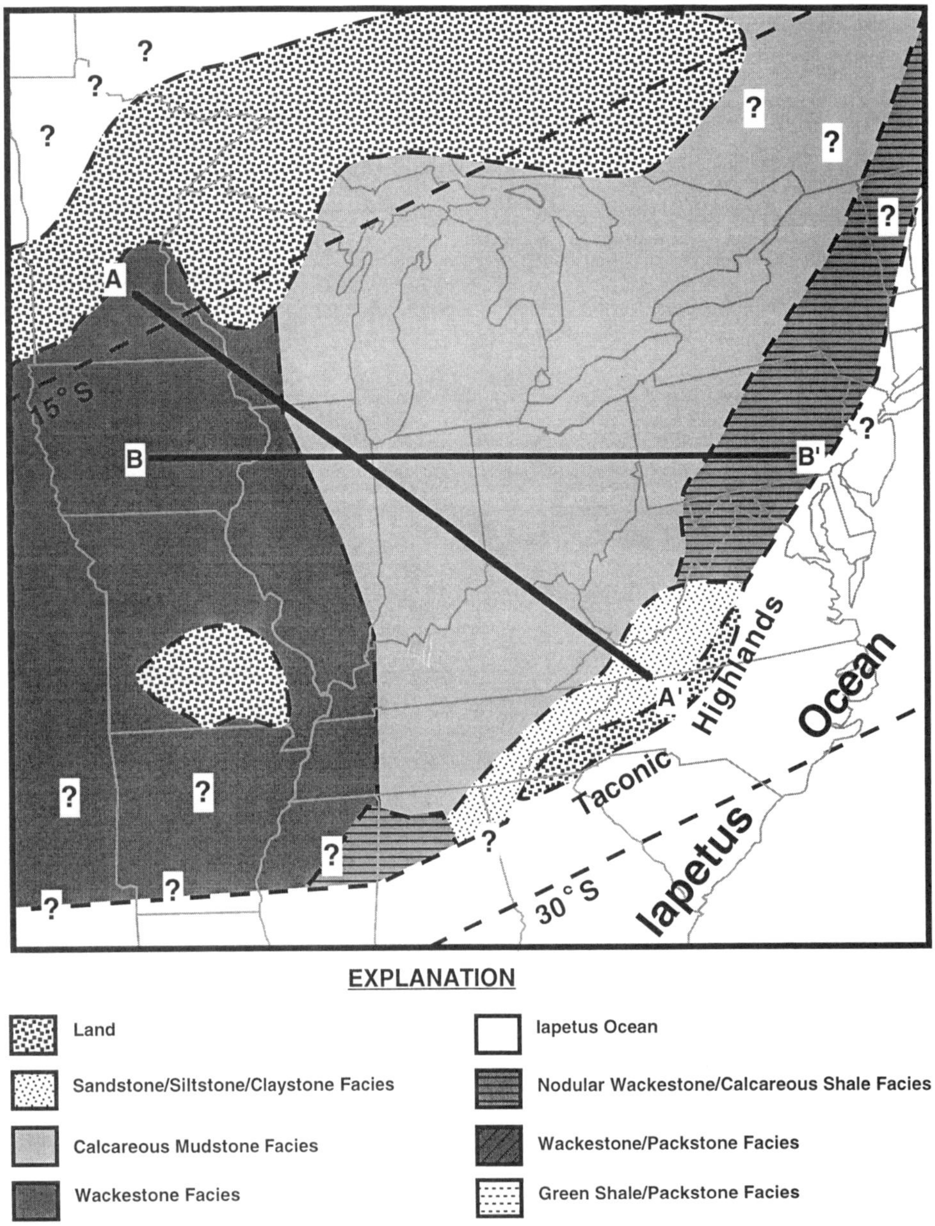

Figure 3. Sub-Deicke (latest Turinian) lithofacies distribution, not corrected for tectonic shortening.

Deicke distribution area by the Wackestone/Packstone Facies in post-Millbrig time. There is no perceptible distributional change in the Nodular Wackestone/Calcareous Shale Facies in Central Alabama from pre-Deicke to post-Millbrig time. However, the wackestone in the pre-Deicke Nodular Wackestone/Calcareous Shale Facies in the north-central Appalachians grades into a sparsely fossiliferous mudstone by post-Millbrig time. The area of the Sandstone/Siltstone/Claystone Facies does not change much from pre-Deicke to post-Millbrig time. The most noteworthy change that occurs within this broadly defined facies is from a finer-grained clastic facies in pre-Deicke time to a coarser-grained clastic facies in post-Millbrig time.

DEPOSITIONAL ENVIRONMENTS

The depositional environment of each lithofacies is interpreted for both pre-Deicke and post-Millbrig time. These interpretations are based upon the dominant lithologies and, to a lesser extent, the general faunal aspect of each lithofacies.

The Upper Mississippi Valley region is interpreted to have been a shallow to very shallow subtidal environment during pre-Deicke time (cf. Keith, 1988, fig. 2). Mud cracks in the Wackestone Facies indicate occasional subaerial exposure. However, they are uncommon everywhere and are not present on the bedding plane immediately below the Deicke. The Mud-

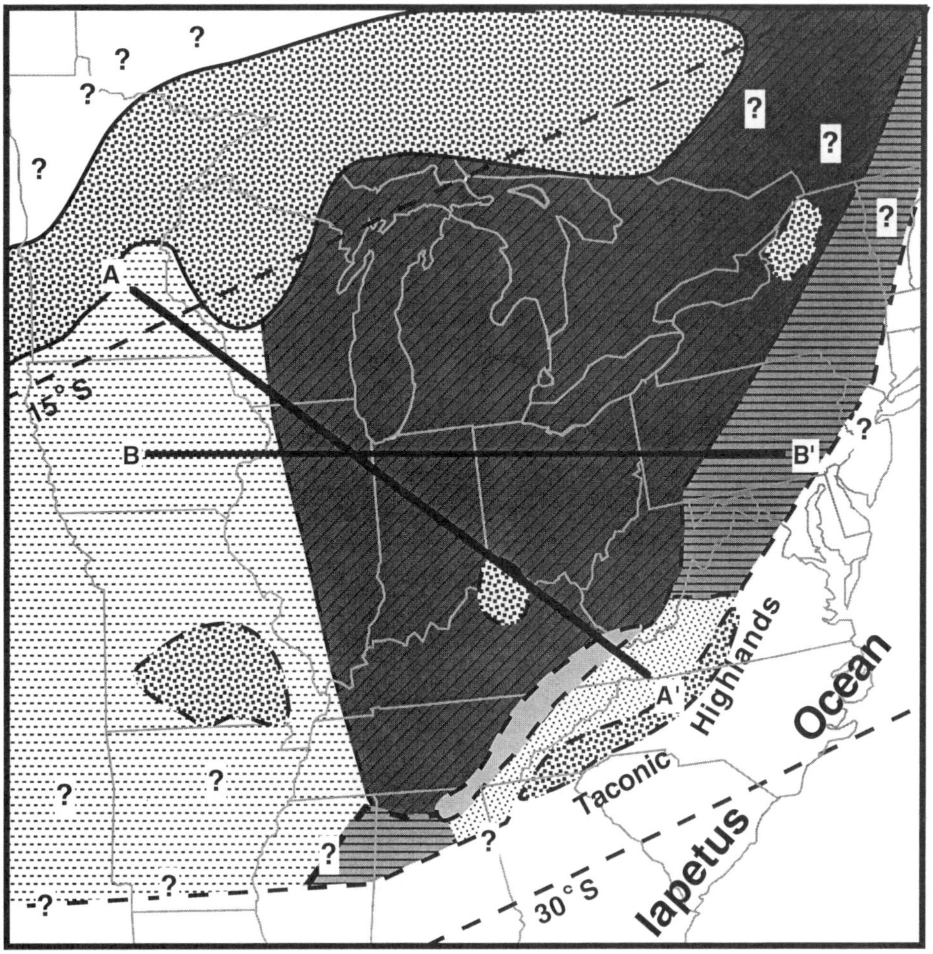

Figure 4. Supra-Millbrig (earliest Chatfieldian) lithofacies distribution, not corrected for tectonic shortening. For explanation of patterns, see Figure 3.

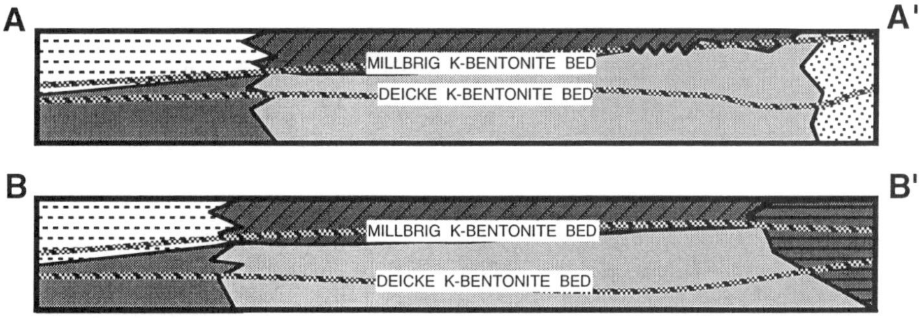

Figure 5. Schematic cross sections, A-A′ and B-B′ from Figures 3 and 4, that illustrate the distribution of lithofacies in the stratigraphic interval around the Deicke and Millbrig K-bentonite beds. For explanation of patterns, see Figure 3.

stone Facies is indicative of a shallow-water, periodically subaerially exposed environment. Abundant mud cracks and algal structures occur in the fenestral mudstone that makes up this facies. The north-central part of the area (Manitoulin Island) of this lithofacies is a rocky shore environment where abundant quartz sand derived from the underlying and adjacent Precambrian Lorraine Quartzite occurs in the mudstone. The Nodular Wackestone/Calcareous Shale Facies is a subtidal slope to "deeper" subtidal deposit. This interpretation is based on the rubbly and wavy nature of the limestone beds. Reworking of material that was deposited on this slope is particularly evident in central Alabama, where the nodular wackestone beds have incorporated biotite grains from the Millbrig K-bentonite Bed. The Sandstone/Siltstone/Claystone Facies represents clastic influx into a shallow basin from a southeastern source area (Read, 1980). These sediments thicken from northwest to southeast, and in general they are finer grained and more calcareous on the western and northern margins of the area covered by this facies. The geometry and lithology of this facies support the interpretation that this is the molasse filling (Van Houten, 1973; Miall, 1981) of the Sevier Basin, the southeasternmost subbasin in the series of basins that make up the Taconic foreland basin region.

The facies distribution in the Upper Mississippi Valley and eastern midcontinent areas changed substantially from pre-Deicke time to post-Millbrig time (Fig. 5). The Wackestone Facies was replaced by a Green Shale/Packstone Facies in the Upper Mississippi Valley. There is no evidence for subaerial exposure in this post-Millbrig facies in the Upper Mississippi Valley, indicating slightly deeper water than in pre-Deicke time. The abundant green shale represents a large supply of very fine grained clastic material, which is interpreted as deposited in a muddy subtidal setting with a patchy distribution of shell-rich "storm" beds preserved as packstone lenses.

The vast shallow-water and intermittently subaerially exposed mud flat that characterized much of the eastern midcontinent in pre-Deicke time was confined to a north-east–trending belt in the southern Appalachians by post-Millbrig time. The algal-laminated fenestral limestone that occupied much of the eastern midcontinent region was replaced by wackestone and packstone in post-Millbrig time. This change also supports the idea of an increase in water depth of the shallow epicontinental sea that covered much of eastern North America (Leslie and Bergström, 1995).

NET ROCK ACCUMULATION RATE

The precise interval of time between deposition of the Deicke and Millbrig K-bentonite beds is difficult to assess within the limits of absolute age dating techniques. Tucker and McKerrow (1995), who used isotope dilution mass spectometry single-grain zircon $^{206}Pb/^{238}U$ measurements, reported an age difference of 1.4 m.y. Kunk and Sutter (1984, table 2, sample 31D = Deicke, sample 30B = Millbrig), using the $^{40}Ar/^{39}Ar$ age spectrum technique, gave an older age for the Millbrig (preferred age 454.4 Ma) than for the Deicke (preferred age 454.6 Ma). However, the Deicke is stratigraphically lower and the older of the two beds. The isotope ages in both studies overlap when their stated errors are considered, which suggests that further studies are needed to get a reliable age estimate for this stratigraphic interval. We believe that it is probably less than 0.5 m.y., but the precise figure is not important for the present discussion.

The occurrence of two K-bentonite beds that are stratigraphically close together and correlatable over a large part of eastern North America provides a unique opportunity to examine regional differences in net rock accumulation rates within a well-defined general time interval. Figure 6 shows the locations of 45 sections that contain both Deicke and Millbrig K-bentonite beds. The stratigraphic thickness between the Deicke and Millbrig K-bentonite beds was determined at each of these sections (Table 1).

Figure 7 illustrates the net rock accumulation between the Deicke and Millbrig K-bentonite beds throughout eastern North America. The distribution of localities is constrained to the north by the difficulty of recognizing the Deicke and Millbrig with confidence from core data in the Illinois and Michigan basins and by the apparent absence of the Millbrig in northern Michigan, southern Ontario, and New York. In these areas, the Millbrig is missing owing to nondeposition, is too thin to be recognized in core or in the field, or was deposited but eroded away. A similar problem with the identification of Deicke and Millbrig in core exists to the west (Kansas, Nebraska), where the stratigraphic interval of interest is not exposed. Also, the distribution of sample sites is constrained to the east (eastern Georgia, South Carolina, North Carolina, eastern Virginia, eastern Maryland) by the absence of latest Turinian and earliest Chatfieldian sedimentary rocks.

Net rock accumulation is variable throughout the study area (Fig. 7). The range in the sections that we sampled is from as little as 0.37 m in the Upper Mississippi Valley to as much as 14.62 m in southwestern Virginia. If one assumes that the stratigraphic interval between these K-bentonite beds corresponds to 1.4 m.y., the net rock accumulation rate for these selected areas was 0.026 cm/1,000 yr and 1.044 cm/1,000 yr, respectively. If the span was 0.5 m.y. (well within the stated errors of Kunk and Sutter, 1984; and Tucker and McKerrow, 1995), the corresponding values are 0.074 cm/1,000 yr and 2.924 cm/1,000 yr, respectively. Net rock accumulation is subject to many postdepositional controls, including compaction, erosion, and dissolution. Therefore, the thicknesses reported in Table 1 and shown in Figure 7 are not direct records of subsidence rates or of sedimentation rates that reflect accommodation space. Rather, they indicate the sum of all controls on deposition, and all processes that have acted upon the sediment, since its deposition.

DISCUSSION

Some interesting correlations are apparent between net rock accumulation (Fig. 7) and lithofacies distribution (Figs. 3 and 4). The area of greatest net rock accumulation is in southwestern Virginia. This area represents the Sandstone/Siltstone/Claystone Facies in both pre-Deicke and post-Millbrig time. The Sandstone/Siltstone/Claystone Facies is the thick molasse that represents the filling of the Sevier Basin. The correlation between the area of greatest net rock accumulation and this clastic facies illustrates increased accommodation space effected by loading of the basin as a result of clastic deposition, along with tectonic loading of the plate margin and voluminous sediment supply to fill it.

There is also a correlation between the Nodular Wackestone/Calcareous Shale Facies and a relative decrease in net rock accumulation. This facies represents the deepest-water environment of the study area, so that deposition was most likely continuous from pre-Deicke to post-Millbrig time. The

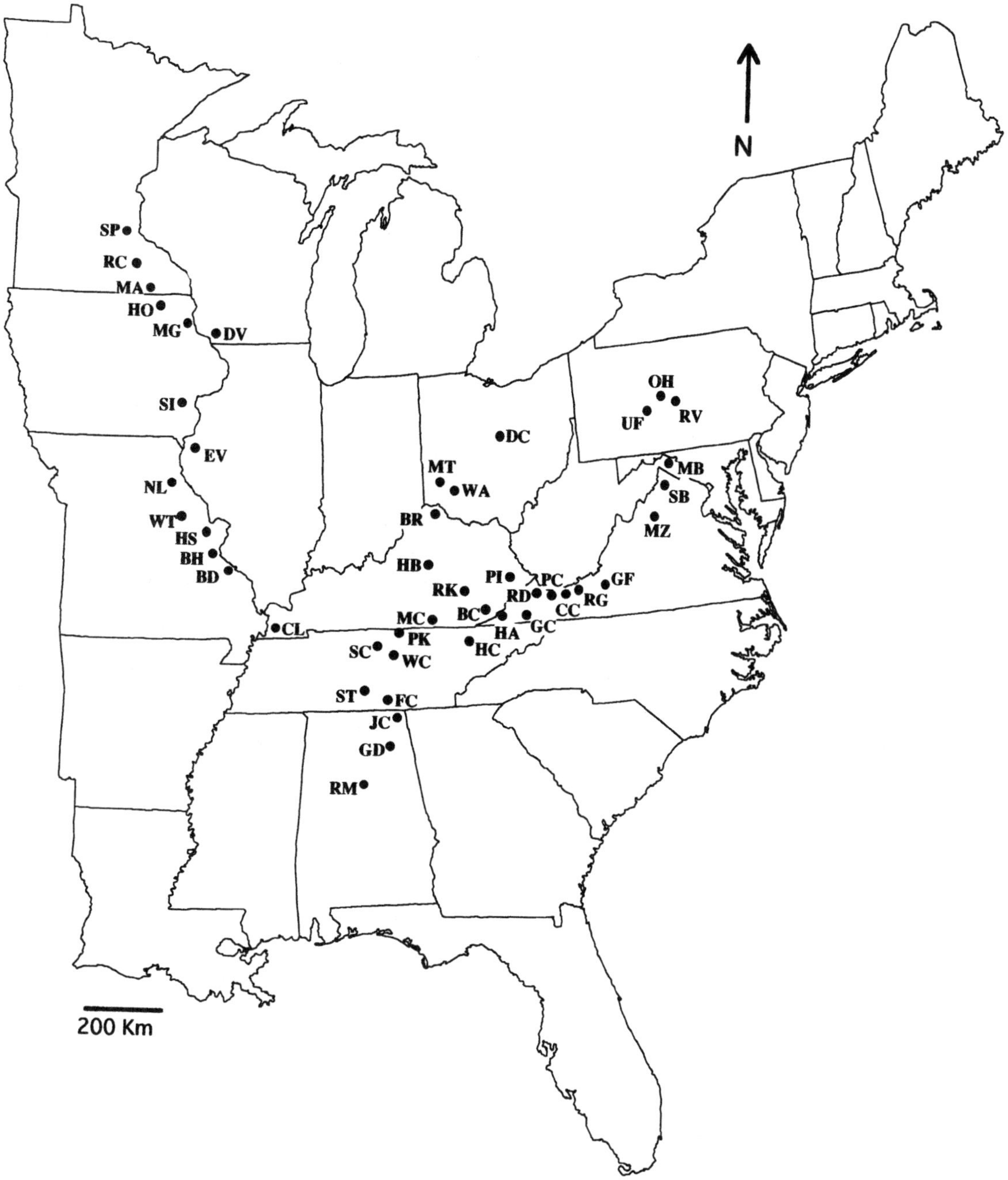

Figure 6. Locations of the 45 sections used to calculate stratigraphic thickness between the Deicke and Millbrig K-bentonite beds. See Table 1 for locality abbreviations.

relatively low net rock accumulation rate in this facies is attributed to two factors: low sediment supply and postdepositional compaction.

Most of the sediment that filled the Sevier Basin was derived from the tectonic highlands south and southeast of the basin (Read, 1980). The bulk of this sediment was deposited in the Sevier Basin during pre-Deicke time. It was not until just before the deposition of the Millbrig K-bentonite Bed that deposition of Martinsburg flysch began in northern Virginia, and the facies changed to a sequence dominated by calcareous shale and sparsely fossiliferous mudstone. The relatively low net rock accumulation rates may represent slow deposition in the deeper-water settings as sediment derived from the tectonic highlands was being trapped during the filling of the Sevier Basin.

The tristate area of Ohio-Indiana-Kentucky shows a gradual lowering of net rock accumulation in the Mudstone Facies. In this area the Millbrig is commonly absent, in most cases probably because it was eroded after deposition. The Mudstone Facies was laid down in very shallow water and periodically was exposed subaerially. Little sediment could accumulate during

TABLE 1. STRATIGRAPHIC THICKNESS BETWEEN THE DEICKE AND MILLBRIG K-BENTONITE BEDS

State	Abbreviation	Locality	Meters Between Deicke and Millbrig	References*
AL	RM	Birmingham	2.60	1
AL	GD	Gadsen	3.97	†
AL	JC	Jackson Co.	7.16	2
IA	HO	Hanover	0.53	3
IA	SI	Lousia Co.	1.18	†
IA	MG	McGregor	0.35	†
IL	EV	Elderville	2.00	3
KY	BC	Bell Co.	12.31	2
KY	BR	Black River Mine	2.71	4
KY	CL	Carlisle Co.	6.69	2
KY	HB	High Bridge	7.51	4
KY	MC	McCreary Co.	10.64	2
KY	PI	Pike Co.	9.53	2
KY	RK	Rockcastle Co.	9.14	2
MN	MA	Mabel	1.50	3
MN	RC	Rochester	2.56	3
MN	SP	St. Paul	2.17	3
MO	BH	Barnhart	2.38	3
MO	BD	Bloomsdale	1.75	3
MO	HS	House Springs	1.57	3
MO	NL	New London	1.90	3
MO	WT	Warrenton	1.80	3
OH	DC	Delaware Co.	8.23	5
OH	MT	Middletown Core	6.10	6
OH	WA	Warren Co. Core	4.27	7
PA	OH	Oak Hall	7.50	8
PA	RV	Reedsville	3.80	8
PA	UF	Tyrone	6.35	†
TN	FC	Franklin Co.	6.88	2
TN	HC	Hines Creek Quarry	8.07	4
TN	PK	Picket Co.	10.38	2
TN	ST	Singleton	5.65	†
TN	SC	South Carthage	5.60	†
TN	WC	White Co.	9.68	2
VA	CC	Cove Creek	9.42	4
VA	GC	Gate City	8.03	4
VA	GF	Goodwins Ferry	14.62	4
VA	HA	Hagan	13.64	†
VA	MZ	Mauzy	2.75	†
VA	RG	Rocky Gap	12.50	4
VA	RD	Rosedale	9.70	4
VA	SB	Strasburg	3.30	†
VA	PC	Tazewell	8.64	†
WI	DV	Dickeyville	0.37	3
WV`	MB	Martinsburg	2.61	†

*1 = Haynes, 1994; 2 = Huff and Kolata, 1990; 3 = Kolata et al., 1986; 4 = Haynes, 1992; 5 = Stith, 1986; 6 = Leslie, 1995; 7 = Shrake et al., 1990; 8 = McVey, 1993.
†See Appendix for locality information.

times of subaerial exposure, and some of the sediment deposited between Deicke and Millbrig time was likely eroded.

The lowest rate of net rock accumulation is displayed in the Upper Mississippi Valley area. Commonly less than1 m of wackestone is present between Deicke K-bentonite Bed and the green shale and packstone beds of the Spechts Ferry Formation. The Upper Mississippi Valley area is the farthest inboard region from the continental margin of Laurentia in this study, and, in general, areas closer to the center of a continent have lower subsidence rates than marginal basins (Sloss, 1988, figs. 1–10). This, together with extremely slow sedimentation rate and postdepositional compaction of the clay, is responsible for the small net rock accumulation.

CONCLUSIONS

Regional tracing of K-bentonite beds that are relatively close in stratigraphic position provides a unique framework in which to examine lithofacies distributions, changes of lithofacies distributions through time, and net rock accumulation rates. The Deicke and Millbrig K-bentonite beds provide such a framework in the latest Turinian and earliest Chatfieldian strata of eastern North America.

Four distinct lithofacies are present in pre-Deicke rocks in eastern North America. They are the Wackestone Facies, Mudstone Facies, Nodular Wackestone/Calcareous Shale Facies, and Sandstone/Siltstone/Claystone Facies. By post-Millbrig time the Wackestone Facies was replaced by the Green Shale/Packstone Facies and the Mudstone Facies was replaced by the Wackestone/Packstone Facies. This lithofacies change was associated with an increase in water depth in the epicontinental sea that covered much of eastern North America at this time. The areas of the other lithofacies changed little between pre-Deicke and post-Millbrig time.

Net rock accumulation between the Deicke and Millbrig K-bentonite beds in the sections that we sampled varies from as little as 0.37 m in the Upper Mississippi Valley to as much as 14.62 m in southwestern Virginia. The area of greatest net rock accumulation corresponds to the Sandstone/Siltstone/Claystone Facies and represents molasse-filling of the Sevier Basin. Net rock accumulation in the deeper-water setting of the Nodular Wackestone/Calcareous Shale Facies is relatively low because of low sedimentation rate caused in part by trapping of clastic sediments in the Sevier Basin. Low net rock accumulation in central Kentucky is probably due to subaerial exposure of the sequence and possible erosion of sediment that accumulated after deposition of the Deicke K-bentonite Bed and before deposition of the Millbrig K-bentonite Bed. The smallest accumulation in the Upper Mississippi Valley is due to compaction of the green shale that is the dominant lithology above the Deicke K-bentonite Bed and below the Millbrig K-bentonite Bed and to the generally lower subsidence rates in this more cratonic setting.

The present study provides an illustration of how K-bentonite beds can be used as time-planes for a variety of regional studies of depositional patterns. As far as we are aware, this is the first study of this type in the Lower Paleozoic. Because such ash layers are both relatively common and widespread geo-

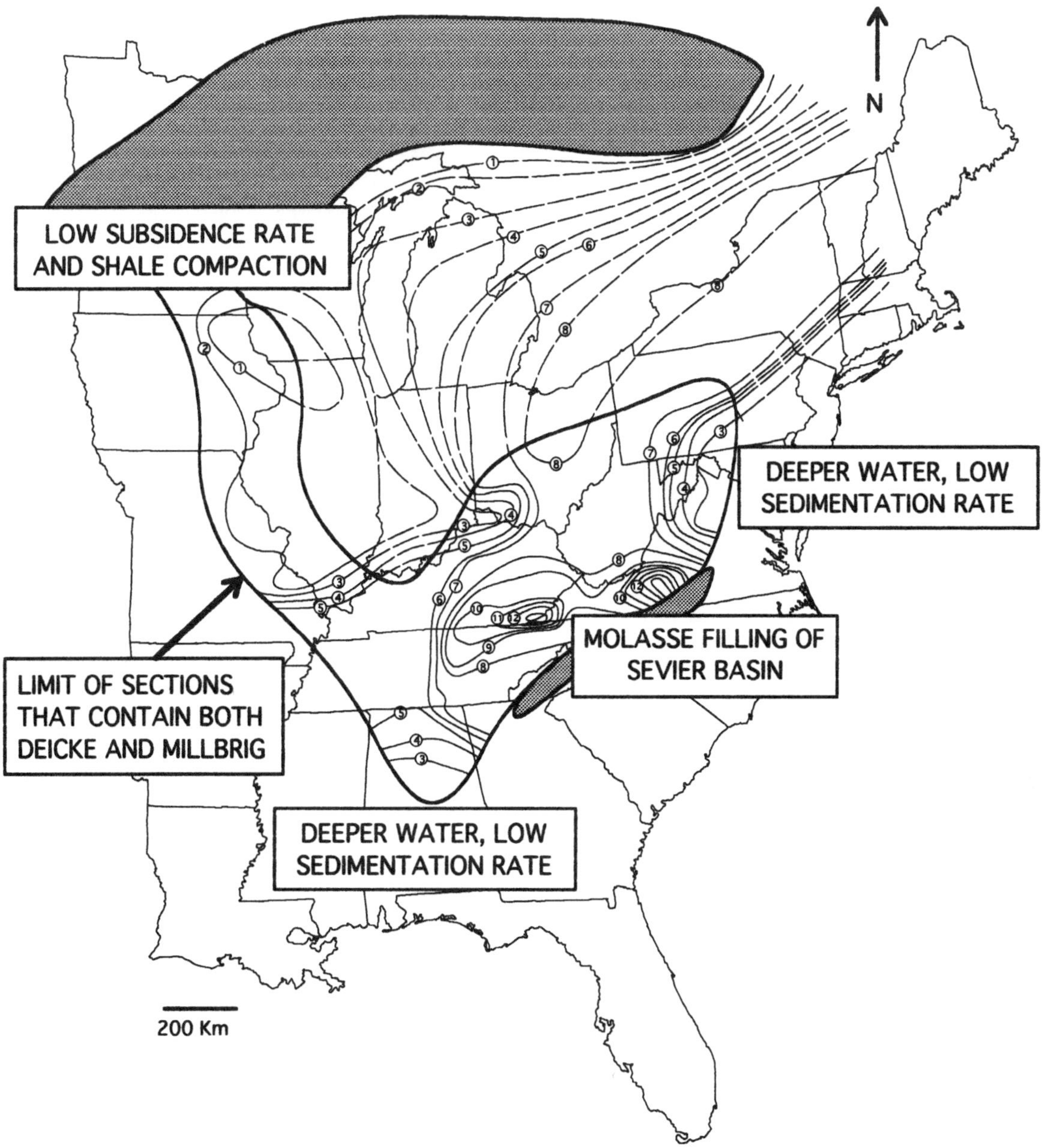

Figure 7. Net rock accumulation between the Deicke and Millbrig K-bentonite beds. Contour interval is 1 m.

graphically in several portions of the North American stratigraphic column, great potential exists for similar investigations in, for instance, the Devonian and Cretaceous. Very critical for such studies is that individual K-bentonite beds can be traced with confidence regionally, and for a long time this has been a serious problem. However, the recent development of trace element fingerprinting methods, particularly those based on fluid inclusions (see, for instance Delano et al., 1990), offer great potential to overcome this problem and make possible much wider use of K-bentontites for the kind of research carried out in the present study.

ACKNOWLEDGMENTS

This work was completed as part of a dissertation at The Ohio State University by the senior author. We thank P. MacKenzie for field assistance. D. R. Kolata provided helpful suggestions and improved an early version of this manuscript. R. L. Ethington and S. C. Finney provided insightful reviews. This work was supported by Shell Oil Company through The Ohio State University Department of Geological Sciences Shell Doctoral Fellowship and by grants from The Friends of Orton Hall, Sigma Xi, and the Geological Society of America (to SAL) and the National Science Foundation (EAR-9205981, to SMB).

APPENDIX: LOCALITY INFORMATION FOR SECTIONS MEASURED IN THIS STUDY

McGregor, Iowa. Near localities 23 and 24 of Kolata et al. (1986). New road cut on the east side of Highway 18. Section begins 0.16 km north of mile marker 302, just north of the Burke Grain Co.

Southeastern Iowa cores. Composite section based on Hutchinson No. 1 and A. Jordan No. 1 cores. Hutchinson No. 1 drilled by the Natural Gas Pipeline Corporation of America (NGCPA) in Lousia County, Iowa, NW1/4, NE1/4, NW1/4, Sec. 21, T74N, R4W. A. Jordan No. 1 drilled by NGCPA in Louisa County, Iowa, SSW1/4, NW1/4, SW1/4, Sec. 32, T75N, R4W.

Gadsden, Alabama. Road cut through Big Ridge approximately 16 km north of Gadsden, Etowah County, Alabama, 0.15 km south of mile marker 194 on Interstate 59 (NE1/2, Sec. 27, T10S, R6E, Kenner 7.5-minute Quadrangle). Section referred to as Stop 2 in Wilson et al. (1971), I-59 Road cut in Kunk and Sutter (1984), and as the Gadsden Road Cut Section in Hall (1986). Section containing the Deicke and Millbrig exposed on the east side of the northbound lane.

Singleton, Tennessee. Originally described by Wilson (1949, p. 75). Both the Deicke and Millbrig exposed on the north side of Highway 130. Exposure of Carters Formation on a natural hillside outcrop along side of the road with the Hermitage Formation capping the hill immediately to the east of Woods Bluff on the Shelbyville 7.5-minute Quadrangle.

South Carthage, Tennessee. Three exposures along Highway 53 between Interstate 40 and Highway 70N provide a continuous section from the lower Carters Formation through the lower Hermitage Formation. Lower part of the section exposed on the east side of Highway 53 2.5 km north of Interstate 40 and continues up-section to top of the lower Carters Formation (Deicke K-bentonite) along Bently Harris Way. Interval from top of the lower Carters through the lowest Hermitage Formation also exposed along the west side of Highway 53, 4.3 km south of Highway 70N at South Carthage, Tennessee (Locality 8 of Huff and Kolata, 1990). Top part of the upper Carters Formation and the lower part of Hermitage Formation also well exposed in the cut at the back of a parking lot on the west side of Highway 53 immediately north of Interstate 40 Gordonsville exit (exit 258).

Plum Creek, Tazewell, Virginia. Section measured by Rosenkrans (1936), described in detail as Geologic section 65 in Hergenroder (1966), and illustrated by Haynes (1994). Exposure about 4.8 km southwest of Tazewell, Tazewell South 7.5-minute Quadrangle.

Mauzy, Virginia. Section graphically illustrated in Rosenkrans (1933), on Route 608 approximately 0.5 km southeast of Route 11 at elevation marker 1086, Tenth Legion 7.5-minute Quadrangle.

Strasburg, Virginia. Type section of the Oranda Formation described by Cooper and Cooper (1946) and Rader and Read (1989). Section exposed in a road cut on northeast side of Highway 55, 0.64 km west of Strasburg.

Martinsburg, West Virginia. Section measured at the Capitol Cement Quarry, a large cement quarry south of Route 9 and east of Route 11, Martinsburg 7.5-minute Quadrangle.

Union Furnace, Pennsylvania. Road cut exposure on both sides of Pennsylvania State Highway 453, 8.8 km southeast of Tyrone (Thompson, 1963).

REFERENCES CITED

Cooper, B. N., and Cooper, G. A., 1946, Lower Middle Ordovician stratigraphy of the Shenandoah Valley, Virginia: Geological Society of America Bulletin, v. 57, p. 35–113.Delano, J., Schirnick, C., Bock, B., Kidd, W. S. F., Heizler, M. T., Putman, G. W., De Long, S. E., and Ohr, M., 1990, Petrology and geochemistry of Ordovician K-bentonites in New York State: Constraints on the nature of a volcanic arc: Journal of Geology, v. 98, p. 157–170.

Dunham, J. B., 1962, Classification of carbonate rocks according to depositional texture, *in* Ham, W. E., ed., Classification of carbonate rocks: American Association of Petroleum Geologists Memoir 1, p. 108–121.

Hall, J. C., 1986, Conodonts and conodont biostratigraphy of the Middle Ordovician in the western overthrust region and Sequatchie Valley of the southern Appalachians [Ph.D. thesis]: Columbus, The Ohio State University, 346 p.

Haynes, J. T., 1992, Reinterpretation of Rocklandian (Upper Ordovician) K-bentonite stratigraphy in southwest Virginia, southeast West Virginia, and northeast Tennessee: With a discussion of the conglomeratic sandstones in the Bays and Moccasin formations: Virginia Division of Mineral Resources Publication 126, 58 p.

Haynes, J. T., 1994, The Ordovician Deicke and Millbrig K-bentonite beds of the Cincinnati Arch and southern Valley and Ridge Province: Geological Society of America Special Paper 290, 80 p.

Hergenroder, J. D., 1966, The Bays Formation (Middle Ordovician) and related rocks of the southern Appalachians [Ph.D. thesis]: Blacksburg, Virginia Polytechnic Institute and State University, 323 p.

Huff, W. D., 1983, Correlation of Middle Ordovician K-bentonites based on chemical fingerprinting: Journal of Geology, v. 91, p. 657–669.

Huff, W. D., and Kolata, D. R., 1990, Correlation of the Ordovician Deicke and Millbrig K-bentonites between the Mississippi Valley and the southern Appalachians: American Association of Petroleum Geologists Bulletin, v. 74, p. 1736–1747.

Huff, W. D., Bergström, S. M., and Kolata, D.R., 1992, Gigantic Ordovician volcanic ash fall in North America and Europe: Biological, tectonomagmatic, and event-stratigraphic significance: Geology, v. 20, p. 875–878.

Johnson, J. G., 1993, Gigantic Ordovician volcanic ash fall in North America and Europe: Biological, tectomagmatic, and event-stratigraphic significance): Comment: Geology, v. 21, p. 383.

Keith, B. D., 1988, Regional facies of the Upper Ordovician Series of eastern North America, in Keith, B. D., ed., The Trenton Group (Upper Ordovician Series) of eastern North America: American Association of Petroleum Geologists Studies in Geology, v. 29, p. 1–16.

Kolata, D. R., Frost, J. K., and Huff, W. D., 1986, K-bentonites of the Ordovician Decorah Subgroup, upper Mississippi Valley: Correlation by chemical fingerprinting: Illinois Department of Energy and Natural Resources, State Geological Survey Division Circular 537, 30 p.

Kolata, D. R., Huff, W. D., and Bergström, S. M., 1996, Ordovician K-bentonites of eastern North America: Geological Society of America Special Paper 313, 84 p.

Kunk, M. J., and Sutter, J. F., 1984, $^{40}Ar/^{39}Ar$ age spectrum dating of biotite from Middle Ordovician bentonites, eastern North America, *in* Bruton, D. L. ed., Aspects of the Ordovician System: Palaeontological Contributions from the University of Oslo 295, p. 11–22.

Leslie, S. A., 1995, Upper Middle Ordovician conodont biofacies distribution patterns in eastern North America and northwestern Europe: Evaluations using the Deicke, Millbrig and Kinnekulle K-bentonite beds as time planes [Ph.D. thesis]: Columbus, The Ohio State University, 450 p.

Leslie, S. A., and Bergström, S. M., 1995. Revision of the North American late Middle Ordovician standard stage classification and timing of the Trenton transgression based on K-bentonite bed correlation, *in* Cooper, J. D. et al., eds., Ordovician odyssey: Short papers for the Seventh International Symposium on the Ordovician System, Pacific Section of the Society for Sedimentary Geologists (SEPM), Book 77, p. 49–54.

McVey, D., 1993, Timing of the Blount and Martinsburg foreland basin development during the Taconic Orogeny based on the Deicke and Millbrig K-bentonite marker horizons [M.S. thesis]: Cincinnati, Ohio, University of Cincinnati, 133 p.

Miall, A. D., ed., 1981, Sedimentation and tectonics in alluvial basins: The Geological Association of Canada Special Paper 23, 272 p.

Rader, E. K., and Read, J. F., 1989, Early Paleozoic continental shelf to basin transition, northern Virginia, *in* Sedimentation and stratigraphy of carbonate rock sequences, Volume 1, Field Trip Guidebook T221: 29th International Geological Congress, American Geophysical Union, Washington, D.C., p. T221:1–T221:9.

Read, J. F., 1980, Carbonate ramp-to-basin transition and foreland basin evolution, Middle Ordovician, Virginia Appalachians: American Association of Petroleum Geologists Bulletin, v. 64, p. 1575–1612.

Rosenkrans, R. R., 1933, Bentonite in northern Virginia: Journal of the Washington Academy of Sciences, v. 23, p. 413–419.

Rosenkrans, R. R., 1936, Stratigraphy of Ordovician bentonite beds in southwestern Virginia: Virginia Geological Survey Bulletin, v. 46-I, p. 85–111.

Samson, S. D., Patchett, P. J, Roddick, J. C., and Parrish, R. R., 1989, Origin and tectonic setting of Ordovician bentonites in North America: Isotopic and age constraints: Geological Society of America Bulletin, v. 101, p. 1175–1181.

Shrake, D. L., Wolfe, P. J., Richard, B. H., Swinford, E. M., Wickstrom, L. H., Potter, P. E., and Sitler, G. W., 1990, Lithologic and geophysical description of a continuously cored hole in Warren County, Ohio, including description of the Middle Run Formation (Precambrian?) and a seismic profile across core site: Ohio Division of Geological Survey Information Circular 56, 11 p.

Sloss, L. L., 1988, Tectonic evolution of the craton in Phanerozoic time, in Sloss, L. L., ed., Sedimentary cover—North American craton: U.S.: Boulder, Colorado, Geological Society of America, Geology of North America, v. D-2, p. 25–51.

Stith, D. A., 1986, Supplemental core investigations for high-calcium limestones in western Ohio and discussion of natural gas and stratigraphic relationships in the Middle to Upper Ordovician rocks of southwestern Ohio: Ohio Division of Geological Survey Report of Investigations 132, 17 p.

Sweet, W. C., 1979, Conodonts and conodont biostratigraphy of post-Tyrone Ordovician rocks of the Cincinnati Region: U.S. Geological Survey Professional Paper 1066G, p. 1–25.

Sweet, W. C., 1984, Graphic correlation of upper Middle and Upper Ordovician rocks, North American midcontinent Province, USA, in Bruton, D. L, ed., Aspects of the Ordovician System: Palaeontological Contributions from the University of Oslo, p. 23–35.

Sweet, W. C., 1995, Graphic assembly of a conodont-based composite standard for the Ordovician System of North America, in Mann, K. O., and Lane, H. R., eds., Graphic correlation: SEPM (Society for Sedimentary Geology) Special Publication 53, p. 139–150.

Templeton, J. L., and Willman, H. B, 1963, Champlainian Series (Middle Ordovician) in Illinois: Illinois State Geological Survey Bulletin 898, 260 p.

Thompson, R. R., 1963, Lithostratigraphy of the Middle Ordovician Salona and Coburn Formations in central Pennsylvania: Pennsylvania Geological Survey Bulletin, Fourth Series, G 38, 154 p.

Tucker, R. D., and McKerrow, W.S., 1995, Early Paleozoic chronology: A review in light of new U-Pb zircon ages from Newfoundland and Britain: Canadian Journal of Earth Sciences, v. 32, p. 368–379.

Van Houten, F. B, 1973, Meaning of molasse: Geological Society of America Bulletin, v. 84, p. 1973–1976.

Wilson, A. O., McKinney, F. K., and Neathery, T. L., 1971, Stop 2 Interstate 59 cut, in Drahovsal, J. A., and Neathery, T. L., eds., The Middle and Upper Ordovician of the Alabama Appalachians: Alabama Geological Society Guidebook 9, p. 179–183.

Wilson, C. W., Jr., 1949, Pre-Chattanooga stratigraphy in central Tennessee: Tennessee Division of Geology Bulletin 56, p. 1–407.

Manuscript Accepted by the Society February 27, 1997

Printed in U.S.A.

Geological Society of America
Special Paper 321
1997

Comparison of Lower Silurian shores and shelves in North America and Siberia

Markes E. Johnson
Department of Geosciences, Williams College, Williamstown, Massachusetts 01267
Yuri I. Tesakov
United Institute for Geology, Geophysics, and Mineralogy, Siberian Branch of the Russian Academy of Sciences, 3 University Avenue, Novosibirsk 630090, Russia
Nikolai N. Predtetchensky
All Russian Geological Research Institute, Sredny Prospect 74, St. Petersburg 199026, Russia
B. Gudveig Baarli
Department of Geosciences, Williams College, Williamstown, Massachusetts 01267

ABSTRACT

Siberia and Laurentia were independent Silurian continents that accumulated extensive carbonate deposits on broad epicontinental shelves attached to limited land areas. Siberian Anabar was a domed lowland, but Taconia in eastern North America was an elongate highland thrust up as a result of oceanic foreshortening in the Proto-Atlantic. Both lands shed quartzitic detritus into coastal environments, permitting accurate map reconstructions of their shorelines. The greater topographic relief of Taconia sheltered the inboard carbonate platform from the most severe storms, but Anabar stood in the direct path of hurricanes that swept across the flanking carbonate shelf and stimulated deposition of cyclic storm beds.

During the middle Aeronian and early Telychian, Siberia was imprinted by two global highstands in sea level that can be well correlated to Laurentia and other paleocontinents, such as Avalonia and Baltica, on the basis of stratigraphic discontinuities in beds with rich concentrations of the brachiopod *Pentamerus oblongus*. An earlier highstand recorded by widespread distribution of the brachiopod *Virgiana decussata* in North America is masked by a prolonged Rhuddanian deepening event on Siberia. Thicker Silurian deposits in Siberia as opposed to North America and especially Siberia's extensive Rhuddanian shales and interbedded fine limestones, attributed to a deep-shelf setting, may be explained by comparatively low hypsography.

INTRODUCTION

The intent of this collaborative project is to compare aspects of shore and shelf sequences preserved in the Lower Silurian strata of North America and Siberia. During the early Silurian, North America (Laurentia) and Siberia were distinct continents (Ziegler et al., 1977; Scotese and McKerrow, 1990) with vast carbonate-dominated shelves bordered by comparatively limited land areas (Kaljo et al., 1991). One of the principal lands of Laurentia was Taconia, which extended as a narrow highland from Quebec to central Alabama (Ziegler et al., 1977). The Anabar Dome in north-central Siberia also formed a discrete Silurian land mass (Tesakov, 1981). How are the former shorelines of these lands marked in the stratigraphic record, and how did they change through Early Silurian time? In what ways are the carbonate cycles that accumulated on

Johnson, M. E., Tesakov, Y. I., Predtetchensky, N. N., and Baarli, B. G., 1997, Comparison of Lower Silurian shores and shelves in North America and Siberia, *in* Klapper, G., Murphy, M. A., and Talent, J. A., eds., Paleozoic Sequence Stratigraphy, Biostratigraphy, and Biogeography: Studies in Honor of J. Granville ("Jess") Johnson: Boulder, Colorado, Geological Society of America Special Paper 321.

adjacent shelves differentiated stratigraphically? What is their possible interregional and intercontinental correlation? The answers to these questions underscore similarities and differences in the bathymetric configuration of marine shelves occupying independent paleocontinents. Evaluation of the roles played by eustasy and regional tectonics in regulating relative sea-level change is one application for these comparative data (M. E. Johnson et al., 1991; M. E. Johnson and McKerrow, 1991). Differences in shelf development may also relate to more subtle variations among Silurian continents in terms of their hypsography, geographic position, and climate.

It is apparent from standard references by Nikiforova and Andreeva (1961) and Berry and Boucot (1970) that the richly fossiliferous Lower Silurian strata of Siberia and North America are suitable subjects for comparison. Much overlap in shelly faunas occurs in the carbonate and mixed carbonate-clastic rocks of these former shelves. Experience with tabulate coral faunas (Sokolov and Tesakov, 1963; Tesakov, 1978) and brachiopod faunas (M. E. Johnson and Campbell, 1980; M. E. Johnson, 1981, 1989) was directly available to the participants of this study, as was a collective knowledge of regional stratigraphies. Silurian outcrop distribution and facies relationships throughout 11 districts of the Siberian platform were treated by Tesakov et al. (1979) prior to formulation of this project. Likewise, 22 regions belonging to Silurian Laurentia were summarized by M. E. Johnson (1987). This background allowed for the planning and realization of joint field excursions to some of the most accessible and instructive Silurian sections in North America and Siberia over the project's two-year duration.

BACKGROUND AND PLAN OF JOINT STUDIES

In July 1993 project goals were defined and an agreement on a field program was concluded as a result of consultations in St. Petersburg and Novosibirsk. The project was expected to entail analysis of detailed stratigraphic schemes for the Silurian of the North American and Siberian platforms, their correlation and comparison with the standard stratigraphic scale, the determination of global events, production of maps based on these events, and a comparative analysis of the history of platform sedimentation.

Tesakov and Predtetchensky visited the United States and Canada during the summer of 1994. Joint fieldwork followed the trail of the Niagaran escarpment, focusing on the Lower Silurian (Rhudannian, Aeronian, and Telychian stages) of New York State, Ontario, and Michigan (Fig. 1). Localities reviewed in the Upper Michigan Peninsula, Manitoulin Island, and the Bruce Peninsula on the margin of the Michigan Basin correspond to those previously detailed by M. E. Johnson and Campbell (1980), M. E. Johnson (1981), and Colville and M. E. Johnson (1982). Sites visited in New York State conform to the transect crossing the Appalachian Trough along the Mohawk River valley to Niagara Falls, including sections interpreted by Walker (1975), Rickard (1975), Brett et al. (1990), Goodman and Brett (1994), and LoDuca and Brett (1994). Detailed stratigraphic logging was undertaken at a fresh road cut in Frankfort Gorge near Utica, New York, providing previously unpublished data for this report.

Knowledge of Arctic Siberia's Silurian stratigraphy accrued as the result of systematic exploration of river systems begun in 1873 by A. L. Chekanovskiy (Barr, 1991). Sections exposed on the Maymecha and Gorbiachin rivers, as well as cores from bore holes in the Norilsk Region (Fig. 2), were demonstrated to M. E. Johnson during his visit to Siberia in the summer of 1995. The Gorbiachin River sections were first explored by Yu. I. Tesakov in 1959 (Sokolov and Tesakov, 1963), and a comprehensive redescription of the sections in 1973 provided a detailed record through the Ruddanian, Aeronian, and part of the Telychian stages (Tesakov et al., 1980). The Maymecha River was explored and its Aeronian sections described by the same team in 1986, although the data remained unpublished until now.

Generalizations herein regarding the comparison of Silurian shorelines were drawn from the Frankfort Gorge and South Moyer Creek sections in New York State and Maymecha River sections in Arctic Siberia, representing margins of Taconia and Anabar on their respective paleocontinents. Conclusions relevant to associated carbonate shelves were derived from sections in the northern Michigan Basin and the northwestern Siberian platform. Further interpretations were reached through construction of onshore-offshore profiles and their comparison.

METHODS

Lithology, sedimentary structures, and fossil content must be studied and recorded on a bed-by-bed basis through stratigraphic sections from one locality to another in order to adequately test for cyclicity of an interregional or ultimately an intercontinental nature. Despite no communication until late in their independent research programs on ancient carbonate platforms, the organizers of this project quickly established a working relationship based on the primacy of this formula. Stratigraphic logs for all sections described in this report follow a format modified from the Shell Oil Company standard legend for rock types, texture, sedimentary structures, and fossils (Fig. 3). Interpretations were enhanced through use of polished rock samples and petrographic thin sections.

Descriptive stratigraphy leads to the successful integration of interregional and intercontinental data only when members of the research team share similar interpretations of incremental change in cyclic strata. Critical to the development of our project was a shared approach to the understanding of shelf bathymetry and the relative positions of normal (fair-weather) wave base and storm base with respect to interpretive models. Despite assignment of different numerical codes to keep track of shelf bathymetry, it was found that approaches in Russian and the United States to the study of Silurian platforms are compatible. Figure 4 illustrates the basic facies model elaborated by Tesakov et al. (1979, 1980, 1986), using a 10-point subdivision of bathy-

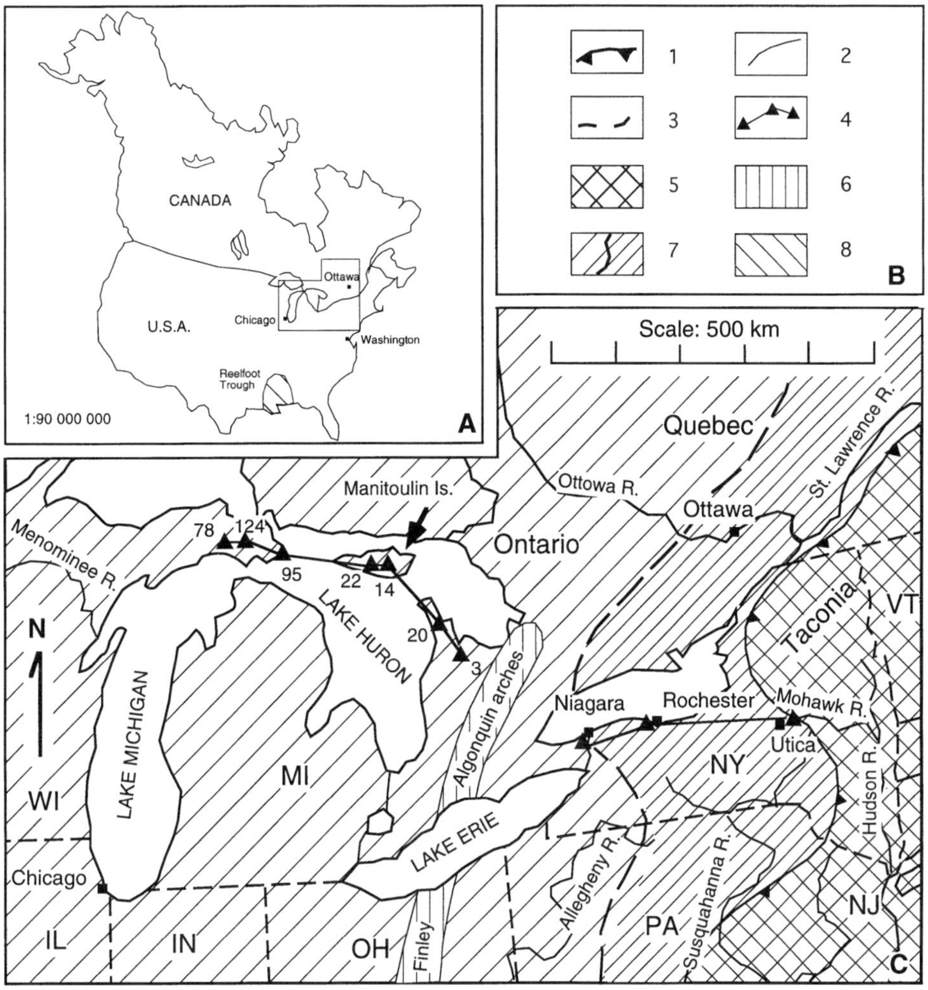

Figure 1. Lower Silurian facies of North America. A, Map of Canada and the United States with inset showing part of the continent covered. B, Key to map symbols. (1) basin margin during maximum Aeronian transgression, (2) facies boundaries, (3) basin boundary during the early Rhuddanian, (4) line of stratigraphic profile with triangles indicating study localities, (5) Silurian land, (6) lagoon or shoal, (7) shallow shelf, (8) deep shelf. C, Combined Rhuddanian and Aeronian paleogeographic map of the Great Lakes Region and adjacent eastern areas.

metric environments. This figure also serves as a key to the lithological and paleontological symbols used to construct our onshore-offshore panels.

Appended to the 10-point subdivision is the scheme of benthic assemblage zones 0–6, as originated by Boucot (1975, p. 49) and widely used by Western geologists. Based on this correlation, BA 1 corresponds to the lagoon, back-shoal, and mid-shoal environments; BA 2 is the fore-shoal to upper shallow shelf; BA 3 and BA 4 relate to middle and lower shallow shelf positions; and BA 5 and BA 6 equate with the upper and lower deep shelf, respectively. The overlap of these numerical codes was tested by comparing the occurrence of key brachiopod genera as routinely assigned to different environments on stratigraphic logs compiled by geologists in Siberia, Europe, and North America (Table 1). This scheme takes into consideration the overlapping membership of Silurian brachiopods in different communities, as originally defined by Ziegler et al. (1968). *Eocoelia*, as a dominant element in the 50% range, occurs in BA 2 = Nos. 4 and 5 (upper shallow shelf to fore-shoal). As a minor element in the range of 10% or less, however, it may also occur in BA 3–4 = Nos. 4 and 3 (lower to middle shallow shelf).

The bathymetric model (Fig. 4) affixes normal (fair-weather) wave base to the boundary between the upper and middle shallow shelf (BA 2/BA 3 = Nos. 4/5) and assigns maximum storm base to a position within the lower shallow shelf (BA 4 = No. 3). This arrangement accommodates evidence on the derivation of shelly tempestites (M. E. Johnson, 1989), in which *Pentamerus oblongus* is interpreted to have lived on the seafloor across a range of water depths affected by storm turbulence that often scattered and fragmented shells at the shallower end but only suffocated populations in

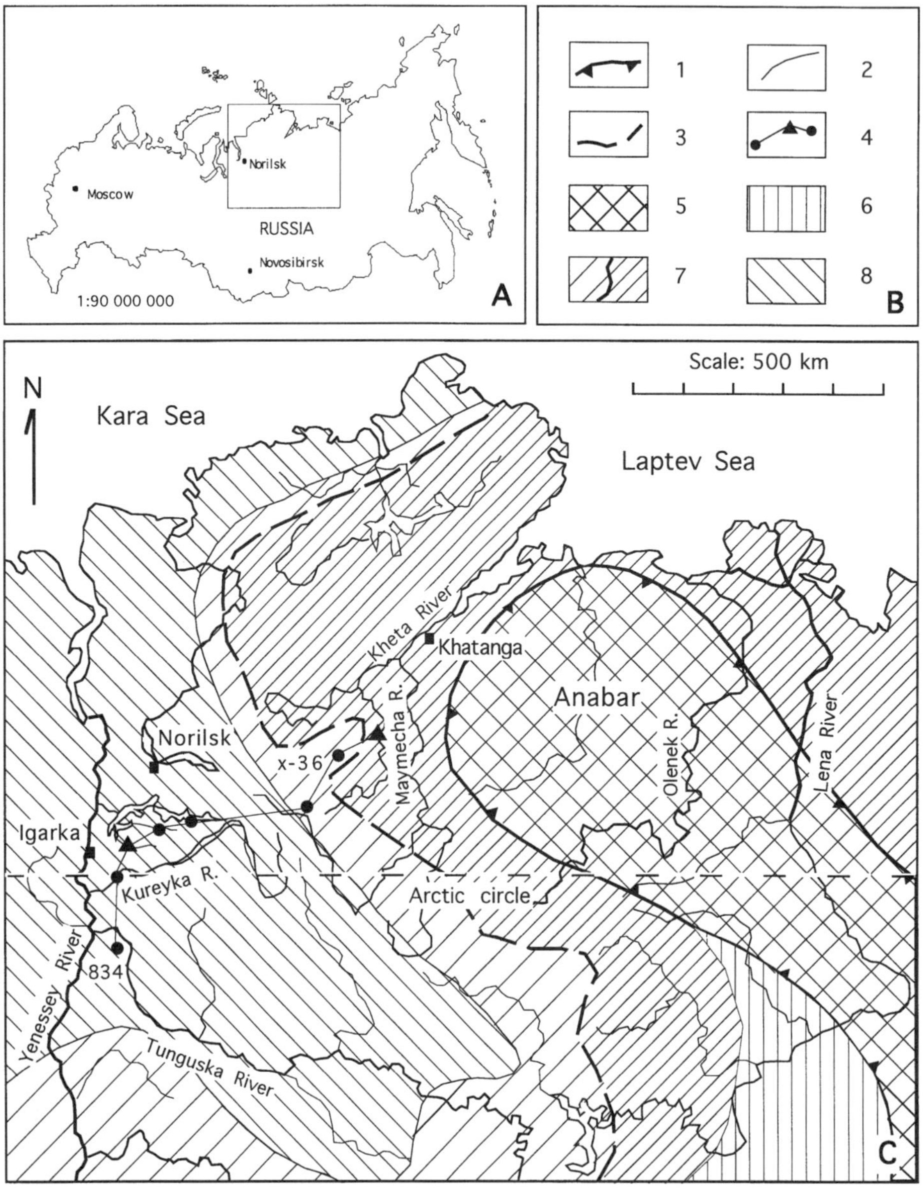

Figure 2. Lower Silurian facies of Siberia. A, Map of Russia with inset showing the part of the country covered. B, Key to map symbols (see descriptions with Fig. 1). C, Combined Rhuddanian and Aeronian paleogeographic map of northern Siberia.

growth position at the deeper end as a consequence of sediment fallout in the absence of physical agitation.

Another component important to a research program of this scope is reference to a common chronostratigraphic scale. Global stratotype sections and points (GSSPs) for the Silurian series and stages defined by authority of the Subcommission on Silurian Stratigraphy (under the International Commission on Stratigraphy and International Union of Geological Sciences) are summarized by Holland and Bassett (1989). These boundaries, particularly within the Llandovery Series, provided us with at least a starting point for intercontinental time-rock correlations.

SILURIAN SHORE DEPOSITS

Frankfort Gorge and South Moyer Creek (Appalachian Trough)

Over 18 m of Lower Silurian strata are well exposed in a road cut (Fig. 5A) on New York State Highway 171 near the confluence of South Moyer Creek and Moyer Creek 6 km west of the town of Frankfort and 10 km southeast of Utica, New York. This is one of the most easterly Silurian sections in the greater Mohawk River valley (Fig. 1). The stream section in South Moyer Creek (Fig. 5B) was logged in detail by Walker

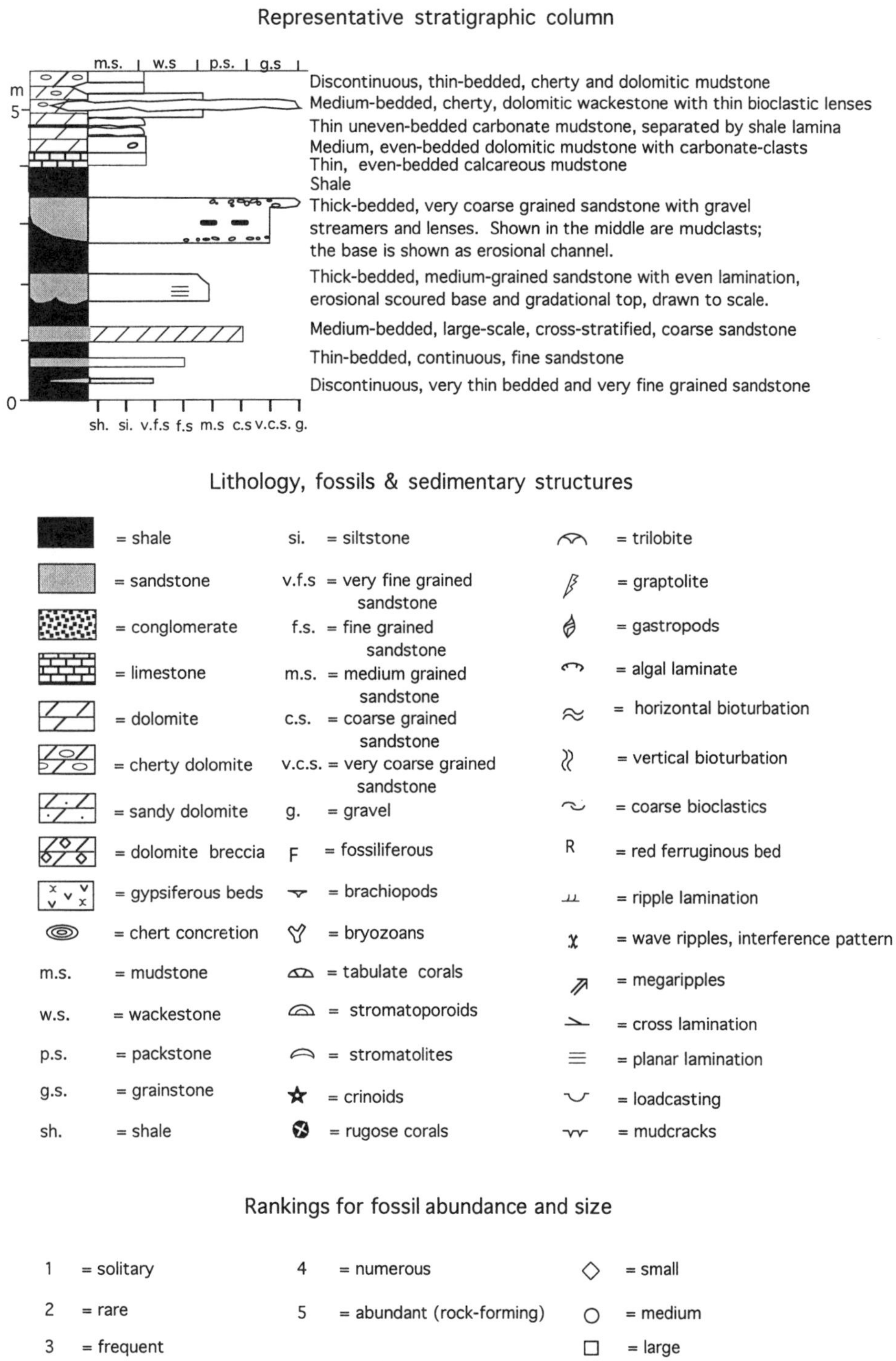

Figure 3. Key to the stratigraphic logs compiled for this chapter and the symbols denoting various lithologies, fossils, and sedimentary structures as well as rankings for fossil abundance and size used in Appendix 2.

(1975), but exposure in the bluffs outcropping along Moyer Creek in Frankfort Gorge was much improved by blasting in 1993 for a highway expansion project.

The Silurian Oneida Formation is represented in the road cut by a 4-meter-thick pebble conglomerate that sits unconformably on the Ordovician Frankfort Shale (Fig. 6A). Rip-up clasts derived from the Ordovician shale are common at the base of the formation, but they are insignificantly represented in thin sections (1.5%). The main component present in the conglomerate is common quartz (70 to 80%). A distinct bimodality

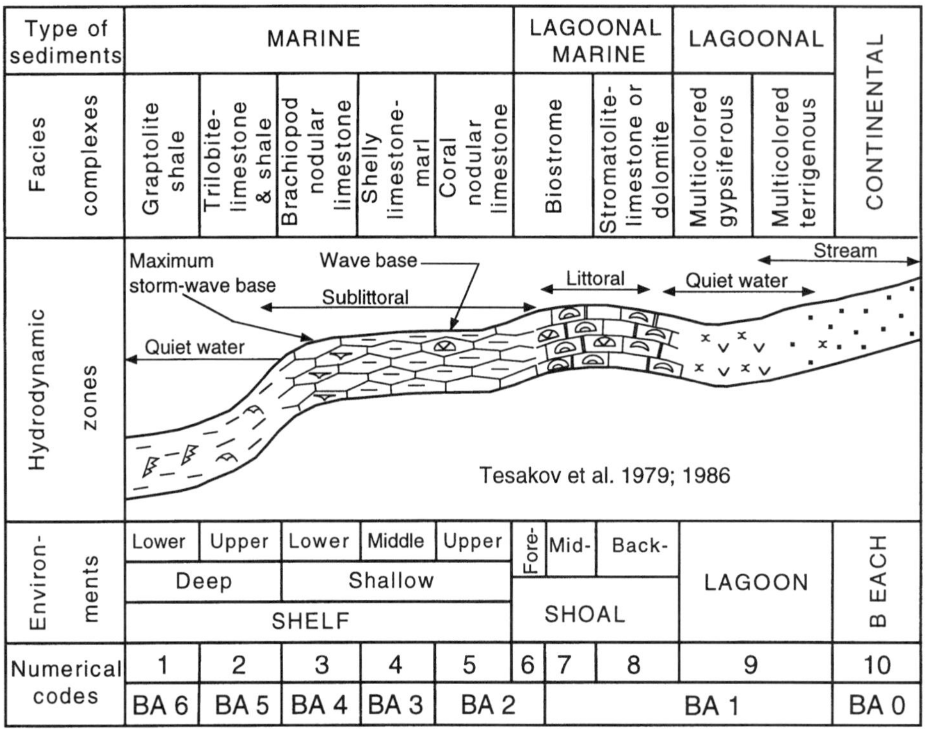

Figure 4. Facies model adopted for this study, with BA 0–6 indicating appropriate Benthic Assemblage zones.

in quartz populations occurs, with the sand fraction consisting of subangular common quartz and the pebbles consisting of vein quartz (5 to 15%), composite quartz from metaquartzite (3.5 to 8%), or chert (<1%). The metaquartzite and chert clasts tend to be angular. Very little matrix is present, as the sand grains are in direct contact with boundaries mostly dissolved by pressure solution. The amount of common quartz increases upward through the unit, and the proportion of metaquartzite increases relative to vein quartz in the pebble fraction. At the top of the formation, vein quartz is more common in the sand-size fraction. As noted by Folk (1960) in his study of the Silurian Tuscarora Sandstone in West Virginia, vacuoles tend to be more common in the pebble fraction than in the sand fraction. The bimodality of the Oneida Formation at this locality appears to be related to derivation from two different source rocks. Hydrothermal quartz veins in the Ordovician Frankfort Shale were the source for most of the pebbles. A fine-grained source rock, such as older sandstone or granite, was the probable source for the sand-size fraction.

Large-scale tabular cross beds are conspicuous, starting 1.45 m above the base of the Silurian. Repetitious beds with a pattern of small-scale, coarsening upward sedimentation (quartzite sand to very coarse quartz pebbles) are found in the upper 2 m of the Oneida. Shale and marine fossils are absent

TABLE 1. ASSIGNMENT OF BRACHIOPODS TO BENTHIC ASSEMBLAGE ZONES (RHUDDANIAN AND AERONIAN STAGES)

BA 6 (1)	BA 5 (2)	BA 4 (3)	BA 3 (4)	BA 2 (5 and 6)
Major components				
Glassia	*Clorinda*	*Stricklandia*	*Pentamerus*	*Eocoelia*
Skenidoides	*Skenidoides*	*Kulumbella*	*Virgiana*	
	Dicoelosia	*Protatrypa*	*Brachyprion*	
	Leangella	*Eospirifer*		
	Eoplectodonta	*Eoplectodonta*		
	Glassia	*Hyattidina*		
Minor components				
	Stricklandia	*Pentamerus*	*Eocoelia*	*Pentamerus*
	Pentamerus	*Eocoelia*		

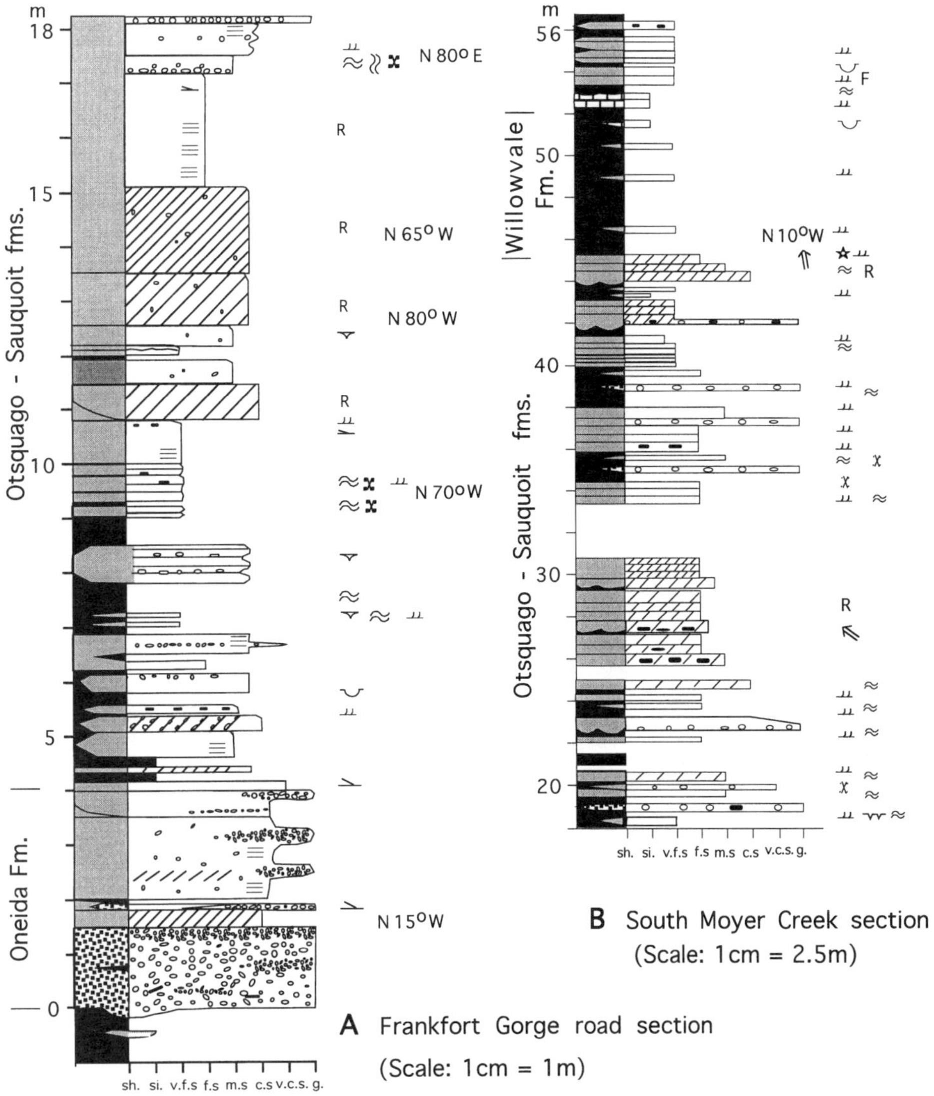

Figure 5. Stratigraphic sections from the Lower Silurian (Aeronian and Telychian stages) near Utica, NewYork. A, Road section along Frankfort Gorge. B, Stream section on South Moyer Creek.

from this part of the section. All features are consistent with channel development in a braided stream that crossed Ordovician bedrock from the present southeast to northwest.

Interfingered strata belonging to the Otsquago Formation (sandstone) and Sauquoit Formation (sandstone and shale) comformably overlie the Oneida Formation and complete the road exposure. The first part of this package, from 4 to 9 m (Figs. 5A and 6A), shows a sharp lithological change with the introduction of interbedded sandstone and shale containing occasional gravel streamers. These beds are laterally discontinuous. Some ripple marks are preserved on sandstone bedding surfaces. Rare brachiopods and some trace-fossil trails first appear 7 m above the base of the section. The fossils and the general fining-upward signature of this interval indicate a marine transgression, although rip-up clasts and quartzite gravel suggest episodic pulses of high energy from flood runoff or storms redepositing and reworking relict river material as the transgression advanced. Subtidal deposition in a brackish estuarine environment was most likely during deposition of this interval.

The succeeding interval, from 9 to 12.5 m (Fig. 5A), is characterized by a gradual coarsening in sediment size through beds that are laterally continuous and separated from one another by thin shale partings. Shale rip-up clasts are common, but quartzite gravel is rare. Interference ripples and symmetrical ripples with rounded crests 7 cm apart are well represented, as are trace-fossil trails. A prominent channel, filled with sandstone that is tabular cross-bedded and hematitic, crosses the outcrop at the 11-m level. A modest shallowing event, still within the upper subtidal range, is suggested by sheeted sands truncated by a channel.

From 12.5 to 18 m, strata follow a gradual fining-upward pattern. Coarse-grained, well-sorted, ferruginous sandstones

Figure 6. Views of the Lower Silurian near Utica, New York. A, Road section along Frankfort Gorge, showing the basal conglomerate of the Oneida Formation and overlying interbedded shales and sandstones of the Otsquago-Sauquoit formations. B, Asymmetrical ripple marks in the Otsquago-Saquoit sequence of South Moyer Creek (wave length = 65 cm).

with tabular cross beds give way to medium-grained, planar laminated sandstones. This sequence is consistent with sedimentation in a rapidly deposited lower tidal flat. Transition to a middle tidal flat is indicated in the uppermost meter of the road cut, where bedding surfaces exhibit interference ripples, asymmetrical ripples having crests 7 cm apart, and frequent trace-fossil trails.

The nearby succession in South Moyer Creek (Fig. 5B) is exceptional for its many extensive exposures of bedding planes that follow behind a series of low waterfalls in broad steps. The road and creek sections overlap and show closely parallel development, although the interval from about 1.25 to 8 m in the road section is not exposed in South Moyer Creek. The section illustrated in Figure 5B begins at the same lithological transition where the road section ends. This transition in tidal-flat facies occurs after the initial 10 m of section in the streambed (not shown in Fig. 5B). Interfingering of thin sandstones and shales is characteristic of the Otsquago-Sauquoit succession through the next 27 m, except for 3 m of tabular cross-bedded sandstone in the interval from 26 to 29 m. A major change in lithology occurs at the 45-m horizon, where the shales of the Willowvale Formation take over (Fig. 5B). The overall progression through this sequence signifies a transgression from tidal flats and tidal channels to shallow, subtidal conditions.

Examples of megaripples are spectacularly exposed on bedding planes at about the 27-m and 45-m levels (Fig. 5B). The latter example (Fig. 6B) exhibits asymmetrical ripples having a maximum wave length of 65 cm and an amplitude of 8 cm. The direction of movement (perpendicular to wave crests and down the steep side of troughs) is on a bearing N10°W. Other bedding planes, especially those exposed between the 34-m and 36-m levels, show excellent examples of interference ripples. Maximum wave lengths measured on interference ripples from this interval are 18 cm on the primary axis and 12 cm on the secondary axis, with a maximum amplitude of 2 cm. The scale of these Silurian megaripples and interference ripples is comparable to structures commonly observed on contemporary intertidal sand flats in Nova Scotia's Bay of Fundy.

Trace fossils are commonly preserved on the top surfaces of sandstone strata throughout much of the Otsquago-Sauquoit sequence (Fig. 5). Meandering trails belong to the monospecific ichnotaxon *Archaeonassa fossulata* Fenton and Fenton 1937 (Fig. 7A). This ichnotaxon was redescribed by Buckman (1994), who argued that its trace-fossil trails are typical of gastropods living in intertidal regimes. Another trace fossil sometimes associated with the *Archaeonassa* trails exhibits trackways marked by individual push-up mounds (Fig. 7B). The tracks were made by an arthropod that carried its body above the substrate on legs that pushed into the substrate and displaced a small mound of sediment backward through the pressure of traction. Figures 7C and 7D illustrate several superimposed trackways made across a set of interference ripples. The tracks may be attributed to the ichnogenus *Diplichnites* (R. K. Pickerill, personal communication, 1995).

The exact chronostratigraphic placement of the Oneida, Otsquago, and Sauquoit formations is problematic, owing to a lack of index fossils. Traditionally, these units form part of the middle Clinton Group assigned to the Upper Llandovery Series (Rickard, 1975). The brachiopod *Eocoelia sulcata* occurs in the lower part of the Willowvale Formation on South Moyer Creek (Walker, 1975). Based on lateral facies relationships with formations farther west in New York State, the basal part of the Oneida Formation is considered coeval with the Wallington limestone, bearing *Pentamerus oblongus*, and the Willowvale Formation is interpreted as coeval with the Williamson shale,

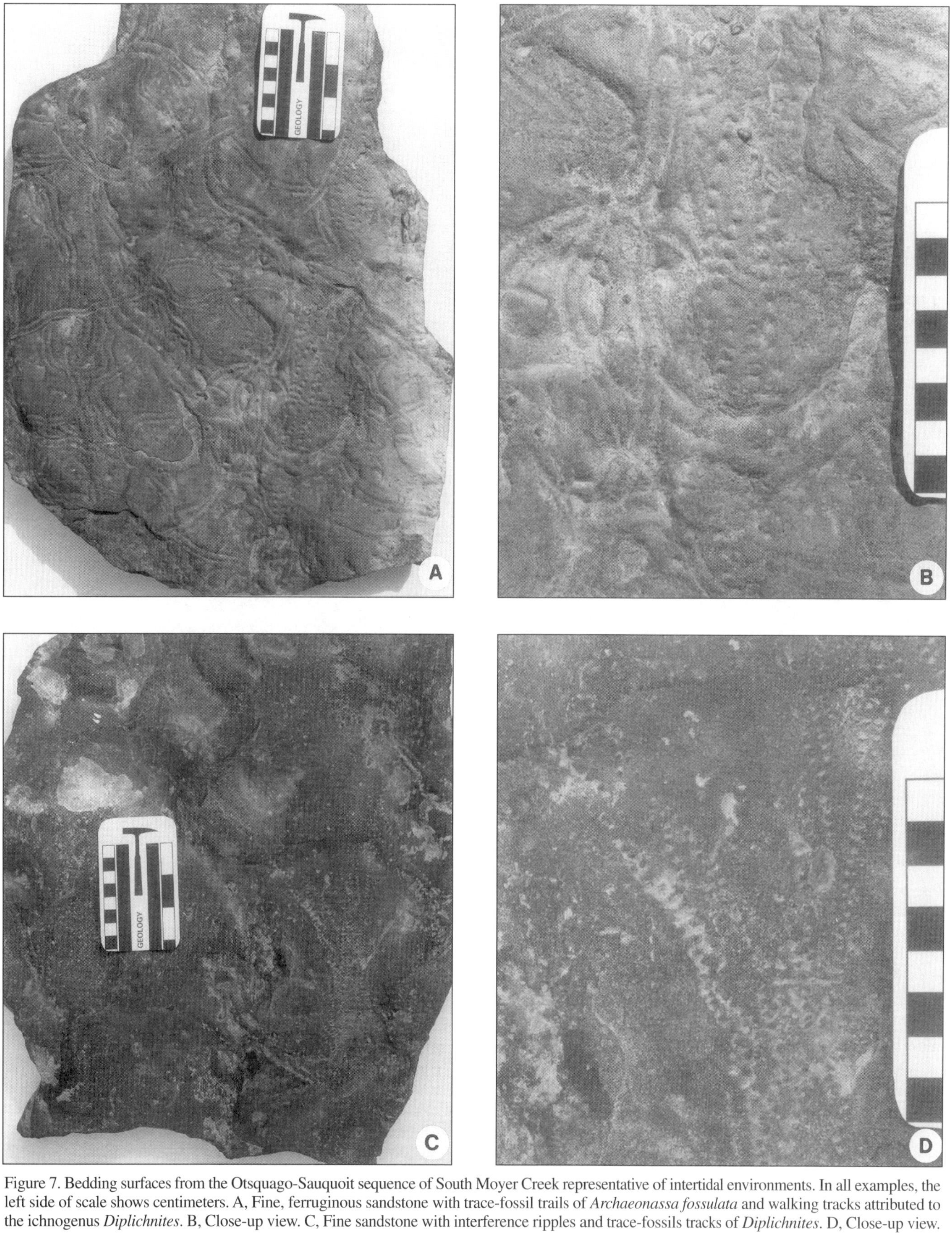

Figure 7. Bedding surfaces from the Otsquago-Sauquoit sequence of South Moyer Creek representative of intertidal environments. In all examples, the left side of scale shows centimeters. A, Fine, ferruginous sandstone with trace-fossil trails of *Archaeonassa fossulata* and walking tracks attributed to the ichnogenus *Diplichnites*. B, Close-up view. C, Fine sandstone with interference ripples and trace-fossils tracks of *Diplichnites*. D, Close-up view.

bearing *Costistricklandia lirata* (Brett et al., 1990). These brachiopods suggest that the basal Oneida occurs in the middle Aeronian Stage and the Willowvale belongs to the upper Telychian or lower Sheinwoodian stages. Two rises in relative sea level along this part of the Taconian shoreline may be inferred to have occurred during late Aeronian–early Telychian time and during late Telychian–early Sheinwoodian time.

Maymecha River (Maymecha District)

Stratigraphic sections through the Kuranakh and Orachin formations, straddling the Ordovician-Silurian boundary, are exposed on both banks of the Maymecha River near its confluence with Kuranakh Creek approximately 100 km south of Khatanga (Fig. 2). Dolomite strata dip gently 7° to 8° northwest beneath a thick cover of Permian basalt dissected by stream erosion. Situated 300 to 400 m upstream from the mouth of the Kuranakh, the more complete exposure on the right bank of the Maymecha River is 11 m thick (Fig. 8A). Much of this section consists of strata with laminations of probable algal origin and trace fossils that are biostratigraphically insignificant (Fig. 9A). Based on tabulate corals, brachiopods, and other macrofossils, however, units 11 and 13 at the top of the Orachin Formation are known to be Early Silurian in age. Reddish-colored marly dolomites and dolomitic marls at the base of the Kuranakh Formation are regarded as Ordovician. A detailed description of all layers in this section is provided in Appendix 1.

Identification of the Ordovician-Silurian boundary is problematic. One interpretatation would set it at the top of Unit 6 (Fig. 9A), based on fine-grained, vertically burrowed dolomite thinly draped by about 0.5 cm of quartzite granules. The *Skolithus*-type burrows suggest that Unit 6 was deposited in very shallow water. The quartizite-sand drape is laterally traceable for tens of meters along the entire length of outcrop exposing this horizon.

Another possible position for the Ordovician-Silurian boundary in this section is the junction between units 9 and 10 where laminated dolomites of probable algal origin, forming the base of Unit 10, fill 25 cm of relief among small knolls eroded in the top of Unit 9 (Fig. 8B). No quartzite clasts are present at this boundary, and the topographic relief is localized over a lateral distance of only a few meters. Some erosional activity on a hard substrate clearly took place at this level, and algal colonization of the resulting surface occurred in relatively shallow water. The localized nature of this feature, however, makes it inappropriate as a systemic boundary.

A unit-by-unit correlation of beds across to the opposite (left) bank of the Maymecha River is indicated in part by Fig. 9B. Most striking is the absence of units 6 through 9, presumably cut out by more extensive downcutting at the Ordovician-Silurian boundary. On the left bank, pockets of quartzite sand 2 to 3 cm thick occur in small channels eroded in reddish, silty dolomite at the top of Unit 5. They are directly overlain by laminar dolomite from the lower part of Unit 10. Thus, about 1 m of strata present on the right bank is missing from exposures on the left bank. This narrow interval encompasses both unconformities observed on the right bank.

Figure 8. Views of the Upper Ordovician and Lower Silurian (Aeronian Stage) on the Maymecha River in Arctic Siberia. A, Left bank of the Maymecha River; arrow points to likely Ordovician-Silurian boundary. B, Intraformational erosion surface less than 1 m above the Ordovician-Silurian boundary indicated in Figure 8A.

Lower in the left-bank section is a well-developed channel filled with quartzite pebbles (Fig. 9B). The channel is more than 5 m wide and forms a depression 50 to 70 cm deep. The topmost dolomite and middle chert-concretion layers of Unit 4 are fully incised by the channel; the basal layer of Unit 4 with its undulating laminations is only partially eroded. Based on maximum erosion, vertically as well as laterally, this lower unconformity cuts out a lesser volume of strata and is more likely Ordovician in age. Thus, the Ordovician-Silurian boundary is identified to

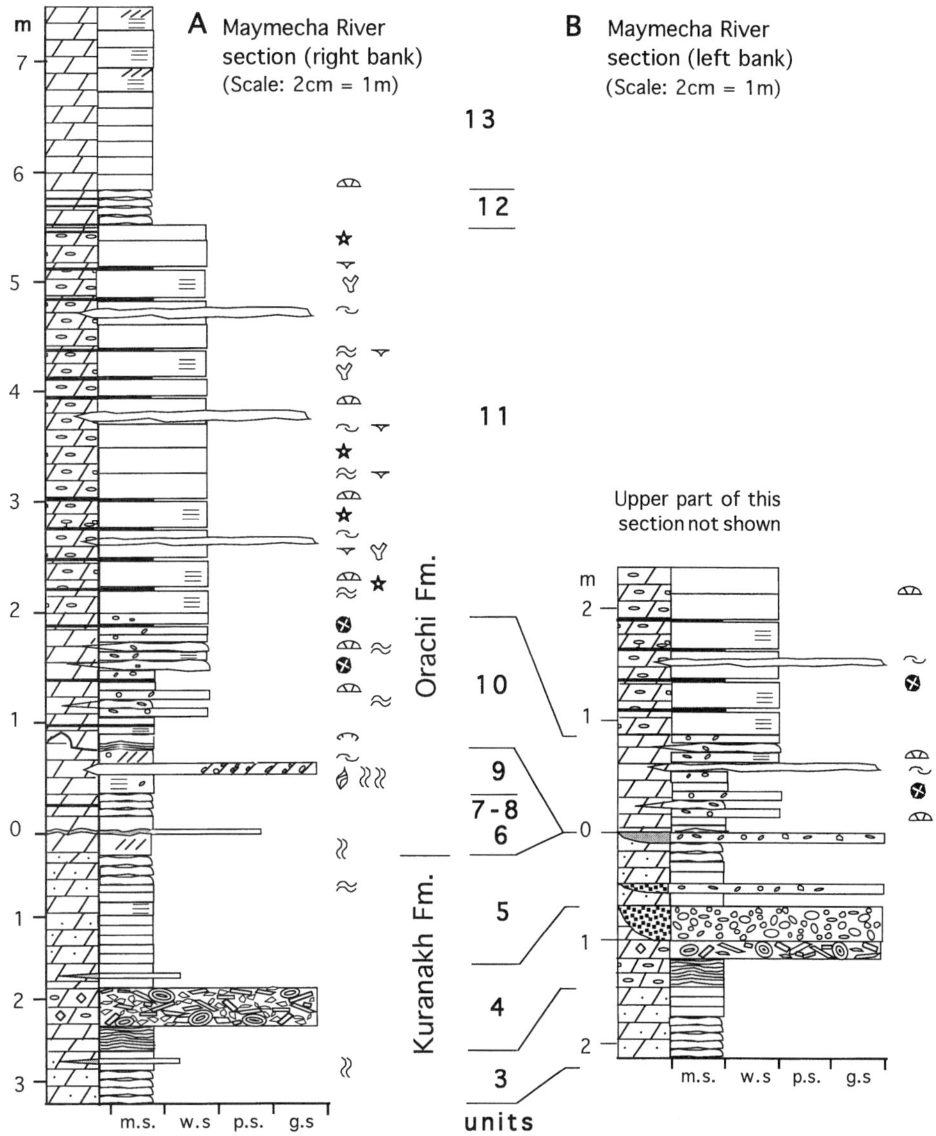

Figure 9. Stratigraphic sections from the Upper Ordovician and Lower Silurian (Aeronian Stage) on the Maymecha River in Arctic Siberia. A, Right bank of the Maymecha River. B, Left bank of the Maymecha River.

within about 1 m, considering stratigraphic variations in the two bank exposures. What is more significant, however, is the importation of exotic material from Precambrian formations located in Anabar, as represented by the presence of quartzite clasts at two or more horizons on or near the systemic boundary.

Silurian strata on the left bank are more accessible for fossil collecting than those on the right bank, and the following occurrences are noted. Bed 11 contains *Mesofavosites dualis* and *Favosites gothlandicus*. The highest diversity of species is found in the upper part of bed 13, including tabulate corals (*Multisolenia tortuosa*, *Halysites catenularius*, *Parastriatopora rhizoides*, *Favosites gothlandicus*, *Mesofavosites dualis*, *Calamopora alveolaris*, and *Propora tubulata*); stromatoporoids (*Clathrodictyon boreale* and *Labechia venusta*); a gastropod (*Straparollus alacer*); and brachiopods (*Pentamerus oblongus*, *Strophomena kulumbensis*, *S. pectenoides*, and *Coolinia gracilis*). *Pentamerus oblongus* confirms a middle to late Aeronian age. The succession from channeled sands to algal carbonates to strata with corals and brachiopods signifies a gradual marine transgression across Ordovician bedrock.

SILURIAN SHELF DEPOSITS

Northern Great Lakes (Michigan Basin)

Silurian carbonates dominated the craton of the Great Lakes region beyond the clastic veil shed by Taconia in the Appalachian Trough (Fig. 1). With the exception of the Cabot Head Shale, all other Lower Silurian strata in this region are dolomites. Differing nomenclatures for these strata are employed on opposite sides of the U.S.-Canadian boundary, but they form a readily correlated band of outcrops stretching across the northern rim of the Michigan Basin from the upper peninsula of Michigan to Ontario's Manitoulin Island and Bruce Peninsula (Colville and M. E. Johnson, 1982, Fig. 5). Only the basal Silurian Manitoulin Dolomite and the succeeding Cabot Head Shale are recognized by name as the same entities in Michigan and Ontario.

Joint fieldwork in 1995 took us to the Hendricks, Marble Bluff, and Trout Lake sections in northern Michigan (localities 78, 95, and 124 in M. E. Johnson and Campbell, 1980); the type Fossil Hill locality and the Blue-Jay, Lake Saugigansing, Kagawong, Cup and Saucer, Jerusalem Hill, Red Rock, Ice Lake, and Indian Hill sections on Manitoulin Island (localities 1, 3, 12–14, 17, 20, 22, and 25 in M. E. Johnson, 1981); and the Inglis Falls and Lion's Head sections on the Bruce Peninsula (localities 3 and 21 in Colville and M. E. Johnson, 1982).

The key to time-rock correlations through this part of the Michigan Basin is related to the distribution of four stratigraphically distinct intervals rich in pentamerid brachiopods (i.e., the dominant element accounting for more than 75% of the fauna). These brachiopods are often found preserved in growth position, allowing for their confident assignment to a bathymetric position in BA 3. The oldest interval bears abundant *Virgiana decussata*: In northern Michigan the host unit is called the Lime Island Formation (M. E. Johnson and Campbell, 1980), but in Ontario it is named the Dyer Bay Formation (M. E. Johnson, 1981; Colville and M. E. Johnson, 1982). The second and third intervals are rich in *Pentamerus oblongus*: In Michigan and Ontario the Schoolcraft and Fossil Hill formations include these intervals, respectively. The fourth interval is characterized by *Pentameroides subrectus*: The middle part of the Cordell Formation is the host unit in Michigan; the top of the Fossil Hill Formation is the host unit on Manitoulin Island.

Stratigraphically arrayed among these four pentamerid-bearing beds are intervals dominated by tabulate corals and stromatoporoids (attributed to BA 2) and algal-laminated dolomites with occasional dessication cracks (attributed to BA 1). Four successive highstands in fluctuating sea levels are related to the recurrence of brachiopod-dominated communities in the *Virgiana-Pentamerus-Pentameroides* succession. With the co-occurrence of rarer specimens of *Eocoelia, Stricklandia,* and *Costistricklandia,* it is possible to constrain the age of the last three events as middle Aeronian, early Telychian, and late Telychian to early Sheinwoodian, respectively. The first highstand, or *Virgiana* event, is much more difficult to constrain biostratigraphically but could be latest Rhuddanian to earliest Aeronian in age.

The entire Lower Silurian package is approximately 160 m thick in northern Michigan, little more than 95 m thick on Manitoulin Island, and only 50 m thick on the Bruce Peninsula. Entire beds, such as the *Pentameroides* interval, pinch out to the east on the Algonquin Arch between the Michigan Basin and Appalachian Trough (Colville and M. E. Johnson, 1982). Carbonates are less common on the opposite side of the Algonquin Arch, but the Wallington limestone in New York State may correlate with one of the two *Pentamerus oblongus* intervals in the Michigan Basin.

Gorbiachin River (Igarka District)

Located approximately 80 km west of Igarka, a 262.5-m-thick sequence is represented by the Chamba and Ugiyuk formations in the valley of the Gorbiachin River. Lithostratigraphic units are conformable and well exposed in an east-dipping monoclinal structure (Figs. 10 and 11). The Chamba Formation is divided into two subformations (Tesakov et al., 1980, Fig. 6). The lower unit (intervals 1–6: 74 m thick) is represented by graptolitic shales with rare intercalations of limestone; the upper unit (intervals 7–11: 53.5 m thick) consists of alternating thin intervals of shales and limestones separated by thicker shales (Figs. 11A–11C). Likewise, the Ugiyuk Formation is divided into two subformations (Tesakov et al., 1980, Fig. 6). The lower unit (intervals 12–82: 72 m thick) exhibits distinct cyclic components, where uneven and nodular limestones are interbedded with shale and marl (Fig. 10B). The volume of organo-detrital material increases upward through this part of the formation, as does overall richness in macrofossils. The upper subformation (intervals 83–117: 63 m thick) is characterized by uneven, nodular, and polydetritic limestones showing rare intercalcations with marl. The diversity and total volume of macrofossils decreases through this part of the formation. The overlying sequence belongs to the incompletely exposed Olen Formation. Dominated by marls, this formation (intervals 119–128: 34.5 m thick) marks a major change in sedimentation from the Ugiyuk Formation (Fig. 10C).

Appendix 2 lists mostly key brachiopods from the Ugiyuk Formation, as distilled from the massive biostratigraphic tabulation of Tesakov et al. (1980, Fig. 10) for the full Gorbiachin succession of 128 lithological units. The original summary involved a team of specialists who identified all macrofossils belonging to each interval and then came to a consensus regarding the dominant species. Bathymetric correlation with a benthic association based on these data alone is misleading, however, since the original ranking was performed whether the overall density of fossils in any given rock unit was high or low. Thus, care was taken to amend these data with new census information giving frequency and size rankings for all species belonging to *Eocoelia*, *Pentamerus*, and *Kulumbella* (rankings defined in Fig. 3). *Pentamerus oblongus* may be the "dominant" fossil in a given rock interval by default. It does not follow that a BA 3 assignment is warranted, particularly if individuals are small and relatively scarce. The BA 3 designation is earned by medium to large specimens of *P. oblongus* that are frequent to abundant (rock forming) and dominate numerically over associated species of *Eocoelia* and *Kulumbella*. Likewise, the BA 4 designation is earned by medium to large specimens of *Kulumbella*

Figure 10. Views of the Lower Silurian (Rhuddanian to Telychian stages) on the Gorbiachin River, Arctic Siberia. A, Ugiyuk and Lower Olen formations (Aeronian to Telychian stages) to the left and Chamba Formation (Rhuddanian Stage) to the right, looking upstream. B, Interbedded marls and resistant limestone units from the Lower Ugiyuk Formation, with limestone beds forming natural groins in the river. C, Boundary (arrow) between the Upper Ugiyuk Formation (left) and marls of the overlying Olen Formation (right), approximately equivalent to the Aeronian-Telychian boundary.

kulumbensis that outnumber associated species of *P. oblongus* and *Eoceolia memisphaerica* (an endemic variant distinct from *E. hemisphaerica*).

Sufficient evidence is available to delimit a bathymetric profile for the entire Ugiyuk Formation. Interval 80, near the boundary between the lower and upper subformations, constitutes the nadir of shallow-water deposition based on a strong representation of *E. memisphaerica* (BA 2). Rich concentrations of robust *P. oblongus* (BA 3) occur above this turning point in the Upper Ugiyuk (intervals 83, 84, 86, and 94) as well as below

Figure 11. Views of the Chamba Formation (left bank of the Gorbiachin River). A, General view of left bank, showing interbedded shales and thin limestones. B, Close-up view of intercalated fine limestones and shales from the middle part of the Chamba Formation (left side of scale in centimeters). C, Bedding surface showing abundant graptolites, *Glyptograptus tamariscus* (upper side of scale in centimeters).

in the Lower Ugiyuk (intervals 63–71, 78, and 79). *Clorinda undata* (BA 5) occurs at the top of the Upper Ugiyuk (intervals 111, 113, 115, and 117) and the base of the overlying Olen (interval 119). Likewise, *C. undata* occurs toward the base of the Lower Ugiyuk (interval 22). The sharpest change in relative sea level, however, is indicated by intervals 28–30, with a strong representation of *K. kulumbensis* (BA 4) bracketed by beds immediately above and below bearing *P. oblongus* (BA 3).

Considering the position of this formation between the Olen and Chamba Formations, the Gorbiachin River sequence may be generalized as a continuous shallowing-upward sequence to the midpoint of the Ugiyuk, followed by a deepening-upward cycle through the Olen. This bathymetrical profile, however, is superimposed on a complex pattern of small-scale lithological cycles. According to the bathymetric model (Fig. 4), normal (fair-weather) wave base affected the Gorbiachin seafloor only for a relatively short period of time when concentrations of *Eocoelia* were swept into their characteristic limestone lenses. Storm wave base, however, potentially affected the local seafloor over a much longer time, spanning the stratigraphic interval of *P. oblongus* through much of the Ugiyuk Formation.

Few brachiopods in the Ugiyuk Formation are preserved articulated and in growth position. Lenses of *Meristima lacrima* (intervals 76 and 66), observed with as many as 40 articulated individuals clustered in growth position, are unusual. *P. oblongus* occurs only rarely as articulated specimens (intervals 67 and 79) but never in crowded growth position. *P. oblongus* is one of the most ubiquitous brachiopods in the Ugiyuk Formation, occurring in shales, marls, nodular limestones, and limestone. Some intervals in this formation closely resemble *Pentamerus* tempestite deposits from the Lower Silurian of southern Norway (M. E. Johnson, 1989). Limestone from interval 82 and marl from interval 63, for example, consist of whole but unbroken (disarticulated) valves heaped together. This would suggest low levels of post mortem exposure and transportation on the seabed. Other limestone units, such as interval 79, reflect a more condensed jumble of broken shells, indicating exposure and transportation under more vigorous bottom conditions. The varying fabrics shown by these *Pentamerus* marls and limestones are suggestive of storm cycles. Storm activity would not only winnow away marl and fine lime but also physically concentrate shelly materials. A firmer substrate than marl may also have further stimulated the growth of pentamerid populations.

The regularity of these cycles is evident from the photo in Figure 10B, where three natural groins formed by equally spaced limestone or nodular limestone beds from the Lower Ugiyuk Formation are shown projecting into the Gorbiachin River. The intervening, less resistant marls, have been scoured out by the river's flow. Periodic removal of fine lime and marl from the shallow shelf affected the adjacent upper deep shelf, where much of the material apparently settled out. Although the Chamba Formation is older than the Ugiyuk Formation, the two units may be understood as related facies of shallow and deeper shelf environments (Fig. 4). Interbedded fine limestone (light) and shale (dark) from the Upper Chamba Formation are shown in Fig. 11B. The shale probably represents normal, background sedimentation, whereas the fine limestone may have been introduced by storms sweeping across a coeval shallow shelf from Anabar.

The oldest regional stage applied to the Silurian of Siberia, the Moyerokanian Stage, embraces the entire Chamba Formation. The next regional stage, the Khaastyrian Stage, includes the entire Ugiyuk Formation. The boundary between these two regional stages probably does not correlate with the global Rhuddanian-Aeronian boundary but may interface with the lower Aeronian (Tesakov et al., 1979). In terms of graptolite biostratigraphy, the age of the Chamba Formation is poorly constrained. The most prolific graptolite is an endemic species, *Glyptograptus tamariscus* (Fig. 11C). The brachiopod *Zygospiraella duboisi*, which is restricted to the Rhuddanian of Siberia, the Baltic region, southern Norway, the British Isles, and Nova Scotia (Nikiforova and Andreeva, 1961, p. 237; Cocks, 1978, p. 159; Thomsen and Baarli, 1982, p. 65), does occur in the middle of the Chamba Formation (intervals 7–8). The "possibly conspecific" *Zygospiraella planoconvexa* also occurs throughout the basal Silurian Manitoulin Dolomite on Manitoulin Island, Ontario (Copper, 1982, p. 695).

Occurring 10.5 m above the base of the Ugiyuk Formation (interval 16), the brachiopod *Virgiana moyeroensis* is consistent with an early Aeronian age. The wide range of *Pentamerus oblongus* (intervals 11–94) is most consistent with a later Aeronian age. The presence of *Monograptus sedgwickii* (interval 70) above the deepening represented by the BA 4 incursion of *Kulumbella kulumbensis* (intervals 28–30) strongly supports a mid-Aeronian age for this event. The last regional stage in the Lower Silurian of Siberia is the Agidian Stage, which includes the Olen Formation. The boundary of the Khaastyrian and Agydian stages is believed to correlate with the boundary between the Aeronian and Telychian stages (Tesakov et al., 1979). Except for the fact that the Olen Formation begins well above the last stratigraphic occurrence of *Pentamerus oblongus*, there is no other constraint on the age of the deepening event represented by the BA 5 incursion of *Clorinda undata* beginning at the Ugiyuk-Olen formational boundary (Fig. 10C). Regional correlation, however, does suggest an early Telychian age for this event. It should be noted that occurrence of *Pentameroides subrectus* has not been documented anywhere in Siberia, but the water depth in the Gorbiachin area was clearly beyond the brachiopod's preferred limits during Telychian time.

COMPARISON OF ONSHORE-OFFSHORE PROFILES AND PALEOGEOGRAPHY

A profile cutting 750 km across the Lower Silurian of Siberia (Fig. 12) connects stratigraphic sections interpreted from the Gorbiachin River and Maymecha River areas. It also incorporates data from additional sections not detailed in this chapter. The cross section closely resembles the onshore-offshore model developed in Figure 4. Despite fluctuations in sea level, the Gorbiachin sections consistently retain a more offshore signature than the Maymecha sections. The Maymecha sections also are much thinner than the Gorbiachin sections, mainly because they represent only Aeronian deposition as opposed to Rhuddanian and Aeronian deposition. This is a direct consequence of gradual marine onlap against the margin of Anabar lands.

Similar profiles covering a combined distance of 900 km were previously published for the Niagaran escarpment cutting along the Mohawk River valley of New York State to the Niagara Falls region (Brett et al., 1990, p. 203–204; Goodman and Brett, 1994, p. 165; LoDuca and Brett, 1994, p. 164) and for the escarpment's continuation across the northern rim of the Michigan Basin (Colville and M. E. Johnson, 1982, p. 969). The maximum thickness of the North American sections is about half that of the Siberian sections, particularly on the offshore shelves. Thicknesses in the onshore direction, however, are greater on the flank of Taconia than Anabar (Figs. 5 and 6, 8 and 9).

In conjunction with additional data from other studies, the North American and Siberian profiles were used to reconstruct the Rhuddanian-Aeronian paleogeographies shown for part of Laurentia (Fig. 1) and Siberia (Fig. 2). In North America, research by Smith (1970) and Cotter (1983, 1990) details the changing position of Taconia's early Silurian shoreline to the present south in New York State and Pennsylvania (Fig. 1). To the present south and east in Siberia, additional work by Tesakov et al. (1992) on the Tunguska synclinorium documents the changing location of Anabar's early Silurian shoreline (Fig. 2).

A significant difference between the Siberian and Laurentian profiles and their associated paleogeographies is that the former are continuous across a single bathymetric basin (Figs. 2 and 12), whereas the latter are bisected by the Algonquin and Finlay arches structurally separating the Appalachian Trough from the Michigan Basin (Fig. 1). Although some of the North American basin and arches delimiting Silurian strata are post-Silurian in origin (M. E. Johnson, 1987), others were structurally and topographically active during Silurian time. Orogenic phases, such as the Taconician and the Uralian, are popularly understood as the result of interplate tectonics. According to Beaumont et al. (1988), however, lithospheric flexures may be modeled to extend well beyond interplate margins to affect lithospheric upwarp and downwarp on "stable" cratons. Eastward pinch out of Lower Silurian strata in the Michigan Basin (Colville and M. E. Johnson, 1982) implies that episodic uplift of the Algonquin Arch was a coeval process.

Another significant difference concerns the observation that much of Silurian Laurentia was a comparatively shallow platform accumulating a relatively condensed sequence of carbonate strata in contrast to the extensive and thicker, deep-shelf deposits of mixed shales and carbonates in Siberia. No deep-shelf deposits are plotted on the map in Figure 1C, and perhaps the only comparable setting on Laurentia with an intracratonic deep shelf was the Reelfoot Trough situated under the Mississippi embayment (Fig. 1A). This feature, which is interpreted as a failed rift dating from the Precambrian (Reed, 1993, p. 629), may have influenced deposition off the adjacent Taconic foreland during the Early Silurian. Thick Lower Silurian shales with interbedded sandstones and limestones, including a succession of brachiopods in the *Stricklandia lens lens* to *Costistricklandia lirata* lineage, are only briefly described as yet from the Gadsden area of Alabama (Bolton, 1990; Bolton and Driese, 1990).

Storm cycles, indicated by amalgamated sandstones with traces of swaley cross-stratification, form part of the Gadsden sequence in Alabama. Storm activity also is registered by the nearshore Silurian sands and carbonates of central Pennsylvania (Cotter, 1990). Elsewhere on the Laurentian platform, carbonates with common preservation of pentamerid brachiopods in crowded growth position are not as suggestive of persistent storm activity, except for the fact that large populations of articulated shells are usually truncated by scour surfaces.

No major tectonic features have an implied relationship with the Silurian geography of Arctic Siberia, a fact underscored by the low topography of Anabar. Subaerial erosion of Anabar was comparatively light, judging by the low volume of quartzitic detritus incorporated at or near the position of the Ordovician-Silurian boundary. Taconia, in contrast, shed a massive volume of quartzitic sand onto the Silurian coastal plain (Fig. 6A; Folk, 1960; Smith, 1970; Cotter, 1983). High gradients on streams dissecting Taconia were undoubtedly influenced by orogenic uplift associated with closure of the Proto-Atlantic ocean between Laurentia and Baltica.

The inboard, cratonic shores of Silurian Taconia and Anabar both show signs of strong tidal activity. The small-scale knolls preserved as an intraformational erosion surface in strata of the Maymecha River area (Fig. 8B) are similar in shape and size to tidal runnels developed on a Pleistocene abrasion platform in Western Australia (M. E. Johnson et al., 1995). The abundant interference ripples and asymmetrical megaripples, now exposed in the steam bed of South Moyer Creek (Fig. 6B), developed on sand flats that remained locally active over a longer interval of time than the channel deposits in the Maymecha River area of Siberia.

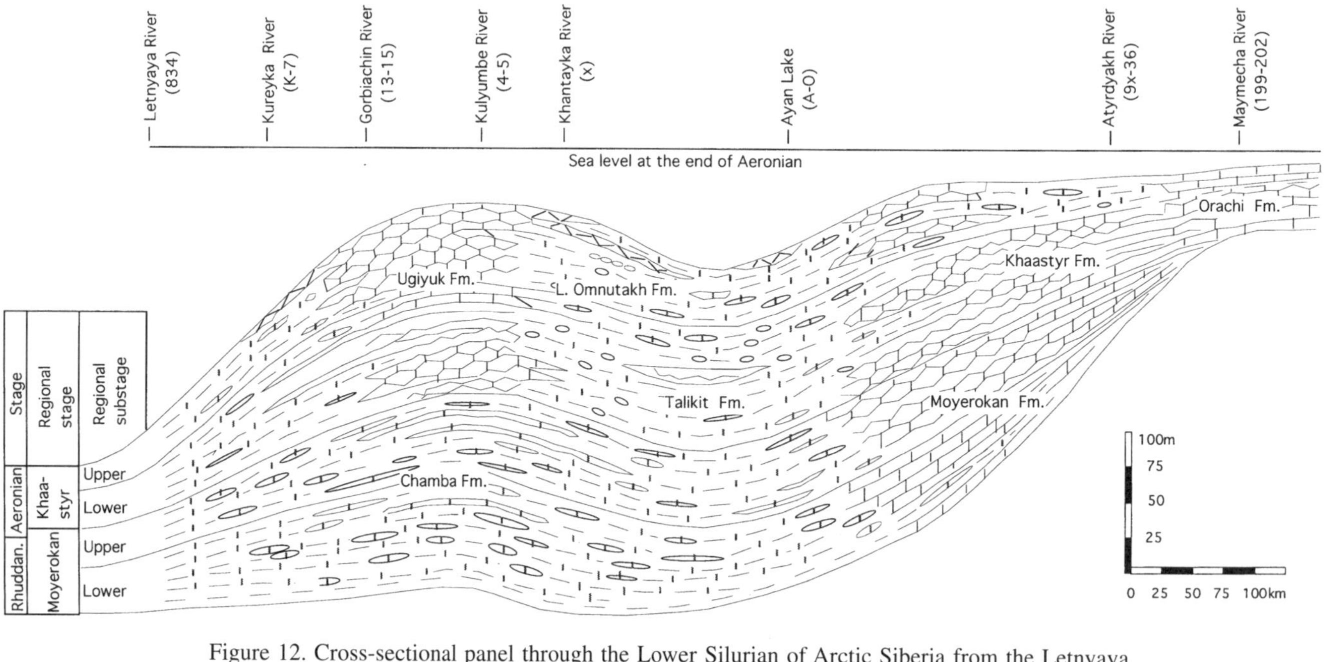

Figure 12. Cross-sectional panel through the Lower Silurian of Arctic Siberia from the Letnyaya River to the Maymecha River (see transect line in Fig. 2).

The most significant comparison between the onshore-offshore profiles of Silurian Laurentia and Siberia concerns preservation of coeval events recording sea-level fluctuations. A Mid-Aeronian deepening event is represented by the widespread occurrence of the first *Pentamerus* interval in the Michigan Basin and an interval with robust *Kulumbella* in the Lower Ugiyuk Formation of Siberia's Gorbiachin River area. An early Telychian deepening event is represented by the second *Pentamerus* interval in the Michigan Basin and a sustained *Clorinda* interval across the Ugiyuk-Olen formational boundary. The mid-Aeronian event is also implied by deepening trends recorded in the onshore sections from the Maymecha River area in Siberia and the Utica area in New York State. The mid-Aeronian and early Telychian events are not as widely recognized in Arctic Siberia as in North America because of the more extensive area of deep-shelf environments. A relatively minor change in sea level on the order of 50 to 60 m (or less) is not expected to have had much effect on the marine benthos living below 150 m. Indeed, development of the Chamba Formation (Fig. 12) and its shallowing-upward pattern through the Rhuddanian may be interpreted to mask the more focused *Virgiana* event recorded in the Michigan Basin and elsewhere in North America.

DISCUSSION

The strongest relationship between Silurian Siberia and North America concerns sea-level fluctuations regarded as global events. That relative sea-level changes recorded by strata show some combination of local and regional tectonics as well as eustatic patterns is logical. It is a difficult task, however, to distinguish the various causal agents. Jess Johnson was an original thinker and contributor on this subject, both in general and pertaining specifically to Devonian strata. His paper, "Timing and coordination of orogenic, epeirogenic, and eustatic events" (J. G. Johnson, 1971), was seminal, and his work with other colleagues on Devonian eustasy (J. G. Johnson et al., 1985) unfolded at about the same time as, and provided encouragement for, parallel Silurian studies.

The concept of eustasy has its detractors. Regarding Paleozoic basins and arches in eastern North America, for example, Beaumont et al. (1988, p. 415) concluded that: "With only minor exceptions around the Ordovician-Silurian boundary, the major features of the observed stratigraphy can be explained by flexurally raising and lowering the surface of the lithosphere against a background of constant sea level." A more accommodating viewpoint is that shifting depocenters in the Silurian Appalachian Basin of Pennsylvania and New York State were affected by plate tectonics but also overprinted by third-order cycles attributed to eustatic sea-level changes (Cotter, 1983; Brett et al., 1990). Although the Illinois Basin might possibly have been affected by flexures during the Silurian (Beaumont et al., 1988, p. 397–398), what about basins even more remote from Taconia, such as the Williston Basin? What about basins on other paleocontinents that record coeval patterns of changing sea level?

These data from Silurian Siberia complement information on coeval sea-level cycles recorded on several paleocontinents: Laurentia, Avalonia, Baltica, South China, and Australian Gondwana (M. E. Johnson et al., 1991; M. E. Johnson and McKerrow, 1991). The tectonic histories of paleocontinents may be closely linked, as are Laurentia, Avalonia, and Baltica through the later Taconic and associated Caledonian orogenies. A striking unit-for-unit resemblance exists between the coeval Chamba-Ugiyuk-Olen succession in Siberia and the Solvik-Rytteråker-Vik succession in Norway's Oslo region, which both exhibit shallowing-upward strata followed by deepening-upward strata. Baarli (1990) related the relative sea-level changes in the Norwegian sequence to a migrating flexure in a foreland basin, but her data also include evidence for overprinting by eustatic cycles. Siberia, however, followed an independent tectonic history, and there is no compelling evidence that a flexure should have affected the Gorbiachin area at exactly the same time.

Under certain circumstances, regional tectonics may be used to reinforce the notion of eustasy, rather than compete against it. Rong and Johnson (1996) showed that in China's Guizhou Province, for example, mid-Aeronian strata overstepped 63 m of paleotopography on an Ordovician escarpment undergoing long-term uplift. The coeval event in Siberia probably was the deposition of strata rich in BA 4 *Kulumbella kulumbensis*, reflecting a minor reversal in the overall shallowing-upward succession of the Lower Ugiyuk Formation. Essentially the same pattern occurred in the Olso region, where the coeval event probably was the deposition of strata in the Upper Solvik Formation rich in BA 4 *Stricklandia lens progressa* (Baarli, 1990, p. 152–154). A series of repetitive glacial deposits through the Lower Silurian of Brazil (Grahn and Caputo, 1992) provides a plausible eustatic agent for these widespread paleoecological events. The argument that sea level was constant through the Early Silurian and that all relative changes in sea level were due to local and regional tectonism is thus unsupportable.

Why should the initial Silurian strata of Siberia, as well as Norwegian Baltica, be so much thicker and represent deeper-water settings than the more condensed carbonates of the North American platform? In terms of their paleolatitudes, Laurentia, Baltica, and Siberia were all well within the appropriate climatic range of carbonate deposition (Scotese and McKerrow, 1990; Kaljo et al., 1991). The Manitoulin Dolomite was being deposited across the shallow shelf area of the Michigan Basin while the Chamba and Solvik shales were being deposited in Siberia and Norway. All three formations may be assigned to the Rhuddanian Stage on the basis of the brachiopods *Zygospiraella duboisi* and *Z. planoconvexa*, but this determination leaves much room for adjustment within the stage. One possibility is that the continents of Siberia and Baltica were at a slightly lower hypsographic level than Laurentia. Not all continents shared the same hypsographic exposure through the late Tertiary or share it today (Bond, 1978). If this were the case at the start of the Silurian, then the lowest continents would be the first to be affected by a glacial-eustatic rise in sea level, and they might accumulate a thicker sedimentary cover over deep shelves.

An interesting peculiarity of the Gorbiachin River area concerns discrete packages of alternating marls and limestones in the Aeronian Ugiyuk Formation. The packages have no counterpart on the Silurian shallow shelf of the Michigan Basin. Interpretation of these small-scale cycles as storm beds is supported by the paleogeographic analysis of Marsaglia and Klein (1983, p. 131), which proposed the most likely hurricane pathways for the Middle Silurian. One such pathway moved from east to west along the N 5° latitude, but shifted gradually northward to directly cross the Annabar lowlands at 25°N. Another proposed pathway moved east to west along the S 5° latitude but was strongly deflected south by Baltica and Laurentia. Storm beds are preserved in the Silurian Appalachian Basin, but the higher topography of Taconia may have shielded the Laurentian craton from the full brunt of hurricanes in the Southern Hemisphere. The alternating marls and limestones in Siberia do not represent individual hurricanes. Rather, the marls may have been deposited during episodes of lesser storm activity and the limestones amalgamated during episodes of more frequent storm activity, possibly related to periods of global warming associated with the waxing and waning of glaciers on South American Gondwana.

CONCLUSIONS

North America and Siberia are the repositories of the two most extensive carbonate platforms that developed in epicontinental seas during Silurian time (Kaljo et al., 1991). Their architecture exhibits some of the same diagnostic time lines of global sea-level events associated with stratigraphic discontinuities in the *Pentamerus oblongus* community of Ziegler et al. (1968, p. 15) or Benthic Assemblage 3 of Boucot (1975, p. 49). Mid-Aeronian and early Telychian highstands in sea level are defined by layers of *P. oblongus* that contain frequent material in growth position and are laterally persistent over a distance of more than 200 km around the Michigan Basin of North America. Near the break between shallow and deep shelves in the Gorbiachin River area of Silurian Siberia, the same two highstands are registered by layers with BA 4 to BA 5 faunas interspersed by layers bearing high concentrations of *P. oblongus*.

During Aeronian and Telychian times, large parts of the Siberian platform were consistently deeper than the North American platform. This condition may have had its origins in a slight hypsographic inequity between the two paleocontinents already evident during Rhuddanian times, when the Siberian platform adjacent to Anabar accumulated thick deposits of interbedded shales and fine-grained limestone. This contrasts with the extensive biohermal limestones accumulated at the same time on Manitoulin Island in North America. A late Rhuddanian to early Aeronian highstand in sea level registered by the pentamerid brachiopod *Virgiana decussata* in North America is masked by persistent deep-water conditions through all of the Rhuddanian and earliest Aeronian over much of Arctic Siberia.

Both of the extensively flooded paleocontinents retained well-defined land areas that shed pebbles derived from quartz veins and quartzitic sand derived from exposed Precambrian formations via river systems to Silurian shores. Taconia in North America was an elongate mountain chain owing its considerable topographic relief to the foreshortening of the Proto-Atlantic ocean. Anabar in Siberia is a dome-shaped geological structure that already existed in low relief during the Silurian Period.

Rhuddanian intercalations of shale and fine limestone followed by Aeronian marls and amalgamated limestones imply the repetition of episodic storm cycles in Silurian Siberia: These have no obvious counterpart on the carbonate platform of Silurian Laurentia. The Aeronian storm beds of Siberia often are composed of *P. oblongus* debris very similar in fabric to carbonate tempestites previously described from the Lower Silurian of southern Norway (M. E. Johnson, 1989). Differences in continental position and the relative topography of land masses in fixed hurricane pathways (Marsaglia and Klein, 1983) appear to account for the presence or absence of these Silurian storm beds.

ACKNOWLEDGMENTS

The undertaking of this project shows that the comparative study of depositional systems on an intercontinental basis benefits from the good will and cooperation of international teams dedicated to the basic fundamentals of stratigraphy. Serious oversights and inaccuracies would undoubtedly have found their way into this study if it were enacted unilaterally or strictly as a literature review. Global studies require global cooperation.

The exchange visits and joint fieldwork in the United States, Canada, and Russia that stimulated this contribution were supported by the National Geographic Society (grant #5251-94), the Global Studies Project at Williams College under sponsorship of the Keck Foundation, the Office of the President at Williams College, and the Russian Foundation for Fundamental Research (grant # 95-05-155564 under the project title "Global events in Silurian sedimentary platform basins"). We are especially grateful for the encouragement and support of administrators at our home institutions: Francis C. Oakley, President of Williams College, and Alexei E. Kontorovich, Vice Director of the United Institute for Geology, Geophysics, and Mineralogy at Novosibirsk. Fieldwork conducted in Arctic Siberia in July 1995 benefited substantially from housing and transport arranged by the Polar Expedition in Talnakh and the Norilsk Surveying Expedition in Norilsk. R. K. Pickerill (University of New Brunswick) is thanked for advice regarding trace fossils. An early draft of this chapter was reviewed by A. M. Ziegler (University of Chicago) and B. J. Witzke (Iowa Geological Survey Bureau), whose comments helped improve the final version.

APPENDIX 1: EXPOSURE P-86201 ON RIGHT BANK OF MAYMECHA RIVER IN ARTIC SIBERIA (SEE FIGS. 8 AND 9)

Orochin Formation

Unit 13: Dolomite, light-gray, coarse-grained with some fine-grained concentrations, flaggy beds (7 to 20 cm thick), alternating in the upper part with more massive, horizontally to obliquely laminated, vuggy dolomite showing frequent stylolites; tabulate corals (Silurian) in the lower part. . . . . 1.8 m

Unit 12: Dolomite, gray, unevenly bedded flagstones (2 to 5 cm thick) with thin lenses of gray-green marl bearing trace-fossil burrows (2 to 3 mm in diameter). . . . . 0.3 m

Unit 11: Dolomite, light gray, fine-grained, medium-bedded flagstones (15 to 30 cm thick) with laminations, large chert concretions; abundant small tabulate corals, brachiopods, bryozoans, and crinoids (Silurian) occurring in lenses; trace fossils of "shoe-string" burrows along laminations; the flagstones are separated by green clay partings . . . . 3.5 m

Unit 10: Dolomite, light gray, fine-grained, thinly bedded flagstones (5 to 10 cm thick); lower part shows regular laminations (probably algal in origin); upper part shows regular laminations intercalated with discontinuous and plicated laminations . . . . 1.0 to 1.2 m

Unit 9: Dolomite, light gray, fine-grained, medium-bedded (10 to 20 cm thick) with cross laminations having sets that are 2 to 3 cm and rarely 10 cm thick, includes elongate (5 to 10 cm) pieces of brecciated limestone; empty molds and some casts of gastropods; some interbeds with bioturbated horizonal laminations (burrows 2 to 3 mm in diameter); the upper surface of this bed is locally uneven, with knolls up to 25 cm of relief (Fig. 9B) . . . . 0.2 to 0.5 m

Unit 8: Dolomite, light greenish gray, fine-grained, unevenly bedded with intercalations of gray-green marl (1 to 2 mm thick). . . 0.2 m

Unit 7: Dolomite, violet-brown to spotted green, fine-grained, numerous vertical burrows (2 mm in diameter); the lower surface of this bed is ocherous . . . . 0.15 m

Kuranakh Formation

Unit 6: Dolomite, light greenish gray to violet-brown, fine-grained, obliquely laminated; numerous vertical trace-fossil burrows in the upper part (2 to 3 mm in diameter); irregular concretions of pyrite (1 to 1.5 cm in diameter); quartzite granules form a drape deposit (about 0.5 cm thick) filling channels of gentle relief at the top of this bed . . . . 0.3 m

Unit 5: Dolomite marl, brown-red with lilac-gray intercalations, horizontally laminated; middle 30 cm is composed of flaggy dolomites (up to 10 cm thick) with numerous horizonal burrows; upper 20 cm is composed of silty dolomite of violet-brown to red-brown color, showing undulating laminations . . . . 1.05 m

Unit 4: Algal dolomite of complex structure; lower part (30 cm thick) shows undulating beds with thin laminations of light-gray color; middle part consists of a dolomite breccia with numerous concretions of epigenetic gray chert (up to 20 cm in diameter); upper part (30 cm thick) is composed of massive, light-gray, fine-grained dolomite with lenses (up to 5 mm thick) of fine-grained sandy material . . . . 1.0 m

Unit 3: Red beds of complex structure; basal part (20 cm thick) consists of intercalations of lilac-gray dolomitic marl and light-gray, thin flagstones of marly dolomite with lenselike horizontal laminations; middle part includes red-brown dolomite; upper part (20 cm thick) is composed of marly dolomite of lilac-gray color with abundant trace-fossil burrows . . . . 0.5 m

Unit 2: Dolomite, light-gray, with thin horizonal laminations; white chert concretions of irregular form (possibly algal in origin) occur in the middle and upper parts of the stratum; upper part includes fine-grained, uncemented sand . . . . 0.15 m

Unit 1: Marly dolomite, lilac-brown, thin bedded, obliquely laminated, with intercalations of flaggy beds (up to 8 cm thick); asymmetrical ripple marks. . . . . 0.5 m

APPENDIX 2. FAUNAL DOMINANCE, GORBIACHIN SECTION, SIBERIA

Interval	Lithology	Thickness	Dominant fossils	Rank	Benthic zone
119	Marl	1.10m	*Clorinda undata*		BA 5
118	Limestone	0.80m	*Hesperorthis rubeli*		
117	Marl	1.70m	*Clorinda undata*		BA 5
116	Limestone	0.90m	*Hesperorthis rubeli*		
115	Marl	1.35m	*Clorinda undata*		BA 5
114	Limestone	0.65m	*Favosites gothlandicus*		
113	Limestone	1.10m	*Clorinda undata*		BA 4/5
			Kulumbella biconvexa		
112	Limestone	0.50m	*Favosites gothlandicus*		
111	Marl	0.95m	*Favosites gothlandicus*		BA 4/5
			Clorinda undata		
110	Limestone	0.45m	*Favosites gothlandicus*		
109	Nodular limestone	2.00m	*Strophomena columbensis*		
108	Limestone	2.30m	*Mesofavosites dualis*		
107	Nodular limestone	1.85m	*Favosites gothlandicus*		
106-104	Limestone	6.15m	*Favosites gothlandicus*		
103	Limestone	0.50m	*Calamopora alveolaris*		
102-100	Limestone	3.65m	*Favosites gothlandicus*		
99-95	Limestone	11.00m	*S. solutus*		
94	Limestone	7.20m	*Pararaphistoma qualteriatum*		BA 3
			Eocoelia memispaerica	[2]	
			Pentamerus oblongus	[3]	
			Kulumbella kulumbensis	[1]	
93	Marl & limestone	1.25m	*Holopea pervetusta*		
			Kulumbella kulumbensis	[2]	
			Clorinda undata		
92	Limestone	1.50m	*Hormotoma gracilis*		BA 3
			Pentamerus oblongus	②	
91	Marl & limestone	0.60m	*Hormotema gracilis*		
			Eocoelia hemisphaerica		
			Kulumbella kulumbensis		
			Clorinda undata		
90	Limestone	1.50m	*Pentamerus oblongus*	②	BA 3
89	Marl & limestone	0.70m	nothing		
88	Limestone	4.80m	*Pentamerus oblongus*	②	BA 3
			Eocoelia memisphaerica	[1]	
			Kulumbella kulumbensis	①	
87	Marl	0.50m	nothing		
86	Limestone	1.40m	*Pentamerus oblongus*	[3]	BA 3
85	Marl	0.70m	*Murchisonia sp.*		
			Pentamerus oblongus	②	
84	Limestone	5.50m	*Eocoelia memispahaerica*	③	BA 3
			Pentamerus oblongus	④	
83	Limestone	0.90m	*Pentamerus oblongus*	[4]	BA 3
82	Limestone	1.30m	*Eocoelia memisphaerica*	[3]	BA 2/3
			Pentamerus oblongus	③	
81	Limestone	1.40m	*Favosites gothlandicus*		BA 2
			Eocoelia memisphaerica	[2]	
			Pentamerus oblongus	②	

APPENDIX 2. FAUNAL DOMINANCE, GORBIACHIN SECTION, SIBERIA (continued, p. 2)

Interval	Lithology	Thickness	Dominant fossils	Rank	Benthic zone
80	Marl	1.40m	*Meifodia recta*		BA 2
			Eocoelia memisphaerica	④	
			Kulumbella kulumbensis	①	
79	Limestone	0.80m	Crinoidea	[3]	BA 3
			Pentamerus oblongus	③	
78	Marl	1.50m	*Eocoelia memisphaerica*	[3]	BA 2/3
			Pentamerus oblongus	①	
			Kulumbella kulumbensis	①	
			Kulumbella biconvexa		
77	Limestone	1.20m	*Plectostroma tenuiplanum*		
			Pentamerus oblongus	①	
76	Marl	1.90m	Trilobita		
			Eocoelia memisphaerica	②	
75	Limestone	1.30m	Trilobita		
			Eocoelia memisphaerica	②	
			Pentamerus oblongus	◇1	
74	Limestone & marl	1.55m	*Dalmanella neocrassa*		BA 2
			Eocoelia memisphaerica	③	
			Kulumbella kulumbensis	[2]	
73	Limestone	0.85m	*Lebechia obrouchevi*		BA2
			Eocoelia memisphaerica	[1]	
72	Marl	1.10m	*Pentamerus oblongus*	◇2	
71	Limestone	1.10m	*Dalmanella neocrasa*		BA 3
			Pentamerus oblongus	[4]	
			Kulumbella kulumbensis	①	
70	Shale	1.70m	*Meristima lacrima*		
			Pentamerus oblongus	◇2	
			Kulumbella kulumbensis	①	
69	Limestone	0.95m	*Calamopora alveolaris*	[4]	BA 3
			Pentamerus oblongus	[4]	
			Kulumbella kulumbensis	①	
68	Shale	1.20m	*Strophomena kulumbensis*		
67	Limestone	1.20m	*Pentamerus oblongus*	[4]	BA 3
			Kulumbella kulumbensis	①	
			Eocoelia memisphaerica	①	
66	Marl	1.80m	*Meristima lacrima*		
65	Limestone	1.35m	*Dalmanella neocrassa*		BA 3
			Pentamerus oblongus	③	
64	Marl & limestone	1.05m	*Kulumbella kulumbensis*		
63	Marl	1.20m	*Pentamerus oblongus*	[3]	BA 3
62	Limestone	0.90m	*Favosites gothlandicus*		
			Pentamerus oblongus	②	
61	Marl	0.90m	*Meifodia rectus*	[4]	
			Pentamerus oblongus	②	
			Eocoelia memisphaerica		
60	Shale	0.50m	*Favosites gothlandicus*		
59	Limestone	0.70m	*Favosites gothlandicus*		
			Eocoelia memisphaerica		
58	Nodular limestone	1.00m	*Eoplectodonta transversalis*		BA 4
57	Limestone	0.55m	*Kulumbella kulumbensis*	②	
			Pentamerus oblongus	②	
56	Marl	1.50m	*Pentamerus oblongus*	②	

APPENDIX 2. FAUNAL DOMINANCE, GORBIACHIN SECTION, SIBERIA (continued, p. 3)

Interval	Lithology	Thickness	Dominant fossils	Rank	Benthic zone
55	Limestone	1.45m	*Favosites gothlandicus*		
			Kulumbella kulumbensis	①	
54	Limestone	1.70m	unidentified brachiopods		
53	Limestone	0.30m	nothing		
52	Limestone	2.00m	*Meifodia recta*		
51	Limestone	0.60m	*Favosites gothlandicus*		
50	Limestone	1.00m	*Meifodia recta*		
49	Limestone	0.35m	Crinoidea		
48	Limestone	0.85m	Unidentified brachiopods		
47	Nodular limestone	0.85m	*Pentamerus oblongus*	②	
46	Shale	1.10m	*Septatrypa pentagonalis*		
45	Limestone	0.50m	*Clorinda undata*		BA5
44	Limestone	0.80m	Unidentified brachiopods		
43-42	Marl	1.50m	Unidentified brachiopod		
41	Nodular limestone	0.35m	*Kulumbella kulumbensis*		
40	Marl	0.80m	*Multisolenia tortuosa*		BA 3
			Kulumbella kulumbensis	[2]	
			Kulumbella biconvexa	[1]	
			Pentamerus oblongus	③	
39	Marl	0.50m	*Pentamerus oblongus*	②	
38	Shale	1.00m	*Pentamerus oblongus*		
			Clorinda undata		
37	Marl	0.50m	*Catenipora escharoides*		
			Clorinda undata		
36	Limestone	0.30m	*Leptostrophia andreevae*		BA 3
			Kulumbella kulumbensis	②	
			Pentamerus oblongus	②	
			Eocoelia memishpaerica		
35	Limestone	0.75m	*Favosites gothlandicus*		BA 3
			Pentamerus oblongus	②	
34-33	Shale	1.40m	rugosa		
32	Nodular limestone	0.95m	rugosa		
31	Limestone	0.80m	*Favosites gothlandicus*		BA 3
			Pentamerus oblongus	[2]	
30-29	Marl	1.20m	*Kulumbella kulumbensis*	[4]	BA 4
			Clorinda undata		
			Pentamerus oblongus		
28	Nodular limestone	0.50m	*Kulumbella kulumbensis*	[3]	BA 4
			Clorinda undata		
			Pentamerus oblongus		
27	Limestone	0.70m	*Catenipora escharoides*		BA 3
			Pentamerus oblongus	[4]	
26	Limestone	0.50m	*Favosites gothlandicus*		BA 3
			Kulumbella kulumbensis	②	
25	Marl	1.50m	unidentified brachiopods		
24	Nodular limestone	0.50m	*Favosites gothlandicus*		
23	Nodular limestone	0.80m	*Favosites gothlandicus*		BA 3
			Pentamerus oblongus	②	
22	Nodular limestone	1.20m	*Clorinda undata*		BA 4/5
21	Shale	0.75m	*Favosites gothlandicus*		
20	Limestone & marl	1.20m	*Favosites gothlandicus*		BA 3
			Pentamerus oblongus		

APPENDIX 2. FAUNAL DOMINANCE, GORBIACHIN SECTION, SIBERIA (continued, p. 4)

Interval	Lithology	Thickness	Dominant fossils	Rank	Benthic zone
19	Limestone	1.00m	*Favosites gothlandicus*		BA 3
			Pentamerus oblongus	[1]	
18	Nodular limestone	1.00m	*Favosites gothlandicus*		BA 4/5
			Kulumbella biconvexa	[1]	
17	Shale	2.20m	*Clorinda undata*	◇2	BA 5
			Kulumbella kulumbensis	[1]	
			Kulumbella biconvexa	[1]	
16	Nodular limestone	1.20m	*Protatrypa alia*	◇3	BA 4/5
			Meifodia recta		
			Virgiana moyeroensis	②	
15	Limestone	2.10m	*Favosites gothlandicus*		
			Multisolenia tortuosa		
			Catenipora eschaoides		
14	Nodular limestone	0.90m	*Favosites gothlandicus*		
			Pentamerus oblongus	[1]	
13	Limestone & shale	4.00m	*Meifodia recta*	④	BA 5
			Favosites gothlandicus		
12	Limestone	2.45m	*Favosites gothlandicus*		
			Catenipora escharoides		
			Protatrypa alia	◇4	
			Eocoelia memisphaerica	◇1	
11	Nodular limestone	1.00m	*Favosites gothlandicus*		BA 4/5
			Catenipora escharoides		
			Pentamerus oblongus	[3]	
			Kulumbella biconvexa		
			Clorinda undata	◇1	
10	Shale	2.00m	*Glyptograptus tamariscus*		BA 5
			Eocoelia memispaherica	◇1	
			Favosites gothlandicus		

REFERENCES CITED

Baarli, B. G., 1990, Peripheral bulge of a foreland basin in the Oslo region during the Early Silurian: Palaeogeography, Palaeoclimatology, Palaeoecology, v. 78, p. 149–161.

Barr, W., 1991, Aleksandr Lavrent'yevich Chekanovskiy, pioneer geologist and explorer of north central Siberia, 1873–76: Earth Sciences History, v. 10, p. 106–129.

Berry, W. B. N., and Boucot, A. J., 1970, Correlation of the North American Silurian rocks: Geological Society of America Special Paper 102, 289 p.

Beaumont, C., Quinlan, G., and Hamilton, J., 1988, Orogeny and stratigraphy: Numerical models of the Paleozoic in the eastern interior of North America: Tectonics, v. 7, p. 389–416.

Bolton, J. C., 1990, Sedimentological data indicate greater range of water depths for *Costistricklandia lirata* in the southern Appalachians: Palaios, v. 5, p. 371–374.

Bolton, J. C., and Driese, S. G., 1990, The determination of substrate conditions from the orientations of solitary rugose corals: Palaios, v. 5, p. 479–483.

Bond, G., 1978, Evidence for Late Tertiary uplift of Africa relative to North America, South America, Australia, and Europe: Journal of Geology, v. 86, p. 47–65.

Boucot, A. J., 1975, Evolution and extinction rate controls: Amsterdam, Elsevier, 427 p.

Brett, C. E., Goodman, W. M., and LoDuca, S. T., 1990, Sequences, cycles, and basin dynamics in the Silurian of the Appalachian foreland basin: Sedimentary Geology, v. 69, p. 191–244.

Buckman, J. O., 1994, *Archaeonassa* Fenton and Fenton 1937 reviewed: Ichnos, v. 3, p. 185–192.

Cocks, L. R. M., 1978, A review of British Lower Palaeozoic brachipods including a synoptic revision of Davidson's monograph: London, Palaeontolographical Society, 256 p.

Colville, V. R., and Johnson, M. E., 1982, Correlation of sea-level curves for the Lower Silurian of the Bruce Peninsula and Lake Timiskaming District (Ontario): Canadian Journal of Earth Sciences, v. 19, p. 962–974.

Copper, P., 1982, Early Silurian atrypoids from Manitoulin Island and Bruce Peninsula, Ontario: Journal of Paleontology, v. 56, p. 680–702.

Cotter, E., 1983, Shelf, paralic, and fluvial environments and eustatic sea-level fluctuations in the origin of the Tuscarora Formation (Lower Silurian) of central Pennsylvania: Journal of Sedimentary Petrology, v. 53, p. 25–49.

Cotter, E., 1990, Storm effects on siliciclastic and carbonate shelf sediments in the medial Silurian succession of Pennsylvania: Sedimentary Geology, v. 69, p. 254–258.

Folk, R. L., 1960, Petrography and origin of the Tuscarora, Rose Hill, and Keefer Formations, Lower and Middle Silurian of eastern West Virginia: Journal of Sedimentary Petrology, v. 30, p. 1–58.

Goodman, W. M., and Brett, C. E., 1994, Roles of eustasy and tectonics in development of Silurian stratigraphic architecture of the Appalachian foreland basin, *in* Dennison, J. M., and Ettensohn, F. R., eds., Tectonic and eustatic controls on sedimentary cycles: SEPM (Society for Sedimentary Geology) Concepts in Sedimentology and Paleontology, v. 4, p. 147–169.

Grahn, Y., and Caputo, M. V., 1992, Early Silurian glaciations in Brazil: Palaeogeography, Palaeoclimatology, Palaeoecology, v. 99, p. 9–15.

Holland, C. H., and Bassett, M. G., eds., 1989, A global standard for the Silurian system: National Museum of Wales Geological Series 9, 325 p.

Johnson, J. G., 1971, Timing and coordination of orogenic, epeirogenic, and eustatic events: Geological Society of America Bulletin, v. 82, p. 3263–3298.

Johnson, J. G., Klapper, G., and Sandberg, C. A., 1985, Devonian eustatic fluctuations in Euramerica: Geological Society of America Bulletin, v. 96, p. 567–587.

Johnson, M. E., 1981, Correlation of Lower Silurian strata from the Michigan Upper Peninsula to Manitoulin Island: Canadian Journal of Earth Sciences, v. 18, p. 869–883.

Johnson, M. E., 1987, Extent and bathymetry of North American platform seas in the early Silurian: Paleoceanography, v. 2, p. 185–211.

Johnson, M. E., 1989, Tempestites recorded as variable *Pentamerus* layers in the Lower Silurian of southern Norway: Journal of Paleontology, v. 63, p. 195–205.

Johnson, M. E., and Campbell, G. T., 1980, Recurrent carbonate environments in the Lower Silurian of northern Michigan and their inter-regional correlation: Journal of Paleontology, v. 54, p. 1041–1057.

Johnson, M. E., and McKerrow, W. S., 1991, Sea level and faunal changes during the latest Llandovery and earliest Ludlow (Silurian): Historical Biology, v. 5, p. 153–169.

Johnson, M. E., Kaljo, D., and Rong, J.-Y., 1991, Silurian eustasy, *in* Bassett, M. G., Lane, P. D., and Edwards, D., eds., The Murchison Symposium: Proceedings of an International Conference on the Silurian System: Special Papers in Palaeontology, v. 44, p. 145–163.

Johnson, M. E., Baarli, B. G., and Scott, J. H., Jr., 1995, Colonization and reef growth on a Late Pleistocene rocky shore and abrasion platform in Western Australia: Lethaia, v. 28, p. 85–98.

Kaljo, D., Nestor, H., and Einasto, R., 1991, Aspects of Silurian carbonate platform sedimentation, *in* Bassett, M. G., Lane, P. D., and Edwards, D., eds., The Murchison Symposium: Proceedings of an International Conference on the Silurian System: Special Papers in Palaeontology, v. 44, p. 205–224.

LoDuca, S. T., and Brett, C. E., 1994, Revised stratigraphic and facies relationships of the lower part of the Clinton Group (middle Llandoverian) of western New York State, *in* Landing, E., ed., Studies in stratigraphy and paleontology in honor of Donald W. Fisher: New York State Museum Bulletin 481, p. 161–182.

Marsaglia, K. M., and Klein, G. D., 1983, The paleogeography of Paleozoic and Mesozoic storm depositional systems: Journal of Geology, v. 91, p. 117–142.

Nikiforova, O. I., and Andreeva, O. N., 1961, Ordovician and Silurian stratigraphy of the Siberian Platform and its paleontological basis (Brachiopoda): Nauchno-Issledov, Geological Institute (VSEGEI), Trudy, v. 56, 412 p., 56 plates.

Reed, J. C., Jr., 1993, Development of the southern reaches of Laurentia: A view from the conterminous United States, *in* Reed, J. C., Jr., et al., eds., Precambrian: Conterminous U.S.: Boulder, Colorado, Geological Society of America, Geology of North America, v. C-2, p. 622–636.

Rickard, L. V., 1975, Correlation of the Silurian and Devonian rocks of New York State: New York State Museum and Science Service Map and Chart Series 24, 16 p., 4 plates.

Rong, J.-Y., and Johnson, M. E., 1996, A stepped karst unconformity as an early Silurian rocky shoreline in Guizhou Province (South China): Palaeogeography, Palaeoclimatology, Palaeoecology, v. 121, p. 115–129.

Scotese, C. R. and McKerrow, W. S., 1990, Revised world maps and introduction, *in* McKerrow, W. S. and Scotese, C. R., eds., Palaeozoic palaeogeography and biogeography: Geological Society of London Memoir 12, p. 1–21.

Smith, N. D., 1970, The braided stream depositional environment: Comparison of the Platte River with some Silurian clastic rocks, north-central Appalachians: Geological Society of America Bulletin, v. 81, p. 2993–3014.

Sokolov, B. S., and Tesakov, Yu. I., 1963, Paleozoic tabulates of Siberia: Moscow and Leningrad, Moscow-Leningrad Publishers, USSR Academy of Sciences, 188 p. [in Russian].

Tesakov, Yu. I., 1978, Tabulata—Population, biocoenosic and biostratigraphic analysis: Transactions of the Institute of Geology and Geophysics, USSR Academy of Sciences, Siberian Branch, v. 409, 262 p. [in Russian].

Tesakov, Yu. I., 1981, Development of ecosystems in old platform sedimentary basins, *in* The problems of geological processes evolution: Novosibirsk, Nauka, p. 186–199. [in Russian].

Tesakov, Yu. I., Predtechensky, N. N., Bazarova, L. S., et al., 1979, Silurian of the Siberian platform: New regional and local stratigraphic divisions: USSR Academy of Sciences, Siberian Branch, Novosibirsk, Nauka, 96 p. [in Russian].

Tesakov, Yu. I., Predtechensky, N. N., Bazarova, L. S., et al., 1980, The reference sections of the Gorbiachin River, *in* Sokolov, B. S., Obut, A. M., Tesakov, Yu. I., and Predtchensky, N. N., eds., Silurian of the Siberian Platform: Reference sections of the NW Siberian platform: USSR Academy of Sciences, Siberian Branch, Novosibirsk, Nauka, p. 7–55 [in Russian].

Tesakov, Yu. I., Predtechensky, N. N., Khromych, V. G., et al., 1986, Fauna and flora of the polar areas of the Siberian platform: USSR Academy of Sciences, Siberian Branch, Novosibirsk, Nauka, 216 p. [in Russian]

Tesakov, Yu. I., Predtchensky, N. N., Khromych, V. G., et al., 1992, Silurian sections and fauna of the northern Tunguska Synclinorium: Transactions of the Institute of Geology and Geophysics, Russian Academy of Sciences, Siberian Branch, v. 789, 192 p. [in Russian].

Thomsen, E., and Baarli, B. G., 1982, Brachiopods of the Lower Llandovery Saelabonn and Solvik formations of the Ringerike, Asker and Oslo districts, *in* Worsley, D., ed., Subcommission on Silurian Stratigraphy Field Meeting, Oslo Region 1992: Paleontological Contributions from the University of Oslo 278, p. 63–77.

Walker, J. P., 1975, Sedimentology of the Medina and Clinton groups (Lower and Middle Silurian) of western and central New York [M.S. thesis]: Urbana-Champaign, University of Illinois, 105 p.

Ziegler, A. M., Cocks, L. R. M., and Bambach, R. K., 1968, The composition and structure of Lower Silurian marine communities: Lethaia, v. 1, p. 1–27.

Ziegler, A. M., Hansen, K. S., Johnson, M. E., Kelly, M. A., Scotese, C. R., and Van der Voo, R., 1977, Silurian continental distributions, paleogeography, climatology, and biogeography: Tectonophysics, v. 40, p. 13–51.

Manuscript Accepted by the Society February 27, 1997

Printed in U.S.A.

Geological Society of America
Special Paper 321
1997

Wenlock (Silurian) depositional sequences, eustatic events, and biotic change on the southern shelf of North America

James E. Barrick
Department of Geosciences, Texas Tech University, Lubbock, Texas 79409-1053

ABSTRACT

Wenlock strata on the southern cratonic shelf of North America comprise a thin veneer of carbonates that accumulated as three depositional sequences, the magnitude and extent of which are related to sea-level events. Depositional sequences and associated biotic events were dated and correlated using conodont range events for graphic correlation and a newly modified version of the Silurian Composite Standard. Similarities of southern North America depositional sequences with patterns of sea-level change and oceanic episodes determined elsewhere indicate that eustasy was the primary control. The magnitude and timing of eustatic events inferred here for the Wenlock, though, differ from most previously proposed global sea-level curves for the series.

Erosion during a major late Telychian (latest Llandovery) fall in sea level produced a regional unconformity. A major early Sheinwoodian rise in sea level rapidly flooded the southern shelf, such that only thin transgressive shales formed before deep-water carbonates appeared with the offshore *Dapsilodus obliquicostatus* conodont biofacies and associated acrotretid brachiopod and foraminiferal faunas. Overlying Sheinwoodian carbonate lithofacies and microfaunal biofacies record a shallowing-upward trend on the shelf. A late Sheinwoodian acme in *D. praecipuus* is attributed to a brief episode of altered oceanic circulation over the outer shelf. The smaller early Homerian sea-level rise briefly reversed the shallowing trend in lithofacies and biofacies and introduced greater quantities of fine terrigenous clastics into the spectrum of carbonate environments. During the late Homerian, a major influx of fine terrigenous clastics, possibly related to a sea-level rise, spread across the inner and outer shelf regions, producing a succession of silty marlstones that extends into the Ludlow.

INTRODUCTION

Wenlock strata comprise a relatively thin package of shelf carbonates that stretches discontinuously across the southern cratonic margin of North America. Within this relatively complete depositional package is a series of biotic and depositional events that represent the combined influence of local, regional, and global changes. The proper sequencing and correlation of events within the region depends heavily on the careful application of conodont biostratigraphy, which also provides the best means by which to compare southern North America events with patterns of eustatic sea-level change and other global events. By a combination of lithofacies analysis and biostratigraphy, Silurian eustatic events can be interpreted with the same care and accuracy that J. G. Johnson and his coworkers have used to resolve Devonian eustatic events (Johnson et al., 1985; Johnson and Klapper, 1992).

In this chapter, I review the framework of biostratigraphic

Barrick, J. E., 1997, Wenlock (Silurian) depositional sequences, eustatic events, and biotic change on the southern shelf of North America, *in* Klapper, G., Murphy, M. A., and Talent, J. A., eds., Paleozoic Sequence Stratigraphy, Biostratigraphy, and Biogeography: Studies in Honor of J. Granville ("Jess") Johnson: Boulder, Colorado, Geological Society of America Special Paper 321.

correlation that has been developed to correlate lithostratigraphic units and depositional and biotic events of Wenlock age in southern North America. The evaluation of biostratigraphic ranges of conodonts in the region forces a reconsideration of a portion of the Silurian Composite Standard of Kleffner (1989, 1995), which remains the best means by which to compare the chronology of regional events with proposed global events. Using a modified biostratigraphic framework based largely on graphic correlation, the chronological sequence of biotic and depositional events is summarized and interpreted with reference to models of regional and global sea-level change.

LITHOSTRATIGRAPHIC FRAMEWORK

Shelf strata of Wenlock age are exposed in discontinuous outcrop areas in southern Oklahoma, northern Arkansas, central and western Tennessee, and along the Mississippi River in southeastern Missouri and southwestern Illinois (Fig. 1). More extensive areas of Wenlock strata lie buried in the subsurface of the Illinois Basin in Illinois, the Anadarko Basin of Oklahoma, and the Permian Basin region of west Texas and eastern New Mexico. The rarity of fossil material from basinal sediments of the Ouachita Facies has hindered the recognition of Wenlock-age units, but shales and cherts of this age have been identified in both the southeastern Oklahoma (Barrick and Haywa-Branch, 1994; BKR on Fig. 1) and west Texas (Noble, 1994; M on Fig. 1) parts of the belt. Each area possesses its own series of lithostratigraphic subdivisions and names. The correlation of lithostratigraphic units shown in Figure 2 is the result of the work accomplished by a number of workers, discussed below, and revisions based on the correlation framework discussed later in this chapter.

In southern Oklahoma and the Anadarko Basin, Wenlock strata comprise the Clarita Formation and basal beds of the Henryhouse Formation of the Hunton Group. The lithostratigraphy, biostratigraphy, and paleoecology of these units in surface and subsurface sections have been described in a number of papers (e.g., Amsden, 1960, 1968, 1976; Amsden et al., 1980; Amsden and Barrick, 1988). The Clarita Formation includes two members, the lower thin Prices Falls Shale, and the upper Fitzhugh Member, a thicker carbonate unit. Conodont faunas from surface sections of the Clarita were described by Barrick and Klapper (1976) and Barrick (1977). Additional conodont faunas have been obtained from subsurface sections during a restudy of the Hunton Group in the Anadarko Basin (Amsden and Barrick, 1993). Amsden (1960, 1976) and Bar-

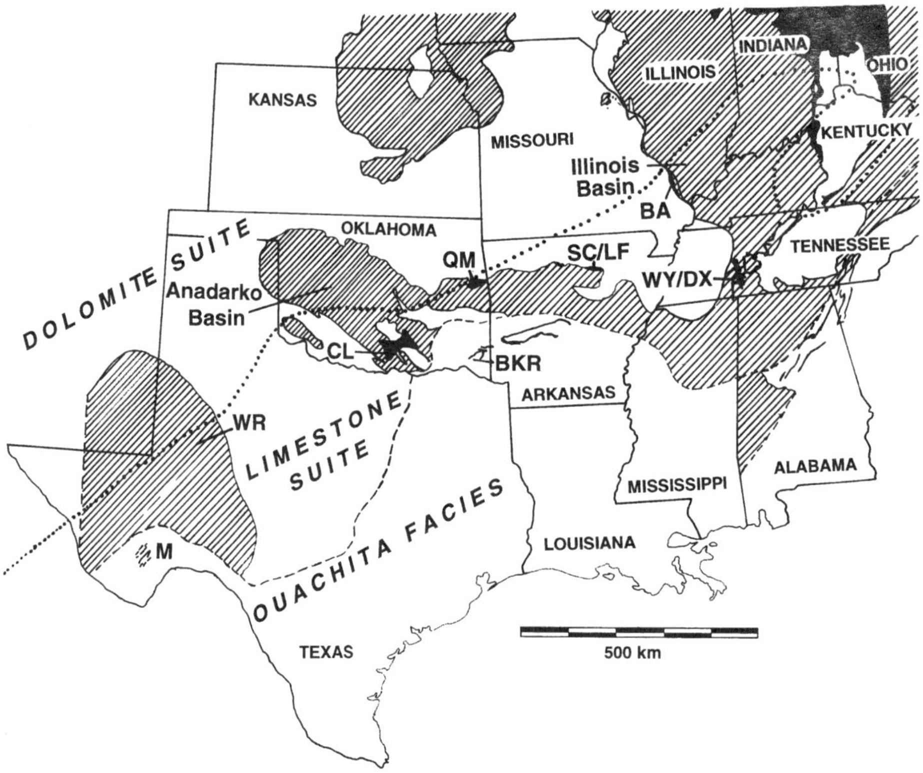

Figure 1. Distribution of Silurian strata in the southern United States and areas of Wenlock strata discussed in this paper. Black—outcrops; shaded—subsurface extent. M—Marathon uplift; WR—Wristen Group; CL—Clarita and Henryhouse Formations; QM—Quarry Mountain Formation; BKR—Black Knob Ridge; SC/LF—St. Clair and Lafferty limestones; WY/DX—Wayne and Dixon Formations; BA—Bainbridge Group. Modified from Berry and Boucot (1970, pl. 1), which includes a discussion of the dolomite and limestone suites.

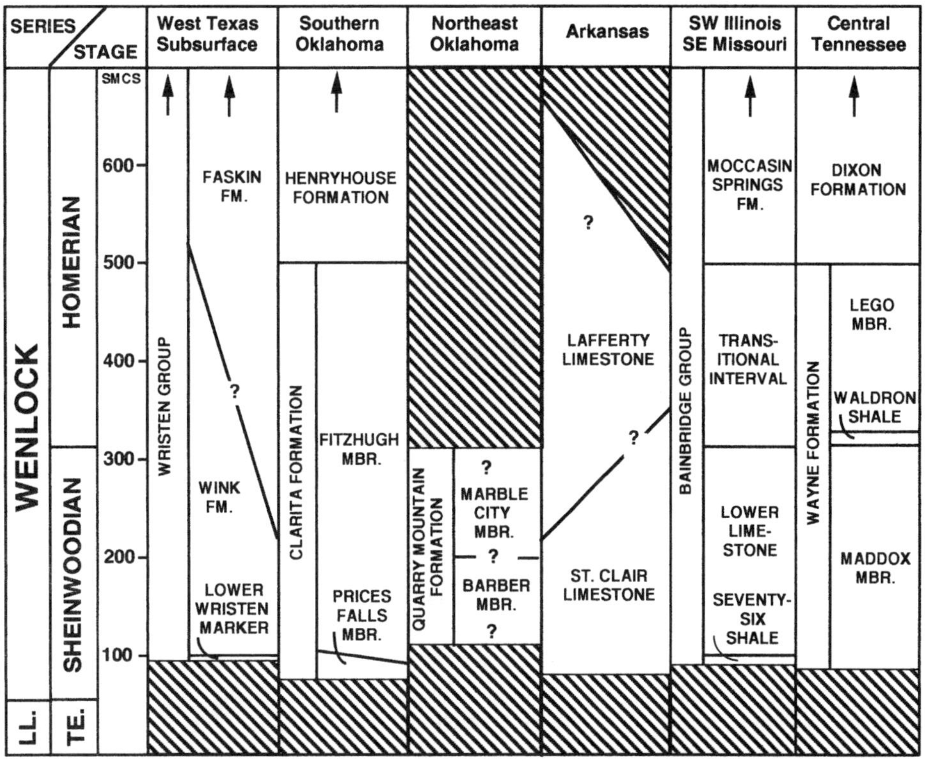

Figure 2. Chart of lithostratigraphic units of Wenlock age in the southern United States. References to individual columns are given in the text. Correlations are based on conodonts and the Southern Mid-continent Composite Standard (SMCS) discussed in the text. An arrow at top of a column indicates that the unit is known to range into the Ludlow. LL.—Llandovery Series; TE.—Telychian Stage.

rick et al. (1990) summarize lithofacies, biofacies, and biostratigraphic information for the Henryhouse. Conodont faunas of the Henryhouse Formation are currently under study. A small area of Wenlock strata, the Quarry Mountain Formation, crops out in northeastern Oklahoma on the flank of the Ozark Dome. Amsden (1978) reviewed the lithofacies and macrofauna of the two members of the Quarry Mountain Formation, the lower Barber and upper Marble City Members. The small conodont fauna from the Quarry Mountain is reported in Barrick (1978).

Craig (1969, 1993) provides the most comprehensive summary of the lithofacies and conodont biostratigraphy of the St. Clair Limestone and lower Lafferty Limestone in northern Arkansas. Amsden (1968) discussed the brachiopod fauna and biostratigraphy of the St. Clair. Conodont biofacies of the St. Clair are described more fully in Craig (1968) and Barrick (1978). Because so little biostratigraphic data are available for the Lafferty (Craig, 1993), it is difficult to determine whether it is Wenlock or younger in age.

The Bainbridge Group comprises a thick succession of carbonates and argillaceous carbonates that ranges in age from the Wenlock to possibly as high as the Prídolí (Droste and Shaver, 1987). The Bainbridge crops out along both sides of the Mississippi River in southeastern Missouri and southwestern Illinois. It extends into the subsurface of the Illinois Basin, where it locally contains reefs and associated facies (Droste and Shaver, 1987). In this chapter, only the outcropping sections of the Bainbridge are included because too few biostratigraphic data are available for the subsurface sections. The base of the Bainbridge Group is a thin shale, the Seventy-Six Shale. The relatively thin sequence of carbonates lacking argillaceous interbeds overlying the Seventy-Six Shale has been assigned to the St. Clair Limestone, borrowing the name from the Arkansas succession. Because the "St. Clair" of the Bainbridge is not lithologically similar nor contiguous with the St. Clair of Arkansas, the informal term "lower Bainbridge limestone," is used in this chapter. The lower Bainbridge limestone grades into a transitional interval of interbedded argillaceous and nonargillaceous carbonates. This transitional interval grades into the dominantly argillaceous carbonates and shales of the Moccasin Springs Formation, which forms the upper part of the Bainbridge Group. Conodont faunas from the Seventy-Six Shale were described by Satterfield and Thompson (1975), and those of the lower Bainbridge limestone have been described by Cooper (1976). Barrick (1978) includes additional information on conodont faunas and biofacies from the lower Bainbridge limestone and the overlying transitional interval, both of which are Wenlock in age.

Barrick (1983) summarized the lithofacies and conodont biostratigraphy and biofacies of the Wayne Formation in central and southern Tennessee. The Wayne Formation consists of a series of members that span the Wenlock, in ascending order:

Maddox Member (Osgood and Laurel Members of some authors), Waldron Shale, Lego Member. The lower part of the overlying Dixon Formation is latest Wenlock in age, but conodonts are known only from the base of the unit.

In the subsurface of West Texas, the Wristen Group includes strata of Wenlock age, but the upper parts of the unit may range as high as the Prídolí (Barrick, 1995). Detailed biostratigraphic data are sparse, most coming from one core, the Mobil Pegasus in Midland County, Texas (Barrick et al., 1993). The Pegasus core lies offshore of a major Wristen carbonate shelf margin assigned to the Fasken Formation, which is characterized by a complex series of organic buildups and related facies (Ruppel, 1993; Ruppel and Holtz, 1994). Although, on lithostratigraphic evidence, the lower part of the Fasken carbonate platform is clearly equivalent to the Wink Formation in the Pegasus core, no faunal data from the Fasken are available to establish a biostratigraphic correlation.

BIOSTRATIGRAPHIC FRAMEWORK

Because graptolites are rare to absent in the predominantly carbonate sections that constitute the Wenlock Series in the southern United States, one must rely primarily on conodonts to effect regional and global correlations. The Wenlock zonation proposed by Barrick and Klapper (1976) works well where the four successive species of *Kockelella* can be recovered, but this is rarely the case in strata outside of Oklahoma. Aldridge and Schönlaub (1989) suggested that three apparently successive species of *Ozarkodina* (*O. sagitta rhenana, O. sagitta sagitta,* and *O. bohemica*) could offer some promise for a revised Wenlock zonation for the British succession. Although the first and third species occur with some regularity in the southern United States, *O. sagitta sagitta* has only been reported questionably once from Tennessee (Barrick, 1983). Regardless of the system employed, the use of conodont zones for Wenlock strata allows only four to five intervals of time to be recognized, which is too coarse a time frame upon which to locate depositional and biotic events. Even more troubling are the difficulties with relating conodont zones to the boundaries of the Wenlock Series and the Sheinwoodian and Homerian stages (Aldridge and Schönlaub, 1989).

Kleffner (1989; 1995) has developed a system of Silurian chronostratigraphy based on graphic correlation of conodont and graptolite range events, using data from sections from Europe and North America, including several from the southern United States. The Silurian Composite Standard (SCS) includes the correlated positions of the Wenlock Series and its stages, as well as some proposed global events (e.g., the Ireviken event of Aldridge et al., 1993). Besides including a vast array of biostratigraphic information composited in a systematic, semiquantitative manner, the graphic correlation technique permits a far greater degree of biostratigraphic and, it is to be hoped, chronostratigraphic, resolution to be obtained with rich data. Even when relatively few taxon range events occur, graphic correlation permits a reasonable estimate of chronostratigraphic relations to be made (see Kleffner, 1995, for further discussion).

Direct application of the most recent version of the SCS of Kleffner (1995) to sections from the southern United States, however, is problematic. Kleffner's graphs (1995, figs. 3 and 4) correlating sections of the Clarita Formation (Fitzhugh Member) from Oklahoma show results that are far different from previous correlations. Barrick and Klapper (1976) placed the base of the Fitzhugh Member at most sections (Highway 77, Ca2, Haragan Creek) in the lowermost part of their lower Wenlock *ranuliformis* Zone, just above the top of the *amorphognathoides* Zone. The underlying Prices Falls Member bears an *amorphognathoides* Zone fauna at all Clarita sections. In contrast, Kleffner's correlations of these sections place the base of the Fitzhugh well above the top of the *amorphognathoides* Zone. The base of the Fitzhugh is correlated so high in the SCS that a considerable unconformity appears to exist between the Prices Falls and Fitzhugh Members, and strata of Sheinwoodian age are absent. Because local and regional stratigraphic relations argue for more or less continuous deposition from the *amorphognathoides* Zone into the *ranuliformis* Zone (for example, the occurrence of the *amorphognathoides* Zone in the base of the Fitzhugh at section C1; Barrick and Klapper, 1976), it appears that Kleffner's correlations of the Oklahoma sections are erroneous.

In each of Kleffner's graphs, it should be noted, the line of correlation (LOC) is based on the positions of points occurring well above the base of the Fitzhugh, and the line is extrapolated to the base of the member. Slight changes in the positions of the lower points used to define the LOC will have major effects on the interpreted age of the base of the Fitzhugh. A provisional graphic correlation composite standard for the Oklahoma Wenlock sections, which I prepared in an attempt to provide greater time resolution for detailed biofacies studies, is unpublished. This provisional Southern Midcontinent Composite Standard (SMCS) uses the Highway 77 section of Barrick and Klapper (1976) as a standard reference section, is based thus far on only three rounds of correlation, and incorporates the interpretation of more or less continuous deposition from the *amorphognathoides* Zone into the *ranuliformis* Zone. It has recently been updated to include data from all Clarita sections, the more complete sections from Tennessee (Barrick, 1983), and unpublished range data from sections from the Oklahoma subsurface; the St. Clair Limestone, Arkansas; the lower part of the Bainbridge Group, southwestern Illinois and southeastern Missouri (Barrick, 1978); and the Mobil Pegasus core in the subsurface of west Texas (Barrick et al., 1993).

When the provisional SMCS is plotted against the SCS of Kleffner (1995), the reason for the different correlations of the base of the Fitzhugh is obvious (Fig. 3). There is exceptional agreement for the first appearance datums of taxa in the upper part of the Wenlock (*Kockelella variabilis, K. stauros, K. patula,* and *O. bohemica*). An LOC generated by these points intersects the top of the *amorphognathoides* Zone (last appear-

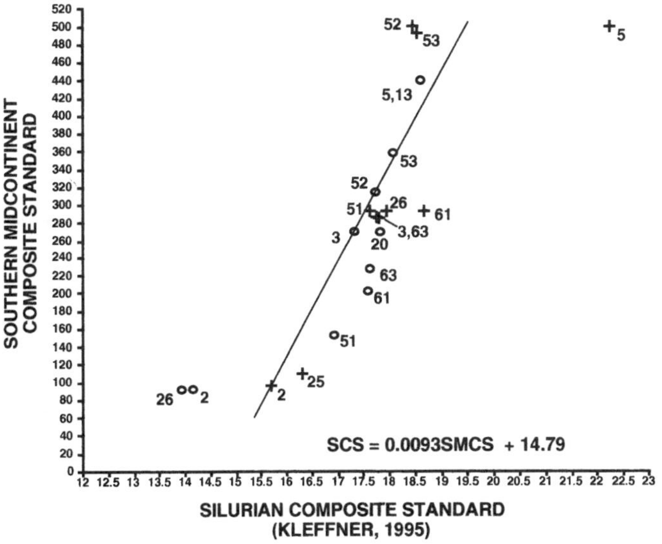

Figure 3. Graph of ranges of selected species of conodonts in the Southern Midcontinent Composite Standard (SMCS) plotted against ranges in the Silurian Composite Standard (SCS) of Kleffner (1995). Species represented by numbers are indicated on Figure 4. o—first appearance datum; +—last appearance datum. Asterisk (*) indicates datums whose positions in the SCS were determined by sections also used in the SMCS; these points were not used to draw the Line of Correlation (LOC). Formula of the LOC between the two standards is given on the lower right of the graph.

ance datum) on the SCS, just below where the base of the Fitzhugh Member lies in the SMCS, which would indicate that only a slight break exists at the base of the Fitzhugh Member at the Highway 77 section. The graph shows that taxa in the lower part of the Wenlock (*Ozarkodina sagitta rhenana, Kockelella walliseri, K. amsdeni*, and *K. absidata*) appear much lower in the SMCS than the SCS of Kleffner. Based on the graphic correlation of the provisional SMCS with the SCS of Kleffner, the first appearances of these taxa need to be revised significantly downward in the SCS. Other sections in the SCS of Kleffner for which the first appearances of these species were used to determine the LOC need to be reexamined.

Kleffner (1995) shows the base of the Homerian Stage (first appearance datum of *Cyrtograptus lundgreni*) at 16.15 in the SCS, a short interval above the base of the Wenlock at 15.30. Even with the recommended range revisions, the first appearance datums of *Ozarkodina sagitta rhenana* (16.21) and *Kockelella walliseri* (16.67) occur in the Homerian on the SCS. This is surprising, considering that both species are known from only the Sheinwoodian in Great Britain (Aldridge and Schönlaub, 1989) and from comparable strata in the Baltic region (Jeppsson et al., 1994). Also, the first occurrence of *O. sagitta sagitta,* which appears in upper Sheinwoodian strata in Britain, lies at 17.68, well above the base of the Homerian on the SCS. The position of the first appearance datum of *C. lundgreni* and other Wenlock graptolites in the SCS is based on correlation of faunas from sections in central Nevada that lack these conodont species (Klapper and Murphy, 1975; Berry and Murphy, 1975). Klapper et al. (1995), following the guidelines of Alan Shaw, emphasize the importance of *not* drawing the LOC such that ranges of well-known species overlap when they have not been observed to overlap in any actual section. Because the position of the base of the Homerian as correlated by Kleffner (1995) forces the overlap of species that do not occur together and are separated by significant intervals of strata in the British succession, some miscorrelation of at least the Wenlock parts of the Nevada sections with the SCS must have occurred.

Short of recorrelating the graptolite-bearing sections in the SCS, a reasonable estimate of the position of the base of the Homerian can be obtained by inspecting the ranges of conodont species on the SCS and comparing them to the British sections. *Ozarkodina sagitta sagitta* (17.67) appears in uppermost Sheinwoodian strata and *O. bohemica* (17.72) occurs in lower Homerian strata in Britain (Aldridge and Schönlaub, 1989). Jeppsson et al. (1994) provide additional integration of graptolite and conodont data from the Baltic region and place the base of the Homerian above the first occurrence of *Kockelella patula* (17.30 on the SCS) and *K. absidata* (17.30, revised herein; 17.81 before revision). *Ozarkodina bohemica* appears in strata below the Mulde beds in Gotland, which have *Gothograptus nassa* Zone (late Homerian) graptolites (Jeppsson et al., 1994). Using this information, the base of the Homerian is estimated to lie at about 17.70 on the SCS or at 313 on the SMCS, for the purposes of this chapter. The true position of the base of the Homerian on the SCS, however, can only be determined by using the graphic correlation technique directly.

In contrast to the problems with the Homerian, the position of the base of the Wenlock Series (and Sheinwoodian Stage) appears to be well established at 15.30 on the SCS (Kleffner, 1995), which correlates with 55 on the SMSC. Kleffner (1995) shows the base of the Ludlow to lie at 21.27 on the SCS, which falls within the *crassa* Zone. This point lies at 697 on the SMSC, well above the highest level (510) for which I have reliable data. No direct conodont control for this boundary exists in Britain, but the conodont succession from the Baltic (Jeppsson et al., 1994) fits well with this interpretation.

The differences in interpretation regarding the age of the lower Fitzhugh Member of the Clarita and the position of the base of the Homerian in the SCS are not an indication of any failings in the graphic correlation method. The composite standard that is developed by graphically compositing ranges of taxa from numerous sections does not represent a final result but, like a zonation, is a hypothesis about the ranges of species that will be tested and revised over time. The variable quality of biostratigraphic data and any errors introduced when the LOC is determined will cause uncertainties to arise about the placement of some points. However, when subsequent graphs are plotted, as in this chapter, anomalies in the placement and sequence of range events become obvious and can be reevaluated and corrected as necessary.

For Wenlock strata in the southern North America region, a

series of range events appears to be more reliable than others in correlating sections, whether graphic correlation or zones based on ranges are used. These events are shown on Figure 4 with reference to their position in the SMCS and to the SCS. Above the last occurrence of *Pterospathodus amorphognathoides* (99 SMCS) and below the first appearance of *Ozarkodina sagitta rhenana* (153) is a short interval of time that is not characterized by the presence of any particular species. The successive first occurrences of *O. s. rhenana, Kockelella walliseri* (202), *K. amsdeni* (228), and *K. patula* (270) or *K. absidata* (270) permit detailed subdivision of the upper part of the Sheinwoodian. In the southern midcontinent region the first four of these species do not range into the Homerian, as placed here (17.7 SCS; 313 SMCS). Although it may be an artifact of the compositing process, the first four of these Sheinwoodian species disappear at nearly the same stratigraphic level, just below the top of the Sheinwoodian (Fig. 4).

The first appearance datum of *Ozarkodina bohemica* (315 SMCS) lies just above the base of the Homerian, largely because this datum was used to help position the base of the Homerian in the SMCS. The successive first appearance datums of *Kockelella stauros* (358) and *K. variabilis* (440) or *K. crassa* (*"Ozarkodina" crassa*; 440) allow subdivision of the lower part of the Homerian. Only a few samples higher than the first occurrence of *K. crassa* have been analyzed, all from the lower part of the Henryhouse Formation in Oklahoma. The presence of *Ancoradella ploeckensis*, which appears in the lower Ludlow (22.3 in SCS) in two Henryhouse sections, helps to locate the top of the Wenlock in these sections.

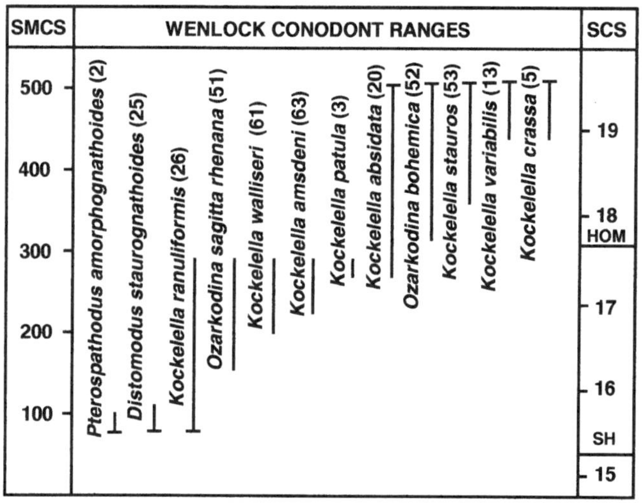

Figure 4. Ranges of selected species of conodonts scaled relative to their position in the Southern Midcontinent Composite Standard (SMCS). Positions of datums are shown where they correlate with the Silurian Composite Standard (SCS) of Kleffner (1995) using the LOC on Figure 3. Numbers refer to points on Figure 3. Cross-bars on range lines indicate that the species is known to range either lower (base of line) or higher (top of line) in other areas. SH—Sheinwoodian Stage; HOM—Homerian Stage.

Depositional and biotic events preserved in shelf strata in southern North America were correlated in one of two ways. When sufficient data existed, the graphic correlation method was used with the SMCS as the standard. Unless otherwise noted, numerical time values refer to position in the Southern Midcontinent Composite Standard. Sections bearing fewer biostratigraphic events were correlated using the ranges of species directly, as in zonal biostratigraphy. Although the numerical values given to sedimentological and biological events using the graphic correlation method implies a high degree of accuracy and precision, the true accuracy and precision of the correlations are difficult to estimate. However, it is certain that correlations derived using the graphic correlation method are more tightly constrained than those based on zones.

WENLOCK DEPOSITIONAL SEQUENCES AND MICROFAUNAL EVENTS

Wenlock strata in the southern United States comprise three depositional sequences, each possessing a characteristic series of lithofacies and microfaunal biofacies. The lowest, the Sheinwoodian sequence (84 to 310 SMCS), includes most of the Sheinwoodian Stage. The second, the lower Homerian sequence (310 to 500), begins near the base of the Homerian Stage and includes the lower part of the stage. The third, the upper Homerian/lower Ludlow sequence (500 to 700?) includes the upper part of the Homerian and ranges into the Ludlow Series.

Sheinwoodian depositional sequence

Lower Wenlock shelf strata in the southern United States overlie upper Llandovery units unconformably. The duration of the hiatus separating the two sequences appears to be relatively short. Although the upper Llandovery shallow-water carbonate sequence shows widely variable thickness and evidence of karstification, at no place was it completely removed before deposition of Wenlock sediments. Diagnostic faunas occur at the top of the upper Llandovery section only in Oklahoma and Arkansas. Conodonts of the *celloni* Zone occur at the top of the Cochrane Formation in surface and subsurface sections in Oklahoma (Barrick et al., 1990) and in the upper limestone unit ("Brassfield") of the Cason Shale in Arkansas (Craig, 1993). The oldest beds of the Sheinwoodian sequence bear faunas of the *amorphognathoides* Zone, and the hiatus probably includes parts of both zones. Only in the basinal Missouri Mountain Shale of the Ouachita Facies at Black Knob Ridge in southern Oklahoma (BKR, Fig. 1) does deposition appear to have been continuous during this time interval (Barrick and Haywa-Branch, 1994). In this succession, a thick interval of shale and sandstone (20 m) that represents a series of gravity-flow deposits contains transported conodont faunas that occur in the expected order of succession from the *celloni* into the *amorphognathoides* Zone. Barrick and Haywa-Branch (1994) suggested that these beds were deposited during the time of erosion and dissolution of Llandovery carbonates. In Arkansas, a clastic-rich section (Craig, 1993), was deposited at the same time on the adjacent shelf region.

The lithological character of the lowest parts of the Sheinwoodian sequence varies geographically. In many sections a thin shale that may contain abundant glauconite and conodonts of the *amorphognathoides* Zone lies at the base. The thin basal shale is recognized in the subsurface of west Texas (lower Wristen Marker), Oklahoma (Prices Falls Shale), along the Mississippi River (Seventy-Six Shale), and in the most southern sections of the Maddox Member in Tennessee. This shale represents a lag deposit of erosional residuum that was reworked into a marine shale during the initial phases of transgression. Some late Llandovery conodonts and probably all of the glauconite were derived from weathering and erosion of the underlying carbonates. The presence of local topography on the erosional surface can be best seen in a quarry on the Lawrence Uplift in Oklahoma, as illustrated by Amsden (1957, pl. 1A).

Skeletal wackestones to mudstones, containing little terrigenous clastic material, rest with a sharp contact on the thin shale. Carbonate lithofacies and macrofaunal biofacies indicate offshore, relatively to moderately deep shelf environments (Amsden et al., 1980; Barrick, 1983). These strata are characterized by sparse ostracodes, trilobite fragments, nautiloids, and pelmatozoan debris and generally lack typical shallow-water suspension-feeding benthic organisms such as stromatoporoids, corals, and bryozoans. Little change in lithofacies occurs in most sections in the lower part of the carbonate section. In parts of the southern Oklahoma outcrop belt (Coal and Pontotoc Counties, sections C1 and P1; Amsden, 1960), coarse-grained, unabraded pelmatozoan packstones and interbedded wackestones are overlain by grainstones with abundant brachiopods and trilobites.

In the Batesville District of Arkansas, the St. Clair Limestone rests unconformably on the "button shale" of the Cason Shale (Craig, 1993). The basal beds of the St. Clair are ostracode wackestones; the base of this interval is heavily mineralized and rests with a sharp contact on the Cason Shale. Coarse-grained unabraded pelmatozoan packstones and grainstones with brachiopods and trilobites appear less than a meter above the base and characterize the major part of the St. Clair. They grade upward into the silty ostracode wackestones and mudstones of the Lafferty Limestone. At the Love Hollow Quarry the transition occurs about 5 m about the base of the formation. At other sections, the pelmatozoan grainstones are thicker, ranging up to 30 m. In the Ozark Region of northeastern Oklahoma (Quarry Mountain Formation) and the Buffalo River area of Arkansas (St. Clair Limestone), a basal shale is absent, and pelmatozoan packstones and grainstones rest directly on upper Llandovery carbonates (Craig, 1975). Amsden (1978) compared the grain composition of the packstones and grainstones in the Clarita, St. Clair, and Quarry Mountain Formations. He showed that the grainstones of these formations formed in a spectrum of moderate-energy, subtidal environments, the most shallow of which formed the Quarry Mountain.

The basal shale thickens and becomes calcareous with the addition of skeletal debris across the Tennessee outcrop belt from southwest to northeast, apparently in response to an increased influx of terrigenous clastics from the east (Barrick, 1983). These thicker calcareous shales grade more into argillaceous carbonates that contain bryozoan debris and rare corals, suggesting a somewhat shallower position on the shelf to the northeast. The Maddox Member displays a distinct upward-shallowing trend. The quantity of fine terrigenous detritus decreases upward, and abraded skeletal packstones and grainstones appear, especially in sections to the northeast.

It is not possible to date the age of the base of the transgressive shale precisely. All faunas from the shale can be assigned to the *amorphognathoides* Zone, and the presence of older Llandovery taxa is attributed to reworking. The age of the base of the shale can be estimated by extending the LOC downward from the lowest reliable point (the last appearance datum of *P. amorphognathoides*), but one needs to make the questionable assumption that the basal beds represent the same rate of rock accumulation as the overlying carbonate section. The lowest value obtained is 84 SMCS, 15.47 on the SCS, which indicates that the shale represents only the upper part of the *amorphognathoides* Zone (top = SMCS 99). The initial transgression began during the early Sheinwoodian, because the base of the Wenlock lies lower in the *amorphognathoides* Zone at the stratotype section (15.30 SCS; Kleffner, 1995).

At most sections, the appearance of carbonates coincides with a level near the top of the *amorphognathoides* Zone, around 100 SMCS. The basal bed of carbonate overlying the thin basal shale lag generally lacks *Pterospathodus amorphognathoides* but still contains *Distomodus staurognathoides*, which ranges slightly higher in the composite standard (111 SMCS). Reworking of older conodont elements from the shale into the basal carbonate bed has occurred at some sections, but the selective occurrence of *D. staurognathoides* in the carbonates cannot be explained by reworking. In contrast, the basal carbonate beds of the Fitzhugh Member of the Clarita in Coal County, Oklahoma (sections C1 and C2; 96 SMCS; Barrick and Klapper, 1976), and the St. Clair in the Batesville district of Arkansas (84) contain an *amorphognathoides* Zone fauna (Craig, 1969). In the central Tennessee sections, where no sharp distinction exists between the basal calcareous shale and overlying argillaceous carbonates, the last occurrence of *P. amorphognathoides* in the highest shale is often followed by the last occurrence of *D. staurognathoides* in the overlying argillaceous carbonates (100–105).

The pelmatozoan grainstone facies appears in the Arkansas region at the end of the *amorphognathoides* Zone (96 to 100). At the Love Hollow Quarry, the transition from the pelmatozoan grainstones to the wackstones of the Lafferty occurs at about 230. The time of the transition is not known for other sections but was probably somewhat later. In Oklahoma, pelmatozoan grainstones at sections C1 and P1 (Amsden, 1960) appear at 130 and 232, respectively. It is not possible to date the pelmatozoan grainstones of northeastern Oklahoma, but they appear to belong to the Sheinwoodian sequence based on the brachiopod correlations of Amsden (1978). The shift from the skeletal packstones of Facies II to the shallower water abraded skeletal packstones and grainstones of Facies III in Tennessee (Barrick, 1983) occurs at a level about 225, in every section with sufficient data to draw an LOC. Based on lithologic correlations, the Fasken carbonate platform began to grow at about this time in west Texas (Barrick, 1995).

Abundant conodonts, agglutinated foraminifers, and acrotretid brachiopods are the most prominent elements of the microfauna of the Sheinwoodian sequence. Within the upper *amorphognathoides* Zone (84 to 99), two possible conodont biofacies can be distinguished in shale and carbonate samples. The more offshore sections contain a diverse conodont biofacies, containing up to 22 species that occur in abundances of hundreds of elements per kilogram. No single species is strongly dominant, but a variable core of four or five species constitutes the bulk of the fauna. Species most commonly forming the core are *Panderodus unicostatus, P.* sp. cf. *P. recurvatus* (see Barrick, 1983), *P. langkawiensis, Walliserodus sancticlairi, Distomodus staurognathoides,* and *Decoriconus fragilis*. All species are most abundant in this biofacies, although the abundance of individual species varies strongly from sample to sample. In the shallower-water calcareous shale at the base of the Maddox Member in central Tennessee, a second shallower-water biofacies occurs. This biofacies is characterized by the strong dominance of *Panderodus unicostatus*, which often constitutes over 60% of the elements. *Ozarkodina polinclinata*, *P.* sp. cf. *P. recurvatus*, and *D. staurognathoides* are the other common species. Samples of this biofacies have moderate abundance, tens of elements per kilogram, and moderate diversity, usually five to 12 species. Most samples of the *amorphognathoides* Zone that contain elements of *Dapsilodus* bear only one species, *D. praecipuus.* In Arkansas elements of *D. obliquicostatus,* a species characteristic of the overlying carbonates, becomes moderately common in the uppermost sample of the zone.

Foraminiferal tests are moderately common in samples with both conodont biofacies of the *amorphognathoides* Zone. Tests of *Ammodiscus* are most prominent, with the greatest number of *Hyperammina* and *Thurammina* occurring with the more offshore conodont fauna. Acrotretid brachiopods are rare to most typically absent in this interval. A few specimens of *Artiotreta parva* occur in the skeletal wackestones in Oklahoma and Arkansas, but none have been recovered from the thin basal shale nor from the calcareous shales in central Tennessee.

The most significant turnover in conodont faunas in the Silurian occurs during the early Sheinwoodian, at the transition from the *amorphognathoides* Zone to the *ranuliformis* Zone (99 to 100 SMCS; Jeppsson, 1987; Aldridge, 1988). In the southern United States, this change occurs either at the transition from the basal shale to carbonates or in the lowest part of the carbonate section. Most of the younger conodont faunas are dominated by coniform species. Species rare in older strata—*Dapsilodus obliquicostatus* and *Pseudooneotodus bicornis*—and species common in the *amorphognathoides* Zone—*Panderodus unicostatus, Walliserodus sancticlairi,* and *Decoriconus fragilis*—constitute more than 95% of the elements. Of the many characteristic pectiniform-bearing species of the underlying *amorphognathoides* Zone, only three species range higher. *Distomodus staurognathoides* ranges only a short interval higher (111). *Kockelella ranuliformis* is one of the two common pectiniform-bearing species in the lower Sheinwoodian, the other being the long-ranging species *Ozarkodina excavata excavata*.

Barrick (1983) identified three major intergrading normal marine conodont biofacies in the Tennessee region that define an offshore to nearshore, deep-water to shallow-water gradient. The offshore *obliquicostatus* Biofacies is strongly dominated by *Dapsilodus obliquicostatus,* but all species attain their greatest absolute abundance in this biofacies, often thousands of elements per kilogram. Samples from this biofacies have the greatest diversity and are those most likely to contain conodont species used to subdivide the Sheinwoodian. The most onshore biofacies, the *unicostatus* Biofacies, is dominated by *Panderodus unicostatus*; *D. obliquicostatus* is uncommon. *Walliserodus sancticlairi* is the second most common species, and other species are uncommon to absent. Diversity can be reduced to one, the nominate species, and abundance is low, usually tens of elements or fewer per kilogram. In the intermediate biofacies, the Mixed Biofacies, no one species strongly dominates. A core of three to five of the following species constitutes the bulk of the fauna: *D. obliquicostatus, Decoriconus fragilis, P. unicostatus, Pseudooneotodus bicornis, Ozarkodina excavata*, and *W. sancticlairi*. Diversity and abundance are moderate, with five to 10 species comprising tens to hundreds of elements per kilogram.

In strata immediately overlying the *amorphognathoides* Zone, the *obliquicostatus* Biofacies and Mixed Biofacies occur. The *obliquicostatus* Biofacies stretches across most of the southern North America shelf, being absent only in sections in northeastern Oklahoma and central Tennessee and northward. The greatest dominance of *Dapsilodus obliquicostatus* occurs in the Pegasus core from the subsurface of west Texas and sections in southern Oklahoma. In these areas the *obliquicostatus* Biofacies persists through the Sheinwoodian. Minor faunal variations occur within the Sheinwoodian *obliquicostatus* Biofacies, as described by Amsden et al. (1980) for the Fitzhugh Member of the Clarita Formation in Oklahoma.

In Tennessee, the *obliquicostatus* Biofacies is restricted to the most southern sections, and all other sections contain the Mixed Biofacies (Barrick, 1983). The latter replaces the *obliquicostatus* Biofacies by 150 to 180, and the *unicostatus* Biofacies replaces the Mixed Biofacies, appearing earliest in the northern sections (150 to 180), and in some southern sections (e.g., Clifton and Iron City; Barrick, 1983) by 280. Although the lithofacies show a similar shift to shallower-water conditions, the change in conodont biofacies does not correspond exactly with the lithofacies transitions (Barrick, 1983).

In the Mississippi River sections of the lower Bainbridge limestone, where lithofacies shifts are more subtle, the Mixed Biofacies appears as low as 150 in the northernmost section and spreads southward to occupy the unit at all sections by 220. Even in the Droke core, which contains the Fitzhugh Member in the subsurface of Oklahoma (Kingfisher County; Barrick, 1978), the conodont biofacies shifts from the *obliquicostatus* to the Mixed Biofacies by 225, and then to the *unicostatus* Biofacies at 275, in the apparent absence of corresponding lithofacies transitions.

The *obliquicostatus* and Mixed Biofacies occur in the pelmatozoan packstones and grainstones of the St. Clair in Arkansas. In contrast to the high conodont abundance in the similar lithofacies in the Clarita in Oklahoma, abundance drops to a few tens of elements per kilogram in the St. Clair. As the typical St. Clair pelmatozoan grainstone lithofacies grades into the wackestones of the Lafferty at Love Hollow, the *unicostatus* Biofacies appears at 230, just before all conodonts disappear from that section. In northeastern Oklahoma, the sparse conodont faunas of the lower part of the Quarry Mountain Formation, the Barber Member, can tentatively be referred to the Mixed Biofacies. The upper Marble City Member contains a sparse *unicostatus* Biofacies fauna.

Superimposed on the general pattern of shifting conodont biofacies in the Sheinwoodian sequence is a strong acme of *Dapsilodus praecipuus*. Although relatively common in offshore late Llandovery conodont faunas (Barrick and Haywa-Branch, 1994), *D. praecipuus* is generally rare in Sheinwoodian faunas. During the interval of 255 to 300, *D. praecipuus* rises to dominance in sections otherwise characterized by the *obliquicostatus* and Mixed Biofacies. In samples where *D. praecipuus* is abundant, the absolute abundance of *D. obliquicostatus* falls dramatically. It appears that *D. obliquicostatus* was being replaced by *D. praecipuus* in the conodont association. The acme is most pronounced and longest in duration in the Mobil Pegasus core of west Texas, where two subpeaks of abundance are poorly defined. In Oklahoma and southern Tennessee, the two subpeaks are more clearly resolved, the lower at about 255 to 260, and the upper at about 280 to 300. In most sections, only one peak in the abundance of *D. praecipuus* is present, usually in the 280 to 300 interval. The acme of *D. praecipuus* does not coincide with any lithofacies or other biofacies events in the region. The species appears to occupy an ecological position farther offshore in southern North America than even *D. obliquicostatus,* which Barrick (1983) suggested may have been an indicator of oceanic water conditions. Conodont samples from Sheinwoodian carbonates associated with graptolitic shales in the Carnic Alps, Austria, are strongly dominated by *D. praecipuus, D. obliquicostatus*, and the less common species *D. sparsus*, almost to the exclusion of other conodont taxa (Oberbuchach 1, sample 91a, Schönlaub, 1980, p. 27; Cellon, samples 12C and 14E, Walliser, 1964). The brief interval of abundant *D. praecipuus* in the Sheinwoodian can be interpreted to represent a change in marine circulation patterns, such that more offshore or deeper oceanic waters briefly spread across the outer shelf of southern North America.

In the upper part of the Sheinwoodian sequence several species become extinct in southern North America: *Kockelella walliseri* (294), *K. amsdeni* (294), *K. patula* (287), and *Ozarkodina sagitta rhenana* (289). These taxa either range substantially higher elsewhere (*K. walliseri*) or are a part of a lineage that gave rise to younger species. Their disappearance lies near the top of the Sheinwoodian sequence (310) and may be related either to the overall shallowing in the region or to other local ecological effects.

The acrotretid brachiopod *Artiotreta parva* attains maximum abundance in the lower part of the Sheinwoodian sequence. The species reaches abundances exceeding thousands of valves per kilogram in the more offshore skeletal wackestones bearing the *obliquicostatus* Biofacies in the Mobil Pegasus core of west Texas, in southern Oklahoma (Amsden et al., 1980), and in the lower Bainbridge limestone along the Mississippi River. In these areas, the greatest abundance occurs low in the Sheinwoodian, in the interval from 100 to 200, and decreases in younger beds. Before the top of the Sheinwoodian sequence is reached, *A. parva* has disappeared, regardless of lithofacies and conodont biofacies. Where pelmatozoan packstones and grainstones occur in southern Oklahoma and Arkansas, *A. parva* is uncommon, even where the *obliquicostatus* Biofacies is well developed. In Tennessee, *A. parva* occurs in more moderate numbers and is restricted to the lower Sheinwoodian interval of 100 to 200. Also present in the more offshore Sheinwoodian strata are a species of *Opsiconidion* and fragments of linguloid brachiopods, both of which range into Homerian strata.

Agglutinated foraminifers occur in tremendous numbers in the Sheinwoodian carbonates, a feature noted by many workers (Ireland, 1939; Dunn, 1942; Amsden, 1960). The greatest diversity of foraminifers appears to occur in the lower part of the Sheinwoodian sequence. Although this is not clear in sections of the Clarita in southern Oklahoma (Amsden et al., 1980), Zuravel (1986) showed that species-level diversity and test abundance decreases up-section and from south to north in the Maddox Member of the Wayne Formation in Tennessee. The decrease in diversity and abundance he attributed to the higher energy levels of the environment, which affected the distribution of the foraminifers and decreased the potential for preservation of tests.

Lithofacies and biofacies evidences indicate that just after the beginning of the Sheinwoodian the entire southern margin of the North America craton was inundated by a rapid rise in sea level. This event is represented in the more offshore sections by the thin basal shale, a lag deposit that formed prior to initiation of carbonate deposition in an offshore deeper-water shelf. In more shoreward settings the basal shale-carbonate transition was more gradual. Gradual shallowing through the Sheinwoodian is shown best in the lithofacies and biofacies in more shoreward sections in central Tennessee and less well by the development of offshore skeletal sands in Arkansas and parts of Oklahoma. In southern Oklahoma, and the Mississippi Valley outcrop belt, deeper-water carbonate environments were apparently little affected.

Lower Homerian depositional sequence

The base of the lower Homerian depositional sequence is best delineated in Tennessee by the appearance of the Waldron Shale Member; it represents a brief pulse of terrigenous clastics that spread southwestward across the outcrop belt. The age of the Waldron is better determined in the more southern sections of the Wayne Formation, where the biostratigraphic data are richer. The base of the Waldron falls in a narrow time interval, 300 to 320, which corresponds approximately with the base of the Homerian (310) used in this chapter. The Waldron clastic pulse terminated the shallowing-upward succession of the underlying Maddox Member. Conodont biofacies change to more offshore biofacies: Mixed to *obliquicostatus* in the southern sections and *unicostatus* to Mixed in central Tennessee (Barrick, 1983). The biofacies facies shift is less apparent in the base of the overlying Lego Member. Carbonate lithofacies in the base of the Lego Member, however, show a more distinct shift in the offshore direction. However, a very short distance above the base of the Lego, the shallower-water lithofacies and conodont biofacies that occur in the upper part of the Maddox Member again dominate the section.

At approximately the same time the Waldron clastic pulse appeared in Tennessee, a silty argillaceous lithofacies (marlstone) appeared in the Bainbridge along the Mississippi River. The first occurrence of the reddish-colored marlstone interbedded with cleaner carbonates typical of the underlying lower Bainbridge limestone defines the interval transitional to the Moccasin Springs. The time of appearance of these marlstones varies from section to section, but most fall in the 300 to 325 range, coincident with the Waldron Shale. No obvious shift in conodont biofacies occurs with the appearance of the marlstone, but the *obliquicostatus* Biofacies predominates over the Mixed Biofacies.

Similar silty, argillaceous carbonates appear in the upper part of the Fitzhugh Member of the Clarita Formation in southern Oklahoma. In a few sections the marlstone beds are interbedded with cleaner carbonates like those of the underlying Sheinwoodian sequence, but more typically a homogenous upper subunit of ostracode silty marlstones is present (Amsden et al., 1980). The time of appearance of the silty marlstones in Oklahoma is also coincident with the Waldron Shale, approximately 300 to 325. The *obliquicostatus* Biofacies continues from the Sheinwoodian carbonates into the ostracode silty marlstones. No comparable change in lithofacies or biofacies marks the correlated base of the Homerian in the Wink Formation in the Mobil Pegasus core of west Texas.

Problems in extrapolating LOCs through the tops of Clarita sections of pelmatozoan grainstones make it difficult to confidently determine whether the facies persisted from the Sheinwoodian into the Homerian. The correlated position of the top of the section at C1 (Amsden, 1960) is 384, suggesting an early Homerian age for the uppermost beds. At section P1 (Amsden, 1960), the distinctive biomicrites of the *Placotriplesia* beds cap the pelmatozoan grainstone section at about 330 to 340, barely into the Homerian. Both of the sections are unconformably overlain by the basal beds of the Henryhouse Formation, which at section Pl contain a Ludlow *siluricus* Zone fauna. Although the oldest beds of the Lafferty Limestone at Love Hollow appear in the Sheinwoodian, Craig (1993) suggests that the pelmatozoan grainstone facies may have persisted into the Ludlow. This is based on the presence of a single specimen of *Polygnathoides siluricus* recovered from the top of the thick section of St. Clair (30 m) at the nearby Batesville Stone Quarry. My collections from the same locality are undiagnostic, and it is possible that some leaked conodont elements may have been recovered. My tentative conclusion is that the pelmatozoan grainstone facies in southern Oklahoma and Arkansas as well as those in northeastern Oklahoma, from which the unique triplesid-dominated brachiopod faunas were described by Amsden (1968; 1978; see also Boucot, 1975, p. 310), are mostly Sheinwoodian in age.

Although the three basic conodont biofacies present in Sheinwoodian strata continue into the lower Homerian depositional sequence, changes in species composition occur, probably in response to the increased quantity of terrigenous clastics, rather than any shift in water depth. *Walliserodus sancticlairi* and its descendant, *Walliserodus* sp. A (Barrick, 1983), major components of Sheinwoodian faunas, become uncommon in Homerian strata. *Belodella silurica,* which occurs rarely in Sheinwoodian units, is common to abundant in the Homerian marlstones. Similarly, *Panderodus recurvatus*, although never common, occurs in a far greater number of samples in Homerian than does it or its possible ancestor, *Panderodus* sp. cf. *P. recurvatus* (Barrick, 1983), in Sheinwoodian strata. Amsden et al. (1980) noted that in the *obliquicostatus* Biofacies in the ostracode silty marlstones of the upper Fitzhugh, *Panderodus unicostatus* and *Ozarkodina excavata excavata* become somewhat more abundant, and *Pseudooneotodus bicornis* becomes much less common. These changes in relative abundance also occur in the Mixed and *unicostatus* Biofacies.

Little information exists on foraminiferal distributions in the lower Homerian sequence. Amsden et al. (1980) show that most species of foraminifers occurring in the Sheinwoodian arthropod micrites extend into the ostracode silty marlstones in southern Oklahoma. I have seen no specimens of the acrotretid brachiopod *Artiotreta parva* in the lower Homerian sequence and only a few examples of *Opsiconidion* and lingulid fragments.

The lower Homerian depositional sequence represents an extremely brief rise in sea level that was accompanied by the introduction of fine terrigenous clastics across the southern North American shelf. In the more nearshore settings, shallow-water carbonate environments and conodont biofacies returned almost immediately, suggesting that the magnitude of the rise was small and easily overcome by sediment deposition. In offshore settings, no corresponding shift in carbonate facies or conodont biofacies occurs, although the presence of the increased quantity of fine terrigenous detritus modified lithofacies and biofacies compositions.

Upper Homerian/lower Ludlow sequence

The interbedded marlstones and clean limestones of the lower Homerian sequence are overlain by marlstones that may be interbedded with calcareous shales. The change in lithology is sufficiently distinct that it is used as a formational boundary: Oklahoma—base of Henryhouse Formation; Illinois Basin—Moccasin Springs Formation; Tennessee—Dixon Formation. In each of these areas, *Kockelella variabilis* and *Kockelella crassa* appear just below the formational contact, and the base of the marlstone/shale formation lies at 500 or within the *crassa* Zone. Barrick and Klapper (1976), Barrick (1983), and Aldridge and Schönlaub (1989) tentatively equated the base of the *crassa* Zone with the base of the Ludlow, but Kleffner's most recent Silurian Composite Standard (1995) places the base of the Ludlow within the *crassa* Zone. The age of the top of this depositional sequence is not known with certainty because of a lack of biostratigraphic control. At the Highway 77 section in Oklahoma, the sequence extends through the Ludlow *ploeckensis* Zone and perhaps as high as the *siluricus* Zone.

The greatly increased quantity of terrigenous clays and silts in upper Homerian beds makes interpretation of the depth-significance of the lithofacies difficult. In Oklahoma, the contact between the two sequences and the Clarita/Henryhouse boundary is sharp, appearing as a surface within a single bed (Amsden, 1957). The *obliquicostatus* Biofacies passes from the upper Clarita into the Henryhouse. However, the conodont elements of the upper sequence are more badly broken and abraded and have a more heterogeneous pattern of abundance from sample to sample. In the Bainbridge Group, the base of the Moccasin Springs Formation is generally placed where nonmarly beds disappear from the section (e.g., Thompson, 1993). This transition is less sharp than in Oklahoma, but conodont biofacies pass unchanged through it. In Tennessee, the contact between the Lego Member of the Wayne Formation and the overlying Dixon Formation is gradational in the south but sharp in the north where Dixon marlstones rest directly on grainstones of the uppermost Lego. In both areas, conodont biofacies pass across the contact with little change except for a decrease in quality of preservation. At the correlated position of the base of this sequence in the Pegasus core in west Texas, dark siliceous shales that yield no conodonts appear in the section.

Because little information is available, it is difficult to interpret the nature of the transition from the lower Homerian sequence to the upper Homerian/lower Ludlow sequence. No good lithologic evidence exists to suggest a change in water depth, although many workers have used the appearance of fine terrigenous clastics to infer deepening water conditions. The persistence of conodont biofacies across the sequence boundary suggests little change in ecological factors controlling conodont distribution. However, the increased breakage and abrasion of conodont elements in more offshore settings point to greater current and wave energy in the environment. The evidence does indicate a significant change in depositional processes between the two sequences, but additional work will be required to interpret it in terms of sea-level or tectonic events.

DISCUSSION

Summary of Wenlock sea-level changes in southern North America

The nature of the three Wenlock depositional sequences described above suggests the following overall pattern of sea-level change on the southern shelf of North America (Fig. 5):

1. A latest Llandovery fall in sea level produced a regional unconformity across the entire shelf.
2. An early Sheinwoodian rapid rise in sea level flooded the shelf. The rise in sea level occurred so quickly that only a thin basal shale records the transgressive episode in outer shelf settings. Deep-water, condensed, offshore carbonate facies rest directly on the shale on the outer shelf regions. On the inner shelf regions, shallower-water carbonates overlie thicker calcareous shales.
3. Shallowing of the shelf began shortly after maximum transgression, partly by carbonate deposition filling the accommodation space at a rate faster than that of shelf subsidence and by a fall in sea level. The lack of an unconformity at the top of the Sheinwoodian sequence and persistence of deep-water facies in outer shelf locations suggest that sea level did not fall significantly until near the end of the stage.
4. A second rise in sea level occurred near the beginning of the Homerian. In Tennessee, the Waldron Shale forms the base of the lower Homerian sequence. Carbonate lithofacies and conodont biofacies document a brief deepening event that was immediately followed by the return of shallower-water lithofacies and conodont biofacies. During sea-level rise, an increased quantity of fine terrigenous detritus spread across the outer shelf. The origin of the detritus is uncertain, but it probably came from foreland basins adjacent to the Taconic orogen.
5. Starting in the late Homerian and extending into the early Ludlow, a flood of terrigenous clays and silts blanketed the entire shelf, strongly modifying the pattern of carbonate lithofacies. No unconformity exists at this level, nor do conodont biofacies shift. The flood of clastics may be attributed either to a sea-level rise or to some event in the orogen or foreland region that permitted an even greater quantity of terrigenous detritus to spread westward across the southern shelf region.

Comparisons with other regions

Over the last 20 years numerous papers describing regional to global sea-level events in the Silurian have been published,

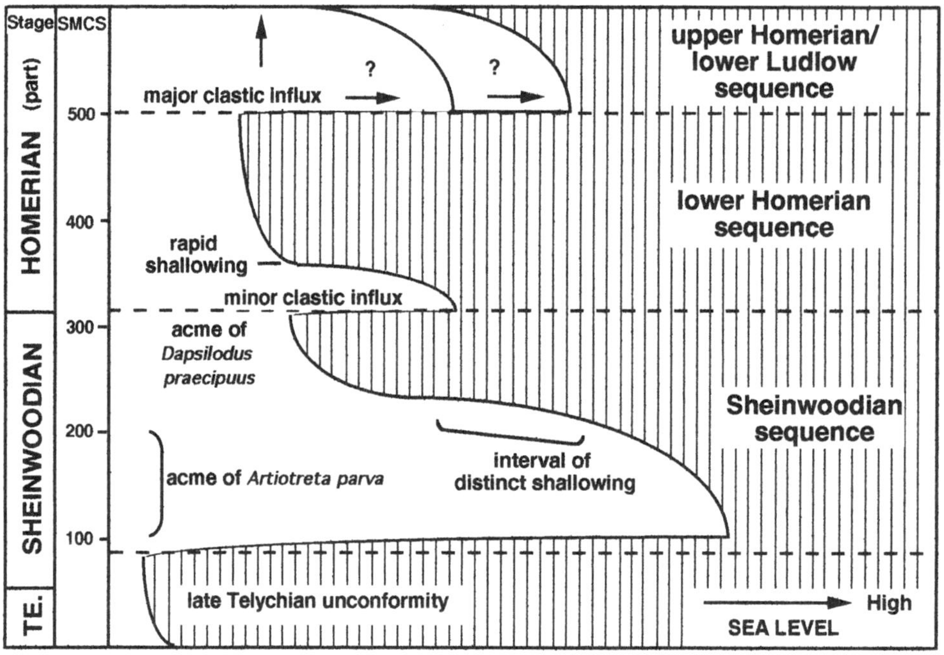

Figure 5. Transgressive-regressive cycles interpreted for Wenlock strata on the southern shelf of North America and their relation to sea-level events. Dashed lines indicate the boundaries of the three depositional sequences. Depositional sequences and events shown on the figure are scaled relative to the Southern Midcontinent Composite Standard (SMCS). Upper Homerian/lower Ludlow sequence continues for an undetermined interval into the Ludlow (vertical arrow at top of figure), the base of which lies at 697 SMCS. Relative magnitude of sea-level change for this sequence is uncertain. TE.—Telychian Stage.

with varying degrees of biostratigraphic documentation. The literature is already so vast that a complete review, even for the poorly known Wenlock, is not feasible here. Instead, this chapter will emphasize a smaller number of more recent papers that detail North American events and summarize previous literature covering other parts of the globe. Comparison of depositional events and inferred changes in sea level in southern North America with other regions is hampered by a lack of precision in biostratigraphic correlation. To adequately test whether events in disjunct regions are coeval, more than a similarity in pattern of timing and magnitude is required.

Droste and Shaver (1987) proposed a sea-level curve for Wenlock strata in the Great Lakes region that includes the Michigan Basin, Appalachian Basin, Illinois Basin, and the interbasin areas that separate them (Fig. 6). Their curve is presented showing information from the stratigraphic column in Indiana; this is similar to the column in Tennessee. They show a major rise in sea level occurring at the Llandovery/Wenlock boundary (within the *amorphognathoides* Zone). Sea level gradually falls to just below the Waldron Shale, where a second major rise in sea level occurs. Sea level gradually falls again, until the base of the Mississinewa Shale, an extremely argillaceous and silty carbonate like the marlstone of the upper Homerian/lower Ludlow sequence in Tennessee. Droste and Shaver (1987) show the base of the marlstone sequence to be a deepening event of a magnitude similar to that of the earlier rises in sea level. Although Droste and Shaver (1987) indicate that the base of the Mississinewa lies well within the Ludlow, my interpretation of the position of the base of the Ludlow, based on Kleffner (1995), places the base of the Mississinewa in the upper Homerian.

Ross and Ross (1996) present a synthetic sea-level curve for North America that appears to be based heavily on depositional patterns in the Illinois Basin region (see also Moore et al., 1994). Their sea-level curve, which is scaled in meters of water depth, shows a major rise in sea level of over 60 m starting in the latest Llandovery and continuing into the early Sheinwoodian (Fig. 6). A mid-Sheinwoodian fall (25 m), is followed by a somewhat smaller late Sheinwoodian rise (20 m). Sea level declines gradually from the latest Sheinwoodian through the Homerian. The trend in falling sea level reaches its lowest point at approximately the Wenlock/Ludlow boundary.

The Ross and Ross sea-level curve matches that for the southern shelf region if slight adjustments in the ages of events in their curve are made. Because the *amorphognathoides* Zone spans the Llandovery/Wenlock boundary at the stratotype, most workers show any event occurring in that zone as potentially spanning the boundary. My correlations indicate that the transgression is restricted to the Wenlock. The mid-Sheinwoodian rise in sea level of Ross and Ross (1996) may be the same as the early Homerian rise inferred for the southern shelf region, but I lack sufficient biostratigraphic data to confirm or reject the

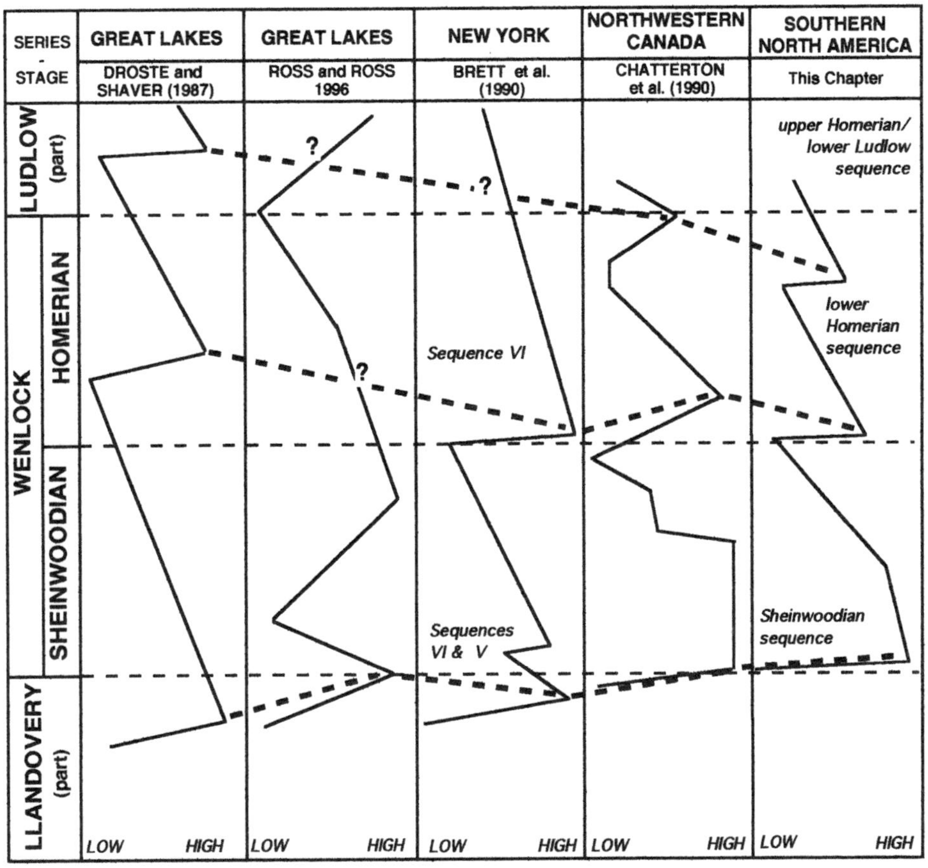

Figure 6. Latest Llandovery to earliest Ludlow sea-level curves from different regions of North America correlated with the Wenlock sea-level curve from the southern United Stated proposed here. Heavy dashed lines indicate suggested matches of highs in sea level. See text for further discussion.

match. The low stand of sea level at the end of the Wenlock on the Ross and Ross chart lies at the base of the Moccasin Springs Formation and has been discussed above. Their interpretation of a sea-level fall is influenced by the presence of an unconformity lying between the Joliet and Sugar Run Formations in the Chicago area.

Brett et al. (1990) and Goodman and Brett (1994) subdivided the Silurian succession in the Appalachian foreland basin into a series of sequences, which they interpret to represent third-order sea-level cycles (Fig. 6). The base of Sequence IV of Brett et al. (1990) is marked by a widespread angular unconformity that Goodman and Brett (1994) correlate with the major late Telychian (C_5) fall in sea level of Johnson (1987). Sequence IV (basal upper Clinton) comprises a thin section of dominantly argillaceous lithologies that contain an *amorphognathoides* Zone fauna. The base of Sequence V (upper Clinton) of Brett et al. (1990) is a sharp planar disconformity that they interpret to be an erosional surface, although there is little evidence of regional beveling. Goodman and Brett (1994) interpret the disconformity as recording a major early Wenlock fall in sea level and Sequence V as representing the subsequent rise. In some areas, a deepening-upward succession occurs that is interpreted as a transgressive systems tract. Overlying units preserve a general shallowing trend through the late Wenlock, parts of which form a high-stand systems tract. Sequence VI (Lockport Group and equivalents) is bounded at its base by a sharp unconformity that truncates underlying units. After a major fall in sea level, a subtle, but significant, transgression produced Sequence VI (Brett et al., 1990). Lithofacies in Sequence VI indicate a shallowing-upward trend, modified in part by the shifting position of the basin depocenter. In the Decrew Dolomite, which lies at the top of Sequence V, *sagitta* Zone conodonts are present, which Brett et al. (1990) and Goodman and Brett (1994) use to give the lower part of Sequence VI a Homerian age.

Kleffner (1991), using conodont faunas from the Niagara Gorge area, showed that the Williamson Shale and Rockaway Dolomite Member of the Irondequoit Limestone (Sequence V), as well as the base of the upper member of the Irondequoit (Sequence VI), all contain an *amorphognathoides* Zone fauna (see also Goodman and Brett, 1994, fig. 10). Based on graphic correlation (Kleffner, 1991), all of these units lie in the upper part of the zone and are Sheinwoodian in age. No biostratigraphic evidence exists for a substantial hiatus between Sequences IV and V. The change in lithofacies described by

Brett et al. (1990) between the two sequences is like that seen in the basal part of coeval Wenlock sections on the southern shelf of North America. Sequence IV probably represents only the basal portion of the early Sheinwoodian rise in sea level that produced the overlying Sequence V. The physical evidence for an early Sheinwoodian eustatic fall in sea level proposed by Brett et al. (1990) and Goodman and Brett (1994) at the base of Sequence V appears to be a result of the sedimentologic response to local tectonic effects in the active foreland basin.

The base of the Lockport Group, the base of Sequence VI, Kleffner (1991) dates as occurring within the upper part of his *sagitta* Chronozone, which ranges from the top of the Sheinwoodian through the lower Homerian, as used in this chapter. The base of Sequence VI appears to correspond with the lower Homerian sequence of the southern North American shelf. Kleffner (1991) shows that the argillaceous dolostones of the Eramosa Formation and Goodman and Brett (1994, fig. 6) show that the upper part of the underlying Goat Island Dolomite, which lie in the upper part of Sequence VI, contain conodonts of the *crassa* Zone, which spans the Wenlock/Ludlow boundary (updated in Kleffner, 1994, 1995). The marlstones of the lower part of the upper Homerian/lower Ludlow sequence on the southern shelf appear to correlate with the upper part of Sequence VI of Brett et al. (1990) and Goodman and Brett (1994).

The three Wenlock depositional sequences recognized on the southern shelf of North America can be matched generally with equivalent sequences in the Great Lakes region and Appalachian Basin. All three areas, though, lie close to the Taconic Orogen, and the possibility of regional tectonic effects cannot be discounted. In western North America, the likelihood of similar tectonic effects is less likely; similarity in the pattern of sequences would more strongly support eustatic control.

Unfortunately, no detailed synthesis of Wenlock sea-level events exists for the carbonate shelf, slope, and basin sediments of the Cordilleran region in the Utah-Nevada area. Sheehan and Boucot (1991) present a comprehensive summary of the Silurian paleogeography of the western United States, but it is difficult to infer details of sea-level changes from the information they present. An important consideration in interpreting local sea-level changes is the collapse of the Cordilleran carbonate shelf in the late Llandovery (Johnson and Potter, 1975). This event coincides with the second transgressive-regressive cycle of Johnson and Murphy (1984). Ross and Ross (1996; see also Moore et al., 1994) miscorrelated the late Llandovery unconformity in central Nevada and subsequent deepening event in the Roberts Mountains Formation and equivalents with the late Llandovery fall and early Wenlock rise in sea level discussed here. Murphy et al. (1979) clearly show that *celloni* Zone conodonts (Telychian C_5) occur in the base of the units overlying the unconformity. Wenlock conodonts appear substantially higher in the section (e.g., Klapper and Murphy, 1975). Lenz (1982) could only weakly resolve a mid-Wenlock regression in central and southern Nevada. Johnson and Murphy (1984) and Sheehan and Boucot (1991) describe deposition during the Wenlock as consisting of regressive carbonates and associated westward progradation of shelf carbonates over upper slope deposits.

Chatterton et al. (1990) derived a history of sea-level change during the Wenlock for the central Mackenzie Mountains in northwestern Canada through lithofacies analyses (Fig. 6). Their sea-level curve is remarkably like that described from the southern shelf of North American, indicating the following: (1) A sharp rise in sea level at the base of the Wenlock and high stand, (2) a fall in sea level during the late Sheinwoodian, (3) a smaller rise in the early Homerian (associated with *Monograptus testis*), (4) a fall in the later Homerian, and (5) an insignificant rise just below the Wenlock/Ludlow boundary. The pattern of sea-level change in the Mackenzie Mountains is similar to the more broadly drawn sea-level curve derived by Lenz (1982) based on seven sections in western North America and the Canadian Arctic.

Because the pattern of sea-level change during the Wenlock on the southern shelf of North America can be recognized not only in midwestern and eastern North America but also in western Canada, I conclude that these were primarily eustatic events. Some discrepancies do exist, however, between our results and Silurian global sea-level curves (Fig. 7). Johnson (1987) described a late Telychian high stand in sea level for North America and illustrated its magnitude in terms of regional coverage and estimated water depth. The high stand is shown as lying in the conodont succession at the boundary of the *Icriodella inconstans* Zone (=*celloni* Zone) and the *amorphognathoides* Zone. Johnson et al. (1991) show this high stand as a global event, lying in the *Costistricklandia* Biozone, which spans the Llandovery/Wenlock boundary. Johnson and McKerrow (1991) discuss the late Telychian sea-level rise in more detail and state that deepening commenced during the *Monograptus griestoniensis* Zone and reached its peak in the *M. crenulata* Zone before the end of the Llandovery. Shallowing followed the start of the Wenlock in the *Cyrtograptus centrifugus* Zone.

I believe that this late Telychian sea-level high stand is the same as my early Sheinwoodian high stand (Fig. 7). Differences in the interpretation of the biostratigraphy of shelly faunas relative to conodont and graptolite zones are probably the cause of differences in correlation. In particular, brachiopod zonations using the stricklandiide, *Eocoelia,* and pentameride lineages are not able to distinguish the latest Telychian from the early Sheinwoodian (Bassett, 1989). The "late Telychian sea-level high stand" of Johnson et al. (1991) lies at the level of the major unconformity that stretches across the southern shelf of North America and, by our correlations, at the same stratigraphic position elsewhere in North America. The unconformity should also be present in the shallow shelf carbonate successions in the northern midcontinent region (Iowa, Michigan, etc.), but physical evidence for an unconformity may be obscured by the pervasive dolomitization in the area. Most of the biostratigraphic evidence presented in Johnson (1987), Johnson and McKerrow (1991), and Johnson et al. (1991) for a latest Telychian age can be interpreted just as easily to show

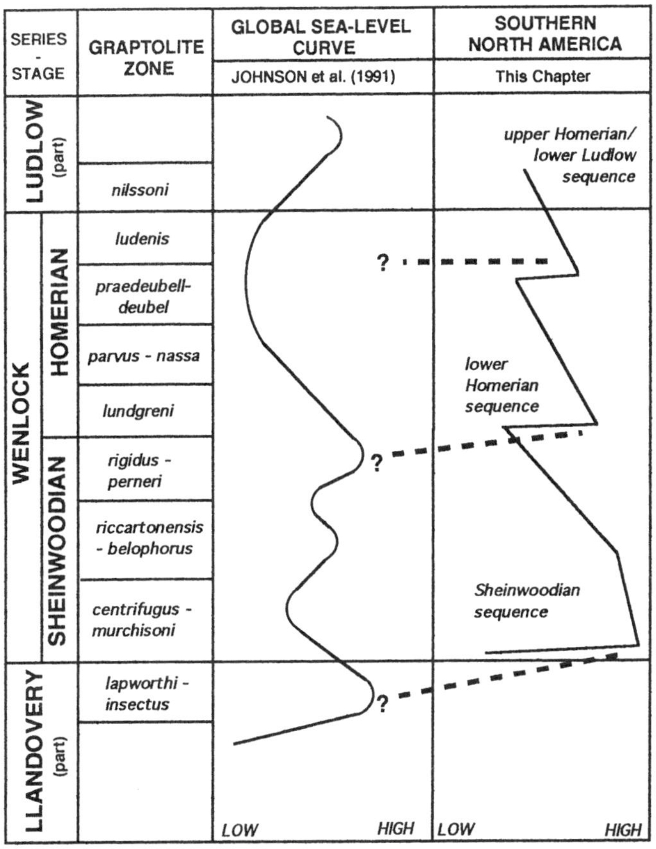

Figure 7. Comparison of the Wenlock global sea-level curve of Johnson et al. (1991) with the Wenlock sea-level curve proposed here for the southern United States. Graptolite zonation is that proposed for use by the Subcommission on Silurian Stratigraphy (1995). Heavy dashed lines indicate suggested matches of highs in relative sea level. See text for further discussion.

that the point of maximum rise of sea level occurred during the earliest Sheinwoodian. In some cases, for example, Australia, where graptolite data instead of shelly faunas are used for dating, the transgressive phase peaks in the early Wenlock (Johnson and McKerrow, 1991, p. 158). In China, as well, five of six localities on the Yangtze Platform demonstrate trends culminating in a deep-water peak in latest Llandovery or earliest Wenlock time (Johnson et al., 1985, p. 1392).

The Sheinwoodian fall in sea level shown by Johnson et al. (1991) and Johnson and McKerrow (1991) is also probably placed too low in the stratigraphic column. They show the low stand of sea level as occurring near the boundary between the *C. centrifugus* and *C. murchisoni* graptolite zones and at the base of the *Ozarkodina sagitta rhenana* Zone. In southern North America, shallowing does start to occur above the first occurrence of *O. sagitta rhenana*, but this species first appears well above the base of the Wenlock (Fig. 4). Shallowing occurs through the entire Sheinwoodian in southern North America; a late Sheinwoodian high stand like that shown by Johnson et al. (1991) and Johnson and McKerrow (1991) cannot be supported. It is probable, though, that the rise of sea level that I interpret as occurring near the base of the Homerian is the same as their late Sheinwoodian rise (Fig. 7).

Johnson and McKerrow (1991) summarized the evidence for a global rise in sea level near the base of the Ludlow. Data from several regions show a shallowing trend in the upper part of the Homerian, followed by the appearance of deeper-water facies near its upper boundary. The slight differences in age of the deepening event from region to region appear to be partly a function of an imprecise definition of the base of the Ludlow in the absence of graptolite faunas. The upper Homerian/lower Ludlow depositional sequence in southern North America falls during the span of time in which the Wenlock/Ludlow boundary lies and probably is also a result of deepening water, although the local evidence is equivocal. To ascertain whether the deepening occurred before or after the Wenlock/Ludlow boundary will require better biostratigraphic correlations than are available now.

Jeppsson (1990), Aldridge et al. (1993), and Jeppsson et al. (1995) propose that a series of oceanic episodes and events that are not necessarily related to sea-level changes controlled the global pattern of lithologic and biotic change during the Silurian (Fig. 8). Primo episodes were characterized by less carbonate deposition in shallow shelf settings and increased diversity and abundance in marine microplankton and communities dependent on plankton productivity; the oceanic state was one with cold bottom waters and high nutrient supply. During secundo episodes, carbonate environments with rich and diverse benthic faunas expanded in shallow shelf settings; the oceanic state was one with salinity-dense bottom waters and diminished nutrient supply. The primary reference for resolving primo and secundo episodes is the sedimentary succession on the island of Gotland, Sweden, in the Baltic Sea. Important comparative sections occur in nearby Estonia (Jeppsson and Männik, 1993; Männik and Viira, 1993).

The later part of the Ireviken Event correlates exactly with the transgressive base of the Sheinwoodian depositional sequence, based on the extinction level of *Pterospathodus amorphognathoides* (Fig. 8). The Vattenfallet Secundo Episode corresponds with the lower part of the Sheinwoodian sequence, but its upper boundary has not been identified in southern North America. The series of short-lived episodes and events that Jeppsson et al. (1995) infer for the upper part of the Sheinwoodian (Sanda Primo Episode, Boge Event, Lansa Secundo Episode, and Allekvia Primo Episode) have not been recognized in southern North America, nor do they appear to be present in Great Britain (Jeppsson et al., 1995, fig. 2). These authors indicated that the Waldron Shale of Tennessee could correspond to the late Sheinwoodian Allekvia Primo Episode, but my correlations suggest that the Waldron is younger, early Homerian in age.

The transgressive base of the lower Homerian depositional sequence in southern North America is best correlated with the early Homerian Valleviken Event (Fig. 8), based on the appearance of *Ozarkodina bohemica* just above both events. The

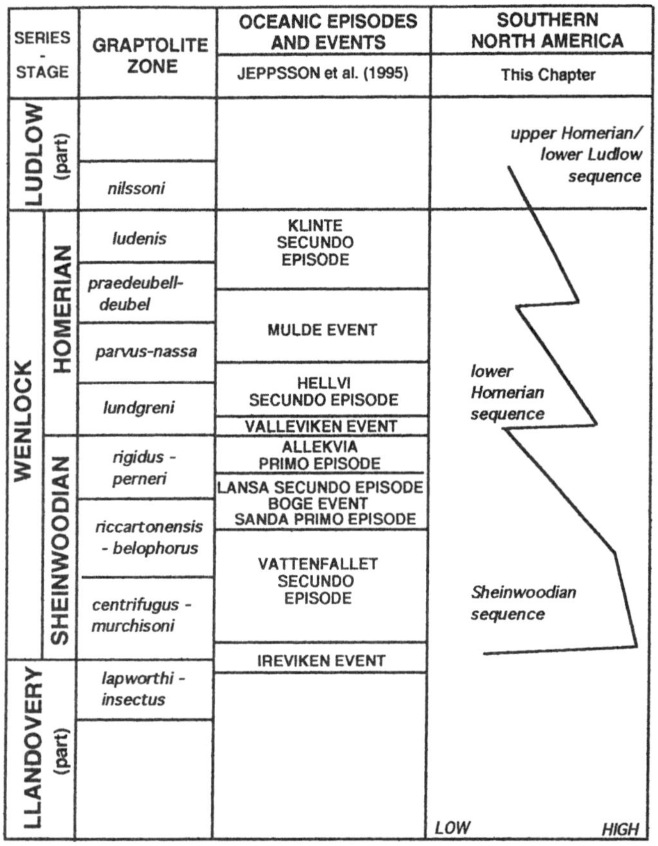

Figure 8. Comparison of Wenlock oceanic events and episodes of Jeppsson et al. (1995) with the Wenlock sea-level curve proposed here for the southern United States. Graptolite zonation is that proposed for use by the Subcommission on Silurian Stratigraphy (1995). See text for further discussion.

Hellvi Secundo Episode corresponds to the main part of the lower Homerian depositional sequence. Jeppsson et al. (1995, fig. 2) show that the Hellvi Secundo Episode is represented in Great Britain by the lower part of the Much Wenlock Limestone. Goodman and Brett (1994) correlated the basal part of the Much Wenlock Limestone with their Sequence VI (Lockport), which I correlate with the lower Homerian depositional sequence in southern North America.

By stratigraphic position, the Mulde Event appears to correlate with the basal transgressive part of the upper Homerian/lower Ludlow depositional sequence in southern North America (Fig. 8). No direct biostratigraphic evidence is available to evaluate this correlation. The Klinte Secundo Episode, the last of the Wenlock episodes, is the same age as at least the lower part of the upper Homerian/lower Ludlow depositional sequence. Both the Klinte Episode and the upper Homerian/lower Ludlow depositional sequence extend an unknown duration of time into the Ludlow.

Each of the two lower Wenlock depositional sequences in southern North America, inferred to have been a result of a major rise in sea level, corresponds directly to an Event–Secundo Episode pair. If the model of Jeppsson et al. (1995) is correct, then it is possible that major rises in eustatic sea level could have been a cause of these major shifts in oceanic state and global climatic setting. It has not been possible to recognize definite faunal or lithologic evidence for primo events in southern North America. Perhaps the incursion of abundant *Dapsilodus praecipuus* across the southern shelf late in the Sheinwoodian is related to a primo event, for it occurs in the approximate stratigraphic position of the two late Sheinwoodian primo events (Sanda and Allekvia). The primo events, if they are global in distribution, may represent a type of change in oceanic circulation patterns not strongly coupled with eustatic sea-level changes.

A number of significant biological events, especially extinctions, coincide with major changes in sea level; attempts have been made to determine the causal factors (Chatterton at al., 1990; Copper and Brunton, 1991; Jeppsson, 1990; Johnson and McKerrow, 1991; Kaljo and Märss, 1991; Quinby-Hunt and Berry, 1991; Koren and Urbanek, 1994, and many other papers). The major event affecting conodont faunas during the Wenlock is the rapid turnover, largely by extinction, of taxa in the transition from the *amorphognathoides* Zone to the *ranuliformis* Zone during the Ireviken event (Jeppsson, 1987; Aldridge, 1988; Männik and Viira, 1993; Aldridge et al., 1993). Most of the series of eight datum points constituting the Ireviken event reported by Aldridge et al. (1993) and Jeppsson and Männik (1993) from the Baltic area have not been not distinguished in southern North America. Only three of their datum points (2—extinction of *Ozarkodina polinclinata*, 3—extinction of *Pterospathodus amorphognathoides*, and 7—extinction of *Distomodus staurognathoides*) occur in the proper order in southern North America. These events may represent true last occurrences, whereas some of the other events are probably local terminations of ranges due to local ecological factors, a possibility noted by Aldridge et al. (1993). Our information sheds no additional light on the nature of the changes responsible for the rapid series of extinctions.

SUMMARY

The Sheinwoodian and lower Homerian depositional sequences in the southern United States correspond to major depositional sequences identified elsewhere in North America and to major secundo episodes in northern Europe. The geographically widespread nature of the Sheinwoodian and lower Homerian events indicates that major eustatic sea-level events are a likely cause. Future work should concentrate on the identification of coeval events in other geographic regions and on the detailed examination of faunal changes that may occur in association with these events.

The upper Homerian/lower Ludlow depositional sequence in the southern United States cannot be as confidently recognized outside of the region, nor can an eustatic origin for the sequence be convincingly demonstrated. Additional studies directed toward a better understanding of regional and global events during the latest Wenlock and Ludlow will be required to interpret the significance of this sequence.

Better resolution of Wenlock global sea-level events and presumably related biotic crises awaits a more precise system of biostratigraphic correlation. The current lack of precise integration of graptolite, conodont, shelly fossils, and other groups into a single biostratigraphic scheme will continue to hinder determination of the synchronism of similar events in different regions. The current poor status of our ability to correlate Silurian chronostratigrapic boundaries and events is clearly demonstrated by the discussion of Wenlock events given here. I doubt that any series of coordinated zonations will be able to provide the degree of resolution and widespread applicability required to accomplish the task. For this reason, I will continue to use the graphic correlation method, following the lead of Kleffner (1989; 1995), to develop a better system of biostratigraphy for the Silurian.

ACKNOWLEDGMENTS

Gilbert Klapper (University of Iowa) and Thomas W. Amsden (Oklahoma Geological Survey) were instrumental in introducing me to the Silurian conodont faunas of the southern United States and encouraging my continuing research in this area. Special thanks are due to Mark A. Kleffner (Ohio State University–Lima), who shared his most recent unpublished version of the Silurian Composite Standard on which so many of the correlations in this paper are based. Funding for this research has come from sources acknowledged in previous papers. ARCO Oil and Gas Company provided the funds that permitted the most recent results to be published. Mark A. Kleffner and Stephen C. Ruppel (Bureau of Economic Geology, Austin, Texas) carefully reviewed the manuscript and contributed several important improvements.

REFERENCES CITED

Aldridge, R. J., 1988, Extinction and survival in the Conodonta, *in* Larwood, G. P., ed., Extinction and survival in the fossil record: Oxford, Clarendon Press, Systematics Association Special Volume 34, p. 231–256.

Aldridge, R. J., and Schönlaub, H. P., 1989, Conodonts, *in* Holland, C. H., and Bassett, M. G., eds., A global standard for the Silurian System: National Museum of Wales Geological Series 9, p. 274–279.

Aldridge, R. J., Jeppsson, L., and Dorning, K. J., 1993, Early Silurian oceanic episodes and events: Geological Society of London Journal, v. 150, p. 510–513.

Amsden, T. W., 1957, Stratigraphy and paleontology of the Hunton Group in the Arbuckle Mountain region. Part 1: Stratigraphy of the Hunton Group: Oklahoma Geological Survey Circular 44, 57 p.

Amsden, T. W., 1960, Stratigraphy and paleontology of the Hunton Group in the Arbuckle Mountain region. Part 6: Hunton stratigraphy: Oklahoma Geological Survey Bulletin 84, 311 p.

Amsden, T. W., 1968, Articulate brachiopods of the St. Clair Limestone (Silurian), Arkansas, and the Clarita Formation (Silurian), Oklahoma: Paleontological Society Memoir 1 (Journal of Paleontology, v. 42, supplement), 117 p.

Amsden, T. W., 1976, Hunton Group (Late Ordovician, Silurian, and EarlyDevonian) in the Anadarko Basin of Oklahoma: Oklahoma Geological Survey Bulletin 121, 214 p.

Amsden, T. W., 1978, Articulate brachiopods of the Quarry Mountain Formation (Silurian), eastern Oklahoma: Oklahoma Geological Survey Bulletin 125, 75 p.

Amsden, T. W., and Barrick, J. E., 1988, Late Ordovician through Early Devonian annotated correlation chart and brachiopod range charts for the southern Midcontinent region, U.S.A., with a discussion of Silurian and Devonian conodont faunas: Oklahoma Geological Survey Bulletin 143, 66 p.

Amsden, T. W., and Barrick, J. E., 1993, Pre-Woodford subcrop map and stratigraphic sections of the Anadarko Basin: Oklahoma Geological Survey Map GM-34, 2 plates and appendix, 20 p.

Amsden, T. W., Toomey, D. F., and Barrick, J. E., 1980, Paleoenvironment of Fitzhugh Member of Clarita Formation (Silurian, Wenlockian), southern Oklahoma: Oklahoma Geological Survey Circular 83, 54 p.

Barrick, J. E., 1977, Multielement simple-cone conodonts from the Clarita Formation (Silurian) Arbuckle Mountains, Oklahoma: Geologica et Palaeontologica, v. 11, p. 47–68.

Barrick, J. E., 1978, Wenlockian (Silurian) depositional environments and conodont biofacies, south-central United States [Ph.D. thesis]: Iowa City, University of Iowa, 273 p.

Barrick, J. E., 1983, Wenlockian (Silurian) conodont biostratigraphy, biofacies and carbonate lithofacies, Wayne Formation, central Tennessee: Journal of Paleontology, v. 57, p. 208–239.

Barrick, J. E., 1995, Biostratigraphy of uppermost Ordovician through Devonian depositional sequences in the Permian Basin, west Texas and southeastern New Mexico, *in* Pausé, P. H., and Candelaria, M. P., eds., Carbonate facies and sequence stratigraphy: Practical application of carbonate models: Permian Basin Section—SEPM (Society for Sedimentary Geology) Publication 95-36, p. 207–216.

Barrick, J. E., and Haywa-Branch, J. N., 1994, Conodont biostratigraphy of the Missouri Mountain Shale (Silurian–Early Devonian?) and the Arkansas Novaculite (Devonian), Black Knob Ridge, Atoka County, Oklahoma, *in* Suneson, N. H., and Hemish, L. A., eds., Geology and resources of the eastern Ouachita Mountains frontal belt and southeastern Arkoma Basin, Oklahoma: Oklahoma Geological Survey Guidebook 29, p. 161–177.

Barrick, J. E., and Klapper, G., 1976, Multielement Silurian (late Llandoverian–Wenlockian) conodonts of the Clarita Formation, Arbuckle Mountains, Oklahoma, and phylogeny of *Kockelella*: Geologica et Palaeontologica, v. 10, p. 59–100.

Barrick, J. E., Klapper, G., and Amsden, T. W., 1990, Late Ordovician–Early Devonian conodont succession in the Hunton Group, Arbuckle Mountains and Anadarko Basin, Oklahoma, *in* Ritter, S. M., ed., Early to middle Paleozoic conodont biostratigraphy of the Arbuckle Mountains, southern Oklahoma: Oklahoma Geological Survey Guidebook 27, p. 55–92.

Barrick, J. E., Finney, S. C., and Haywa-Branch, J. N, 1993, Revision of ages of the Fusselman, Wristen, and Thirtyone Formations (Late Ordovician–Early Devonian) in the subsurface of West Texas based on conodonts and graptolites: Texas Journal of Science, v. 45, p. 231–247.

Bassett, M. G., 1989, Brachiopods, *in* Holland, C. H., and Bassett, M. G., eds., A global standard for the Silurian System: National Museum of Wales Geological Series 9, p. 232–241.

Berry, W. B. N., and Boucot, A, J., 1970, Correlation of the North American Silurian rocks: Geological Society of America Special Paper 102, 289 p.

Berry, W. B. N., and Murphy, M. A., 1975, Silurian and Devonian graptolites of central Nevada: University of California Publications in Geological Sciences, v. 110, 109 p.

Boucot, A. J., 1975, Evolution and extinction rate controls: Amsterdam, Elsevier, Developments in Paleontology and Stratigraphy 1, 427 p.

Brett, C. E., Goodman, W. M., and LoDuca, S. T., 1990, Sequences, cycles, and basin dynamics in the Silurian of the Appalachian Foreland Basin:

Sedimentary Geology, v. 69, p. 191–244.

Chatterton, B. D. E., Edgecombe, G. D., and Tuffnell, P. A., 1990, Extinction and migration in Silurian trilobites and conodonts of northwestern Canada: Geological Society of London Journal, v. 147, p. 703–715.

Cooper, B. J., 1976, Multielement conodonts from the St. Clair Limestone (Silurian) of southern Illinois: Journal of Paleontology, v. 50, p. 205–217.

Copper, P., and Brunton, F., 1991, A global review of Silurian reefs, *in* Bassett, M. G., Lane, P. D., and Edwards, D., eds., The Murchison Symposium: Proceedings of an international conference on the Silurian System: Special Papers in Palaeontology 44, p. 225–259.

Craig, W. W., 1968, The stratigraphy and paleontology of Ordovician and Silurian strata, Batesville district, Independence and Izard counties, Arkansas [Ph.D. thesis]: Austin, University of Texas, 395 p.

Craig, W. W., 1969, Lithic and conodont succession of Silurian strata, Batesville district, Arkansas: Geological Society of America Bulletin, v. 80, p. 1621–1628.

Craig, W. W., 1975, Stratigraphy and conodont faunas of the Cason Shale and the Kimmswick and Fernville Limestones of northern Arkansas, *in* Wise, A. O., and Headrick, K., eds., Contributions to the geology of Arkansas, Volume 2: Little Rock, Arkansas Geological Commission, p. 61–95.

Craig, W. W., 1993, Stratigraphy of the Cason Shale, the Brassfield, St. Clair, and Lafferty Limestones, and the Penters Chert (Upper Ordovician–Lower Devonian) in the Arkansas Ozarks, *in* Johnson, K. S., ed., Hunton Group core workshop and field trip: Oklahoma Geological Survey Special Publication 93-4, p. 135–147.

Droste, J. B., and Shaver, R. H., 1987, Paleoceanography of Silurian seaways in the midwestern basins and arches region: Paleoceanography, v. 2, p. 213–227.

Dunn, P. H., 1942, Silurian Foraminifera of the Mississippi basin: Journal of Paleontology, v. 16, p. 317–342.

Goodman, W. M., and Brett, C. E., 1994, Roles of eustasy and tectonics in development of Silurian stratigraphic architecture of the Appalachian foreland basin, *in* Dennison, J. M., and Ettensohn, F. R., eds., Tectonic and eustatic controls on sedimentary cycles: SEPM (Society for Sedimentary Geology) Concepts in Sedimentology and Paleontology, v. 4, p. 147–169.

Ireland, H. A., 1939, Devonian and Silurian Foraminifera from Oklahoma: Journal of Paleontology, v. 13, p. 190–202.

Jeppsson, L., 1987, Lithological and conodont distributional evidence for episodes of anomalous oceanic conditions during the Silurian, *in* Aldridge, R. J., ed., Palaeobiology of conodonts: Chichester, United Kingdom, Ellis Horwood Limited, p. 129–145.

Jeppsson, L., 1990, An oceanic model for lithological faunal changes tested on the Silurian record: Geological Society of London Journal, v. 147, p. 663–674.

Jeppsson, L., and Männik, P., 1993, High-resolution correlations between Gotland and Estonia near the base of the Wenlock: Terra Nova, v. 5, p. 348–358.

Jeppsson, L., Viira, V., and Männik, P., 1994, Silurian conodont-based correlation between Gotland (Sweden) and Saaremaa (Estonia): Geological Magazine, v. 131, p. 201–218.

Jeppsson, L., Aldridge, R. J., and Dorning, K. J., 1995, Wenlock (Silurian) oceanic episodes and events: Geological Society of London Journal, v. 152, p. 487–498.

Johnson, J. G., and Klapper, J., 1992, North American Midcontinent Devonian T-R cycles, *in* Chaplin, J. R., and Barrick, J. E., eds., Special papers in paleontology and stratigraphy: A tribute to Thomas W. Amsden: Oklahoma Geological Survey Bulletin 145, p. 127–135.

Johnson, J. G., and Murphy, M. A., 1984, Time-rock model for Siluro-Devonian continental shelf, western United States: Geological Society of America Bulletin, v. 95, p. 1349–1359.

Johnson, J. G., and Potter, E. C., 1975, Silurian (Llandovery) downdropping of the western margin of North America: Geology, v. 3, p. 331–334.

Johnson, J. G., Klapper, G., and Sandberg, C. A., 1985, Devonian eustatic fluctuations in Euramerica: Geological Society of America Bulletin, v. 96, p. 567–587.

Johnson, M. E., 1987, Extent and bathymetry of North American platform seas in the Early Silurian: Paleoceanography, v. 2, p. 185–211.

Johnson, M. E., and McKerrow, W. S., 1991, Sea-level and faunal changes during the latest Llandovery and earliest Ludlow (Silurian): Historical Biology, v. 5, p. 153–169.

Johnson, M. E., Rong Jia-Yu, and Yang Xue-Chang, 1985, Intercontinental correlation by sea-level events in the Early Silurian of North America and China: Geological Society of America Bulletin, v. 96, p. 1384–1397.

Johnson, M. E., Kaljo, D., and Rong Jia-Yu, 1991, Silurian eustasy, *in* Bassett, M. G., Lane, P. D., and Edwards, D., eds., The Murchison Symposium: Proceedings of an international conference on the Silurian System: Special Papers in Palaeontology 44, p. 145–163.

Kaljo, D., and Märss, T., 1991, Pattern of some Silurian bioevents: Historical Biology, v. 5, p. 145–152.

Klapper, G., and Murphy, M. A., 1975, Silurian–Lower Devonian conodont sequence in the Roberts Mountain Formation of central Nevada: University of California Publications in Geological Sciences, v. 111, 62 p.

Klapper, G., Kirchgasser, W. T., and Baesemann, J. F., 1995, Graphic correlation of a Frasnian (Upper Devonian) composite standard, *in* Mann, K. O., and Lane, H. R., eds., Graphic correlation: SEPM (Society for Sedimentary Geology) Special Publication 53, p. 177–184.

Kleffner, M. A., 1989, A conodont-based Silurian chronostratigraphy: Geological Society of America Bulletin, v. 101, p. 904–912.

Kleffner, M. A., 1991, Conodont biostratigraphy of the upper part of the Clinton Group and the Lockport Group (Silurian) in the Niagara Gorge region, New York and Ontario: Journal of Paleontology, v. 65, p. 500–511.

Kleffner, M. A., 1994, Conodont biostratigraphy and depositional history of strata comprising the Niagaran sequence (Silurian) in the northern part of the Cincinnati Arch region, west-central Ohio, and evolution of *Kockelella walliseri* (Helfrich): Journal of Paleontology, v. 68, p. 141–153.

Kleffner, M. A., 1995, A conodont- and graptolite-based Silurian chronostratigraphy, *in* Mann, K. O., and Lane, H. R., eds., Graphic correlation: SEPM (Society for Sedimentary Geology) Special Publication 53, p. 159–176.

Koren, T. N., and Urbanek, A., 1994, Adaptive radiation of monograptids after the Late Wenlock crisis: Acta Palaeontologica Polonica, v. 39, p. 137–167.

Lenz, A. C., 1982, Ordovician to Devonian sea-level changes in western and northern Canada: Canadian Journal of Earth Sciences, v. 19, p. 1919–1932.

Männik, P., and Viira, V., 1993, Events in the conodont history during the Silurian in Estonia: Proceedings of the Estonian Academy of Sciences, v. 42, p. 58–69.

Moore, G. T., Jacobson, S. R., Ross, C. A., and Hayashida, D. N., 1994, A paleoclimate simulation of the Wenlockian (late Early Silurian) world using a general circulation model with implications for early land plant paleoecology: Palaeogeography, Palaeoclimatology, Palaeoecology, v. 110, p. 115–144.

Murphy, M. A., Dunham, J. B., Berry, W. B. N., and Matti, J. C., 1979, Late Llandovery unconformity in central Nevada: Brigham Young University Studies, v. 26, p. 21–36.

Noble, P. J., 1994, Silurian radiolarian zonation for the Caballos Novaculite, Marathon uplift, west Texas: Bulletins of American Paleontology, v. 106, 55 p.

Quinby-Hunt, M. S., and Berry, W. B. N., 1991, Late Wenlock (Middle Silurian) global bioevent: Possible chemical cause for mass graptolite mortalities: Historical Biology, v. 5, p. 171–181.

Ross, C. A., and Ross, J. R. P., 1996, Silurian sea level fluctuations, *in* Witzke, B. J., Ludvigsen, G. A., and Day, J., eds., Paleozoic sequence stratigraphy: Views from the North American Craton: Geological Society of America Special Paper 306, p. 187–192.

Ruppel, S. C., 1993, Depositional and diagenetic character of Hunton-equivalent rocks in the Permian Basin of west Texas, *in* Johnson, K. S., ed., Hunton Group core workshop and field trip: Oklahoma Geological Survey Special Publication 93-4, p. 91–106.

Ruppel, S. C., and Holtz, M. H., 1994, Depositional and diagenetic facies patterns and reservoir development in Silurian and Devonian rocks in the Permian Basin: The University of Texas, Bureau of Economic Geology Report of Investigations 216, 89 p.

Satterfield, I. R., and Thompson, T. L., 1975, Seventy-Six Shale, a new member of the Bainbridge Formation (Silurian) in southeastern Missouri: Missouri Geological Survey, Department of Natural Resources Report of Investigation 57, p. 109–119.

Schönlaub, H. P., 1980, Carnic Alps, with contributions by H. Jaeger and others, *in* Schönlaub, H. P., ed., Second European conodont symposium (ECOSII), Guidebook, Abstracts: Vienna, Abhandlungen der Geologischen Bundesansalt 53, p. 5–57.

Sheehan, P. M., and Boucot, A. J., 1991, Silurian paleogeography of the western United States, *in* Cooper, J. D., and Stevens, C. H., eds., Paleozoic paleogeography of the western United States—II: Society of Economic Paleontologists and Mineralogists, Pacific Section , v. 67, p. 51–82.

Subcommission on Silurian Stratigraphy, 1995, Correlation chart: Silurian Times, no. 3, p. 8.

Thompson, T. L., 1993, Paleozoic succession in Missouri. Part 3: Silurian and Devonian Systems: Missouri Department of Natural Resources, Division of Geology and Land Survey Report of Investigations 70, 228 p.

Walliser, O. H., 1964, Conodonten des Silurs (Conodonts of the Silurian): Abhandlungen des Hessischen Landesamtes für Bodenforschung, v. 41, 106 p.

Zuravel, D. L., 1986, Wenlockian (Silurian) agglutinated Foraminifera from the Wayne Formation, Tennessee [M.S. thesis]: Lubbock, Texas Tech University, 130 p.

MANUSCRIPT ACCEPTED BY THE SOCIETY FEBRUARY 27, 1997

Printed in U.S.A.

Geological Society of America
Special Paper 321
1997

Sedimentation and stratigraphic architecture of a Middle Devonian (late Givetian) transgressive-regressive carbonate-evaporite cycle, Coralville Formation, Iowa area

Brian J. Witzke and Bill J. Bunker
Iowa Department of Natural Resources, Geological Survey Bureau, Iowa City, Iowa 52242

ABSTRACT

Depositional facies of Coralville strata in the Iowa area characterize late stages of the Taghanic marine onlap and offlap along the margins of the Transcontinental Arch in the cratonic interior of North America. The Coralville Formation forms a discrete transgressive-regressive depositional cycle bounded by subaerial unconformity surfaces in the shoreward areas and submarine discontinuity surfaces in more offshore areas. Distinctive sedimentary and biotic associations are identified in inner- and middle-shelf settings of the continental interior. (1) During transgression, the inner-shelf region was characterized by muddy, relatively quiet-water carbonate sedimentation, with low diversity, probably euryhaline shelly faunas. Regression was marked by the progradation of peritidal and evaporitic sediments and by the loss of shelly faunas across the inner shelf. (2) Current energy was frictionally dissipated across the distal inner shelf and shelf margin during sea-level highstand, and peritidal facies prograded to the shelf margin during regression. (3) The middle-shelf region was characterized by subtidal carbonate deposition throughout the Coralville cycle. Relatively diverse normal-marine shelly faunas typify the middle-shelf region, but lower diversity faunas may suggest salinity stresses during late stages of sea-level fall. The limited biotope across the inner shelf formed an effective environmental filter that restricted faunal migrations between eastern and western North America during sea-level highstands. Migrations ceased altogether during sea-level lowstands, as inner-shelf areas became emergent.

INTRODUCTION

The onlap of Devonian seas into the continental interior of North America was a topic that challenged and fascinated the probing mind of Jess Johnson. He evaluated the geologic and biologic implications and consequences of the waxing and waning of Devonian epeiric seas in a series of influential papers (e.g., Johnson, 1970, 1971, 1974, 1990; Johnson et al., 1985; Johnson and Klapper, 1992). In particular, the Taghanic onlap of the Middle Devonian (late Givetian) was a focus of his formative thinking on large-scale processes associated with continental onlap.

Johnson (1971, p. 3273) recognized the Taghanic onlap as the "largest single transgressive event of the Kaskaskia Sequence." He characterized it not as a single marine incursion onto the continent but as a multiphased onlap marked by three or more separate transgressive pulses (Johnson et al., 1985; Johnson et al., 1989; Johnson and Klapper, 1992). The early phase of Taghanic onlap, defined within the context of the Givetian conodont zonal chronology, was initiated during the Middle *varcus* Subzone, with subsequent transgressive deepening events near the base of the *hermanni* Zone and within the *disparilis* Zone (Johnson and Klapper, 1992). The deepest depositional environments associated with Taghanic onlap in eastern and central North America are recognized within the *her-*

Witzke, B. J., and Bunker, B. J., 1997, Sedimentation and stratigraphic architecture of a Middle Devonian (late Givetian) transgressive-regressive carbonate-evaporite cycle, Coralville Formation, Iowa area, *in* Klapper, G., Murphy, M. A., and Talent, J. A., eds., Paleozoic Sequence Stratigraphy, Biostratigraphy, and Biogeography: Studies in Honor of J. Granville ("Jess") Johnson: Boulder, Colorado, Geological Society of America Special Paper 321.

manni and *disparilis* Zones (e.g., Geneseo black shale of New York [Johnson et al., 1985, p. 576], Rapid Member in Iowa [Witzke and Bunker, 1994]).

Johnson (1970, 1990) emphasized the efficacy of continental onlap in explaining the dispersal, evolution, and biogeography of benthic marine organisms. He suggested that "the major factor [for the dispersal of benthic invertebrates] is the extent to which marine seaways or broad, onlapping epeiric seas encroached on the main areas of the continental platforms" (Johnson, 1970, p. 2077). In his Taghanic example, the onlap and flooding of the Transcontinental Arch, the so-called continental backbone of North America, played a primary role in brachiopod dispersal and the resultant loss of endemism that characterized much of the earlier Devonian. "True cosmopolitanism among the brachiopods on both sides of North America may have only been achieved when shallow-water marine seas were truly widespread across, and on both sides of, the continental backbone" (Johnson, 1970, p. 2089).

In particular, Johnson (1970, p. 2078) proposed marine connections between "Iowa and the Williston Basin" during the Taghanic onlap to explain benthic dispersal patterns, but he suggested that such connections "probably encompassed a limited biotope restricting free movement of many elements of the megafauna" across the continental backbone. In this sense, Johnson envisioned the shallow carbonate-platform environments in the Iowa region as having acted as a sort of environmental filter within the late Givetian epeiric sea that permitted some, but not all, invertebrate groups to migrate between western and eastern North America. The margins of these interior carbonate platforms would thereby mark major biofacies boundaries that would "wax and wane in conjunction with changes in relative sea level" (Johnson, 1990, p. 930).

Johnson et al. (1985) interpreted synchronous changes in sedimentation in disjunct regions of the Devonian Euramerican continent to be a response to eustatic changes in sea level. They subdivided the Devonian into 14 third-order transgressive-regressive (T-R) depositional cycles and used this framework to propose a "correlation hypothesis of cratonic rocks in Iowa . . ." and other areas of the continental interior (Johnson et al., 1985, p. 579). Their T-R cycle IIa corresponded to the Taghanic onlap. Johnson (1970, p. 2083) had observed earlier that "in the Iowa Basin and adjacent areas of Nebraska, Kansas, and Missouri, Taghanic carbonate onlap is represented principally by the Cedar Valley [Group]." They proposed three deepening events within the Cedar Valley Group of Iowa (Johnson et al., 1985; Johnson and Klapper, 1992). Of particular note for the purposes of this chapter was their proposal of a "deepening pulse . . . at the base of the Coralville Member [Formation of this report], . . . a prominent discontinuity surface" (Johnson et al., 1985, p. 580). Johnson's proposals for Cedar Valley deposition were largely borne out by the stratigraphic investigations of Witzke et al. (1989) in the Iowa Basin area.

Depositional facies and stratigraphy of the Coralville Formation, the second formation of the Cedar Valley Group, are considered in this chapter. Although sedimentation of the Cedar Valley Group has been addressed to varying degrees in previous studies (e.g., Kettenbrink, 1973; Witzke et al., 1989), the interval of the Coralville Formation was selected for further elaboration within a memorial volume for Jess Johnson for several reasons. (1) The Coralville is a widely recognizable unconformity-bounded unit deposited during the later part of Johnson's Taghanic onlap. Following initial marine onlap, the Coralville records a major marine offlap from the continental interior at the end of T-R cycle IIa of Johnson et al. (1985). (2) Characterization of Taghanic sedimentation in the Iowa area is needed to constrain the progression of sedimentary environments in the cratonic seas that ultimately flooded and then receded from the "continental backbone." The biotic response of shallow-water benthic faunas to the rise and fall of sea level in the interior epeiric seas, as hypothesized by Johnson (1970, 1990), is considered within the context of Coralville sedimentation. (3) The Coralville Formation represents the least complicated of three major transgressive-regressive cycles developed within the Cedar Valley Group. Its facies architecture can serve as a simple stratigraphic model for the general characterization of Devonian transgressive-regressive cycles in carbonate platform settings of the expansive subtropical epeiric seaways of the North American interior.

GEOLOGIC SETTING

Devonian sedimentation was initiated in portions of eastern Iowa during the late Eifelian (lower Wapsipinicon Group). The locus of maximum sediment accumulation shifted to central Iowa during deposition of the Cedar Valley Group (late Givetian–early Frasnian) to form the Iowa Basin (Bunker et al., 1988). This basin was not a bathymetric basin but was a region of thicker stratigraphic accumulation where shallow-water and mudflat sedimentation kept pace with subsidence (Witzke et al., 1989). A general northwestward onlap of formational units within the Cedar Valley Group onto the margins of the Transcontinental Arch ("continental backbone") is recognized in western Iowa (Fig. 1). Following an erosional hiatus, strata of the Lime Creek Formation (upper Frasnian) further onlapped the Cedar Valley edge. The eroded surface that developed at the top of the Lime Creek was subsequently onlapped to the northwest by progressively younger Famennian depositional cycles (Witzke and Bunker, 1996).

All Givetian and Frasnian stratigraphic units are consistently characterized by thinner but deeper-water depositional facies in southeastern Iowa when compared with coeval strata in central Iowa. A relatively narrow geographic zone along the profile of Cedar Valley and Lime Creek strata is characterized by an abrupt shift in depofacies and a marked southeastward thinning of strata; this zone is labeled the "shelf break" on Figure 1. This intracratonic shelf margin serves as a consistent and convenient boundary that separates two broad groupings of carbonate depofacies in the Iowa region as delimited by Witzke and Bunker (1996): (1) an "inner-shelf" region of shallow-marine,

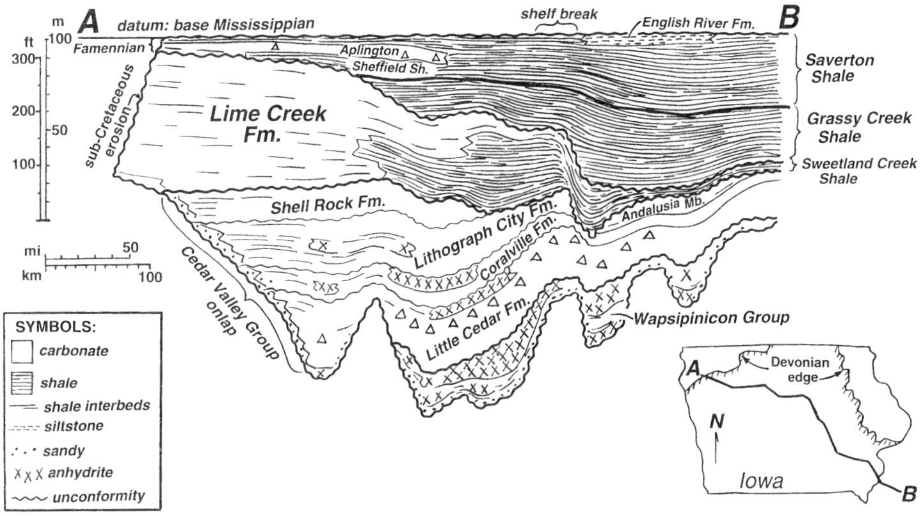

Figure 1. Generalized northwest-southeast cross section of Devonian strata across Iowa. Cedar Valley Group and Lime Creek Formation onlap to the northwest. Region where strata of the Lime Creek Formation and Cedar Valley Group show marked southeastward thinning is labeled as "shelf break"; this intracratonic shelf break divides a broad "inner shelf" area to the northwest and a "middle shelf" area to the southeast. Famennian shale units (Grassy Creek, Saverton, Sheffield) onlap the unconformity at the top of Lime Creek–Sweetland Creek interval. (abbreviations: Fm., Formation; Mb., Member).

peritidal, and mudflat/evaporite facies and (2) a deeper "middle-shelf" region marked by subtidal carbonate facies punctuated by submarine discontinuity surfaces (peritidal facies are absent). These cratonic facies groupings are analogous to the "middle- and inner-shelf rocks" described by Johnson et al. (1989) for the Devonian of the western United States. Outboard from the Iowa region, Givetian and Frasnian carbonates give way to deeper-water basinal dark shale facies of the "outer shelf," which extend to the margins of the continent (Witzke, 1987; Witzke and Bunker, 1996). This latter facies tract was termed the "outer-shelf-basin" by Johnson et al. (1989).

Stratigraphic nomenclature of the Cedar Valley Group in Iowa is shown on Figure 2. Because of significant geographic variation in lithofacies, a revised usage of member names was introduced to distinguish large-scale facies groupings of the inner and middle shelves within each formation of the Cedar Valley Group (Witzke et al., 1989; see Fig. 2). Each formation within the group forms a T-R depositional cycle, marked by initial transgressive deepening and regional flooding succeeded by general shallowing-upward patterns of sedimentation. These large-scale cycles (formations) are bounded by subaerial erosion surfaces on the inner shelf and by submarine discontinuity surfaces (or hardgrounds) on the middle shelf. However, the formation-level T-R cycles do not display simple progressive deepening and shallowing patterns of sedimentation but can be subdivided into a more complex series of smaller-scale depositional cycles, or subcycles (as schematically illustrated on the sea-level curve, Fig. 2). Goodwin and Mauriello (1994) recognized stacked patterns of smaller-scale allocycles within a portion of the Cedar Valley Group, whose hierarchy may be an expression of orbital forcing. The boundaries of the major Cedar Valley T-R cycles and subcycles (Witzke et al., 1989; Witzke and Bunker, 1996) correspond directly to the cycle boundaries and transgressive events delimited by Johnson et al. (1985), providing general confirmation of their "cratonic hypothesis."

GENERAL STRATIGRAPHY OF THE CORALVILLE FORMATION

Distribution and general lithostratigraphy of Coralville Formation

The Coralville Formation was selected to characterize one of the large-scale unconformity-bounded T-R cycles within the Cedar Valley Group. The development of stratigraphic nomenclature within the Cedar Valley Group and Coralville Formation is summarized elsewhere (Kettenbrink, 1973; Witzke et al., 1985; Witzke et al., 1989; Bunker and Witzke, 1989, 1992). Coralville strata are recognized within the Cedar Valley outcrop belt that stretches from northwestern Illinois and southeastern Iowa to northern Iowa and southeastern Minnesota (Fig. 3). Coralville strata extend in the subsurface westward to eastern Nebraska (Day, 1990) and southward into northern Missouri (Woodruff, 1990) and central Illinois (toward the northern margin of the Sangamon Arch). Coralville strata were previously thought to be generally absent in the Cedar Valley outcrop area of central Missouri along the northern flank of the Ozark Uplift (Witzke et al., 1989; Day, 1996), but Coralville equivalents have recently been recognized in that area based on conodont faunas (Day, this volume). Near the southern erosional limits of Cedar

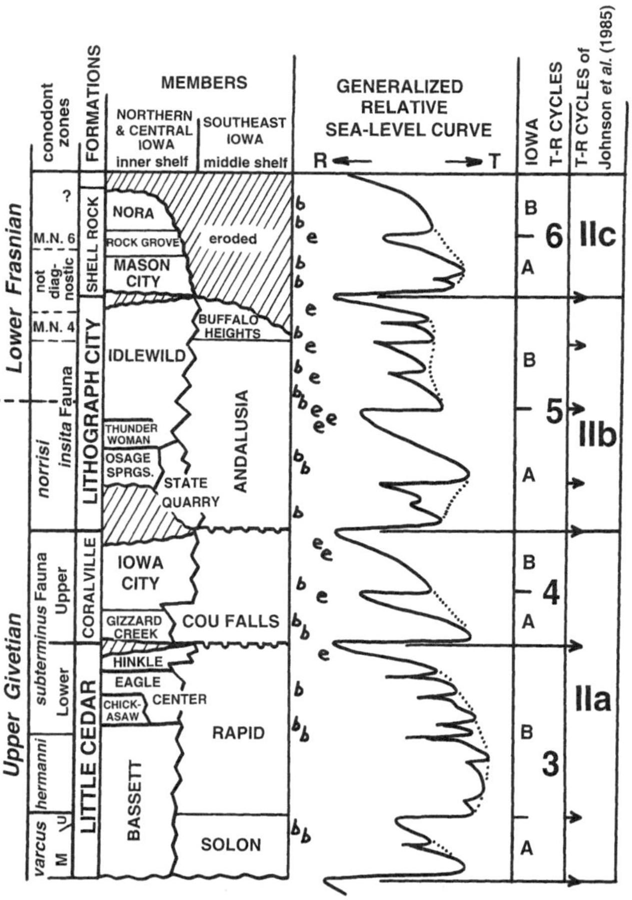

Figure 2. Stratigraphic nomenclature of Cedar Valley Group and interpreted relative sea-level curve for Cedar Valley deposition. Sea-level curve patterned largely after Witzke et al. (1989) and Witzke and Bunker (1994, 1996). Curve primarily illustrates interpretations of fourth- and/or fifth-order changes in sea level; dotted segments generalize lower-order patterns (third- and fourth-order). Right-pointing arrows show position of deepening events associated with T-R cycles of Johnson et al. (1985) and Johnson and Klapper (1992). Conodont zones follow Witzke et al. (1989) and Klapper and Johnson (1990); Frasnian zones after Montagne Noire (M.N.) succession of Klapper (1989). Stratigraphic positions of biostromal units (b) and evaporites (e) are shown. (Additional abbreviations: R, regression; T, transgression.)

Valley strata in central Missouri, strata correlative with the Lithograph City Formation locally and disconformably overlap Little Cedar equivalents (Witzke et al., 1985, p. 49). In general, the depositional or erosional edge of the Coralville Formation apparently stretches across portions of northern and central Missouri into central Illinois. The Coralville edge is overstepped southward by upper Cedar Valley strata.

Both the Gizzard Creek and Iowa City members of the Coralville Formation (Fig. 2) are restricted to inner-shelf areas of the Iowa Basin. These stratigraphic intervals were informally recognized in Minnesota and northern Iowa by Kohls (1961) and Mossler (1978), who described a "biogenic" and "sparsely fossiliferous dolostone" interval below a "pelletal calcilutite facies" within the "Coralville Member" (which at that time also encompassed strata now assigned to the Lithograph City Formation). These intervals were given formal member status by Witzke et al. (1989). The Gizzard Creek Member is a vuggy and fossiliferous dolomite, containing molds of brachiopods and crinoid debris, that ranges 4 to 7 m in thickness. The Iowa City Member, dominated by peritidal, supratidal, and lagoonal carbonate facies (7 to 14 m thick; 23 to 46 ft), is not recognized east of its depositional edge in southeastern Iowa and northeastern Missouri (Fig. 3). This edge delineates the "inner-shelf margin" for Coralville sedimentation. Evaporite lithologies (anhydrite, gypsum) are recognized within the Iowa City Member in the subsurface of central Iowa (Fig. 3), and extensive collapse breccias suggest the former presence of evaporites across areas of northern Iowa and Minnesota.

The Cou Falls Member is primarily an argillaceous skeletal limestone unit (wackestone to packstone fabrics), coral rich in part. This member makes up the entire Coralville Formation across middle-shelf regions of southeastern Iowa and adjacent Illinois, where it averages about 7 m in thickness (one-third the thickness of the formation in northern Iowa). The lower portion of the Cou Falls extends northwestward beyond the inner-shelf margin of the underlying Hinkle Member, which caps the underlying T-R cycle. The Cou Falls Member makes up the lower half of the Coralville Formation in distal areas of the inner shelf, equivalent to the interval of the Gizzard Creek Member to the north and west.

The northwestern edge of the Coralville Formation parallels the southeastern margin of the Transcontinental Arch, where only limited subsurface information is available to characterize the formation. This edge is apparently overstepped by upper Cedar Valley strata in northwest Iowa (Fig. 1). Cedar Valley strata further lap onto the Transcontinental Arch in the area of the Nebraska Sag (Bunker et al., 1988), a region where eroded Devonian strata almost trend across the arch, connecting the southern Williston Basin (southern South Dakota) with the Iowa Basin (across northeast Nebraska). It is hypothesized that late Givetian and Frasnian seaway connections were likely established across the Transcontinental Arch in the region of the Nebraska Sag and possibly central Minnesota during sea-level highstands. Although Cedar Valley strata are presently eroded (sub-Pennsylvanian) north and east of the study area, eastward seaway connections were undoubtedly developed across northern Illinois during the Middle and Late Devonian.

Age and correlation of Coralville Formation

The age of the Coralville Formation has been generally constrained by recent reevaluation of the conodont and brachiopod faunas. Earlier correlations were hampered by lithostratigraphic misconceptions that equated strata of the Coralville and Lithograph City formations. Conodonts are known in relatively low abundance from strata of the Cou Falls and Gizzard Creek members but are apparently absent in the Iowa City Member.

The Coralville conodont fauna is characterized by occur-

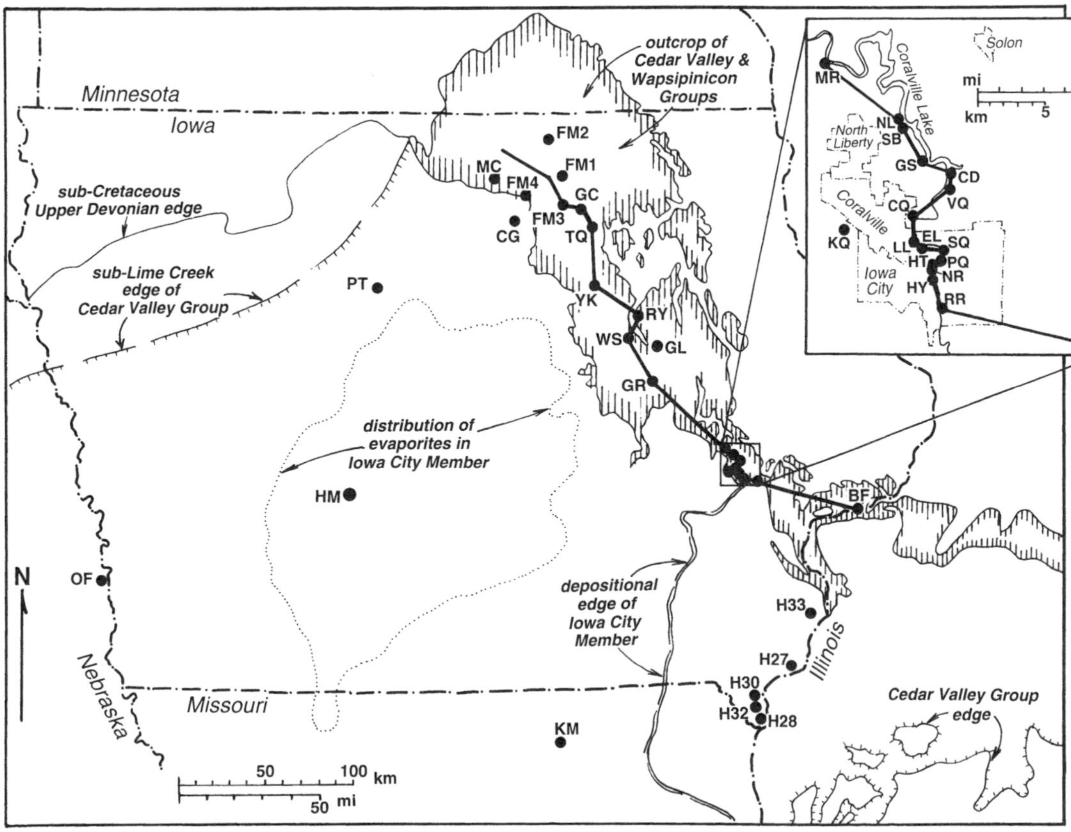

Figure 3. Locality map for Coralville Formation in Iowa area. Site locations and references for each locality are given in the Appendix. Edge of Cedar Valley Group is hachured; edge of Coralville evaporite facies is dotted; outcrop of Eifelian through lower Frasnian strata is vertically ruled. Depositional edge of Iowa City Member forms dividing line between inner-shelf facies to the northwest and middle-shelf facies to the southeast. Inset map shows locations in Johnson County, Iowa, the type area for the Coralville Formation.

rences of *Icriodus subterminus*, *Polygnathus angustidiscus*, and *Mehlina gradata*, but additional elements have been recovered, including undescribed species of *Polygnathus* (*P.* sp. A, *P.* sp. B of Witzke et al., 1985, *P.* n. sp. A of Rogers, 1990), *P. xylus xylus*, *Ozarkodina* sp. aff. *O. semialternans*, and *Schmidtognathus*? *gracilis* (Klug, 1982b; Bunker and Klapper, 1984; Witzke et al., 1985; Witzke et al., 1989; Woodruff, 1990; Rogers, 1990; Day, 1992). This association makes up the cratonic Upper *subterminus* Fauna as proposed by Witzke et al. (1985, 1989). The fauna correlates with a portion of the *disparilis* Zone of the standard (outer shelf) conodont zonation, most likely the upper part. The brachiopod zonal correlation scheme for the Cedar Valley Group proposed by Day (1989a, 1992, 1994) also supports correlation of Coralville strata with the upper *disparilis* Zone. Overlying strata of the lower Lithograph City Formation contain conodonts of the *insita* Fauna and *norrisi* Zone (Fig. 2).

Consideration of conodont- and brachiopod-based biostratigraphies, placed within the sea-level framework of Johnson et al. (1985), leads to some suggestions about interregional correlations across portions of North America. Eastward correlations to the Michigan Basin are not yet resolved, but the recovery of *Palmatolepis* (*Klapperina*) *disparalvea*, an indicator for the *disparilis* Zone, in basal beds of the type Petoskey Formation of western Michigan is of note (Bunker and Witzke, 1992). Like the Coralville Formation in Iowa, the Petoskey Formation overlies a prominent T-R cycle boundary (erosional surface on peritidal carbonates of the Charlevoix Formation). In addition, the Petoskey and Coralville formations contain shared brachiopod (Day, 1992) and coral (Sloss, 1939) taxa. However, the position of the Petoskey within the *disparilis* Zone is not known, and it may correlate instead with strata of the upper Little Cedar Formation in the Iowa area (Lower *subterminus* Fauna). If the latter suggestion is true, the Coralville would then presumably correlate with the Whiskey Creek–Thunder Bay interval in the Michigan Basin.

Farther to the east, the Geneseo Shale Member of the Genesee Formation in New York spans the *disparilis* Zone (Kirchgasser et al., 1989), and the Coralville Formation must therefore correlate with some part of the Geneseo. Baird et al. (1989) described a T-R cycle boundary (discontinuity) within the Geneseo above the Fir Tree Limestone Submember. The overlying Hubbard Quarry Shale Submember (Upper *disparilis* Zone below lower Penn Yan strata of the *norrisi* Zone) probably correlates with part or all of the Coralville T-R cycle in Iowa.

In contrast to the poorly resolved status of the eastward correlations, the Coralville T-R cycle of Iowa can be directly correlated with strata in the Williston Basin (Manitoba) and the northern Elk Point Basin (northern Alberta, southwestern District of Mackenzie) on the opposite side of the Transcontinental Arch. As in the Coralville of Iowa, the first appearance of the Upper *subterminus* Fauna is marked in the Canadian sequences above a regional disconformity. Coralville equivalents include the "argillaceous limestone beds" of the lower Point Wilkins Member, Souris River Formation in Manitoba (Witzke et al., 1989; Day et al., 1996), and the Slave Point Formation of the Elk Point Basin (Day et al., 1996). Johnson et al. (1989) outlined further correlation of *disparilis* Zone strata in the western United States.

BASAL TRANSGRESSIVE SURFACE

Basal Coralville strata sharply and disconformably overlie strata of the upper Little Cedar Formation throughout the region, except along the western and northwestern margin where Coralville strata apparently onlap Ordovician units (Fig. 1). In northern Iowa and across most of the inner-shelf area, fossiliferous strata of the basal Gizzard Creek Member record general marine transgression across an irregular surface developed on the Hinkle Member. The Hinkle includes supratidal facies and evidence of subaerial exposure during its deposition ("birdseye" fenestrae). Further lowering of base levels during late regressive phases of the Little Cedar T-R cycle culminated in the development of a regional subaerial erosion surface across the inner shelf with local relief to about 1 m. This surface was transgressed by the advancing Coralville sea. Initial deepening left no apparent sedimentary record, and the basal deposits of the Gizzard Creek Member are marked by subtidal skeletal mudstone fabrics locally with reworked lithoclasts of underlying Hinkle lithologies.

Inner-shelf-margin facies of the upper Little Cedar include cross-bedded echinoderm grainstones and packstones, but peritidal facies are absent. Meteoric diagenesis in this area resulted from the influx of meteoric waters sourced from the regional subaerial exposure surface on the adjacent inner shelf and "periodic subaerial exposure of the grainstone shoals on the inner shelf-margin" (Plocher and Ludvigson, 1989, p. 20). Basal Coralville strata in the inner-shelf-margin area sharply overlie a slightly irregular to burrowed surface developed on an underlying complex of shoal and lagoonal facies.

The Little Cedar–Coralville contact in middle-shelf areas forms "the best defined and most easily recognized stratigraphic boundary within the Cedar Valley" (Kettenbrink, 1973, p. 63). The contact is marked by a prominent burrowed discontinuity surface, commonly stylolitic, separating argillaceous skeletal mudstones to wackestones below and fine skeletal packstones above. Prominent vertical and horizontal burrows, most with diameters of 1 to 3 cm, are developed at the contact and are typically infilled with basal Coralville packstone lithologies. These burrow networks commonly penetrate 10 to 15 cm into the top of the Little Cedar Formation, and vertical penetrations of 30 to 35 cm are known at some localities. The sharp boundaries of these large burrows, some showing grain truncation, and the open burrow network required for subsequent sediment infilling indicate that the burrowing occurred in semilithified sediment.

The burrowed discontinuity at some localities is further expressed as a submarine hardground surface with an irregular sculpted surface that truncates skeletal grains. The hardground surface is locally encrusted with bryozoans and spirorbids (e.g., Loc. KQ, Appendix); millimeter-sized borings (*Trypanites*) are scattered to common. The hardground is phosphatic (apatite-impregnated) and pyritic at some localities (e.g., Loc. H33, Appendix).

Basal Coralville sediments above the burrowed discontinuity/hardground surface are primarily coral-bearing fine skeletal packstones that are largely indistinguishable from overlying Cou Falls strata but that locally incorporate small lithoclasts (1 cm or less) of underlying Rapid lithologies. In places where burrow networks are not well developed, basal Coralville strata locally include lenses of whole-valve brachiopod packstone (*Neatrypa waterlooensis*).

The basal Coralville surface in middle-shelf areas does not show unequivocal evidence of subaerial exposure; the surface bounds subtidal open-marine facies both above and below the formational contact. The surface is interpreted to have formed entirely in a subtidal marine setting. The sea-level fall at the close of the Little Cedar T-R cycle (Iowa T-R cycle 3; Fig. 2) was apparently of insufficient magnitude to produce subaerial exposure on the middle shelf. The prominent burrowed discontinuity or hardground surface is interpreted as a surface of nondeposition marking the initiation of Coralville transgression (a transgressive discontinuity). Submarine cementation (lithification) probably accompanied development of the discontinuity surface. The surface was colonized by infaunal burrowing organisms and secondarily by borers and encrusters. Burrow networks were infilled, and the surface was buried as the environments deepened and sediments once again began to accumulate.

INNER-SHELF FACIES OF THE LOWER CORALVILLE

Lower Coralville biostromal facies of the distal inner shelf

Lower Coralville strata of the Cou Falls Member are well exposed in Johnson County, Iowa (see inset Fig. 3 for general location). This area occupies the distal portion of the inner shelf, or approximately the outer 50 to 75 km of the inner shelf, and is sharply bounded at the shelf break (Fig. 4). These strata are dominated by richly fossiliferous and commonly biostromal skeletal packstones. Previous informal subdivisions of these strata include a lower "*Cranaena* beds" interval and an upper "*Idiostroma* beds" interval (see Figure 4; Calvin, 1897; Stainbrook, 1941; Kettenbrink, 1973).

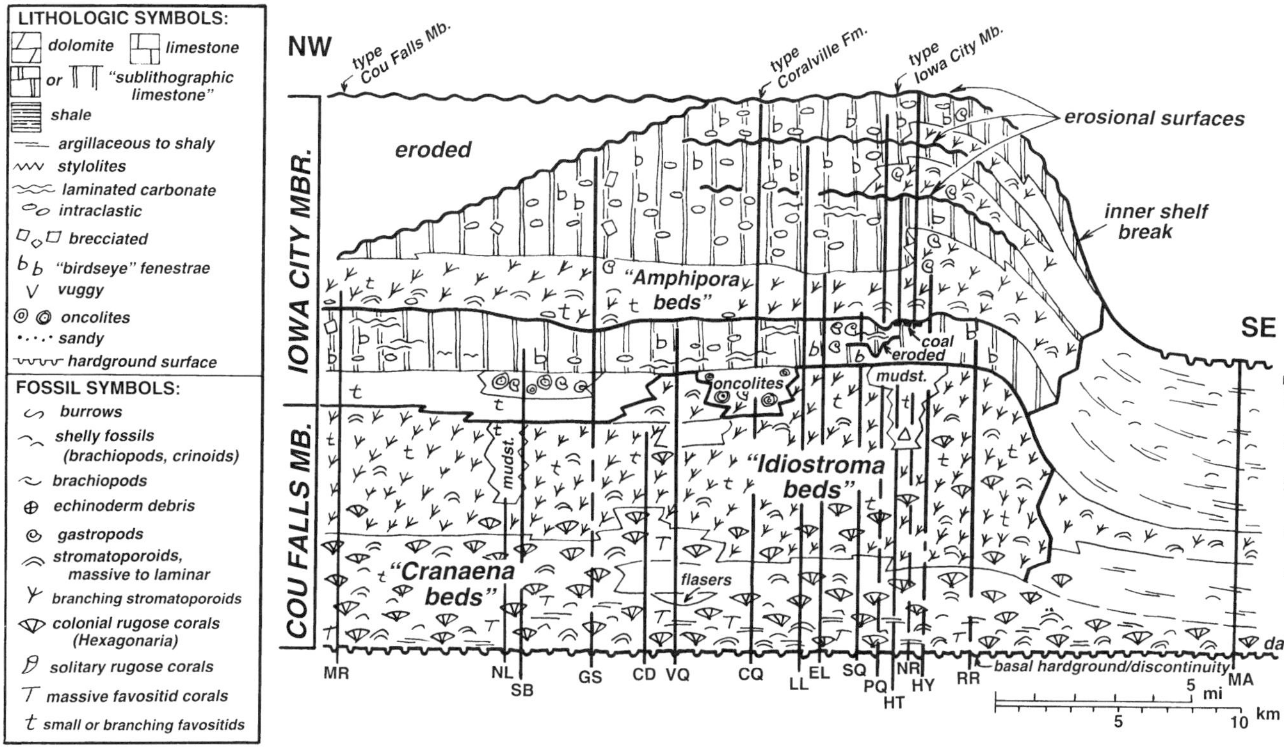

Figure 4. Cross section of Coralville Formation in the type area, Johnson County, Iowa. Vertical bars show extent of measured sections. See Figure 3 inset and Appendix for localities. Inner-shelf break marked by stratigraphic thinning and offshore loss of peritidal facies. (Abbreviations: mudst., mudstone; Mb. or Mbr., Member.) Coralville section along cross section line is essentially all limestone.

"Cranaena *beds."* Lower Cou Falls strata of the "*Cranaena* beds" in Johnson County include coral- and stromatoporoid-rich biostromal beds. Much of the interval is biostromal, although lateral variations of the coralline component are noted in the area, and not all beds are biostromal (intervals to 1 m in thickness may be only sparsely coralline). The overall lithology of the interval is a fine skeletal packstone, but a spectrum of skeletal wackstone to grainstone is observed. The fine grains are typically broken and abraded, volumetrically dominated by coral, stromatoporoid, brachiopod, and crinoid debris (Kettenbrink, 1973). The overall broken and abraded character of the grains suggests significant bottom current activity during deposition. However, the abundance of thin dark shaly carbonaceous (bituminous) partings (Kettenbrink, 1973, p. 43), especially in the lower beds, implies that quiet-water sedimentation, dysoxic in part, was also part of the general deposition. This co-occurrence of environmentally disparate sediment types is here regarded to reflect overall quiet subtidal sedimentation for the "*Cranaena* beds," with recurring episodes of vigorous bottom current activity probably generated by storm events. Such environmental alternations may further reflect small-scale sea-level fluctuations.

Sedimentary structures are rare in these strata, but the discontinuous carbonaceous partings subdivide thin horizontal layers of packstone that are amalgamated into larger bedforms. Faintly cross-bedded units and graded submarine channel-fills are locally observed (Kettenbrink, 1973, p. 43). Although rare, the presence of shallow channel fills (8.5 m × 35 cm) displaying flaserlike to hummocky cross-stratification (Loc. VQ, Fig. 4) provides further evidence of probable storm current activity during deposition.

Strata of the "*Cranaena* beds" typically contain densely packed to encrusting biostromal accumulations of corals and stromatoporoids. The corals and stromatoporoids within each biostrome bed are variably oriented, some in life position but many overturned or rotated. These are commonly encrusted with later generations of stromatoporoid and coral growth. Overturned corals are interpreted as evidence of episodic storm current activity, with subsequent encrustations reflecting quieter sedimentation.

Coral faunas are characterized by colonial rugosans (*Hexagonaria thomasi*), solitary rugosans (*Tortophyllum milleri*, *Tabulophyllum traversensis*, others [see Pitrat, 1962]), and tabulates (*Favosites* sp., less common alveolitids and pachyporids). *Favosites* and *Hexagonaria* colonies display general tabular to bulbous or domal forms (Trier, 1993), and are commonly 10 to 30 cm in diameter (locally reaching diameters of 50 to 100 cm). Massive stromatoporoids are of similar form and size, but their taxonomic identity has not been determined (these are commonly informally referred to "*Stromatopora*," but may include "*Trupetostoma*" according to Bils [1983]).

Favositid and stromatoporoid encrustations are common on colony surfaces.

The relatively diverse biota of the "*Cranaena* beds" is interpreted to be a stenohaline marine assemblage. Trilobites are scattered to common in the lower beds (*Crassiproetus searighti*, *Dechenella* sp., *Mystrocephala raripustulosa*; Hickerson, 1992, 1994). The molluscan fauna is dominated by rostroconchs (*Conocardium* sp.), but scattered gastropods, bivalves (*Paracyclas*), and straight-shelled nautiloids are known. Bryozoan debris is present, including fenestellids and branching cystodictyonids. Crinoid debris is common throughout; coarse columnals and stems are locally prominent. The general absence of taxonomically identifiable echinoderm fossils precludes a meaningful assessment of echinoderm diversity; a cup of *Botryocrinus* sp. (see Calhoun, 1983) and an abundance and variety of isolated plates indicate that inadunate and camerate crinoid taxa were significant components of the benthic community. Calcareous foraminifera and calcareous algae are rare in strata of the "*Cranaena* beds," but become more common in the overlying "*Idiostroma* beds" (Kettenbrink, 1973).

The relatively diverse brachiopod fauna of the "*Cranaena* beds" contains upwards of 20 species (Day, 1989a, 1989b, 1992, 1994). The full fauna is not listed here, but species of atrypids (*Pseudoatrypa*, *Independatrypa*, *Seratrypa*), spiriferids (*Orthospirifer*, *Tylothyris*, *Cyrtina*), *Athyris*, *Pentamerella*, *Cranaena*, *Schizophoria*, and *Strophodonta* are characteristic. Brachiopods are generally more abundant and diverse in the lower 1 m of the interval than in upper strata. Additional fauna includes conodonts (see earlier) and rare ptyctodont placoderm tritors.

Facies of the "*Cranaena* beds" were deposited in the most offshore, most marine, and deepest environments of Coralville sedimentation on the inner shelf, probably coincident with maximum transgression in the Coralville T-R cycle. Biostromal and skeletal-rich sedimentation is interpreted to have been generally above storm wavebase. Distal inner-shelf sediments were winnowed and abraded by storm and other currents, but these facies are replaced shoreward by more mud-rich sediments deposited in quieter and more restricted settings. This leads to the suggestion that significant wave and current energy decreased shoreward across the distal inner shelf, analogous to the shelf-margin high-energy zone envisioned by Johnson (1990, p. 927). The biotic association that includes diverse brachiopods, echinoderms, bryozoans, trilobites, and rugose corals is interpreted to be a stenohaline normal-marine assemblage. Storm, tidal, and wave currents must have effectively bathed the inner-shelf margin with marine waters from the open seaway to the east, but the shoreward dissipation of such currents limited marine circulation across more proximal areas of the inner shelf.

"Idiostroma *beds*." Upper Cou Falls strata of the distal inner shelf are characterized by biostromes rich in broken segments (generally 1 to 3 cm) of branching stromatoporoids, primarily *Idiostroma*. The lowest occurrence of these branching stromatoporoids marks the base of the interval, but the contact between the "*Idiostroma*" and "*Cranaena* beds" shows some facies intergradation across Johnson County (Fig. 4). In general, the two intervals are faunally and lithologically gradational. The "*Idiostroma* beds" contain lithologies similar to those of the "*Cranaena* beds," primarily fine abraded-grain coral- and stromatoporoid-rich packstones, but there are notable variations in the abundance and composition of the skeletal grains between the two units. In addition, carbonaceous partings characteristic of lower strata are absent. Upper strata of the "*Idiostroma* beds" show lateral facies between stromatoporoid-rich packstones and more sparsely skeletal, slightly argillaceous mudstones and wackestones and oncolitic facies (Figs. 4, 5).

Biotic changes are evident both laterally and vertically within the interval. Branching sticklike stromatoporoids (*Idiostroma* and *Stachyodes* [Trier, 1993]) are scattered to abundant throughout most of the interval, showing a general decrease in grain abundance upward (Bils, 1983). They are locally absent within upper mudstone facies. Massive tabular and hemispherical stromatoporoids ("*Stromatopora*") are scattered to common. The largest forms reach 30 to 50 cm in diameter (lower beds), but most range between 3 and 10 cm, especially in the upper beds. Massive stromatoporoids and favositid corals commonly encrust other coralline elements. Favositid (*Favosites*) morphologies show variation with respect to occurrences in the underlying "*Cranaena* beds," with bulbous, hemispherical, and branching forms noted (Bils, 1983; Trier, 1993).

Colonial rugose corals (*Hexagonaria*) are less common and generally smaller that in underlying "*Cranaena* beds" strata, but they remain a significant part of the biostromal fauna in the lower "*Idiostroma* beds." *Hexagonaria* disappears both vertically and proximally within the interval and is known from upper beds primarily in distal settings marginal to the shelf break (Figs. 4, 5). Solitary rugose corals are scattered through the interval but are most common in the lower beds. Massive corals and stromatoporoids are commonly overturned or rotated from life position, and fragments of branching stromatoporoids are concentrated in abraded packstone layers and lenses. Tabulate and rugose corals disappear near the proximal limits of the "*Idiostroma* beds" (Loc. GR, Fig. 6), where the interval is characterized by packed to biostromal stromatoporoid accumulations, including abundant *Idiostroma* and massive irregular to spherical forms.

Trilobites and conodonts are not known from the "*Idiostroma* beds," and bryozoans are essentially absent. Echinoderm debris generally decreases upward and proximally within the interval and is scarce to absent in upper beds and proximal facies. Rostroconch molluscs (*Conocardium*) are scattered in the interval. Brachiopods are less common and less diverse than in the underlying "*Cranaena* beds," and the brachiopod faunas show a general upward and proximal decrease in abundance and diversity. Lower beds include spiriferids (*Elita*), atrypids, athyrids, *Pentamerella*, *Productella*, and *Cranaena* (Day, 1992; Witzke, 1984), but brachiopods are rare to absent in upper strata (primarily *Athyris* only).

Calcareous algae and foraminifera are more abundant in

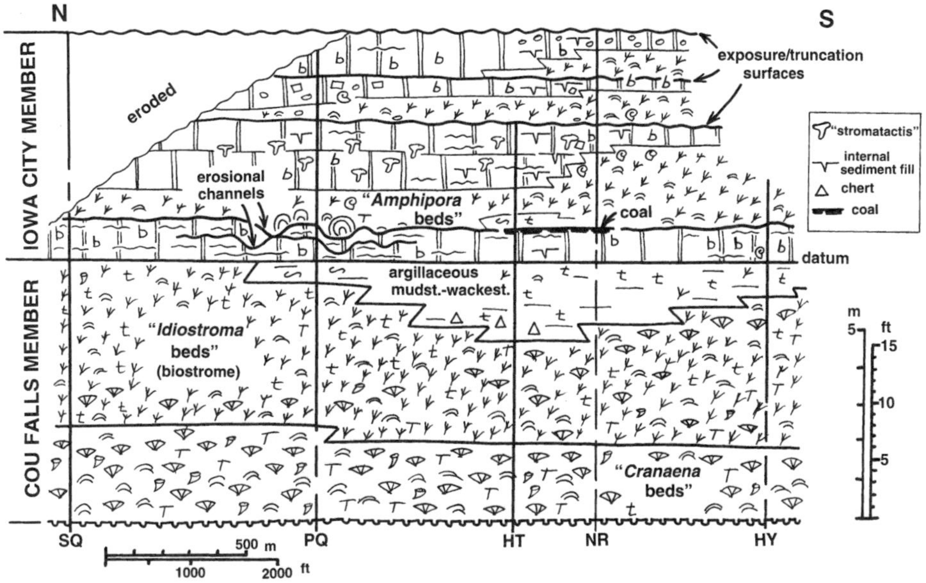

Figure 5. Cross section of Coralville Formation in the University of Iowa campus area, Iowa City, Iowa. See Figure 3 inset and Appendix for localities. Symbols and abbreviations as in Figure 4 unless noted. Several erosional surfaces are shown within the Iowa City Member. A slightly argillaceous mudstone-wackestone (mudst.-wackest.) facies has limited geographic extent.

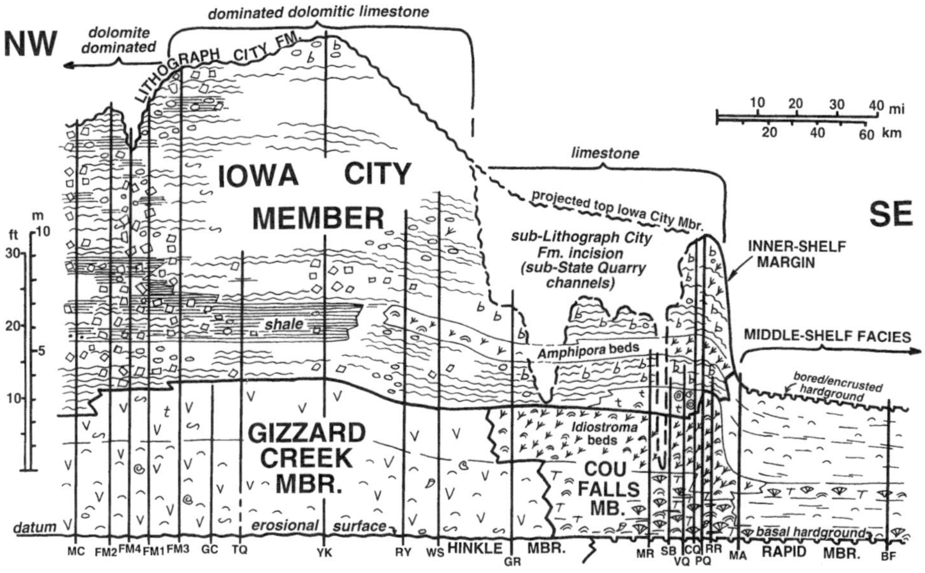

Figure 6. Northwest-southeast cross section of Coralville Formation, north-central to east-central Iowa. See Figure 3 for cross section line and localities; site locations given in Appendix. Symbols and abbreviations as in Figure 4. Sub–Lithograph City erosional incision of Coralville strata is most pronounced in the distal regions of the inner shelf area and is schematically illustrated. Facies contain more shale and dolomite to the northwest.

the "*Idiostroma* beds" than in the underlying interval. Foraminifera include simple monolocular bisphaerid and lagenid-like forms, uniserial eonodosarids, encrusting forms (*Evlania*), and parathuraminids. Algae include dasyclads (*Vermiporella*, calcispheres) and reds (*Parachaetetes*), and blue-green cyanobacteria filaments (*Girvanella*) are noteworthy (Kettenbrink, 1973; Kettenbrink and Toomey, 1975).

The upward and proximal decrease of stenohaline faunal elements and a coeval increase in carbonate mud suggest increasing salinity stress and decreasing bottom current activity during deposition of the "*Idiostroma* beds." Upward changes in coral and stromatoporoid growth forms within the Cou Falls Member were interpreted by Trier (1993) to be an ecophenotypic response to shallowing trends accompanied by decreasing wave and current energy and increasing light intensity. Kettenbrink (1973, p. 158) proposed that the skeletal packstones of the Cou Falls interval "were deposited in progressively shallower and more restricted marine waters, perhaps with greater salinity fluctuation." Bottom currents that impinged along the inner-shelf margin were apparently dissipated across the distal reaches of the inner shelf, where abraded-grain biostromal facies were developed. As the environments shallowed, circulation became less effective onto the inner shelf, and salinity stresses were imposed. The shallowing trend culminated in the spread of supratidal facies in overlying strata of the lower Iowa City Member.

Lower Coralville Marine and Restricted-Marine Subtidal Facies of the Proximal Inner Shelf

Deposition of biostromal and abraded grain facies characterized the distal region of the inner shelf during maximum transgressive phases of the Coralville T-R cycle. However, coeval facies across the proximal and central regions of the inner shelf are considerably less skeletal and more mud rich, and the contained faunas are notably less diverse. These latter facies are included in the Gizzard Creek Member, a vuggy fossil-moldic dolomite interval (Witzke et al., 1989) (Fig. 6). Sedimentary fabrics indicate primary deposition of Gizzard Creek sediments as skeletal mudstones to wackestones. Subsequent regional dolomitization of these carbonate sediments probably occurred during later regressive portions of the Coralville T-R cycle, possibly within meteoric-marine mixing-zone environments that accompanied seaway withdrawal across the inner shelf.

Skeletal grains (molds) are scattered to common in lower Gizzard Creek strata, but these become progressively less common upward in the member. Brachiopods and crinoid debris are characteristic, but these decrease in abundance upward and are commonly absent in upper strata. Bivalves and gastropods are scattered through the member, and bryozoans are occasionally seen in lower strata. Branching to domal stromatoporoids, branching favositids, and solitary rugose corals are locally observed in the distal regions of the member, generally transitional with biostromal strata of the Cou Falls Member (e.g., Loc. KM, Fig. 7). Small favositid corals are locally seen in upper strata in more proximal settings (Loc. FM3, Fig. 6). Sparsely fossiliferous mudstone lithologies commonly display faintly burrowed fabrics, primarily simple horizontal burrow traces. Onychodontid crossopterygian fish teeth have also been noted (Loc. RY, Fig. 6).

Brachiopods of the Gizzard Creek Member are generally represented by only a few species, primarily an athyrid, *Athyris vitatta*, and an atrypid, *Independatrypa randalia* (Day, 1992, 1994). A few additional brachiopods are recognized in more distal portions of the inner shelf, including a new species of *Tecno-*

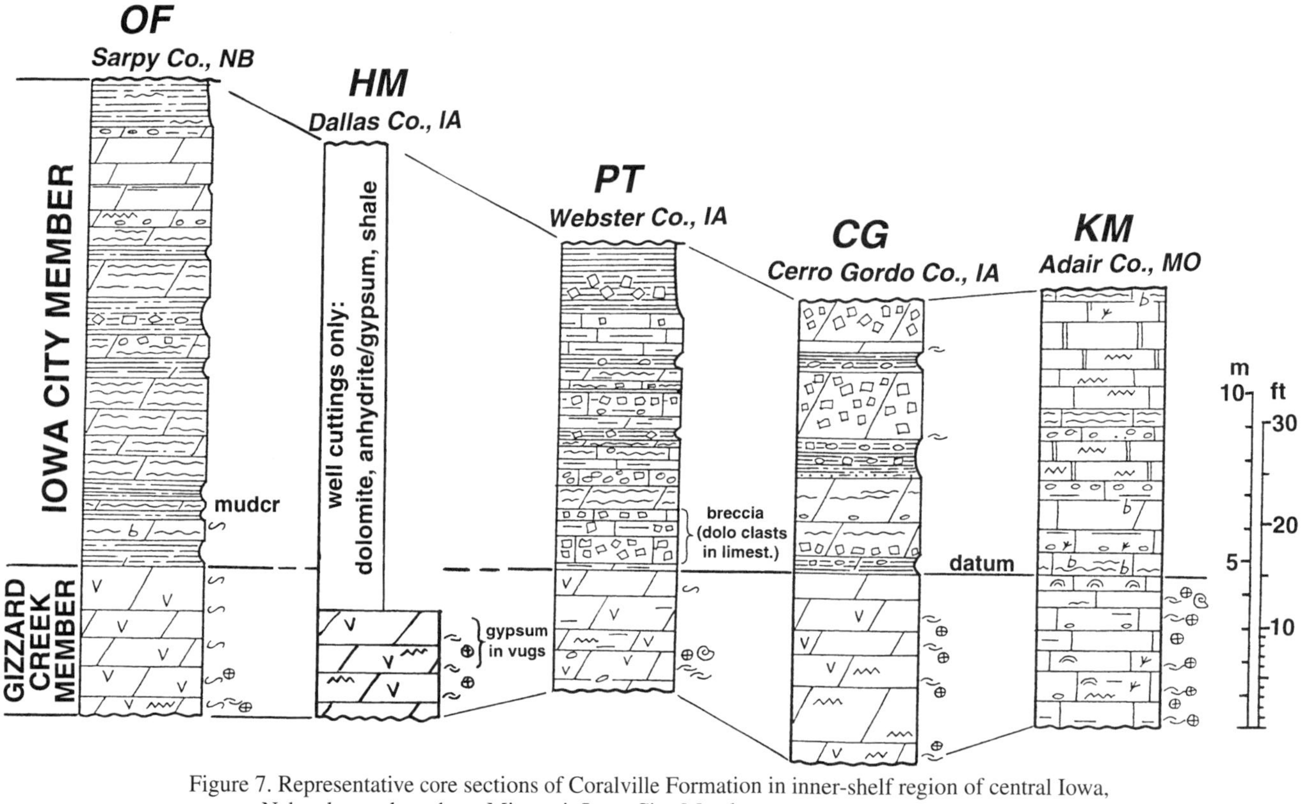

Figure 7. Representative core sections of Coralville Formation in inner-shelf region of central Iowa, eastern Nebraska, and northern Missouri. Iowa City Member represented only by well cuttings in Hummel core (Locality HM). See Figure 3 and Appendix for localities. Symbols as in Figure 4. (Abbreviations: mudcr, mudcracks; dolo, dolomite; limest., limestone.)

cyrtina (Loc. GL; see Day, this volume) and *Seratrypa rotunda*. Brachiopod molds occur as isolated valves or whole shells in the mudstone to wackestone lithologies in lower to middle Gizzard Creek strata and are locally abundant within thin skeletal-rich horizons. Brachiopods are commonly absent in upper Gizzard Creek strata or are represented by occasional *Athyris* only. The diversity of the Gizzard Creek brachiopod fauna is dramatically reduced in comparison to coeval strata of the lower Cou Falls Member, suggesting increasing environmental restrictions in a shoreward direction. *Independatrypa* and *Athyris* are interpreted to be more euryhaline than other contemporary brachiopod taxa that required more open-marine salinities (of the middle-shelf and inner-shelf margin).

Conodonts are not abundant but include a typical Upper *subterminus* Fauna, recovered in lower Gizzard Creek strata. The general upward decline in abundance and diversity of brachiopods, echinoderm debris, and conodonts within the member is interpreted to reflect increasing stress associated with progressive shallowing. Environmental restriction was likely related to increasing salinity stress in a shoreward direction. Continued shallowing and environmental restriction culminated in upper Coralville supratidal and evaporitic sedimentation across the inner shelf.

Mud-rich sedimentation dominated in the Gizzard Creek facies, indicating relatively quiet depositional environments. In contrast to the distal grain-rich facies, a shoreward decrease in bottom current energy is indicated. Currents must have been weakened across the distal inner-shelf region, decreasing shoreward circulation and promoting euryhaline conditions (Witzke et al., 1989). However, episodic storm currents probably influenced Gizzard Creek sedimentation, especially during phases of maximum transgression. Hummocky stratification is observed in lower strata near the distal limits of Gizzard Creek sedimentation (Loc. GR, Fig. 6). Skeletal-rich lenses in more proximal settings may have formed by episodic storm current activity, although current energies must have been reduced compared to those impinging across the grain-rich distal areas.

INNER-SHELF FACIES OF THE UPPER CORALVILLE

Upper Coralville restricted-marine and peritidal facies of the distal inner shelf

A complex of restricted marine and peritidal facies prograded across the inner shelf during the regressive phase of the Coralville T-R cycle. These facies are included within the Iowa City Member (Witzke et al., 1989). Iowa City strata are characterized by limestone facies in distal reaches of the inner shelf, but the facies become more dolomitic and evaporitic in a proximal (shoreward) direction. Distal Iowa City strata display a sequence of facies that reflects overall depositional shallowing but is punctuated by a minor transgressive episode (a T-R subcycle) in the middle part.

Lower mudstone and oncolite beds. Transitional mudstone and oncolitic facies are identified in portions of Johnson County that separate characteristic biostromal lithologies of the upper "*Idiostroma* beds" and peritidal "sublithographic" limestones of the lower Iowa City Member (Figs. 4, 5). Although this transitional interval is laterally equivalent to biostromal units of the upper Cou Falls Member (Figs. 4, 5), it has generally been included with upper Coralville (Iowa City Member) facies because of its pelleted to mud-rich (calcilutitic) character (Kettenbrink, 1973). Decreasing current energy and shallowing conditions during the final stages of *Idiostroma*-rich biostromal sedimentation culminated in the local accumulation of restricted mud-rich facies near the inner-shelf margin (Fig. 4). These beds are commonly slightly argillaceous and contain a low-diversity fauna of scattered *Idiostroma*, small hemispherical stromatoporoids, and small branching favositids (Fig. 5). They are locally burrowed, and scattered chert nodules are noted.

These restricted facies locally include oncolitic beds, marked by an abundance of small oncolites (commonly 1 to 4 cm) or oncolitic encrustations (Locs. CQ, GS, SB; Fig. 4). The oncolitic facies contain common branching favositids, scattered *Idiostroma*, abundant large gastropods ("*Straparollus*" and rare high-spired forms), calcareous algae, foraminifers, and ostracodes. Subspherical oncolites are locally common, completely encrusting gastropods or other skeletal grains. Oncolitic laminae are bound by cynaobacterial filaments and are commonly encrusted with spirorbid worm tubes and foraminifera (Kettenbrink, 1973). Bottom currents were insufficient to significantly winnow the carbonate mud, but episodic bottom agitation was needed to move the oncolites and allow coatings to develop on all sides. These shallow restricted facies were buried beneath peritidal sediments as shallowing progressed.

Lower peritidal facies. The spectrum of environments that spans intertidal and supratidal settings is grouped under the label "peritidal," including primarily carbonate mudflat facies in the Iowa City Member. A 1- to 2-m-thick interval of dense pelmicrite in the lower Iowa City Member, beneath the "*Amphipora* beds," is persistent across the distal region of the inner shelf (Figs. 4, 5). These dense unfossiliferous beds commonly display discontinuous laminations and "birdseye" structures, and the homogeneous muddy lithologies typically break with conchoidal fracture (these have historically been termed "sublithographic" limestone). The "sublithographic" limestones are composed of muddy pelleted lime sediment (pelmicrite) with scattered to common small intraclasts.

"Birdseye" structures comprise small spar-filled bubble-like voids and larger coalesced planar fenestral voids aligned with laminae. Such spar-filled voids (1 to 6 mm) are interpreted to have formed in supratidal mudflats, the result of the release of gas bubbles and periodic desiccation in subaerial settings. The lower "sublithographic" interval displays at least two erosional surfaces (top and middle part) with local relief to 50 cm in channeled cut-outs (Loc. PQ, Fig. 5). These likely represent part of a network of storm-flood and tidal channels that characterizes the distal regions (channeled belt) of the extensive mudflat environments. It is conceivable that some erosional surfaces (Figs. 4, 5) may mark small-scale eustatic cycle boundaries. Peritidal deposition was terminated across the entire distal region of the inner shelf by a minor marine transgression that initiated deposition of the overlying "*Amphipora* beds."

A thin (0.6 to 3.8 cm) spore-rich coal is locally developed at or near the top of the peritidal interval at a few localities in Johnson County (Locs. HT, NR; Fig. 4; Dow [1960] also noted it at Locs. KQ, VQ, CQ). Algal spores make up the bulk of the coal, and trilete miospores are also abundant (Sanders, 1968; Klug, 1990). The coal has been variably labeled a "cannel coal" (Dow, 1960) or "boghead coal" (Sanders, 1968) to indicate its spore-rich composition. Scolecodonts have been noted in the coal (Dow, 1960), suggesting marginal marine connections likely related to burial during initial transgression of the overlying "*Amphipora* beds." The depositional environment for the boghead coal is speculative, although Witzke (1984, p. 26) suggested an algal-rich ponded setting protected from oxygenating water currents. Thin coals are also known within regressive portions of other T-R cycles of the Cedar Valley Group associated with peritidal and lowland settings, including a paper coal (cuticle coal) from Lithograph City strata (Klug and Schabilion, 1992).

"Amphipora beds." A 0.6- to 2-m-thick limestone unit containing segments of the branching stromatoporoid, "*Amphipora*," is recognized throughout the distal region of the inner shelf (Calvin, 1897; Kettenbrink and Heckel, 1975; Witzke, 1984). It includes marginal- and restricted-marine facies, biostromal in part. The unit sharply to disconformably overlies the "sublithographic" limestone interval, and it marks the base of a smaller-scale T-R cycle (subcycle 4B, Fig. 2) within the larger Coralville cycle. The lithologies typically comprise an *Amphipora*-rich wackestone to packstone with a pelmicrite matrix; *Amphipora* grains commonly show a thin oncolitic coating (Kettenbrink, 1973). Calcareous algae and foraminifers are also abundant.

The "*Amphipora* beds" show lateral variation in fossil content and abundance. Although generally dominated by *Amphipora* (branches 5 to 15 mm), additional stromatoporoids are commonly represented in lesser numbers, including *Idiostroma* and domal to globular forms (usually < 8 cm, rarely to 25 cm). Small branching favositid corals and brachiopods (*Athyris*) are occasionally seen. Near the inner-shelf margin, upper strata are observed to interfinger with birdseye-bearing sublithographic limestones (Loc. NR, Figs. 4, 5), indicating that deposition occurred in part adjacent to peritidal mudflats. Large gastropods (to 8 cm), some with oncolitic coatings, are particularly abundant near the peritidal margins.

Interesting local faunal variations are observed where basal strata infill small erosional channels incised in the underlying "sublithographic" unit (Loc. PQ, Fig. 5). Very large massive blocky to hemispherical stromatoporoids, the largest observed anywhere within the Coralville Formation, are confined within these channels (up to 70 × 45 cm). These variably occur in life position or are slumped and rotated within the channel. A thin horizon with common atrypid brachiopods (*Independatrypa*) is recognized above the massive stromatoporoids.

Strata of the "*Amphipora* beds" extend onto the inner shelf up to 130 km shoreward of the shelf margin (Fig. 6). The more shoreward facies of the "*Amphipora* beds" interfinger with peritidal facies and locally include intraclastic and ostracode-rich beds. Large leperditiid ostracodes (2 to 10 mm) are locally common (Loc. GR).

The "*Amphipora* beds" represent a transgression of restricted- and marginal-marine facies onto the inner shelf. Kettenbrink (1973, p. 124) suggested that *Amphipora* was "likely to have tolerated salinity and temperature extremes greater than most other marine organisms" and flourished in restricted shallow environments. *Amphipora*-rich environments interfingered with and were ultimately buried by peritidal and supratidal mudflat facies as the shallow marginal euryhaline seaway withdrew from the inner shelf.

Upper peritidal facies. Strata of the upper Iowa City Member (to 4 m thick) above the "*Amphipora* beds" are dominated by sublithographic limestones (pelmicrites), commonly faintly to prominently laminated. These strata are intraclastic to varying degrees, and coarse intraclastic beds are locally developed. "Birdseye" structures are well developed in many beds, and mudcracked and irregular erosional surfaces are displayed in the interval (Figs. 4, 5). These features suggest extensive sedimentation in supratidal mudflats with subaerial desiccation and episodic erosion. Large domal stromatolites (up to 1 m wide, 40 cm high) are observed above the "*Amphipora* beds" near its northern extent (Loc. RY).

Supratidal mudflat facies are observed to interfinger with *Amphipora*-bearing restricted-marine units at the shelf edge (Loc. NR; Figs. 4, 5), where lithologies and faunas similar to the underlying "*Amphipora* beds" are identified. These units are bounded by erosionally truncated exposure surfaces, locally capped with vadose pisoliths, which delineate small-scale T-R packages at the shelf edge. The upper peritidal interval contains a restricted biota including calcareous algae (and common calcispheres), *Girvanella*, foraminifera, and ostracodes.

These strata commonly display an interconnected fenestral network of geopetal void fills (internal sediment fills) and spar-filled "stromatactis" cavities (Kettenbrink, 1973). This network likely resulted from dissolution and desiccation in subaerial and vadose environments as base levels were lowered during phases of regional regression. Sediment fills variably include muddy, intraclastic, and ostracode-rich sediments. Void-filling cements primarily grew in vadose and freshwater phreatic diagenetic environments that accompanied the Coralville regression, but later-stage marine cements were precipitated within a microkarstic network during subsequent marine transgression of the Lithograph City Formation (Ludvigson, 1989; Plocher et al., 1992).

The upper peritidal facies records the final regressive phases of sedimentation of the Coralville T-R cycle in the proximal areas of the inner shelf. Minor erosional episodes interrupted deposition during the phased regression. A major erosional surface caps the sequence, which developed during the sea-level lowstand that terminated the T-R cycle across the inner shelf. This surface includes prominent erosional channels (schematically shown on Figure 6), in places incised through the entire thickness of the Coralville Formation. Erosional channels are recognized on the proximal inner shelf up to 85 km from the shelf margin (Loc. GR). The depth of the sub–Lithograph City incision provides unambiguous evidence that base levels must have been lowered at least 25 m with respect to earlier peritidal deposition on the inner shelf (Witzke and Bunker, 1994). The erosional channels were infilled with sediments of the State Quarry Member during subsequent transgressive phases of T-R Cycle IIb of Johnson et al. (1985).

Upper Coralville peritidal and hypersaline facies of the proximal inner shelf

Strata of the Iowa City Member in central and proximal areas of the inner shelf are characterized by dolomite, evaporite, and shale facies (Fig. 6). This tract of facies, overlying strata of the Gizzard Creek Member, contrasts with the limestone-dominated facies of the distal inner shelf.

Dolomitic facies. Carbonate strata of the Iowa City Member become progressively more dolomitic in a shoreward direction (Fig. 6). A compositional spectrum of limestone, dolomitic limestone, and dolomite mineralogies characterizes the inner-shelf facies, but dolomite becomes the dominant carbonate lithology in the proximal areas. Replacive fabrics indicate that most if not all of the dolomite is a diagenetic replacement of primary limestone deposits. Dolomitization of Coralville strata may have occurred in mixing-zone diagenetic environments that accompanied the regressive phases of the Coralville cycle and/or may relate to brine reflux processes associated with hypersaline environments of the inner shelf.

Limestone and dolomitic limestone strata in the central region of the inner shelf (Fig. 6) preserve primary depositional features similar to those seen in more distal areas. Laminated, intraclastic, mudcracked, and "birdseye"-bearing intervals resemble peritidal facies closer to the shelf edge. Primary depositional features in dolomite strata are more difficult to recognize because of diagenetic overprinting, but many dolomites faithfully preserve similar laminated and intraclastic fabrics. Dissimilar lithologies include extensively brecciated dolomites and unfossiliferous argillaceous mudstone fabrics (limestone and dolomite). The generally featureless mudstones locally show varying degrees of brecciation, and burrow traces are locally noted.

Carbonate facies across the broad central and proximal areas of the inner shelf are interpreted to have been deposited in a mosaic of peritidal and shallow quiet-water euryhaline and hypersaline environments. Broad mudflat environments apparently interfingered with shallow lagoons and restricted embayments along the retreating margins of the regressing epicontinental seaway. Marine organisms were largely excluded from these environments, but the presence of calcareous algae (calcispheres), ostracodes, pellets, and burrowers in some beds provides evidence of a restricted-marine biota. In a distal direction, additional biota, including gastropods, brachiopods (*Athyris*), and *Amphipora*, are locally seen in a few beds. This biotic distribution likely reflects an overall decrease in salinity stress distally on the inner shelf.

Evaporite facies. Carbonate strata interbed with evaporite units in the Iowa City Member of central Iowa (Fig. 3), but, unfortunately, these strata are known only from well cuttings. Anhydrite is the dominant evaporite mineral, but gypsum is also recognized. Halite and sylvite have not been identified. A cored interval of Gizzard Creek strata includes secondary gypsum void fills (Loc. HM, Fig. 7), which likely precipitated from descending brines associated with overlying Iowa City evaporite deposition.

Outside the region of preserved evaporite lithologies in proximal regions of the inner shelf, the former presence of evaporite deposits is suggested by (1) geographically extensive breccias (Figs. 6, 7) of probable collapse origin and (2) moldic and replacive mineral fabrics in carbonate rocks, including molds of gypsum crystals and nodular anhydrite. Most breccias are composed of angular clasts of dolomite (varying from less than 1 cm to greater than 1 m in diameter) in a shaley to dolomitic mudstone matrix. Less commonly, dolomite breccia clasts occur in a limestone matrix (Loc. PT, Fig. 7), indicating dolomitization of the source beds prior to their brecciation and incorporation into a nondolomitized matrix. Breccia clasts are commonly lithologically indistinguishable from many nonbrecciated beds of laminated dolomite seen within the Iowa City Member.

Dissolution of interbedded and nodular evaporite lithologies would result in collapse and brecciation of associated carbonate strata. The angular and poorly sorted fabrics of the breccias are inconsistent with lateral transport; the breccias are interpreted to largely result from in situ stratal collapse. Evaporite dissolution would be expected to occur in undersaturated freshwater phreatic environments, which would migrate through the shelf deposits during regression (sea-level fall and subaerial exposure).

Evaporite occurrences provide clear evidence of hypersaline environments on the inner shelf during deposition of the Iowa City Member. Evaporite deposition is interpreted to have occurred in two general settings: (1) subaqueous precipitation within shallow hypersaline embayments and salinas and (2) nodular precipitation within sediments of extensive supratidal sabkha mudflats.

Shale facies. Although Iowa City carbonate strata in distal regions of the inner shelf are relatively free of argillaceous impurities, shale units become increasingly significant in a proximal direction (Fig. 6). These units are typified by gray and green-gray clay shales, dolomitic and silty to varying degrees and locally sandy in part. Many shales occur as centimeter to decimenter-thick units within intervals of laminated to brecciated dolomite. However, a regionally prominent shale interval (to 2 m thick) is recognized in northern Iowa at a stratigraphic position roughly equivalent to the "*Amphipora* beds" in the distal shelf area (Fig. 6). Additional thick shales (1 to 3 m) are identified in central Iowa and Nebraska (Fig. 7).

Many shale units are variably intraclastic to brecciated, containing clasts of rounded to angular dolomite. Carbonate intraclasts indicate penecontemporaneous sediment reworking, possibly in peritidal settings. Brecciated carbonate clasts within shaley intervals may have formed by solution collapse of interbedded carbonate, shale, and evaporite strata. The shales are largely featureless and unfossiliferous, but laminated and burrowed fabrics are known.

The greatest abundance of shale, as well as the highest proportion of quartz silt and sand, is recognized at northern and western localities (e.g., Loc. MC, Fig. 6; Locs. OF, PT, Fig. 7). The general proximal increase in siliciclastic detritus identifies source areas along the trend of the Transcontinental Arch. Although evaporite deposition indicates an arid climate for the region, minor stream influx from the arch delivered clay and silt to various depositional settings on the inner shelf. This likely included discharge into shallow restricted lagoons and embayments as well as flood discharge across coastal lowland regions and mudflat environments.

SUBTIDAL MARINE FACIES OF THE MIDDLE SHELF

Middle-shelf facies of the Cou Falls Member are characterized by limestone and dolomitic limestone beds of skeletal wackestone to packstone. The beds are variably argillaceous with scattered thin argillaceous and shale partings. These facies include a lower, partly coralline interval that correlates with Gizzard Creek–Cou Falls strata of the inner shelf and an upper interval that correlates with strata of the Iowa City Member (Figs. 6, 8).

Lower Cou Falls Interval

The lower interval commonly includes accumulations of corals, locally packed in biostromal units (Loc. H33, Fig. 8). However, unlike Cou Falls strata in distal inner-shelf areas, coral biostromes have limited geographic extent on the middle shelf, and corals are absent through much of the sequence. *Hexagonaria* typically dominates most coral-bearing associations, but additional elements include solitary rugosans, massive to branching *Favosites*, rare *Alveolites*, and tabular to hemispherical stromatoporoids. Branching stromatoporoids (*Idiostroma*) are identified only near the inner-shelf margin (Loc. MA, Figs. 4, 8). The coral-bearing units occur in a lithologic and faunal association that is largely indistinguishable from the "*Cranaena* beds" of the distal inner shelf. Like the "*Cranaena* beds, " overturned corals and the abraded fine skeletal packstone matrix indicate episodic bottom current activity during deposition.

Coral-bearing units interbed with noncoralline strata in the lower interval. The noncoralline units include fine to coarse skeletal packstone, wackestone, and mudstone lithologies not seen in inner-shelf associations. Lenses of coarse crinoidal packstone and whole-valve brachiopod packstone stringers are recognized within finer-grained wackestone to packstone intervals. These lenses and stringers are interpreted to have been generated by episodic storm currents (tempestites). Portions of the lower interval include abraded grain lithologies similar to the "*Cranaena* beds" on the inner shelf, but overall quieter-water environments characterized much of middle shelf sedimentation. Argillaceous mud-rich units, including sparsely fossiliferous burrowed lithologies, were clearly deposited in less agitated environments. In addition, hardground surfaces and phosphatic horizons punctuate the interval (Fig. 8); these are interpreted to have formed during episodes of sediment starvation associated with minor transgressive pulses. Development of these surfaces provides further support for overall lower rates of sedimention on the middle shelf.

The lower interval includes a faunal association largely identical to that seen in the coralline "*Cranaena* beds," but additional associations also occur. Day (1989a, 1989b, 1994) documented a moderately diverse brachiopod association, the *Strophodonta* Fauna, from these strata sharing many taxa with the "*Cranaena* beds" but apparently lacking *Pentamerella*. Brachiopod stringers in the interval are generally dominated by valves of atrypids (*Pseudoatrypa*, *Independatrypa*) and *Strophodonta*. Coarse crinoidal packstone lithologies suggest the presence of echinoderm-dominated faunas on the middle shelf. A sparse bryozoan fauna occurs in the "*Cranaena* beds," but bryozoans are notably more common in some facies of the middle shelf (branching cystodictyonids and fenestellids). Tentaculites are locally noted with the crinoidal and bryozoan-rich lithologies, but tentaculites are not known from inner-shelf facies.

Upper Cou Falls interval

Upper strata across the middle shelf are characterized by argillaceous skeletal wackestone to packstone lithologies, primarily crinoidal to brachiopodal. Unlike lower strata, corals are generally absent, although rare small favositids are known (Loc. H27, Fig. 8). Lenses and stringers of brachiopod-rich and crinoidal packstones are common through much of the interval. Atrypids (*Pseudoatrypa*, *Independatrypa*) are characteristic, but *Athyris* is common to dominant in some beds (the *Athyris-Cyrtina* and *Pseudoatrypa-Athyris* Faunas of Day, 1989a, 1994). There is a general decline in brachiopod diversity within the interval, with upward loss of *Strophodonta* and spiriferids (*Orthospirifer*, *Cyrtina*).

The upper interval can be subdivided into two units at the base of a shaley to very argillaceous bed (dashed line within upper interval, Fig. 8). This boundary is characterized by an abrupt change in lithology and is locally expressed as a submarine discontinuity or hardground surface (Loc. BF, Fig. 8). It is interpreted to mark the base of a minor transgressive pulse, possibly the same

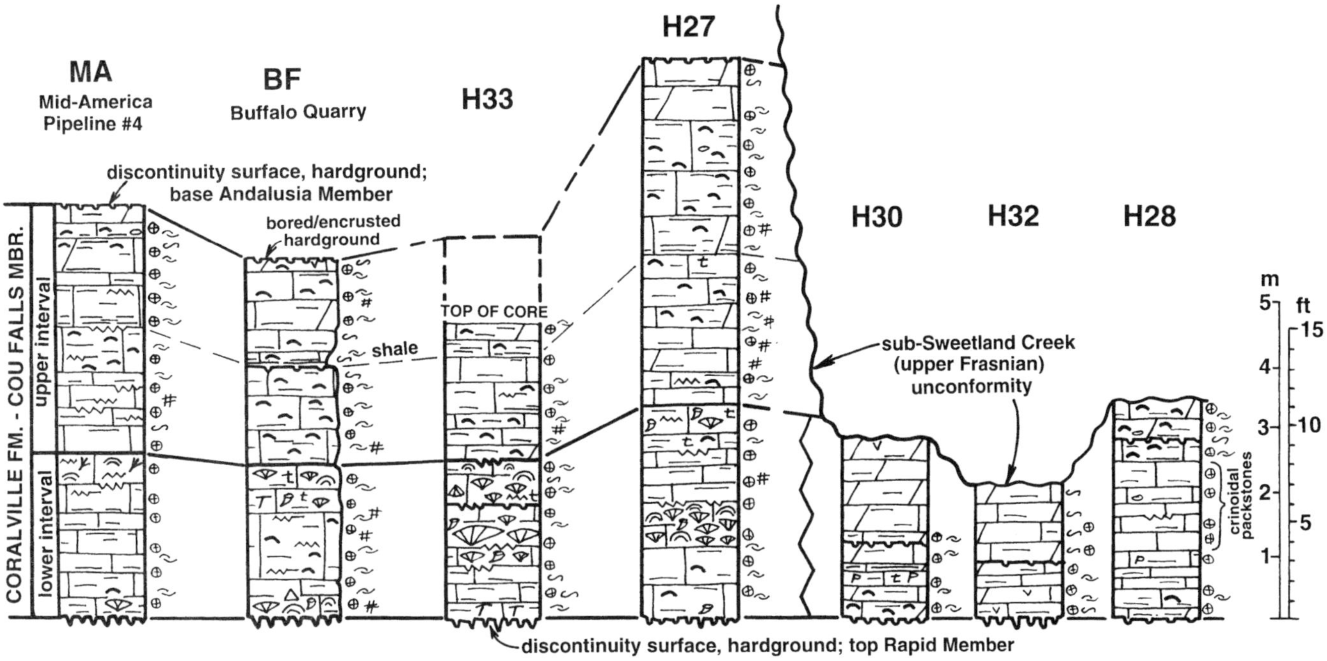

Figure 8. Representative core sections of Coralville Formation in middle-shelf region of southeastern Iowa. Depositional facies are entirely of subtidal origin. Coralville strata are truncated southward beneath the sub–Sweetland Creek unconformity. See Figure 3 and Appendix for localities. Symbols as in Figures 4 and 5. (Additional symbols: #, bryozoans; P, phosphatic grains.)

transgression recorded by the "*Amphipora* beds" on the inner shelf. The lower unit of the upper interval, which contains the most bryozoan-rich facies of the Coralville Formation, includes fenestellids, cystodictyonids (and other cryptostomes), and trepostomes. Rare tentaculites occur. The upper unit is dominated by crinoid- and brachiopod-rich wackestones and packstones (with local grainstones near the top), but sparsely fossiliferous burrowed mudstones and fine skeletal wackestones also occur.

Middle-shelf faunas are interpreted to be largely stenohaline associations, dominated by normal-marine crinoid, brachiopod, and bryozoan elements. However, the upward decline in faunal diversity within the upper interval likely indicates a general increase in benthic stress associated with overall seaway shallowing. The increased dominance of *Athyris* in the interval, which is the most eurytopic brachiopod taxon known from inner-shelf facies, and the upward loss of characteristic stenohaline bryozoan faunas are interpreted to reflect increasing salinity stress. The effectiveness of marine circulation across the middle shelf may have declined as environments shallowed, resulting in more euryhaline conditions.

The impact of storm currents on the middle-shelf environments is suggested by the incorporation of skeletal stringers and abraded-grain packstones in the sequence. However, the abundance of carbonate mud and argillaceous material in the interval indicates that current winnowing was not pervasive and that benthic conditions were less agitated than along the distal inner shelf. The source of the clay influx onto the middle shelf for this and other T-R cycles of the Cedar Valley Group is not known with certainty but likely represents a distal influx of suspended sediment transported from regions of the Transcontinental Arch and/or Canadian Shield.

The absence of peritidal facies and other evidence of subaerial exposure in the interval supports an entirely subtidal depositional setting on the middle shelf during regional regression of the Coralville cycle. The top of this depositional cycle on the middle shelf is marked by a burrowed submarine discontinuity or hardground surface, which probably formed during an episode of sediment starvation associated with transgression of the subsequent depositional cycle. This surface locally displays borings (*Trypanites*) and encrusting auloporid corals (Loc. BF, Fig. 8).

STRATIGRAPHIC AND DEPOSITIONAL MODEL

The lateral and vertical stacking of lithofacies, biofacies, and depofacies within the Coralville Formation, as described above, can be summarized and integrated into a regional depositional model that can serve to characterize general sedimentation patterns in the cratonic interior of North America during the later part of the Taghanic onlap. The progression of facies through time forms a large-scale transgressive-regressive cycle of deposition that is considered to be a consequence of the rise and fall of sea level.

Initial marine transgression

Marine transgression flooded the subaerial unconformity surface that had developed across the inner shelf during the previous sea-level lowstand. The retrogradation of inner-shelf environments during initial transgression, when rates of coastal onlap exceeded sediment progradation, resulted in general sediment starvation across the inner and middle shelves. Burrowing, boring, and encrusting organisms populated the sediment-starved surface on the middle shelf during early stages of transgression.

Sea-level highstand

Carbonate sediments began to accumulate across the expanse of epicontinental shelves only after the rate of transgression had slowed. Of note, the basal deposits of the Coralville T-R cycle are among the deepest-water facies of the entire cycle of deposition in both inner- and middle-shelf settings, suggesting that an equilibrium between coastal onlap and sedimentation rates was achieved only as maximum transgression was approached. This recurring pattern of Paleozoic carbonate sedimentation in the Iowa area (Witzke and Bunker, 1996) reflects both overall slow rates of sedimentation and the extremely low slope of the transgressive surface across the inner shelf and onto the margins of the Transcontinental Arch. The lower the slope, the greater the coastal onlap for any given relative increase in sea level. The general absence of early transgressive deposits leaves a gap in the sedimentary record of many cratonic T-R cycles, including the Coralville cycle.

Broad-scale sedimentary patterns became established on the epicontinental shelves when sea-level rise began to slow and absolute sea-level highstand was approached. These initial patterns include the deepest water and most offshore sedimentary facies preserved in the Coralville cycle at any given locality (as seen in the lower "*Cranaena* beds," lower Gizzard Creek, and the "lower interval"). Three major depositional systems were established during general sea-level highstand (maximum transgression through early regression) on the epicontinental shelves (Fig. 9). "Highstand" is used in the absolute sense of Witzke and Bunker (1996).

Middle shelf. Middle-shelf environments are dominated by argillaceous carbonates with skeletal wackestone to packstone fabrics. These preserve brachiopod, echinoderm, and coralline biofacies typified by the *Strophodonta* Fauna and *Hexagonaria*-stromatoporoid associations. The benthic environments primarily occupied a relatively quiet-water setting of normal-marine salinity. However, high energy conditions, probably storm-generated currents, episodically impinged across the bottom to produce grain abrasion and transport.

Shelf margin. The-inner shelf margin and adjacent distal regions of the inner shelf are characterized by stromatoporoid-coral biostromal facies (Fig. 9). Fine abraded-grain packstones are the dominant lithology. The high degree of grain abrasion indicates relatively high-energy depositional conditions. Storm, wave, and tidal currents frictionally dissipated much of their energy across the breadth of this depositional system (the shelf-margin high-energy zone of Johnson, 1990, p. 927). Normal-marine faunas in lower strata indicate a generally stenohaline environment, but vertical and shoreward loss of certain stenohaline elements suggests increasing euryhalinity across the shelf-margin area. The deeper parts of this depositional system, and the adjacent middle shelf, were not continuously agitated (episodic storm currents). The preservation of carbonaceous laminae in these areas may indicate occasional impingement of oxygen-deficient water from the stratified water mass in the adjacent epicontinental seaway (Witzke, 1987).

Inner shelf. Sea-level highstand across most of the inner shelf was characterized by deposition of skeletal mudstone and wackestone (Fig. 9). These facies show a significant decline in biotic diversity over that seen in the shelf-margin area and are characterized by low-diversity high-dominance associations of brachiopods (especially *Independatrypa*), scattered echinoderms, and burrowers. Benthic conditions were slightly restricted with respect to the shelf margin, probably due to increasing euryhalinity in the shallow marginal sea. The dominance of carbonate mud in this depositional system indicates relatively low-energy bottom conditions (the onshore low-energy zone of Johnson, 1990). This inner-shelf depositional system incorporates the only biofacies capable of breaching the Transcontinental Arch and thereby defines the "limiting biotope" for benthic dispersal originally envisioned by Johnson (1970, p. 2078).

Falling Sea Level

Highstand depositional systems underwent significant sedimentary reorganization as water depths shallowed, particularly on the inner shelf. These shallowing depositional systems have commonly been included under the label of "highstand systems tract" in the Exxon sequence stratigraphic nomenclature (Van Wagoner et al., 1988). However, Witzke and Bunker (1996) interpreted recurring patterns of shallowing Paleozoic sedimentation in cratonic areas to have been associated with falling sea level, not with sea-level highstand. Therefore, the genetic Exxon terminology is not adopted to describe regional regression of major T-R cycles in the Iowa area. Three general depositional systems characterize shallowing sedimentary patterns associated with falling sea level across the inner and middle shelves (Fig. 9). These encompass strata of the Iowa City Member and the "upper interval" of the Cou Falls Member.

Middle shelf. Shallowing middle-shelf environments are dominated by argillaceous carbonates with skeletal wackestone to packstone fabrics, similar to depofacies developed during earlier sea-level highstand. However, important biotic changes accompanied shallowing on the middle shelf, including general upward loss of rugose corals and bryozoans and a significant decline in brachiopod diversity. These changes suggest increasing environmental restriction associated with shallowing, possibly due to decreasing effectiveness of the marine circulation. Increasing euryhaline stresses may have characterized late regression on the middle shelf, as discussed earlier.

Shelf margin. Areas of the distal inner shelf and shelf margin are characterized by a mosaic of supratidal, peritidal, and

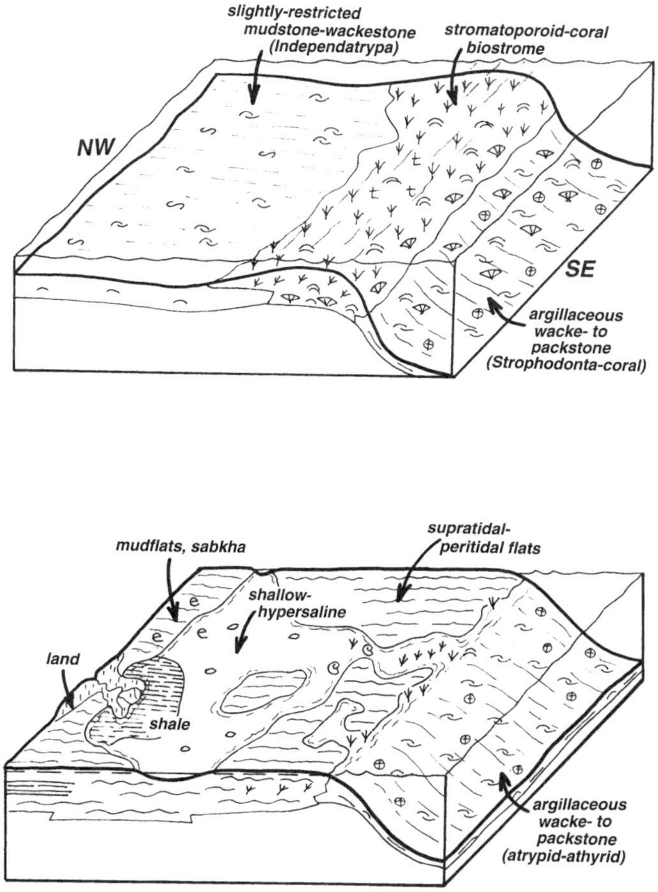

Figure 9. Schematic representation of depositional facies and biofacies, Coralville Formation, Iowa area. Foreground edge of block diagrams corresponds to cross section line shown in Figure 6 (approximately 300 km long); vertically exaggerated. Upper diagram illustrates widespread subtidal deposition of middle Cou Falls (lower "*Idiostroma* beds") and middle Gizzard Creek facies. Lower diagram schematically depicts peritidal and shallow-restricted facies across the inner shelf (middle Iowa City Member) with coeval subtidal facies on the middle shelf (upper Cou Falls Member). Symbols as in Figure 4. (Abbreviation: e, evaporite deposition.)

restricted lagoonal facies that prograded as the seaway shallowed. These facies extended to the margin of the inner shelf but do not extend farther basinward onto the middle shelf. Multiple episodes of subaerial exposure and erosion punctuated deposition, including significant erosional channeling following the final seaway withdrawal stages of deposition. A minor transgressive pulse across the distal shelf is marked by incursion of shallow restricted-marine environments with faunas of branching stromatoporoids and ostracodes.

Inner shelf. The remainder of the inner shelf was marked by the progradation of a mosaic of peritidal, supratidal, sabkha, and shallow hypersaline depofacies during sea-level fall (Fig. 9). The lithofacies are characterized by mixed associations of carbonates, evaporites, and shales. Normal-marine biotas are absent, and benthic faunal dispersal would have been impossible across the Transcontinental Arch during deposition of these facies as shorelines retreated across the inner shelf. An episode of regional subaerial exposure and erosion preceded transgression of the succeeding Lithograph City cycle (T-R Cycle 5; Fig. 2). Meteoric phreatic diagenesis progressed through the inner-shelf facies as the seaway withdrew. Regional stratal relationships suggest the development of small-scale depositional cycles across the inner shelf, delimiting sequences of punctuated aggredational cycles (PACs) and/or parasequence sets.

STABLE-ISOTOPE PALEOHYDROLOGY OF THE CEDAR VALLEY CYCLES

Stable carbon and oxygen isotopic data from carbonates of the Cedar Valley Group provide information about the paleogeochemistry of marine and meteoric hydrologic systems associated with late Givetian and early Frasnian T-R cycles of the North American interior. For each cycle on the inner shelf, the progression from marine phreatic to freshwater phreatic systems left an isotopic record retained within the carbonate grains, muds, and associated cements. A brief summary of the results obtained from Cedar Valley carbonate samples is presented here to characterize the marine and freshwater geochemical signatures contained within individual T-R cycles. Marine and meteoric calcite compositions are estimated for each depositional cycle.

Diagenetically unaltered fabric-retentive brachiopods (nonluminescent under cathodoluminoscope) from open-marine carbonate units were used to estimate original marine calcite compositions. The most enriched $\delta^{13}C$ and $\delta^{18}O$ values of covariant trends were accepted to reflect original marine compositions (Popp et al., 1986; Veizer et al., 1986; Lohmann and Walker, 1989). Meteoric calcite lines were estimated from covariant plots of isotopic values derived from blocky equant calcite cements (which display oscillatory luminescence under cathodoluminoscope). The meteoric calcite line displays invariant $\delta^{18}O$ values and highly variable soil-derived light $\delta^{13}C$ reflective of variable rock-water interaction in meteoric phreatic environments (Lohmann, 1988).

Estimated marine calcite values are noted in ascending stratigraphic order for the Little Cedar (Plocher and Ludvigson, 1989; Plocher et al., 1990), Coralville (Plocher et al., 1992; Ludvigson, 1992), and Lithograph City formations (Metzger et al., 1994) as follows: $\delta^{13}C$ (‰ PDB) +1, +2, +1.5; $\delta^{18}O$ (‰ PDB) −6.5, −4, −5.5. Meteoric calcite lines interpreted for the same formations define $\delta^{18}O$ (‰ PDB) values of −8.0, −8.0, and −7.0, respectively. In general, early meteoric phreatic diagenesis integrates the mean annual $\delta^{18}O$ of local precipitation and mean annual temperature, providing an indicator of paleoclimate (Hays and Grossman, 1991; Swart et al., 1993).

The essential points derived from the stable isotopic data from Cedar Valley carbonates include the following: (1) Meteoric calcite lines change very little through the Little Cedar, Coralville, and Lithograph City cycles, suggesting little variation in climate during the interval. (2) Increased $\delta^{13}C$ and $\delta^{18}O$

marine calcite values for the Coralville Formation might record evaporative influences related to greater restriction of open-marine currents during the Coralville cycle or alternatively could reflect secular change in global marine carbonate values.

CONCLUSIONS

The distribution of Coralville lithofacies and biofacies constrains interpretations of regional sedimentary and biotic patterns associated with a transgressive-regressive cycle during the late stages of Taghanic onlap within the North American continental interior. An examination of this region allows aspects of Jess Johnson's biofacies models and hypotheses of cratonic onlap to be evaluated. Several general conclusions pertinent to Johnson's proposals are drawn from interpretations of Coralville deposition.

1. The Taghanic onlap of Johnson (1970), which coincides with T-R cycle IIa of Johnson et al. (1985), is a multiphased onlap characterized by smaller-order cyclic patterns of transgression and regression. A discrete unconformity-bounded T-R cycle marks the final stage of Taghanic onlap in inner-shelf areas of the continental interior, typified by the Coralville Formation. This final cycle of the Taghanic is recognized at other North American localities, lending support (but not conclusive evidence) for a eustatic origin.

2. The early phases of transgression and coastal onlap in the Coralville cycle, as well as other Devonian cycles of the Iowa area, were marked by regional sediment starvation. Sedimentation patterns were first established across the inner and middle shelf areas during later stages of transgression as maximum water depths were approached.

3. Maximum diversity of benthic faunas, which occurs in basal Coralville strata, coincides with maximum transgression of the Coralville cycle. This conforms with the predictions of Johnson's (1990) biofacies model.

4. Interregional migration of benthic faunal elements across the "continental backbone" of North America was possible only during periods of sea-level highstand within particular T-R cycles. Breaching of the Transcontinental Arch coincided with maximum transgression (Stage 4 biofacies model of Johnson, 1990). Subsequent regression led to progradation of peritidal and nonmarine facies across inner-shelf areas that bordered the arch and the reestablishment of migratory barriers between eastern and western North America.

5. Johnson (1970) proposed that the shallow seas that periodically transgressed the Transcontinental Arch during the Taghanic T-R cycles encompassed a limited biotope that formed an environmental filter for faunal migration. Only those elements of the biota that were adapted for life in the inner-shelf environments, or whose larva could survive transport across such environments, were in a position to migrate across the arch. The inner-shelf margin thereby marked an effective biofacies boundary across which deeper-water and more open-marine faunal elements endemic to middle-shelf environments could not migrate. This pattern is generally supported by the Coralville example. Elements of the inner-shelf Coralville brachiopod fauna, including *Tecnocyrtina*, *Cranaena*, and *Athyris*, migrated across the arch (Day, 1996), but wholesale migration of the middle-shelf benthic fauna did not occur. The inner-shelf conodont fauna of the Coralville Formation is recognized on both sides of the Transcontinental Arch, indicating free migration of these taxa.

6. Salinity stresses are interpreted to have formed the primary environmental filter to faunal migration across the inner shelf during the Taghanic onlap in the Coralville example, although additional depth-dependent factors acted to further limit the biotope. The low-energy muddy environments that characterized the inner shelf at maximum sea-level highstand were inhabited by a eurytopic biota that was apparently adapted to increased salinity stress. Elevated salinities would have produced an anti-estuarine thermohaline circulation pattern across the inner shelf, but middle and outer shelves were probably characterized by a contrasting stratified quasi-estuarine circulation system of normal-marine salinity (Witzke, 1987). Subsequent shallowing of Coralville environments led to hypersalinity and evaporite precipitation on the inner shelf, and the inner-shelf shelly fauna was decimated at that time. Elevated salinities likely posed environmental stress and modified circulation patterns on the middle shelf during late regressive phases of the Coralville cycle as the inner-shelf region became emergent. The basinward shift of euryhaline conditions late in the T-R cycle is marked by a decline of shelly benthic diversity, loss of specific stenohaline faunal elements, and increased dominance of more eurytopic taxa on the middle shelf (Stage 5 biofacies model of Johnson, 1990).

7. The absolute magnitudes of sea-level change associated with transgression and regression of the Coralville cycle are difficult to constrain. However, based on the limits of storm currents and light penetration in modern seas, it seems likely that water depths never exceeded more than about 30 to 60 m on the middle shelf (see discussion by Brett et al., 1993). Post-Coralville and pre–Lithograph City erosional downcutting of at least 25 m along the inner-shelf margin provides minimum constraints on the magnitude of base-level lowering associated with late regression and seaway withdrawal from the inner-shelf margin.

ACKNOWLEDGMENTS

We humbly acknowledge the significant role that Jess Johnson's ideas about continental onlap, transgression and regression, and faunal change have played in our interpretations of Devonian sediments and faunas in the Iowa area. We also thank a number of individuals and groups for their assistance in various aspects of our studies of Cedar Valley strata, including fieldwork, paleontologic studies, petrographic investigations, and logistical support. The help of Jed Day, Curt Klug, Mike Woodruff, Greg Ludvigson, Orrin Plocher, Ron Metzger, Luis González, Brian Glenister, Phil Heckel, Gilbert Klapper, numerous quarry operators, the Geological Survey Bureau, and the Army Corps of Engineers (Coralville Lake) is gratefully acknowledged. Stable isotopic data was attained at K. C. Lohmann's University of Michigan Stable Isotope Laboratory, and we gratefully acknowledge the ongoing interest and support of the Michigan group. Isotopic sampling was facilitated by a grant from the Iowa Science Foundation (ISF-91-25). Helpful review comments and suggestions were provided by Gil Klapper, Pete Goodwin, Jed Day, and Greg Ludvigson.

APPENDIX: LOCALITY REGISTER FOR CORALVILLE FORMATION SECTIONS

All localities are shown on Figure 3. Localities are arranged in general order from northwest to southeast (see Figs. 4, 5, 6) except for localities illustrated on Figures 7 and 8. Cores are reposited with the Iowa Department of Natural Resources, Geological Survey Bureau, Iowa City, Iowa, unless noted. Core depths given in feet (ft) and meters (m). All localities are in Iowa unless noted.

MC. Mason City core, Northwest Portland Cement Company; NW sec. 33, T897N, R20W, Cerro Gordo Co. (136.7–194.1 ft; 41.7–59.2 m); Witzke and Bunker (1984a, 1985).

FM2. Floyd-Mitchell Project core 2; SE NE NE SE sec. 23, T99N, R17W, Mitchell Co. (78.2–138.2 ft; 23.8–42.1 m); Witzke and Bunker (1985).

FM4. Floyd-Mitchell Project core 4; NE SW NW SE sec. 31, T96N, R18W, Floyd Co. (158.7–211.1 ft; 48.4–64.4 m); Witzke and Bunker (1985); Witzke et al. (1989).

FM1. Floyd-Mitchell Project core 1; SE NE NE sec. 33, T97N, R16 W, Floyd Co. (65.8–127.5 ft; 20.1–38.9 m); Witzke and Bunker (1985); Witzke et al. (1989).

FM3. Floyd-Mitchell Project core 3; SW NW SW NW sec. 22, T95N, R16W, Floyd Co. (144.1–210 ft; 43.9–64 m); Witzke and Bunker (1985).

GC. Type section of Gizzard Creek Member, Maurice Bunn Quarry; SW NE NW SW sec. 23, T95N, R15W, Floyd Co.; Witzke et al. (1989).

TQ. Tracy Quarry, Nashua; N 1/2 SW NE sec. 29, T94N, R14W, Chickasaw Co.; Witzke et al. (1989).

YK. Yokum Quarry, lower quarry section and core log (Iowa Department of Transportation log); NW SW sec. 4, T90N, R14W, Black Hawk Co.; Anderson (1972); Bunker et al. (1986); Witzke et al. (1989).

RY. Raymond (Pints) Quarry (abandoned); SW sec. 36, T89N, R12W, Black Hawk Co.; Kettenbrink (1972); Anderson and Garvin (1984). Nearby Peske Quarry (active); NE sec. 2, T88N, R12W.

WS. Waterloo–South Quarry; SW NW NW sec. 18, T97N, R12W, Black Hawk Co.; Anderson and Garvin (1984); Witzke et al. (1989).

GL. Glory Quarry; NW SE NW sec. 36, T87N, R11W, Black Hawk Co.; Anderson and Garvin (1984); Day (1992).

GR. Garrison Quarry; NW NE sec. 33, T85N, R11W, Benton Co.; Witzke et al. (1989); Witzke et al. (1992).

MR. Mid-River Marina Quarry, type section of Cou Falls Member; NE NW NE SE sec. 27, T81N, R7W, Johnson Co.; Witzke et al. (1989); Plocher et al. (1989).

NL. North Liberty east, west shore of Coralville Lake; NW NE NE NW sec. 8, T80N, R6W, Johnson Co.

SB. West shore of Coralville Lake, opposite Sugar Bottoms; SW NW SE NE sec. 8, T80N, R6W, Johnson Co.

GS. Gooseberry North area, southwest shoreline of Coralville Lake; NE sec. 21, T81N, R6W, Johnson Co.

CD. Coralville Dam, east face above spillway; NE SW NE sec. 22, T80N, R6W, Johnson Co.

VQ. Vogel Quarry; NE NW NE sec. 27, T80N, R6W, Johnson Co.; Kettenbrink (1973).

CQ. Conklin Quarry, River Products Co., type section of Coralville Formation; NW NW sec. 33, T80N, R6W, Johnson Co.; Kettenbrink (1973); Witzke and Bunker (1984c); Witzke et al. (1989); Bunker and Witzke (1992).

KQ. Klein Quarry, River Products Co.; NE NE sec. 2, T79N, R7W, Johnson Co.; Michael and Welp (1957); Rogers (1990).

LL. Lovers Leap section, east bank of Iowa River; SW SW NW sec. 4, T79N, R6W, Johnson Co.

EL. Elks Club property, exposures along north bank of Iowa River; NW NW NW SE sec. 4, T79N, R6W, Johnson Co.

SQ. Old Sanders Quarry, North Dubuque Street, Iowa City; NE NE SE SW sec. 3, T79N, R6W; Calvin (1897); Witzke and Bunker (1984b).

PQ. Old Public Quarry area, below President's house along Iowa River, University of Iowa campus; SW NE NW sec. 10, T79N, R6W, Johnson Co.; Witzke and Bunker (1984b).

HT. Old Hutchinson Quarry, opposite Art Building on North Riverside Drive, University of Iowa campus; SE SE SE NE sec. 9, T79N, R6W, Johnson Co.; Calvin (1897); Witzke and Bunker (1984b); Witzke et al. (1989).

NR. Nursing Building road-cut section, near intersection of Riverside Drive and Newton Road, University of Iowa campus; NW SE NE SE sec. 9, T79N, R6W, Johnson Co.; Witzke and Bunker (1984b).

HY. Hydraulics Annex, old quarry area along Riverside Drive across from Hydraulics Building, University of Iowa campus; SW SE SE SE sec. 9, T79N, R6W, Johnson Co.; Witzke and Bunker (1984b).

RR. Railroad cut, south edge of Iowa City; NW NE SE NW sec. 22, T79N, R6W, Johnson Co.

MA. Mid-America Pipeline, Iowa City terminal, core 4; NE NW SE SE sec. 27, T79N, R5W, Johnson Co. (208.7–229.6 ft; 63.6–70 m).

BF. Buffalo Quarry, Davenport Cement Company; NE NW NE sec. 23, T77N, R2E, Scott Co.; Witzke and Bunker (1985); Day (1992).

Following localities for Figure 7:

H33. New Jersey Lead-Zinc core H-33; SW NW NE sec. 15, T71N, R2W, DesMoines Co. (241.5–256.7 ft; 73.6–78.3 m); Klug (1982a).

H27. New Jersey Lead-Zinc core H-27; NE NE NW sec. 21, T68N, R3W, Lee Co. (412–441 ft; 125.6–134.5 m); Klug (1982a).

H30. New Jersey Lead-Zinc core H-30; NE NW NW sec. 8, T66N, R5W, Lee Co.

H32. New Jersey Lead-Zinc core H-32: NE NW SW sec. 32, T66N, R5W, Lee Co. (639.2–646.1 ft; 199.4–197 m).

H28. New Jersey Lead-Zinc core H-28; SE SE sec. 22, T65N, R5W, Lee Co. (537.4-556.7 ft; 163.8–169.7 m); Klug (1982a).

Following localities for Figure 8:

OF. Offutt Air Force Base, Strategic Air Command; SE SE NW sec. 11, T13N, R13E, Sarpy Co., Nebraska (1,162.5–1,225.5 ft; 354.4–373.6 m); unpublished core log of Jed Day; core stored at University of Nebraska, Conservation and Survey Division, Lincoln, Nebraska.

HM. Hummell core 4; NW SW SW sec. 18, T79N, R28W, Dallas Co. (1,140–1,150.5 ft; 347.6–350.8 m); Klug (1982b).

PT. Peterson core 1; NE NE NW sec. 10, T90N, R27W, Webster Co. (704.5–748 ft; 214.8–228 m); Klug (1982b); Witzke et al. (1989).

CG. Cerro Gordo Project core CG-1; NE NW SW sec. 15, T94N, R19W, Cerro Gordo Co. (342.5–389 ft; 104.4–118.6 m); unpublished core log of Jed Day.

KM. Cominco core KM-1; SW SE SE NW sec. 33, T64N, R17W, Adair Co., Missouri (1,012–1,058.4 ft; 308.5–322.7 m); Woodruff (1990); core stored at Missouri Department of Natural Resources, Geological Survey, Rolla, Missouri.

REFERENCES CITED

Anderson, W. I., 1972, The general geology of Black Hawk County and adjacent areas: Geological Society of Iowa Guidebook 26, 55 p.

Anderson, W. I., and Garvin, P. L., 1984, The Cedar Valley Formation (Devonian), Black Hawk and Buchanan counties: Carbonate facies and mineralization: Geological Society of Iowa Guidebook 42, 47 p.

Baird, G. C., Brett, C. E., and Kirchgasser, W. T., 1989, Genesis of black shale–roofed discontinuities in the Devonian Genesee Formation, western New York State, *in* McMillan, N. J., Embry, A. F., and Glass, D. J., eds., Devonian of the world: Calgary, Canadian Society of Petroleum Geologists Memoir 14, v. 2, p. 357–375.

Bils, J., 1983, Morphometric trends in tabulates corals, Coralville Member, Cedar Valley Limestone (Devonian), Johnson County, Iowa [M.S. thesis]: Iowa City, University of Iowa, 67 p.

Brett, C. E., Boucot, A. J., and Jones, B., 1993, Absolute depths of Silurian benthic assemblages: Lethaia, v. 26, p. 25–40.

Bunker, B. J., and Klapper, G., 1984, Conodont zones and correlation of the Cedar Valley-State Quarry interval of eastern Iowa: Geological Society of Iowa Guidebook 41, p. 15–18.

Bunker, B. J., and Witzke, B. J., 1989, The Coralville and Lithograph City formations of the Coralville Lake area, *in* Plocher, O. W., ed., Geologic reconnaissance of the Coralville Lake area: Geological Society of Iowa Guidebook 51, p. 37–46.

Bunker, B. J., and Witzke, B. J., 1992, An upper Middle through lower Upper Devonian lithostratigraphic and conodont biostratigraphic framework of the Midcontinent carbonate shelf area, Iowa, *in* Day, J., and Bunker, B. J., eds., The stratigraphy, paleontology, depositional and diagenetic history of the Middle-Upper Devonian Cedar Valley Group of central and eastern Iowa: Iowa Department of Natural Resources, Geological Survey Bureau, Guidebook Series 16, p. 3–25.

Bunker, B. J., Witzke, B. J., and Day, J., 1986, Upper Cedar Valley stratigraphy, north-cental Iowa, Lithograph City Formation: Geological Society of Iowa Guidebook 44, 41 p.

Bunker, B. J., Witzke, B. J., Watney, W. L., and Ludvigson, G. A., 1988, Phanerozoic history of the central Midcontinent, *in* Sloss, L. L., ed., Sedimentary cover—North American Craton: U.S.: Boulder, Colorado, Geological Society of America, Geology of North America, v. D-2, p. 243–260.

Calhoun, S. H., 1983, Crinoids of the Devonian Cedar Valley Limestone in Iowa [M.S. thesis]: Iowa City, University of Iowa, 151 p.

Calvin, S., 1897, Geology of Johnson County: Iowa Geological Survey Annual Report, v. 7, p. 33–104.

Day, J., 1989a, The brachiopod succession of the late Givetian-Frasnian of Iowa, *in* McMillan, N. J., Embry, A. F., and Glass, D. J., eds., Devonian of the world: Calgary, Canadian Society of Petroleum Geologists Memoir 14, v. 3, p. 303–326.

Day, J., 1989b, Middle-Late Devonian (late Givetian-early Frasnian) brachiopod faunas and biostratigaphy of the Coralville and Lithograph City formations of eastern Iowa, *in* Plocher, O. W., ed., Geologic reconnaissance of the Coralville Lake area: Geological Society of Iowa Guidebook 51, p. 47–60.

Day, J., 1990, Cyclic deposition and stratigraphy of the Middle and Upper Devonian (Eifelian-Frasnian) strata in the subsurface of southeastern Nebraska: Geological Society of America Abstracts with Programs, v. 22, no. 5, p. 7.

Day, J., 1992, Middle-Upper Devonian (late Givetian–early Frasnian) brachiopod sequence in the Cedar Valley Group of central and eastern Iowa, *in* Day, J., and Bunker, B. J., eds., The stratigraphy, paleontology, depositional and diagenetic history of the Middle-Upper Devonian Cedar Valley Group of central and eastern Iowa: Iowa Department of Natural Resources, Geological Survey Bureau, Guidebook Series 16, p. 53–105.

Day, J., 1994, Late Middle and early Upper Devonian brachiopod faunas of southeastern Iowa and northwestern Illinois, *in* Hickerson, W., and Anderson, R. C., coordinators, Bunker, B. J., ed., Paleozoic stratigraphy of the Quad-Cities region, east-central Iowa, northwestern Illinois: Geological Society of Iowa Guidebook 59, p. 65–83.

Day, J., 1996, Faunal signatures of Middle-Upper Devonian depositional sequences and sea level fluctuations in the Iowa Basin: U.S. midcontinent, *in* Witzke, B. J., Ludvigson, G. A., and Day, J., eds., Paleozoic sequence stratigraphy: Views from the North American Craton: Geological Society of America Special Paper 306, p. 277–300.

Day, J., Uyeno, T., Norris, W., Witzke, B. J., and Bunker, B. J., 1996, Middle-Upper Devonian relative sea level histories of central and western North American interior basins, *in* Witzke, B. J., Ludvigson, G. A., and Day, J. E., eds., Paleozoic sequence stratigraphy: Views from the North American Craton: Geological Society of America Special Paper 306, p. 259–275.

Dow, V. E., 1960, Some notes on the occurrence of a coal seam in the Cedar Valley Formation of Johnson County, Iowa: Proceedings Iowa Academy of Sciences, v. 67, p. 253–259.

Goodwin, P. W., and Mauriello, D. J., 1994, Orbital forcing of a hierarchy of carbonate allocycles: Devonian Cedar Valley Group, Iowa: Geological Society of America Abstracts with Programs, v. 26, no. 5, p. 17.

Hays, P. D., and Grossman, E. L., 1991, Oxygen isotopes in meteoric calcite cements as indicators of continental climate: Geology, v. 19, p. 441–444.

Hickerson, W., 1992, Trilobites from the late Givetian Solon Member, Little Cedar Formation of eastern Iowa and northwestern Illinois, *in* Day, J., and Bunker, B. J., eds., The stratigraphy, paleontology, depositional and diagenetic history of the Middle-Upper Devonian Cedar Valley Group of central and eastern Iowa: Iowa Department of Natural Resources, Geological Survey Bureau, Guidebook Series 16, p. 123–139.

Hickerson, W., 1994, Silurian, Devonian, and Pennsylvanian trilobites from Rock Island County, Illinois: Geological Society of Iowa Guidebook 59, p. 23–29.

Johnson, J. G., 1970, Taghanic onlap and the end of North American Devonian provinciality: Geological Society of America Bulletin, v. 81, p. 2077–2105.

Johnson, J. G., 1971, Timing and coordination of orogenic, epeirogenic, and eustatic events: Geological Society of America Bulletin, v. 82, p. 3263–3298.

Johnson, J. G., 1974, Extinction of perched faunas: Geology, v. 2, p. 479–482.

Johnson, J. G., 1990, Lower and Middle Devonian brachiopod-dominated communities of Nevada, and their position in a biofacies-province-realm model: Journal of Paleontology, v. 64, p. 902–941.

Johnson, J. G., and Klapper, G., 1992, North American midcontinent Devonian T-R cycles, *in* Chaplin, J. R., and Barrick, J. E., eds., Special papers in paleontology and stratigraphy: A tribute to Thomas W. Amsden: Oklahoma Geological Survey Bulletin 145, p. 127–135.

Johnson, J. G., Klapper, G., and Sandberg, C. A., 1985, Devonian eustatic fluctuations in Euramerica: Geological Society of America Bulletin, v. 96, p. 567–587.

Johnson, J. G., Sandberg, C. A., and Poole, F. G., 1989, Early and Middle Devonian paleogeography of the western United States, *in* McMillan, N. J., Embry, A. F., and Glass, D. J., eds., Devonian of the world: Calgary, Canadian Society of Petroleum Geologists Memoir 14, v. 1, p. 161–182.

Kettenbrink, E. C., 1972, The Cedar Valley Formation at Pint's Quarry, *in* Anderson, W. I., General geology in the vicinity of northern Iowa: Guidebook for the 36th Annual Tri-State Geological Field Conference, Cedar Falls, Iowa, p. 35–41.

Kettenbrink, E. C., 1973, Depositional and post-depositional history of the Devonian Cedar Valley Formation, east-central Iowa [Ph.D. thesis]: Iowa City, University of Iowa, 191 p.

Kettenbrink, E. C., and Heckel, P. H., 1975, Cedar Valley Limestone in east-central Iowa, *in* Heckel, P. H., Kettenbrink, E. C., and Klapper, G., Devonian limestone facies—Cedar Valley and State Quarry limestones in the Iowa City region: Guidebook for the 39th Annual Tri-State Geological Field Conference, Iowa City, Iowa, p. 2–5.

Kettenbrink, E. C., and Toomey, D. F., 1975, Distribution and paleoecological implications of calcareous foraminifera in the Devonian Cedar Valley

Formation of Iowa: Journal of Foraminiferal Research, v. 5, p. 176–187.

Kirchgasser, W. T., Baird, G. C., and Brett, C. E., 1989, Regional placement of Middle/Upper Devonian (Givetian-Frasnian) boundary in western New York State, *in* McMillan, N. J., Embry, A. F., and Glass, D. J., eds., Devonian of the world: Calgary, Canadian Society of Petroleum Geologists Memoir 14, v. 3, p. 113–118.

Klapper, G., 1989, The Montagne Noire Frasnian (Upper Devonian) conodont succession, *in* McMillan, N. J., Embry, A. F., and Glass, D. J., eds., Devonian of the world: Calgary, Canadian Society of Petroleum Geologists Memoir 14, v. 3, p. 449–468.

Klapper, G., and Johnson, J. G., 1990, Revision of Middle Devonian conodont zones: Journal of Paleontology, v. 64, p. 934–941 [Appendix 3 *in* Johnson, 1990].

Klug, C. R., 1982a, The subsurface Devonian lithostratigraphy of southeastern Iowa: Iowa Geological Survey Open File Report 82-1, 33 p.

Klug, C. R., 1982b, Devonian stratigraphy and conodont biostratigraphy from portions of two cores in central Iowa: Iowa Geological Survey Open File Report 82-2, 56 p.

Klug, C. R., 1990, Miospores and conodonts of the Cedar Valley Group (Middle to Upper Devonian) of Iowa [Ph.D. thesis]: Iowa City, University of Iowa, 280 p.

Klug, C. R., and Schabilion, J. T., 1992, A Devonian paper coal (cutinite) in the Lithograph City Formation at Garrison, Iowa, *in* Day, J., and Bunker, B. J., eds., The stratigraphy, paleontology, depositional and diagenetic history of the Middle-Upper Devonian Cedar Valley Group of central and eastern Iowa: Iowa Department of Natural Resources, Geological Survey Bureau, Guidebook Series 16, p. 147–159.

Kohls, D. W., 1961, Lithostratigraphy of the Cedar Valley Formation in Minnesota and northern Iowa [Ph.D. thesis]: Minneapolis, University of Minnesota, 200 p.

Lohmann, K. C., 1988, Geochemical patterns of meteoric diagenetic systems and their application to studies of paleokarst, *in* James, N. P., and Choquette, P. W., eds., Paleokarst: New York, Springer-Verlag, p. 58–80.

Lohmann, K. C., and Walker, J. C. G., 1989, The $\delta^{18}O$ record of Phanerozoic abiotic marine calcite cements: Geophysical Research Letters, v. 16, p. 319–322.

Ludvigson, G. A., 1989, Observations on the cathodoluminescence petrography of the Coralville Formation near Iowa City: Implications for regional diagenetic patterns in the Devonian Cedar Valley Group, *in* Plocher, O. W., ed., Geologic reconnaissance of the Coralville Lake area: Geological Society of Iowa Guidebook 51, p. 63–81.

Ludvigson, G. A., 1992, Paleo-ocean chemistry of Devonian seas in Iowa: Cedar Falls, Final Project Report to Iowa Science Foundation on ISF-91-25, University of Northern Iowa, 10 p.

Metzger, R. A., Ludvigson, G. A., Witzke, B. J., González, L. A., and Day, J. E., 1994, Stable isotopic chemostratigraphy of the Middle Devonian Cedar Valley Group: Initial data from the uppermost Givetian Lithograph City Formation: Geological Society of America Abstracts with Programs, v. 26, no. 5, p. 55.

Michael, R. D., and Welp, T. L., 1957, The Devonian section at the Klein Quarry, Johnson County, Iowa: Proceedings Iowa Academy of Science, v. 64, p. 443–447.

Mossler, J. H., 1978, Cedar Valley Formation (Devonian) of Minnesota and northern Iowa: Minnesota Geological Survey Report of Investigations 18, 44 p.

Pitrat, C. W., 1962, Devonian corals from the Cedar Valley Limestone of Iowa: Journal of Paleontology, v. 36, p. 1155–1162.

Plocher, O. W., and Ludvigson, G. A., 1989, Facies of the upper Little Cedar Formation, *in* Plocher, O. W., ed., Geologic reconnaissance of the Coralville Lake area: Geological Society of Iowa Guidebook 51, p. 11–25.

Plocher, O. W., Bunker, B. J., and Witzke, B. J., 1989, Stop 4—Mid-River Marina Quarry, *in* Plocher, O. W., ed., Geologic reconnaissance of the Coralville Lake area: Geological Society of Iowa Guidebook 51, p. 113–118.

Plocher, O. W., Ludvigson, G. A., and González, L. A., 1990, Secular variation of carbon-oxygen isotopes in successive carbonate T-R cycles, Middle Devonian (Givetian) of the mid-continent, U.S.A.: Geological Society of America Abstracts with Programs, v. 22, no. 7, p. A115.

Plocher, O. W., Ludvigson, G. A., Witzke, B. J., González, L. A., and Day, J. E., 1992, Stable isotopic systematics of the Coralville T-R cycle, *in* Day, J., and Bunker, B. J., eds., The stratigraphy, paleontology, depositional and diagenetic history of the Middle-Upper Devonian Cedar Valley Group of central and eastern Iowa: Iowa Department of Natural Resources, Geological Survey Bureau, Guidebook Series 16, p. 27–34.

Popp, B. N., Anderson, T. F., and Sandberg, P. A., 1986, Brachiopods as indicators of original isotopic compositions in some Paleozoic limestones: Geological Society of America Bulletin, v. 97, p. 1262–1269.

Rogers, F. S., 1990, Stratigraphy, depositional history, and conodonts of the Little Cedar and lower Coralville formations of the Cedar Valley Group (Middle Devonian) of Iowa [Ph.D. thesis]: Iowa City, University of Iowa, 97 p.

Sanders, R. B., 1968, Devonian spores of the Cedar Valley coal of Iowa, U.S.A.: Journal of Palynology, v. 2, p. 16–32.

Sloss, L. L., 1939, Devonian rugose corals from the Traverse beds of Michigan: Journal of Paleontology, v. 13, p. 52–73.

Stainbrook, M. A., 1941, Biotic analysis of Owen's Cedar Valley Limestone: Pan-American Geologist, v. 75, p. 321–327.

Swart, P. K., McKenzie, J. A., and Lohmann, K. C., 1993, Introduction, *in* Swart, P. K., Lohmann, K. C., McKenzie, J. A., and Savin, S., eds., Climate change in continental isotopic records: American Geophysical Union Geophysical Monograph 78, p. xi–xiii.

Trier, S. M., 1993, Trends in growth form of stromatoporoids and corals, Coralville Formation (Devonian), Iowa [M.S. thesis]: Iowa City, University of Iowa, 123 p.

Van Wagoner, J. C., Posamentier, H. W., Mitchum, R. M., Jr., Vail, P. R., Sarg, J. F., Loutit, T. S., and Hardenbol, J., 1988, An overview of the fundamentals of sequence stratigraphy and key definitions, *in* Wilgus, C. K., et al., eds., Sea-level changes: An integrated approach: Society of Economic Paleontologists and Mineralogists Special Publication 42, p. 39–45.

Veizer, J., Fritz, P., and Jones, B., 1986, Geochemistry of brachiopods: Oxygen and carbon isotopic records of Paleozoic oceans: Geochimica et Cosmochimica Acta, v. 50, p. 1679–1696.

Witzke, B. J., 1984, The Devonian stratigraphic sequence exposed in Iowa City, *in* Witzke, B. J., ed., Geology of the University of Iowa campus area, Iowa City: A walking tour along the Iowa River: Iowa Geological Survey Guidebook 7, p. 18–34.

Witzke, B. J., 1987, Models for circulation patterns in epicontinental seas applied to Paleozoic facies of North American craton: Paleoceanography, v. 2, p. 229–248.

Witzke, B. J., and Bunker, B. J., 1984a, Devonian stratigraphy of north-central Iowa: Iowa Geological Survey Open File Report 84-2, p. 107–149.

Witzke, B. J., and Bunker, B. J., 1984b, Tour guide and stop descriptions, *in* Witzke, B. J., ed., Geology of the University of Iowa campus area, Iowa City: A walking tour along the Iowa River: Iowa Geological Survey Guidebook 7, p. 65–76.

Witzke, B. J., and Bunker, B. J., 1984c, The underburden: The Devonian stratigraphic sequence at Conklin Quarry, *in* Bunker, B. J., and Hallberg, G. R., eds., Underburden-overburden: An examination of Paleozoic and Quaternary strata at the Conklin Quarry near Iowa City: Geological Society of Iowa Guidebook 41, p. 2–10.

Witzke, B. J., and Bunker, B. J., 1985, Stratigraphic framework for the Devonian aquifers in Floyd and Mitchell counties, Iowa: Iowa Geological Survey Open File Report 85-2, p. 21–32.

Witzke, B. J., and Bunker, B. J., 1994, Classic geological exposures, old and new, Coralville Lake and spillway: Geological Society of Iowa Guidebook 60, 68 p.

Witzke, B. J., and Bunker, B. J., 1996, Relative sea-level changes during Middle Ordovician through Mississippian deposition in the Iowa area, North American craton, *in* Witzke, B. J., Ludvigson, G. A., and Day, J. E., eds., Paleozoic sequence stratigraphy: Views from the North American Craton: Geological Society of America Special Paper 306, p. 307–330.

Witzke, B. J., Bunker, B. J., and Klapper, G., 1985, Devonian stratigraphy in the

Quad Cities area, eastern Iowa–northwestern Illinois, *in* Hammer, W. R., Anderson, R. C., and Schroeder, D. A., Devonian and Pennsylvanian stratigraphy of the Quad-Cities region, Illinois-Iowa: Society of Economic Paleontologists and Mineralogists, Great Lakes Section, 15th Annual Field Conference, Guidebook, p. 19–64.

Witzke, B. J., Bunker, B. J., and Rogers, F. S., 1989, Eifelian through lower Frasnian stratigraphy and deposition in the Iowa area, central Midcontinent, U.S.A., *in* McMillan, N. J., Embry, A. F., and Glass, D. J., eds., Devonian of the world: Calgary, Canadian Society of Petroleum Geologists Memoir 14, v. 1, p. 221–250.

Witzke, B. J., Bunker, B. J., and Day, J. E., 1992, Stop 1. Garrison Quarry, Benton County, Vulcan Materials Corp., *in* Day, J., and Bunker, B. J., eds., The stratigraphy, paleontology, depositional and diagenetic history of the Middle-Upper Devonian Cedar Valley Group of central and eastern Iowa: Iowa Department of Natural Resources, Geological Survey Bureau, Guidebook Series 16, p. 163–172.

Woodruff, M. L., 1990, Middle and Upper Devonian (Givetian-Famennian) conodont biostratigraphy and lithostratigraphy in the subsurface of northeastern Missouri [M.S. thesis]: Iowa City, University of Iowa, 239 p.

Manuscript Accepted by the Society February 27, 1997

Printed in U.S.A.

Geological Society of America
Special Paper 321
1997

Sequence stratigraphy of the Middle to Upper Devonian Guilmette Formation, southern Egan and Schell Creek ranges, Nevada

Todd A. LaMaskin and Maya Elrick
Department of Earth and Planetary Sciences, University of New Mexico, Albuquerque, New Mexico 87131

ABSTRACT

The upper Middle to Upper Devonian Guilmette Formation (~800 m thick) of eastern Nevada was deposited on a low-energy, westward-deepening carbonate platform. Five depositional facies are recognized, including tidal-flat, restricted shallow subtidal, shallow subtidal, intermediate subtidal, and deep subtidal. Facies are arranged into meter-scale, upward-shallowing peritidal and subtidal cycles that have average periods of between ~30 to165 k.y. Intermediate through deep subtidal facies are also present as thick, noncyclic intervals. The mechanism that best explains the presence of subtidal cycles, transgressive-prone cycles, and exposure-capped cycles and of systematic cycle stacking patterns is high-frequency (10^4 to 10^5 yr), glacio-eustatic sea-level fluctuations. Noncyclic intermediate and deep subtidal intervals represent missed sea-level oscillations when the seafloor lay too deep to record the effects of high-frequency fluctuations.

Eleven, fourth- to third-order depositional sequences are recognized from deepening and shallowing trends in depositional facies, changes in cycle stacking patterns, and subaerial exposure features. *Catch-up style* sequences deepen to intermediate to deep subtidal water depths during transgression and maximum flooding, indicating that sedimentation rates lagged behind accommodation space gains. *Keep-up style* sequences deepen only to shallow subtidal water depths, indicating that sedimentation rates kept pace with accommodation space gains throughout sequence development.

Combining sequence stratigraphic interpretations and conodont biostratigraphy permits correlation across the study area and correlation with previously published Devonian sea-level curves. Sequences 1, 2, 3, and 4 correlate with T-R cycles IIa-1, IIa-2, IIb, and IIc, respectively. Sequence stratigraphic relationships suggest that initial deepening of T-R cycle IId may be represented by maximum flooding zones of Sequences 5, 6, or 7. Sequences 8, 9, and 10 are interpreted to represent regression at the end of T-R cycle IId.

Sequence-scale facies patterns reflect second-order accommodation space changes related to the Kaskaskia supersequence. In particular, catch-up Sequences 1 through 7 represent the second-order transgressive systems tract; thick intervals of deep to intermediate subtidal facies in Sequences 3 and 4 are interpreted to represent the second-order maximum flooding zone. Keep-up sequences 8 through 10 record the second-order highstand systems tract.

LaMaskin, T. A., and Elrick, M., 1997, Sequence stratigraphy of the Middle to Upper Devonian Guilmette Formation, southern Egan and Schell Creek ranges, Nevada, *in* Klapper, G., Murphy, M. A., and Talent, J. A., eds., Paleozoic Sequence Stratigraphy, Biostratigraphy, and Biogeography: Studies in Honor of J. Granville ("Jess") Johnson: Boulder, Colorado, Geological Society of America Special Paper 321.

INTRODUCTION

Shallow-marine carbonates are characterized by a hierarchy of stratigraphic cyclicity ranging from meter-scale, upward-shallowing cycles (or parasequences) to depositional sequences that are tens to hundreds of meters thick. The detailed evolution of entire carbonate platforms on a 10^5 to 10^7 year timescale is best understood through recognition of how meter-scale cycles are organized into depositional sequences and how those sequences change through time (e.g., Elrick and Read, 1991; Goldhammer et al., 1993; Montañez and Osleger, 1993).

This study focuses on the late Middle to Upper Devonian (Givetian-Famennian) Guilmette Formation of eastern Nevada, which is composed of shallow- through deep-subtidal carbonates that are arranged into meter-scale cycles and depositional sequences. Previous workers have attempted to divide the thick and relatively homogeneous Guilmette Formation into regionally correlative, mappable members (Kellogg, 1963; Ackman, 1991). Results from this study suggest that these previous divisions, which were based largely on weathering profiles, are not genetic in origin; consequently they are of little use in regional correlations or for evaluating the controls on Middle-Upper Devonian carbonate platform evolution.

The main objectives of this chapter are to (1) describe and interpret the depositional environments represented by the Guilmette Formation, (2) describe and interpret the origin of meter-scale, upward-shallowing cycles that developed across the study area, (3) illustrate how stratal stacking patterns at individual stratigraphic sections combined with conodont biostratigraphy can be used to identify and correlate depositional sequences, and (4) evaluate the relationship of Guilmette depositional sequences to the Devonian eustatic sea-level curve of Johnson et al. (1985, 1991).

GEOLOGIC AND STRATIGRAPHIC SETTING

The Middle-Upper Devonian deposits of the eastern Great Basin were deposited along a westward-deepening, carbonate platform that was ~300 km wide and ~1500 km long, extending from southern California to Alberta, Canada (Fig. 1; Johnson and Sandberg, 1977; Sandberg et al., 1989; Johnson et al., 1991). The partially emergent Transcontinental Arch lay to the east of the platform, and oceanic deposits lay to the west. The study area in eastern Nevada represents deposition along the central platform region (i.e., inner shelf of Johnson and Murphy, 1984; Johnson et al., 1991).

The Guilmette Formation and time-equivalent units overlie a 3- to-7-km-thick succession of passive-margin carbonates and siliciclastics of latest Precambrian through Middle Devonian age (Stewart and Poole, 1974). The upper part of the Guilmette Formation (middle Frasnian–lower Famennian) is temporally equivalent to the lower Pilot Shale (Fig. 2), which is interpreted to represent the initial sedimentary response to the latest Devo-

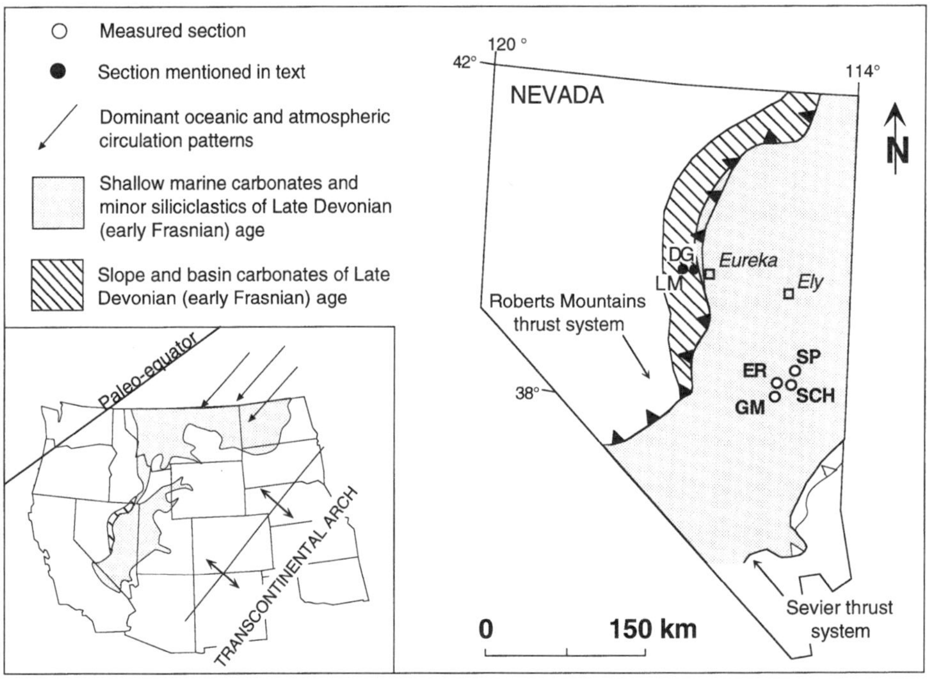

Figure 1. Map of study area and location of measured sections. Inset illustrates the extent of early Frasnian (Upper Devonian) shallow and deep marine deposits (modified from Sandberg et al., 1989) and dominant atmospheric and oceanic circulation patterns (modified from Witzke and Heckel, 1989). DG = Devils Gate; ER= southern Egan Range; GM = Gap Mountain; LM = Lone Mountain; SCH = southern Schell Creek Range. SP = Sidehill Pass, Schell Creek Range.

nian–Early Mississippian Antler orogeny (Sandberg and Poole, 1977; Sandberg et al., 1989; Goebel, 1991). The Guilmette Formation is overlain by the Upper Devonian West Range Limestone in eastern Nevada and parts of western Utah and the Upper Devonian to Lower Mississippian Pilot Shale in central Nevada to western Utah (Fig. 2). Unconformably overlying these Upper Devonian–Lower Mississippian deposits, is a succession up to 3,000 m thick of Mississippian siliciclastic strata composed of submarine-fan to fluvial-deltaic deposits, which filled the Antler foreland basin (Poole, 1974; Harbaugh and Dickinson, 1981).

DEPOSITIONAL FACIES

Four stratigraphic sections were measured (Fig. 1) and described on a bed-by-bed scale, resulting in the recognition of five depositional facies (or facies assemblages). In order of increasing water depths they are: tidal-flat, restricted shallow subtidal, shallow subtidal, intermediate subtidal, and deep subtidal facies (Table 1). Depositional facies are composed of 15 subfacies representing subenvironments that are defined on the basis of grain types, sedimentary structures, fossil content, and vertical facies relationships. Subfacies are arranged into meter-scale, upward-shallowing peritidal cycles capped by tidal-flat facies and subtidal cycles composed wholly of subtidal facies. Shallow, intermediate, and deep subtidal facies also are present as noncyclic intervals. Detailed description of subfacies are presented in Table 1 and in LaMaskin (1995); brief environmental interpretations are given below.

Presently, the distance between the Egan Range (Sections ER and GM) and Schell Creek Range (Sections SCH and SP) is

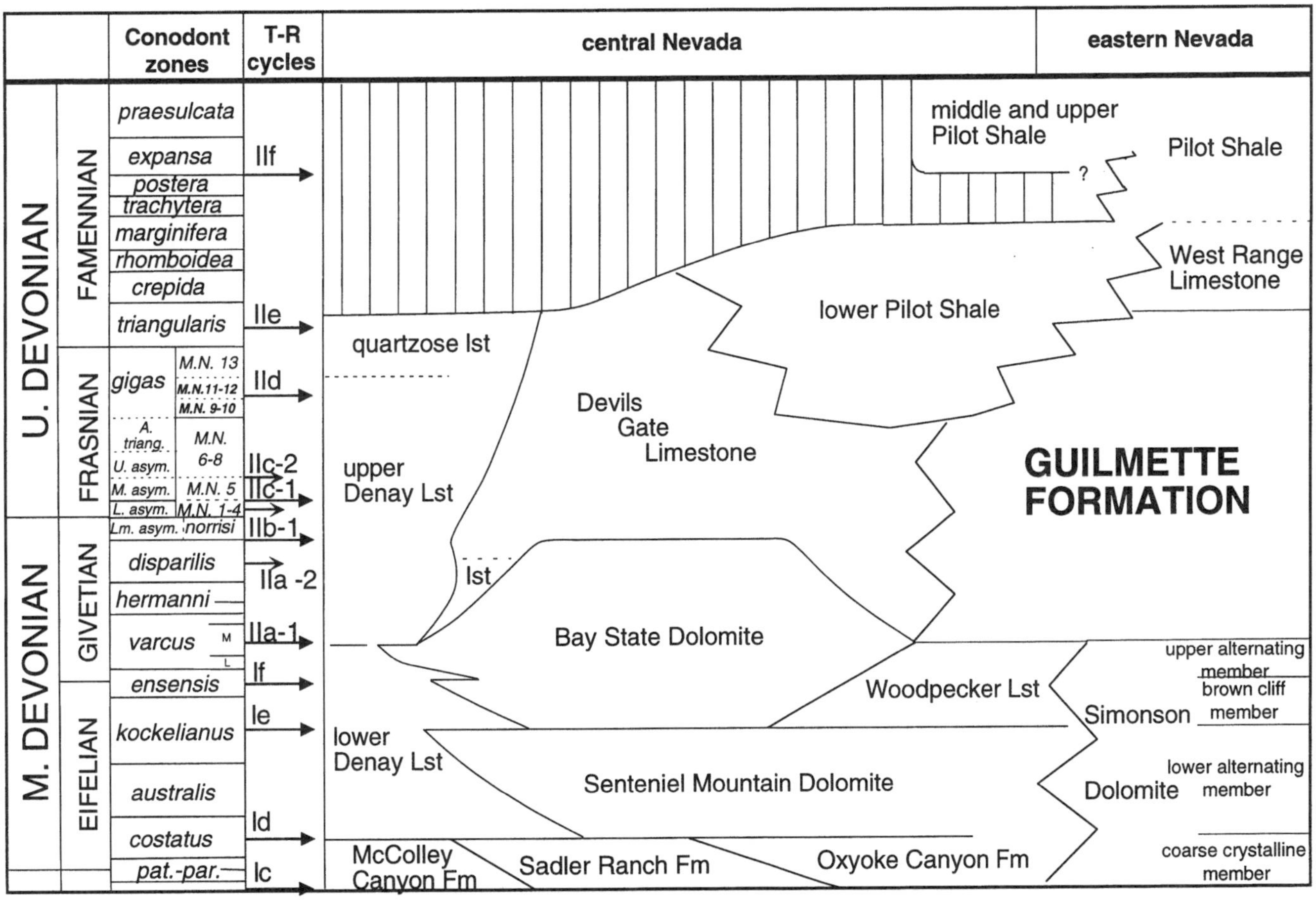

Figure 2. Chronostratigraphic diagram for Middle and Upper Devonian strata. Vertical pattern indicates depositional hiatus or unconformity. Dashed lines indicate vertical or lateral variation in nomenclature. T-R cycles are transgressive-regressive cycles (or sequences) of Johnson et al. (1985, 1991) with modifications by Day (1994) and Day et al. (1996). Long arrows indicate major transgressive starts; short arrows indicate intra–T-R cycle transgressive starts. M.N. = Montagne Noire conodont zonation of Klapper (1989). Correlation of M.N. Zones with standard Frasnian conodont zones is based on comparisons of Johnson (1990), Johnson and Klapper (1992), Johnson et al. (1985), and Day (1994). Dashed lines between conodont zones indicate uncertain correlations between the two independent Frasnian zonations. Thicknesses of Frasnian conodont zones are not related to time. Modified after Johnson and Murphy (1984), Johnson et al. (1996), and K. A. Giles (personal communication, 1994).

TABLE 1. DESCRIPTION OF GUILMETTE FORMATION SUBFACIES

TIDAL-FLAT FACIES

(1) Laminated Dolomite Subfacies

Occurrence: Caps to peritidal cycles (0.02 – 2 m thick) and as transgressive units at the bases of cycles. Thick laminites commonly grade up into thin laminites.

Bedding and Lithology: Graded thin (<1 – 3 mm) to thick (3 – 15 mm) laminae composed of medium- to fine-crystalline dolomite. Rare undolomitized beds are composed of pellets, peloids, lime mud, and fine skeletal debris. Commonly contains cauliflower-shaped, calcified-evaporite nodules (<1 – 3 cm) and rare calcite pseudomorphs after gypsum.

Sedimentary Structures: Thin laminated dolomite: flat-pebble conglomerates and rare mudcracks. Thick laminated dolomite: ripple cross-laminae, shallow scours, flat-pebble conglomerates, and rare mudcracks.

Biota: Thin laminites: abraded ostracode fragments and fine skeletal debris, rare laterally linked hemispheroids (LLH) stromatolites. Thick laminites: abraded ostracode fragments, fine skeletal debris, rare gastropods and LLH stromatolites.

Color: Weathered: yellow; fresh: light gray.

(2) Dolomitic Siltstone Subfacies

Occurrence: A single 3-m-thick bed lying ~3 m below the West Range Limestone.

Bedding and Lithology: Thin-bedded, dolomitic quartz siltstone to fine sandstone. Coarse silt- to fine sand-sized quartz, angular to sub-angular, well sorted. Dolomite occurs as cement and detrital framework grains. Plagioclase feldspar composes <1% of framework grains.

Sedimentary Structures: Abundant thin planar laminae.

Biota: Barren.

Color: Weathered and fresh surface: light tan to brown.

RESTRICTED SHALLOW SUBTIDAL FACIES

(3) Massive Dolomite Subfacies

Occurrence: Caps to shallow subtidal cycles (0.05 – 4 m thick), units within peritidal cycles, and as transgressive units at base of peritidal and subtidal cycle.

Bedding and Lithology: Medium-bedded, fine- to medium-crystalline dolomite. Rare undolomitized beds composed of skeletal pellet wackestone or ostracode pellet packstone. Rare calcite pseudomorphs after gypsum.

Sedimentary Structures: Rare laminae; most primary sedimentary structures destroyed by bioturbation.

Biota: Abraded and whole ostracodes, whole gastropods, fine skeletal debris, and bioturbation.

Color: Weathered: yellow; fresh: light gray.

(4) Massive Quartz Arenite Subfacies

Occurrence: Rare beds (1.5 – 15 cm thick) interbedded with laminated dolomite subfacies and interbedded peloidal grainstone to mudstone subfacies.

Bedding and Lithology: Thin- to thick-bedded quartz arenite composed of fine- to medium sand-sized grains. Moderately well to well sorted, well rounded. Dolomite cemented.

Sedimentary Structures: Massive.

Biota: Barren.

Color: Fresh and weathered surfaces: tan.

(5) Stromatolite Boundstone Subfacies

Occurrence: Base of the yellow, slope-forming interval (YSFI); rare in the remainder of the formation as units within peritidal cycles. LLH stromatolites commonly overlie laminated dolomite subfacies and grade upward into digitate stromatolites.

Bedding and Lithology: Medium to thick beds of LLH stromatolites (10 – 30 cm diameter, 5 cm thick) overlain by upward-branching, unlinked digitate stromatolites (2 – 5 cm in diameter, as much as 15 cm tall). Interhead matrix of digitate stromatolites composed of peloidal intraclast grainstone. Stromatolites commonly nucleate on scoured surfaces and intraclasts.

Sedimentary Structures: Stromatolites composed of graded pelletal/peloidal laminae (<1 – 3 mm) that thin laterally and exceed angle of repose, scalloped and truncated laminae, and fenestrae. Matrix: imbricated intraclasts.

Biota: Rare gastropods.

Color: Weathered: dark gray; matrix yellow; fresh: dark gray.

(6) Interbedded Peloidal Grainstone to Mudstone Subfacies

Occurrence: Caps to shallow subtidal cycles (0.05 – 1.5 m thick) units within peritidal and subtidal cycles, and rare transgressive units at the cycle bases. Interbedded with burrow-mottled peloidal grainstone/mudstone subfacies.

Bedding and Lithology: Wavy thin beds (~1 cm) composed of multiple graded laminae of peloidal grainstone to ostracode mudstone, interbedded with thin beds of fine- to medium-crystalline dolomite.

Sedimentary Structures: Laterally discontinuous thin beds and isolated lenses draped by fine-crystalline dolomite, graded laminae, ripple-cross laminae, and centimeter-scale scours.

Biota: Abraded ostracodes, rare whole gastropods, rare abraded brachiopods, and fine skeletal debris.

Color: Weathered: bands of light gray and yellow; fresh: bands of gray and yellow gray.

(7) Burrow-mottled Peloidal Grainstone/Mudstone Subfacies

Occurrence: Caps to shallow subtidal cycles (0.05 – 2 m thick), units within peritidal and subtidal cycles, and rare tgransgressive units at cycle bases. Intercalated with interbedded peloidal grainstone to mudstone subfacies.

Bedding and Lithology: Isolated lenses (<1 – 5 cm thick) of peloidal grainstone/mudstone within fine- to medium-crystalline dolomite; e.g., bioturbated interbedded peloidal grainstone to mudstone subfacies.

Sedimentary Structures: None.

Biota: Abraded ostracodes and brachiopods, whole gastropods and brachiopods, fine skeletal debris, and *Thalassinoides* burrows.

Color: Weathered: mottled gray and yellow; fresh: mottled gray and yellow gray.

(8) *Amphipora* Wackestone/Grainstone Subfacies

Occurrence: Beds (<1 – 3 m thick) at the base or middle of peritidal cycles and in middle of subtidal cycles; rare caps to shallow subtidal cycles.

Bedding and Lithology: Thin- to thick-bedded *Amphipora* skeletal pelletal wackestone/grainstone containing variable amounts of peloids and intraclasts in discrete lenses. Rare calcite pseudomorphs after gypsum. Locally dolomitized.

Sedimentary Structures: *Amphipora* grains lie parallel to bedding and are commonly current aligned in packstone and rare grainstone subfacies.

Biota: Abundant *Amphipora,* minor gastropods, brachiopods, calcispheres, rare reworked subspherical and nodular stromatoporoids, and bioturbation.

Color: Weathered: dark blue-gray with *Amphipora* grains weathering white; fresh: dark gray. Dolomitized beds are brown.

TABLE 1. DESCRIPTION OF GUILMETTE FORMATION SUBFACIES (continued)

SHALLOW SUBTIDAL FACIES

(9) Peloidal Skeletal Packstone/Grainstone Subfacies

Occurrence: Units (<1 – 11 m thick) at the base or in the middle of peritidal and subtidal cycles.

Bedding and Lithology: Thin- to thick-bedded, peloidal skeletal packstone/grainstone, intraclastic peloidal packstone/grainstone, and rare coated-grain packstone/grainstone. Ovoid peloids (0.1 – 2 mm) composed of lime mud, fossiliferous wackestone, and micritized skeletal grains, and commonly form grapestone lumps. Intraclasts (0.2 – 2.5 cm) composed of peloidal skeletal packstone. Locally dolomitized.

Sedimentary Structures: None.

Biota: Abraded and whole brachiopods, crinoids, *Amphipora, Stachyoides* ?, gastropods, *Thamnopora* coralis, calcispheres, reworked and in-growth-position sub-spherical and nodular stromatoporoids, fine skeletal debris.

Color: Weathered: light gray; fresh: light gray. Dolomitized beds are brown.

Subfacies variations: *Thamnopora* packstone with peloidal skeletal packstone matrix is present in all measured sections as beds within peritidal and subtidal cycles (<0.5 – 8 m), traced laterally into peloidal skeletal packstone subfacies.

(10) Skeletal Pellet Packstone Subfacies

Occurrence: Beds (<12 m thick) at the base or middle of peritidal and subtidal cycles; rare caps to shallow subtidal cycles. Present as matrix to stromatoporoid floatstone subfacies.

Bedding and Lithology: Thin- to thick-bedded, fine skeletal pellet packstone/wackestone. Locally dolomitized.

Sedimentary Structures: None due to bioturbation.

Biota: Brachiopods, gastropods, *Thamnopora, Amphipora, Stachyoides* ?, calcispheres, bioturbation, and fine skeletal debris.

Color: Weathered: dark blue-gray; fresh: dark gray. Dolomitized beds are brown

(11) Stromatoporoid Floatstone/Rudstone Subfacies

Occurrence: Beds (<1 – 10 m thick) at the base or middle of peritidal and subtidal cycles.

Bedding and Lithology: Medium- to thick-bedded stromatoporoid floatstone/rudstone containing abundant sub-spherical, nodular, bulbous, and domal stromatoporoids (<3 – 70 cm in diameter), and minor tabular to lamellar stromatoporoids. Smaller stromatoporoids are commonly reworked; larger ones are in-growth-position. Matrix composed of fine skeletal pellet packstone with variable amounts of peloids and intraclasts in discrete lenses. Locally dolomitized.

Sedimentary Structures: None, due to bioturbation.

Biota: Diverse biota including stromatoporoids, *Amphipora, Stachyoides* ?, gastropods, brachiopods, calcispheres, crinoids, rugose and thamnoporid corals, fine skeletal debris, and bioturbation.

Color: Weathered: dark blue-gray; fresh: dark gray. Dolomitized beds are brown.

INTERMEDIATE SUBTIDAL FACIES

(12) Tabular Stromatoporoid Boundstone Subfacies

Occurrence: Caps to deep subtidal cycles (<1 – 1.5 m thick), within noncyclic intervals; rare occurrences within peritidal and shallow subtidal cycles.

Bedding and Lithology: Medium- to thick-bedded tabular stromatoporoid boundstone/rudstone and tabular-lamellar stromatoporoid boundstone with lime mudstone, or fine- to medium-crystalline dolomite matrix. Stromatoporoids (2 – 10 cm thick and 10 – 50 cm wide), oriented parallel to bedding.

Sedimentary Structures: None.

Biota: Brachiopods, crinoids, rugose and thamnoporid corals, encrusting algae, fine skeletal debris, minor sub-spherical and nodular stromatoporoids, and rare colonial corals (*Disphyllum*?).

Color: Weathered: matrix weathers light gray to yellow, stromatoporoids light gray; fresh: dark gray.

(13) Nodular-bedded Skeletal Wackestone Subfacies

Occurrence: Noncyclic, recessive intervals (<30 m thick), and as caps to deep subtidal cycles (<4 m); rare within peritidal and subtidal cycles.

Bedding and Lithology: Thin, nodular-beddded, brachiopod-pellet wackestone/mudstone, crinoidal-pellet wackestone/mudstone, coral-pellet wackestone/mudstone, and pellet wackestone/mudstone. Common laterally discontinuous thin lenses of skeletal intraclast packstone. Micron-size pyrite crystals, abundant clay, silt, and fine sand-sized quartz in insoluble residues.

Sedimentary Structures: Graded skeletal-intraclast packstone lenses and shallow scours.

Biota: Brachiopods, crinoids, rugose and thamnoporid corals, tentaculites, encrusting algae, branching and meandering horizontal burrow networks. Conodonts include: *Mesotaxis* sp., *Polyganthus* sp., *Icriodus* sp., *Playfordia* sp., *Pandorinellina* sp., *Palmatolepis* sp., and *Mehlina* sp.

Color: Weathered: light gray; fresh: light to medium gray.

DEEP SUBTIDAL FACIES

(14) Graded Peloidal Grainstone to Packstone

Occurrence: Rare occurrences as noncyclic intervals (<7 m thick), and rarely within peritidal and subtidal cycles (<2 m thick). Consistently interbedded with intermediate subtidal facies.

Bedding and Lithology: Graded laminae (~1.5 – 15 cm thick) of peloidal grainstone to mudstone; minor interbedded flat-pebble conglomerates with clasts composed of adjacent rock type.

Sedimentary Structures: Graded thin laminae, ripple-cross laminae, shallow scours.

Biota: None.

Color: Weathered: light gray; fresh: very dark gray.

(15) Interbedded Lime Mudstone and Argiullite Subfacies

Occurrence: Recessive intervals (<2 – 8 m thick) associated with noncyclic intermediate subtidal facies; rare occurrences within peritidal and subtidal cycles.

Bedding and Lithology: Thin- to medium-bedded, laminated lime mudstone interbedded with thin-bedded, laminated argillaceous lime mudstone (argillite). Commonly grades upward into bioturbated equivalent.

Sedimentary Structures: Lime mudstone: sparse thin laminae; argillaceous lime mudstone: abundant thin laminae.

Biota: Tentaculites. Conodonts include *Mesotaxis* sp., *Ancryodella* sp., *Polygnathus* sp.

Color: Weathered: dark gray; fresh: dark gray to black.

approximately 15 km (Fig. 1). Due to Basin and Range extension, the two ranges presumabably lay closer to one another during the Devonian; however, the amount of extension between the ranges in the study area has not been investigated. It is unlikely that 300% extension has occurred as is interpreted by Gans (1987) for the Schell Creek–Snake Range region to the north. Because of uncertainties in the amount of shortening, present or actual distances between sections are shown in the present study. The use of present distances does not alter facies or sequence stratigraphic interpretations discussed below.

Tidal-flat facies

Tidal-flat facies include thick and thin laminated dolomite and dolomitic siltstone subfacies (Table 1).

(1) *Thick and thin laminated dolomite subfacies* are present at each stratigraphic section and form caps to peritidal cycles and thin transgressive bases to some peritidal cycles. This subfacies is interpreted to represent intertidal to supratidal laminites. Millimeter- to centimeter-thick laminae represent traction deposition followed by suspension settling of grains from storm and tidal currents. Rare mudcracks attest to intermittent subaerial exposure and a paucity of skeletal and trace fossils indicates restricted conditions. The conformable upward transition from thick laminae (centimeter-scale) to thin laminae (millimeter-scale) represents landward thinning of storm and tidal-generated sediment sheets and is the result of seaward progradation of supratidal flats over intertidal flats (Hardie and Shinn, 1986). Laminae displaced by calcified evaporite nodules indicate postdepositional growth of evaporites from hypersaline groundwaters. Laterally discontinuous flat-pebble conglomerate beds represent reworking of semilithified subtidal and intertidal deposits by storm currents.

(2) *Dolomitic siltstone subfacies* are present as a 3-m-thick bed at the top of the Guilmette Formation at Section SCH. Deposition in a tidal-flat environment is indicated by the intercalation with laminated dolomite subfacies, thin planar laminae, and the absence of skeletal and trace fossils. The silt-size quartz grains are interpreted as eolian in origin and were subsequently reworked in the marine environment (Kukal and Saadallah, 1973).

Restricted shallow subtidal facies

Restricted shallow subtidal facies include massive dolomite, massive quartz arenite, stromatolite boundstone, interbedded peloidal grainstone to mudstone, burrow-mottled peloidal grainstone/mudstone, and *Amphipora* wackestone/grainstone subfacies (Table 1). These subfacies are present in all measured sections as units within peritidal and subtidal cycles.

(3) *Massive dolomite subfacies* were deposited in restricted, shallow subtidal environments. Evidence for this interpretation includes low-diversity skeletal fauna suggestive of restricted conditions, presence of bioturbation, lack of subaerial-exposure features, and the stratigraphic position below tidal-flat laminites.

(4) *Massive quartz arenite subfacies* are present as two <1.5-m-thick beds at the top of the formation at Section SCH. This subfacies is interpreted to represent restricted shallow subtidal deposition as indicated by the conformable stratigraphic position below tidal flat facies, grain size and texture, and lack of skeletal fossils.

(5) *Stromatolite boundstone subfacies* are present in a stratigraphic interval termed the "yellow, slope-forming interval" (YSFI) throughout eastern Nevada (Hurtubise, 1989) but are rare in the remainder of the formation. This subfacies is interpreted to have been deposited in restricted, shallow subtidal environments with energy levels ranging from low to high. This interpretation is indicated by the absence of skeletal fauna, matrix composed of mudstone to peloid intraclastic grainstone, and the stratigraphic position below tidal-flat facies.

(6) *Interbedded peloidal grainstone to mudstone subfacies* were deposited in restricted, moderate energy, shallow subtidal environments as indicated by low-diversity skeletal fauna, lack of bioturbation and subaerial exposure features, and the stratigraphic position below tidal-flat facies. Graded laminae and ripple-cross laminae draped by dolomudstone layers indicate episodic traction sedimentation followed by suspension settling related to storms and tides (Reineck and Singh, 1980; Demicco, 1983). Reineck and Singh (1980) described similar thin-bedded siliciclastic sediments from shallow subtidal and intertidal regions of the North Sea and ascribed their formation to ebb and flood tidal currents and to slack-water conditions.

(7) *Burrow-mottled peloidal grainstone/mudstone subfacies* commonly underlies and contains a slightly more diverse skeletal fauna than the closely associated interbedded peloidal grainstone to mudstone subfacies. This stratigraphic relationship suggests that burrow-mottled subfacies were deposited in slightly less restricted and/or slightly deeper waters than the interbedded peloidal grainstone to mudstone subfacies.

(8) *Amphipora wackestone/grainstone subfacies* were deposited in restricted, low- to moderate-energy, shallow subtidal environments. Evidence for this interpretation includes a monospecific skeletal fauna, presence of bioturbation, absence of subaerial exposure features, and the stratigraphic position below tidal-flat facies. Rare current-aligned *Amphipora* grainstone beds represent locally transported deposits, whereas beds with randomly oriented *Amphipora* represent in situ accumulations. *Amphipora*-rich strata are characteristic of many Middle to Upper Devonian deposits worldwide and are commonly interpreted as representing restricted shallow subtidal environments (Read, 1973; Stearn, 1982; Dorobek, 1991; Wendte, 1992).

Shallow subtidal facies

Shallow subtidal facies include peloidal skeletal packstone/grainstone, skeletal pellet packstone, and stromatoporoid floatstone/rudstone subfacies (Table 1). These subfacies are present in each measured section as units within peritidal and subtidal cycles and rarely as noncyclic intervals. Shallow subtidal facies are differentiated from restricted shallow subtidal facies by the abundance and diversity of open-marine skeletal fauna and bioturbation and by the concurrent paucity of evaporites and subaerial exposure features.

(9) *Peloidal skeletal packstone/grainstone subfacies* represent deposition in moderate energy, open-marine, subtidal shoals as indicated by the packstone/grainstone textures, abundant open-marine skeletal fauna, and the absence of subaerial exposure features. Interbedded peloid intraclast layers reflect reworking of early-cemented substrates by episodic storm events. Packstone/grainstone beds directly overlain by tidal-flat facies represent fringing shoal deposits, whereas packstone/grainstone beds overlain by restricted shallow subtidal facies represent offshore subtidal shoals that were subsequently overlain by prograding backshoal deposits.

(10) *Skeletal pellet packstone subfacies* were deposited in low-energy, open-marine, shallow subtidal environments. Evidence for this interpretation includes abundant open-marine skeletal fauna and bioturbation, the absence of subaerial exposure features, and the fine grain size. Modern bioturbated skeletal pellet muds are common in shallow subtidal, offshore areas of the Great Bahama Bank (Hardie and Shinn, 1986) and western portions of the Trucial Coast in the Persian Gulf (Purser and Evans, 1973).

(11) *Stromatoporoid floatstone/rudstone subfacies* represent open-marine, shallow subtidal biostromes. Stromatoporoid biostromes are composed of subspherical, nodular, bulbous, and domal stromatoporoids (terminology of Abbott, 1973); individual beds may be continuous for hundreds of meters along strike. A predominance of fine-grained (pellet wackestone/packstone) matrix and the absence of wave- or current-generated features suggest that the biostromes did not act as wave-resistant barriers; that is, the buildups do not represent reef complexes (*sensu stricto*) such as those described in Upper Devonian carbonates of Alberta (Kobluk, 1978; Stearn, 1982). Stromatoporoid biostromes similar to those observed in the study area are present in the Upper Devonian of Alberta and are interpreted by Wendte (1992) to have developed on the platform interior in low-energy, subtidal environments.

Power (1984) reported what may be stromatoporoid and coral reefs in the Devils Gate Formation in the Roberts Mountains of central Nevada; however, the textures and faunal associations she described are not observed in the Egan or Schell Creek Ranges.

Intermediate subtidal facies

Intermediate subtidal facies include tabular stromatoporoid floatstone/boundstone and nodular-bedded skeletal wackestone subfacies (Table 1). These subfacies are present in each measured section as noncyclic intervals and less commonly within subtidal and peritidal cycles. Intermediate subtidal facies are differentiated from shallow subtidal facies by the greater amount of micrite matrix, a higher diversity of open-marine fauna (brachiopods, corals, and crinoids), and a change in stromatoporoid morphology from spherical to tabular and lamellar forms.

(12) *Tabular stromatoporoid floatstone/boundstone subfacies* were deposited in low-energy, open-marine subtidal waters as low-relief biostromes. This interpretation is supported by the abundance of typical open-marine skeletal fauna such as corals and crinoids, abundance of lime mudstone matrix, and tabular stromatoporoid morphologies. The alternative interpretation that tabular and lamellar stromatoporoid morphologies reflect high-energy conditions (e.g., Fischbuch, 1968; Noble and Ferguson, 1971) is not supported by the presence of abundant micrite matrix, which implies minimal winnowing, and the abundance of delicate coral forms.

Reso (1963) described a coral-stromatoporoid reef in the Guilmette Formation on Mount Irish near Alamo, Nevada. Reconnaissance examination of the Mount Irish section reveals that the strongly dolomitized and recrystallized "reef" is composed of tabular stromatoporoid boundstone with sparse corals (*Disphyllum?* sp.) and probably represents a biostrome similar to those in the Guilmette Formation in the study area.

(13) *Nodular-bedded skeletal wackestone subfacies* were deposited in low-energy, open-marine, subtidal environments below fair-weather wave base. Evidence for this interpretation includes abundant open-marine fauna dominated by brachiopods, crinoids, and corals; bioturbation; and abundant micrite matrix. Laterally discontinuous lenses of abraded skeletal intraclastic packstone represent reworking and deposition of semilithified subtidal deposits by episodic storm events (e.g., Markello and Read, 1981).

Deep subtidal facies

Deep subtidal facies include graded peloid grainstone to packstone and interbedded lime mudstone and argillite subfacies (Table 1). These subfacies are present as noncyclic intervals interbedded with intermediate subtidal facies and less commonly in subtidal cycles.

(14) *Graded peloid grainstone to packstone subfacies* represent deposition in deep subtidal waters as storm deposits. Evidence for storm deposition includes graded laminae and beds, shallow scours, and interbedded flat-pebble conglomerates. The lack of bioturbation or skeletal fossils suggests low oxygen levels and/or rapid sedimentation rates.

(15) *Interbedded lime mudstone and argillite subfacies* were deposited in quiet, poorly oxygenated waters below storm wave base. Individual, millimeter-thick, graded laminae in limestone and argillite layers represent deposition from discrete distal storm events and/or from dilute turbidity currents. Bundling of laminae into centimeter-thick carbonate-rich and carbonate-poor couplets likely represents periodic fluctuations in detrital carbonate influx or changes in siliciclastic fluvial influx into the marine basin (Elrick et al., 1991; Elrick and Hinnov, 1996).

MICROKARST FEATURES

Microkarst features include downward-tapering cracks, erosion surfaces, and dolomite breccias. These features affect tidal-flat facies and some shallow subtidal facies.

Downward-tapering cracks are observed in tidal-flat facies. They range in width from a few millimeters to 5 cm, are up to 1 m long, and are infilled with subtidal deposits from overlying units (Fig. 3). Larger cracks commonly branch into shorter (1 to 10 cm) and thinner (<2 cm) solution-enhanced horizontal cracks.

Erosion surfaces are smooth to irregular surfaces with 2 to 50 cm of relief that affect laminated dolomite subfacies. Erosion is indicated by truncation of primary laminae in the host rock. The surfaces commonly pass laterally into overhanging walls and pinnacles. Transgressive portions of overlying units may onlap delicately sculptured surfaces and contain reworked clasts derived from the underlying eroded unit.

Dolomite breccias are monomictic and polymictic. Monomict breccia clasts are composed of dolomitized tidal-flat facies in a matrix of fine-crystalline dolomite with abundant quartz grains. Dolomitized polymict breccias clasts composed of tidal flat and shallow subtidal facies are less common. Both breccia types are associated with downward-tapering cracks and erosion surfaces and may pass along strike into nonbrecciated tidal flat or subtidal units. Breccia clasts commonly are reworked into overlying subtidal units.

Downward-tapering cracks, erosion surfaces, and dolomite breccias are interpreted to have developed during prolonged periods of subaerial exposure when downward-percolating, undersaturated fluids moved through partially lithified tidal-flat deposits (e.g., Heckel, 1983; Goldstein et al., 1991). During initial periods of subaerial exposure, vertical fractures developed in response to shrinkage and expansion during desiccation (Goldstein et al., 1991). These fractures likely acted as conduits for undersaturated fluids, resulting in solution enlargement of the vertical and horizontal cracks as well as along bedding planes. Continued exposure and dissolution resulted in the development of pinnacles and overhanging walls. Dolomite breccias formed when downward-tapering cracks coalesced and whole beds collapsed.

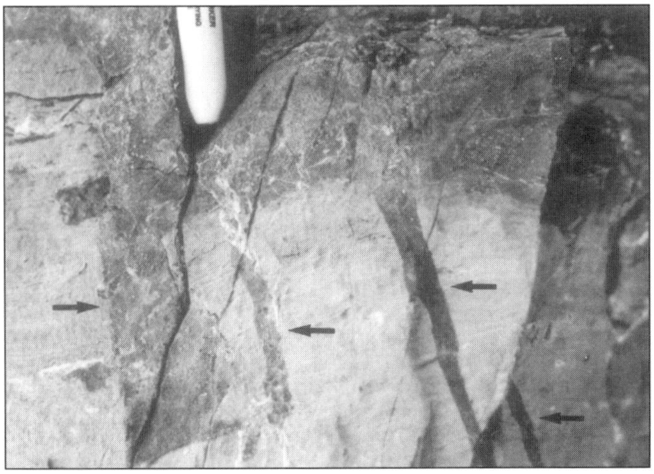

Figure 3. Photograph of downward-tapering cracks in laminated dolomite subfacies that caps a peritidal cycle at Section SCH. Arrows point to cracks that are filled with dark-gray subtidal facies of overlying cycle. Note that laminae are truncated by crack. Base of pen for scale.

Several lines of evidence suggest that microkarsting occurred prior to deposition of the immediately overlying beds; these include infiltration of subtidal sediment from the overlying units into newly formed cavities, mantling of relief on erosion surfaces by overlying deposits, and reworking of breccia clasts into overlying units. Early lithification of tidal flat deposits is indicated by angular breccia clasts displaying fitted fabrics, truncation of laminae, and a lack of soft-sediment deformation features.

METER-SCALE, UPWARD-SHALLOWING CYCLES

Tidal-flat through deep subtidal facies are arranged into meter-scale, upward-shallowing cycles (Fig. 4). Peritidal cycles (84% of all cycles) are composed of an upward-shallowing succession of shallow and restricted subtidal facies and are capped by tidal-flat facies; intermediate or deep subtidal facies are rare (Figs. 4, 5). Progressive upward shallowing within these cycles is reflected by an upsection decrease in open-marine skeletal fauna (brachiopods, crinoids, corals, stromatoporoids), size of skeletal grains, and degree of bioturbation and by an increase in evidence of former evaporite minerals.

Two types of subtidal cycles (16% of all cycles) are recognized; *shallow subtidal cycles* are composed of shallow and restricted subtidal facies (Fig. 4B), whereas *deep subtidal cycles* are composed of an upward-shallowing succession of deep to intermediate subtidal facies (Fig. 4C).

Over 85% of the peritidal and subtidal cycles are regressive-prone; that is, the entire cycle thickness records upward shallowing or regression. Transgressive-prone cycles are characterized by basal, upward-deepening trends followed by typical upward-shallowing trends (Figs. 4A, B). The average duration of all cycles is between ~30 to 165 k.y. based on timescales of Palmer (1983) and Harland et al. (1989) and are therefore fifth- to fourth-order in scale (Vail et al., 1977).

The three mechanisms commonly cited to explain the origin of meter-scale, carbonate cycles are (1) autogenic variations in sediment production (Ginsburg, 1971; Pratt and James, 1986), (2) fault-induced subsidence (Cisne, 1986; Hardie et al., 1991), and (3) high-frequency, glacio-eustatic fluctuations in sea level (e.g., Goodwin and Anderson, 1985; Heckel, 1986; Goldhammer et al., 1987). The degree of control that each mechanism has on the development and characteristics of cycles has been the subject of much debate, particularly because unequivocal evidence supporting one mechanism over another has yet to be clearly demonstrated.

The mechanism that can explain all the features observed in the Guilmette cycles is high-frequency (10^4 to 10^5 yr), glacio-eustatic, sea-level oscillations. Transgressive-prone cycles indicate that subtidal sediment was being produced during relative sea-level rises; that is, sedimentation lag times were absent to minimal during flooding, precluding the Ginsburg autogenic model. Exposure-capped cycles represent times when high-frequency sea level fell below the platform surface, resulting in microkarsting of the deposits ("missed beats"; Goldhammer et

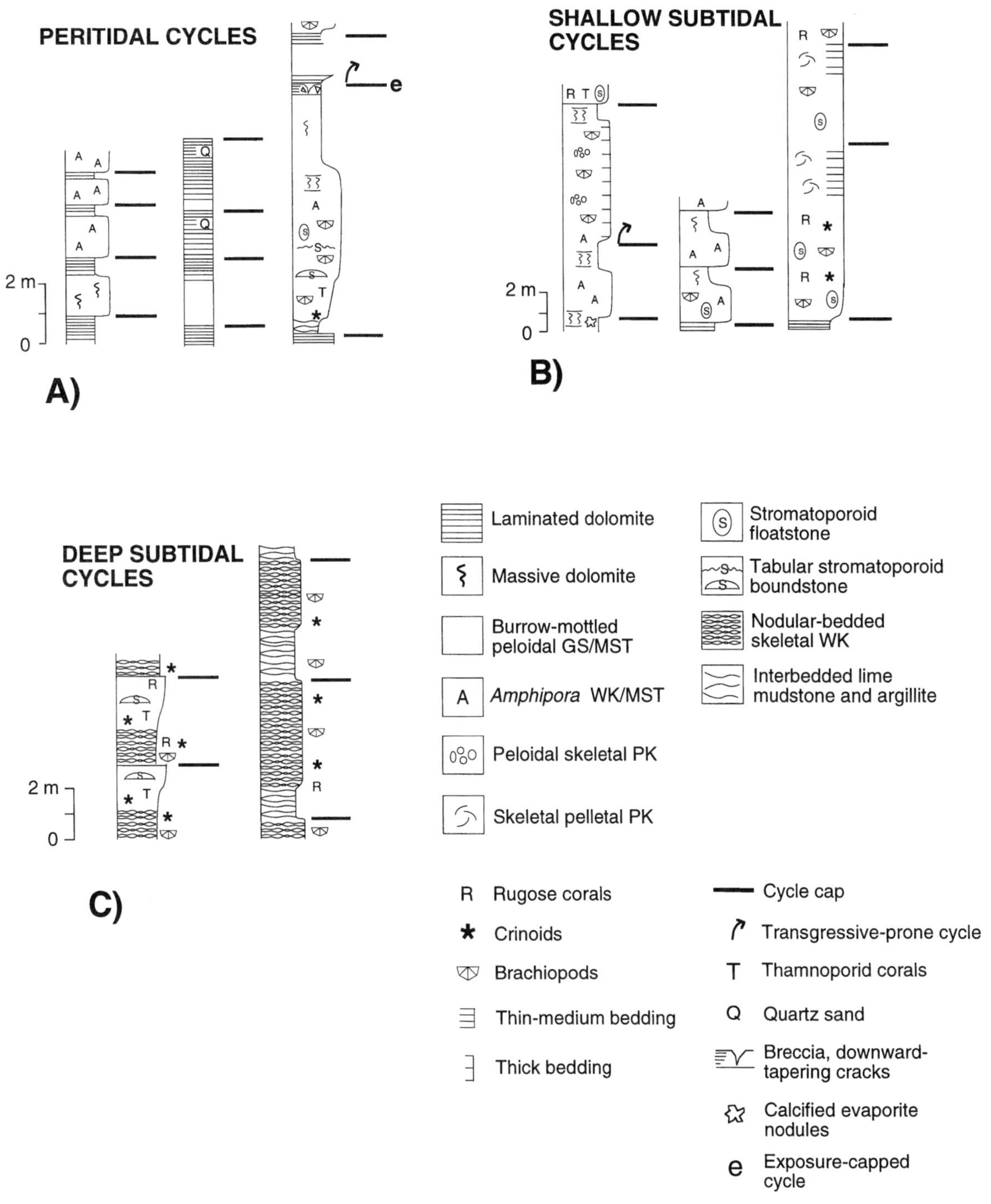

Figure 4. Partial stratigraphic columns illustrating typical peritidal, shallow subtidal, and deep subtidal cycles. A, Peritidal cycles from Sections GM and SCH are capped by tidal-flat facies and rarely deepen to deep subtidal facies. Peritidal cycles can be exposure capped and transgressive prone. B, Shallow subtidal cycles from Sections SCH and SP are composed of an upward-shallowing succession of restricted to shallow subtidal facies; cycles may be transgressive-prone. C, Deep subtidal cycles in Sections SCH and ER are capped by intermediate subtidal facies. MST = lime mudstone; WK = wackestone; PK = packstone; GS = grainstone.

Figure 5. Photograph of typical peritidal cycle in Section SCH with burrow-mottled peloid grainstone/mudstone subfacies gradationally overlain by laminated dolomite. Top of cycle indicated by arrow and black line. Cycle cap is abruptly overlain by dark-gray, shallow subtidal facies of the overlying cycle.

al., 1990; Elrick, 1995); cycle-capping exposure features also preclude the autogenic model. Subtidal cycles reflect incomplete shallowing and infilling of accommodation space before the ensuing high-frequency rise in sea level. Noncyclic subtidal intervals are interpreted to represent the results of the seafloor, lying too deep to record the effects of high-frequency sea-level changes (subtidal missed beats)

Glacial-eustasy as a cycle-generating mechanism is also supported by the evidence of continental glaciation in South America and Africa during the Givetian(?) and Famennian (Rocha-Campos, 1981a, 1981b; Caputo and Crowell, 1985; Veevers and Powell, 1987). The average cycle periods of ~30 to 165 k.y. fall within the Milankovitch range; however, Algeo and Wilkinson (1988) caution that because of the relatively narrow range of long-term accumulation and subsidence rates, any 1- to 20-m-thick stratigraphic interval will lie within the Milankovitch time range. As such, average cycle periods lying within the Milankovitch band, alone, are insufficient evidence for interpreting a glacio-eustatic origin.

PLATFORM DEPOSITIONAL MODEL

Interpretation of previous work in the Devonian of the Great Basin region indicates that the Guilmette Formation and temporal equivalents represent a westward-deepening, gently sloping platform; the study area lies along the central platform. The nature of the platform edge (reef/shoal rimmed or distally steepened) and gradient along its slope is not known because the pertinent sections are obscured as a result of the Late Devonian–Early Mississippian Antler orogeny and younger tectonism or the result of cover beneath upper Cenozoic alluvium.

The gently dipping platform morphology was modified in the middle Frasnian by the formation of the deeper water Pilot Basin in central Nevada (Fig. 2; Sandberg and Poole, 1977; Sandberg et al., 1989); this has been interpreted as reflecting the initial affects of the Antler orogeny. Specifically, the basin has been interpreted as the result of increased subsidence before development of the Mississippian foreland basin (Poole, 1974; Sandberg and Poole, 1977). Alternatively, the basin has been interpreted as a fault-bound, back-bulge basin that formed in response to flexural loading of the continental margin by the eastward-migrating Roberts Mountains allochthon (Goebel, 1991; Giles and Dickinson, 1995). Deposits along the eastern edge of the Pilot Basin pass eastward into shallow to intermediate subtidal environments of the Guilmette Formation, whereas deposits along the western edge of the basin probably pass into deeper-water deposits of the Devils Gate Limestone (Fig. 2; Sandberg et al., 1989).

Vertical facies relationships within meter-scale cycles and depositional sequences in the Guilmette Formation permit the interpretation of lateral paleoenvironmental relationships across the study area. During times of maximum flooding, tidal-flat environments passed seaward into restricted shallow subtidal environments characterized by low faunal diversity and generally low-energy conditions, which were episodically interrupted by storm events (Fig. 6). Restricted shallow subtidal environments passed seaward into a broad belt of shallow subtidal facies with stromatoporoid biostromes, peloidal-skeletal packstone shoals, and muddy skeletal sands, which contained more diverse skeletal assemblages (Fig. 6). Seaward of shallow subtidal environments, water depths increased to intermediate subtidal environments as did faunal diversities and abundances. Intermediate subtidal environments graded seaward into deep subtidal waters characterized by poorly oxygenated, substorm wave base conditions (Fig. 6). During times of regional regression, the central platform region was dominated by tidal-flat and restricted shallow subtidal environments; shallow subtidal through deep subtidal environments lay farther to the west.

The Guilmette Formation is dominated by fine-grained sediments; wave- and current-generated features such as cross-bedding, scours, and winnowed grainstones are rare. This suggests

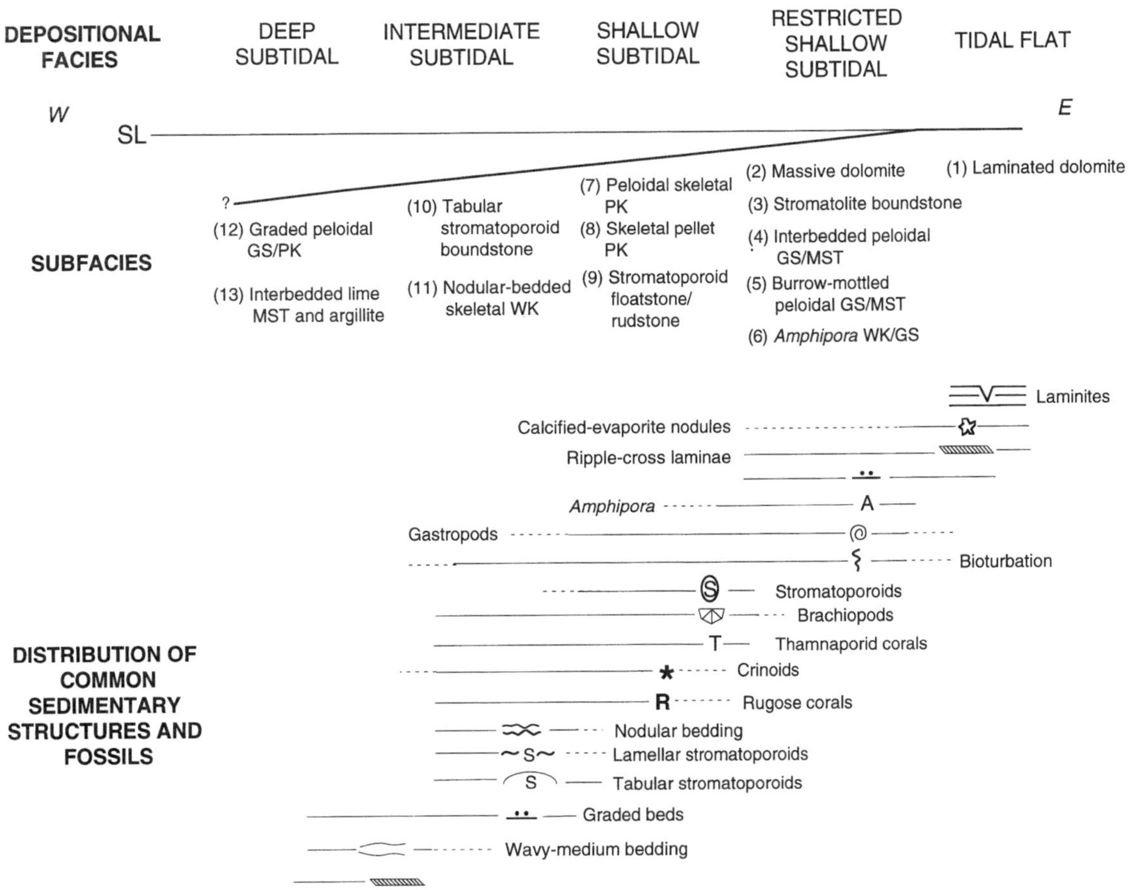

Figure 6. Interpreted lateral distribution of subfacies across the upper Middle to Upper Devonian platform inferred from vertical facies relationships in the Guilmette Formation in the study area. The lateral distribution represents interpreted conditions during maximum flooding when intermediate through deep subtidal environments were deposited along the central platform.

either that fair-weather waves were weak as a result of the wave-dampening across the broad shallow platform or that wave energy was reduced by platform margin reefs and/or shoals. The low-energy conditions along the central platform resulted in abundant bioturbating organisms, which thoroughly homogenized the majority of shallow subtidal deposits. Similar low-energy conditions were interpreted for Upper Devonian tidal-flat through shallow-subtidal deposits in Idaho, Wyoming, and Montana (Dorobek, 1991), and western Utah (Larsen, 1989), suggesting regional low-energy conditions during Late Devonian time.

DEPOSITIONAL SEQUENCES

Methods used to identify depositional sequences in structurally isolated outcrops typical of the Great Basin region are different from those originally used to define seismic-scale sequences, because onlapping and downlapping stratal geometries are difficult to recognize in strata deposited along very low paleoslopes and in laterally discontinuous outcrops. The Guilmette Formation sequences were identified from vertical changes in depositional facies, changes in cycle stacking patterns, and from the stratigraphic distribution of subaerial exposure features (e.g., Montañez and Osleger, 1993; Elrick, 1996). The stratal stacking patterns at individual sections were correlated between sections, then were independently verified using conodont biostratigraphy. Eleven transgressive-regressive sequences (Sequences 1 through 11) are recognized. According to the timescales of Palmer (1983) and Harland et al. (1989), the sequences are ~0.5 to 3.2 m.y. in duration; therefore, they are fourth- to third-order in scale (Vail et al., 1977).

The bed-by-bed resolution available from this outcrop study reveals that sequence and systems tract boundaries are generally gradational over meters to tens of meters (Montañez and Osleger, 1993). These gradational boundaries differ from those interpreted from seismic studies because typical seismic data have the spatial resolution of a few tens of meters; thus boundaries between sequences or systems tracts appear as surfaces rather than stratigraphic intervals or zones.

The first method used to identify depositional sequences was to construct a paleobathymetric curve of a complete section (composite of Sections SCH and ER). To construct a curve that can be interpreted at the sequence scale, the five depositional facies were combined to form three major facies groups; tidal-flat and restricted shallow subtidal facies compose the first group, shallow subtidal facies forms a group on its own, and intermediate and deep subtidal facies are combined to form the third group. The paleobathymetric curve shown in Figure 7 illustrates the 11 transgressive-regressive sequences identified in the Guilmette Formation. Sequence 1 begins in the Fox Mountain Member of the Simonson Dolomite, and Sequence 11 ends ~ 3 m below the contact with the overlying West Range Limestone.

The transgressive or upward-deepening portion of sequences (transgressive systems tract; TST) ends in the maximum flooding zone (MFZ), which records the deepest water depths attained for each sequence. The regressive or upward-shallowing portion of sequences (highstand systems tract; HST) ends at the sequence boundary zone (SBZ), which records the effects of minimum long-term accommodation. Lowstand systems tracts are not recognized along the central platform region.

Changes in cycle stacking patterns also aided in sequence identification and correlation. Cycle stacking patterns are best illustrated using Fischer plots (Fig. 8), which are a graphical means

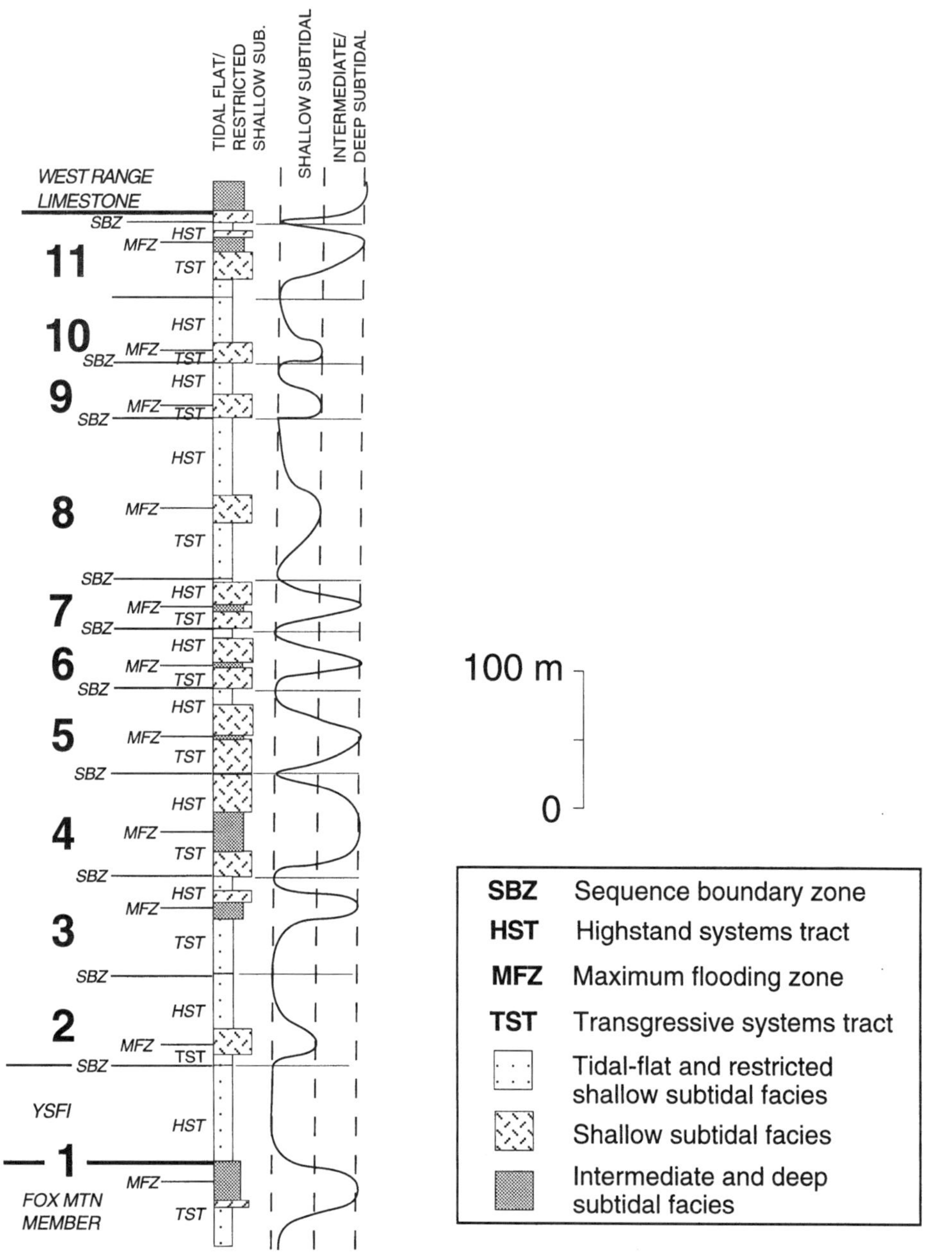

Figure 7. Relative paleobathymetric curve for composites of Sections SCH and ER (includes only the basal 60 m of Section ER) illustrating Sequences 1 through 11.

of representing cumulative departure from mean cycle thickness for a given cyclic stratigraphic section (Read and Goldhammer, 1988; Sadler et al., 1993). Successions of thicker-than-average cycles form rising limbs on Fischer plots and are interpreted as the result of increased accommodation space during long-term relative sea-level rises. Successions of thinner-than-average cycles form falling limbs on the plots and are interpreted to reflect decreased accommodation space during long-term relative falls. The wave forms of Fischer plots composed of peritidal cycles have been interpreted to represent long-term variations in accommodation space and can be a valuable tool for correlating between isolated stratigraphic sections and between sections that lack sufficient biostratigraphic control; when single plots are combined with cycle subfacies data, they provide information for sequence identification in single sections (e.g., Read and Goldhammer, 1988; Goldhammer et al., 1993; Montañez and Osleger, 1993).

Fischer plots were constructed for the nearly complete Section SCH and plotted with histograms of percentage subfacies per cycle (Fig. 8); the result was then compared to Fischer plots from partial Sections ER, GM, and SP to aid in sequence identification and correlation.

Two types of depositional sequences are recognized, utilizing the methods outlined above. *Catch-up sequences* (Sequences 1, 3, 4, 5, 6, 7, and 11) are characterized by TSTs that contain intermediate to deep subtidal facies either as noncyclic intervals or within individual cycles. The presence of these deeper water facies indicates that sedimentation rates along the central platform did not keep pace with the combined effects of fifth- through third-order accommodation gains (catch-up style sedimentation of Schlager, 1981). *Keep-up sequences* (Sequences 2, 8, 9, and 10) deepen only to shallow subtidal water depths during TST development; this implies that, in general, sedimentation rates along the central platform kept pace with combined fifth- through third-order accommodation gains; that is, the platform remained aggraded throughout sequence development (keep-up style sedimentation of Schlager, 1981).

The following sections illustrate how sequences were recognized and correlated across the study area using the methods outlined above. For brevity, the discussion focuses on Sequences 3, 4, 5, and 6; however, the same methods used to identify these specific sequences were applied to the remaining sequences. Figure 9 illustrates the lateral and vertical facies

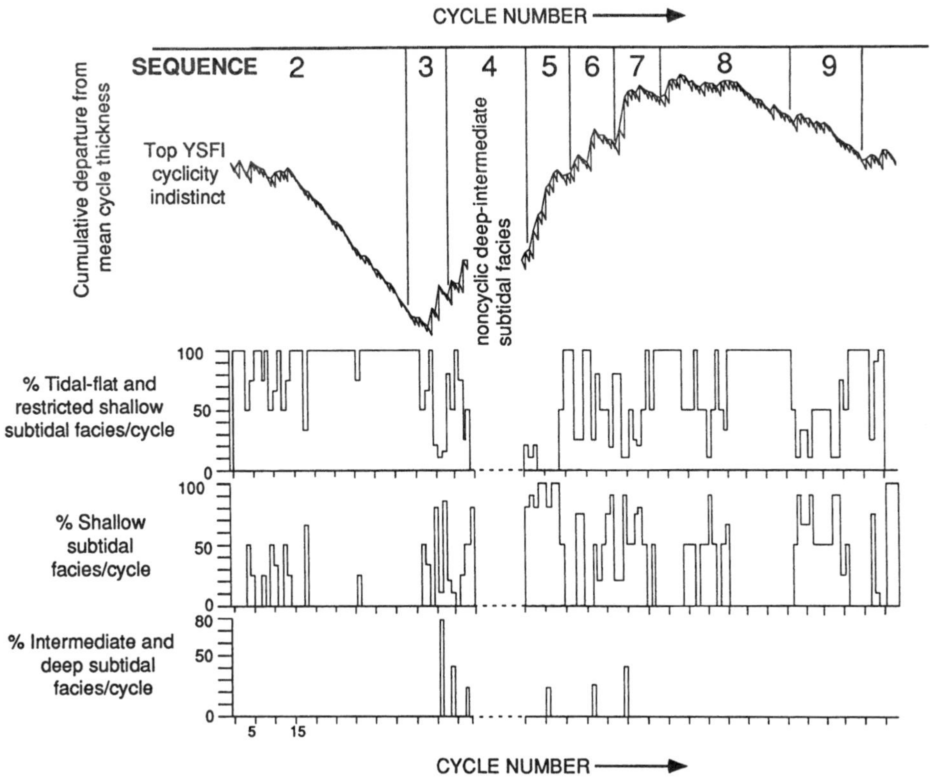

Figure 8. Fischer plot of Section SCH and histograms of percent subfacies per cycle. Plot begins at the top of the yellow, slope-forming interval (YSFI) and ends at the base of Sequence 10, as a result of poor exposure of cycles. In general, falling limbs of the Fischer plot are coincident with thinner-than-average cycles dominated by tidal-flat and restricted shallow subtidal facies, indicating a loss in third-order accommodation space. Rising limbs are associated with thicker-than average cycles, which are dominated by shallow subtidal facies, indicating a gain in third-order accommodation space. Note that the horizontal width of sequences on the plot is related to number of cycles per sequence rather than to sequence thickness.

relationships, cycle stacking patterns, and systems tracts of Sequences 3, 4, 5, and 6.

Depositional Sequence 3

The description of Sequence 3 begins with MFZ 3, which represents the maximum water depths attained along the central platform during Sequence 3 development. MFZ 3 is composed of a distinctive succession (<6 m thick) of noncyclic, graded peloidal grainstone to packstone subfacies (deep subtidal facies); MFZ 3 at Section SP is covered by Quaternary alluvium (Fig. 9). Overlying MFZ 3, peritidal and shallow subtidal cycles, which are dominated by *Amphipora* wackestone/grainstone (restricted shallow subtidal facies) and tidal-flat facies, suggest a loss in third-order accommodation space (HST 3). Third-order accommodation space minimum (sequence boundary zone; SBZ 3) is defined by a succession of 2 to 3 peritidal cycles, which across the study area are immediately overlain by a relatively thick, subtidal-facies dominated cycle (Fig. 9).

Depositional Sequence 4

Upward-deepening facies trends related to TST 4 are variably expressed at each of the sections (Fig. 9). At Section GM, upward-deepening is reflected by shallow subtidal cycles overlain by a deep subtidal cycle, which is, in turn, overlain by an ~18-m-thick interval of noncyclic, nodular-bedded skeletal wackestone subfacies (intermediate subtidal facies). An increase in third-order accommodation space at Section ER is indicated by the underlying SBZ 3 overlain by a thick, shallow subtidal cycle followed by nodular-bedded skeletal wackestones (intermediate subtidal facies). At Section SCH, upward-deepening is expressed as relatively thick peritidal cycles, two of which deepen to intermediate and or deep subtidal water depths, overlain by noncyclic nodular-bedded skeletal wackstone. TST 4 is partially exposed at Section SP and is composed of stromatoporoid boundstone (shallow subtidal facies) overlain by noncyclic nodular-bedded skeletal wackestones (intermediate subtidal facies). Within the noncyclic intervals at Sections GM and ER, a <3-m-thick interval of interbedded lime mudstone and argillite (deep subtidal facies) is interpreted as MFZ 4. At Section SCH, MFZ 4 is defined as the thinnest-bedded interval within a succession of noncyclic, nodular-bedded skeletal wackestone (intermediate subtidal facies). MFZ 4 at Section SP is covered.

Upward-shallowing facies trends or third-order accommodation space loss (HST 4) at each section is indicated by deep subtidal cycles gradationally overlain by shallow subtidal and/or peritidal cycles. Minimum accommodation space related to SBZ 4 is defined by a succession of thinner-than-average peritidal cycles at each section (Fig. 9). At Section ER, prolonged subaerial exposure at the top of SBZ 4 is indicated by the occurrence of a monomict breccia and an associated erosion surface with 5 to 10 cm of relief.

Depositional Sequence 5

Upward-deepening (TST 5) is indicated at Sections GM and SCH by a 10-m-thick cycle dominated by shallow subtidal facies immediately overlying relatively thin peritidal cycles of SBZ 4. At Sections ER and SP, TST 5 is represented by a thick succession of noncyclic stromatoporoid floatstone (shallow subtidal facies). Maximum water depths (MFZ 5) are reflected at Sections GM, SCH, and SP by a 3- to 5-m-thick interval of nodular-bedded, skeletal wackestone (intermediate subtidal facies) contained within a single cycle. MFZ 5 is covered at Section ER.

Loss of third-order accommodation space (HST 5) is indicated at each section by upward-shallowing facies trends. At Sections GM and ER, HST 5 is composed of a succession of peritidal cycles that become progressively more restricted in nature upsection (Fig. 9). At Sections SCH and SP, upward-shallowing is indicated by cycles with bases composed of stromatoporoid floatstone/rudstone subfacies (shallow subtidal facies) overlain by peritidal cycles dominated by restricted shallow subtidal facies.

SBZ 5 at Sections GM, SCH, and SP is defined as a succession of thin peritidal cycles dominated by restricted shallow subtidal facies immediately overlain by a relatively thick cycle composed dominantly of shallow to intermediate subtidal facies. The top of Section ER is faulted out; however, an 8-m-thick interval of amalgamated tidal-flat laminites suggests that these deposits represent a minimum in long-term accommodation space (Fig. 9).

BIOSTRATIGRAPHIC CORRELATIONS

Nineteen samples from the four measured sections (Fig. 10) were collected for conodonts to verify correlations based on sequence stratigraphic techniques and to evaluate how Middle to

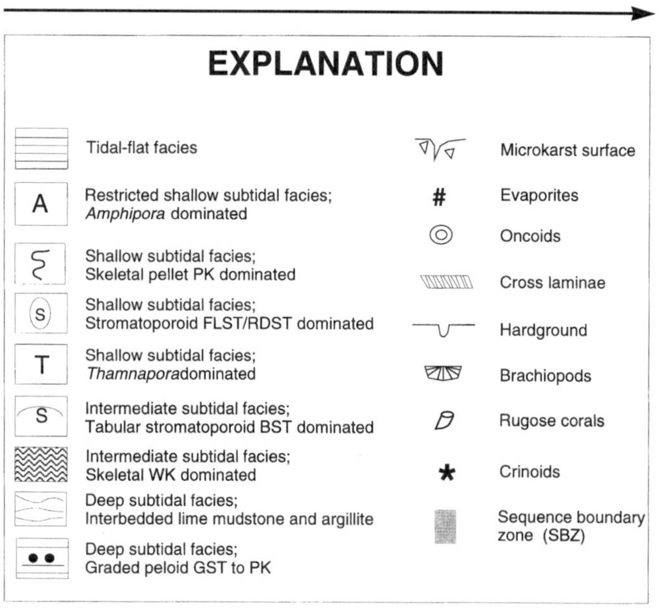

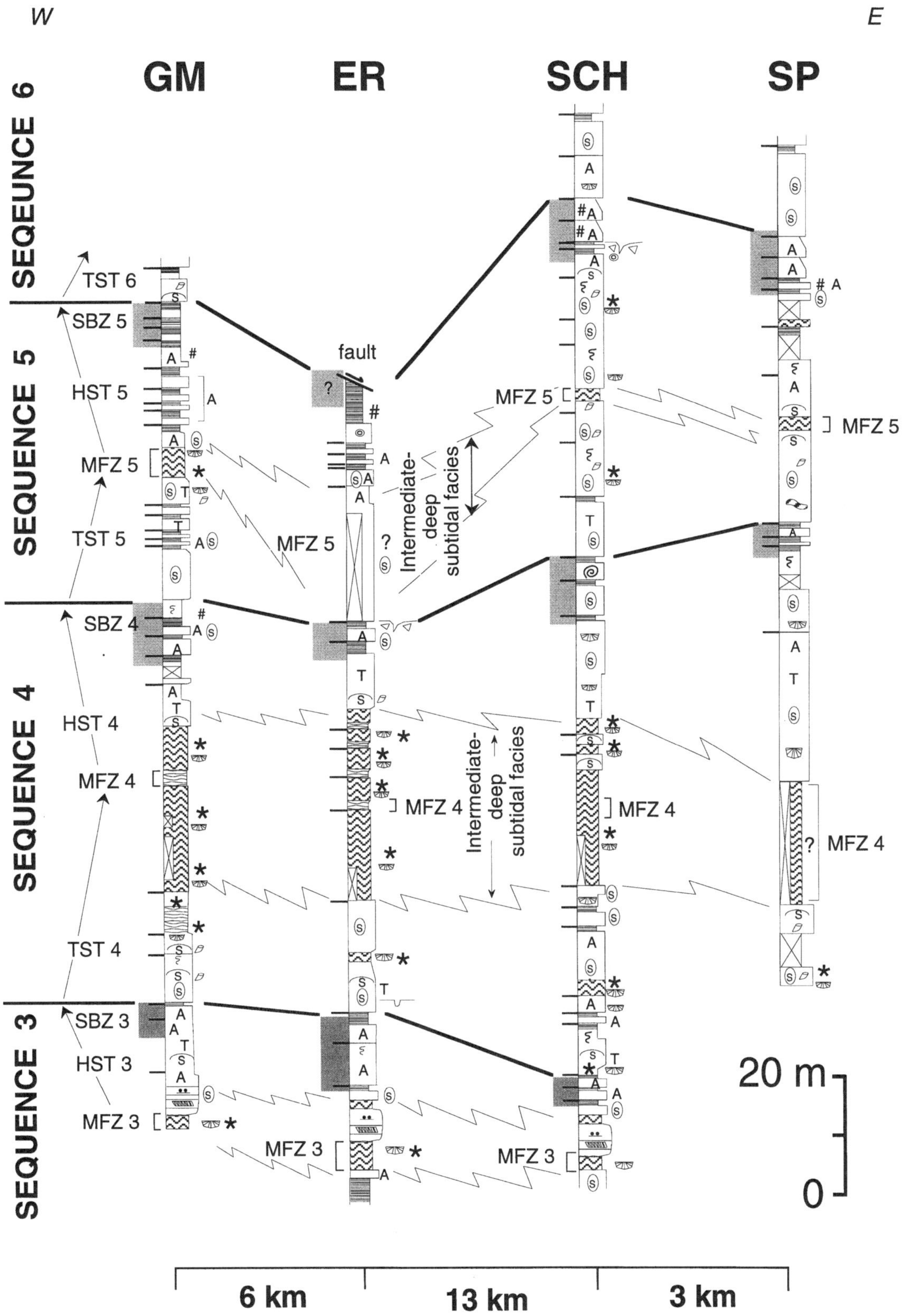

Figure 9. Partial stratigraphic columns of Sequences 3 through 6 illustrating systems tracts and boundaries; see text for detailed description of sequences. Note that present distances between sections are shown because of uncertainties in the amount of Basin and Range extension within the study area. FLST/RDST = floatstone/rudstone; BST = boundstone. See Figure 4 caption and Figure 7 for additional abbreviations.

Upper Devonian sequences of the Guilmette Formation correlate with the third-order Devonian eustatic sea-level curve of Johnson et al. (1985; 1991) (Fig. 11). Results from the conodont data are summarized in Table 2.

Conodont samples that yield a Givetian age were assigned to the standard conodont zonation used by Klapper (1977), Ziegler and Klapper (1982), and Johnson (1990). Two different conodont zonation schemes exist for the Frasnian. Most Devonian studies from western North America employ the standard conodont zonation originally proposed by Ziegler (1971) and used by Johnson et al. (1985; 1991) and Sandberg et al. (1989). More recent work in the western United States (Johnson and Klapper, 1992; Johnson et al., 1996), the midcontinent, and western Canada (Day, 1994; 1996; Day et al., 1996) has successfully employed the Montagne Noire (M.N.) zonal scheme proposed by Klapper (1989). Frasnian samples from this study are zoned according to the M.N. zonation to facilitate correlation with the work of Day (1994; 1996) and Day et al. (1996). Figure 2 illustrates the relationship of conodont zonations to the Middle to Upper Devonian stratigraphy in the eastern Great Basin.

Resultant data from the conodonts collected for this study do not bracket the specific upper and lower zonal limits of the Guilmette sequences; however, by integrating the conodont identifications with sequence stratigraphic interpretations, the sequences can be evaluated in terms of the Devonian eustatic events outlined by Johnson et al. (1985; 1991) and modified by Day (1994). In particular, Sequences 1, 2, 3, and 4 appear to correlate with their T-R cycles IIa-1, IIa-2, IIb, and IIc, respectively (Fig. 11). To date, biostratigraphically diagnostic conodonts have not been recovered from Sequences 5, 6, 8, 9, 10, and 11, and specific stratigraphic intervals representative of T-R cycle IId have not been biostratigraphically identified in the study area. Conodonts indicative of T-R cycle IIe have been recovered from the uppermost Guilmette Formation and the basal West Range Limestone.

The following sections discuss the relationships of the Guilmette sequences to specific characteristics of the Devonian sea-level curve of Johnson et al. (1985; 1991) (Fig. 11).

T-R cycle IIa

The upper Simonson Dolomite and lower Guilmette Formation were postulated to represent the Taghanic onlap by Johnson and Murphy (1984) and Johnson et al. (1991). Hurtubise (1989) first described the significant open-marine character of the Fox Mountain Member of the upper Simonson Dolomite; however, specific biostratigraphic evidence indicating the Middle *varcus* Subzone (age of initial Taghanic onlap; Johnson et al., 1985) for these units has not been demonstrated. In this study, the conodont species *Icriodus brevis* and *Polygnathus xylus xylus* (Table 2, ER-57) collected from the upper, open-marine part of the Fox Mountain Member (TST 1 and MFZ 1) indicate a Givetian age probably within the *varcus* Zone, although these conodonts range throughout the zone and thus do not permit a finer identification (Klapper, personal communication, 1995, and comment in Table 2). From this approximate determination and, more especially, the significant open-marine character of the upper Fox Mountain Member, this part of the sequence may be interpreted to represent the Taghanic onlap along the central platform (T-R cycle IIa-1) (Fig. 10).

The catch-up style of Sequence 1 indicates that the initial Taghanic transgression was rapid enough to significantly backstep environments, or that suppressed sedimentation rates during the transgression resulted in extensive retrogradation. TST 1 and MFZ 1 are relatively thick with respect to overlying TSTs, suggesting that sedimentation rates during transgression were not suppressed; rather, the transgression was sufficiently rapid to generate catch-up style sedimentation patterns.

Highstand shallowing in Sequence 1 is represented by the yellow, slope-forming interval (YSFI), which has been recognized throughout eastern Nevada and western Utah (Fig. 10; Hurtubise, 1989). The upper portion of the YSFI is composed of peritidal cycles capped by laminated, dolomitic siltstone containing abundant mudcracks and represents a significant shallowing event following the initial Taghanic onlap. Diagnostic faunas are absent from this environmentally restricted unit; however, conodonts collected from the underlying Fox Mountain Member (MFZ 1) and the overlying MFZ 2 allow us to postulate a zonal span for the YSFI ranging from the *hermanni* Zone to the *dispar-*

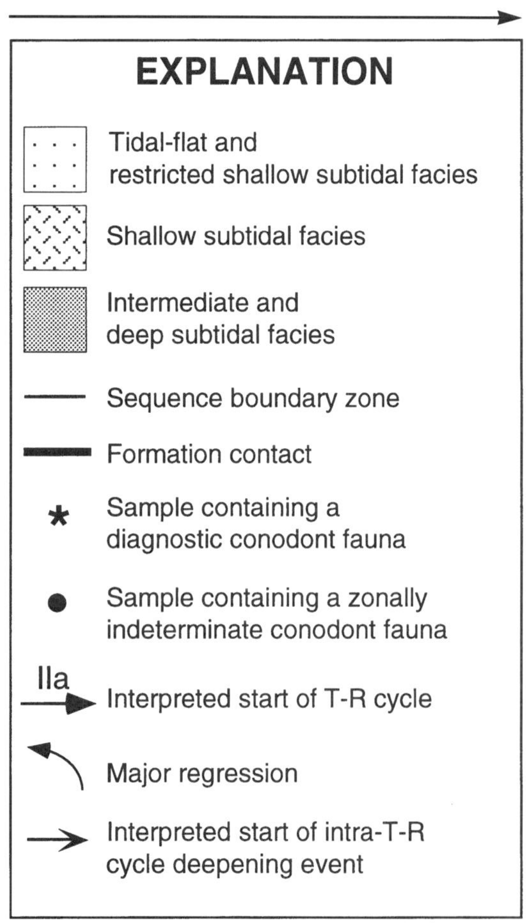

ilis Zone (Fig. 10). In the midcontinent and in western Canada, Day et al. (1996) and Day (personal communication, 1994) recognize significant shallowing from the Upper *hermanni* Zone to the Lower *disparilis* Zone and interpret this as the result of T-R cycle IIa-1 regression. These stratigraphic and biostratigraphic relationships suggest that the YSFI represents T-R cycle IIa-1 regression (late Givetian) (Fig. 10).

The relationship of MFZ 2 lying above probable *varcus* Zone strata and below strata of T-R cycle IIb (MFZ 3) suggests that Sequence 2 deepening is in the *disparilis* Zone. This deepening is tentatively correlated with upper *disparilis* Zone IIa-2 deepening, as recognized in the midcontinent (Fig. 10; Day, 1994; Day et al., 1996).

HST 2 represents a major regression in the study area as indicated by successions of thinner-than-average, exposure-capped peritidal cycles (Fig. 8). The exposure-capped cycles indicate multiple periods of subaerial exposure during HST 2. This regression is recorded in upper T-R cycle IIa strata along the outer platform in central Nevada (Johnson et al., 1985, fig. 2). In addition, Day et al. (1996) recognize widespread regression, karstification, and evaporite deposition in late *disparilis* time (upper T-R cycle IIa-2) in the midcontinent and western Canada.

T-R cycle IIb

The stratigraphically lowest occurrence of nodular-bedded skeletal wackestone (intermediate subtidal facies) in Sequence 3 contains the conodont species *Playfordia primitiva*, suggesting that the deposits lie in the range of M.N. Zones 2 through 4 (T-R cycle IIb; Fig. 10). This is the first catch-up style sequence since the Taghanic onlap (Sequence 1) and represents significant deepening or backstepping of environments. The poorly developed HST at the top of Sequence 3 is consistent with a geographically widespread, minor regression between T-R cycles IIb and IIc (Johnson et al., 1985).

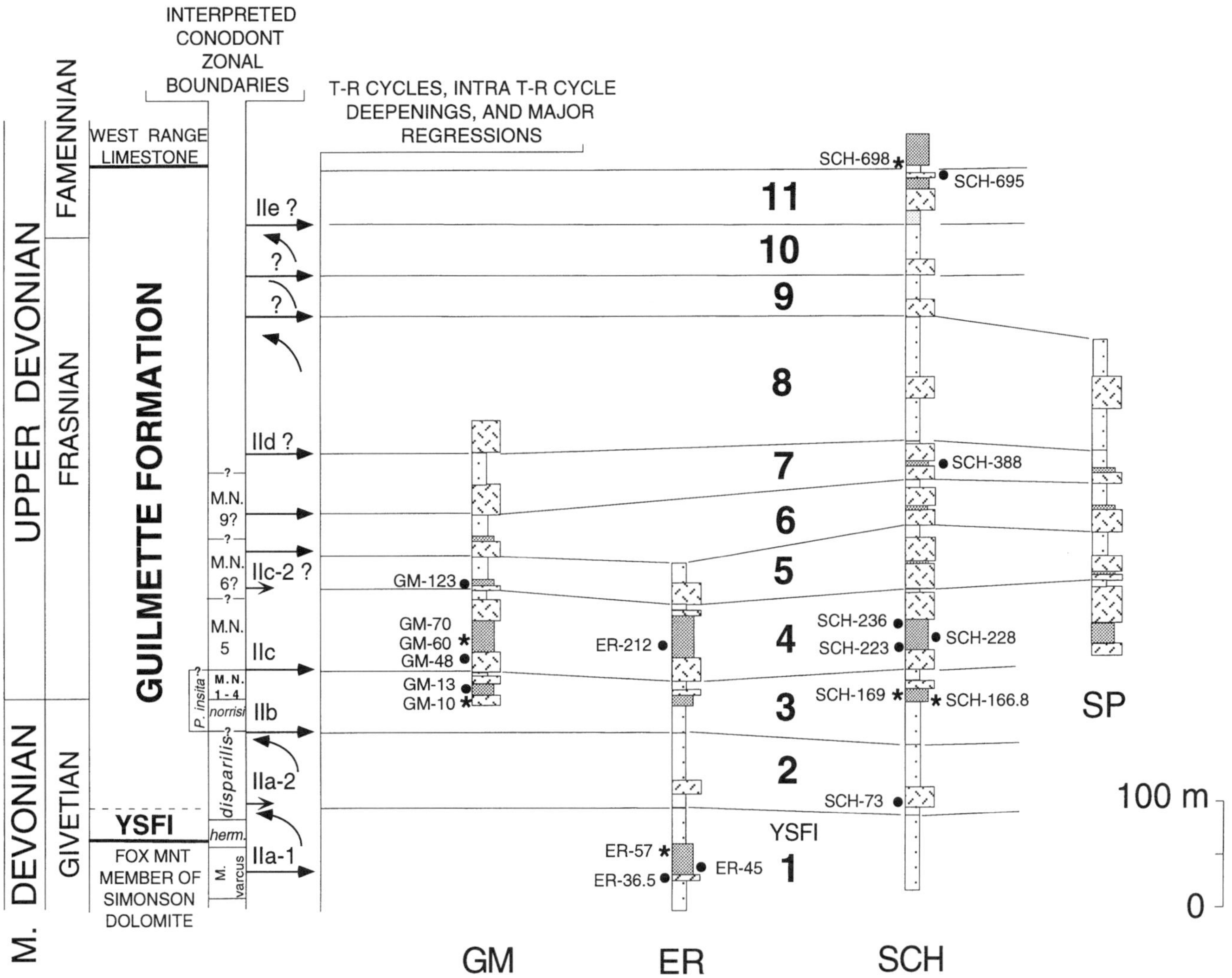

Figure 10. Interpreted correlation of sequences and correlations with T-R cycles of Johnson et al. (1985; 1991), Day (1994), and Day et al. (1996) based on sequence stratigraphic and biostratigraphic analysis. Numbers and symbols on side of columns refer to position of conodont samples (Table 2).

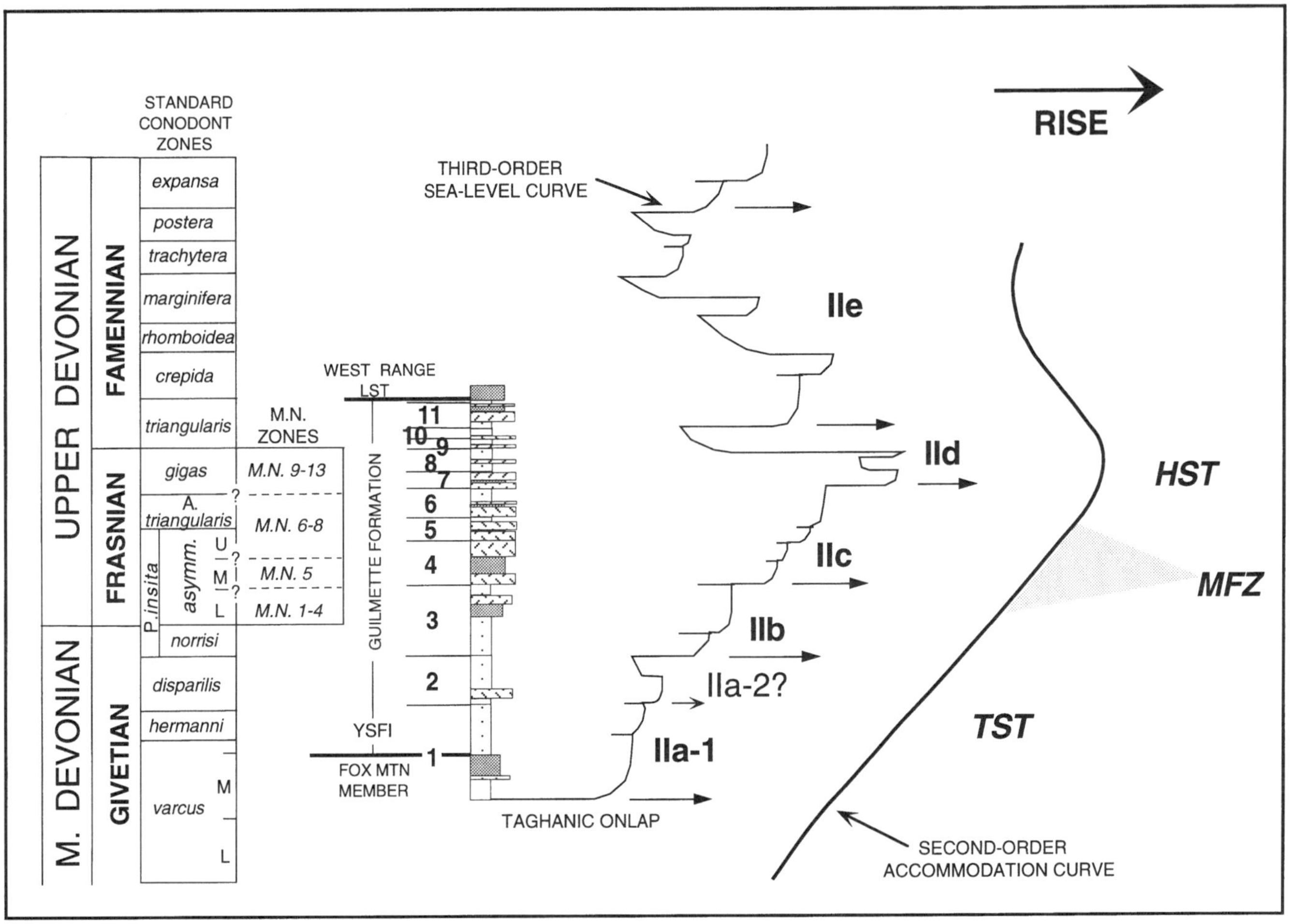

Figure 11. Interpreted correlation of Guilmette Sequences 1 through 11 with the eustatic sea-level curve of Johnson et al. (1985, 1991) and as modified by Day (1994). Note that the actual thickness of the stratigraphic column has been modified to fit the biostratigraphically controlled eustatic curve. The second-order accommodation curve represents the envelope of third-order highstand positions and reflects the changes in accommodation space associated with the Kaskaskia sequence of Sloss (1963). The second-order TST is composed of catch-up style sequences (excluding Sequence 2); second-order MFZ is composed of catch-up sequences with relatively thick intervals of deep to intermediate subtidal facies indicating maximum rates of second-order accommodation space gain. The second-order HST is composed of stacked, keep-up style Sequences 8 through 10. Deepening related to Sequence 11 may record renewed catch-up style sedimentation related to the ensuing second-order accommodation space gain or may reflect increased subsidence related to the Antler orogeny. Symbols explained in Figure 7.

T-R cycle IIc

MFZ 4 and TST 4 yield conodonts including *Mesotaxis asymmetrica*, *Ancyrodella africana*, *Mesotaxis johnsoni*, and *Ancyrodella gigas* form 1; *M. johnsoni* (Table 2, GM-70) indicates that the MFZ 4 lies in M.N. Zones 5 through 7 (T-R cycle IIc; Fig. 10). TST 4/MFZ 4 is composed of 20 to 30 m of weakly cyclic to noncyclic intermediate to deep subtidal facies and is recognized throughout the study area as a thick, recessive-weathering interval. The open-marine character of this thick, deep-water interval suggests that this transgressive event was as significant and sustained as the initial Taghanic onlap (T-R cycle IIa-1).

Johnson et al. (1985) did not recognize a T-R cycle IIc deepening in western North America; consequently, it was not used in the construction of that part of the Devonian sea-level curve. However, in subsequent papers, Johnson and Sandberg (1989) and Sandberg et al. (1989) suggested that T-R cycle IIc was represented by the lower Guilmette Formation (Sequences 1 and 2 of this study). In particular, Sandberg et al. (1989, p. 191) suggested that the upper portions of the YSFI were representative of T-R cycle IIc on the central platform. The results of this study demonstrate that the YSFI is older than T-R cycle IIc.

Results from this study indicate that T-R cycle IIc deepening (TST 4/MFZ 4) resulted in the deposition of thick, noncyclic intermediate to deep subtidal facies, and that the magnitude of T-R cycle IIc deepening along the central platform has been previously underestimated. Future sedimentologic and biostratigraphic studies of T-R cycle IIc strata in the Great Basin will lead to a better understanding of the nature of this event.

Shallowing at the top of Sequence 4 (HST 4) is weakly developed and is marked by three to four, thin peritidal cycles above a thick succession of intermediate to deep subtidal facies. This minor shallowing is poorly resolved on the Fischer plot (Fig. 8) and is interpreted to correlate with geographically widespread, weak regressive trends (Johnson et al., 1985; 1991).

T-R Cycles IId-IIe

Sequences 5, 6, and 7 are catch-up style sequences; however, unlike catch-up Sequences 1, 3, and 4, deepening to intermediate or deep subtidal water depths is limited to the basal portions of a single cycle. This suggests that although deepening was pronounced enough to backstep environments, sedimentation quickly filled accommodation space (catch-up followed by keep-up sedimentation).

The zonally indeterminate conodonts *Polygnathus aspelundi* and *Ozarkodina postera* from MFZ 7 (Table 2, SCH-388) constrain a zonal span no finer than M.N. Zones 7 through 13 (upper T-R cycle IIc through IId; Fig. 10). Because Sequences 5, 6, and 7 lie above MFZ 4, which is interpreted as representing T-R cycle IIc transgression, then either MFZ 5, MFZ 6, or MFZ 7 may represent T-R cycle IId transgression. Additional conodont sampling is necessary to refine this interpretation.

According to Johnson et al. (1985; 1991), T-R cycle IId ends with a major eustatic sea-level fall, and they note that this regression is reflected by *Amphipora*-rich deposits overlain by shallow subtidal, quartz sandstones and supratidal laminites (Johnson and Sandberg, 1989; Johnson et al., 1991). In the study area, *Amphipora* wackestone-grainstone subfacies overlain by quartz sand-bearing, tidal-flat facies form Sequences 8, 9, and 10 and are interpreted to represent regression related to the end of T-R cycle IId (Fig. 10).

Sequence 11 records the first significant deepening to open-marine conditions (MFZ 11; nodular-bedded skeletal wackestone subfacies) since Sequence 7. Lower Famennian but zonally indeterminate conodonts were recovered from the top of this interval (Table 2, SCH-695). However, at Section SCH (698) immediately above Sequence 11, *Palmatolepis wolskae* was recovered from the basal West Range Limestone (Fig. 10), indicating the Middle-Upper *crepida* Zone of the lower Famennian (T-R cycle IIe). Consequently, Sequence 11 may represent the base of T-R cycle IIe.

The timing of sedimentation influenced by Antler-age deformation in the study area is not known; however, regional sedimentation patterns in the overlying West Range Limestone suggest influence by Antler-related tectonism (Goebel, 1991). As a consequence, Sequence 11 deepening (MFZ 11) may represent tectonically induced subsidence related to Antler deformation. Additional data from conodont biostratigraphy and regional stratigraphy are needed to evaluate when tectonically influenced deposition first affected the central platform region.

Upper HST 11 is composed of dolomitic siltstone subfacies and is interpreted to represent the northern extent of the upper Guilmette Formation siliciclastic facies belt (Hurtubise, 1989). The abundance of siliciclastic material increases to the south (Seaman Range, Hurtubise, 1989) and east (Burbank Hills, Biller, 1976; Larsen et al., 1989); however, quartz sand-bearing beds are absent in ranges directly to the north and west (Niebuhr, 1980; Hurtubise, 1989). The overlying West Range Limestone also contains abundant fine sand- and silt-size quartz grains, suggesting that siliciclastic deposition may reflect local and/or regional uplift and a shift in drainage patterns due to Antler tectonism or a climate change.

DISCUSSION

Results from the combined sequence stratigraphic and biostratigraphic correlations across the study area indicate that four and perhaps five of the Guilmette sequences (Sequences 1, 2, 3, 4, 11?) represent several T-R cycles previously recognized by Johnson et al. (1985; 1991) and Day (1994). The other six (or seven) sequences (Sequences 5 through 10 and perhaps 11?) represent smaller-scale, transgressive-regressive events within T-R cycles (fourth-order in scale). These smaller-scale sequences would be less evident in outer platform, slope, and basinal environments where Johnson et al. (1985) originally defined their curve because in these regions the seafloor lay too deep to respond to fourth-order sea-level oscillations. Along the shallow platform of the study area, slight fluctuations in water depth resulted in significant changes in current, wave, and circulation patterns, producing relatively distinct sedimentologic changes.

Stacking patterns of sequence-scale facies trends provide information about changes in long-term or second-order accommodation space changes related to the Kaskaskia sequence of Sloss (1963). With the exception of Sequence 2, Sequences 1 through 7 record catch-up style sedimentation during fourth- to third-order transgressions and maximum flooding; that is, during late TSTs and MFZs sedimentation rates did not keep pace with accommodation space gains, and intermediate to deep subtidal water depths were attained along the central platform (defining the second-order TST; Fig. 11). In particular, Sequences 3 and 4 contain the thickest successions (10 to 40 m) of intermediate to deep subtidal facies, suggesting that the rate of long-term accommodation space gain was greatest during these intervals (defining the second-order MFZ; Fig. 11). Overlying Sequences 8 through 10 reflect keep-up style sedimentation; that is, water depths in this region did not exceed shallow subtidal water depths (~10 to 20 m), and the central platform remained nearly aggraded to sea level throughout sequence development because sedimentation rates kept pace with accommodation space gains. These facies patterns suggest decreased rates of long-term accommodation space gain (second-order HST; Fig. 11). Transgression and maximum flooding during Sequence 11 record renewed catch-up style sedimentation, indicating that sedimentation rates lagged behind accommodation

TABLE 2. SUMMARY OF CONODONT DATA FROM GUILMETTE FORMATION

Sample	Species Identified	Conodont Zone	Comments
ER-36.5	*Polygnathus* sp. indet.	Zonally indeterminate	
ER-45	Indet. ramiform element	Indeterminate	
ER-57	*Icriodus brevis* Stauffer *Polygnathus xylus xylus* Stauffer	*Varcus* Zone (unrestricted)	An immediately adjacent zone cannot be excluded.
ER-212	Barren of conodonts		
GM-10	*Playfordia primitiva* (Bischoff and Ziegler) *Pandorinellina insita* (Stauffer) *Polygnathus angustidiscus* Youngquist *Polygnathus* sp. *Schmidtognathus*? sp. indet. Indet. ramiform elements	M.N. Zones 2–4 (within T-R IIb)	On evidence of range of *Playfordia primitiva.*
GM-48	*Polygnathus angustidiscus* Indet. ramiform elements	Zonally indeterminate	Identified species ranges from *disparilis* Zone (u. Givetian) through the Frasnian.
GM-60	*Ancyrodella africana* Garcia-López *Ancyrodella gigas* Youngquist form 1 of Klapper, 1989 *Mesotaxis asymmetrica* (Bischoff and Ziegler) *Polygnathus alatus* Huddle *Polygnathus* sp. *Icriodus subterminus* Youngquist Indet. ramiform elements	Upper part of M.N. Zone 4 – Zone 6 (T-R IIb—IIc)	On evidence of combined range of first two species.
GM-70	*Mesotaxis johnsoni* Klapper et al. *Ancyrodella gigas* form 1 *Polygnathus* sp. *Icriodus subterminus*	M.N. Zones 5–7 (= in the range from base of T-R IIc to within IIc)	On evidence of first species, which is most common in Zone 5 but is known to range higher.
GM-123	*Ozarkodina?* sp. indet. (Pa element) Indet. ramiform elements	Indeterminate	
SCH-73	*Icriodus subterminus* Pa element of *Prioniodina* sp. Indet. ramiform elements	Indeterminate	*I. subterminus* ranges from *disparilis* Zone (u. Givetian) through the Frasnian.
SCH-166.8	*Playfordia primitiva* *Polygnathus* sp. indet. *Pandorinellina insita?* Indet. ramiform elements	M.N. Zones 2–4 (within T-R IIb)	On evidence of range of *Playfordia primitiva.*
SCH-169	*Playfordia primitiva*	As preceding sample	One well-preserved specimen.
SCH-223	*Icriodus subterminus* Indet. ramiform elements	Indeterminate	See range of species under SCH-73
SCH-228	Indet. fragment	Indeterminate	
SCH-236	Indet. ramiform elements	Indeterminate	

TABLE 2. SUMMARY OF CONODONT DATA FROM GUILMETTE FORMATION (continued)

Sample	Species Identified	Conodont Zone	Comments
SCH-388	*Polygnathus angustidiscus* *Polygnathus aspelundi* Savage and Funai *Polygnathus* spp. *Ozarkodina postera* Klapper and Lane Indet. ramiform elements	Zonally Indeterminate	*O. postera* ranges in the Frasnian Composite Standard from M.N. Zones 7 to 13. *P. aspelundi* ranges from Zones 8 to 11, but its range is not well constrained.
SCH-695	*Icriodus cornutus* Sannemann *Icriodus* or *Pelekysgnathus* coniform element *Mehlina gradata* Youngquist *Apatognathus?* sp.	Lower Famennian, but zonally indeterminate	*I. cornutus* is a lower Famennian species but is not zonally diagnostic.
SCH-698	*Palmatolepis wolskae* Ovnatanova *Palmatolepis* sp. indet. *Polygnathus* sp. indet. *Icriodus cornutus* *Icriodus* or *Pelekysgnathus* *coniform element* *Apatognathus* sp.	Lower Famennian, Middle to Upper *crepida* Zone (lower part of T-R IIe)	On evidence of reported range of first species.

space gains or perhaps reflecting increased subsidence related to the Antler orogeny.

CONCLUSIONS

1. The late Middle to Upper Devonian Guilmette Formation in the southern Egan and Schell Creek Ranges of Nevada was deposited along a low-energy, westward-deepening carbonate platform. Fifteen subfacies are recognized along the central platform and are grouped into five depositional facies representing tidal flat, restricted shallow subtidal, shallow subtidal, intermediate subtidal, and deep subtidal environments. Tidal-flat through deep subtidal facies are arranged into meter-scale, upward-shallowing peritidal and subtidal cycles, which have average cycle periods between ~30 to165 k.y.

2. Eleven, fourth- to third-order depositional sequences are identified from shallowing and deepening trends of depositional facies, changes in cycle stacking patterns, and the stratigraphic distribution of subaerial exposure features. *Catch-up style* sequences are characterized by TSTs and/or MFZs composed of intermediate to deep subtidal facies; these facies patterns indicate that sedimentation rates lagged behind accommodation space gains. *Keep-up style* sequences deepen only to shallow subtidal water depths, indicating that sedimentation rates kept pace with accommodation space gains.

3. Conodont biostratigraphy and sequence stratigraphy permit correlation between measured sections and correlation with T-R cycles of the Johnson et al. (1985; 1991) and Day (1994) eustatic sea-level curve. The base of Sequence 1 (TST 1; Fox Mountain Member of the Simonson Dolomite) is inferred to represent the initial Taghanic onlap (Middle *varcus* Subzone; T-R cycle IIa-1). Pronounced shallowing recorded in the YSFI is tentatively interpreted to represent regression at the end of T-R cycle IIa-1 (Upper *hermanni*–Lower *disparilis* zones). Deepening and shallowing in Sequence 2 are tentatively correlated with lower and upper T-R cycle IIa-2, respectively (Upper *disparilis* Zone). Sequence 3 represents T-R cycle IIb (M.N. Zones 2 through 4) and T-R cycle IIc is represented by Sequence 4 (M.N. Zone 5 through 7). The available conodont biostratigraphy does not resolve T-R cycle IId; however, sequence stratigraphic relationships suggest that initial deepening of T-R cycle IId may be represented by MFZs of Sequences 5, 6, or 7. Keep-up Sequences 8, 9, and 10 are interpreted to represent regression at the end of T-R cycle IId. Sequence 11 records the first significant deepening to open-marine conditions since Sequence 7. Conodonts recovered from this interval indicate an Early Fammenian age, and conodonts from the immediately overlying West Range Limestone indicate the Middle to Upper *crepida* Zone of the Lower Famennian (T-R cycle IIe). These biostratigraphic and sequence stratigraphic relationships suggest that Sequence 11 may represent the base of T-R cycle IIe.

4. Sequence-scale facies patterns suggest that changes in second-order accommodation space related to the Kaskaskia sequence of Sloss (1963) controlled sequence-scale stacking patterns. In particular, catch-up Sequences 1 through 7 (excluding Sequence 2) record increased rates of second-order accommodation space gain (second-order TST). The accumulation of thick successions of deep to intermediate subtidal facies during TSTs and MFZs of Sequences 3 and 4 reflects the effects of second-order maximum flooding. Sequences 8 through 10 record keep-up style sedimentation reflecting decreased rates of second-order accommodation space gain (second-order HST).

ACKNOWLEDGMENTS

This chapter represents a portion of the senior author's masters thesis at the University of New Mexico (UNM). Funding was provided by the Student Resources Allocations Committee at UNM, Alumni Scholarship Fund, Jean-Luc Miossec Scholarship Fund, and PRF grant #25593-G8 from the American Chemical Society. Many thanks go to Johan Forsman, Justin Sieberg, and Kathy Pierce for their help in the field. The manuscript benefited from reviews by Gary Smith, Barry Kues, Daniel Larsen, Cathy Ratcliff, Eric Kirby, Katherine Giles, Jed Day, Cal Stevens, and Gilbert Klapper.

REFERENCES CITED

Abbott, B. M., 1973, Terminology of stromatoporoid shapes: Journal of Paleontology, v. 47, p. 805–806.

Ackman, B. W., 1991, Stratigraphy of the Guilmette Formation, Worthington Mountains and Schell Creek Range, southeastern Nevada [M.S. thesis]: Golden, Colorado School of Mines, 207 p.

Algeo, T. J., and Wilkinson, B. W., 1988, Periodicity of mesoscale Phanerozoic sedimentary cycles and the role of Milankovitch orbital modulation: Journal of Geology, v. 96, p. 313–322.

Biller, E. J., 1976, Stratigraphy and petroleum possibilities of lower Upper Devonian (Frasnian and lower Famennian) strata, southwestern Utah: U.S. Geological Survey Open File Report 76–343, 105 p.

Caputo, M. V., and Crowell, J. C., 1985, Migration of glacial centers across Gondwana during Paleozoic Era: Geological Society of America Bulletin, v. 96, p. 1020–1036.

Cisne, J. L., 1986, Earthquakes recorded stratigraphically on carbonate platforms: Nature, v. 323, p. 320–322.

Day, J., 1994, Late Middle and early Upper Devonian brachiopod faunas of southeastern Iowa and northwestern Illinois, in Bunker, B. J., ed., Paleozoic stratigraphy of the Quad Cities region, East-Central Iowa, northwestern Illinois: Geological Society of Iowa Guidebook 59, p 65–83.

Day, J., 1996, Faunal signatures of Middle-Upper Devonian depositional sequences and sealevel fluctuations in the Iowa Basin: U.S. Midcontinent, in Witzke, B. J., Ludvigson, G. A., and Day, J., eds., Paleozoic sequence stratigraphy: Views from the North American Craton: Geological Society of America Special Paper 306, p. 277–300.

Day, J., Uyeno, T., Norris, W., Witzke, B., and Bunker, B., 1996, Middle-Upper Devonian relative sea-level histories of central and western North American interior basins, in Witzke, B. J., Ludvigson, G. A., and Day, J., eds., Paleozoic sequence stratigraphy: Views from the North American Craton: Geological Society of America Special Paper 360, p. 259–275.

Demicco, R. V., 1983, Wavy and lenticular-bedded ribbon rocks of the Upper Cambrian Conococheague Limestone, central Appalachians: Journal of Sedimentary Petrology, v. 53, p. 1121–1132.

Dorobek, S. L., 1991, Cyclic platform carbonates of the Devonian Jefferson Formation, southwestern Montana, in Cooper, J. D. and Stevens, C. H., eds., Paleozoic paleogeography of the western United StatesÑII: Society of Economic Paleontologists and Mineralogists, Pacific Section, v. 67, p. 509–526.

Elrick, M., 1995, Cyclostratigraphy of Middle Devonian carbonates, eastern Great Basin: Journal of Sedimentary Research, v. B65, p. 61–79.

Elrick, M., 1996, Sequence stratigraphy and platform evolution of Lower to Middle Devonian carbonates, eastern Great Basin; Geological Society of America Bulletin, v. 108, p. 392–416.

Elrick, M. and Hinnov, L. A., 1996, Millenial-scale climate origin for stratification in Cambrian and Devonian deep-water rhythmites, western U.S.A.: Palaeogeography, Palaeoclimatology, Palaeoecology, v. 123, p. 353–372.

Elrick, M., and Read, J. F., 1991, Cyclic ramp-to-basin carbonate deposits, Lower Mississippian, Wyoming and Montana: A combined field and computer modeling study: Journal of Sedimentary Petrology, v. 61, p. 1194–1224.

Elrick, M., Read, J. F., and Çoruh, C., 1991, Short-term paleoclimatic fluctuations expressed in Lower Mississippian ramp-slope deposits, southwestern Montana: Geology, v. 19, p. 799–802.

Fischbuch, N. R., 1968, Stratigraphy, Devonian Swann Hills reef complexes of central Alberta: Bulletin of Canadian Petroleum Geology, v. 16, p. 444–587.

Gans, P. B., 1987, An open-system, two-layer curstal stretching model for the eastern Great Basin: Tectonics, v. 6, p. 1–12.

Giles, K. A., and Dickinson, W. R., 1995, The interplay of eustasy and lithospheric flexure in forming stratigraphic sequences in foreland settings: An example from the Antler foreland, Nevada and Utah, in Dorobek, S. L. and Ross, G. M., eds., Stratigraphic evolution of foreland basins: SEPM (Society of Sedimentary Geology) Special Publication 52, p. 187–212.

Ginsburg, R. N., 1971, Landward movement of carbonate mud: new model for regressive cycles in carbonates [abs.]: American Association of Petroleum Geologists Bulletin, v. 55, p. 340.

Goebel, K. A., 1991, Paleogeographic setting of Late Devonian to Early Mississippian transition from passive to collisional margin, Antler Foreland, eastern Nevada and western Utah, in Cooper, J. D., and Stevens, C. H., eds., Paleozoic paleogeography of the western United StatesÑII: Society of Economic Paleontologists and Mineralogists, Pacific Section, v. 67, p. 401–418.

Goldhammer, R. K., Dunn, P. A., and Hardie, L. A., 1987, High-frequency, glacio-eustatic sea-level oscillations with Milankovitch characteristics recorded in Middle Triassic platform carbonates in northern Italy: American Journal of Science, v. 287, p. 853–892.

Goldhammer, R. K., Dunn, P. A., and Hardie, L. A., 1990, Depositional cycles, composite sea-level changes, cycle stacking patterns, and the hierarchy of stratigraphic forcing: Examples from platform carbonates of the Alpine Triassic: Geological Society of America Bulletin, v. 102, p. 535–562.

Goldhammer, R. K., Lehman, P. J., and Dunn, P. A., 1993, The origin of high-frequency platform carbonate cycles and third-order sequences (Lower Ordovician El Paso Group, west Texas): Constraints from outcrop data and stratigraphic modeling: Journal of Sedimentary Petrology, v. 63, p. 318–360.

Goldstein, R. H., Anderson, J. E., and Bowman, M. W., 1991, Diagenetic responses to sea-level change: Integration of field, stable isotope, paleosol, paleokarst, fluid inclusion, and cement stratigraphy research to determine history and magnitude of sea-level fluctuation, in Franseen, E. K., Watney, W. L., Kendall, G. St. C., and Ross, W., eds., Sedimentary modeling: Computer simulations and methods for improved parameter definition: Kansas Geological Survey Bulletin 233, p. 139–162.

Goodwin, P. W., and Anderson., E. J., 1985, Punctuated aggradational cycles: A general hypothesis of episodic stratigraphic accumulation: Journal of Geology, v. 93, p. 515–533.

Harbaugh, D. W., and Dickinson, W. R., 1981, Depositional facies of Mississippian clastics, Antler foreland basin, central Diamond Mountains, Nevada: Journal of Sedimentary Petrology, v. 51, p. 1223–1234.

Hardie, L. A., and Shinn, E. A., 1986, Carbonate depositional environments, modern and ancient. 3: Tidal Flats: Colorado School of Mines Quarterly, v. 81, 74 p.

Hardie, L. A., Dunn, P. A., and Goldhammer, R. K., 1991, Field and modeling studies of Cambrian carbonate cycles, Virginia Appalachians — Discussion: Journal of Sedimentary Petrology, v. 61, p. 636–646.

Harland, W. B., Armstrong, R.L., Cox, A. V., Craig, L. E., Smith, A. G., and Smith D. G., 1989, A geologic time scale: Cambridge, Cambridge University Press, 263 p.

Heckel, P. H., 1983, Diagenetic model for carbonate rocks in Midcontinent Pennsylvanian eustatic cyclothems: Journal of Sedimentary Petrology, v. 53, p. 733–759.

Heckel, P. H., 1986, Sea-level curve for Pennsylvanian eustatic marine transgressive-regressive depositional cycles along midcontinent outcrop belt, North America: Geology, v. 14, p. 330–334.

Hurtubise, D. O., 1989, Stratigraphy and structure of the Seaman Range, Nevada, with an emphasis on the Devonian System: [Ph.D. thesis]: Golden, Colorado School of Mines, 443 p.

Johnson, J. G., 1990, Lower and Middle Devonian brachiopod dominated communuties of Nevada, and their position in a biofacies-province-realm model, with a section on revision of Middle Devonian conodont zones, by G. Klapper and J. G., Johnson: Journal of Paleontology, v. 64, p. 902–941.

Johnson, J. G., and Klapper, G., 1992, North American Midcontinent Devonian T-R cycles, in Chaplin, J. R., and Barrick, J. E., eds., Special Papers in Paleontology and Stratigraphy: A tribute to Thomas W. Amsden: Oklahoma Geological Survey Bulletin 145, p. 127–135.

Johnson, J. G., and Murphy, M. A., 1984, Time-rock model for Siluro-Devonian continental shelf, western United States: Geological Society of America Bulletin, v. 95, p. 1349–1359.

Johnson, J. G., and Sandberg, C. A., 1977, Lower and Middle Devonian continental-shelf rocks of the western United States, in Murphy, M. A., Berry, W. B. N., and Sandberg, C. A., eds., Western North America: Devonian: University of California, Riverside Campus Museum Contribution 4, p. 121–143.

Johnson, J. G., and Sandberg, C. A., 1989, Devonian eustatic events in the western United States and their biostratigraphic responses, in McMillan, N. J., Embry, A. F., and Glass,D. J., eds., Devonian of the world: Calgary, Canadian Society of Petroleum Geologists Memoir 14, v. 3, p. 171–178. [imprint 1988]

Johnson, J. G., Klapper, G., and Sandberg, C. A., 1985, Devonian eustatic fluctuations in Euramerica: Geological Society of America Bulletin, v. 96, p. 567–587.

Johnson, J. G., Sandberg, C. A., and Poole, F. G., 1991, Devonian lithofacies of western United States, in Cooper, J. D. and Stevens, C. H., eds., Paleozoic paleogeography of the western United StatesÑII: Society of Economic Paleontologists and Mineralogists, Pacific Section, v. 67, p. 83–105.

Johnson, J. G., Klapper, G., and Elrick, M., 1996, Devonian transgressive-regressive cycles and biostratigraphy, northern Antelope Range, Nevada: Establishment of reference horizons for global cycles: Palaios, v. 11, p. 3–14.

Kellogg, H. E.,1963, Paleozoic stratigraphy of the southern Egan Range, Nevada: Geological Society of America Bulletin, v. 74, p. 685–708.

Klapper, G., 1977, Lower and Middle Devonian conodont sequence in central Nevada, with contributions by D. B. Johnson, in Murphy, M. A., Berry, W. B. N., and Sandberg, C. A., eds., Western North America: Devonian: University of California, Riverside Campus Musem Contribution 4, p. 33–54.

Klapper, G., 1989, The Montagne Noire Frasnian (Upper Devonian) conodont succession, in McMillan, N. J., Embry, A. F., and Glass, D. J., eds., Devonian of the world: Calgary, Canadian Society of Petroleum Geologists Memoir 14, v. 3, p. 449–468. [imprint 1988]

Kobluk, D. R., 1978, Reef stromatoporoid morphologies and dynamic populations: Application of field data to a model and the reconstruction of an Upper Devonian reef: Bulletin of Canadian Petroleum Geology, v. 26, p. 218–236.

Kukal, Z., and Saadallah, A., 1973, Aeolian admixtures in the sediments of the northern Persian Gulf, in Purser, B. H., ed., The Persian Gulf: Holocene carbonate sediementation and diagenesis in a shallow epicontinental sea: New York, Springer-Verlag, p. 155–122.

LaMaskin, T. A., 1995, Cyclostratigraphy and sequence stratigraphy of the Middle-Upper Devonian Guilmette Formation, southern Egan and Schell Creek ranges, Nevada [M.S. thesis]: Albuquerque, University of New Mexico, 204 p.

Larsen, B. R., 1989, Facies, depositional environments and stratigraphy of the upper Devonian upper member of the Guilmette Formation, west-central Utah [M.S. thesis]: Salt Lake City, University of Utah, 170 p.

Larsen, B. R., Chan, M. A., and Bereskin, S. R., 1989, Cyclic stratigraphy of the upper member of the Guilmette Formation (uppermost Givetian, Frasnian), west-central Utah, in McMillan, N. J., Embry, A. F., and Glass, D. J., eds., Devonian of the world: Calgary, Canadian Society of Petroleum Geologists Memoir 14, v. 2, p. 569–579. [imprint 1988]

Markello, J. R., and Read, J. F., 1981, Carbonate ramp-to-deeper shale shelf transitions of an Upper Cambrian intrashelf basin, Nolichucky Formation, southwest Virginia Appalachians: Sedimentology, v. 28, p. 573–597.

Montañez, I. P., and Osleger, D. A., 1993, Parasequence stacking patterns, third-order accommodation events, and sequence stratigraphy of Middle to Upper Cambrian platform carbonates, Bonanza King Formation, southern Great Basin, in Loucks, B., and Sarg, J. F., eds., Recent advances and applications of carbonate sequence stratigraphy: American Association of Petroleum Geologists Memoir 57, p. 305–326.

Niebhur, W. W., Jr., 1980, Biostratigraphy and paleoecology of the Guilmette Formation (Devonian) of eastern Nevada [Ph.D. thesis]: Berkeley, University of California, 181 p.

Noble, J. P. A., and Ferguson, R. D., 1971, Facies and faunal relations at edge of early Middle Devonian carbonate shelf, South Nahanni River area, N.W.T.: Bulletin of Canadian Petroleum Geology, v. 19, p. 570–588.

Palmer, A. R., compiler, 1983, The Decade of North American Geology: 1983 geologic time scale: Geology, v. 11, p. 503–504.

Poole, F. G., 1974, Flysh deposits of Antler foreland basin, western United States, in Dickinson, W. R., ed., Tectonics and sedimentation: Society of Economic Paleontologists and Mineralogists Special Publication 22, p. 58–83.

Power, J. D., 1984, The Devils Gate Limestone of the northern Roberts Mountains, Nevada [M.S. thesis]: Riverside, University of California, 98 p.

Pratt, B. R., and James, N. P., 1986, The St. George Group (Lower Ordovician) of western Newfoundland: Tidal-flat island model for carbonate sedimentation in shallow epiric seas: Sedimentology, v. 33, p. 313–343.

Purser, B. H., and Evans, G., 1973, Regional sedimentation along the Trucial Coast, SE Persian Gulf, in Purser, B. H., ed., The Persian Gulf: Holocene carbonate sedimentation and diagenesis in a shallow epicontinental sea: New York, Springer-Verlag, p. 1–10.

Read, J. F., 1973, Carbonate cycles, Pillara Formation (Devonian), Canning Basin, Western Australia: Bulletin of Canadian Petroleum Geology, v. 21, p. 38–51.

Read, J. F., and Goldhammer, R. K., 1988, Use of Fischer plots to define third-order sea-level curves in Ordovician peritidal cyclic carbonates, Appalachians: Geology, v. 16, p. 895–899.

Reineck, H. E., and Singh, I. B., 1980, Depositional sedimentary environments: New York, Springer-Verlag, 439 p.

Reso, A., 1963, Composite columnar section of exposed Paleozoic and Cenozoic rocks in the Pahranagat Range, Lincoln County, Nevada: Geological Society of America Bulletin, v. 74, p. 901–918.

Rocha-Campos, A. C., 1981a, Late Devonian Curara Formation, Amazon basin, Brazil, in Hambrey, M. J., and Harland, W. B., eds., Earth's pre-Pleistocene glacial record: Cambridge, Cambridge University Press, p. 888–891.

Rocha-Campos, A. C., 1981b, Middle-Late Devonian Cabeças Formation, Parnaíba Basin, Brazil, in Hambrey, M. J. and Harland, W. B., eds., Earth's pre-Pleistocene glacial record: Cambridge, Cambridge University Press, p. 892–895.

Sadler, P. M., Osleger, D. A., and Montañez, I. P., 1993, On the labeling, length, and objective basis of Fischer plots: Journal of Sedimentary Petrology, v. 63, p. 360–368.

Sandberg, C. A., and Poole, F. G., 1977, Conodont biostratigraphy and depositional complexes of Upper Devonian cratonic-platform and continental-shelf rocks in the western United States, in Murphy, M. A., Berry, W. B. N., and Sandberg, C. A., eds., Western North America: Devonian: University of California, Riverside Campus Museum Contribution 4, p. 144–182.

Sandberg, C. A., Poole, F. G., and Johnson, J. G., 1989, Upper Devonian of western United States, in McMillan, N. J., Embry, A. F., and Glass, D. J., eds., Devonian of the world: Calgary, Canadian Society of Petroleum Geologists, Memoir 14, v. 1, p. 183–220. [imprint 1988]

Schlager, W., 1981, The paradox of drowned reefs and carbonate platforms: Geological Society of America Bulletin, v. 92, p. 197–211.

Sloss, L. L., 1963, Sequences in the cratonic interior of North America: Geological Society of America Bulletin, v. 74, p. 93–114.

Stearn, C. W., 1982, The shapes of Paleozoic and modern reef builders: A critical review: Paleobiology, v. 8, p. 228–241.

Stewart, J. H., and Poole, F. G., 1974, Lower Paleozoic and uppermost Precambrian Cordilleran miogeocline, Great Basin, western United States, in Dickinson, W. R., ed., Tectonics and sedimentation: Society of Economic Paleontologists and Mineralogists Special Publication 22, p. 28–57.

Vail, P. R., Mitchum, R. M., and Thompson, S., III, 1977, Seismic stratigraphy and global changes of sea level. Part 4: Global cycles of relative changes of sea level, in Payton, C. E., ed., Seismic stratigraphy—Applications to hydrocarbon exploration: American Association of Petroleum Geologists Memoir 26, p. 83–97.

Veevers, J. J., and Powell, C. M., 1987, Late Paleozoic glacial episodes in Gondwanaland reflected in transgressive-regressive depositional sequences in Euramerica: Geological Society of America Bulletin, v. 98, p. 475–487.

Wendte, J., 1992, Platform evolution and its control on reef inception and localization, in Wendte, J., Stoakes, F. A., and Campbell, C. V., eds., Devonian-Early Mississippian carbonates of western Canada sedimentary basin: A sequence stratigraphic framework: SEPM (Society for Sedimentary Geology) Short Course 28, p. 41–87.

Witzke, B. J., and Heckel, P. H., 1989, Paleoclimatic indicators and inferred Devonian paleolatitudes of Euramerica, in McMillan, N. J., Embry, A. F., and Glass, D. J., eds., Devonian of the world: Calgary, Canadian Society of Petroleum Geologists, Memoir 14, v. 1, p. 49–63. [imprint 1988]

Ziegler, W., 1971, Conodont stratigraphy of the European Devonian, in Sweet, W. C., and Bergström, S. M., eds., Symposium on conodont biostratigraphy: Geological Society of America Memoir 127, p. 227–284.

Ziegler, W. and Klapper, G., 1982, The disparilis conodont zone, the proposed level for the Middle-Upper Devonian boundary: Courier Forschungsinstitut Senckenberg, v. 55, p. 463–491.

MANUSCRIPT ACCEPTED BY THE SOCIETY FEBRUARY 27, 1997

Geological Society of America
Special Paper 321
1997

Graphic correlation of Frasnian (Upper Devonian) sequences in Montagne Noire, France, and western Canada

Gilbert Klapper
Department of Geology, University of Iowa, Iowa City, Iowa 52242

ABSTRACT

Through graphic correlation (Shaw's method) of Frasnian sections in the Montagne Noire, southwestern France, with those in the Alberta Rockies and Hay-Trout Rivers area of western Canada, the zonal and/or informal conodont faunal sequences developed for the three areas can be aligned with some precision for the first time. All of the zones are scaled to a Frasnian composite standard (CS) that derives from graphic correlation of over 70 sections in North America, Europe (including European Russia), and Western Australia. Graphic correlation tests the opposing hypotheses of isochrony versus diachrony of biostratigraphic events based on zonally defining conodont species. A substantial number of bases of zonally defining species are shown to be diachronous in the present study. This indicates that the Frasnian CS developed through graphic correlation will eventually be preferable to the synthesis of Frasnian biostratigraphic data through use of a zonation.

INTRODUCTION

Practitioners of graphic correlation (beginning with Shaw, 1964; see also the papers collected in Mann and Lane, 1995) have long recognized that the method offers a far greater resolution than is attainable with zonal biostratigraphy. Furthermore, graphic correlation is not burdened with the assumption deeply imbedded in zonal correlation charts that zonally defining species are everywhere synchronous. As graphic correlation makes evident, the majority of species fall off the line of correlation (LOC) and are thus diachronous to the extent they depart from the line. That there is error in positioning the LOC will not change this empirical observation, only the details. In discussing the philosophical conflicts between practitioners of zonal methods and graphic correlation, MacLeod (1995, p. 227) emphasized that "the simple exercise of constructing a two-dimensional plot of coeval biostratigraphic data from different sections should be sufficient to convince anyone that intra-datum diachrony is routinely encountered. . ." Graphic correlation thus challenges the necessary general assumption underpinning zonal methods, namely that biostratigraphic events used to define zonal boundaries are isochronous. Clearly, their isochronous nature should not be assumed but rather must be demonstrated, to paraphrase Shaw (1969, p. 1097) writing on another subject—the species problem in paleontology. In terms of graphic correlation, what I mean by "demonstrated" is that the points representing the bases and tops of species ranges must lie on or very close to the LOC to be considered isochronous or nearly so. And, of course, these cannot be the points used to fix the exact position of the LOC.

Graphic correlation of Frasnian sections in the Montagne Noire, southwestern France, began in the late 1980s in a joint proj-

Klapper, G., 1997, Graphic correlation of Frasnian (Upper Devonian) sequences in Montagne Noire, France, and western Canada, *in* Klapper, G., Murphy, M. A., and Talent, J. A., eds., Paleozoic Sequence Stratigraphy, Biostratigraphy, and Biogeography: Studies in Honor of J. Granville ("Jess") Johnson: Boulder, Colorado, Geological Society of America Special Paper 321.

ect with John F. Baesemann. Initially, the Montagne Noire sections formed the principal basis of a Frasnian composite standard (CS). William T. Kirchgasser joined the research in 1990 as sections were added from western New York and Western Australia in which key beds and goniatites were also used. Western Canadian, midcontinent United States, and southern Russian Platform sections were also correlated, and a preliminary account indicated the status of knowledge of the Frasnian CS as of early 1994 (Klapper et al., 1995). Subsequently, results from graphic correlation of seven cores in the Timan-Pechora Basin of Russia further refined ranges of widespread conodont species in the CS and added ranges of a number of endemic species (Klapper et al., 1996).

Prior to the availability of results from graphic correlation, however, the sequences in the Montagne Noire (Klapper, 1989) and the Alberta Rockies (Klapper and Lane, 1989) formed the bases for separate Frasnian conodont zonations. At that time, detailed correlation between the two sequences was largely unknown. Now through graphic correlation, the alignment of the Montagne Noire sequence with that in the Alberta Rockies can be presented as a hypothesis for which there is considerable support. Since the Frasnian Montagne Noire (MN) zonation is applicable in sections as far removed as those in Western Australia and the Timan-Pechora Basin (Klapper et al., 1996), continued use of the Alberta zonation except for local descriptive purposes is unnecessary.

The sequences in the Montagne Noire and Alberta Rockies are dominated by the *Palmatolepis* conodont biofacies. The *Polygnathus* biofacies (as these terms for biofacies are used by Klapper and Lane, 1985), however, characterizes the upper part of the Alberta sequence, as it does throughout the Hay-Trout Rivers sequence, southwestern Northwest Territories. A hypothesis toward resolving the relationship between these contrasting, mostly mutually exclusive biofacies resulted from graphic correlation in the paper of Klapper et al. (1995, figs. 7, 8). That preliminary account of the development of the Frasnian CS focused mainly on the methods used to locate the line of correlation in representative graphs. Nevertheless, the basis for correlation of the Montagne Noire sequence with those in the Alberta Rockies and Hay-Trout Rivers area of western Canada was laid by the graphs of the preliminary paper. The resulting alignment of the zones and faunal intervals of the three sequences are portrayed in this chapter for the first time.

As the Montagne Noire sections provided the initial basis for the CS, some details of three of the key graphs for those sections are briefly discussed under the heading "Montagne Noire Sections." Resulting ranges of species, especially their bases, established in the CS through graphic correlation of the Montagne Noire sections then effected a correlation hypothesis for four of the MN sections with two representative sections in western Canada (discussed in "Alignment of Montagne Noire and Canadian Sequences"). Observed ranges, scaled in terms of projected composite standard values, for all the species in each of the six sections are given in Appendix 1.

In the early rounds of graphing for the Frasnian CS, the tendency was to position the LOC exactly on the bases of zonally defining species for the purpose of having the CS mirror the MN zonation. In later rounds (third through fifth), the LOC was determined largely by the method of "splitting" the bases and tops (discussed in Klapper et al., 1995, p. 178–179). In this method, all points on the graph are considered, but for convenience two points are chosen to position ("fix") each segment of the LOC exactly. The bases and tops of such points cannot properly be a part of a test of the degree of diachrony, for they will necessarily have the same value as the maximum bases and tops in the CS. In the section "Diachrony of Frasnian (MN) Zonal Species," certain points are stated not to have been used to "fix" the LOC, that is, only in the special sense of "fix" just defined. Thus, the degree of divergence of these points from the LOC is a measure of diachrony at individual sections. To reiterate, however, all points influence the positioning of the LOC in the method of splitting the bases and tops.

MONTAGNE NOIRE SECTIONS

The graphing of La Serre Trench D with the Frasnian CS has been outlined by Klapper et al. (1995, p. 178–179, figs. 1, 2). In the fifth-round plot, the bases of two species, *Palmatolepis ljaschenkoae* (no. 66) and *P. winchelli* (no. 82) were used to fix the LOC. Figure 1 represents only the lower part of the graph in order to show the relation of the points more effectively than in the previous paper (the base of *P. winchelli* is only shown in the previously published graph). The LOC for La Serre D is one of the most tightly controlled of all the sections in the CS, a nearly optimum separation of bases and tops (the interested reader should refer to the original graph to evaluate this opinion in the context of the entire section [Klapper et al., 1995, fig. 1]). Note that many of the bases in Figure 1 form a narrow channel close to the LOC, indicating very little diachrony between occurrences in La Serre D and that of the Composite Standard. This is equally true of species of *Ancyrodella* and *Ancyrognathus* as for species of *Palmatolepis* (all labeled points are identified in Appendix 1).

Only the bases of *Palmatolepis winchelli* (no. 82) and *P. bogartensis* (no. 50), two zonally defining late Frasnian species, were used to fix the fifth segment of the LOC for the graph of the Upper Coumiac Quarry (Fig. 2; the whole graph is shown in Klapper et al., 1995, fig. 5). This was determined in an early round of graphing prior to extending the graph into the Famennian, from the base of *P. triangularis* (no. 81) to a higher position. Thus, the LOC fixed on two Frasnian species was not influenced by the Famennian part of the graph, even though the bases of a remarkable number of significant Famennian species are almost

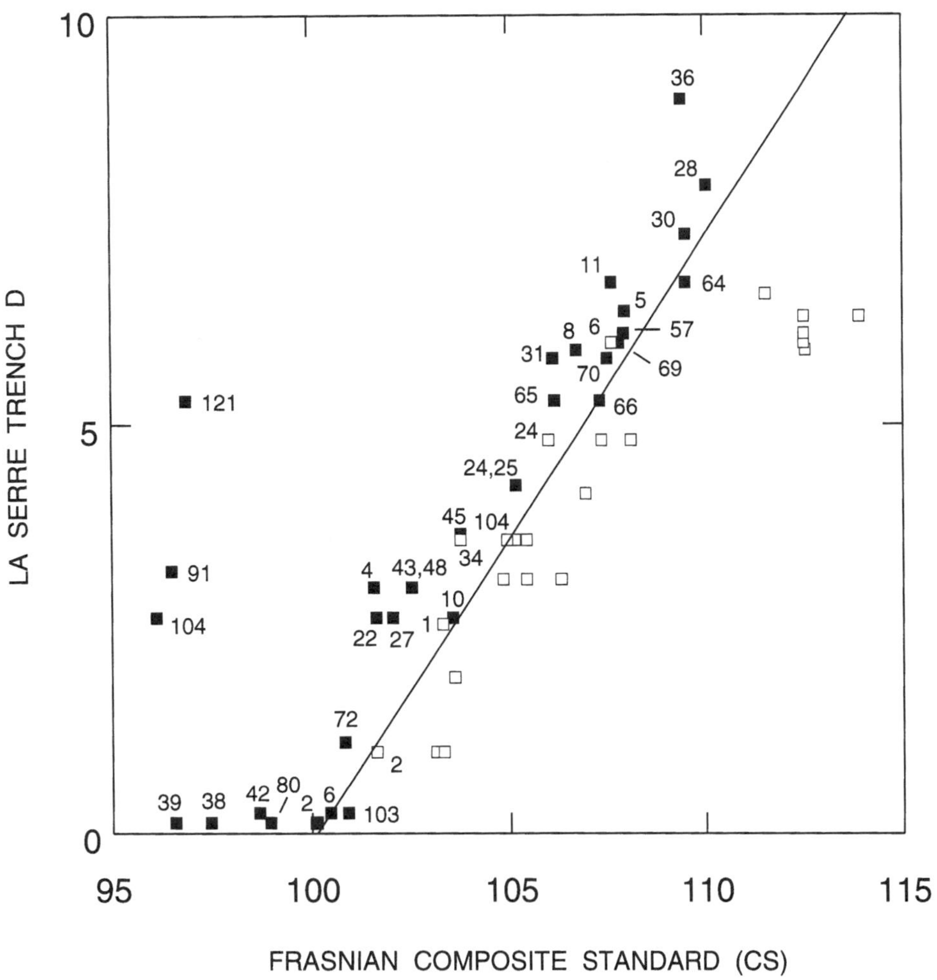

Figure 1. Detail of fifth-round plot of La Serre Trench D against the Frasnian CS, representing the lower left quarter of the full graph (Klapper et al., 1995, fig. 1). Numbers represent species indicated in Appendix 1. Solid squares are bases; open squares are tops. Base of Frasnian is at 96.9 composite standard units (CSU, Table 1). Scale on y-axis in meters above section base; scale on x-axis in CSU in Figures 1 through 3.

exactly intersected (briefly noted by Klapper et al., 1995, p. 181). Among others are the bases of *Palmatolepis triangularis* (no. 81), *P. clarki* (no. 54), *P. minuta minuta* (no. 68), *P. quadrantinodosalobata* (no. 73), *P. termini* (no. 79), and *P. glabra prima* (no. 60). All are defining or characterizing species for lower Famennian zones.

The base of Bed 32a at the Upper Coumiac Quarry defines the boundary stratotype (GSSP) for the base of the Famennian Stage (Klapper et al., 1994). Beds 32a and 32b have abundant *Palmatolepis triangularis* (no. 81) and represent the Lower *triangularis* Zone at Coumiac. There is one specimen of the species in the underlying highest bed of Frasnian Zone 13 (Klapper et al., 1994, p. 436, Bed 31g, plus one questionably identified specimen), but as this occurrence may represent stratigraphic leak it is not considered in the graphing. As mentioned above, the LOC (Fig. 2) almost exactly intersects the base of *P. triangularis* in Bed 32a (the base is 0.01 composite standard units to the left of the LOC). That there is no change in slope necessitated for the LOC between the Frasnian and the lower Famennian, nor a terrace developed between Beds 31g and 32a, further confirms the adequacy of the Upper Coumiac Quarry as the boundary stratotype for the definition of the base of the Famennian (as earlier noted in Klapper et al., 1995, p. 181).

The LOC for Causses-et-Veyran South (CVS) consists of three segments: the first fixed on *Palmatolepis hassi* form 2 (no. 61) and *P. winchelli* (no. 82), the second at a slightly different

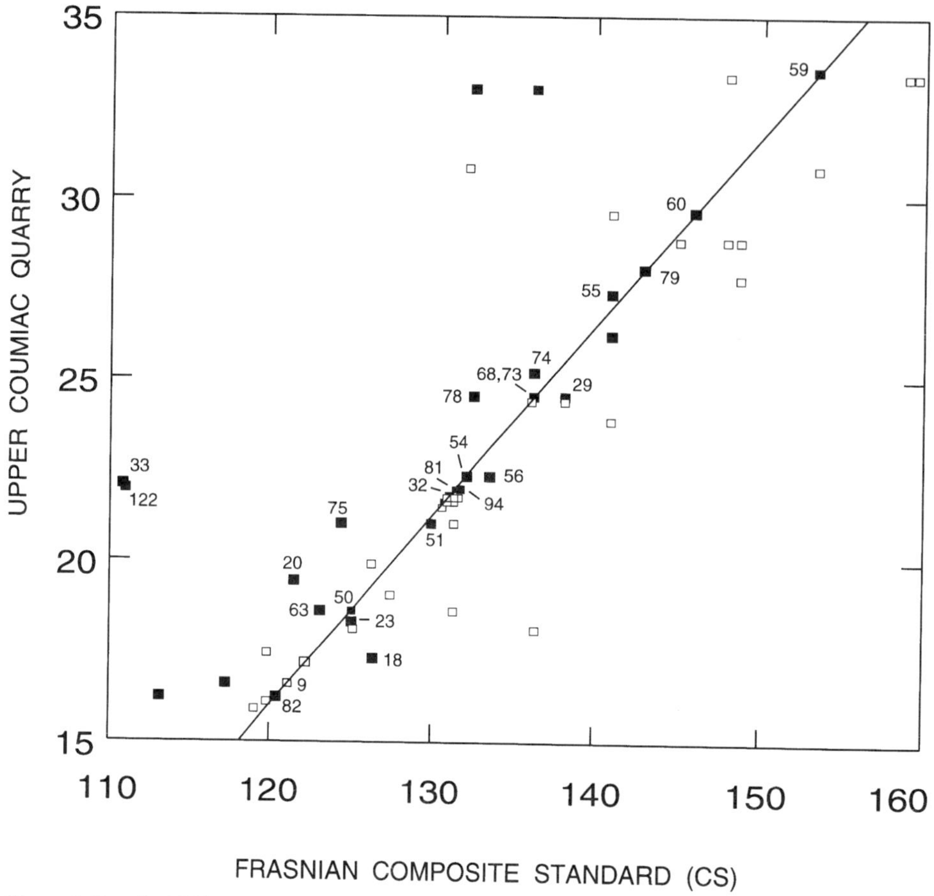

Figure 2. Detail of fifth-round plot of the Upper Coumiac Quarry against the Frasnian CS, representing only the fifth segment of the LOC and about the upper right two-thirds of the full graph (Klapper et al., 1995, fig. 5). Fixed on species no. 82 and no. 50, LOC intersects a number of other species (see text). Significantly, there is an almost exact intersection with the base of *Palmatolepis triangularis* (no. 81), at the position of the boundary stratotype (GSSP = base of Bed 32a) for the base of the Famennian (131.4 CSU). Further explanation in Figure 1.

slope fixed on no. 82 and *P. bogartensis* (no. 50), and the third fixed on no. 50 and *P. triangularis* (no. 81). A number of other bases and tops are either on or extremely close to the LOC for CVS (Fig. 3), for example, *Ancyrognathus triangularis* (no. 30), *Ancyrodella ioides* (no. 9), and *Ancyrognathus ubiquitus* (no. 32).

DIACHRONY OF FRASNIAN (MN) ZONAL SPECIES

The principal purpose of the following discussion is to attempt to quantify the degree of diachrony of the zonally defining species of the Frasnian (MN) zonation. To this end, ranges of zonally defining species, as well as some significant characterizing species (Murphy, 1977), are analyzed in terms of the correlation hypothesis represented in Figure 4 involving four Montagne Noire and two western Canadian sections. This analysis focuses not only specifically on the variation in the observed lowest occurrence of zonally defining species at these six sections but also more broadly on such variation in all the sections in the Frasnian CS. As the MN zonation (Klapper, 1989, 1990) was defined in terms of first occurrences only, a seemingly prevailing bias in Paleozoic biostratigraphy, the presentation here emphasizes bases. But graphic correlation demonstrates that tops must be considered as well. In my view, graphic correlation will eventually replace the commonly used biostratigraphic zonations not only because of demonstrated diachrony of defining species but also because of the greater degree of refinement attainable. Nevertheless, for ease of presentation, the foregoing analysis is partitioned in terms of the Frasnian MN zonal sequence.

Zone 4 through 8

Palmatolepis transitans (no. 80), the defining species for the base of MN Zone 4 (Klapper, 1989; Table 1 herein), first occurs

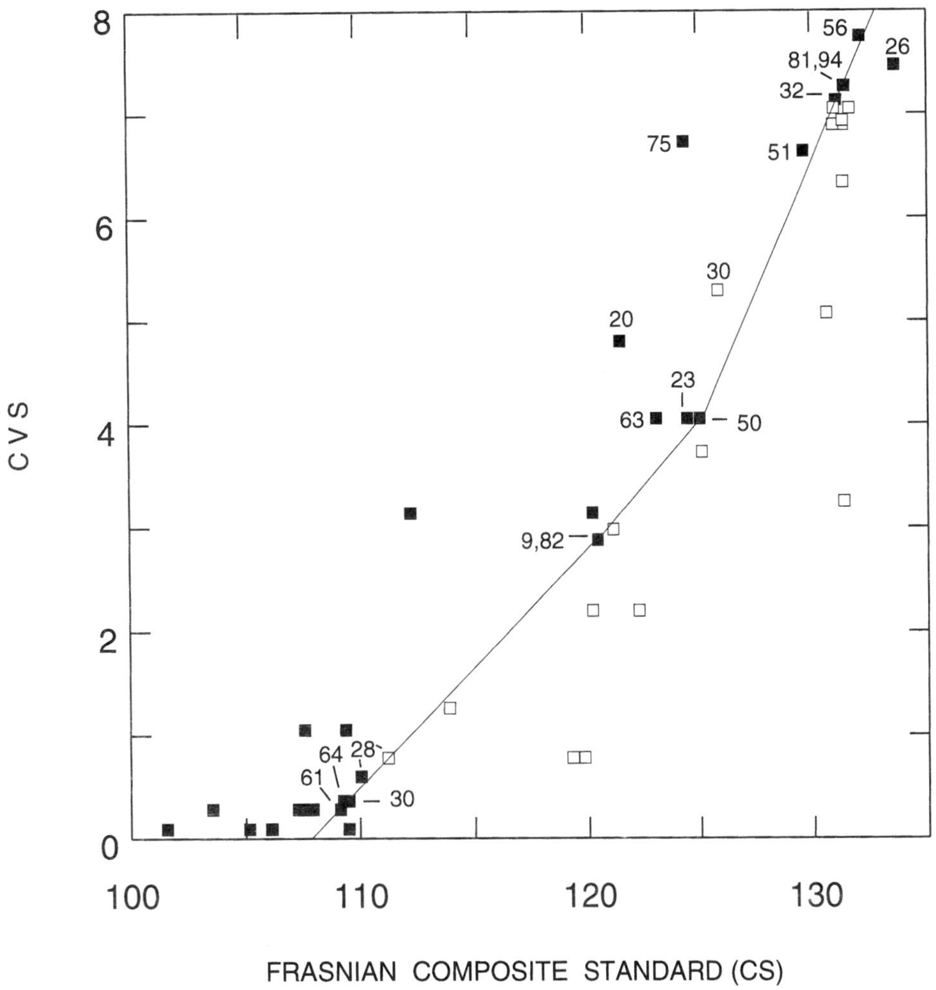

Figure 3. Fifth-round plot of Causses-et-Veyran South (CVS) against the Frasnian CS, representing the full graph. Base of Famennian at 131.4 CSU. Further explanation in Figure 1.

within Zone 4 at La Serre Trench D in the Montagne Noire and Luscar Mountain in the Alberta Rockies. At the Upper Coumiac Quarry, however, the lowest sample collected in the section projects by graphic correlation to a position within Zone 5 (Fig. 4). The lowest occurrences of the species in the CS are in the New York regional composite and Timanites Hill in the Canning Basin. Of the 17 bases in the Frasnian CS (Appendix 1), 11 are within Zone 4, five are within Zone 5, and one is within Zone 6. Thus, the observed entry of *Palmatolepis transitans* is zonally diachronous in terms of the Frasnian MN zonation, the variation amounting to 2.2 composite standard units (CSU).

The base of *Palmatolepis punctata* (no. 72), the defining species for the base of MN Zone 5 (Table 1), is at the same position (within 0.1 CSU) at CPS-H, La Serre D, and Coumiac, that is, in the upper half of Zone 5 according to the results of graphic correlation (Fig. 4). In the graphs for these sections (Klapper et al., 1995, figs. 1, 2, 5), the species was not used to fix the LOC. The lowest occurrence at Pic de Bissous, Montagne Noire, establishes the lowest known base in the CS, and this lowest range was used to fix the LOC in the New York regional composite (Klapper et al., 1995, figs. 3, 6). Of the 19 bases in the CS (Appendix 1; the species occurs in 20 sections, but one is an isolated top), 10 are within Zone 5, six are within Zone 6, and three Alberta bases are within Zone 7. In contrast to the upper Zone 5 position in the three Montagne Noire sections, *Palmatolepis punctata* enters the Hay River sequence at a higher position that is graphically equivalent to almost the top of Zone 6, and it first occurs at the base of Zone 7 at Luscar Mountain (Fig. 4). Biofacies differences may account for the higher entries in Alberta, but lack of sampling of Zone 5 at Hay River may be a factor in the Northwest Territories sequence. The observed entry of *Palmatolepis punctata* is zonally diachronous in terms of the MN zonation, that is, slightly over 3 CSU.

Mesotaxis johnsoni (no. 41) has its base within Zone 5 at Luscar Mountain but is within Zone 6 at Hay River just above the lowest sample graphed. The lowest occurrence in the Fras-

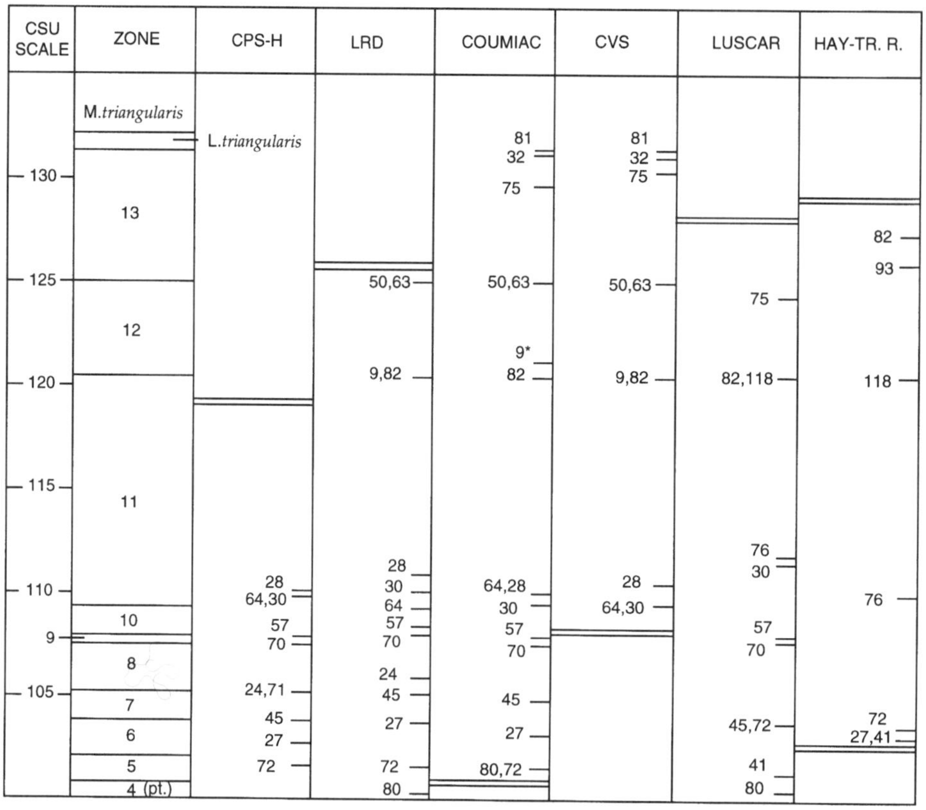

Figure 4. Graphic correlation of four Montagne Noire sections (columns 3 through 6; CPS-H = Col du Puech de la Suque, the standard reference section for the Frasnian CS; LRD = La Serre Trench D; Coumiac = Upper Coumiac Quarry; CVS = Causses-et-Veyran South) with two in western Canada (column 7, Luscar Mountain, Alberta Rockies; column 8, Hay-Tr. R. = Hay-Trout Rivers area of southwestern Northwest Territories). This diagram represents the correlation hypothesis cited near the end of the "Introduction." Vertical scale of chart in composite standard units (CSU, column 1). Second column in part represents Frasnian numbered zones first recognized in the Montagne Noire sequence (Klapper, 1989; 1990) and later replicated to varying degrees of completeness in sections in North America, Australia, and Russia. Lower three Frasnian zones and part of Zone 4 not shown; base of Famennian at 131.4 CSU. L. and M. *triangularis* = Lower and Middle *triangularis* Zones of the lower Famennian (Ziegler, 1962). Double lines in columns indicate sampling limits. Numbers in columns 3 through 6 are oberved bases (and one top marked by an asterisk) in the individual sections given in terms of their projected position in the CS and represent the species listed in Appendix 1. Only in some instances do they coincide with the lowest bases in the CS, and most of these are mentioned in the text.

nian CS is in the Timan-Pechora region (Klapper et al., 1996, table 2 and fig. 5, core 2060), slightly above the base of Zone 5. Of the 12 bases in the CS (Appendix 1; the species does not occur in the Montagne Noire), three are within Zone 5, eight are within Zone 6, and one is within Zone 7. The observed entry of *Mesotaxis johnsoni* is zonally diachronous, equal to 3.2 CSU.

Ancyrognathus primus (no. 27), the defining species for the base of MN Zone 6 (Klapper, 1990, p. 1000; Table 1 herein), enters at different levels within this zone in four of the sections shown in Figure 4 (three in the Montagne Noire and one at Hay River). Of the 11 bases in the Frasnian CS (Appendix 1), all are within Zone 6, ranging from 102.1 (Pic de Bissous, Montagne Noire) to 103.7 (La Serre Trench D), a relatively narrow range in terms of CSU values.

The standard reference section for the Frasnian CS, CPS-H, establishes the lowest base of *Ozarkodina nonaginta* (no. 45), the defining species for the base of Zone 7 (Table 1; 103.8 CSU). Its entry at La Serre Trench D is slightly below the top of Zone 7 (105.1 CSU) and at a slightly lower level within the same zone at Coumiac (Fig. 4). The base of the species is one of three points used to fix the LOC at Luscar Mountain (Klapper et al., 1995, fig. 7), thus coincident with that at CPS-H. Of the 12 bases in the Frasnian CS (Appendix 1), all are within Zone 7, a relatively narrow range of 1.3 CSU.

As with the base of Zone 7, CPS-H establishes the lowest base of *Palmatolepis* aff. *P. proversa* (no. 71), the defining species for the base of Zone 8 (Table 1). The new species is unknown in the Alberta Rockies but does occur in the Alberta subsurface in a core (B. A. Shell Berland River), where it was used to fix the LOC. There are two bases in the Frasnian CS as a result of Montagne Noire sections, a base within Zone 8 at 105.7 CSU at Horse Spring in the Canning Basin, Western Australia, and the Alberta core. Consequently, the species has too few occurrences in the CS as yet to form a confident conclusion about the degree of diachrony. *Ancyrognathus barba* (no. 24) enters with the new species at CPS-H and higher in the zone at La Serre Trench D (Fig. 4). There are four bases in the CS, all in the Montagne Noire and all low in Zone 8.

Zones 9 and 10

The lowest base in the CS of *Palmatolepis proversa* (no. 70), the defining species for the base of Zone 9 (Table 1), is at CPS-H. This base was used to fix the lower segment of the LOC at Coumiac (Klapper et al., 1995, fig. 5), but *P. proversa* enters within Zone 9 at Luscar Mountain and low within Zone 10 at La Serre Trench D (Fig. 4). Of the 16 bases in the Frasnian CS (Appendix 1), seven are within Zone 9, five are within Zone 10, and four are low within Zone 11. The observed entry of *P. proversa* is zonally diachronous in terms of the MN zonation, the variation amounting to 2.7 CSU.

The base in the CS of *Palmatolepis domanicensis* (no. 57), the defining species for the base of Zone 10 (Table 1), is also established by CPS-H. Remarkably, the base of *P. domanicensis* is intersected by the LOC, fixed on other species, in the graphs for Coumiac (the unlabeled point just above that of *P. proversa*) and Luscar Mountain (Klapper et al., 1995, figs. 5, 7). Of the 15 bases in the CS (Appendix 1; the sixteenth section has only an isolated top), 10 are within Zone 10 and the rest are low in Zone 11. The variation in observed first occurrences at these 15 sections is 2.4 CSU.

Zone 11

La Serre Trench D establishes the lowest base of *Palmatolepis jamieae* (no. 64), the defining species for the base of Zone 11. Bases of this species at five other sections in the Montagne Noire are restricted to very low in the zone (three are shown in Fig. 4), whereas that at Horse Spring in the Canning Basin, Western Australia, is 110.7 CSU, less than 0.7 CSU above the highest base in the French sections. The variation in observed first occurrences of the seven bases in the CS is a relatively narrow range of 1.4 CSU, all low within Zone 11. The base of *P. jamieae* was not used to fix the LOC in the graphs in which it occurs.

TABLE 1. SCALE IN COMPOSITE STANDARD UNITS OF A FRASNIAN CS*

Scale in CSU	Base of Zone	Defining Species for Base
131.4	Lower *triangularis*	*Palmatolepis triangularis*
125.0	13	*P. bogartensis*
120.4	12	*P. winchelli* (= *P. subrecta*)
109.3	11	*P. jamieae*
107.9	10	*P. domanicensis*
107.5	9	*P. proversa*
105.2	8	*P.* aff. *P. proversa*
103.8	7	*Ozarkodina nonaginta*
102.1	6	*Ancyrognathus primus*
100.8	5	*Palmatolepis punctata*
99.0	4	*P. transitans*
98.4	3	*Ancyrodella rugosa*
97.5	2	*A. rotundiloba* late form
96.9	1	*A. rotundiloba* early form

*Values in first column for bases of 13 Frasnian zones and base of Lower *triangularis* Zone, coincident with base of Famennian Stage, are those of lowest bases in the CS.

Similarly, the first occurrences of *Ancyrognathus seddoni* (no. 28) are within a narrow range low in Zone 11 in the Montagne Noire, less than 0.9 CSU at five sections (four of which are shown in Fig. 4). These are the only occurrences in the CS, and the species was used to fix a segment of the LOC only in the graph of Coumiac (Klapper et al., 1995, fig. 5). Although the base of *Ancyrognathus triangularis* (no. 30) is within an even narrower range (109.5 to 110.1 CSU) in six Montagne Noire sections, it enters at a somewhat higher level low in Zone 11 at Luscar Mountain (111.5, Fig. 4). The base of this species was not used to fix the LOC in any graphs. Of nine bases in the CS, all are within Zone 11, the variation amounting to 5.9 CSU.

Palmatolepis semichatovae (no. 76) does not occur in the Montagne Noire, but its base projects to a position low in Zone 11 in western Canada, western New York State, the Canning Basin, the Russian Platform (Aristov, 1988), and Timan-Pechora (Klapper et al., 1996). Of 12 bases in the Frasnian CS, the lowest was originally established at Horse Spring, Canning Basin, which was then used to fix the LOC in the Hay-Trout River regional composite (110.1 CSU). The base at Luscar Mountain (111.9) is somewhat higher (Fig. 4). The variation in the level of entries at the 12 sections in the CS (Appendix 1; the thirteenth section has only an isolated top) is 2.2 CSU, all low within Zone 11.

Zones 12 and 13

The base of *Palmatolepis winchelli* (no. 82), the defining species for the base of Zone 12, was set in the initial graphing of

La Serre Trench D against CPS-H (Klapper et al., 1995, fig. 2), by projection from the former section. This has remained the lowest occurrence in the Frasnian CS and was used to fix the LOC at Coumiac and CVS (Fig. 4). The entry of *P. winchelli* is 0.1 CSU higher at Luscar Mountain, where the base of *Polygnathus samueli* (no. 118) fixes the seond segment of the LOC (Klapper et al., 1995, fig. 7). The latter species was also used to fix the LOC in the Hay-Trout Rivers regional composite, in which *P. winchelli* does not occur until a position graphically equivalent to Zone 13 (Fig. 4). An explanation relating to biofacies differences probably accounts for the much higher entry, as this is a sequence dominated by the *Polygnathus* biofacies (Klapper and Lane, 1985). The variation in the levels of entry of *Palmatolepis winchelli* at the 13 sections in the CS (Appendix 1; the fourteenth section has an isolated top) is 9.2 CSU, the greatest degree of diachrony among the species discussed.

The base of *Ancyrodella ioides* (no. 9) exactly coincides with that of *Palmatolepis winchelli* in La Serre Trench D and CVS at the base of Zone 12, whereas the top is fixed by an isolated occurrence at Coumiac (Fig. 4). The base of the species at Horse Spring, Canning Basin (Klapper et al., 1995, fig. 4), is 0.03 CSU higher than its base in the Montagne Noire. There are only these three bases in the CS, an insufficient test.

The base of *Palmatolepis bogartensis* (no. 50), the defining species for the base of Zone 13, was set by the initial graphing of La Serre D against CPS-H (Klapper et al., 1995, fig. 2), as for *P. winchelli*. In the fifth-round graph of the section, the base of *P. bogartensis* falls less than 0.1 CSU to the left of the LOC fixed on other species (Klapper et al., 1995, p. 178; the caption for their fig. 1 incorrectly indicates the opposite relationship, although it is correctly stated in the text). The base of *P. bogartensis* was used to fix segments of the LOC in the graphs for Coumiac and CVS (Figs. 2 through 4). Of six bases in the CS (Appendix 1, two sections have isolated tops), all are within Zone 13, the variation amounting to 2.1 CSU

Palmatolepis hassi (no. 63) enters in exactly the same samples as *P. bogartensis* at the base of Zone 13 in three Montagne Noire sections (Fig. 4), but it occurs in the higher part of Zone 12 in the second New York regional composite (which extends from within Zone 11 through Zone 13; Kirchgasser and Klapper, 1992). Of six bases in the CS, the variation is 2.4 CSU.

In the sections analyzed, *Palmatolepis rhenana* (no. 75) has its lowest base in the upper part of Zone 12 at Luscar Mountain, but its base is in the upper part of Zone 13 in the Montagne Noire (Coumiac and CVS, Fig. 4). Of six bases in the CS, the variation in the levels of entry is 6.0 CSU, a considerable degree of diachrony. The earliest occurrence of the species has been used to define the base of the Late *rhenana* Zone of Ziegler and Sandberg (1990).

Ancyrognathus ubiquitus (no. 32) is at a position just below the top of Zone 13 and the coincident base of the Famennian at Coumiac and CVS (Fig. 4); it falls exactly on the LOC fixed on other species in the first section and less than 0.1 CSU to the left of the LOC in the second. There are only these two occurrences in the sections graphed as yet for the Frasnian CS.

Base of Famennian Stage

The lowest occurrence of *Palmatolepis triangularis* at Sessacker Trench II in the Rhenish Slate Mountains (Ziegler, 1962) establishes its base in the Composite Standard. This species defines the base of the Lower *triangularis* Zone, which was used to define the boundary stratotype (GSSP) for the base of the Famennian Stage at the Upper Coumiac Quarry (Klapper et al., 1994). As mentioned under "Montagne Noire Sections," the LOC at Coumiac fixed on other species almost exactly intersects the base of *P. triangularis* (no. 81). The same base was then used to fix the third segment of the LOC at CVS (Figs. 3, 4). The variation in four entries in the CS is 0.3 CSU (Sessacker, Coumiac, CVS, and Bagan well 3 in the Timan-Pechora Basin; Klapper et al., 1996). The lowest occurrence at Horse Spring in the Canning Basin, however, is 3.3 CSU higher than the lowest base in the CS owing to a sparsely sampled, dolomitic interval above the highest sample of Frasnian Zone 13.

Conclusion

The foregoing analysis of the amount of variation in the entries of zonally defining and characterizing species of the Frasnian MN zonation, determined through graphic correlation, demonstrates that a substantial number of entries are diachronous to a considerable degree in different sections of the CS. This conclusion applies to some, though not all, of the species of *Palmatolepis*. In some instances, species of *Ancyrognathus*, *Ancyrodella*, and *Ozarkodina* are less diachronous than some of the zonally defining *Palmatolepis*.

ALIGNMENT OF MONTAGNE NOIRE AND CANADIAN SEQUENCES

Through graphic correlation of the Montagne Noire sections with those in the Alberta Rockies and the Hay-Trout Rivers regional composite (Fig. 4), the zonal and/or informal faunal sequences developed for the three areas can be aligned with some precision for the first time. This is summarized in Figure 5, a diagram scaled to the composite standard.

The sequence in the Alberta Rockies was based largely on two sections, Luscar Mountain and Mount Haultain (Klapper and Lane, 1989), but was augmented by unpublished studies of over 100 cores obtained by Amoco Canada Ltd. from the central Alberta subsurface. Only the topmost part of Alberta Zone 1 is

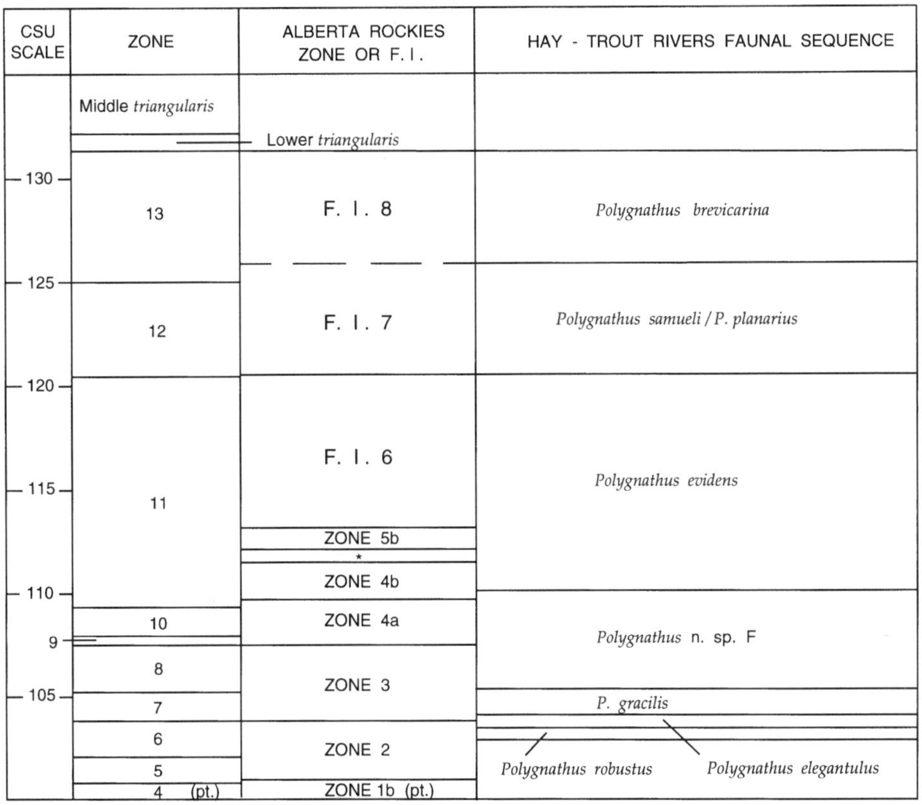

Figure 5. Graphic correlation of Frasnian zones (column 2, first recognized in the Montagne Noire, see explanation of Fig. 4) with the zones and sequence of faunal intervals (F. I.) developed for the Alberta Rockies (column 3, Klapper and Lane, 1989). * = Zone 5a. Fourth column is informal faunal sequence described for the Hay and Trout Rivers, southwestern Northwest Territories (Klapper and Lane, 1985). Vertical scale of chart in composite standard units (CSU, column 1).

developed in the outcrop at Luscar Mountain, shown as 1b in Figure 5. The full zone is recognized in the subsurface where it was subdivided into 1a and 1b and is approximately equivalent to Frasnian (MN) Zones 1 through 4.

Close correlation between columns two and three in Figure 5 exists at several levels. The base of Alberta Zone 3 coincides exactly with the base of Frasnian (MN) Zone 7, as the zonally defining species *Ozarkodina nonaginta* was used to fix the LOC in the graph for Luscar Mountain. The entry of *Palmatolepis proversa* defines both the bases of MN Zone 9 and Alberta Zone 4a and was used to fix the LOC at Mount Haultain. The lowest *P. domanicensis* coincides exactly with that in the Montagne Noire, as it is intersected by the LOC at Luscar, low in Alberta Zone 4a.

Higher in the Alberta sequence, the base of faunal interval 7 corresponds closely with that of MN Zone 12, as the entry of *Polygnathus samueli* and *Palmatolepis winchelli* at Luscar Mountain is only very slightly higher than that of the latter species in the Montagne Noire, as mentioned earlier. An occurrence of the Lower *triangularis* Zone at Cinquefoil Mountain (Wang and Geldsetzer, 1995), 25 km southeast of Mount Haultain, demonstrates the position of the base of the Famennian in this part of the Alberta Rockies and is indicated on Figure 5, although the Cinquefoil section has not been graphed.

Alignment of the informal faunal sequence outlined for the Hay and Trout Rivers, southwestern Northwest Territories (Klapper and Lane, 1985), with those depicted in columns two and three of Figure 5 resulted from a fifth-round plot of the regional composite for these sections (Klapper et al., 1995, fig. 8). Since the bases of *Palmatolepis semichatovae* (no. 76) and *Polygnathus samueli* (118) were used to fix the LOC in this graph, they coincide with the lowest bases in the Frasnian CS (Fig. 4 and Appendix 1).

The entry of *Polygnathus brevicarina* (no. 93) at Trout River corresponds to a CS position low in Zone 13. The lowest occurrence of the species was used to define the base of Faunal Interval 8 (Klapper and Lane, 1989), but it does not occur at either Luscar Mountain or Mount Haultain (in the Ancient Wall), the only sections in the Alberta Rockies that we have graphed. However, *P. brevicarina* is known at several sections in the Rockies, including the nearby southeast Thornton Creek section of the Ancient Wall (unpublished collection C-1C of M. T. Whalen). Here it is associated with *Palmatolepis boogaardi* in a sample about 23 m above the top of the Mount Hawk Formation, indicating a correlation with the upper part of Zone 13. Nevertheless, the position of the base of F. I. 8 in column three of Figure 5 is a largely unsupported projection of the base of *P. brevicarina* from Trout River. The highest graphed occurrence of the species projects in the CS to just below the top of Zone 13 at Sweetland Creek, Iowa. The conodont sequence across the Frasnian/Famennian boundary at Trout River has been discussed by Orchard (1989).

CONCLUSIONS

Graphic correlation of sections in the Montagne Noire with those in the Alberta Rockies and Hay-Trout Rivers area results in precise alignment of the zones and informal faunal intervals developed earlier for the three sequences. A substantial expansion in knowledge of ranges of species in the Frasnian composite standard has resulted from the correlation of the French and Canadian sections. The maximum bases and tops of all the species in the sections in the Frasnian CS provide information for locating the line of correlation in subsequent graphs. Splitting bases and tops to position the LOC, or by whatever alternative method it is determined, is no more than a correlation hypothesis. Nonetheless, the dispersion of bases and tops in relation to the LOC provides a quantitative estimate of the degree to which the occurrence of each species in each section graphed is diachronous. The analysis demonstrates that many, though not all, of the zonally defining species of the Frasnian MN zonation are diachronous to a considerable degree in different sections of the CS.

ACKNOWLEDGMENTS

I am greatly indebted to my colleagues, John F. Baesemann and William T. Kirchgasser, for their extensive participation in the research on graphic correlation. Heartfelt thanks are also extended to H. Richard Lane and Allen R. Ormiston for continued encouragement of my research for many years and for originally collecting much of the data from Luscar Mountain and Mount Haultain (all of the geologists involved in this collecting are cited in Klapper and Lane, 1989). Ross A. McLean generously provided the lithostratigraphic composite that was essential to developing the Hay-Trout Rivers regional composite. The guidance for all my research in the Montagne Noire has been expertly and unstintingly given by Raimund Feist.

Peter M. Sadler, Lucy E. Edwards, and Michael A. Murphy provided thoughtful and highly constructive reviews of an earlier version of the manuscript. I am of course responsible for all remaining errors of interpretation.

From 1966 until July 1994, Jess Johnson was the dominant force in providing the crucial and original ideas for our numerous joint research projects, which first centered on the Devonian of Nevada but later extended elsewhere. Moreover, his inspirational thinking stimulated various aspects of almost all of my research projects, especially those that turned out to be worthwhile. I do not wish to minimize the importance of our professional relationship, but his friendship was of far more significance.

APPENDIX 1: COMPOSITE STANDARD RANGES

The Frasnian Composite Standard currently includes 32 sections in Europe, North America, and Australia. New York regional composite I and the Hay-Trout Rivers regional composite consist of 20 and 26 individual sections, respectively. Additionally, three exclusively Famennian sections were used as a basis for correlation with the three Frasnian sections (Coumiac, CVS, and Horse Spring in Western Australia) that extend into the lower Famennian.

There are 156 Frasnian and 129 Famennian conodont species in the CS; additionally, three species cross the stage boundary. Twenty-three species and 16 goniatite genera are largely known in the New York and Australian sequences that also form an integral part of the Frasnian CS.

Observed ranges in composite standard units (CSU) are given below for 127 conodont species in each of five Montagne Noire and the two western Canadian sections (Luscar Mountain in the Alberta Rockies and Hay-Trout Rivers in the southwestern Northwest Territories) in which they occur. The French sections are Col du Puech de la Suque H (CPS-H), which is the standard reference section for the Frasnian CS; Causses-et-Veyran South (CVS); Upper Coumiac Quarry (Coumiac); La Serre trench D (LRD); and Pic de Bissous (VS-E, graph shown in Klapper et al., 1995, Fig. 3). The maximum base and top in the CS is at the bottom of the entry for each species, the number in parentheses being the number of sections in the CS in which they occur.

Numbers at beginning of species entries are those used in Figures 1 through 4 and in the text. Reference to original author(s) of species, and to some later revisions that have clarified the understanding of the species, are cited. Abbreviation: tr. w. = transitional with.

APPENDIX 1

Species (number used in figures)	Section	Base CSU	Top CSU
1. *Ancyrodella africana*	CPS-H	100.2	
García-López, 1981	LRD		103.6
	Luscar		101.0
	VS-E	99.1	101.9
	CS (13 sections)	99.1	103.6
2. *A. africana* (tr. w.) *A. gigas*	LRD	100.3	101.5
Klapper, 1985	VS-E	100.1	100.4
	CS(5)	100.1	101.6
3. *A. alata* s.s.	CPS-H	98.9	100.0
Glenister and Klapper, 1966	VS-E	98.9	99.9
	CS(11)	98.3	100.4
4. *A. curvata* early form	CPS-H	103.8	117.5
(Branson and Mehl, 1934)	CVS	108.3	111.3
Klapper, 1989	Coumiac	105.9	119.8
	LRD	104.2	108.7
	Luscar	103.8	111.5
	VS-E	102.0	103.0
	CS(18)	101.6	119.8
5. *A. curvata* late form	CPS-H	108.4	119.0
as no. 4	CVS	109.2	131.0
	Coumiac	108.3	131.0
	LRD	108.8	127.1
	Luscar	108.0	112.4
	CS(13)	108.0	131.6
6. *A. gigas*	CPS-H	101.7	106.3
Youngquist, 1947	Coumiac	101.6	107.0
form 1 of Klapper, 1989	Hay-Trout Rivers	103.5	103.6
	LRD	100.5	108.3
	Luscar	100.6	102.2
	VS-E	100.6	103.8
	CS(21)	100.5	108.3
7. *A. gigas* form 2	CPS-H	102.5	
as no. 6	CS(5)	102.5	103.5
8. *A. gigas* form 3	CPS-H	106.7	107.7
as no. 6	LRD	108.2	108.2
	CS(5)	106.7	112.5
9. *A. ioides*	CVS	120.4	120.8
Ziegler, 1958	Coumiac		121.1
	LRD	120.4	
	CS(4)	120.4	121.1
10. *A. lobata*	CPS-H	103.8	110.0
Branson and Mehl, 1934	CVS	109.2	
	Coumiac	103.6	108.3
	Hay-Trout Rivers	103.6	105.1
	LRD	103.7	108.7
	Luscar	105.7	106.0
	CS(12)	103.6	123.0
11. *A. nodosa*	CPS-H	109.2	119.0
Ulrich and Bassler, 1926	CVS	112.5	121.9
	Coumiac	109.5	125.0
	LRD	109.3	121.1
	Luscar	111.5	124.1
	CS(15)	107.6	131.3
12. *A. nodosa* (tr. w.) *A. ioides*	CPS-H	119.1	
Klapper, 1989	Coumiac	120.4	
	LRD	120.4	
	Luscar	113.1	122.2
	CS(10)	113.1	130.9
13. *A. pramosica*	CPS-H	99.9	100.0
Perri and Spalletta, 1981	VS-E	99.1	99.9
	CS(4)	99.1	100.0
14. *A. recta*	CPS-H	98.2	98.5
Kralick, 1994	VS-E	98.2	98.8
	CS(4)	98.2	98.8
15. *A. rotundiloba* early form	CPS-H	96.9	97.2
(Bryant, 1921)	VS-E	96.9	97.5
Klapper, 1985	CS(5)	96.9	98.2
16. *A. rotundiloba* late form	CPS-H	97.5	98.3
(Bryant, 1921)	VS-E	97.5	98.7
Klapper, 1985; Kralick, 1994	CS(7)	97.5	99.6
17. *A. rugosa*	VS-E	98.4	99.1
Branson and Mehl, 1934	CS(6)	98.4	100.3
18. *A. subrotunda*	Coumiac	122.5	
Miller and Youngquist, 1947	CS(3)	122.5	130.7
19. *A. triangulata*	CPS-H	98.5	100.0
Kralick, 1994	VS-E	98.5	98.9
	CS(3)	98.5	100.0
20. *Ancyrognathus amana*	CVS	126.5	127.0
Müller and Müller, 1957	Coumiac	126.6	130.5
	CS(6)	121.5	130.6
21. *A. amplicavus*	CPS-H	103.8	
Klapper et al., 1996	CS(2)	103.8	107.9
22. *A. ancyrognathoideus*	CPS-H	102.8	
(Ziegler, 1958)	Coumiac	103.6	104.1
	Hay-Trout Rivers	103.1	104.8
	LRD	103.7	104.4
	Luscar		103.8
	VS-E	102.0	103.0
	CS(14)	101.6	104.8
23. *A. asymmetricus*	CVS	125.0	131.0
(Ulrich and Bassler, 1926)	Coumiac	124.4	131.0
	LRD	126.6	127.1
	CS(3)	124.4	131.0

APPENDIX 1. (continued - page 2)

Species (number used in figures)	Section	Base CSU	Top CSU
24. *A. barba*	CPS-H	105.2	
Sandberg et al., 1992	LRD	105.9	106.7
	CS(4)	105.2	106.7
25. *A. coeni*	CPS-H	105.2	112.7
Klapper, 1990	CVS	108.3	111.3
	Coumiac	106.1	111.4
	LRD	105.9	112.2
	CS(8)	105.2	112.7
26. *A. cryptus*	CVS	131.8	132.1
Ziegler, 1962	CS(2)	131.8	133.6
27. *A. primus*	CPS-H	102.8	
Ji, 1986	Coumiac	103.1	104.1
	Hay-Trout Rivers	103.1	104.8
	LRD	103.7	104.4
	VS-E	102.0	
	CS(11)	102.0	105.4
28. *A. seddoni*	CPS-H	110.1	111.2
Klapper, 1990	CVS	110.5	111.3
	Coumiac	110.1	110.9
	LRD	110.9	
	CS(5)	110.1	111.3
29. *A. sinelaminus*	Coumiac	136.3	142.5
(Branson and Mehl, 1934)	CS(2)	136.3	148.8
30. *A. triangularis*	CPS-H	109.8	119.0
Youngquist, 1945	CVS	109.5	127.5
	Coumiac	109.5	125.8
	LRD	110.1	124.8
	Luscar	111.5	122.2
	CS(11)	109.5	127.4
31. *A. tsiensi*	CPS-H	106.1	113.9
Mouravieff, 1982	CVS	108.3	113.4
	Coumiac	108.2	111.4
	LRD	108.0	108.7
	CS(7)	106.1	113.9
32. *A. ubiquitus*	CVS	131.2	
Sandberg et al., 1988	Coumiac	131.1	
	CS(2)	131.1	
33. *A.* n. sp. (Coumiac)	Coumiac	131.5	
Klapper et al., 1994	CS(1)	131.5	
34. *A.* n. sp. L	CPS-H	103.8	103.8
Klapper, 1990	LRD		105.0
	VS-E	104.2	
	CS(4)	103.8	105.0
35. *A.? sp.* 1	Coumiac		104.8
Klapper, 1989	VS-E	103.8	103.8
	CS(3)	103.8	104.8?
36. *Icriodus alternatus*	CPS-H	110.1	119.0
Branson and Mehl, 1934	CVS	112.5	132.1
	Coumiac	117.0	148.3
	LRD	112.3	127.1
	Luscar	112.5	122.2
	CS(11)	109.4	148.3
37. *I. subterminus*	Hay-Trout Rivers	103.5	110.1
Youngquist, 1947	Luscar	101.3	122.1
	CS(9)	96.5	128.8
38. *I. symmetricus*	CPS-H	101.7	117.5
Branson and Mehl, 1934	CVS	108.3	123.8
	Coumiac	101.6	124.0
	LRD	100.3	122.7
	Luscar	101.7	122.2
	VS-E	97.5	103.8
	CS(25)	97.5	125.1
39. *Mesotaxis asymmetrica*	CPS-H	96.6	100.0
(Bischoff and Ziegler, 1957)	Coumiac	101.6	101.6
	LRD	100.3	102.7
	Luscar		100.5
	VS-E	96.9	101.2
	CS(20)	96.6	103.6
40. *M. bogoslovskyi*	CPS-H		100.0
Ovnatanova and Kuz'min, 1991	VS-E	99.6	99.7
	CS(8)	99.3	100.4
41. *M. johnsoni*	Hay-Trout Rivers		103.0
Klapper et al., 1996	Luscar	101.3	102.7
	CS(12)	101.0	105.4
42. *M. ovalis*	CPS-H		101.6
(Ziegler and Klapper, *in*	LRD	100.5	101.5
Ziegler et al., 1964)	VS-E	99.6	101.2
	CS(12)	98.7	103.1
43. *Ozarkodina bidentatiformis*	CPS-H	102.5	103.8
(Pham, 1979)	Coumiac	103.6	105.2
	LRD	104.2	105.0
	VS-E	102.5	103.8
	CS(7)	102.5	105.4
44. *O. dissimilis*	Hay-Trout Rivers	127.1	128.4
Klapper and Lane, 1985	CS(2)	123.5	128.4
45. *O. nonaginta*	CPS-H	103.8	105.1
Klapper et al., 1996	Coumiac	104.8	105.3
	LRD	105.1	105.8
	Luscar	103.8	105.5
	VS-E	103.8	103.8
	CS(12)	103.8	106.9
46. *O. postera*	Hay-Trout Rivers	104.1	110.7
Klapper and Lane, 1985	Luscar	111.9	125.4
	CS(4)	104.1	128.8

APPENDIX 1 (continued - page 3)

Species (number used in figures)	Section	Base CSU	Top CSU
47. *O. sannemanni*	VS-E		96.7
(Bischoff and Ziegler, 1957)	CS(3)		98.4
48. *O. trepta*	CPS-H	102.5	102.7
(Ziegler, 1958)	Coumiac		104.1
	LRD	104.2	104.4
	VS-E	103.0	103.0
	CS(8)	102.5	106.3
49. *Palmatolepis amplificata*	Luscar		111.5
Klapper et al., 1996	CS(5)	108.8	112.5
50. *P. bogartensis*	CVS	125.0	131.0
(Stauffer, 1938)	Coumiac	125.0	131.0
Klapper and Foster, 1993	LRD	125.0	127.1
	CS(8)	125.0	131.0
51. *P. boogaardi*	CVS	130.2	130.7
Klapper and Foster, 1993	Coumiac	129.6	130.7
	CS(5)	129.6	131.3
52. *P.* cf. *P. boogaardi*	Hay-Trout Rivers		127.3
Klapper et al., 1996	CS(?)	124.4	127.3
53. *P. brevis*	Luscar		111.5
as no. 52	CS(5)	110.3	111.5
54. *P. clarki*	Coumiac	132.1	136.0
Ziegler, 1962	CS(3)	132.1	138.2
55. *P. crepida*	Coumiac	141.7	153.1
Sannemann, 1955b	CS(3)	141.0	153.1
56. *P. delicatula*	CVS	132.4	
Branson and Mehl, 1934	Coumiac	132.1	136.0
	CS(3)	132.1	136.1
57. *P. domanicensis*	CPS-H	107.9	109.3
Ovnatanova, 1976	CVS	109.2	
	Coumiac	107.9	108.3
	LRD	108.4	108.7
	Luscar	108.0	112.1
	CS(16)	107.9	112.5
58. *P. ederi*	CPS-H	109.5	
Ziegler and Sandberg, 1990	CS(5)	109.0	111.9
59. *P. glabra acuta*	Coumiac	153.5	
Helms, 1963	CS(4)	153.4	158.3
60. *P. glabra prima*	Coumiac	146.0	153.1
Ziegler and Huddle, 1969	CS(3)	146.0	158.9
61. *P. hassi* form 2	CPS-H	109.2	109.4
Klapper, 1989	CVS	109.2	
	Coumiac	109.5	
	CS(4)	109.2	109.8

Species (number used in figures)	Section	Base CSU	Top CSU
62. *P. hassi* form 3	CPS-H	117.2	119.0
as no. 61	Coumiac	121.1	122.8
	LRD	118.3	119.4
	CS(4)	117.2	122.8
63. *P. hassi* s.s.	CVS	125.0	
Müller and Müller, 1957	Coumiac	125.0	127.5
Klapper and Foster, 1993	LRD	124.0	
	CS(6)	123.1	127.5
64. *P. jamieae*	CPS-H	109.8	116.3
Ziegler and Sandberg, 1990	CVS	109.5	117.5
= *P.* sp. B of Klapper and	Coumiac	110.1	120.2
Foster, 1986	LRD	109.3	118.2
	CS(7)	109.3	120.2
65. *P. kireevae*	CPS-H	107.8	117.1
Ovnatanova, 1976	CVS	108.3	117.5
Klapper et al., 1996	Coumiac	107.9	122.3
	LRD	107.3	117.7
	Luscar	108.0	112.3
	CS(18)	106.2	122.3
66. *P. ljaschenkoae*	CPS-H	107.3	107.8
Ovnatanova, 1976	CVS	109.2	
	Coumiac	107.5	108.3
	LRD	107.3	108.3
	Luscar	108.0	112.4
	CS(14)	107.3	112.4
67. *P. luscarensis*	Luscar	111.2	111.5
Klapper and Foster, 1993	CS(6)	107.9	111.5
68. *P. minuta minuta*	Coumiac	136.3	153.1
Branson and Mehl, 1934	CS(5)	136.2	162.2
69. *P. mucronata*	CPS-H	107.8	
Klapper et al., 1996	LRD	108.3	108.4
	Luscar	111.2	111.5
	CS(7)	107.8	112.5
70. *P. proversa*	CPS-H	107.5	109.4
Ziegler, 1958	CVS	109.2	
	Coumiac	107.5	108.3
	LRD	108.0	109.1
	Luscar	107.7	111.5
	CS(16)	107.5	111.5
71. *P.* aff. *P. proversa*	CPS-H	105.2	105.3
Klapper, 1989	LRD		106.7
	CS(7)	105.2	107.3
72. *P. punctata*	CPS-H	101.7	106.6
(Hinde, 1879)	Coumiac	101.6	106.6
	Hay-Trout Rivers	103.6	104.3
	LRD	101.6	106.7
	Luscar	103.8	107.5
	CS(20)	100.8	108.1
73. *P. quadrantinodosalobata*	Coumiac	136.3	148.3
Sannemann, 1955a	CS(4)	136.2	153.5

APPENDIX 1. (continued - page 4)

Species (number used in figures)	Section	Base CSU	Top CSU
74. *P.* cf. *P regularis*	Coumiac	137.6	144.5
concept followed since	CS(4)	136.2	148.8
Ziegler, 1962			
75. *P. rhenana*	CVS	130.4	130.7
Bischoff, 1956	Coumiac	129.6	130.7
	Luscar	124.3	124.6
	CS(6)	124.3	130.9
76. *P. semichatovae*	Hay-Trout Rivers	110.1	
Ovnatanova, 1976	Luscar	111.9	112.5
	CS(13)	110.1	119.5
77. *P. subperlobata*	Coumiac		152.2
Branson and Mehl, 1934	CS(4)	132.5	155.2
78. *P. tenuipunctata*	Coumiac	136.3	153.1
Sannemann, 1995b	CS(4)	132.6	153.1
79. *P. termini*	Coumiac	143.0	144.5
as no. 78	CS(2)	143.0	145.1
80. *P. transitans*	Coumiac	101.6	103.1
Müller, 1956	LRD	100.3	101.5
	Luscar	100.5	100.6
	VS-E	100.1	
	CS(17)	99.0	103.3
81. *P. triangularis*	CVS	131.4	132.1
Sannemann, 1955a	Coumiac	131.4	135.0
	CS(6)	131.4	141.0
82. *P. winchelli*	CVS	120.4	130.8
(Stauffer, 1938)	Coumiac	120.4	131.0
Klapper and Foster, 1993	Hay-Trout Rivers	127.3	127.5
	LRD	120.4	127.1
	Luscar	120.5	123.2
	CS(14)	120.4	131.3
83. *P.* aff. *P. winchelli*	Luscar	113.1	120.5
Klapper, 1989	CS(4)	113.1	124.4
84. *P.* sp. F	CVS	121.4	
unpublished	CS(2)	120.2	123.2
85. *Pandorinellina insita*	Luscar		112.5
(Stauffer, 1940)	CS(3)	96.8	112.5
86. *Pelekysgnathus planus*	Coumiac	110.6	144.5
Sannemann, 1955b	LRD		118.6
	CS(6)	103.8	148.0
87. *Playfordia primitiva*	CPS-H	98.5	100.0
(Bischoff and Ziegler, 1957)	VS-E	99.1	
	CS(6)	98.4	100.3
88. *Polygnathellus* n. sp.	CPS-H		100.0
unpublished	VS-E	99.1	99.1
	CS(5)	99.1	100.0
89. *Polygnathus aequalis*	Hay-Trout Rivers	103.5	110.1
Klapper and Lane, 1985	CS(9)	103.2	110.1
90. *P. alatus*	CPS-H	98.1	98.1
Huddle, 1934	Coumiac	105.2	111.4
	Hay-Trout Rivers	103.6	110.1
	LRD		105.8
	Luscar	101.0	120.5
	VS-E	98.2	
	CS(15)	96.8	130.6
91. *P. angustidiscus*	Hay-Trout Rivers	104.8	110.1
Youngquist, 1945	LRD	104.5	
	Luscar	112.5	124.6
	CS(13)	96.5	124.6
92. *P. aspelundi*	Hay-Trout Rivers	105.8	110.7
Savage and Funai, 1980	Luscar		112.5
	CS(2)	105.8	112.5
93. *P. brevicarina*	Hay-Trout Rivers	125.9	127.5
Klapper and Lane, 1985	CS(2)	125.9	130.9
94. *P. brevilaminus*	CVS	131.4	131.8
Branson and Mehl, 1934	Coumiac	131.4	145.9
	CS(4)	131.4	145.9
95 *P. brevis*	CVS	121.4	
Miller and Youngquist, 1947	Luscar	112.3	120.2
	CS(5)	112.3	129.5
96. *P. brevis* group	Luscar		121.1
	CS(4)	118.5	121.3
97. *P. collieri*	CPS-H	96.6	96.8
Huddle, 1981	VS-E	97.0	98.0
	CS(4)	96.6	99.7
98. *P. cristatus?*	CPS-H	96.1	96.9
Klapper, 1989	VS-E	96.7	96.9
	CS(4)	96.1	97.3
99. *P. decorosus*	CVS	108.3	129.6
Stauffer, 1938	Coumiac	109.5	129.6
	LRD	122.9	127.1
	CS(8)	108.3	131.3
100. *P.* aff. *P. decorosus*	Coumiac	130.7	
unpublished	CS(2)	130.7	
101. *P. dengleri*	CPS-H	96.1	98.1
Bischoff and Ziegler, 1957	VS-E	96.5	99.1
	CS(6)	96.1	100.3

APPENDIX 1. (continued - page 5)

Species (number used in figures)	Section	Base CSU	Top CSU
102. *P. dengleri* narrow form	CPS-H	99.9	100.0
Klapper, 1989	VS-E	98.1	101.5
	CS(4)	98.1	102.7
103. *P.* aff. *P. dengleri*	CPS-H	101.7	102.7
as no. 102	Coumiac	103.1	104.9
	LRD	100.5	105.0
	VS-E	102.5	103.8
	CS(8)	100.5	105.1
104. *P. dubius*	CPS-H	96.1	102.8
Hinde, 1879	Coumiac	103.1	104.9
	LRD	103.7	105.0
	Luscar	100.5	100.6
	VS-E	96.5	103.8
	CS(14)	96.1	105.0
105. *P. elegantulus*	Hay-Trout Rivers	103.5	103.5
Klapper and Lane, 1985	CS(2)	103.2	103.5
106. *P. ettremae*	Hay-Trout Rivers	120.5	127.3
Pickett, 1972	Luscar	120.5	127.8
	CS(5)	110.1	127.8
107. *P. evidens*	Hay-Trout Rivers	110.1	110.7
Klapper and Lane, 1985	Luscar	108.8	112.5
	CS(5)	108.0	119.9
108. *P. glaber glaber*	Coumiac	139.5	153.1
Ulrich and Bassler, 1926	CS(4)	139.5	159.5
109. *P. gracilis*	Hay-Trout Rivers	104.1	104.3
Klapper and Lane, 1985	CS(1)	104.1	104.3
110. *P. imparilis*	Hay-Trout Rivers	126.1	128.4
as no. 109	Luscar	120.5	125.4
	CS(5)	120.5	130.9
111. *P. morgani*	Hay-Trout Rivers	104.3	104.9
as no. 109	CS(1)	104.3	104.9
112. *P. ordinatus*	CPS-H	96.1	
Bryant, 1921	CS(3)	96.1	97.6
113. *P. pennatus*	CPS-H	96.6	97.8
Hinde, 1879	VS-E	96.9	98.0
	CS(4)	96.6	99.7
114. *P. planarius*	Hay-Trout Rivers	120.6	121.7
Klapper and Lane, 1985	CS(1)	120.6	121.7
115. *P.* cf. *P. planarius*	Luscar	115.4	116.1
Klapper and Lane, 1989	CS(2)	115.4	124.4

Species (number used in figures)	Section	Base CSU	Top CSU
116. *P. politus*	Coumiac	120.0	124.0
Ovnatanova, 1969	Hay-Trout Rivers	103.5	121.7
= *P. pacificus* Savage and	Luscar	105.7	125.4
Funai, 1980	CS(15)	103.5	131.3
117. *P. robustus*	Hay-Trout Rivers	102.9	103.8
Klapper and Lane, 1985	CS(3)	102.6	103.8
118. *P. samueli*	Hay-Trout Rivers	120.5	120.7
as no. 117	Luscar	120.5	122.2
	CS(7)	120.5	123.5
119. *P. timanicus*	Luscar	101.0	102.7
Ovnatanova, 1969	CS(11)	101.0	103.8
120. *P. unicornis*	Hay-Trout Rivers	110.1	126.1
Müller and Müller, 1957	Luscar	112.5	125.4
	CS(9)	110.1	129.5
121. *P. webbi*	CPS-H	97.7	119.0
Stauffer, 1938	CVS	108.3	131.0
= *P. normalis* Miller and	Coumiac	105.9	131.0
Youngquist, 1947	Hay-Trout Rivers	103.2	104.3
Klapper, 1971	LRD	107.3	127.1
	Luscar		112.5
	VS-E	97.5	103.0
	CS(25)	96.9	131.0
122. *P.* n. sp. (Coumiac)	CVS	131.4	132.1
Klapper et al., 1994	Coumiac	131.4	
	CS(2)	131.4	132.1
123. *P.* n. sp. F	Hay-Trout Rivers	105.4	106.3
Klapper and Lane, 1985	CS(1)	105.4	106.3
124. *P.* n. sp. G	Hay-Trout Rivers	107.2	
as no. 123	CS(1)	107.2	
125. *P.* n. sp. R	Luscar	121.1	126.8
Klapper and Lane, 1989	CS(3)	120.5	126.8
126. *Skeletognathus norrisi*	CPS-H	96.4	96.8
(Uyeno, 1967)	VS-E	96.7	96.9
	CS(5)	96.4	98.0
127. *Tortodus variabilis*	CPS-H	96.1	96.8
(Bischoff and Ziegler, 1957)	VS-E	96.5	96.7
	CS(3)	96.1	97.1

REFERENCES CITED

Aristov, V. A., 1988, Devonskie konodonty tsentral′nogo Devonskogo Polya (Russkaya Platforma) (Devonian conodonts of the Central Devonian Field [Russian Platform]): Akademiya Nauk SSSR, Geologicheskiy Institut, Trudy, v. 432, 119 p.

Bischoff, G., 1956, Oberdevonische Conodonten (to Iδ) aus dem Rheinischen Schiefergebirge: Notizblatt des Hessischen Landesamtes für Bodenforschung, v. 84, p. 115–137.

Bischoff, G., and Ziegler, W., 1957, Die Conodontenchronologie des Mitteldevons und des tiefsten Oberdevons: Abhandlungen des Hessischen Landesamtes für Bodenforschung, no. 22, 136 p.

Branson, E. B., and Mehl, M. G., 1934, Conodonts from the Grassy Creek Shale of Missouri: University of Missouri Studies, v. 8, p. 171–259. [imprint 1933]

Bryant, W. L., 1921, The Genesee conodonts, with descriptions of new species: Buffalo Society of Natural Sciences Bulletin, v. 13, p. 1–59.

García-López, S., 1981, Nueva especie de *Ancyrodella* (Conodonta) del Devónico superior de la Cordillera Cantábrica (no. de España): Estudios geológicos, v. 37, p. 263–267.

Glenister, B. F., and Klapper, G., 1966, Upper Devonian conodonts from the Canning Basin, Western Australia: Journal of Paleontology, v. 40, p. 777–842.

Helms, J., 1963, Zur "Phylogenese" und Taxionomie von *Palmatolepis* (Conodontida, Oberdevon): Geologie, v. 12, p. 449–485.

Hinde, G. J, 1879, On conodonts from the Chazy and Cincinnati Group of the Cambro-Silurian, and from the Hamilton and Genesee-Shale divisions of the Devonian, in Canada and the United States: Quarterly Journal of the Geological Society of London, v. 35, p. 351–369.

Huddle, J. W., 1934, Conodonts from the New Albany Shale of Indiana: Bulletins of American Paleontology, v. 21, 137 p.

Huddle, J. W., assisted by Repetski, J. E., 1981, Conodonts from the Genesee Formation in western New York: U.S. Geological Survey Professional Paper 1032-B, 66 p.

Ji, Q., 1986, Conodonts, in Hou, H., Ji, Q., Xian, S., and Wang, J. eds., Middle-Upper Devonian boundary in Maanshan of Xiangzhou, Guangxi: Beijing, Geological Publishing House, p. 18–50, 113–115.

Kirchgasser, W. T., and Klapper, G., 1992, Graphic correlation using litho- and biostratigraphic markers in the uppermost Middle and Upper Devonian (Frasnian) of New York State: Geological Society of America Abstracts with Programs, v. 24, no. 3, p. 32.

Klapper, G., 1971, Sequence within the conodont genus *Polygnathus* in the New York lower Middle Devonian: Geologica et Palaeontologica, v. 5, p. 59–79.

Klapper, G., 1985, Sequence in conodont genus *Ancyrodella* in Lower *asymmetricus* Zone (earliest Frasnian, Upper Devonian) of the Montagne Noire, France: Palaeontolographica Abteilung A, v. 188, p. 19–34.

Klapper, G., 1989, The Montagne Noire Frasnian (Upper Devonian) conodont succession, *in* McMillan, N. J., Embry, A. F., and Glass, D. J., eds., Devonian of the world: Calgary, Canadian Society of Petroleum Geologists Memoir 14, v. 3, p. 449–468. [imprint, 1988]

Klapper, G., 1990, Frasnian species of the Late Devonian conodont genus *Ancyrognathus*: Journal of Paleontology, v. 64, p. 998–1025.

Klapper, G., and Foster, C. T., Jr., 1986, Quantification of outlines in Frasnian (Upper Devonian) platform conodonts: Canadian Journal of Earth Sciences, v. 23, p. 1214–1222.

Klapper, G., and Foster, C. T., Jr., 1993, Shape analysis of Frasnian species of the Late Devonian conodont genus *Palmatolepis*: Paleontological Society Memoir 32 (Journal of Paleontology, v. 67, no. 4, Supplement), 35 p.

Klapper, G., and Lane, H. R., 1985, Upper Devonian (Frasnian) conodonts of the *Polygnathus* biofacies, N.W.T., Canada: Journal of Paleontology, v. 59, p. 904–951.

Klapper, G., and Lane, H. R., 1989, Frasnian (Upper Devonian) conodont sequence at Luscar Mountain and Mount Haultain, Alberta Rocky Mountains, *in* McMillan, N. J., Embry, A. F., and Glass, D. J., eds., Devonian of the world: Calgary, Canadian Society of Petroleum Geologists Memoir 14, v. 3, p. 469–478. [imprint 1988]

Klapper, G., Feist, R., Becker, R. T., and House, M. R., 1994, Definition of the Frasnian/Famennian Stage boundary: Episodes, v. 16, p. 433-441. [imprint, 1993]

Klapper, G., Kirchgasser, W. T., and Baesemann, J. F., 1995, Graphic correlation of a Frasnian (Upper Devonian) composite standard, in Mann, K. O., and Lane, H. R., eds., Graphic correlation: SEPM (Society for Sedimentary Geology) Special Publication 53, p. 177–184.

Klapper, G., Kuz′min, A. V., and Ovnatanova, N. S., 1996, Upper Devonian conodonts from the Timan-Pechora region, Russia, and correlation with a Frasnian composite standard: Journal of Paleontology, v. 70, p. 131–152.

Kralick, J. A., 1994, The conodont genus *Ancyrodella* in the middle Genesee Formation (Lower Upper Devonian, Frasnian), western New York: Journal of Paleontology, v. 68, p. 1384–1395.

MacLeod, N., 1995, Graphic correlation of new Cretaceous/Tertiary (K/T) boundary successions from Denmark, Alabama, Mexico, and the southern Indian Ocean: Implications for a global sediment accumulation model, *in* Mann, K. O., and Lane, H. R., eds., Graphic correlation: SEPM (Society for Sedimentary Geology) Special Publication 53, p. 215–233.

Mann, K. O., and Lane, H. R., editors, 1995, Graphic correlation: SEPM (Society for Sedimentary Geology) Special Publication 53, 263 p.

Miller, A. K., and Youngquist, W., 1947, Conodonts from the type section of the Sweetland Creek Shale in Iowa: Journal of Paleontology, v. 21, p. 501–517.

Mouravieff, A. N., 1982, Conodont stratigraphic scheme of the Frasnian of the Ardennes, *in* Bigey, F., and 14 others, Papers on the Frasnian-Givetian Boundary: Brussels, Geological Survey of Belgium, p. 101–118.

Müller, K. J., 1956, Zur Kenntnis der Conodonten-Fauna des europäischen Devons. 1: Die Gattung *Palmatolepis*: Abhandlungen der Senckenbergischen Naturforschenden Gesellschaft, no. 494, 70 p.

Müller, K. J., and Müller, E. M., 1957, Early Upper Devonian (Independence) conodonts from Iowa, part I: Journal of Paleontology, v. 31, p. 1069–1108.

Murphy, M. A., 1977, On time-stratigraphic units: Journal of Paleontology, v. 51, p. 213–219.

Orchard, M. J., 1989, Conodonts from the Frasnian-Famennian boundary interval in western Canada, *in* McMillan, N. J., Embry, A. F., and Glass, D. J., eds., Devonian of the world: Calgary, Canadian Society of Petroleum Geologists Memoir 14, v. 3, p. 35–52. [imprint 1988]

Ovnatanova, N. S., 1969, Novye verkhnedevonskie konodonty tsentral′nykh rayonov Russkoy Platformy i Timana (New Upper Devonian conodonts from central regions of the Russian Platform and the Timan): Vsesoyuznyy Nauchno-Issledovatel′skiy Geologorazvedochnyy Neftyanoy Institut, Trudy, Vypusk 93, p. 139–141.

Ovnatanova, N. S., 1976. Novye pozdnedevonskie konodonty Russkoy Platformy (New Late Devoniam conodonts of the Russian Platform): Paleontologicheskiy Zhurnal, v. 1976, no. 2, p. 106–115.

Ovnatanova, N. S., and Kuz′min, A. V., 1991, Konodonty tipovykh razrezov Domanikovoy svity na yuzhnom Timane (Conodonts from type sections of the Domanik Formation in southern Timan): Izvestiya Akademii Nauk SSSR, Seriya Geologicheskaya, v. 1991, no. 3, p. 37–50.

Perri, M. C., and Spalletta, C., 1981, *Ancyrodella pramosica* n. sp. from the Lower Frasnian of the Carnic Alps: Società Paleontologica Italiana Bollettino, v. 20, p. 93–98.

Pham, K. N., 1979, Conodonts from limestone pieces of the Late Paleozoic conglomerates in north-west Bulgaria: Geologica Balcanica, v. 9, p. 77–92.

Pickett, J., 1972, Late Devonian (Frasnian) conodonts from Ettrema, New South Wales: Journal and Proceedings, Royal Society of New South

Wales, v. 105, p. 31–37.

Sandberg, C. A., Ziegler, W., Dreesen, R., and Butler, J. L., 1988, Late Frasnian mass extinction: Conodont event stratigraphy, global changes, and possible causes: Courier Forschungsinstitut Senckenberg, no. 102, p. 263–307.

Sandberg, C. A., Ziegler, W., Dreesen, R., and Butler, J. L., 1992, Conodont biochronology, biofacies, taxonomy, and event stratigraphy around Middle Frasnian Lion Mudmound (F2h), Frasnes, Belgium: Courier Forschungsinstitut Senckenberg, no. 150, 87 p.

Sannemann, D., 1995a, Beitrag zur Untergliederung des Oberdevons nach Conodonten: Neues Jahrbuch für Geologie und Paläontologie, Abhandlungen, v. 100, p. 324–331.

Sannemann, D., 1955b, Oberdevonische Conodonten (to IIα): Senckenbergiana Lethaea, v. 36, p. 123–156.

Savage, N. M., and Funai, C. A., 1980, Devonian conodonts of probable early Frasnian age from the Coronados Islands of southeastern Alaska: Journal of Paleontology, v. 54, p. 806–813.

Shaw, A. B., 1964, Time in stratigraphy: New York, McGraw-Hill, 365 p.

Shaw, A. B., 1969, Presidential address: Adam and Eve, paleontology and the non-objective arts: Journal of Paleontology, v. 43, p. 1085–1098.

Stauffer, C. R., 1938, Conodonts of the Olentangy Shale: Journal of Paleontology, v. 12, p. 411–443.

Stauffer, C. R., 1940, Conodonts from the Devonian and associated clays of Minnesota: Journal of Paleontology, v. 14, p. 417–435.

Ulrich, E. O., and Bassler, R. S., 1926. A classification of the toothlike fossils, conodonts, with descriptions of American Devonian and Mississippian species: Proceedings of the U. S. National Museum, v. 68, 63 p.

Uyeno, T. T., 1967, Conodont zonation, Waterways Formation (Upper Devonian), northeastern and central Alberta: Geological Survey of Canada Paper 67–30, 21 p.

Wang, K., and Geldsetzer, H. H. J., 1995, Late Devonian conodonts define the precise horizon of the Frasnian-Famennian boundary at Cinquefoil Mountain, Jasper, Alberta: Canadian Journal of Earth Sciences, v. 32, p. 1825–1834.

Youngquist, W., 1945, Upper Devonian conodonts from the Independence Shale (?) of Iowa: Journal of Paleontology, v. 19, p. 355–367.

Youngquist, W., 1947, A new Upper Devonian conodont fauna from Iowa: Journal of Paleontology, v. 21, p. 95–112.

Ziegler, W., 1958, Conodontenfeinstratigraphische Untersuchungen an der Grenze Mitteldevon/Oberdevon und in der Adorfstufe: Notizblatt des Hessischen Landesamtes für Bodenforschung, v. 87, p. 7–77.

Ziegler, W., 1962, Taxionomie und Phylogenie Oberdevonischer Conodonten und ihre stratigraphische Bedeutung: Abhandlungen des Hessischen Landesamtes für Bodenforschung, no. 38, 166 p.

Ziegler, W., and Huddle, J. W., 1969, Die *Palmatolepis glabra*–Gruppe (Conodonta) nach der Revision der Typen von Ulrich & Bassler durch J. W. Huddle: Fortschritte in der Geologie von Rheinland und Westfalen, v. 16, p. 377–386.

Ziegler, W., and Sandberg, C. A., 1990, The Late Devonian standard conodont zonation: Courier Forschungsinstitut Senckenberg, no. 121, 115 p.

Ziegler, W., Klapper, G., and Lindström, M., 1964, The validity of the name *Polygnathus* (Conodonta, Devonian and Lower Carboniferous): Journal of Paleontology, v. 38, p. 421–423.

MANUSCRIPT ACCEPTED BY THE SOCIETY FEBRUARY 27, 1997

Geological Society of America
Special Paper 321
1997

A new zonation of middle Lochkovian (Lower Devonian) conodonts and evolution of Flajsella *n. gen. (Conodonta)*

J. Ignacio Valenzuela-Ríos
Institut für Geowissenschaften, Technische Universität Braunschweig, Pokelsstr. 4, Braunschweig D-38023, Germany, and Departamento de Geología, Universidad de Valencia, Dr. Moliner, 50, 46100 Burjassot (Valencia), Spain
Michael A. Murphy
Geology Department, University of California, Davis, California 95616

ABSTRACT

We divide the Lochkovian into three parts. The base of the lower part coincides with the Silurian-Devonian boundary; the base of the middle part is taken at the appearance of *Ancyrodelloides omus* α Murphy and Matti; the base of the upper part is taken at the appearance of *Criteriognathus pandora* β (Murphy, Matti, and Walliser). The middle Lochkovian, thus, is characterized by the genus *Ancyrodelloides* and is divided into three zones based on the first evolutionary occurrences of *A. eleanorae* and *A. trigonicus*, respectively, as the criteria for the bases of the intra-middle Lochkovian zones.

The conodont genus *Flajsella* n. gen. includes five species that are confined to the upper two-thirds of the middle Lochkovian and the base of the upper Lochkovian (Lower Devonian) of western North America and Europe. It is an uncommon element of the faunas, but the Pa element of the apparatus is so distinctive that its species make excellent indices for the middle and upper parts of the middle Lochkovian and the base of the upper Lochkovian. The consistent stratigraphic position and morphology of its earliest representative, *Flajsella schulzei* (Bardashev), suggest that the origin of the genus is from the *"Ozarkodina" excavata* group of species in the lower part of the middle Lochkovian. Three additional *Flajsella* with partially or completely overlapping ranges are found in the succeeding strata, *F. sigmostygia* n. sp., *F. stygia* (Flajs), and *F. lanei* (Bardshev). In the upper part of the middle Lochkovian, *Flajsella streptostygia* n. sp. evolved from *F. stygia* and continues into the base of the upper Lochkovian as the sole representative of the genus.

INTRODUCTION

The purposes of this chapter are to introduce the zones *eleanorae-trigonicus* and *trigonicus-pandora* β for the upper part of the *Ancyrodelloides*-bearing strata of the Lochkovian Stage (Lower Devonian); to erect a new genus, *Flajsella*, for the taxa formerly grouped in "*Ozarkodina stygia*" (Flajs); to describe its species of *Flajsella* and their evolution; to present the biostratigraphy of the group based on sequences in Nevada and Spain; and to compare them with sequences described in Alaska (Lane and Ormiston, 1979) and the Carnic Alps (Schönlaub, 1980). For purposes of this discussion, the Lochkovian Stage is informally divided into three parts: lower, middle, and upper.

The study of *Flajsella* n. gen. was undertaken because this distinctive and widely distributed, but not abundant, taxon might offer a way of correlating several levels within the middle Lochkovian sequences in Nevada and Spain (Figs. 1 and 2). The taxon has been subdivided informally by Lane and Ormiston

Valenzuela-Ríos, J. I., and Murphy, M. A., 1997, A new zonation of middle Lochkovian (Lower Devonian) conodonts and evolution of *Flajsella* n. gen. (Conodonta), *in* Klapper, G., Murphy, M. A., and Talent, J. A., eds., Paleozoic Sequence Stratigraphy, Biostratigraphy, and Biogeography: Studies in Honor of J. Granville ("Jess") Johnson: Boulder, Colorado, Geological Society of America Special Paper 321.

(1979) and by Schönlaub (1980). Recently, Bardashev (1989) established subspecific names for four of the morphological variants, which correspond in part to the morphotypes of the earlier authors. Unfortunately, he did not include any data on the ranges of the various subspecies, listing them all from the "*Optima* Zone," which ranges from the middle Lochkovian through the Pragian. We doubt that *Flajsella* ranges as high as the base of the Pragian. In fact, it probably does not range any higher than the lower part of the *pandora* β-*gilberti* biostratigraphic unit of Valenzuela-Ríos (1994a, c).

The range and origin of *Flajsella* have been enigmatic because of its relative rarity in most samples and because the ranges of other Lochkovian taxa with which its range had been compared had not been well worked out. This was a particular problem with respect to species within the genus *Pedavis*. *Pedavis pesavis*, the former zonal index in the upper Lochkovian conodont-based zonal scheme (Klapper and Johnson, 1980), had been interpreted in a very broad way to include what we believe to be several distinct taxa (Valenzuela-Ríos, 1994b, c). For these reasons, the *pesavis* Zone is not defined, nor is it well characterized. Therefore, the ranges and dating of other taxa associated with *P. pesavis* and dependent on it for their own relative ranges have not been accu-

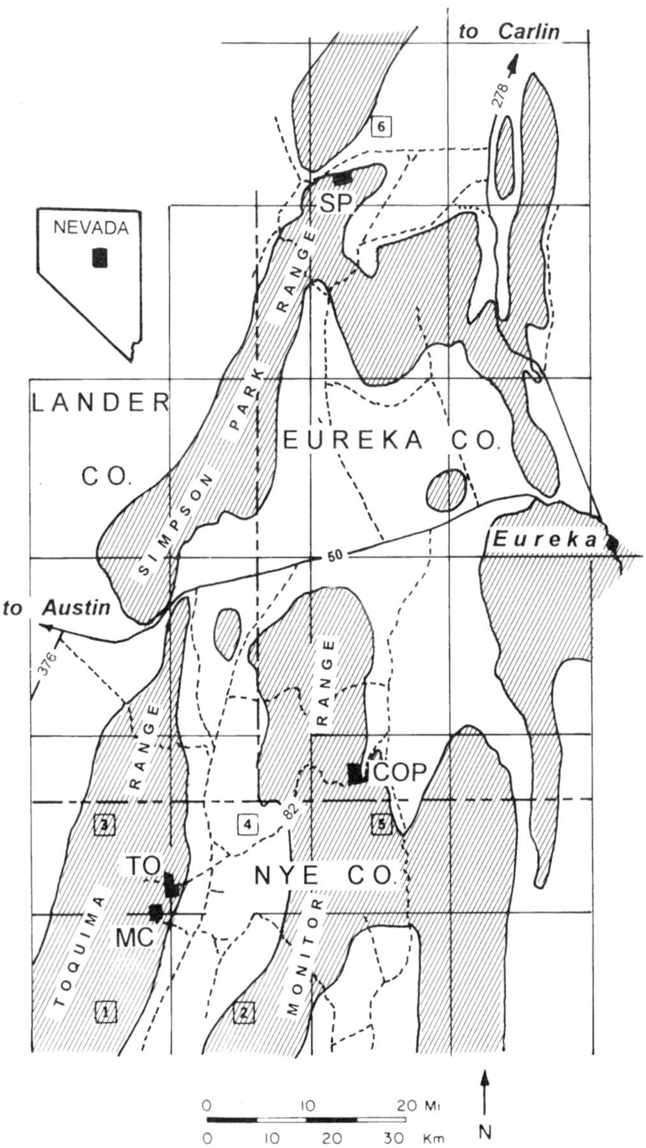

Figure 1. Location map of sections in Nevada. COP indicates the Copenhagen Canyon sections in the Monitor Range; MC, the Mill Canyon section in the Toquima Range; SP, the Coal Canyon sections in the northern Simpson Park Range; TO, the Ikes Canyon section in the Toquima Range.

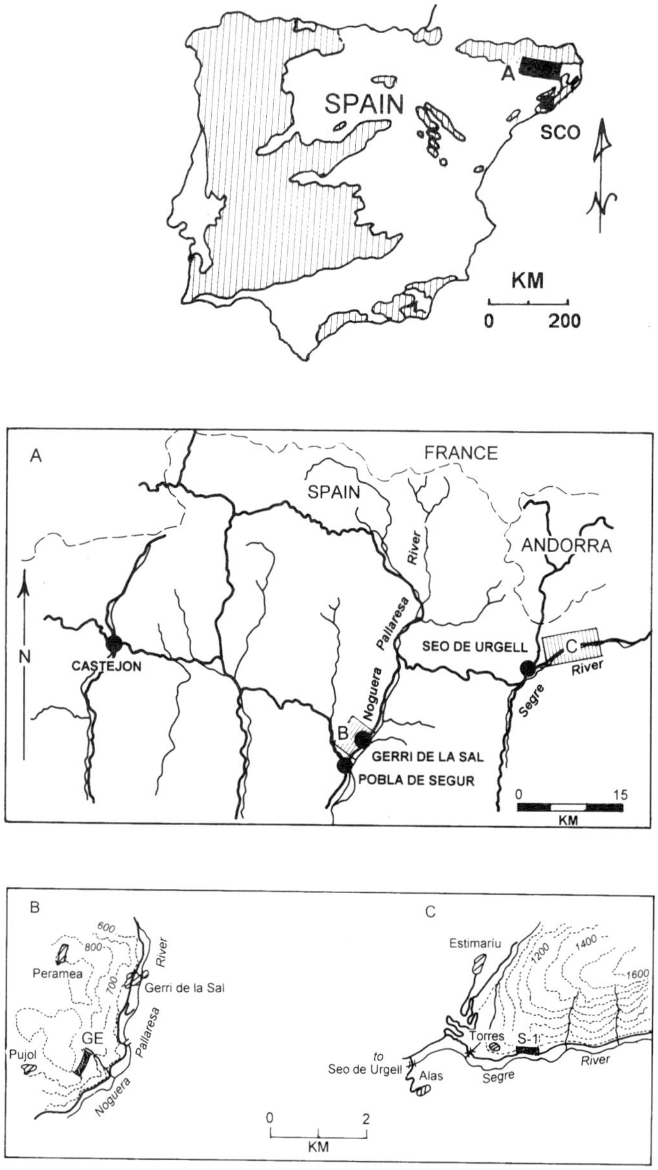

Figure 2. Location map of sections in Spain. A indicates the area of the Spanish Pyrenees localities; SCO, the area of the Catalonian Coastal Ranges; B, the Gerri de la Sal sections (GE); C, the Segre sections (S1). Vertical stripe pattern indicates outcrop of Paleozoic and crystalline rocks.

rately represented. *P. pesavis* was described from a locality in Frankenwald, Germany, that was thought to be a condensed deposit (Bischoff and Sannemann, 1958). The original sample of Bischoff and Sannemann contained many elements of taxa now known to occupy the upper part of the middle Lochkovian (approximately the upper *delta* Zone of Klapper and Johnson, 1980 = the *trigonicus-pandora* β Zone, described below). We now believe *P. pesavis* in a strict sense is restricted stratigraphically to the *trigonicus-pandora* β Zone and that the Frankenwald assemblage mostly represents this fauna. To improve the zonal classification of the upper Lochkovian, a new zonal scheme using widespread and relatively abundant taxa, whose biostratigraphy is established, has been proposed (Valenzuela-Ríos, 1994c). This new zonal scheme uses the *Criteriognathus pandora* lineage (Murphy et al., 1981) and new data concerning the ranges of *Pedavis* species. We further subdivide the Lochkovian in its middle part by describing three new biostratigraphic units (Fig. 3) in addition to those proposed by Valenzuela-Ríos (1994c).

In the text that follows, we have placed some taxa in different categories than previous authors. We have, for example, included the "*Ozarkodina*" *eosteinhornensis-remscheidensis-pandora* plexus of morphologies in *Criteriognathus* Walliser (1972) and excluded the "*Ozarkodina*" *excavata* group of morphologies from *Ozarkodina*. We restrict *Ozarkodina* herein to forms similar to the type species, *O. confluens*, with a tendency for right-handed symmetry in the Pa element.

NEW LOCHKOVIAN ZONATION

Our original purpose in this study was to describe only the succession of species in the genus *Flajsella* and not to set up a new zonal framework for the middle Lochkovian. However, we needed to put the sequence of species within *Flajsella* in an unambiguous stratigraphic scheme that can be globally applied, and we believe that the conodont-based zonation of the Lochkovian derived from the North American Cordillera (Klapper, 1977; Klapper and Johnson, 1980) is not adequate for intercontinental correlation because we do not know the evolutionary history of either *Amydrotaxis praejohnsoni*, the zonal defining species, or *Ancyrodelloides delta*, the characterizing species. Consequently, we do not use the Cordilleran zonation here. Instead, as a substitute hypothesis, we propose several datums that are based upon evolution within the genus *Ancyrodelloides* and that occupy globally consistent stratigraphic positions. This scheme is shown in Figures 3 and 4. (Above, we use the term "globally" to mean on a global scale. This does not mean, however, the zonation is everywhere applicable.)

LOCHKOVIAN STAGE	Division	Global Zone	Cordilleran Zone
	Upper	gilberti	pesavis
		pandora β	? ?
	Middle	trigonicus	delta
		eleanorae	
		omus α	
	Lower	not zoned	eurekaensis / hesperius

Figure 3. Zonation proposed in this chapter compared with that in general use based on the zonation in the Cordilleran region.

We believe that the Lochkovian can be subdivided into three parts: a lower part before the origin of the conodont genus *Ancyrodelloides* s.l. (= lower Lochkovian), a part with *Ancyrodelloides* (= middle Lochkovian), and a part essentially without *Ancyrodelloides* (= upper Lochkovian).

The base of the upper Lochkovian in this scheme is defined by the appearance of *Criteriognathus pandora* β and the following *pandora* β–*gilberti* Zone yields *Ancyrodelloides* of the *trigonicus* group in its lowest part. The middle Lochkovian can be divided into three parts on the basis of the first occurrences of species of two lineages within the genus *Ancyrodelloides* as presently classified: *A. omus* alpha morph, *A. eleanorae*, and *A. trigonicus* (see also Valenzuela-Ríos, 1994b, fig. 26). *Flajsella* is found in the upper two intervals of the middle Lochkovian and the lower part of the upper Lochkovian.

Ancyrodelloides omus α–*Ancyrodelloides eleanorae* Zone

The earliest members of the genus *Ancyrodelloides* (as represented by Murphy and Matti, 1983) belong to the *eleanorae*-group of species and include a series of forms leading from *A. omus* α to *A. eleanorae* (Klapper and Murphy, 1980, figs. 4.1, 4.7–4.9, 4.13, 4.14). Using this classification of *Ancyrodelloides* (the separation of the *eleanorae* group as a separate clade), the *Ancyrodelloides*-bearing interval is approximately equal to the *delta* Zone of the Cordilleran Province zonation (Klapper, 1977; Klapper and Johnson, 1980).

The most widespread and easily recognized *Ancyrodelloides* species found either in the *delta* Zone or the new *omus* α–*eleanorae* Zone is *A. transitans*. However, the lowest occurrence of *A. transitans* occupies an inconsistent position in the sequence of biostratigraphic events in Nevada, Spain, Alaska, and the Carnic Alps. In addition, as for *Am. praejohnsoni* and *A. delta*, we do not know the details of the origin of *A. transitans* and so have chosen *A. omus* α Murphy and Matti, which is a slightly lower datum.

Klapper and Johnson (1980) specified that the *delta* Zone was applicable only to the Cordilleran Province, but it has been adopted without comment for the rest of the world by subsequent authors. We believe *A. omus* α is a better reference datum than the others for this part of the column, because its first occurrence is interpreted as an evolutionary event in the COP IV and SP VII sections in Nevada.

The zone is characterized by the lowest occurrences of

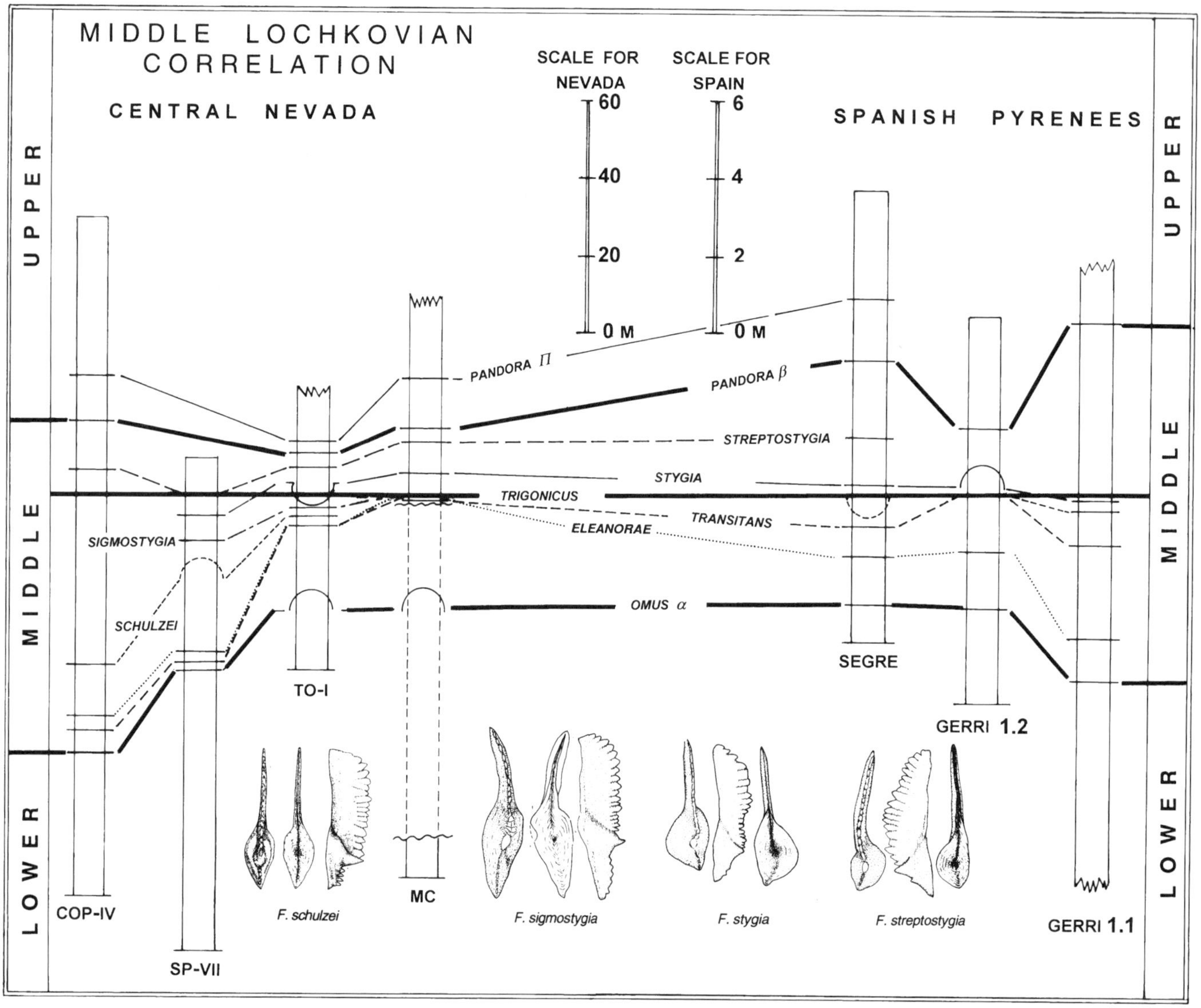

Figure 4. Correlation of middle Lochkovian sections. Abbreviations for the Nevada sections are: COP-IV—section IV, Copenhagen Canyon, Monitor Range; SP-VII—section VII, Coal Canyon, northern Simpson Park Range; TO-I—section I, near Ikes Canyon, Toquima Range; MC—Mill Canyon section, Toquima Range.

Ancyrodelloides transitans, *Ancyrodelloides omus* β, *A. delta*, and *Pedavis breviramus*.

Ancyrodelloides eleanorae–Ancyrodelloides trigonicus Zone

The lower boundary of the *Ancyrodelloides eleanorae–Ancyrodelloides trigonicus* Zone is taken at the appearance of *A. eleanorae*, a distinctive and widespread species that is found in Alaska, Nevada, and Spain. Its appearance is interpreted as an evolutionary event in the lineage *A. omus* α–*A. eleanorae*.

The zone is characterized by the appearance of the genus *Flajsella* and late forms of *A. eleanorae* (Klapper and Murphy, 1980, figs. 4.1, 4.7–4.9, 4.13, 4.14).

Ancyrodelloides trigonicus–Criteriognathus pandora β Zone

The lower boundary of the *Ancyrodelloides trigonicus–Criteriognathus pandora* β Zone is defined by the lowest occurrence of *A. trigonicus* Bischoff and Sannemann. The appearance of *A. trigonicus* is interpreted as an evolutionary event. It has been shown that the species arises from *A. transitans* by its close morphologic and stratigraphc relations (Murphy and Matti, 1983; Murphy and Cebecioglu, 1987; Valenzuela-Ríos, 1994b). It is a distinctive and widespread species (Klapper and Johnson, 1980) whose geographic distribution includes Spain, Nevada, Carnic Alps, and Alaska.

The zone is characterized by the lowest occurrence of *A.*

kutscheri, *Flajsella streptostygia*, *Kimognathus alexii*, and the graptolite *Monograptus hercynicus*.

SYSTEMATIC PALEONTOLOGY

The bulk of the material treated here comes from four sections in central Nevada (Fig. 1) and three in the Spanish Pyrenees (Fig. 2). Additional Spanish material, from a section in the Santa Creu d'Olorda area in the Catalonian Coastal Ranges (García-López et al., 1990; Fig. 2), was made available to us by Susana García-López during a visit by J. I. V-R to Oviedo in 1992. Additional material from central Nevada was shown us by G. Klapper during a visit to Iowa City in 1992. Klapper's collection, labeled IC, corresponds to a section located approximately 50 m south of the IK-IV section in the Toquima Range (Murphy, 1989, fig. 4).

The abbreviations for stratigraphic occurrences used below are:

COP-IV—section IV in Copenhagen Canyon, Monitor Range, central Nevada, United States (Matti et al., 1975).

Ge 1.1 and Ge 1.2—sections 1 and 2 at the Gerri de la Sal outcrop area, central Pyrenees, Spain (Valenzuela-Ríos, 1990, 1994b, c).

MC—section at Gatecliff in Mill Canyon, Toquima Range, central Nevada (Murphy and Matti, 1983).

SCO—Set of sections at Santa Creu d'Olorda, Catalonian Coastal Ranges, Spain (García-López et al., 1990).

Se 1—section at Segre, Spanish central Pyreneees (Valenzuela-Ríos, 1994b, c).

SP-VII—section VII at Coal Canyon, northern Simpson Park Range, Nevada (Berry and Murphy, 1975).

TO-I—section I near Ikes Canyon, Toquima Range, central Nevada (Murphy, 1989, fig. 4).

The shorthand system used in the text to specify localities and material is as follows: Section name/bed (number of specimens). For example, Ge 1.1/17al (2) means two specimens from bed 17a1 of section Gerri 1.1.

We use the descriptive nomenclature of Sweet (1988) where appropriate. The term "carina" is used for a crest formed by a row of denticles. The abbreviations for repository institutions are as follows:

DPO—Department of Paleontology, University of Oviedo, Spain.

MPZ—Museum of Paleontology, University of Zaragoza, Spain.

UCR—University of California, Riverside, California.

All the material from the Pyrenees is deposited at MPZ; the material from SCO is deposited at DPO; the material from Nevada formerly at UCR is now housed at University of California, Davis, California.

Family SPATHOGNATHODONTIDAE Hass, 1959

Genus *Flajsella* new genus

Type species. Spathognathodus stygius Flajs, 1967.

Diagnosis. A spathognathodontid genus whose Pa element, derived from a carminiscaphate ancestor, has a large posterior basal cavity, vestigial posterior blade with small denticles or lacking denticles, and a prominent, nearly conical cusp.

Derivation of the name. In honor of Gerd Flajs, author of the species *Spathognathodus stygius.*

Previous work. Flajs (1967) described the species *Spathognathodus stygius* (=*Flajsella stygia*) from a sample taken in a pipeline tunnel in the Eisenerz region of Austria and suggested that the species came from the Upper Silurian *eosteinhornensis* Zone of Walliser (1964), on the basis of the cooccurrence of Walliser's "H" and "J" conodont apparatuses. The age was soon corrected in two other places in the general region (Schulze, 1968; Pölsler, 1969). Both papers reported its joint occurrence with *Ancyrodelloides transitans*, a Lower Devonian species. It was also reported in the Catalogue of Conodonts (Klapper, 1973, p. 228, 249) from Yukon Territory and Nevada in Lower Devonian rocks.

The specimens illustrated by Flajs (1967) are small; the maximum length of the Pa element is only 0.44 mm. This has caused some problems taxonomically because the ontogenies of the various morphs were unknown, and it was uncertain whether or not the early ontogenies of all the different morphs were so similar as to be indistinguishable or whether they maintained their distinctive character throughout ontogeny. The smallest specimens that we have—one specimen of *F. schulzei*, one of *F. sigmostygia*, and two of *F. stygia*—indicate that they established distinct characteristics at a length of about 0.4 mm.

Lane and Ormiston (1979, p. 57) described a series of *Flajsella* morphs from the Salmontrout River area in Alaska and designated them by the Greek letters. They noted that their beta morph was the oldest, that the alpha and gamma morphs were found together, and that the delta morph was uppermost in their section. Schönlaub (1980, pl. 4) figured a comparable series of morphs from the Carnic Alps and applied the same Greek letters to them, but in a different way—in order of their stratigraphic appearance. Murphy and Matti (1983, p. 10) adopted Schönlaub's notation and pointed out that the post-alpha morphs have overlapping ranges. Bardashev (1989, p. 7) established four subspecies that included the previously described morphotypes of Lane and Ormiston (1979) and Schönlaub (1980), but he grouped them in a third way. Bardashev (1989) indirectly indicated that some overlap of the morph ranges occurs because his synonymies for his subspecies *stygia* s.s., *schulzei*, and *sorokini* all include more than one of Lane and Ormiston's (1979) morphs. However, he did not include any biostratigraphic data in his paper. Wilson (1989, pl. 4 figs. 22–28) recognized the delta morph of Lane and Ormiston (1979) in New South Wales, Australia. García-López et al. (1990) recognized the overlap of the ranges of Lane and Ormiston's morphs beta and gamma in the Catalonian Coastal Ranges of Spain. Rodriguez-Cañero et al. (1990, figs. 2.1, 2.2) figured a specimen of *Flajsella* sp. from northern Morocco that was found in a sample with *Ancyrodelloides transitans*. Bardashev and Ziegler (1992), in a summary of Bardashev's earlier work, gave stratigraphic data that show virtually complete overlap of all of the "subspecies" of Bardashev (1989). Valenzuela-Ríos (1994b) recognized the four morphotypes of Schönlaub in the Spanish Pyrenees and noted that alpha occurs first, then alpha, beta, and gamma overlap, and delta occurs last. He also pointed out that this sequence from the Pyrenees is comparable with that from the Salmontrout River area in Alaska. Figure 5 summarizes the previous nomenclatorial schemes and relates them to one another.

Discussion. The taxa here assigned to *Flajsella* were formerly assigned to *Ozarkodina*. The reorganization of Devonian taxa prompted by the description of the standard Devonian skeletal apparatus types by Klapper and Philip (1971, 1972) suppressed the all-inclusive genus

this paper	Lane & Ormiston 1979	Schönlaub 1980,1985	Murphy & Matti 1983	Bardashev 1989	Valenzuela-Rios 1994b
schulzei	beta	alpha	----	*schulzei* (part)	alpha
sigmostygia	----	----	MC 3C	*schulzei* (part)	----
lanei	alpha	----	----	*lanei*	----
stygia	gamma	beta and gamma	MC 5B	*sorokini*	gamma and beta
streptostygia	delta	gamma and delta	MC 8C	*stygia*	delta

Figure 5. History of morphotype nomenclature and relationship to species now included in *Flajsella* new genus.

Spathognathodus. Most of the taxa in *Spathognathodus* were transferred to *Ozarkodina* by default. The *stygia* morphs never fit well in *Ozarkodina* because the Pa elements are so different. Nothing close to the posterior, very expanded and open basal cavity and high, peglike cusp is present in *Ozarkodina*. Part of the reason for this assignment was the lumping of the three main Early Devonian lineages of the Spathognathodontidae, *Criteriognathus*, *Ozarkodina*, and the members of the *"Ozarkodina" excavata* group, in the same genus in the early studies of Lower Devonian taxa (Fig. 6).

The main reasons that we have introduced a new generic term for the *stygia*-group of taxa are that (1) it clearly is not related to the Pa elements of *Ozarkodina* (type species, *Ozarkodina confluens*) in any feature (Fig. 7), and (2) the Pb elements that were designated by Lane and Ormiston (1979) and that are available for the reconstruction of the *Flajsella* apparatus resemble the Pb elements of the *Criteriognathus* apparatus just as closely as they do those of *Ozarkodina*.

The only previous work on the apparatus of *Flajsella* is the suggestion of Lane and Ormiston (1979, pl. 2, figs. 24–26), who figured two Pb elements that they unreservedly assigned to *F. stygia*. Their conclusion was based on the consistent association of a small number of Pb elements with a large number of Pa elements at two localities in the Salmontrout River Formation, Alaska. The principal difficulty in further reconstruction of the apparatus of *Flajsella* has been the paucity of isolated elements that can be assigned to the genus. After reconstruction of other better-known Lochkovian apparatuses that are present in our collections (e.g. *"Ozarkodina" excavata*, *Amydrotaxis praejohnsoni*) and assignment of other elements that clearly are associated with *Ancyrodelloides* species, the leftover spathognathodontid material in the range of *Flajsella* is scarce. In addition, the residual material does not exhibit uniform characteristics. Our initial opinion is based on the following: (1) these elements cannot be attributed to other taxa, (2) their stratigraphic occurrences are compatible with the range of *Flajsella*, and (3) we recognize some morphologic similarities with the Pa elements. Consequently our suggestions for apparatus reconstructions are speculative.

Our material permits speculation on the partial reconstruction of the apparatus of only three species: *F. sigmostygia* (Pa, Sa, ?Sb, and ?Sc), *F. stygia* (Pa, Pb), and *F. streptostygia* (Pa, Pb, M, Sa or Sb, and Sc). These speculations are described under the systematic descriptions

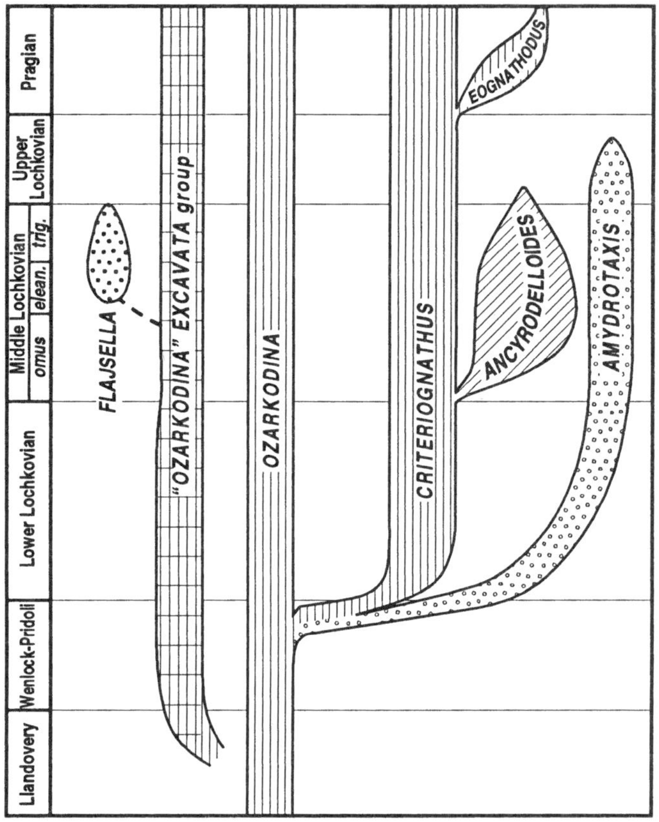

Figure 6. Diagrammatic representation of the evolutionary development of the Devonian genera in the Spathognathodontidae. *Flajsella* is interpreted here as evolving from the *"Ozarkodina" excavata* group.

of the respective species and are based on coincident stratigraphic occurrences with the corresponding Pa elements.

The above classification further restricts Ozarkodina, which in multielement taxonomy included *Criteriognathus* Walliser, *Amydrotaxis* Klapper and Murphy, and much of *Ancyrodelloides* Bischoff and Sannemann, as emended by Murphy and Matti (1983). The classification, as shown in Figure 6, still needs modification, but the emendation and the documentation for it will be presented separately.

Flajsella schulzei (Bardashev, 1989)

Figure 8: 1–4, 7–22

[?] *Spathognathodus seebergensis* SCHULZE, 1968, p. 227, Pl. 17, Fig. 9 [only].

Spathognathodus stygius Flajs. PÖLSLER, 1969, p. 433-434, Pl. 1, Figs. 5, 6 [only].

Ozarkodina stygia (Flajs) alpha morph SCHÖNLAUB, 1980, p. 39, Pl. 4, Figs. 5–8; VALENZUELA-RÍOS, 1994b, p. 66–67, Pl. 6, Figs. 16, 21.

Ozarkodina stygia schulzei BARDASHEV, 1989, p. 7-8, Pl. 1, Fig. 5 [not Fig. 4]; ?BARDASHEV AND ZIEGLER, 1992, p. 15, Pl. 1, Fig. 39 [not Figs. 25, 26].

Ozarkodina stygia (Flajs) beta morphotype of Lane and Ormiston. GARCÍA-LÓPEZ ET AL., 1990, p. 152, Pl. 1, Figs. 3, 4.

Holotype. Specimen figured by Bardashev, 1989, Pl. 1, Fig. 5.

Material. Twenty-six specimens. Ge 1.1/17a1 (2); Ge 1.1/17a2 (2); Ge 1.1/17c (1); Ge 1.2/6g (1); SCO/DC-44 (9); TO-I/17 (3); MC 5/(1?); COP-IV/205 (2); COP-IV/233 (2); IC/8a (3).

Revised diagnosis. A species of *Flajsella* based on a Pa element with anterior and posterior segments of the carina in line in upper view (without offset or angle at the cusp) and a deep, almost symmetrical, spear-shaped posterior basal cavity.

Description. Blade of Pa element straight, with anterior and posterior carinae aligned or with slight deflection of the posterior carina. Anterior carina composed of 10 to 13 denticles of approximately equal height and equal to, or slightly higher than, the cusp. Cusp slightly larger in cross section than denticles and more inflated to almost conical, located in the posterior third of the blade. Posterior carina commonly with two to four denticles appreciably lower than the cusp so that the profile steps down posteriorly, but exceptions are known. Most specimens have an adenticulate space between cusp and posterior carina. Spear-shaped, symmetrical, or slightly asymmetrical basal cavity and platform occupy approximately 50% of Pa element; maximum width is anterior to cusp and about one third of length of element from posterior. Anterior margins at higher angle to blade than posterior ones. Lower profile straight or slightly arched downward at one or both ends.

Discussion. The straight anterior carina with more or less uniform denticles is on a line with the posterior carina in upper view and distinguishes *F. schulzei* from other species of *Flajsella*. No other apparatus elements are recognized.

Range and occurrence. *F. schulzei* is restricted to the upper half of the middle part of the middle Lochkovian (=*eleanorae-trigonicus* Zone). It has been found in central Nevada (TO-I, IC, COP-IV), the Carnic Alps, Spain (Ge 1.1, Ge 1.2, SCO), and Tadzhikistan. In Tadzhikistan, Bardashev and Ziegler (1992, figs. 6, 7) report that it occurs with all other "subspecies" and enters the section above F. stygia.

Flajsella sigmostygia n. sp.

Figure 8: 23–25, 29–33, Figure 9: 5–8, ?9–12

[?] *Ozarkodina stygia* (Flajs). SCHÖNLAUB, 1980, p. 28, Pl. 2, Figs. 18, 19; SCHÖNLAUB, 1985, p. 356, Pl. 2, Figs. 18, 19 [same specimen as preceding].

Ozarkodina stygia (Flajs). MURPHY AND MATTI, 1983, p. 10–12, Pl. 3, Figs. 1, 2, 7, 8.

Ozarkodina stygia schulzei BARDASHEV, 1989, p. 7–8, Pl. 1, Figs. 8, ?9, 10–13; BARDASHEV AND ZIEGLER, 1992, p. 15, Pl. 1, Figs. 25, 26.

[?] *Ozarkodina stygia sorokini* Bardashev. BARDASHEV AND ZIEGLER, 1992, p. 15, Pl. 1, Figs. 28, 29.

Holotype. UCR 8530/1 illustrated by Murphy and Matti, 1983, Pl. 3, Figs. 1, 2, from bed MC 3c of the Bastille Limestone, Mill Canyon section, Toquima Range, Nevada; reillustrated here in Figure 9: 7 and 8.

Derivation of name. Referring to the sigmoidal curvature of the dentical row of the Pa element in upper view.

Material. Thirty-one specimens from TO-I/19 (3); TO-I/20 (4); MC/3C (4); MC/4A (4); MC/5 (6); MC/5A (2); SP-VII/11K (2); SP-VII/11L (4); SP-VII/11Q (1); SP-VII/12 (1).

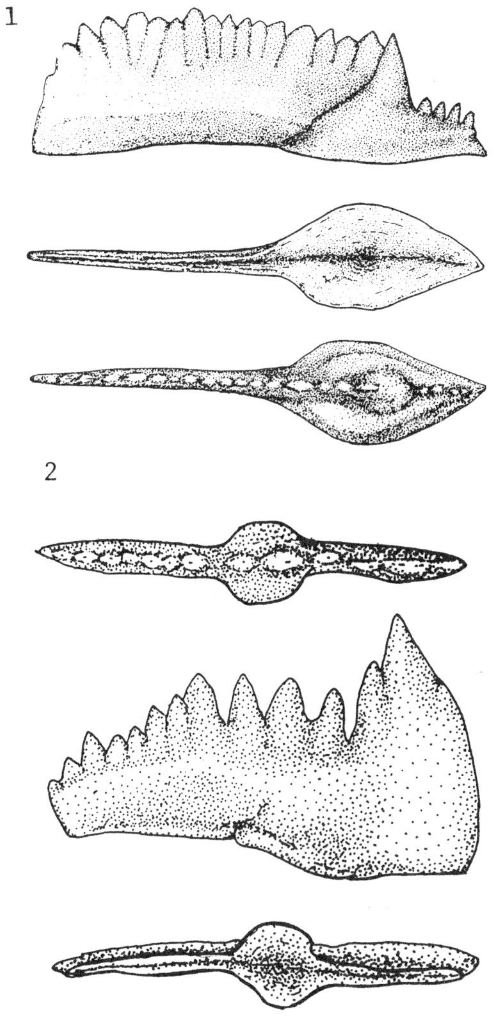

Figure 7. Comparison of *Flajsella* with the type species of *Ozarkodina*. Camera lucida drawings of *Flajsella schulzei* (1) and *Ozarkodina confluens* (2), type species of *Ozarkodina*. Note the different positions and shapes of the basal cavities, the difference in anterior denticles, the prominent cusp and lack of a posterior process in *Flajsella*, and the offset of the anterior carina in *Ozarkodina*.

Diagnosis. A *Flajsella* species whose Pa element has a sigmoidal configuration in upper view; large basal cavity that tapers equally in both directions; posterior lower profile convex.

Description. Blade of Pa element with 19 to 24 denticles that describe a sigmoidal line in upper view; anterior and posterior carinae in line through the cusp. Anterior carina commonly with 14 to 16 denticles in a fan with the highest point near the anterior end and descending slightly to the cusp. Cusp distinctly larger and higher than denticles and located in the posterior third of the Pa element. Posterior carina with generally four, but up to seven, short denticles appreciably lower than anterior denticles. Large, deep, and almost symmetrical basal cavity that occupies two thirds of the length of the blade, generally lenticular in upper view and tapering almost equally in both directions; width is about one third of blade length. Posterior lower margin convex.

Apparatus. Three elements are available in our collections for

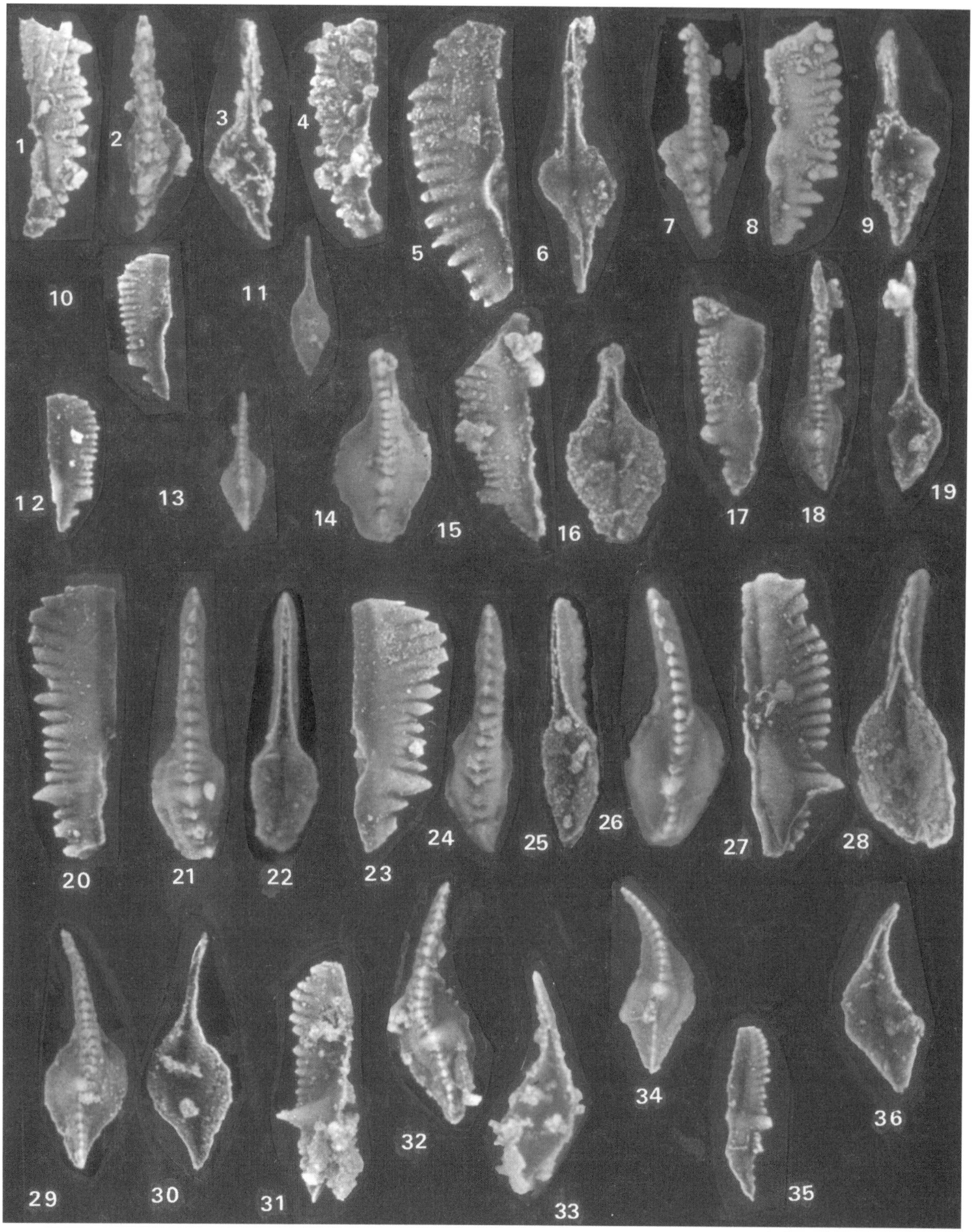

Figure 8. *Flajsella* species and a related form; all figures ×40. 1–4, *Flajsella schulzei* (Bardashev), TO-I 17 I/1. 5, 6, *"Ozarkodina" excavata* (Branson and Mehl), TO-I 17 I/2. 7–22, *F. schulzei*: 7–9, TO-I 17; 10–13, COP-IV 197 II/5; 14–16, TO-I 18 I/6; 17–19, TO-I 18 I/5; 20–22, TO-I 18 I/4. 23–25, *F. sigmostygia* n. sp., TO-I 18 I/3. 26–28, *F. stygia* (Flajs), TO-I 23 I/3. 29–33, *F. sigmostygia* n. sp., TO-I 20 I/49, I/50. 34–36, *F. lanei* (Bardashev), TO-I 23 I/1.

apparatus reconstruction. They are too different among themselves to belong to the same skeletal apparatus. We indicate here only that they are available.

Sa element alate with an acute angle between flattened, processes only slightly turned posteriorly; seven to 10 discrete, elongate denticles; basal cavity narrowly pinched protruding posteriorly (Fig. 9: 9, 10). This element is available, but does not appear to be a strong candidate for the *Flajsella* apparatus because the style of the denticles (discrete and round) is different than in the Flajsella Pa element.

?Sb element digyrate, with a keel originating low on the cusp and reaching to the posterior edge of the basal cavity (Fig. 9: 11).

?Sc element digyrate, with appressed denticles mounted on short, shouldered processes; cusp large, keeled; anterior process longer than the posterior one (Fig. 9: 12).

Discussion. The sigmoidal configuration of the processes of the Pa element in upper view distinguishes *F. sigmostygia* from *F. schulzei* and *F. stygia*. In addition, the basal cavity of *F. sigmostygia* tapers both anteriorly and posteriorly, whereas the anterior margin of the basal cavity of *F. schulzei* bulges abruptly. The convex profile of the posterior lower margin separates *F. sigmostygia* from other species of *Flajsella* except *F. lanei*, whose basal profile is also convex. The difference between the two taxa is that the basal cavity of *F. lanei* extends throughout the entire length of the Pa element. Transitions exist between this taxon and *F. schulzei*, *F. lanei*, and *F. stygia*. Specimens in Figure 8: 23–25 show an almost straight carina, an only slightly convex lower profile, and well-developed posterior denticles similar to *F. schulzei*. The basal cavity, however, is expanded and tapers anteriorly and posteriorly at equal rates. Specimens in Figure 8: 26–28 show the expanded and tapering basal cavity of *F. sigmostygia* but a distinct angle between the carinae at the cusp like *F. stygia*. Specimens in Figure 8: 31–33 approach *F. lanei* in most respects except that the basal cavity does not reach the anterior end as it does in *F. lanei*.

Range and distribution. F. sigmostygia appears in sequence above *F. schulzei* and below *F. stygia* in TO-I. It also occurs in sequence below *F. stygia* in SP-VII (Fig. 4). It is restricted to the middle part of the middle Lochkovian. It has been found at several additional localities in Nevada as well as in the Carnic Alps and Tadzhikistan.

Flajsella lanei (Bardashev, 1989)

Figure 8: 34–36

Ozarkodina stygia (Flajs) alpha morphotype LANE AND ORMISTON, 1979, p. 57-58, Pl. 2, Figs. 20–22.
Ozarkodina stygia lanei BARDASHEV, 1989, p. 10, Pl. 2, Figs. 7–15; BARDASHEV AND ZIEGLER, 1992, p. 15, Pl. 1, Fig. 30.

Holotype. Specimen figured by Bardashev, 1989, Pl. 2, Fig. 7, specimen 1316/136, from southern Tien Shan.

Material. One specimen from TO-I/23.

Revised diagnosis. A *Flajsella* Pa element with a sigmoidal blade in upper view and a basal cavity equal to the length of the element.

Description. Blade of Pa element short, sigmoidal, with small number of denticles and higher, conical cusp; profile variable with posterior part turned up; basal cavity large, sigmoidally lenticular, extending to posterior and anterior ends of element.

Discussion. The large basal cavity extending throughout the whole length of the Pa element separates *F. lanei* from other species of *Flajsella*. *F. lanei* is morphologically close to *F. sigmostygia*. Both have the same overall shape and share the characteristic convex posterior lower margin. The diagnostic basal cavity, smaller numbers of denticles, and smaller size distinguish *F. lanei* from *F. sigmostygia*. See also discussion under *F. sigmostygia*.

Range and distribution. F. lanei has been found in Alaska, Tadzhikistan, and central Nevada, where it occurs together with *F. sigmostygia* in the *eleanorae-trigonicus* Zone, middle Lochkovian. In Tadzhikistan, Bardashev and Ziegler (1992, p. 15) report 20 specimens ("species") from the Shishkat Formation but give the position of only one specimen (fig. 6) in their occurrence tables. In their summary range chart (fig. 7), they show the taxon ranging higher than any ot the other *stygia* "subspecies," contrary to the data in their tables.

Flajsella stygia (Flajs, 1967)

Figure 8: 26–28, Figure 9: 17?, 18?, 26–30

Spathognathodus stygius FLAJS, 1967, p. 204-205, Pl. 5, Figs. 16, 17 [? Figs. 12–15]; PÖLSLER, 1969, p. 433–434, Pl. 1, Figs 1-4.
Spathognathodus seebergensis SCHULZE, 1968, p. 227, Pl. 17, Fig. 4 [? Figs. 10, 11].
Ozarkodina stygia (Flajs) beta morphotype LANE AND ORMISTON, 1979, p. 57-58, Pl. 1, Figs. 45, 46; gamma morphotype LANE AND ORMISTON, 1979, Pl. 2, Figs. 10, 11.
Ozarkodina stygia (Flajs) gamma morph SCHÖNLAUB, 1980, p. 39, Pl. 4, Figs. 23, ?24, 25 [not Figs. 21, 22].
Ozarkodina stygia (Flajs). SCHÖNLAUB, 1980, p. 28, Pl. 2, Figs. ?12, 25, 26 [not Figs. 13, 14, 18, 19]; SCHÖNLAUB, 1985, p. 356, Pl. 2, Figs. ?12; 25, 26 [same specimens as preceding].
Ozarkodina stygia sorokini BARDASHEV, 1989, p. 8–9, Pl. 2, Figs. 1–6; BARDASHEV AND ZIEGLER, 1992, p. 15, Pl. 1, Figs. 22, 23, 31, 32 [? Figs. 36, 37].
Ozarkodina stygia stygia (Flajs). BARDASHEV AND ZIEGLER, 1992, p. 15, Pl. l, Fig. 34 [? Figs. 35, 38].
Ozarkodina stygia (Flajs) beta and delta morphs of Schönlaub. VALENZUELA-RÍOS, 1994b, p. 66–67, Pl. 6, Figs 17–19.

Holotype. Specimen figured by Flajs, 1967, Pl. 5, Fig. 17.

Material. Two hundred and forty-five specimens. Ge 1.1/17c (3); Ge 1.1/20c (1); Se 1/7 (1); Se 1/20 (1); Se 1/23 (2); SCO/DC-44 (3); TO-I/18 (6); TO-I/19 (4) TO-I/23 (1); TO-I/26 (55); MC/5 (25); MC/5A (1); MC/5B (44); MC/6 (13); MC/6B (1); MC/7 (26); MC/11 (1 reworked); COP-IV/287 (1); COP-IV/304 (8); COP-IV/313 (1); IC/8A (1); SP-VII/13A (1); SP-VII/13B (2); SP-VII/13G (5); SP-VII/13J (4); SP-VII/365' (19); SP- VII/14C (6); SP-VII/14E (2); SP-VII/14F (3); SP-VII/370' (4).

Revised diagnosis. A species of *Flajsella* whose Pa element has an angle or offset of the anterior and posterior carinas at the cusp, steeply descending posterior denticles, and asymmetrical platform lobes.

Description. Anterior blade nearly straight except at proximal end, where it is sharply bent to join the cusp. Straight or slightly sigmoidal posterior carina joining the anterior carina at the cusp in an angle or offset. Long anterior carina with 13 to 18 denticles, commonly 15, with a set of 5 to 7 higher and better developed anterior denticles that normally are higher than the cusp; the rest gradually descending toward the cusp. Cusp high and with elliptical cross section, located in the posterior third of the blade. Short posterior carina with one to four (more than 60% of specimens with two to three poorly developed denticles descending steeply from higher cusp; many specimens have a space between the cusp and the denticles of the posterior carina. Large and slightly asymmetrical basal cavity occupies about one half of the length of the Pa element but varies between one half and two thirds; width is approximately one third of blade length; outer lobe is wider than inner. Upper anterior margin is convex, and the highest point of the blade is located in the anterior half of the anterior carina.

Apparatus. The only element available for reconstruction is an angulate Pb element with large, posteriorly inclined cusp. Processes of similar length, each with six well-developed flattened, appressed, keeled denticles. Basal cavity open, expanded, continues as open grooves anteriorly and posteriorly. Arched lower margin, with marked concavity at the basal cavity.

Discussion. Typical *F. stygia* is distinguished by the straight or slightly sigmoidal anterior blade of the Pa element, which is highest near the anterior end, descends gradually to the cusp, and makes a sharp bend at its proximal end to join the cusp. The posterior blade is short, descends steeply from the cusp, and has one to four short denticles with an adenticulate gap just posterior of the cusp. Stratigraphically lower specimens of *F. stygia* have a straight anterior carina, as is characteristic of *F. schulzei*, but these specimens have high anterior denticles that are absent in *F. schulzei*. Some specimens of *F. stygia* have a large basal cavity and a geniculation at the junction of anterior and posterior carinas. They differ from *F. sigmostygia* in having an offset carina and lacking a convex lower posterior margin. In sample SP VII/13G, there are several small specimens with anterior denticles of more or less equal height and with a deep posterior step. They are included in *F. stygia* because of the marked geniculation at the junction point of the anterior and posterior carinas. See also discussion under *F. sigmostygia*.

Range and distribution. F. stygia appears above *F. sigmostygia* within the middle part of the middle Lochkovian and ranges into the upper part of the middle Lochkovian. It has been found in central Nevada (TO-I, MC, COP-IV, SP-VII, IC), Spain (Se 1, Ge 1.1, SCO), Carnic Alps, Alaska, and Tadzhikistan.

Flajsella streptostygia n. sp.

Figure 9: 1–4, ?13–16, ?19–25, 31–37

Ozarkodina stygia (Flajs) delta morphotype LANE AND ORMISTON, 1979, p. 57–58, Pl. 2, Figs. 23, 28.

Ozarkodina stygia (Flajs) gamma morphotype SCHÖNLAUB, 1980, p. 39, Pl. 4, Figs. 21, 22; delta morphotype SCHÖNLAUB, 1980, Pl. 4, Fig. 26.

Ozarkodina stygia (Flajs). MURPHY AND MATTI, 1983, p. 10–12, Text-fig. 2, MC/8c; UYENO, 1991, p. 135, Pl. 1, Fig. 8.

[?] *Ozarkodina stygia stygia* BARDASHEV, 1989, p. 6–7, Pl. 1, Figs. 1, 2.

Holotype. Specimen Sp VII 14L VII/4 illustrated in Figure 9: 32, 33 from Coal Canyon, northern Simpson Park Range, Nevada.

Derivation of the name. Greek, *strepto-*, bent + *stygia.*

Material. One hundred and fifteen specimens. Ge 1.1/17c (1) transitional form with *F. stygia*; Se 1/23 (1); MC/7 (2); MC/8 (30); MC/8b (1); MC/8b+2 (4); COP- IV 8c (2); COP- IV 338 (1); SP-VII/14L (50); SP-VII/15 (14); SP-VII 15A/ (1); TO-I 27 (8).

Diagnosis. A species of *Flajsella* whose Pa element has an curved anterior carina in upper view that meets a straight or curved posterior carina at an angle at the cusp and has a small and markedly asymmetrical basal cavity.

Description. Blade of Pa element kinked at the cusp with strongly curved anterior carina in upper view composed of 15 to 20 denticles that ascend gradually from anterior end to highest denticle, which is located in posterior half of anterior carina and close to cusp. Conical cusp larger than denticles and located on posterior quarter of Pa element. Short and straight, or slightly arched, posterior carina with one to four poorly developed denticles that meets anterior carina at distinct angle at cusp. Small and strongly asymmetrical basal cavity located in posterior quarter of Pa element, one side markedly more convex than the other; maximum width located posterior to cusp and about one-fourth of blade's length from posterior end.

Apparatus. There are four candidates for the Pb element of this species by virtue of their stratigraphic position and availability. The first three are angulate with subequal length processes and a large posteriorly inclined cusp. Anterior process has appressed denticles and an open basal cavity that extends to the distal ends of the processes as a groove. Minor differences in the three elements include the relative height of the cusp with respect to the posterior denticles, the relative size and number of the denticles, and the development of the basal cavity (Fig. 9: 19, 20, 25).

The fourth candidate is a carminate element with a large basal cavity that occupies almost the entire lower surface of the element and a large reclined cusp slightly offset from the posterior process and constituting almost a third of the element. About eight small, appressed denticles

Figure 9. *Flajsella* species, all figures ×40. 1–4, *Flajsella streptostygia* n. sp.: 1, 2, TO-I 27, I/1, top and basal views of Pa element; 3, 4, TO-I 27, I/2, top and lateral views of Pa element. 5–8. *Flajsella sigmostygia* n. sp. MC 3C I/2 and I/1, top and lateral views of Pa elements, respectively, both previously figured in Murphy and Matti (1983, pl. 3). Holotype is MC I/1. 9–12, *?Flajsella sigmostygia* n. sp., possible elements of apparatus, MC 3C: 9, 10, I/40, posterior and anterior views of Sa element; 11, I/44, lateral view of ?Sb element; 12, II/48, lateral view of ?Sc element. 13–16, *?Flajsella streptostygia* n. sp., possible elements of apparatus, SP VII 15 I/26, 27, lateral and basal views of ?Pb elements. 17, 18, *?Flajsella stygia* (Flajs), SP VII 13 I/39, lateral and basal views of Pb element. 19–25, *?Flajsella streptostygia* n. sp., possible elements of apparatus: 19, 20, 25, SP VII 15 I/37, 38, SP VII 14L II/39 lateral views of Pb elements; 21, 22, SP VII 15 I/36, lateral and posterior views of Sb element; 23, SP VII 15 I/34, lateral view of M element; 24, SP VII 15 I/33, posterior view of Sa element. 26–30, *Flajsella stygia*, Pa elements to show typical form: 26, 27, MC 7 I/6, top and lateral views; 28, MC 5 II/29, lateral view; 29, 30, top and lateral views. 31–37, *Flajsella streptostygia* n. sp., Pa elements to show typical profile and curvature of the blade: 31, SP VII 15 I/28, lateral view; 32, 33, holotype, SP VII 14L II/4, top and lateral views; 34, SP VII 14L II/5, top view; 35, 36, SP VII 14L II/5, top and lateral views; 37, SP VII 15 I/29, lateral view.

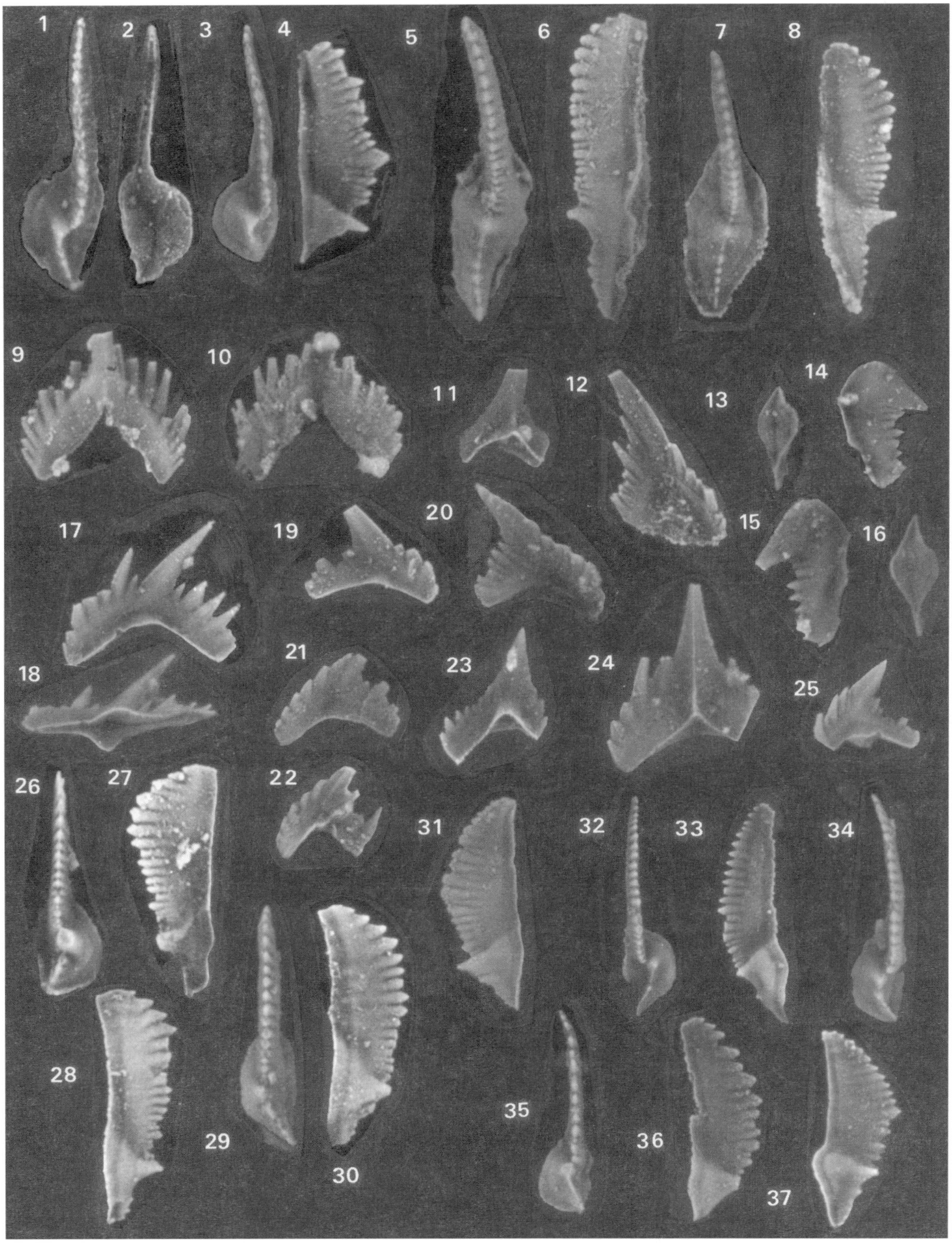
1
2
3
4
5
6
7
8
9
10
11
12
13
14
15
16
17
18
19
20
21
22
23
24
25
26
27
28
29
30
31
32
33
34
35
36
37

occupy the posterior process, but only one poorly developed denticle is present anterior of the cusp, with anterior process barely developed. Lower margin is convex, with the posterior and anterior ends curved up, the anterior one strongly (Fig. 9: 13–16). This element looks like a small Pa element of *Ozarkodina*, especially in the presence of the cusp offset from the posterior process, and so is not given serious consideration here. In fact, it bears a remarkable resemblance to *Spathognathodus primus geniatus* Baranov (1992) from the upper Silurian of northeastern USSR.

The M element is dolabrate with a large, straight, triangular cusp that has a fila along the posterior side. Anterior process short, serrated; posterior process with small, appressed denticles that are larger and more individualized distally. Basal cavity open, expanded (Fig. 9: 23).

The single candidate for the Sa element is a fragmented specimen with a posterior keel that extends from the tip of the cusp to the margin of the basal cavity (Fig. 9: 24).

Discussion. The small basal cavity of *F. streptostygia* distinguishes the Pa element of this species from that of *F. sigmostygia.* The markedly asymmetrical and small basal cavity, arched posterior carina, posterior position of highest denticle on anterior carina, and angular junction of carinae rather than offset distinguish typical *F. streptostygia* from *F. stygia* from which it evolved. Additionally, the posterior carina of many specimens of *F. streptostygia* ends on the lateral side of one lobe of the basal cavity instead of at the posterior tip as it does in *F. stygia*.

Range and distribution. F. streptostygia appears in the upper part of the middle Lochkovian, where it slightly overlaps *F. stygia*, and ranges through the lower part of the upper middle Lochkovian (=upper *eleanorae-trigonicus* to lower *trigonicus-pandora* β Zone). In the Pyrenees, it reaches the lower range of *Ancyrodelloides kutscheri*, and at SP-VII/15 and 15a in Nevada it occurs with *Kimognathus alexeii*. It has been found in Spain (Ge 1.1, Se 1), central Nevada (COP-IV, SP-VII, TO-I, MC), Alaska, Canada, Carnic Alps, and Tadzhikistan.

STRATIGRAPHIC DISTRIBUTION AND EVOLUTION

All morphs of *Flajsella* have a short range in addition to being uncommon. Thus, it is not surprising that the entire sequence of morphs is only rarely discovered. Nevertheless, the partial sequences of taxa are always the same.

The *Flajsella* lineage occurs in the *A. eleanorae-C. pandora* β Zone. It is present in four sections in central Nevada and two in the Spanish Pyrenees (Fig. 4). The TO-I and COP-IV sections have yielded the most complete sequence of *Flajsella*, but SP-VII and MC also furnish important data. In addition, there are several occurrences in the Cortez Range. Of these, the TO-I section is the most instructive. In the lowest occurrence in TO-I, TO-I 17, *F. schulzei* occurs with a morph of the *"Ozarkodina" excavata* lineage that shows the close relationship between the Pa elements of the two taxa. Their profiles are similar except that a cusp is developed above the basal cavity in *F. schulzei* (Fig. 8: 1, 4) and the posterior denticles are stronger in the *"O." excavata* morph (Fig. 8: 5). The basal cavity in *F. schulzei* is deep, open, and spear shaped (Fig. 8: 3, 9, 11), whereas in the *"O." excavata* morph it is shallow and the posterior basal groove is flared but appressed at the posterior end (Fig. 8: 6).

In the next overlying bed (TO-I 18), specimens are larger but show about the same variation. Specimen TO-I 18 I/3 (Fig. 8: 23–25) is intermediate between *F. schulzei* and F. *sigmostygia*. It shows the sigmoidal curvature of the carina and a basal cavity that tapers equally both anteriorly and posteriorly that is typical of *F. sigmostygia*. However, it has the nearly straight lower profile of *F. schulzei*.

Bed TO-I 19 shows the same morphs. Bed TO-I 20 has only *F. sigmostygia* (Figs. 8: 29–33). No *Flajsella* have been found in beds 21 and 22. At TO-I 23, *F. stygia* and *F. lanei* (Figs. 8: 34–36) are present. *F. stygia* (Fig. 8: 26–28) still shows some resemblance to *F. sigmostygia*, but the carina clearly is not on a straight line through the cusp, one of the hallmarks of *F. sigmostygia*. A single level of occurrence of *F. streptostygia* in the TO-I section is present at TO-I 27 (Fig. 9: 1–4).

All other Nevada sections lack one or the other of the *Flajsella* taxa, but all of the remaining sections show support for the general sequence of appearances found in TO-I. The COP-IV section has *F. schulzei*, *F. stygia*, and *F. streptostygia* in sequence. In the MC section, a small unconformity lies immediately below the lowest occurrence of *F. sigmostygia*, eliminating strata that would be expected to bear *F. schulzei*. *F. sigmostygia* is followed by *F. stygia* and *F. streptostygia*, respectively. In SP VII, *F. sigmostygia*, *F. stygia*, and *F. streptostygia* follow one another in sequence.

The sequence of *Flajsella* found in the Road River Formation, Alaska, the other principal North American occurrence of the genus, is similar to that found in Nevada in so far as this can be determined from the literature. In section 1 of Lane and Ormiston (1979), the genus is present in four samples. The lowest reported specimens are not figured but are reported to be the b morph (= *F. stygia*; Lane and Ormiston, 1979, p. 58, table 1a). The next two *Flajsella*-bearing samples in the Road River sequence are reported to have α, β, and γ morphotypes (Lane and Ormiston, 1979, p. 58). In our nomenclature, these are *F. lanei*, *F. stygia*, and *Flajsella* with transitional features between *F. stygia* and *F. streptostygia*. Representatives are figured of *F. lanei* and the transitional form between *F. stygia* and *F. streptostygia* (Lane and Ormiston, 1979, pl. 2, figs. 20–22 and 10, 11, respectively). The highest *Flajsella*-bearing sample has morphotype δ (= *F. streptostygia*; 1979, pl. 2, figs. 23, 28). The only other specimen that is figured is from their section 3 on Linear Ridge. According to Lane and Ormiston (1979, p. 58), the figured specimen from Linear Ridge was believed to represent the oldest morph. We have placed it in *F. stygia* because there is a small deflection of the posterior carina, but it is a juvenile and it is difficult to be sure.

The Alaskan sections, thus, show the same sequence as the upper parts of the SP-VII and TO-I sections of Nevada but also a possible anomaly if the identifications of other taxa are correct (Lane and Ormiston, 1979, table 1a). In section 1 they report *F. stygia* from sample 11 at 50 ft and *"Ozarkodina"* n. sp. F of Klapper and Murphy, (1975) from sample 14 at 60 ft. These two taxa never overlap in any other section, so this report needs to be verified.

In Spain, the Gerri 1.1 section shows the sequence *F.*

schulzei, *F. stygia*, *F. streptostygia*. The relationship of the Nevada and Spanish sections is shown in Figure 4.

Schönlaub (1980, pl. 4) figured a sequence of *Flajsella* morphs from the Rauchkofel Boden section in the Carnic Alps, Austria, that he designated alpha through delta. The earliest morph, alpha (=*F. schulzei*), is from sample 208, which also yielded an early morph of *Ancyrodelloides eleanorae* (= *A. asymmetricus*, Schönlaub, 1980, pl. 4, figs. 3, 4). This would place it just below the occurrence of *A. eleanorae* in Nevada (Fig. 4). Schönlaub's beta morph (partim = *F. stygia*) enters the section with *A. trigonicus* in sample 212. This skips *F. sigmostygia* of the Nevada sequence, but *F. sigmostygia* may be present in Austria, as a poorly preserved example from Oberbuchach (Schönlaub, 1980, pl. 2, figs. 18, 19) is tentatively identified as *F. sigmostygia*. The gamma morph (= *F. streptostygia*) follows in sample 213.

ORIGIN OF *FLAJSELLA*

The Nevadan and Spanish sections have yielded all of the previously described morphotypes of *"Ozarkodina" stygia*. They occur in stratigraphic sequence and are accompanied by other conodonts and well-studied graptolite faunas (Berry and Murphy, 1975; Springer and Murphy, 1994) that control the relative ages of the new genus *Flajsella* and its species.

Flajsella appears to have originated as a branch of either *Criteriognathus* or the *"Ozarkodina" excavata* group of species in the middle Lochkovian between the appearances of *Ancyrodelloides transitans* and *A. trigonicus* and to have maintained a relatively stable morphology until it became extinct without issue in the lowest upper Lochkovian. We favor the latter interpretation, as shown in Figure 6. The earliest species, *Flajsella schulzei*, has the following characteristics: (1) straight or gently arched blade with anterior and posterior carinas aligned; (2) anterior denticles small, appressed except at their tips, numerous, and laterally compressed; (3) cusp prominent, nearly round; (4) posterior denticles small, but well developed; (5) basal cavity large, open, posterior, and nearly symmetrical (Fig. 7).

The spathognathodontid genera that could be the ancestor of *Flajsella* by virtue of their stratigraphic positions include *Ancyrodelloides*, *Amydrotaxis*, *Criteriognathus*, *Ozarkodina*, and the *"Ozarkodina" excavata*-group of species. Of these, *Ancyrodelloides* can be eliminated because of the tendency of the Pa element toward a restricted and centrally located basal cavity; well-developed, shouldered platform; and different style of denticulation. *Amydrotaxis* has the posterior, open basal cavity of *Flajsella*, but the elements are much more robust, the denticulation is very different, and the reconstructed apparatus elements are unique (Klapper and Murphy, 1980; Murphy and Matti, 1983) and have a much greater range than the Pa elements of all *Flajsella* morphs. Two of the remaining candidates, *Criteriognathus* and *Ozarkodina*, are very similar to one another in overall Pa element morphology and have Pb elements that could easily be modified to look like the Pb elements that are available for the reconstruction of the Flajsella apparatus. However, the Pa elements of members of true *Ozarkodina* (type species, *O. confluens*) have high anterior denticles and centrally located basal cavities through most of their history (Fig. 7). This is true for the middle Lochkovian as well, the interval during which *Flajsella* developed. Also, the *Ozarkodina* species do not develop a single, prominent cusp.

A taxon close in morphology to early *Flajsella* among the species of *Criteriognathus* is *C. paucidentata* (Murphy and Matti, 1983, pl. 1, figs. 25–32). This taxon has a number of the characters that would be expected in the ideal ancestor: low, equant, anterior denticles; prominent, round cusp; paucity of denticles behind the cusp; expanded posterior groove that tapers to a point at the posterior extremity. The basal cavity and posterior basal groove, however, are shallow, not deep as in *Flajsella*, and the basal cavity is central rather than posterior. It also has a posterior blade that is inflated at the shoulder. In addition, the early morphs of Flajsella have small but well-developed posterior denticles. It is only later in their history that the posterior denticles are reduced and disappear.

The *"Ozarkodina" excavata* group of species has many similarities with *Flajsella*, and the Pb elements of the apparatuses reconstructed for the two genera may be similar (Klapper and Murphy, 1975; Lane and Ormiston, 1979, respectively). In addition, the group has had a tendency to develop a posteriorly enlarged basal cavity, and it has members that are positioned in the right part of the column to be the ancestor of *Flajsella*. Also, transitional morphologies exist between *"Ozarkodina"* and *Flajsella* (Fig. 8: 1–6). Although we presently favor derivation from an *"Ozarkodina" excavata*–like ancestor, we cannot as yet demonstrate this transition morphometrically.

Bardashev and Ziegler (1992, p. 14) have suggested that *Flajsella* was derived from *Criteriognathus pandora* (Murphy, Matti, and Walliser). This possibility also has merit.

The evolutionary succession of forms within *Flajsella* is based on their stratigraphic distribution as discussed above, as well as morphologic considerations, and is interpreted as a single lineage with a broad range of variation in particular intervals. *F. lanei* may be an exception to this interpretation and represent cladogenesis, but too few elements of the taxon are known to make a reasoned judgment.

ACKNOWLEDGMENTS

S. García-López, Gilbert Klapper, H. Richard Lane, and H.-P. Schönlaub permitted us to study their collections; Fulbright Foundation Fu 8910974/84 was given to J. I. Valenzuela-Ríos while at University of California, Davis (UCD); National Science Foundation, University of California Riverside Intramural Research Fund, and UCD Intramural Research Fund supported various phases of the research; a Humboldt Fellowship was awarded to J. I. Valenzuela-Ríos during the final phases of this work. This chapter had the benefit of critical reviews by J. Barrick and S. Bergström.

REFERENCES CITED

Baranov, V. V., 1992, Novye pozdnesiluriiskie i rannedevonskie konodonty severo-vostoka SSSR (New Late Silurian and Early Devonian conodonts): Paleontologicheskiy Zhurnal, v. 1992, p.129–133.

Bardashev, I. A., 1989, Novye listovidnye konodonty iz nizhnego devona tsentral'nogo Tadzhikistana (New conodonts from Lower Devonian of centrat Tadghikistan): Dushanbe, Akademiya Nauk Tadzhikskoi SSR Institut Geologii, p. 5–14, 211–214.

Bardashev, I. A., and Ziegler, W., 1992, Conodont biostratigraphy of Lower Devonian deposits of the Shishkat section (southern Tien-Shan, Middle Asia): Courier Forschungsinstitut Senckenberg, v. 154, p. 1–29.

Berry, W. B. N., and Murphy, M. A., 1975, Silurian and Devonian graptolites of central Nevada: University of California Publications in Geological Sciences, v. 110, 109 p.

Bischoff, G., and Sannemann, D., 1958, Unterdevonischen Conodonten aus dem Frankenwald: Notizblatt des Hessischen Landesamtes für Bodenforschung, v. 86, p. 87–110.

Flajs, G., 1967, Conodontenstratigraphische Untersuchungen im Raum von Eisenerz, Nördliche Grauwackenzone: Mitteilungen der Geologischen Gesellschaft in Wien, v. 59, p. 157–212.

García-López, S., Julivert, M., Soldevila, J., Truyols-Massoni, M., and Zamarreño, I., 1990, Bioestratigrafía y facies de la sucesión carbonatada del Silúrico Superior y Devónico Inferior de Santa Creu d'Olorda (Cadenas Costeras Catalanas, NE de España): Acta Geològica Hispànica, v. 25, p. 141–168.

Hass, W. H., 1959, Conodonts from the Chappel Limestone of Texas: U.S. Geological Survey Professional Paper 294-J, p. 365–399.

Klapper, G., 1973, *Ozarkodina stygia*, *in* Ziegler, W., ed., Catalogue of Conodonts I: Stuttgart, E. Schweizerbart'sche Verlagsbuchhandlung, p. 228, 249.

Klapper, G., 1977, Lower and Middle Devonian conodont sequence in central Nevada, with contributions by D. B. Johnson, *in* Murphy, M. A., Berry, W. B. N., and Sandberg, C. A., eds., Western North America: Devonian: University of California, Riverside Campus Museum Contribution 4, p. 33–54.

Klapper, G., and Johnson, J. G., 1980, Endemism and dispersal of Devonian conodonts: Journal of Paleontology, v. 54, p. 400–455.

Klapper, G., and Murphy, M. A., 1975, Silurian–Lower Devonian conodont sequence in the Roberts Mountains Formation of central Nevada: University of California Publications in Geological Sciences, v. 111, 62 p. [imprint 1974]

Klapper, G., and Murphy, M. A., 1980, Conodont zonal species from the *delta* and *pesavis* Zones (Lower Devonian) in central Nevada: Neues Jahrbuch für Geologie und Paläontologie, Monatshefte 1980, no. 8, p. 490-504.

Klapper, G., and Philip, G. M., 1971, Devonian conodont apparatuses and their vicarious skeletal elements: Lethaia, v. 4, p. 429–452.

Klapper, G., and Philip, G. M., 1972, Familial classification of reconstructed Devonian conodont apparatuses, *in* Lindström, M., and Ziegler, W., eds., Symposium on conodont taxonomy: Geologica et Palaeontologica, special volume 1, p. 97–114.

Lane, H. R., and Ormiston, A. R., 1979, Siluro-Devonian biostratigraphy of the Salmontrout River area, east-central Alaska: Geologica et Palaeontologica, v. 13, p. 39–96.

Matti, J. C., Murphy, M. A., and Finney, S. C., 1975, Silurian and Lower Devonian basin and basin-slope limestones, Copenhagen Canyon, Nevada: Geological Society of America Special Paper 159, 48 p.

Murphy, M. A., 1989, Lower Pragian Boundary (Lower Devonian) and its application in Nevada: Courier Forschungsinstitut Senckenberg, v. 117p. 61–70.

Murphy, M. A., and Cebecioglu, M. K., 1987, Morphometric study of the genus *Ancyrodelloides* (Lower Devonian, conodonts), central Nevada: Journal of Paleontology, v. 61, p. 583–594.

Murphy, M. A., and Matti, J. C., 1983, Lower Devonian conodonts (*hesperius-kindlei* Zones), central Nevada: University of California Publications in Geological Sciences, v. 123, 82 p. [imprint 1982]

Murphy, M. A., Matti, J. C., and Walliser, O. H., 1981, Biostratigraphy and evolution of the *Ozarkodina remscheidensis–Eognathodus sulcatus* lineage (Lower Devonian) in Germany and central Nevada: Journal of Paleontology, v. 55, p. 747–772.

Pölsler, P., 1969, Conodonten aus dem Devon der Karnischen Alpen (Findenigkofel, Österreich): Jahrbuch der Geologischen Bundesanstalt, v. 112, p. 399–440.

Rodríguez-Cañero, M. R., Maate, A., and Martín-Algarra, A., 1990, Conodontos del paleozoico gomáride (rift septentrional, Marruecos): Nota preliminar: Geogaceta, v. 7, p. 81–84.

Schönlaub, H.-P., 1980, Field Trip A, Carnic Alps: Guidebook and Abstracts, Second European conodont symposium (ECOS II): Abhandlungen der Geologischen Bundesanstalt, v. 35, p. 5–57.

Schönlaub, H.-P., 1985, Devonian conodonts from section Oberbuchach II in the Carnic Alps (Austria): Courier Forschungsinstitut Senckenberg, v. 75, p. 353–374.

Schulze, R., 1968, Die Conodonten aus dem Paläozoikum der mittleren Karawanken (Seeberggebiet): Neues Jahrbuch für Geologie und Paläontologie, Abhandlungen, v. 130, p. 133–245.

Springer, K. B., and Murphy, M. A., 1994, Punctuated stasis and collateral evolution in the Devonian lineage of *Monograptus hercynicus*: Lethaia, v. 27, p. 119–128.

Sweet, W. C., 1988, The Conodonta: Morphology, taxonomy, paleoecology, and evolutionary history of a long-extinct animal phylum: Oxford, Clarendon Press, 212 p.

Uyeno, T. T., 1991, Pre-Famennian Devonian conodont biostratigraphy of selected intervals in the eastern Canadian Cordillera, *in* Orchard, M. J., and McCracken, A. D., eds., Ordovician to Triassic conodont paleontology of the Canadian Cordillera: Geological Survey of Canada Bulletin 417, p. 129–161.

Valenzuela-Ríos, J. I., 1990, Lochkovian conodonts and stratigraphy at Gerrida de la Sal (Pyrenees): Courier Forschungsinstitut Senckenberg, v. 118, p. 53–63.

Valenzuela-Ríos, J. I., 1994a, A new biostratigraphic subdivision of the upper Lochkovian (Lower Devonian) based on conodont sequences from Spain and western North America: Geological Society of America Abstracts with Programs, v. 26, no. 5, p. 66.

Valenzuela-Ríos, J. I., 1994b, Conodontos del Lochkoviense y Praguiense (Devónico inferior) del Pirineo Central español: Memorias del Museo Palaeontológico de la Universidad de Zaragoza, v. 5, 178 p.

Valenzuela-Ríos, J. I., 1994c, The Lower Devonian conodont *Pedavis pesavis* and the *pesavis* Zone: Lethaia, v. 27, p. 199–207.

Walliser, O. H., 1964, Conodonten des Silurs: Abhandlungen des Hessischen Landesamtes für Bodenforschung no. 41, 106 p.

Walliser, O. H., 1972, Conodont apparatuses in the Silurian, *in* Lindström, M., and Ziegler, W., eds., Symposium on conodont taxonomy: Geologica et Palaeontologica, special volume 1, p. 75–80.

Wilson, G. A., 1989, Documentation of conodont assemblages across the Lochkovian-Pragian (Early Devonian) boundary at Wellington, central New South Wales, Australia: Courier Forschungsinstitut Senckenberg, v. 117, p. 117–171.

Manuscript Accepted by the Society February 27, 1997

Printed in U.S.A.

Geological Society of America
Special Paper 321
1997

Middle Devonian (Eifelian) conodonts, and their stratigraphic implications, from the mid-Appalachians of Québec

Tom T. Uyeno
Geological Survey of Canada, 3303 33rd Street, N.W., Calgary, Alberta T2L 2A7, Canada
Pierre J. Lespérance
Département de Géologie, Université de Montréal, Casier Postal 6128, Montréal, Québec H3C 3J7, Canada

ABSTRACT

Middle Devonian, Eifelian-age conodonts are described from Québec for the first time. The faunas are dominated by icriodontids, most of which are endemic to North America; only two species of *Polygnathus*, one of which is represented by two subspecies, are present. The faunas also probably represent the easternmost occurrences of the endemic icriodontid species and belong to the Eastern Faunal Realm of North America (of Sparling). In the near absence of polygnathids, reference to standard conodont zonation is made only indirectly.

The Famine Formation, exposed at and near the townsite of Saint-Georges in the Eastern Townships, has yielded a fauna that is correlated with the *costatus* Zone. The Touladi Formation crops out near Lac Témiscouata, close to the border with the province of New Brunswick; its conodont fauna is correlated with the *australis* to *kockelianus* zones. A new species, *Icriodus lesperancei*, is described.

Mid-Devonian thrusting in the Lac Témiscouata and Saint-Georges areas is presumably responsible for the shallow-water carbonates' being deposited in front of advancing Acadian thrust sheets. The Famine and Touladi formations were subjected to ensuing regional metamorphism during the Acadian Orogeny. This is reflected in the relatively poor preservation and the Color Alteration Index (CAI 4 and 5 of Epstein et al. and Rejebian et al.) of the enclosed conodonts.

INTRODUCTION

Marine Middle Devonian strata in Québec are uncommon and are totally unknown in the Appalachians of the Gaspé Peninsula proper (Figs. 1 and 2). In the Matapédia Valley region, the area between the Gaspé Peninsula and the Témiscouata area, marine Middle Devonian strata have now been identified. (The terms "Témiscouata area" and "Témiscouata" as used herein refer to Témiscouata County, a region the center of which is the town of Cabano.) The uppermost York River Formation in the Ste-Florence area (Berger, 1993) contains the Eifelian brachiopods *Devonochonetes* cf. *D. syrtalis*, *Longispina* sp. (Racheboeuf and Lespérance, 1995), and *Brevispirifer* n. spp. On the northern limb of the same syncline, above the type area of the Four-Mile Brook Member of the York River Formation, east of the Causapscal area, the uppermost York River contains spores, identified by D. C. McGregor, of the *velata-langii* Zone of the early, but not earliest, Eifelian (Richardson and McGregor, 1986). Unfortunately, the physical field relationship in terms of stratigraphy of the brachiopod and spore occurrences is unknown. The Middle Devonian age of the uppermost York River Formation in this area is thus firm, although only one locality is decidedly marine. In the mid-Appalachians of Québec, the Touladi Formation of limestones in the Témiscouata area is now known to be Middle Devonian. The southern segment of the Québec Appalachians, the Eastern Townships, has Middle Devonian strata at both its southern and north-

Uyeno, T. T., and Lespérance, P. J., 1997, Middle Devonian (Eifelian) conodonts, and their stratigraphic implications, from the mid-Appalachians of Québec, *in* Klapper, G., Murphy, M. A., and Talent, J. A., eds., Paleozoic Sequence Stratigraphy, Biostratigraphy, and Biogeography: Studies in Honor of J. Granville ("Jess") Johnson: Boulder, Colorado, Geological Society of America Special Paper 321.

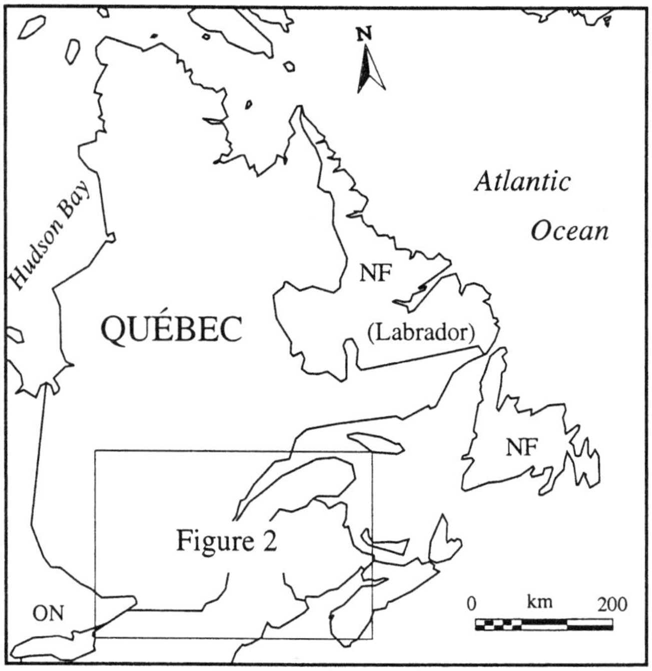

Figure 1. Map of Québec with inset area. NF=Newfoundland, ON=Ontario.

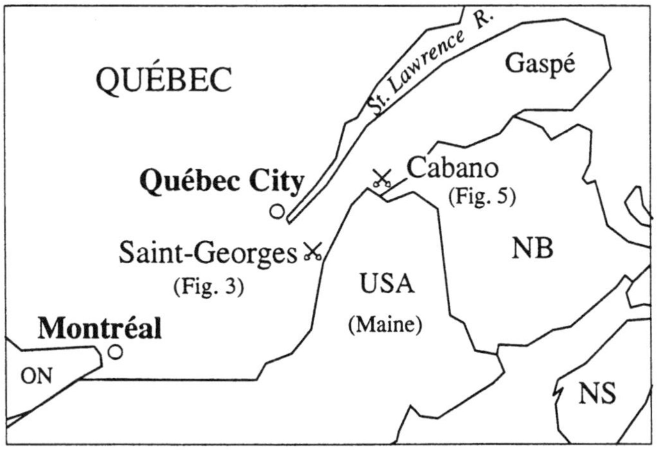

Figure 2. Locality map of southern Québec showing the towns of Saint-Georges and Cabano, St. Lawrence River, and Gaspé. NB=New Brunswick, NS=Nova Scotia, ON=Ontario.

ern extremities. The Mountain Wharf Limestone occurs at Lake Memphremagog (Boucot and Drapeau, 1968), near the Québec-Vermont border, and the "Famine Limestone" occurs southeast of Québec City, close to the Québec-Maine border. Outside the Appalachians, Middle Devonian, Hamilton-age strata occur in a diatreme breccia in Montreal (Boucot et al., 1987). We have no new data on the highly tectonized Mountain Wharf Limestone, and the Famine and Touladi formations are the subject of this study. Middle Devonian conodonts from Québec are described for the first time.

The Connecticut Valley–Gaspé synclinorium is a major structural feature of the Northern Appalachians in Québec and New England and is composed predominantly of slates, in the widest acceptance of the term. These slates are poorly dated, are known under a variety of formational or group names, and, all in all, are poorly understood both stratigraphically and structurally. A review of these slates by Boucot (1970) summarized our deficient stratigraphical knowledge, which this contribution will attempt to redress, but our overall understanding still leaves much to be desired.

Both the Famine and Touladi formations occur on the northwestern limit of the slates, are associated with unconformities, and are adjacent to major thrust faults. These curious coincidences, combined with a change in age assignment of the Touladi by A. J. Boucot in the mid-1960s (Boucot and Johnson, 1967, p. 54, 55; Lajoie et al., 1968, p. 635; Lespérance and Greiner, 1969, p. viii), prompted our investigations. A precise age of these limestones is obviously essential to our understanding of the relationships between the Connecticut Valley–Gaspé synclinorium slates and the limestones.

FAMINE FORMATION

Nomenclature

MacKay (1921, p. 21) introduced the term Famine Series, which was said to comprise a basal conglomerate member, an intermediate shelly (fossiliferous) limestone member, and overlying shale members. Outcrops assigned to the Famine occurred only on the south slope of the Famine River valley in the townsite of Saint-Georges (Fig. 3). Outcrops near Morisset Station, 18 km to the northeast, were only tentatively assigned to the Famine Series by Tolman (1936).

Subsequent mapping of the adjacent areas, at smaller scales (Gorman, 1955) and larger scales to the southwest (St-Julien, 1987) and northeast (Cousineau et al., 1984a; Cousineau et al., 1984b; Cousineau, 1990), has elucidated the distribution of MacKay's Famine. The conglomerate is very restricted in extent and limestones only slightly less so; in none of these publications is the term Famine employed. Usage of the term "Famine," in the past few decades, has been restricted to the paleontological or biostratigraphical literature. The distinctive nature of the Famine (i. e., limestones and limy slates), when compared to the adjacent St. Francis Group turbidites (nonlimy slates and wackes), justifies the use of the term Famine in a lithostratigraphic sense, as was employed by Cousineau (1990) who, however, considered the Famine as a formation within the St. Francis Group. Cousineau and Tremblay (1993), however, considered the Famine (Formation) distinct from the St. Francis (Group).

The "slates" occurring southeastward of the Saint-Georges area are assigned to the St. Francis Group. Although the term St. Juste Group has been used sporadically (Gorman, 1955; Boucot and Drapeau, 1968), St. Francis Group has priority. It ranges in age from Late Silurian to Early Devonian; its precise assignment within the Devonian is unknown, however. St-Julien (1987, p. 59)

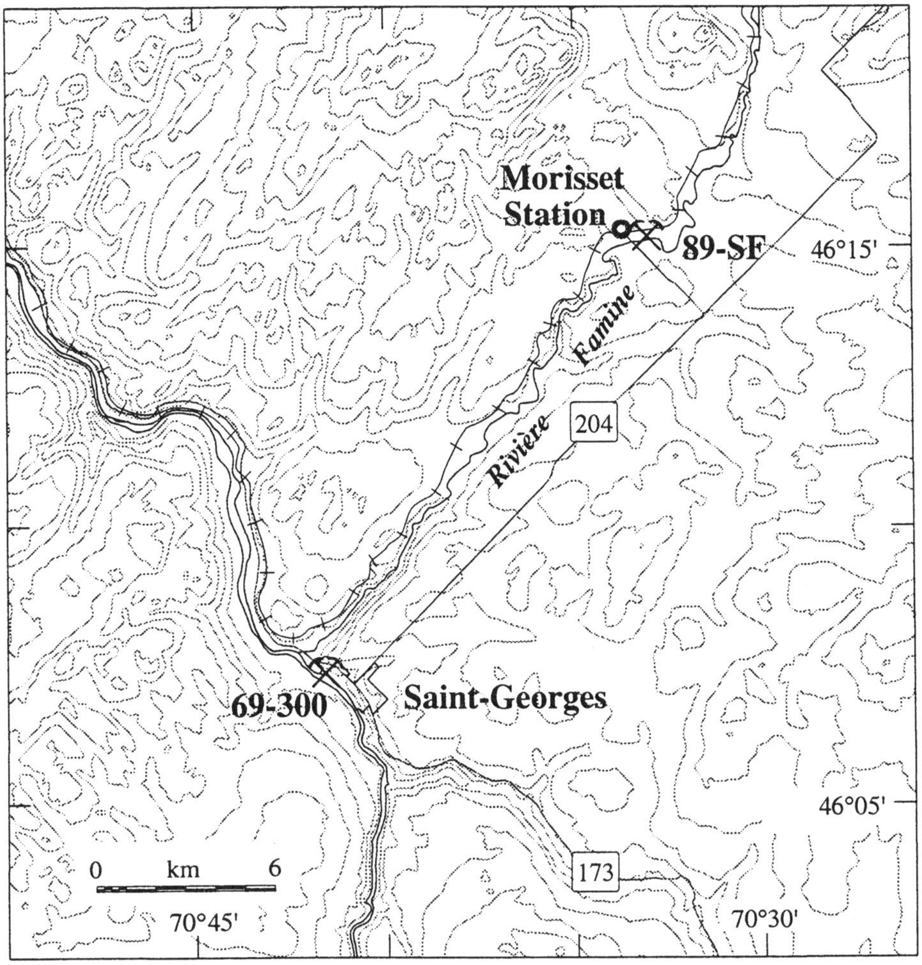

Figure 3. Small-scale locality map of the town of Saint-Georges (section with the 69-300 series field numbers; see Appendix and Fig. 4) and the Morisset Station (89-SF). The location shown of Route 204 within Saint-Georges is that prior to 1990 (see text for additional details).

has correlated most of the slate units (e. g., Fortin, Témiscouata, St. Francis) in Québec within the Connecticut Valley– Gaspé synclinorium; such a correlation can be readily accepted and, in fact, has been implicitly assumed by most workers. When examined biostratigraphically, however, such a correlation is but a broad one (i.e., Early Devonian) and can only be valid if the Famine Formation is excluded from the St. Francis Group. For clarity in this contribution, we herein exclude the Famine from the St. Francis; in any event, these strata are lithologically distinct.

The Saint-Georges area

The well-known section on the main highway (Route 173) in the town of Saint-Georges (Fig. 4; see Appendix for exact location) begins with a small anticline, the core of which exposes the Ordovician Magog Group; from there the section continues eastward. The upper slates are now extensively covered by lawns, but a few outcrops can still be observed. The lower part of the section (the conglomerates and limestones) is much better exposed, although not totally so. The conglomerates have resisted tectonic deformation (stretching and cleavage) far better than the overlying limestones. An angular unconformity is exposed between the gray, non-limy slates, with sparse dolomitic siltstones, of the mid-Ordovician Saint-Victor Formation of the Magog Group (St-Julien, 1987) and the Famine. The basal 9.8 m of the Famine is composed of conglomerate, with subrounded clasts averaging 1 cm in diameter. These conglomerates have been described by MacKay (1921) and Gorman (1955). To the east, overlying these basal beds are 24 m of alternating argillaceous limestones, limy gray slates, and relatively purer limestones (in decreasing order of abundance, with argillaceous content increasing upward) in beds commonly 25 cm thick, but locally ranging from 5 to 90 cm. These limestones are in sharp contact with the underlying conglomerates and are sparingly to extremely fossiliferous; corals are the most conspicuous fossils, brachiopods being less abundant. The uppermost outcrops are almost totally limy slates, with rare argillaceous limestones.

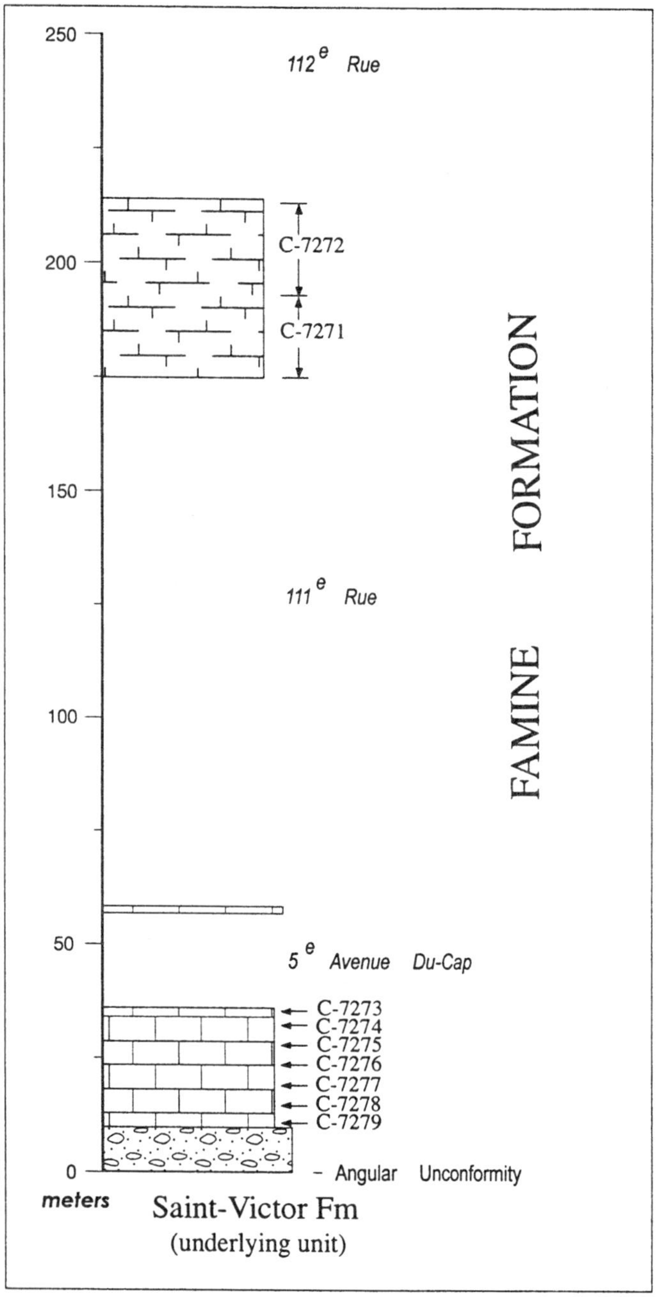

Figure 4. Columnar section of the Famine Formation exposed along Route 173 in Saint-Georges. Sampled intervals and their GSC locality numbers are shown.

There is no evidence of faulting within the section, and the limestones appear totally gradational into the slate sequence, much as was reported by Boucot and Drapeau (1968, p. 37). Sampling for conodonts was directed toward establishing if indeed the limestone sequence was gradational into the limy slate sequence, based on the faunas.

For some 200 m, in the fields immediately adjacent to and north of Route 173, the basal conglomerate and some limestone also crop out on the northern limb of a small anticline exposing the basal part of the Famine. These discontinuous outcrops suggest complicated folding on the northern limb; corals are particularly evident in the few limestone outcrops.

Route 204, which runs northeastward from Saint-Georges, was relocated northwestward within the town-site in 1990. It now bifurcates from Route 173 at the 107e Rue, approximately 0.4 km west of the small anticline exposing the previously described section. From its junction with Route 173, Route 204 now has a northward direction for a short distance, turns east-northeast on a steep incline, and finally runs east-southeast to 120e Rue. The steep incline necessitated extensive excavation, exposing Ordovician strata, the Famine Formation, and St. Francis Group. These outcrops complement nicely the previous section and merit brief description.

The unconformity between the Ordovician (Saint-Victor Formation of the Magog Group) and Famine Formation is well exposed along the new roadcut, at a distance of 1.0 km as measured along the road from the junction of Routes 173 and 204. The basal conglomerate and the overlying limestones are interbedded over a few meters, demonstrating that the unfossiliferous conglomerates are of the same sedimentary sequence as the limestones. The limestones progressively lose their carbonate content up-section, passing successively to very dark gray argillaceous limestones, limy mudstones, and finally limy "slates." Some of the limestones have pyrite concentrations, suggesting euxinic conditions. The Famine Formation thus records the story of a deepening-upward sequence. At the major bend, where Route 204 turns from east-northeast to east-southeast, and 2.0 km from the 173–204 junction, an isolated outcrop of a 30-m-wide fault zone is exposed. This is the La Guadeloupe Fault. Farther along the road, non-limy, gray, turbiditic St. Francis Group strata crop out.

Morisset Station area

Data from the Morisset Station area are sparse, and an attempt to find additional limestones, or other possibly fossiliferous strata, from the limited mapping of Cousineau et al. (1984b), was unsuccessful. There, the strata on the northern bank of the Famine River are more highly sheared than those in the Saint-Georges section, which suggests that the La Guadeloupe Fault is close by. Slates, argillaceous limestones, and purer limestones crop out at this locality; conglomerates are absent. Again, corals are the most evident fossils in the limestones.

Unconformities and faults

According to Tremblay et al. (1989, p. 1934) there is no regionally important angular unconformity between the Ordovician and Siluro-Devonian units in the Eastern Townships. Similarly, no angular unconformity has been recognized between the Ordovician and younger strata in western New England (Hatch, 1982). Whatever interpretation may be given to this contact, the well-exposed unconformity in Saint-Georges between the Ordovician and the Middle Devonian Famine and the equally well-exposed one between the St. Francis Group and the Ordovician, at a locality not far south of Saint-Georges (Labbé and St-Julien, 1989, p. 2276), are certainly significant for the elucidation of the geological history of the region.

The position of a major thrust fault bounding the Connecticut Valley–Gaspé synclinorium in the Eastern Townships has been determined from regional data. St-Julien et al. (1983) named the thrust fault bounding the northwestern edge of the synclinorium, in the northern part of the Eastern Townships, the La Guadeloupe Fault. This thrust fault, deduced from deep seismic data, has been traced northeastward by Cousineau et al. (1984a, 1984b), and Cousineau (1990), and was discussed by Tremblay et al. (1989), Labbé and St-Julien (1989), and Cousineau and Tremblay (1993). In the Saint-Georges area, it separates the Famine Formation from the St. Francis Group.

The deepening-upward nature of the Famine and the regional extent of the La Guadeloupe Fault support a model (Jacobi, 1981) in which a northwestward-advancing thrust sheet (of St. Francis) progessively downwarped the pre-Acadian crust and eventually led to marine deposition of the Famine in Middle Devonian time. Further advancement of the thrust sheet increasingly downwarped the crust, finally overrode the Famine, and signaled the end of our knowledge of mid-Devonian Acadian events in this area.

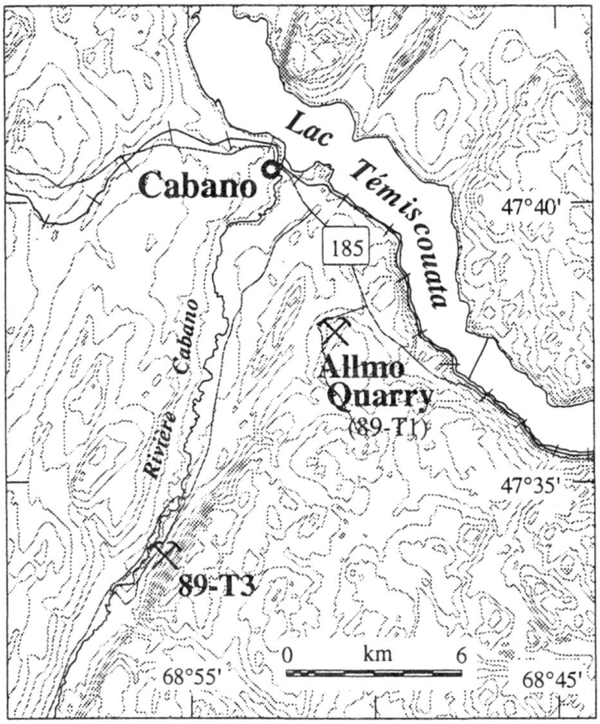

Figure 5. Small-scale locality map of the vicinity of the town of Cabano, showing the sites of the Allmo Quarry (89-T1) and the hillside outcrop (89-T3).

TOULADI FORMATION

The Touladi Formation was erected by Lespérance and Greiner (1969). They did not suggest a type section but noted that the best exposures occurred in the vicinity of Touladi Lake. Thirty years after the field work was accomplished, this is no longer true, and the quarry at our locality 89-T1, south of Témiscouata Lake, is certainly now the best exposure of the formation (Fig. 5). This quarry, now water filled, was short lived; to the best of our knowledge, quarrying was restricted to 1972, and the operator was Allmo Construction. Quarrying exposed both the lower and upper contacts of the Touladi Formation. Herein this site is referred to as the Allmo Quarry.

The thickness of the Touladi Formation ranges from a featheredge to a maximum of 105 m (Lespérance and Greiner, 1969); in the Allmo Quarry, a thickness ranging from 15 to 20 m is exposed. Lespérance and Greiner (1969) suggested that most of the Touladi was composed of fossiliferous, probably bioclastic limestones (i.e., bioclastic wackestones); minor sandstones and conglomerates were, however, noted in scattered incomplete sections.

Regionally, the Touladi Formation rests with very slight angular unconformity on various formations, ranging in age from (possibly) the Ordovician to the latest Silurian. In the area south of Témiscouata Lake, where our samples originate, the Cabano Formation underlies the Touladi near locality 89-T3 (Fig. 5), and the Pointe aux Trembles Formation underlies the Touladi in the Allmo Quarry. The Cabano Formation is now known to extend from the basal Caradocian to the late Llandoverian (thus modifying somewhat the view of David et al., 1985), whereas the Pointe aux Trembles is late Llandoverian. In the vicinity of the Touladi samples, there is no direct evidence for the ages of these two formations, although a Silurian age for the Cabano appears most probable, as the Touladi rests on its upper part.

In the Allmo Quarry, no angular unconformity is evident between the Pointe aux Trembles and Touladi Formations, and the term "paraconformity" best describes this contact. The Middle Devonian Touladi rests alternatively on volcanic conglomeratic sandstones, a few centimeters of mudstones, or 5 cm of sandstones (? paleosols) of the Pointe aux Trembles Formation. The basal Touladi is variable: It may be fine calcarenites, calcarenites with sandstone interbeds, or coral-stromatoporoid wackestones. Touladi wackestones welded to Pointe aux Trembles volcanic sandstones have been observed. Still in the same quarry, the Touladi Formation is in fault contact with the Témiscouata Formation (of "slates," which in this area are predominantly composed of strata containing over 50% carbonates). Faulting and shearing along this contact are evident. This fault was shown in the vicinity of the Allmo Quarry in Lajoie et al. (1968, fig. 2) and was extended laterally in David et al. (1985, fig. 1). It is a major structural feature of the Témiscouata region and in fact extends beyond the region (see below); it has been termed the Témiscouata Fault by Bourque et al. (1993). That this major fault in the Témiscouata area is a thrust fault is obvious from the age of the Témiscouata Formation, which has been correlated with the Oriskany Sandstone of Early Devonian (Siegenian) age (Lajoie et al., 1968; Lespérance and Greiner, 1969) in the Témiscouata area. Data from New Brunswick and

Maine, however, have prompted St. Peter and Boucot (1981) to extend the age of the Témiscouata Formation from the Gedinnian (New Scotland Limestone) to Emsian (Esopus Shale or Schoharie Formation) of the Early Devonian.

In summary, the Touladi Formation in the Lac Témiscouata area occurs above an erosional unconformity cutting into the Silurian. Its restricted distribution is most probably due to overthrusting of the older Témiscouata Formation, although original distributional patchiness cannot be ruled out.

THE AGE AND CORRELATION OF THE FAMINE AND TOULADI FORMATIONS

Regional considerations

Silurian strata are locally present in the Eastern Townships, notably the Cranbourne Formation 16 km northwest of Morisset Station (Tolman, 1936; Cousineau et al. 1984b; Cousineau, 1990; a pertinent summary is in Boucot and Drapeau, 1968, p. 18, 37, map 1607A). These Silurian strata suggest a complex pattern of regressions, erosion, and transgressions, not yet decipherable with the available data. In the Lac Témiscouata region, however, our understanding of the geological history is clearer (Lajoie et al., 1968).

The Témiscouata Fault, bounding the northwestward limit of the Early Devonian "slates," has been traced into northern Maine onto the Rocky Mountain Fault (Roy, 1989, p. 86, fig. 1) and thence suggested to be continuous with the La Guadeloupe Fault. In turn, the Témiscouata Fault continues northeastward (Lajoie et al., 1968), but present interpretations are uncertain as to whether it joins the Causapscal Fault in the Matapédia Valley (Brisebois et al., 1991; Bourque et al., 1993, fig. 23). In any event, these two faults probably belong to the same system, as they are presently interpreted to be separated by only a few kilometers (Bourque et al., 1993).

Both the Famine and Touladi Formations are Middle Devonian, have unconformities as their lower limits, and are bounded southeastward by a major thrust, almost certainly continuous from the Eastern Townships to Lac Témiscouata and, possibly, the Matapédia Valley. The Middle Devonian limestones and limy sequences, succeeding Lower Devonian "slates" that were presumably deposited in deeper water, suggest basin-edge sedimentation, tectonostratigraphically distinct from the northwestward-thrusted slates. Data are too sparse for completely deciphering the Devonian history. Nonetheless, mid-Devonian thrusting in both areas is presumably responsible for the shallow-water carbonates deposited in front of advancing Acadian thrust sheets.

Famine Formation

The "Famine Limestone" has always been considered as Middle Devonian (Onondagan) age (among others: MacKay, 1921; Clark, 1923; Tolman, 1936; Gorman, 1955; Boucot and Johnson, 1967; Boucot and Drapeau, 1968; Oliver, 1971). The latest correlation by Oliver (1976, p. 16) suggested a lower Cazenovia (Seneca, Dundee) age for the Famine, that is, uppermost Onondaga (Eifelian). Brachiopods from the Famine are not numerous, but they do include *Brevispirifer* near *lucasensis* (Boucot and Johnson, 1967, p. 54) and *Brevispirifer* sp. from Morisset Station (Boucot and Drapeau, 1968, p. 31). Famine conodonts from Saint-Georges, initially examined by Uyeno in 1966 (Boucot and Drapeau, 1968, p. 37), also indicated a Middle Devonian age. Since then Middle Devonian conodont faunas of eastern and central parts of North America, many with endemic elements, have come under close scrutiny, and this now allows for more precise dating and correlation. Eifelian-age brachiopods from the Saint-Georges area, illustrated by Boucot (1975, p. 338), are *Athyris* cf. *A. undata*, *Brevispirifer* cf. *B. gregarius*, and *Teichostrophia?* sp.

Table 1 shows the distribution of species within the Famine Formation at both the Morisset Station and Saint-Georges. The Famine conodont fauna is dominated by icriodontid species, with rare presence of polygnathids, making direct correlation with the polygnathid-based standard zonation difficult. There are key species present, the stratigraphic ranges of which do limit the age, however. The species occurring at both localities include *Latericriodus robustus*, *Icriodus orri*, *I.* cf. *I. rectirostratus*, and *I.* cf. *I. stephensoni*. Those taxa that were found only at Morisset Station include *Icriodus angustus angustus*, *I. lesperancei*, and *Polygnathus linguiformis linguiformis*, whereas those restricted to Saint-Georges are *Icriodus* n. sp. D and *I.* n. sp. E → *I. orri*. This collective fauna can be assigned to the *costatus* Zone, the age being constrained by *P. linguiformis linguiformis*, *I. angustus angustus*, and *L. robustus*. The first two of these commence within the *costatus* Zone elsewhere (Klapper and Johnson, 1980, table 7), whereas the last has its highest occurrences reported from the top of the same zone (Klug, 1983; Sparling, 1983, 1988), or the lowest part of the superjacent *australis* Zone (Klapper and Ziegler, 1979, text-fig. 4; Klapper and Johnson, 1980, table 8). D. R. Sparling (personal communication, 1996) is of the opinion that the species was exterminated from the Appalachian Basin at about the end of *costatus* Zone time. *Icriodus* cf. *I. stephensoni* is close to that species originally reported from the Dundee Formation of southwestern Ontario (Uyeno et al., 1982, as *I.* sp. C) and then from north-central Ohio, where it has a relatively long range, from the upper Columbus Limestone through Delaware Limestone (Sparling, 1983).

Regional correlation of the Famine Formation

The nearest location to the Famine Formation of a well-dated *costatus*-age conodont fauna is in west-central New York State. There, the zonal fossils occur in the Onondaga Limestone, in the upper Nedrow to Seneca members interval (Klapper, 1971; Klapper *in* Klapper and Ziegler, 1979; Klapper and Oliver, 1995). Although interval contains *L. robustus*, it lacks the other endemic species, *I. angustus angustus* (refer to Fig. 6).

TABLE 1. DISTRIBUTION OF CONODONTS FROM THE FAMINE FORMATION AT SAINT-GEORGES (69-300) AND THE MORISSET STATION (89-SF)*

Field Number	GSC Locality Number	Weight of Sample (g)	Sample Dissolved (%)	*Latericriodus robustus*	*I. cf. I. rectirostratus* (no spur)	*I. cf. I. rectirostratus* (spur)	*Icriodus orri*	*I. cf. I. stephensoni*	*I. angustus angustus*	*I. lesperancei*	*P. linguiformis linguiformis*	*Icriodus* sp. D	*I. n. sp. E-> I. orri*	Simple Cones	Ramiform Elements	Fish Teeth/Scales
Famine Formation, Morisset Station																
89-SF1A	C-129167	2,945	100	13	28	18	11	5	27	85					X	X
89-SF1B	C-129167	1,858	100	8	19	9	3	5	20	75	3			X		X
89-SF1C	C-129167	2,821	98	3	32	14	2	7	26	92					X	X
Famine Formation, Saint-Georges																
69-300	C-7271	3,835	Unknown	1?	2											
69-301	C-7272	4,230	Unknown												X	
69-302	C-7273	3,850	Unknown	1								3			X	
69-303	C-7274	3,327	Unknown	2	1	4	1						2			X
69-304	C-7275	3,735	Unknown	2	4	3										
69-305	C-7276	4,012	Unknown		1	2						1				
69-306	C-7277	3,540	Unknown	2	3							3			X	
69-307	C-7278	4,380	Unknown		3		1?					7		X		
69-308	C-7279	4,310	Unknown				1cf.	2				3	1			

*Refer to Appendix for stratigraphic information. *I.* = *Icriodus*, *P.* = *Polygnathus*. X = present but not counted.

In southwestern Ontario, the *costatus* Zone encompasses parts of the Dundee Formation and the Anderdon Member of the Lucas Formation (Uyeno et al., 1982; Sparling 1988). (It should be noted that the sampled units on Pelee Island, mapped as Dundee Formation by B. V. Sanford [in Uyeno et al., 1982], were considered as a northward extension of the Marblehead Member of the Columbus Limestone by Sparling [1983, 1988]; a similar conclusion was reached by Birchard and Risk [1990]). The varied fauna from the Pelee Island strata includes *L. robustus*, *I. orri*, *I. angustus angustus*, *P. linguiformis linguiformis*, *P. costatus costatus*, *P. angusticostatus*, and *Tortodus intermedius*.

In north-central Ohio, Sparling (1983, 1988) assigned conodonts from the Marblehead Member and the lowest part of the overlying Venice Member, both of the Columbus Limestone, to the *costatus* Zone. Taxa shared by the upper Columbus Limestone and the Famine Formation include *L. robustus* and *I. orri*. *Icriodus angustus angustus* is also present in the upper Venice Member but in that part assigned to the *australis* Zone.

In south-central Indiana and north-central Kentucky, Klug (1983) assigned a similarly aged fauna from the upper Jeffersonville Limestone to the *Icriodus angustus angustus* Zone. Unlike the Famine Formation, the Jeffersonville fauna includes the index species *Polygnathus costatus costatus* as well.

In the Jaab Lake well of the Moose River Basin, northern Ontario, the correlative of the *costatus* Zone was noted to occur within the Kwataboahegan Formation (in interval between depths of 109.7 m and 146.9 m) (Uyeno and Bultynck, 1993). This was based on limits placed by the lowest occurrence of *I. angustus angustus* (Datum plane 7) and the highest occurrence of *L. robustus* (Datum plane 9).

Touladi Formation

Indication that the Touladi Formation was not of Becraft-Oriskany (Early Devonian) age, as originally determined by A. J. Boucot (Lespérance and Greiner, 1969, p. 77), was first put forward from the study of conodonts by G. Klapper (Lespérance and Greiner, 1969, p. viii). This prompted a reexamination of the brachiopod fauna and led to an assignment to the Eifelian, based on the brachiopod *Brevispirifer* near *lucasensis* Stauffer (Boucot and Johnson, 1967, p. 54) and the conodonts (Boucot and Drapeau, 1968, pp. 19, 37). The Touladi conodonts, initially examined by G. Klapper in 1967, were from the Allmo Quarry and, as noted above, were dated as Eifelian.

Table 2 shows the distribution of species within the Touladi Formation at both the Allmo Quarry (89-T1 and -T2) and in an outcrop on the hill located south-southwest of Cabano (89-T3

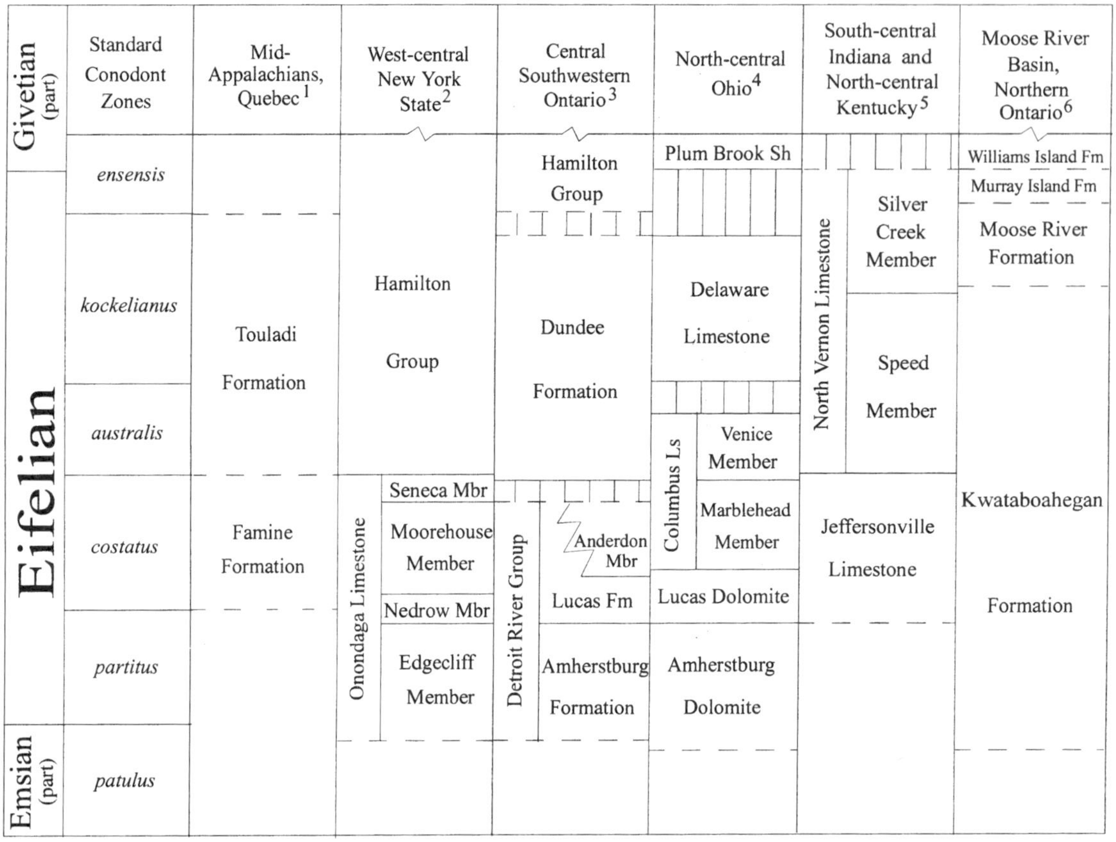

Figure 6. Correlation of the Famine and Touladi formations with stratigraphic sections discussed in the text and with the standard conodont zones and stages. Explanation: [1]After this study and Oliver (1971, 1976); [2]Klapper (1971), Klapper and Ziegler (1979), and Klapper and Oliver (1995); [3]Ferrigno (1971), Orr (1971), Uyeno et al. (1982), and Sparling (1983, 1988); [4]Sparling (1983, 1988, 1995); [5]Klapper and Ziegler (1979) and Klug (1983); [6]Sanford and Norris (1975), and Uyeno and Bultynck (1993).

and -T4). As with the Famine Formation, the Touladi conodont fauna is dominated by icriodontid species. Taxa shared by the two localities include *I. orri*, *I.* n. sp. E → *I. orri*, *I.* cf. *I. rectirostratus*, *P. linguiformis* subsp. indet., and possibly *P. linguiformis linguiformis*. At the Allmo Quarry the fauna also includes *P. linguiformis alveolus*, whereas the hill outcrop locality has an addition of *I. angustus angustus* and *P. pseudofoliatus*. *Icriodus* n. sp. E → *I. orri* is close to *I. orri*, with the main difference being the slightly narrower basal cavity; hence it is closer to *I.* n. sp. E of Weddige (1977). The collective fauna can be assigned to the *australis* and *kockelianus* zones, with the lower and upper constraints provided by *P. pseudofoliatus* and *P. linguiformis alveolus*, respectively (Weddige, 1977; Klapper and Johnson, 1980, Table 8). Previously, the icriodontan species, *I. orri* and *I. angustus angustus*, were thought to have their highest occurrences within the *kockelianus* Zone (Klapper and Johnson, 1980). Those ranges may still hold if the occurrence of both in the younger Lower *varcus* Subzone, reported by Klug (1983), is the result of reworking. This possibility is suggested since there is a hiatus between the Silver Creek Member and Beechwood Member of the North Vernon Limestone (Klug, 1983, figs. 4 and 5). Some of the concerns regarding the dating of the Silver Creek Member were expressed by Sparling (1983, p. 839). At any rate, the Touladi fauna appears to be slightly younger than that of the Famine.

Regional correlation of the Touladi Formation

In west-central New York State, the *australis-kockelianus* zonal interval is encompassed within the lower Hamilton Group (Klapper, 1971; Klapper *in* Klapper and Ziegler, 1979). A diverse polygnathid fauna occurs in the Marcellus Formation, namely in the Cherry Valley Member and the underlying *Werneroceras* bed of the Union Springs Member. The Cherry Valley also contains *Tortodus kockelianus kockelianus*, the name-bearer of the younger of the two zones (refer to Fig. 6).

On the basis of the highest occurrence of *Latericriodus robustus*, Sparling (1983, 1988) placed the upper boundary of the *costatus* Zone in the lowest part of the Dundee Formation at St. Marys, Ontario (Ferrigno, 1971). The diverse conodont fauna of

TABLE 2. DISTRIBUTION OF CONODONTS FROM THE TOULADI FORMATION AT THE ALLMO QUARRY (89-T1) AND AT AN OUTCROP ON SIDE OF HILL (89-T3)*

Field Number	GSC Locality Number	Weight of Sample (g)	Sample Dissolved (%)	*Icriodus orri*	*I.* n. sp. E-> *I. orri*	*I.* cf. *I. rectirostratus* (no spur)	*I.* cf. *I. rectirostratus* (spur)	*P. linguiformis* subsp.	*P. linguiformis linguiformis*	*P. linguiformis alveolus*	*I. angustus angustus*	*P. pseudofoliatus*	*Polygnathus* sp. indet.	*Dvorakia* sp.	Simple Cones	Ramiform Elements	Fish Teeth/Scales
Touladi Formation, Allmo Quarry																	
89-T1A	C-129164	2,824	71	89	57	17									X		X
89-T1B	C-129164	3,465	70	28	10	5											X
89-T2A	C-129164	3,308	97	61	45	16	2	3	1	1				X	X	X	X
89-T2B	C-129164	2,157	99	52	14	14		2	1						X	X	X
	O-96102	1,680	60	6	5	3	1								X		X
A	O-102211	725	45	2	3	1									X		
B	O-102212	2,410	74		2		2									X	
C	O-102213	5,390	40												X		
D	O-102214	9,875	53	3		1									X	X	
E	O-102215	2,630	53					1					1			X	
F	O-102216	1,080	57	1		3											
Touladi Formation, Outcrop on Side of Hill, SSW of Cabano																	
89-T3A	C-129165	2,342	98	14	6	19								X	X		
89-T3B	C-129165	1,388	93	10	4		1								X		
89-T4A	C-129166	1,923	86	13	18	12					2			X	X	X	
89-T4B	C-129166	2,415	92	40	22	20	1	3			1			X	X	X	
89-T4C	C-129166	2,769	98	37	26	9		1		1cf.	4	1		X		X	

*Refer to Appendix for stratigraphic information. *I.* = *Icriodus*, *P.* = *Polygnathus*. X = present but not counted.

the Dundee Formation of southwestern Ontario was assigned to the *costatus* Zone in the lower part and to the *australis* and *kockelianus* zones in the upper (Uyeno et al., 1982). Incidental to this study, the relationship of the Dundee Formation with the underlying Detroit River Group has been a point of discussion in the past. Recently, after a detailed investigation into the stratigraphic relations of the Dundee Formation, Birchard and Risk (1990) concluded that up to 10 m of sediment may have been eroded from the Detroit River strata prior to the deposition of the Dundee limestones.

In north-central Ohio, the higher parts of the Venice Member of the Columbus Limestone have a varied fauna, including *T. kockelianus australis*, the name-bearer of that zone (Sparling, 1983, fig. 6). The assignment of the Delaware Limestone to the *kockelianus* Zone is less certain, since it has not yielded diagnostic taxa.

In south-central Indiana and north-central Kentucky, conodonts from the Speed and Silver Creek members of the North Vernon Limestone include *P. pseudofoliatus*, *P. linguiformis linguiformis*, *I. orri*, *I. angustus angustus*, and *I.* sp. E of Weddige (1977) → *I. orri* (Klug, 1983; Klapper *in* Klapper and Ziegler, 1979). The fauna was assigned to the *Polygnathus pseudofoliatus* Zone, a zone that may correlate with part or all of the sequence from the *australis* to *ensensis* zones (Klug, 1983). The Speed Member was earlier correlated with the *australis* and *kockelianus* zones by Klapper (*in* Klapper and Ziegler, 1979).

The Lower and Middle Devonian conodont faunas in the Jaab Lake well, Moose River Basin in northern Ontario, consist primarily of icriodontid species as in the Touladi Formation (Uyeno and Bultynck, 1993). There, the base of the undivided *australis-kockelianus* zonal interval was placed at Datum plane 9 (the highest occurrence of *L. robustus*) in the upper part of the Kwataboahegan Formation (Uyeno and Bultynck, 1993). The top of the interval was less definite for the following reason. The highest occurrence of *I. angustus angustus*, in the past taken as suggesting a level within the *kockelianus* Zone, was designated as Datum plane 10. This highest occurrence is followed, however, by a 41-m interval of inhospitable bio- and lithofacies, mainly of the Moose River Formation, from which no conodonts were recovered.

THE CONODONT FAUNAS

The conodont faunas of the Famine and Touladi formations are dominated by icriodontid species (Figs. 7 and 8). As a result, correlations are difficult to make with the standard conodont zones, which are primarily based on polygnathids. Further, these icriodontids consist of species that are largely endemic to North America, including *Latericriodus robustus*, *Icriodus orri*, *I. angustus angustus*, *I.* n. sp. E of Weddige transitional to *I. orri* (of Klug, 1983), and *I. lesperancei*, n. sp. The other dominant icriodontan species is *I.* cf. *I. rectirostratus*, which is somewhat similar to a species originally reported from the type Couvinian sections in Belgium (Bultynck, 1970). Sporadic polygnathids do occur, together with simple cones and ramiform elements. The dominance of icriodontids is a common characteristic of shallow-water conodont biofacies. This perhaps reflects the basin-edge setting of the Famine and Touladi sedimentation.

The presence of the endemic species mentioned above probably represents their easternmost known occurrences on the North American continent. The Saint-Georges and Cabano sites were located in an eastward extension of an intracratonic Middle Devonian sea, in a mixed carbonate-terrigenous setting, adjacent to the Acadian Mountains, and located south of 30° south latitude (Heckel and Witzke, 1979, text-fig. 3B). According to Koch (1981, fig. 11), during the early Eifelian the two sites were situated near the junction of the Old World Realm and the two subprovinces (Appohimchi Subprovince and Michigan Basin–Hudson Bay Lowland) of the Eastern Americas Realm. Such a juxtaposition of paleogeographic provinces may provide some clues as to the presence in Nevada of *Latericriodus robustus* in the late Emsian and early Eifelian (Johnson and Klapper, 1981, p. 1241; Kendall et al., 1983, p. 2205), since otherwise the species is known only from eastern North America. In terms of Sparling's (1984) faunal realms during the *costatus* Zone and younger Eifelian time, the Famine and Touladi conodonts can be placed in the Eastern Realm. Here, as in other faunas of this realm, *Icriodus orri* occurs in both the *costatus* and younger zones. Areas included in the Eastern Realm include New York State, southwestern Ontario, north-central and central parts of Ohio, and the eastern Illinois Basin. Uyeno and Bultynck (1993) concluded that the Jaab Lake conodont faunas from the Moose River Basin, northern Ontario, had closest affinity to the Eastern Faunal Realm, with some connection to the northern Michigan Basin.

The Touladi conodonts have fair to good preservation and have a Color Alteration Index (CAI; Epstein et al., 1977) of about 4. The Famine conodonts are black (CAI of 5), are less well preserved, are often contorted and misshapen, and have extraneous material adhering to them. The preservation and color suggest that the Famine strata were subjected to temperatures in the range of 300 to 480° C and to regional metamorphism (Rejebian et al., 1987). This, in turn, is probably a reflection of the tectonic deformation of the Famine Formation in the Saint-Georges and Morisset Station area. Both the Touladi and Famine collections include numerous unidentifiable fragmentary, as well as small (juvenile?), Pa elements of icriodontids that were not counted or listed in the distribution tables. Also present are fish teeth and scales and pyritized(?) ostracodes.

SYSTEMATIC PALEONTOLOGY

We follow the locational notation used in Supplement 2 of Part W of the *Treatise on Invertebrate Paleontology* (Robison, 1981). Only the platform elements are described, illustrated, and counted in this study. The type and figured specimens are deposited in the National Type Collection of Plant and Invertebrate Fossils of the Geological Survey of Canada (GSC), 601 Booth Street, Ottawa, Ontario K1A 0E8.

Family Icriodontidae Müller and Müller, 1957

Genus *Icriodus* Branson and Mehl, 1938

Type species. Icriodus expansus Branson and Mehl, 1938.

***Icriodus angustus* Stewart and Sweet, 1956**
***Icriodus angustus angustus* Stewart and Sweet, 1956**
Figure 7: 1–10

Icriodus angustus STEWART AND SWEET, 1956, p. 267, pl. 33, figs. 4, 5, 11, 15; SPARLING, 1983, p. 850, figs. 12AC, AG; 13AD, AE, AV, AW [synonymy to 1982].
Icriodus angustus angustus Stewart and Sweet. KLUG, 1983, p. 84, figs. 8A–F; UYENO AND BULTYNCK, 1993, p. 19, pl. 2.4, figs. 3–7.

Remarks. The lateral row denticles are round and discrete, whereas the middle row denticles, although similarly discrete, are longitudinally compressed and much smaller than their lateral row counterparts. The posterior margin of the basal cavity may be narrowly rounded (Fig. 7: 2) or pointed (Fig. 7: 9). Two or three denticles on the posterior extension of the middle row are distinctly higher than those on the spindle.

***Icriodus lesperancei* Uyeno, n. sp.**
Figure 7: 11–20

Diagnosis. A species of *Icriodus* the Pa element of which has a narrowly elongate, straight spindle with five to seven round, discrete denticles in the lateral rows. The middle row bears six to eight discrete denticles that may be round to slightly compressed laterally and are nearly aligned, or alternate, with the lateral row denticles. The posterior extension of the middle row bears two or three laterally compressed denticles of height similar to those on the spindle and may be straight to slightly laterally curved, although the latter feature may be an artifact of preservation. In lateral view, the upper margin may be straight to slightly

Figure 7. Magnification 48×; all Pa elements. 1–10, *Icriodus angustus angustus* Stewart and Sweet. 1–3, GSC 112894, GSC loc. C-129166 (T4C), Touladi Fm.; 4, 5, GSC 112895, GSC loc. C-129167 (SF1C), Famine Fm.; 6–8, GSC 112896, GSC loc. C-129167 (SF1A), Famine Fm.; 9, 10, GSC 112897, GSC loc. C-129167 (SF1B), Famine Fm. 11–20: *Icriodus lesperancei* Uyeno, n. sp. 11, 12, GSC 112898, GSC loc. C-129167 (SF1A), Famine Fm.; 13–15, GSC 112899, GSC loc. C-129167 (SF1A), Famine Fm.; 16–18, holotype, GSC 112900, GSC loc. C-129167 (SF1B), Famine Fm.; 19, 20, GSC 112901, GSC loc. C-129167 (SF1C), Famine Fm. 21–25, 31, 32: *Icriodus orri* Klapper and Barrick. 21–23, GSC 112902, GSC loc. C-129164 (T2A), Touladi Fm.; 24, 25, GSC 112903, GSC loc. C-129164 (T1A), Touladi Fm.; 31, 32, GSC 112904, GSC loc. C-129167 (SF1B), Famine Fm. 26–30: *Icriodus* n. sp. E of Weddige (1977) → *Icriodus orri* Klapper and Barrick. 26, 27, GSC 112905, GSC loc. C-129164 (T2A), Touladi Fm.; 28–30, GSC 112906, GSC loc. C-129164 (T1B), Touladi Fm.

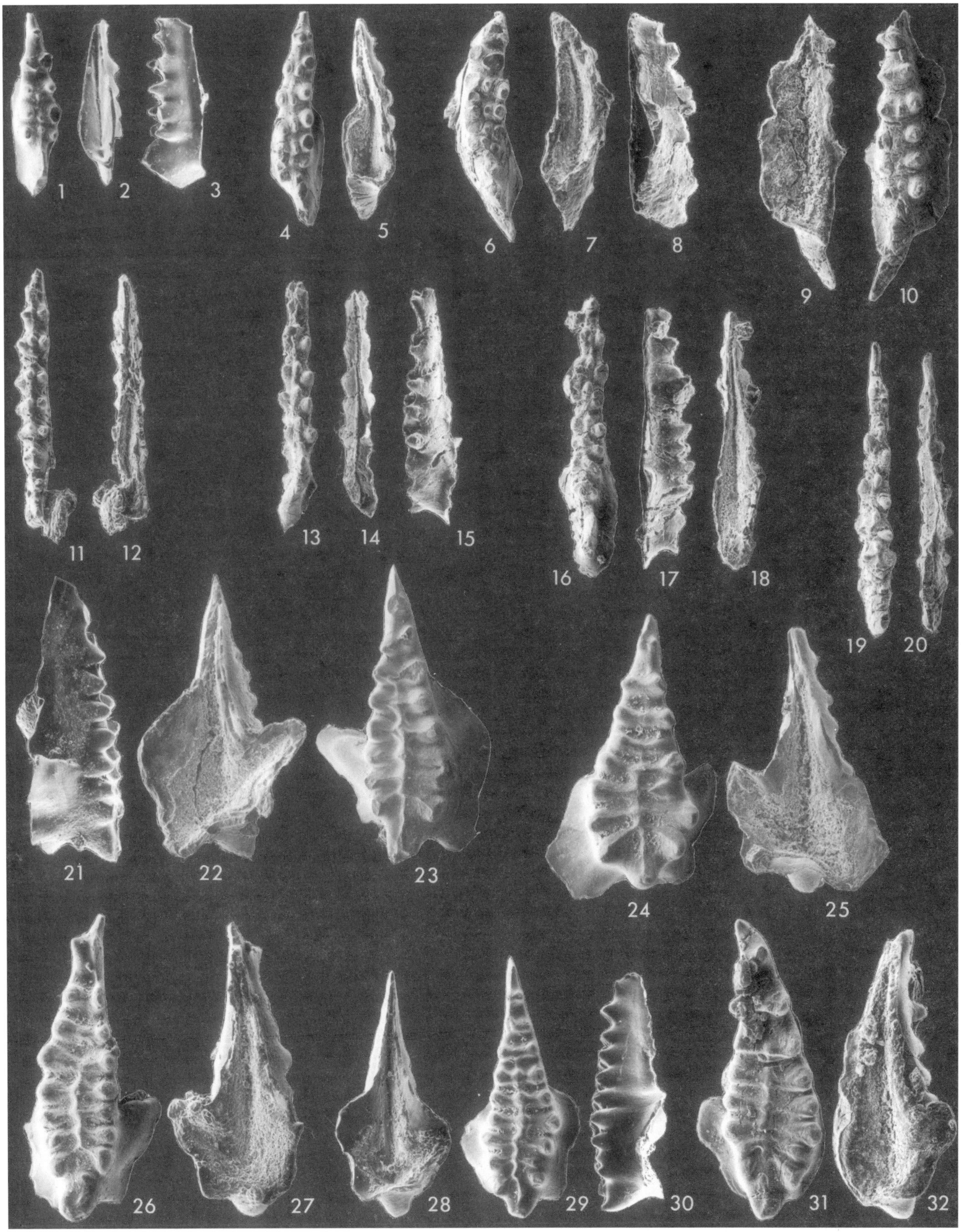
1
2
3
4
5
6
7
8
9
10
11
12
13
14
15
16
17
18
19
20
21
22
23
24
25
26
27
28
29
30
31
32

arched, and the lower margin is straight. The basal cavity outline is narrow and elongate, gradually widening posteriorly, and may have a slight spur developed; its posterior margin may be bluntly pointed or narrowly rounded.

Remarks. The species bears some resemblance to *I. culicellus* (Bultynck) in the character of the spindle but may be distinguished from it, as the latter possesses prominent, high denticles on the middle row extension and a wider basal cavity. It may be distinguished from the early form of *I. obliquimarginatus* Bischoff and Ziegler (sensu Weddige, 1977, p. 295) by its shorter posterior extension with a fewer number of denticles and the narrower basal cavity.

Material. The specimen shown in Figures 7: 16–18 is chosen as holotype (GSC 112900); those shown in Figures 7: 11–15, 19, 20 are paratypes (GSC 112898, 112899, 112901).

Derivation of name. After my distinguished coauthor, Prof. P. J. Lespérance, Université de Montréal, for his numerous contributions to the paleontology of la belle province, Québec.

Icriodus orri Klapper and Barrick, 1983
Figure 7: 21–25, 31, 32

Icriodus orri KLAPPER AND BARRICK, 1983, p. 1230–1231, figs. 9A–U, X–AA, AD-AG; SPARLING, 1995, figs. 4.1–4.5.
Icriodus sp. A KLUG, 1983, p. 87–88, text-figs. 9S–X.

Remarks. As in the Spillville Formation (Klapper and Barrick, 1983), two morphotypes are present, those with and without posterior depression on the upper surface of the spindle. There appears to be an overlap in the features of *I.* n. sp. E of Weddige (1977) and *I. orri* (see Klapper and Barrick, 1983, p. 1231). The main distinguishing criterion is that the posterior part of the basal cavity in Weddige's species is narrower than in *I. orri*, but this feature is difficult to observe in many specimens owing to abrasion. Some Touladi specimens are similar to those referred to as transitional between *I.* n. sp. E and *I. orri* by Klug (1983, figs. 9P-R) from the Speed Member of the North Vernon Limestone (herein identified as I. n. sp. E → *I. orri* in the tables and figure explanations).

The specimens displaying widely flaring basal cavity and, on the upper surface of the spindle, ridges connecting the lateral and middle row denticles approach the Pa element of *I. norfordi* Chatterton (see, e.g., Uyeno, 1990, pl. 19, figs. 23, 24). The latter species does not include those forms with posterior depression, however (Chatterton, 1979, p. 202).

The large (gerontic?) Pa elements of *I. orri* may display a costa on the upper surface of the spur, and in some specimens, one to three small denticles may adorn such costae. In another Pa element, a small antispur with two small denticles on its upper surface can be observed on its outer side.

Icriodus cf. *I. rectirostratus* Bultynck, 1970
Figure 8: 1–6

[cf.] *Icriodus nodosus rectirostratus* BULTYNCK, 1970, p. 107–108, pl. 3, fig. 1; pl. 30, figs. 7, 8.
[cf.] *Icriodus rectirostratus* Bultynck. BULTYNCK AND HOLLARD, 1981, p. 40, pl. 4, figs. 4-6, 9.

Remarks. The spindle of the Québec specimens agrees well with that reported by Bultynck (1970) in its shape and the character of denticulation. Also, the partially fused denticles on the posterior extension of the middle row are similar, numbering two to three, with the posteriormost being the largest; in the Québec collection, however, the posteriormost denticle tends to be inclined posteriorly, whereas the Belgian ones are more erect. The spur on the inner side may be prominent (Fig. 8: 5), or only moderately so (Fig. 8: 2); further, the posterior bowl of the latter specimen is not as wide as that of the former. Neither of the illustrated specimens has a costa on its upper surface, a feature described by Bultynck (1970), although Bultynck and Hollard (1981, pl. 4, fig. 5) illustrated a specimen with only a slightly developed spur and no costa.

Icriodus cf. *I. rectirostratus* differs from the species herein referred to as *I.* n. sp. E of Weddige (1977) → *I. orri* in generally being smaller and having a slimmer spindle, and the posteriormost denticles of the middle row of similar size tend to be discrete. In the latter, conversely, the posteriormost denticle of the middle row is large and bulbous, preceded by a relatively small denticle that is invariably fused onto the large one.

Icriodus cf. *I. stephensoni* Sparling, 1983
Figure 8: 9–13

[cf.] *Icriodus stephensoni* SPARLING, 1983, p. 851–852, figs. 10AA, AM; 11H, J, K, X, Y; 12H–J, AL–AQ; 13I, J, AF–AH, AU.

Remarks. The Famine specimens tend to display lateral row denticles that are laterally elongate, almost touching the median row denticles, whereas in *I. stephensoni* they tend to be more roundish and discrete. The widest part of the spindle is often at about mid-length position and, in *I. stephensoni*, toward the posterior. In both species, the posterior basal cavity outline is roundish with incipient development of spur in some specimens.

Icriodus sp. D
Figure 8: 18–20

Diagnosis. A species of *Icriodus*, the Pa element of which has a narrow, lanceolate spindle with six round, discrete lateral row denticles; the denticles are widely spaced, especially in the anterior half. The six or seven denticles of the middle row are round to laterally compressed and alternate with those of the lateral rows. The posterior extension of the middle row bears three low denticles. In lateral view, the upper margin is flat and rapidly arcs downward at the posterior end; the lower margin slopes gradually posteriorly, so that the lowest part is generally at the extreme posterior end. The basal cavity is lachrymiform and moderately expanded posteriorly.

Remarks. Icriodus sp. D differs from *I. lesperancei* n. sp. in its wider spindle and basal cavity. It appears to be close to *I. culicellus* aff. *monfortensis* Wang from the Moyuela Formation in Celtiberia, Spain (Wang, 1992, figs. 4g, h). The two species are similar in exhibiting the posterior extension of the middle row that is low and arcing downward. The Spanish specimen has fewer lateral row denticles, numbering four. The illustrated specimen from the Famine Formation has a partly broken outer lateral margin in the posterior part of its basal cavity.

Figure 8. Magnification 48×, except 8 at ×33; all Pa elements. 1–6: *Icriodus* cf. *I. rectirostratus* Bultynck. 1–3, GSC 112907, GSC loc. C-7275 (69-304), Famine Fm.; 4–6, GSC 112908, GSC loc. C-129164 (T2A), Touladi Fm. 7: *Polygnathus pseudofoliatus* Wittekindt, GSC 112909, GSC loc. C-129166 (T4C), Touladi Fm. 8: *Polygnathus linguiformis linguiformis* Hinde, GSC 112910, GSC loc. C-129167 (SF1B), Famine Fm. 9–13: *Icriodus* cf. *I. stephensoni* Sparling. 9, 10, GSC 112911, GSC loc. C-129167 (SF1C), Famine Fm.; 11–13, GSC 112912, GSC loc. C-7279 (69-308), Famine Fm. 14, 15, 21–25: *Latericriodus robustus* (Orr). 14, 15, GSC 112913, GSC loc. C-7277 (69-306), Famine Fm.; 21, 22, GSC 112914, GSC loc. C-129167 (SF1C), Famine Fm.; 23–25, GSC 112915, GSC loc. C-7275 (69-304), Famine Fm. 16, 17, *Polygnathus linguiformis alveolus* Weddige, GSC 112916, GSC loc. C-129164 (T2A), Touladi Fm. 18-20, *Icriodus* sp. D, GSC 112917, GSC loc. C-7279 (69-308), Famine Fm. 26, *Polygnathus linguiformis* cf. *linguiformis* Hinde, GSC 112918, GSC loc. C-129166 (T4C), Touladi Fm.

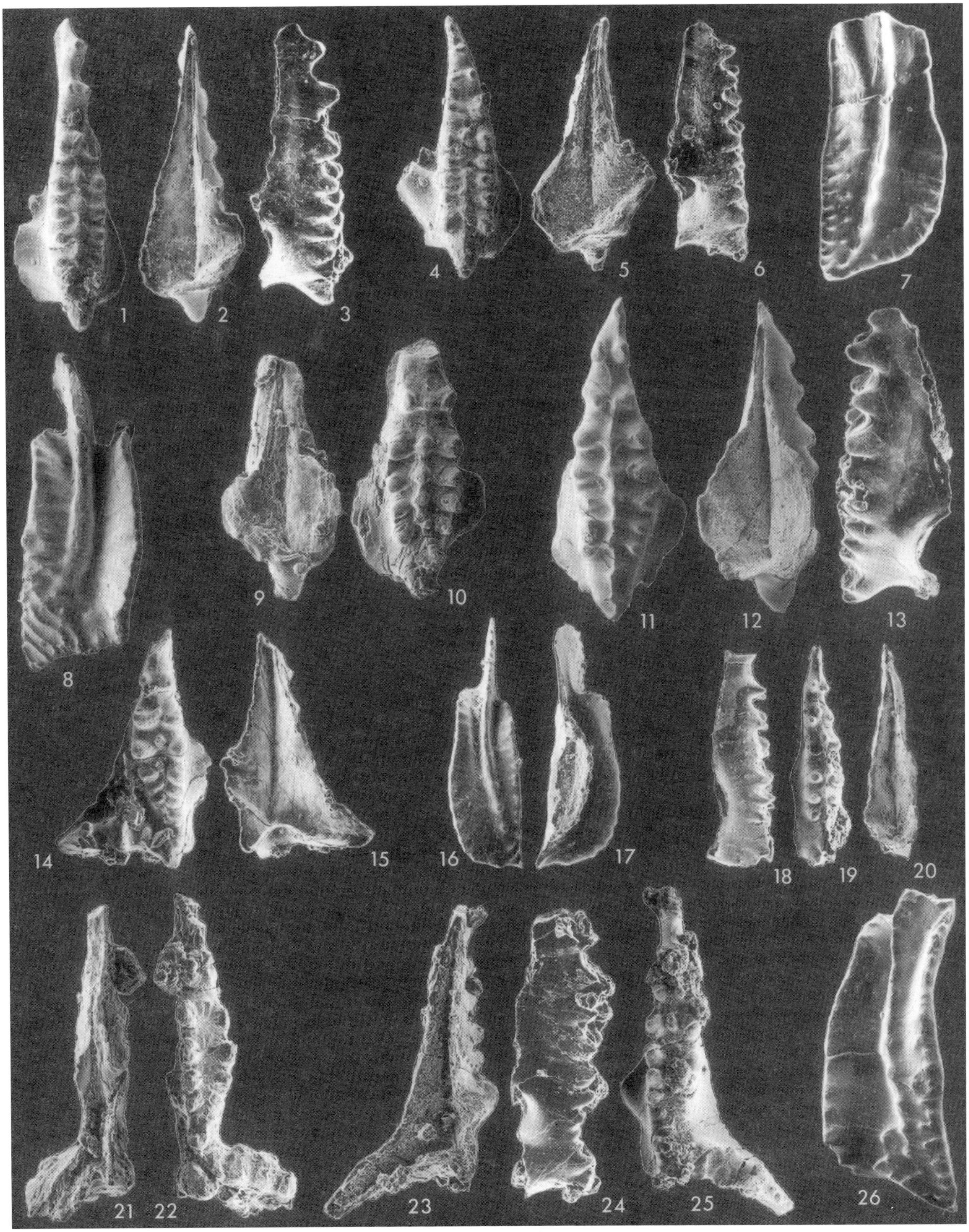
1
2
3
4
5
6
7
8
9
10
11
12
13
14
15
16
17
18
19
20
21
22
23
24
25
26

Genus *Latericriodus* Müller, 1962

Type species. Icriodus latericrescens Branson and Mehl, 1938.

***Latericriodus robustus* (Orr, 1971)**
Figure 8: 14, 15, 21–25

Icriodus latericrescens n. subsp. A. KLAPPER AND ZIEGLER, 1967, p. 75, pl. 8, figs. 7, 9; pl. 9, figs. 1–7.
Icriodus latericrescens robustus ORR, 1971, p. 37, 38, pl. 2, figs. 14–17.
Latericriodus robustus (Orr). UYENO AND BULTYNCK, 1993, p. 22, pl. 2.2, figs. 4–19; pl. 2.4, figs. 1, 2.

Remarks. The denticles of the middle row lie in a depression and are characteristically discrete and roundish. In some larger specimens (Fig. 8: 22), these denticles are obscured toward the posterior of the spindle.

Family Polygnathidae Bassler, 1925

Genus *Polygnathus* Hinde, 1879

Type species. Polygnathus dubius Hinde, 1879.

***Polygnathus linguiformis* Hinde, 1879**
***Polygnathus linguiformis linguiformis* Hinde, 1879**
Figure 8: 8

Polygnathus linguiformis HINDE, 1879, p. 367, pl. 17, fig. 15.
Polygnathus linguiformis linguiformis Hinde gamma forma nova. BULTYNCK, 1970, p. 126, 127, pl. 11, figs. 1–6; pl. 12, figs. 1–6.

Remarks. The gamma morphotype of Bultynck (1970) is considered synonymous with the nominal subspecies (see Weddige, 1977, p. 312–316). One large Pa element displays the characteristic outer flange of the nominate subspecies but only short ridges on its tongue (Figure 8: 26, referred to herein as *P. linguiformis* cf. *linguiformis*).

***Polygnathus linguiformis alveolus* Weddige, 1977**
Figure 8: 16, 17

Polygnathus linguiformis alveolus WEDDIGE, 1977, p. 312–313, pl. 5, figs. 85–87, text-fig. 4/19.

Remarks. The single Pa element displays on the outer side a deep and wide adcarinal trough that is bounded by the outer rim of the platform; the platform, at that point, makes a nearly rectangular bend. The tongue is characteristically minute. It is close to one of the paratypes of Weddige (1977, pl. 5, fig. 86).

***Polygnathus pseudofoliatus* Wittekindt, 1966**
Figure 8: 7

Polygnathus pseudofoliata WITTEKINDT, 1966, p. 637–638, pl. 2, figs. 20–23.
Polygnathus pseudofoliatus Wittekindt. SAVAGE, 1995, p. 546, figs. 3.12–.17; SPARLING, 1995, p. 1136, figs. 2.1–2.8, 2.11–2.19.

Remarks. A single, fragmentary specimen, representing the Pa element, is assignable to the species. It displays the characteristic outline of the outer lateral margin that is anteriorly constricted, followed posteriorly by a rounded expansion. It further displays ridgelike adcarinal margins and has adcarinal troughs that shallow posteriorly. The posterior half of the platform also has transverse ridges, or a series of short, disconnected ridges. This feature is more characteristic of the early forms of the species (see Sparling, 1995, p. 1126).

ACKNOWLEDGMENTS

Pierre-André Bourque in 1969 and Martin Bizzarro in 1989 helped PJL with the fieldwork associated with the sampling; Pierre Cousineau, in 1990, was kind enough to guide that author in the field to demonstrate the stratigraphy of the Saint-Georges area. John Rodgers and Walter Trczienski, in 1991, offered pertinent comments in the field. PJL also wishes to record his indebtedness to the Natural Sciences and Engineering Research Council of Canada for successive operating grants. Dale Sparling and Pierre Bultynck kindly reviewed the chapter and offered several excellent suggestions. The following of the Geological Survey of Canada helped in several ways: Godfrey Nowlan supplied additional conodonts from the Touladi Formation; D. C. McGregor examined two Famine samples for palynomorphs and supplied unpublished information on the York River Formation from western Gaspé, Québec; Jenny Wong took SEM micrographs of conodonts, and Brian Rutley printed the negatives; Brian Ortman, Paul Wozniak, and Denise Then helped in the preparation of locality figures. This is Geological Survey of Canada contribution no. 43395.

APPENDIX: LOCALITY REGISTER

Famine Formation

(1) Saint-Georges area. 1969 collections. All are from Universal Transverse Mercator (UTM) 37027 500904 in the town of Saint-Georges and shown in Figures 3 and 4. See also Table 1 for weights of samples. The outcrops occur on the north side of Route 173 and have been reported in MacKay (1921). The 26.3-m-thick interbedded argillaceous limestones and limestones above the 9.8-m-thick basal conglomerate was arbitrarily divided into the following sample intervals:

Interval	*Field no.*	*GSC loc. no.*
9.8–12.2 m	69-308	C-7279
12.2–16.2 m	69-307	C-7278
16.2–20.2 m	69-306	C-7277
20.2–24.2 m	69-305	C-7276
24.2–28.2 m	69-304	C-7275
28.2–32.2 m	69-303	C-7274
32.2–36.1 m	69-302	C-7273

An isolated outcrop, composed predominantly of slates with little argillaceous limestone, possibly 165 m above the conglomerate (if the same dip of the strata as in the limestones is applicable) was also sampled; its lower half is 69-300 (GSC loc. no. C-7271) and upper half 69-301 (GSC loc. no. C-7272). No palynomorphs were recovered from either sample 69-300 nor 69-301 (D. C. McGregor, personal communication, 1971).

(2) Morisset Station area. 1989 collections: UTM 38125 512300; along north bank of Famine River, approximately 75 m east of bridge at Morisset Station, 18 km northeast of Saint-Georges and the 1969 collections. Locality shown on Figure 3; see also Table 1 for weights of samples. In middle of alternating argillaceous and purer limestones, both cleaved; sample from a purer limestone band, 20 cm thick if presumed bedding is correct (see Cousineauet al., 1984a); locality mentioned by Tolman (1936, p. 16).

GSC loc. no.	*Field no.*
C-129167	89-SF1A
C-129167	89-SF1B
C-129167	89-SF1C

Touladi Formation

(1) Allmo Quarry. UTM 5114 52747: one loose block of calcarenite at northeast end of abandoned Allmo Quarry, 1.3 km west-southwest of Trans-Canada Highway (Route 185), 6.35 km south-southeast of church in Cabano (= fossil locality G28 of Lespérance and Greiner, 1969). Locality shown on Figure 5; see also Table 2 for weights of samples.

GSC loc. no.	*Field no.*
C-129164	89-T1A
C-129164	89-T1B

Same locality as 89-T1; a second calcarenite block:

C-129164	89-T2A
C-129164	89-T2B

Allmo Quarry, collected by Jean David in 1986. Six distinct blocks on the sides of the quarry. Data and conodont slides courtesy of G. S. Nowlan.

GSC loc. no.	*Field no.*
O-102211	A
O-102212	B
O-102213	C
O-102214	D
O-102215	E
O-102216	F

Allmo Quarry, collected by G. S. Nowlan in 1978; one block in quarry, GSC O-96102.

(2) Outcrop on side of hill. UTM 50567 526782: Touladi Formation, outcrop on side of hill, 240 m southeast of secondary gravel road, 13.65 km west-southwest of church in Cabano and 8.84 km southwest of 89-T1 and 89-T2; collection from one 8-cm bed. Locality shown on Figure 5; see Table 2 for weights of samples.

GSC loc. no.	*Field no.*
C-129165	89-T3A
C-129165	89-T3B

Locality as 89-T3; from a single 15-cm-thick bed, 35 cm stratigraphically above 89-T3:

C-129166	89-T4A
C-129166	89-T4B
C-129166	89-T4C

REFERENCES CITED

Bassler, R. S., 1925, Classification and stratigraphic use of the conodonts: Geological Society of America Bulletin, v. 36, p. 218–220.

Berger, J., 1993, Étude structurale et gîtologique de la faille de Sainte-Florence–Gaspésie: Québec, Ministère de l'Énergie et des Ressources MB 93-01, 48 p.

Birchard, M. C., and Risk, M. J., 1990, Stratigraphy of the Middle Devonian Dundee Formation, southwestern Ontario (Grant 340): Ontario Geological Survey Miscellaneous Paper 150, p. 71–86.

Boucot, A. J., 1970, Devonian slate problems in the Northern Appalachians, *in* Shorter contributions to Maine geology: Maine Geological Survey Bulletin 23, p. 42–48.

Boucot, A. J., 1975, Evolution and extinction rate controls: Amsterdam, Elsevier, Developments in Palaeontology and Stratigraphy, v. 1, 427 p.

Boucot, A. J., and Drapeau, G., 1968, Siluro-Devonian rocks of Lake Memphremagog, and their correlatives in the Eastern Townships: Quebec Department of Natural Resources Special Paper 1, 44 p.

Boucot, A. J., and Johnson, J. G., 1967, Paleogeography and correlation of Appalachian Province Lower Devonian sedimentary rocks, *in* Toomey, D. F., ed., Silurian-Devonian rocks of Oklahoma and environs: Tulsa Geological Society Digest, v. 35, p. 10–87.

Boucot, A. J., Brett, C. E., Oliver, W. A., Jr., and Blodgett, R. B., 1987, Devonian faunas of the Sainte-Hélène Island breccia, Montréal, Quebec, Canada: Canadian Journal of Earth Sciences, v. 23, p. 2047–2056. [imprint 1986]

Bourque, P. A., Gosselin, C., Kirkwood, D., Malo, M., and St-Julien, P., 1993, Le Silurien du segment appalachien Gaspésie-Matapédia-Témiscouata; stratigraphie, géologie structurale et paléogéographie: Québec, Ministère de l'Énergie et des Ressources MB 93–25, 115 p.

Branson, E. B., and Mehl, M. G., 1938, The conodont genus *Icriodus* and its stratigraphic distribution: Journal of Paleontology, v. 12, p. 156–166.

Brisebois, D., Lachambre, G., and Piché, G. E., 1991, Carte géologique, Péninsule de Gaspésie (22A, 22B, 22G, 22H et 21O), third edition: Québec, Ministère de l'Énergie et des Ressources Map 2146, scale 1:250,000.

Bultynck, P., 1970, Révision stratigraphique et paléontologique (Brachiopodes et Conodontes) de la Coupe type du Couvinien: Mémoires de l'Institut Géologique de l'Université de Louvain, v. 26, 152 p., 39 plates.

Bultynck, P., and Hollard, H., 1981, Distribution comparée de Conodontes et Goniatites dévoniens des plaines du Dra, du Ma'der et du Tafilalt (Maroc): Aardkundige Mededelingen, Leuven University Press, v. 1, 73 p., 10 plates. [imprint 1980]

Chatterton, B. D. E., 1979, Aspects of late Early and Middle Devonian conodont biostratigraphy of western and northwestern Canada: Geological Association of Canada Special Paper 18, p. 161–231. [imprint 1978]

Clark, T. H., 1923, The Devonian Limestone at St. George, Quebec: Journal of Geology, v. 31, p. 217–225.

Cousineau, P. A., 1990, Le Groupe de Caldwell et le domaine océanique entre Saint-Joseph-de-Beauce et Sainte-Sabine: Québec, Ministère de l'Énergie et des Ressources MM 87-02, 165 p.

Cousineau, P. A., and Tremblay, A., 1993, Acadian deformations in the southwestern Quebec Appalachians, *in* Roy, D. C., and Skehan, J. W., eds., The Acadian orogeny: Recent studies in New England, Maritime Canada, and the autochthonous foreland: Geological Society of America Special Paper 275, p. 85–99.

Cousineau, P. A., Cossette, D., and Marquis, R., 1984a, L'Ordovicien entre St-Georges et Lac Etchemin, région de la Beauce, Québec: Québec, Ministère de l'Énergie et des Ressources DP 84-08, scale 1:50,000, one sheet with marginal notes.

Cousineau, P. A., Robert, C., and Tremblay, A., 1984b, Le Cambro-Ordovicien entre Saint-Joseph-de-Beauce et Saint-Camille-de-Bellechasse: Québec, Ministère de l'Énergie et des Ressources DP 84-55, scale 1:50,000, two sheets with marginal notes.

David, J., Chabot, N., Marcotte, C., Lajoie, J., and Lespérance, P. J., 1985, Stratigraphy and sedimentology of the Cabano, Pointe aux Trembles, and Lac Raymond formations, Témiscouata and Rimouski counties, Quebec: Geological Survey of Canada Paper 85-1B, p. 491–497.

Epstein, A. G., Epstein, J. B., and Harris, L. D., 1977, Conodont color alteration—an index to organic metamorphism: U.S. Geological Survey Professional Paper 995, 27 p.

Ferrigno, K. F., 1971, Environmental influences on the distribution and abundance of conodonts from the Dundee Limestone (Devonian), St. Mary's, Ontario: Canadian Journal of Earth Sciences, v. 8, p. 378–386.

Gorman, W. A., 1955, St. Georges–St. Zacharie area, Beauce and Dorchester counties: Quebec Department of Mines Preliminary Report 314, 5 p. [also published in French].

Hatch, N. L., Jr., 1982, Taconian Line in western New England and its implications to Paleozoic tectonic history, *in* St-Julien, P., and Béland, J., eds., Major structural zones and faults of the northern Appalachians: Geological Association of Canada Special Paper 24, p. 67–85.

Heckel, P. H., and Witzke, B. J., 1979, Devonian world palaeogeography determined from distribution of carbonates and related lithic palaeoclimatic indicators, *in* House, M. R., Scrutton, C. T., and Bassett, M. G., eds., The Devonian System: Special Papers in Palaeontology, no. 23, p. 99–123.

Hinde, G. J., 1879, On conodonts from the Chazy and Cincinnati group of the

Cambro-Silurian, and from the Hamilton and Genesse-Shale divisions of the Devonian in Canada and the United States: Quarterly Journal, Geological Society of London, v. 35, p. 351–369.

Jacobi, R. D., 1981, Peripheral bulge—a causal mechanism for the Lower/Middle Ordovician unconformity along the western margin of the Northern Appalachians: Earth and Planetary Science Letters, v. 56, p. 245–251.

Johnson, D. B., and Klapper, G., 1981, New Early Devonian conodont species of central Nevada: Journal of Paleontology, v. 55, p. 1237–1250.

Kendall, G. W., Johnson, J. G., Brown, J. O., and Klapper, G., 1983, Stratigraphy and facies across Lower Devonian-Middle Devonian boundary, central Nevada: American Association of Petroleum Geologists Bulletin, v. 67, p. 2199–2207.

Klapper, G., 1971, Sequence within the conodont genus *Polygnathus* in the New York lower Middle Devonian: Geologica et Palaeontologica, v. 5, p. 59–79.

Klapper, G., and Barrick, J. E., 1983, Middle Devonian (Eifelian) conodonts from the Spillville Formation in northern Iowa and southern Minnesota: Journal of Paleontology, v. 57, p. 1212–1243.

Klapper, G., and Johnson, J. G., 1980, Endemism and dispersal of Devonian conodonts: Journal of Paleontology, v. 54, p. 400–455.

Klapper, G., and Oliver, W. A., Jr., 1995, The Detroit River Group is Middle Devonian: Discussion on "Early Devonian age of the Detroit River Group, inferred from the Arctic stromatoporoids": Canadian Journal of Earth Sciences, v. 32, p. 1070–1073.

Klapper, G., and Ziegler, W., 1967, Evolutionary developments of the *Icriodus latericrescens* group (Conodonta) in the Devonian of Europe and North America: Palaeontographica, Abteilung A, v. 127, p. 68–83.

Klapper, G., and Ziegler, W., 1979, Devonian conodont biostratigraphy, *in* House, M. R., Scrutton, C. T., and Bassett, M. G., eds., The Devonian System: Special Papers in Palaeontology, no. 23, p. 199–224.

Klug, C. R., 1983, Conodonts and biostratigraphy of the Muscatatuck Group (Middle Devonian), south-central Indiana and north-central Kentucky: Wisconsin Academy of Sciences, Arts and Letters, v. 71, pt. 1, p. 79–112.

Koch, W. F., II, 1981, Brachiopod community paleoecology, paleobiogeography, and depositional topography of the Devonian Onondaga Limestone and correlative strata in eastern North America: Lethaia, v. 14, p. 83–103.

Labbé, J.-V., and St-Julien, P., 1989, Failles de chevauchement acadiennes dans la région de Weedon, Estrie, Québec: Journal Canadien des Sciences de la Terre, v. 26, p. 2268–2277.

Lajoie, J., Lespérance, P. J., and Béland, J., 1968, Silurian stratigraphy and paleogeography of Matapedia-Témiscouata region, Québec: American Association of Petroleum Geologists Bulletin, v. 52, p. 615–640.

Lespérance, P. J., and Greiner, H. R., 1969, Squatec-Cabano area, Rimouski, Rivière-du- Loup, and Témiscouata counties: Quebec Department of Natural Resources Geological Report 128, 111 p. [also published in French].

MacKay, B. R., 1921, Beauceville Map-area, Quebec: Geological Survey of Canada Memoir 127, 105 p. [also published in French, 1923].

Müller, K. J., 1962, Zur systematischen Einteilung der Conodontophorida: Paläontologische Zeitschrift, v. 36, p. 109–117.

Müller, K. J., and Müller, E. M., 1957, Early Upper Devonian (Independence) conodonts from Iowa, Part I: Journal of Paleontology, v. 31, p. 1069–1108.

Oliver, W. A., Jr., 1971, The coral fauna and age of the Famine Limestone in Quebec; *in* Dutro, J. T., Jr., ed., Paleozoic perspectives: A paleontological tribute to G. Arthur Cooper: Smithsonian Contributions to Paleobiology 3, p. 193–201.

Oliver, W. A., Jr., 1976, Noncystimorph colonial rugose corals of the Onesquethaw and lower Cazenovia stages (Lower and Middle Devonian) in New York and adjacent areas: U.S. Geological Survey Professional Paper 869, 156 p.

Orr, R. W., 1971, Conodonts from Middle Devonian strata of the Michigan Basin: Indiana Geological Survey Bulletin 45, 110 p., 6 plates.

Racheboeuf, P. R., and Lespérance, P. J., 1995, Revision of Silurian and Devonian chonetoidean brachiopods from Québec and northern New Brunswick (Canada): Documents des Laboratoires de Géologie Lyon, v. 136, p. 7–99.

Rejebian, V. A., Harris, A. G., and Huebner, J. S., 1987, Conodont color and textural alteration: An index to regional metamorphism, contact metamorphism, and hydrothermal alteration: Geological Society of America Bulletin, v. 99, p. 471–479.

Richardson, J. B., and McGregor, D. C., 1986, Silurian and Devonian spore zones of the Old Red Sandstone continent and adjacent regions: Geological Survey of Canada Bulletin 364, 79 p.

Robison, R. A., ed., 1981, Conodonta: Treatise on invertebrate paleontology, Part W, Miscellanea, Supplement 2: Boulder, Colorado, Geological Society of America (and University of Kansas Press), 202 p.

Roy, D. C., 1989, The Depot Mountain Formation: Transition from syn- to post-taconian basin along the Baie Verte–Brompton Line in northwestern Maine, *in* Tucker, R. D., and Marvinney, R. G., eds.: Studies in Maine geology, v. 2, Structure and Stratigraphy, p. 85–99.

St-Julien, P., 1987, Géologie des régions de Saint-Victor et de Thetford-Mines (moitié est): Québec, Ministère de l'Énergie et des Ressources MM 86-01, 66 p.

St-Julien, P., Slivitsky, A., and Feininger, T., 1983, A deep structural profile across the Appalachians of southern Quebec, *in* Hatcher, R.D., Jr., Williams, H., and Dietz, I., eds., Contributions to the tectonics and geophysics of mountain chains: Geological Society of America Memoir 158, p. 103–111.

St. Peter, C., and Boucot, A. J., 1981, Age and regional significance of brachiopods from the Témiscouata Formation of Madawaska County, New Brunswick: Maritime Sediments and Atlantic Geology, v. 17, p. 88–95.

Sanford, B. V., and Norris, A. W., 1975, Devonian stratigraphy of the Hudson Platform: Geological Survey of Canada Memoir 379, 124 p.

Savage, N. M., 1995, Middle Devonian conodonts from the Wadleigh Limestone, southeastern Alaska: Journal of Paleontology, v. 69, p. 540–555.

Sparling, D. R., 1983, Conodont biostratigraphy and biofacies of lower Middle Devonian limestones, north-central Ohio: Journal of Paleontology, v. 57, p. 825–864.

Sparling, D. R., 1984, Paleoecologic and paleogeographic factors in the distribution of lower Middle Devonian conodonts from north-central Ohio, *in* Clark, D. L., ed., Conodont biofacies and provincialism: Geological Society of America Special Paper 196, p. 113–125.

Sparling, D. R., 1988, Middle Devonian stratigraphy and conodont biostratigraphy, north-central Ohio: Ohio Journal of Science, v. 88, p. 2–18.

Sparling, D. R., 1995, Conodonts from the Middle Devonian Plum Brook Shale of north-central Ohio: Journal of Paleontology, v. 69, p. 1123–1139.

Stewart, G. A., and Sweet, W. C., 1956, Conodonts from the Middle Devonian bone beds of central and west-central Ohio: Journal of Paleontology, v. 30, p. 261–273.

Tolman, C., 1936, Lake Etchemin map-area, Quebec: Geological Survey of Canada Memoir 199, 20 p. [also published in French].

Tremblay, A., St-Julien, P., and Labbé, J.-V., 1989, Mise à l'évidence et cinématique de la faille de La Guadeloupe, Appalaches du sud du Québec: Journal Canadien des Sciences de la Terre, v. 26, p. 1932–1943.

Uyeno, T. T., 1990, Pre-Famennian Devonian conodont biostratigraphy of selected intervals in the eastern Canadian Cordillera, *in* Orchard, M. J., and McCracken, A. D., eds., Ordovician to Triassic conodont paleontology of the Canadian Cordillera: Geological Survey of Canada Bulletin 417, p. 129–161.

Uyeno, T. T., and Bultynck, P., 1993, Lower to Middle Devonian conodonts of the Jaab Lake well, Moose River Basin, northern Ontario: Geological Survey of Canada Bulletin 444, p. 7–35.

Uyeno, T. T., Telford, P. G., and Sanford, B. V., 1982, Devonian conodonts and stratigraphy of southwestern Ontario: Geological Survey of Canada Bulletin 332, 55 p.

Wang Ronghu, 1992, Phylogenie und Evolution der Conodonten aus der Moyuela-Formation (Eifelium) in Keltiberien, Spanien: Neues Jahrbuch für Geologie und Paläontologie, Abhandlungen, v. 184, p. 315–338.

Weddige, K., 1977, Die Conodonten der Eifel-Stufe im Typusgebiet und in benachbarten Faziesgebieten: Senckenbergiana Lethaea, v. 58, p. 271–419.

Wittekindt, H., 1966, Zur Conodontenchronologie des Mitteldevons: Fortschritte in der Geologie von Rheinland und Westfalen, v. 9, p. 621–646. [imprint 1965]

Manuscript Accepted by the Society February 27, 1997

Printed in U.S.A.

Geological Society of America
Special Paper 321
1997

Conodont biostratigraphy of the Java Formation (Upper Devonian) and the Frasnian-Famennian boundary in western New York State

D. Jeffrey Over
Department of Geological Sciences, State University of New York College at Geneseo, Geneseo, New York 14454

ABSTRACT

Conodonts from black shale beds and carbonate intervals in the Java Formation (Pipe Creek Shale, Hanover Shale, and Wiscoy Sandstone) of the West Falls Group enable recognition of global event horizons: Lower Kellwasser interval, Upper Kellwasser interval, and the Frasnian-Famennian boundary. *Palmatolepis winchelli, P.* aff. *P. hassi*, and *Polygnathus samueli* indicate the Pipe Creek Member is in MN Zone 12 and equivalent to the lower Kellwasser interval in Europe. The lower and middle Hanover Shale contain conodonts indicative of MN Zone 13; the uppermost Hanover contains Famennian conodonts of the Lower and Middle *triangularis* zones. The Frasnian-Famennian boundary is resolved across western New York in the upper Hanover Shale Member to a discrete horizon within a black shale bed, marked by a cryptic disconformity. This black shale bed is equivalent to the upper Kellwasser interval. The occurrence of *Palmatolepis triangularis, P. clarki, P. delicatula delicatula*, and *P. subperlobata* in the basal Dunkirk Shale of the Canadaway Group indicates assignment to the Middle *triangularis* Zone or younger. *Ancyrodella buckeyensis* and *Palmatolepis subperlobata* are viable taxa; *Ancyrognathus* sp. A is a new species.

INTRODUCTION

Conodonts from Upper Devonian strata of New York State have received mixed attention from paleontologists. Faunas from lower Frasnian units, featured in some of the first reports on conodonts, have since been described in detail (Hinde, 1879; Bryant, 1921; Ulrich and Bassler, 1926; Huddle, 1981; Klapper, 1981; Kirchgasser, 1994; Kralick, 1994). Upper Frasnian and Famennian conodonts, although referred to the formation level in faunal lists and tables (Hass, 1958; Tesmer, 1963, 1968; Rickard, 1964, 1975; Oliver et al., 1968, 1969; Huddle in Klapper et al., 1971; Klapper, 1981), have not been described or figured. As a result, the recognition of the Frasnian-Famennian stage boundary and correlation of upper Frasnian and Famennian stratigraphic units in New York with other regions have been based largely on lithostratigraphy. The black shale units (Geneseo, Middlesex, Rhinestreet, Pipe Creek, Dunkirk; Fig. 1) are the basis for lithological subdivisions of Upper Devonian strata in New York State (Pepper and de Witt, 1950; Sutton, 1963; Rickard, 1975; Woodrow et al., 1989). However, the precise black shale intervals and intervening units are not definitively correlative to other regions based solely on lithology. The recent use of high resolution conodont stratigraphy and graphical correlation has allowed recognition of global event horizons within the predominantly fine-grained siliciclastic prodeltaic marine facies of the northern Appalachian Basin. Lower and Middle Frasnian marker beds, goniatites, and conodonts have been mapped and correlated (House, 1978; House and Kirchgasser, 1993; Kirchgasser, 1985; Kirchgasser and House, 1981; Kirchgasser et al., 1994; Klapper, 1981; Kralick, 1994; Fig. 1). This chapter describes upper Frasnian and lowest Famennian conodonts and event horizons in western New York State and their zonal relationships.

Frasnian and Famennian marine strata are exposed across

Over, D. J., 1997, Conodont biostratigraphy of the Java Formation (Upper Devonian) and the Frasnian-Famennian boundary in western New York State, *in* Klapper, G., Murphy, M. A., and Talent, J. A., eds., Paleozoic Sequence Stratigraphy, Biostratigraphy, and Biogeography: Studies in Honor of J. Granville ("Jess") Johnson: Boulder, Colorado, Geological Society of America Special Paper 321.

Stage	Group	Formation	Conodont Zone
Famennian	Canadaway	Northeast	?
		Westfield	*rhomboidea*
		Gowanda	*crepida*
		South Wales	
		Dunkirk	*triangularis*
Frasnian	West Falls	Java: Hanover	13
		Java: Pipe Creek	12
		Angola	11
		Rhinestreet	unzoned, 7, 6
	Sonyea	Cashaqua	5
		Middlesex	unzoned
	Genesee	West River	4, 3
		Genundewa	3, 2
		Penn Yan	2, 1, *norrisi*
Giv		Geneseo	*disparilis*

Figure 1. Generalized upper Givetian, Frasnian, and lower Famennian stratigraphy of western New York State and conodont zones of Klapper (1989) and Ziegler and Sandberg (1990). Shaded area is interval considered in this chapter. From Kirchgasser et al. (1994) and Oliver et al. (1969).

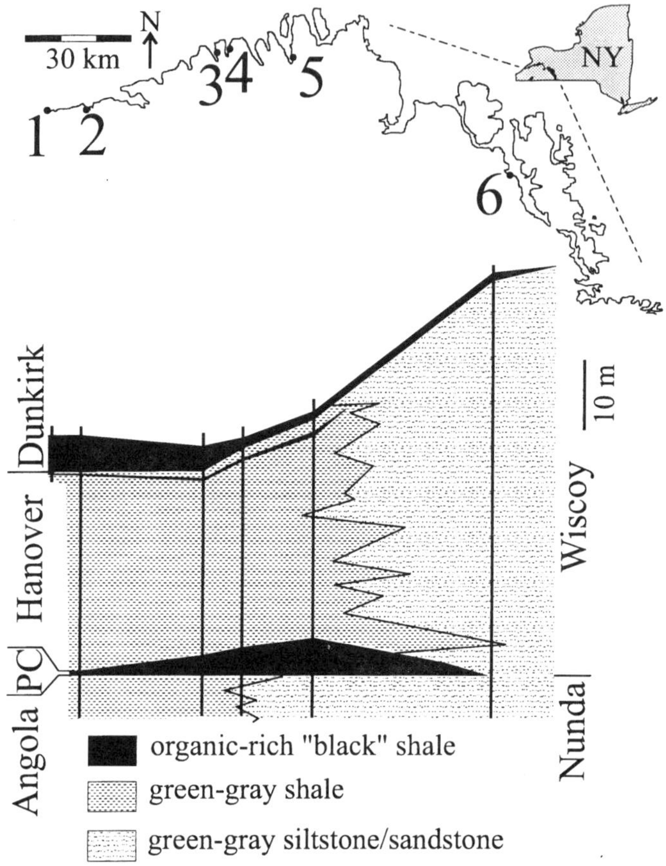

Figure 2. Location of sections, lithostratigraphy, and general east-to-west lithology of Frasnian and lower Famennian strata. The map shows approximate location of the Frasnian-Famennian boundary in New York State; Famennian strata are exposed southwest of line, Frasnian and older strata are exposed to the north. Numbered sections are: 1—Pt. Gratiot; 2—Walnut Creek; 3—Irish Gulf; 4—Pipe Creek; 5—Beaver Meadow Creek; and 6—Arkport Dam. Includes information from Pepper and de Witt (1950).

western New York State (Fig. 2). The Java Formation is the youngest formation of the West Falls Group, which includes the Rhinestreet and Angola formations in deep-water basin/slope facies ("Genesee" and "Portage" of Hall [1843]) and the laterally equivalent Gardeau and Nunda formations in more nearshore settings ("Chemung" facies) (Pepper et al., 1956; de Witt, 1960; House and Kirchgasser, 1993). The Java Formation was proposed by de Witt (1960) to include the Pipe Creek Shale, Hanover Shale, and Wiscoy Sandstone: units that were first established by Chadwick (1923, 1933). The top of the Java Formation and West Falls Group is marked by the overlying Dunkirk Shale of the Canadaway Group, the uppermost regionally persistent black shale interval in western New York State (Fig. 2).

CONODONT ZONATION

Klapper (1989) proposed 13 Frasnian zones based on the first occurrences of conodont species in predominantly carbonate strata exposed in the Montagne Noire (MN) of southern France. The zonation was recognized in other regions and ranges refined through graphical correlation (Klapper, 1990; Klapper and Foster, 1993; Klapper et al., 1995). The Java Formation contains conodonts from the upper two Frasnian zones (Fig. 3). The zonation proposed by Klapper and the zonation scheme originally proposed by Ziegler (1962), most recently updated by Ziegler and Sandberg (1990), are incompatible because of disparate interpretations of certain taxa. The multielement taxonomy and nomenclature of Klapper for Frasnian conodonts is followed herein. The lower Famennian part of the zonation proposed by Ziegler (1962), based on first occurrences of species of *Palmatolepis*, is also used (Fig. 3).

PIPE CREEK SHALE MEMBER

The Pipe Creek Shale Member is a persistent black petroliferous shale unit that is recognized in the subsurface from eastern Kentucky to New York State by a high positive peak on gamma ray logs (de Witt and Roen, 1985). The Pipe Creek represents dysoxic and anoxic benthic conditions that developed as a result of deepening in the Appalachian Basin and deposition of black mud over basin margin and deep-shelf strata of the Angola and Nunda formations. At the type section along Pipe Creek, Town of Colden, the Pipe Creek Member is 4.8 m thick, consisting of dark silty pyritic shale that overlies the Angola Shale (Fig. 4). Along Beaver Meadow Creek, Town of Java, the type section of the Java Formation, the Pipe Creek Shale Member is 6.2 m thick and overlies the Nunda Sandstone. Interbedded and deformed coarse silts and black shale at the base of the Pipe Creek indicate rapid development of black shale facies even as coarse clastic material was transported by storm action or density flows (Fig. 5). The strata likely were deformed during dewatering or by small slumps and plastic flows. Westward, surface exposures of the Pipe Creek thin to less than a meter at Walnut Creek and along the Lake Erie shore. Eastward the Pipe Creek grades into gray shales and disappears near Arkport, Steuben County (Fig. 2, locality 6).

Conodonts of the Pipe Creek Shale were recovered from bedding plane surfaces and consist of isolated elements. Only one or two specimens of each species were collected from the three sections studied (Fig. 4; Table 1; Fig. 6: 1–3). The occurrence of *Palmatolepis winchelli* (Stauffer), *P.* aff. *P. hassi* Müller and Müller, and *Polygnathus samueli* Klapper and Lane below the first occurrence of *Palmatolepis bogartensis* (Stauffer) and *P. juntianensis* Han indicates that the Pipe Creek Member is in MN Zone 12. The underlying Angola Shale contains *P. foliacea*

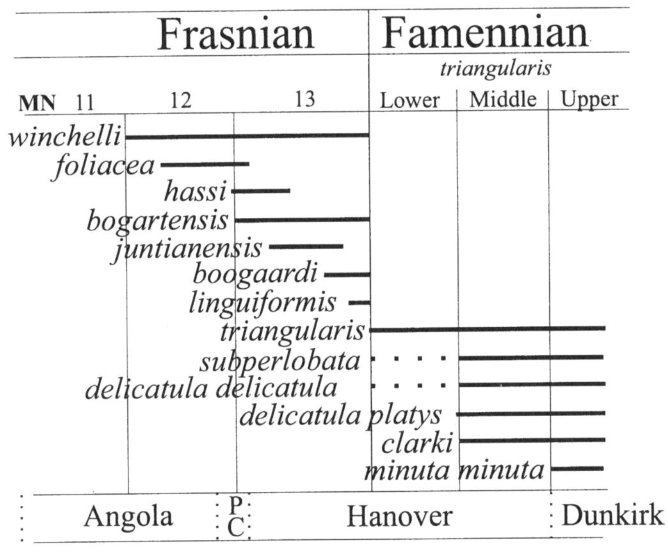

Figure 3. Upper part of Frasnian Montagne Noire (MN) conodont zonation, showing ranges of selected species of *Palmatolepis* based on graphical correlation from Klapper and Foster (1993), and lowermost Famennian conodont zonation (Ziegler, 1962), showing ranges of zonally defining and selected species. PC = Pipe Creek.

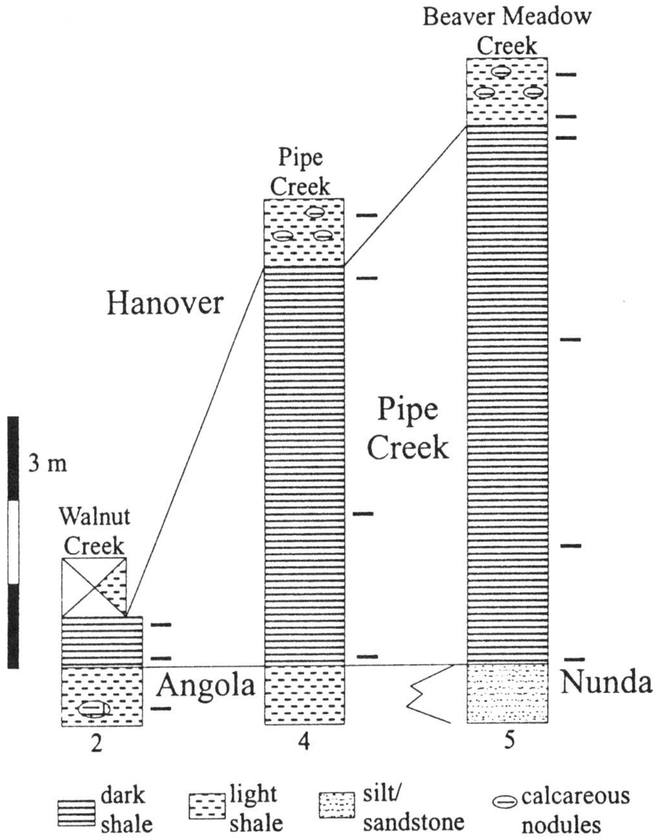

Figure 4. Pipe Creek Shale Member strata and sample position indicated by horizontal dashes. See Figure 2 for map location, Table 1 for fauna list and occurrences, and the Appendix for detailed locality information.

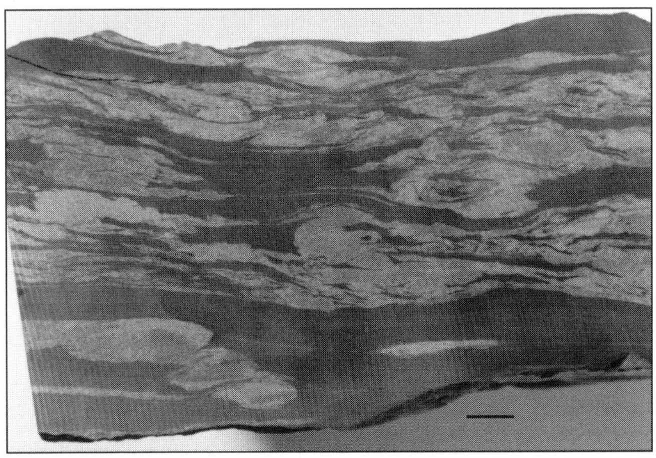

Figure 5. Base of Pipe Creek Shale at Beaver Meadow Creek showing interbedding and deformation of coarse silt and black shale lithologies. Scale bar equals 1 cm.

TABLE 1. DISTRIBUTION OF CONODONT SPECIES IN THE PIPE CREEK SHALE MEMBER OF THE JAVA FORMATION FROM EXPOSURES AT WALNUT CREEK (LOCALITY 2), PIPE CREEK (LOCALITY 4), AND BEAVER MEADOW CREEK (LOCALITY 5)

Locality	Walnut Creek (2)			Pipe Creek (4)			Beaver Meadow Creek (5)		
Sample Interval	WCPC01	WCPC02	WCPC03	PCPC01	PCPC02	PCPC03	BMCPC01	BMCPC02	BMCPC03
Meters above base of Pipe Creek	-0.3	0	0.5	0	1	4.9	0	1.5	4.5
Palmatolepis aff. *P. hassi*				x					
Palmatolepis winchelli			x					x	
Palmatolepis sp. indet.				x	x			x	x
Polyganthus webbi				x					
Polyganthus samueli							x		
Polyganthus sp. indet.	x	x	x	x	x	x	x	x	x
Ancyroganthus sp. indet				x				x	
Ancyrodella curvata (late form)				x					
Ancyrodella sp. indet.									x
Icriodus sp. indet.			x	x					

Youngquist, also indicative of MN Zone 12 (collections of W. T. Kirchgasser). The Pipe Creek Shale corresponds to the lower Kellwasser event, a distinctive interval of organic-rich dark shale and carbonates in western Europe and North Africa characterized by a similar conodont fauna and an ammonoid extinction event in the highest part of MN Zone 12 to lowest MN Zone 13 (Schindler, 1990; Becker et al., 1989; Becker, 1993). The occurrence of *Polygnathus samueli* high in MN Zone 12 in New York State is significantly younger than that indicated by graphical correlation on the Frasnian Composite Standard of Klapper et al. (1995).

HANOVER SHALE MEMBER

The Hanover Shale Member consists of intensely bioturbated green-gray silty pyritic shale interbedded with argillaceous carbonate beds, calcareous nodules, and organic-rich laminated shales (Fig. 7). Siltstones and very fine sandstones become more abundant eastward as the Hanover interfingers with and grades into the Wiscoy Sandstone Member. The Hanover Shale Member is 28 m thick in its type section along Silver Creek in the Town of Hanover. At Beaver Meadow Creek it is 37.5 m thick.

Conodonts were recovered from several carbonate and black shale beds (Tables 2, 3). Cephalopod-bearing concretions in the lower Hanover contain a diverse nektic and low-diversity benthic fauna of acrotretids, ostracodes, tentaculitids, the goniatites *Delphinites cataphractus* and *Sphaeromanitcoceras?*, and the conodonts *Ancyrodella curvata* (Branson and Mehl) (late form), *Ancyrognathus asymmetricus* (Ulrich and Bassler), *Polygnathus decorosus* Stauffer, and *Palmatolepis bogartensis* (Stauffer). The first occurrence of the latter species defines the base of MN Zone 13. The middle of the Hanover contains more siltstone beds and fewer black shale beds, indicative of shallowing and increased substrate oxygenation. At Beaver Meadow Creek (Fig. 7) several carbonate horizons contain crinoid debris and the small solitary rugose coral *Metriophyllum*. A single carbonate horizon yielded abundant conodonts, including *P. bogartensis*, *P. boogaardi* Klapper and Foster, *P. hassi* Müller and Müller, *P. juntianensis*, *P. winchelli*, *Ancyrodella buckeyensis* Stauffer, *Ancyrognathus* aff. *A. altus* Müller and Müller, *A. amana* Müller and Müller, *A. asymmetricus*, *Polygnathus webbi* Stauffer, *P. decorosus*, and *P. brevis* Miller and Youngquist. The fauna is indicative of the middle to upper part of MN Zone 13 and is similar to taxa of the Sweetland Creek Shale and related strata in eastern Iowa (see Müller and Müller, 1957; Johnson and Klapper, 1992). The cooccurrence of *Palmatolepis hassi* and *P. boogaardi*, which do not overlap in range in the Frasnian Composite Standard of Klapper et al. (1995), in conjunction with the occurrence of *Ancyrognathus* aff. *A. altus* suggests a relatively early first occurrence of *P. boogaardi* in New York State.

The upper Hanover Shale is characterized by numerous black shale interbeds, indicative of deepening and a reduction in coarser clastic influx as well as reduced oxygen levels in the substrate. The black shales amalgamate or are truncated westward (Baird and Lash, 1990) as a result of sediment reduction and/or

Figure 6. Conodonts from the Pipe Creek Shale and Hanover Shale members of the Java Formation. BMC = Beaver Meadow Creek; IG = Irish Gulf; PC = Pipe Creek. Specimens are reposited in the New York State Museum (NYSM) unless otherwise indicated. Magnification 37× for specimens 1–5, 7, 13, 52× for specimens 6, 8–12, 14–16. 1: *Polygnathus samueli* Klapper and Lane, 1985, Pa element, NYSM 15636, BMC-PC01. 2, 3: *Palmatolepis* aff. *P. hassi* Müller and Müller, 1957. 2, Pa element, NYSM 15637, PC-PC01; 3, Pa element, NYSM 15638, PC-PC01. 4, 5: *Palmatolepis hassi* Müller and Müller, 1957. 4, Pa element, NYSM 15639, BMC-10; 5, inner lateral view of Pb element, NYSM 15640, BMC-10. 6–9: *Palmatolepis winchelli* (Stauffer, 1938). 6, Pa element, NYSM 15641, IG-03b; 7, Pa element, NYSM 15642, BMC-10; 8, inner lateral view of Pb element, NYSM 15643, BMC-10; 9, inner lateral view of Pb element, NYSM 15644, BMC-10. 10, 11: *Palmatolepis juntianensis* Han, 1987. 10, Pa element, NYSM 15645, BMC-10; 11, Pa element, NYSM 15646, BMC-10. 12: *Palmatolepis* sp. indet., Pa element, NYSM 15647, BMC-10. 13: *Palmatolepis linguiformis* Müller, 1956, Pa element, NYSM 15648, IG-01b. 14, 15: *Palmatolepis bogartensis* (Stauffer, 1938). 14, Pa element, NYSM 15649, IG-01b; 15, Pa element, NYSM 15650, IG-CC. 16: *Palmatolepis boogaardi* Klapper and Foster, 1993, Pa element, NYSM 15651, IG-01b.

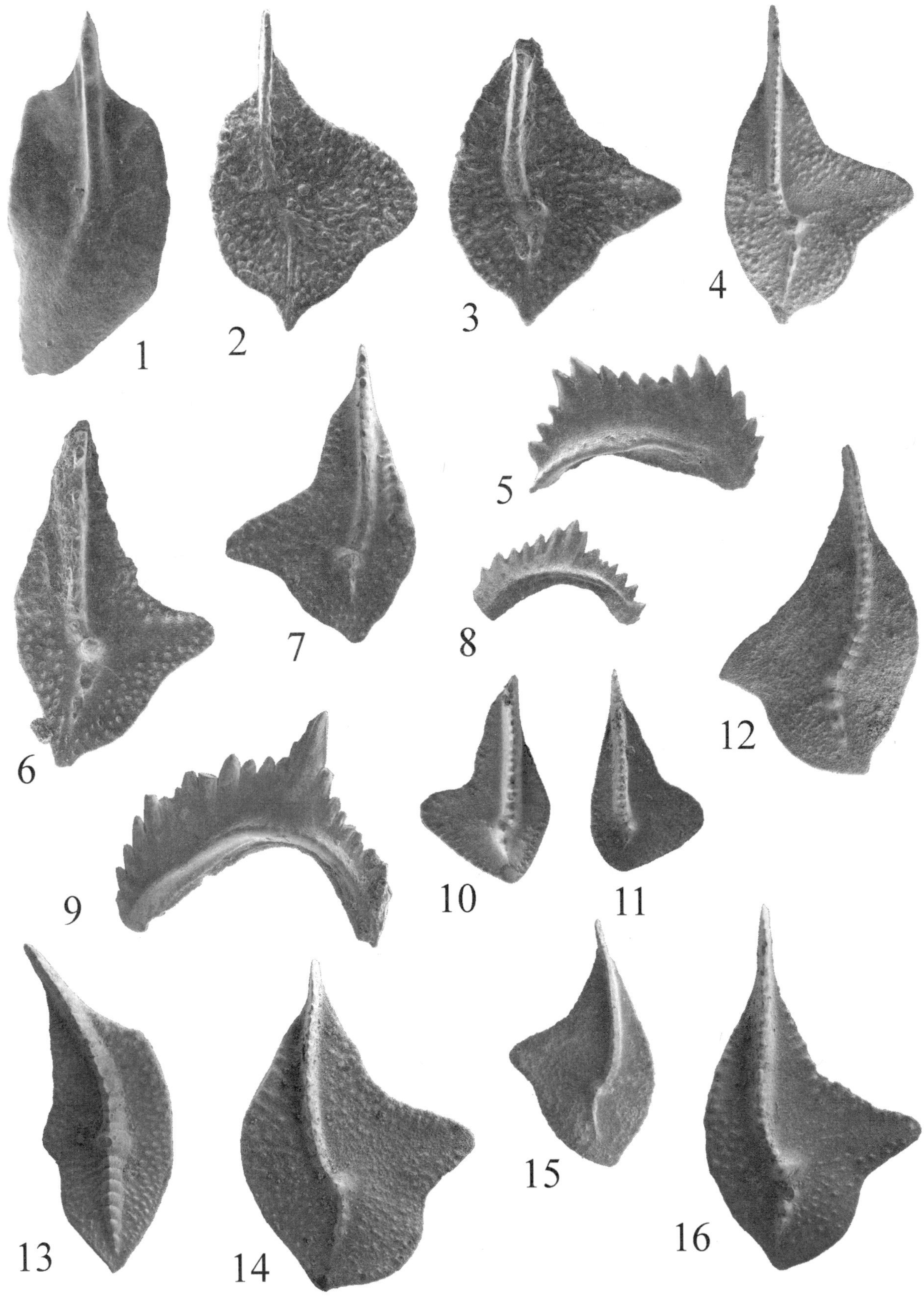
1
2
3
4
5
6
7
8
9
10
11
12
13
14
15
16

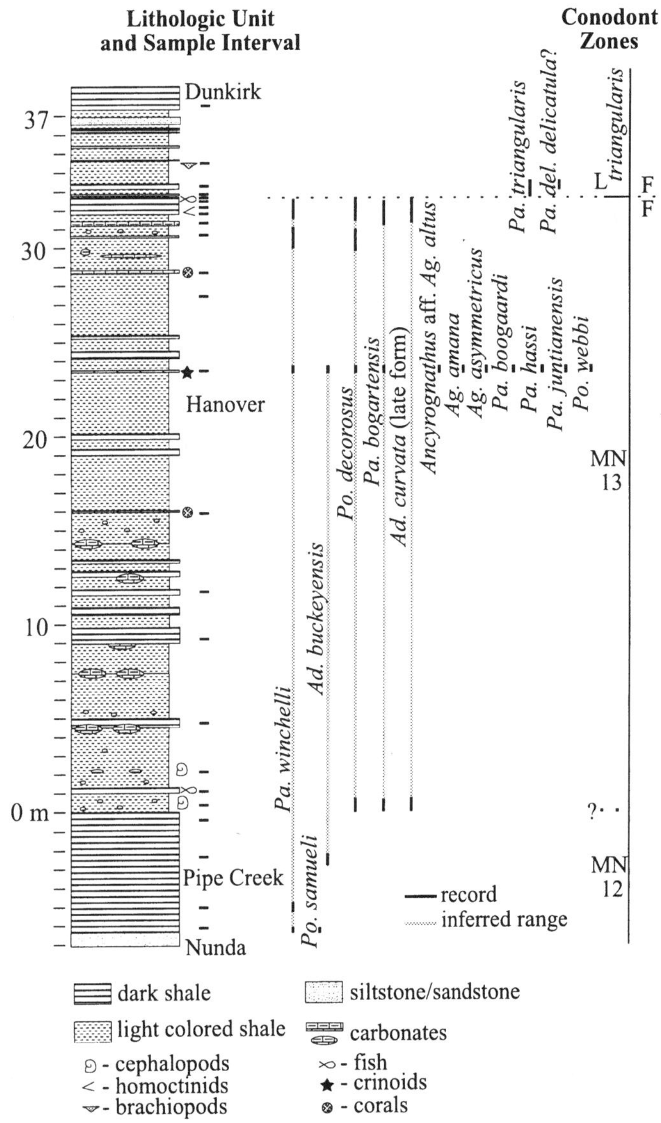

Figure 7. General lithology, conodont occurrences, and conodont zonation of the type section of the Java Formation (Pipe Creek Shale and Hanover Shale members) at Beaver Meadow Creek. See Figure 2 for location, Table 2 for fauna list and occurrences, and the Appendix for detailed locality information. F/F = Frasnian-Famennian boundary; MN = Montagne Noire conodont zones; *Ad.* = *Ancyrodella; Ag.* = *Ancyrognathus; Pa.* = *Palmatolepis; Po.* = *Polygnathus.*

submarine erosion. Black shales and calcareous concretions contain carbonized plant material, gastropods, crinoid debris, pteriacid bivalves, cephalopod aptyci, and an abundant and diverse conodont fauna. A radical change in the conodont fauna marks the Frasnian-Famennian boundary, the precise location resolved in several sections to a distinct horizon. At Irish Gulf (Fig. 8) a calcareous concretion horizon 1.75 m below the Frasnian-Famennian boundary horizon yielded *Polygnathus decorosus*, *P. webbi*, *Palmatolepis winchelli*, *P. bogartensis*, *P. boogaardi*, *P. linguiformis* Müller, and *Ancyrodella buckeyensis*, indicative of the highest part of MN Zone 13. Strata below the Frasnian-Famennian boundary horizon also contain *Palmatolepis triangularis* Sannemann (rare occurrence before Frasnian extinction), *Icriodus alternatus alternatus* Branson and Mehl, *Ancyrodella curvata* (late form), and *Ancyrognathus asymmetricus* (Table 3).

Famennian strata are recognized by the absence of Frasnian species and the predominance of *Palmatolepis triangularis* and *Icriodus alternatus alternatus* characteristic of the Lower *triangularis* Zone. Higher in the Hanover, *Palmatolepis delicatula delicatula* Branson and Mehl, *P. delicatula platys* Ziegler and Sandberg, *P. subperlobata* Branson and Mehl, and *Ancyrognathus* sp. A indicate the Middle *triangularis* Zone. Huddle (in Klapper et al., 1971) reported *Palmatolepis perlobata* Ulrich and Bassler and *P. quadrantinodosalobata* Sannemann from the upper Hanover. Reexamination of the specimens by Klapper (personal communication, 1995), however, indicates they were misidentified and represent specimens of *P. winchelli* in some instances and *P. triangularis* in others.

Black shale beds in the uppermost Hanover, as well as the basal Dunkirk Shale, contain significant disseminated pyrite as well as detrital pyrite in the form of reworked framboids and burrow fills. Reworked pyrite and concentrations of conodonts and other phosphatic material in dark shales indicate subaqueous erosion under oxygen-reducing benthic conditions (Baird and Brett, 1986).

DUNKIRK SHALE

The Hanover Shale Member is overlain by dark petroliferous strata of the Dunkirk Shale, the lowest formation in the Canadaway Group. The Dunkirk is defined by the first occurrence of massive black shale above the green-gray shales and thin black shale interbeds of the Hanover and is equivalent, at least in part, to the lower part of the Huron Shale in Ohio. The type section of the Dunkirk is at Pt. Gratiot, Town of Dunkirk, on the Lake Erie shoreline. The basal Dunkirk yielded *Palmatolepis triangularis*, *P. subperlobata*, *P. delicatula delicatula*, and *P. clarki* Ziegler, characteristic of the Middle *triangularis* Zone. These species range into succeeding zones, but with the exception of juvenile specimens suggestive of *Palmatolepis perlobata* ssp. or *P. tenuipunctata* Sannemann, conodonts indicative of the Upper *triangularis* Zone or younger were not recovered in the present study. *Palmatolepis glabra* (Ulrich and Bassler), *P. perlobata*, *P. quadrantinodosalobata*, *P. subperlobata*, and *Ancyrognathus bifurcatus* (Ulrich and Bassler) were reported by Hass (1958) from the South Wales and Gowanda formations, which overlie the Dunkirk. This fauna is characteristic of a position no lower than the Upper *crepida* Zone.

FRASNIAN-FAMENNIAN BOUNDARY

The Frasnian-Famennian boundary marks one of the major biotic events in the Phanerozoic, recognized by the extinction of shallow-water corals, most stromatoporoids, and numerous species and higher groups of trilobites, cephalopods, brachiopods, and conodonts (Sorauf and Pedder, 1986; Goodfellow

TABLE 2. DISTRIBUTION OF CONODONT SPECIES IN THE HANOVER SHALE MEMBER OF THE JAVA FORMATION FROM EXPOSURES AT BEAVER MEADOW CREEK (LOCALITY 5)

Sample Meters above base of Hanover	BMC-10 23.5	BMC-10E 28.7	BMC-11 30.7	BMC-11B 31.4	BMC-13 31.9	BMC-14 32.45	BMC-15 32.55	BMC-15b 32.65	BMC-16 32.8
Palmatolepis bogartensis	x	?		x		x	?		
Palmatolepis boogaardi	x	?							
Palmatolepis hassi	x								
Palmatolepis aff. *P. hassi*	?								
Palmatolepis juntianensis	x								
Palmatolepis winchelli	x	x	x	x	x	x	x		
Palmatolepis triangularis								x	x
Palmatolepis delicatula delicatula									?
Palmatolepis sp. indet.	x	x	x	x	x	x	x		x
Polygnathus brevis	x								
Polygnathus decorosus	x	x		x			?		
Polygnathus webbi	x	x		x					
Polygnathus sp. indet.	x				x		x	x	
Ancyrodella buckeyensis	x								
Ancyrodella curvata (late form)							?		
Ancyrodella sp. indet.		x			x				
Ancyrognathus aff. *A. altus*	x								
Ancyrognathus amana	x								
Ancyrognathus asymmetricus	x								
Ancyrognathus sp. indet.						x			
Icriodus sp. indet.							x		

et al., 1989; McLaren, 1989; Wang et al., 1991; Becker, 1993, Joachimski and Buggisch, 1993; McGhee, 1996). The mass extinction is associated with sea-level fluctuations and deposition of organic-rich fine-grained sediments in tropical seas. Changes in ocean circulation (Wilde and Berry, 1984; Geldsetzer et al., 1987), oceanic cooling (Copper, 1986), and oceanic warming (Thompson and Newton, 1989; Brand, 1989) have been proposed as extinction mechanisms, possibly triggered by bolide impact (McLaren, 1970; Sandberg et al., 1988; Wang et al., 1991), sea-level change (Joachimski and Buggisch, 1993), or other terrestrial causes such as innovation and diversification of vascular plants (Algeo et al., 1995). The extinction of all taxa was not abrupt (McGhee, 1989; Becker, 1993). Based on conodonts the boundary is recognized by the disappearance of most Frasnian species and the common occurrence of *Palmatolepis triangularis* (Klapper et al., 1994), which first appears in close association with the mass extinction. Although there is a minor disconformity at the boundary horizon in the Hanover Shale, the fine-grained clastic nature of the strata and no major lithologic change suggests that the boundary interval is essentially continuous in some sections (Fig. 9).

The boundary horizon is recognized as a cryptic disconformity within or at the top of a 0.1- to 0.8-m-thick black shale bed in the upper Hanover (Fig. 9). This bed locally contains numerous nowakiids, identified by W. T. Kirchgasser as *Homoctenus*; fish remains, including an articulated specimen of *Coccosteus?*; and numerous cephalopod aptyci. This interval is equivalent to the Upper Kellwasser horizon in Europe, based on similarities of the conodont fauna and organic-rich nature (Schindler, 1990, 1993; Klapper et al., 1994). At Irish Gulf the boundary is in the upper 1 cm of the black shale bed, marked by a concentration of conodonts. *Palmatolepis triangularis* first occurs several centimeters below the boundary horizon in association with typical Frasnian conodonts (Fig. 8). The occurrence of a hummocky cross-bedded siltstone bed within the black shale bed, very thin silt lamina, and the conodont concentration at the boundary horizon indicate a depositional setting above storm wave base within a realm of micro turbidity flows and submarine erosion in an offshore environment with a dysoxic to anoxic substrate. East of Beaver Meadow Creek the boundary horizon has not been resolved to a discrete horizon. Geochemical analysis of this interval did not yield significant concentrations of platinum group elements or microspherules indicative of a bolide impact in association with the boundary horizon (McGhee et al., 1984; Over et al., 1997).

SYSTEMATIC PALEONTOLOGY

Class Conodonti Branson, 1938
Order Prioniodontida Dzik, 1976
Family Icriodontidae Müller and Müller, 1957

Genus *Icriodus* Branson and Mehl, 1938

Type species. Icriodus expansus Branson and Mehl, 1938

TABLE 3. DISTRIBUTION OF CONODONT SPECIES IN THE HANOVER SHALE MEMBER OF THE JAVA FORMATION AND LOWER DUNKIRK FORMATION FROM EXPOSURES AT PT. GRATIOT (LOCALITY 1), WALNUT CREEK (LOCALITY 2), AND IRISH GULF (LOCALITY 3)

Locality	Pt. Gratiot (1)				Walnut Creek (2)				
Sample Interval	D2	D4	D5	D8	WCBpA	WCB04	WCB06	WCB09	WCB12
Meters from base of Dunkirk	-0.4	-0.3–0.2	-0.2–0.0	0.0–0.2	-5	-0.8	-0.6	-0.27– -0.2	0.0–0.2
Palmatolepis bogartensis	?	?	?		?				
Palmatolepis boogaardi		?	?						
Palmatolepis linguiformis									
Palmatolepis winchelli	x		x		x	x			
Palmatolepis triangularis				x			x	x	x
Palmatolepis subperlobata				x				x	
Palmatolepis delicatula delicatula				x					
Palmatolepis delicatula platys				x					
Palmatolepis clarki				?					x
Palmatolepis sp. indet.		x	x	x	x	x	x	x	x
Polygnathus decorosus									
Polygnathus webbi									
Polygnathus sp. indet.	x	x	x	x	x	x		x	
Ancyrodella buckeyensis					?	x			
Ancyrodella curvata (late form)			x						
Ancyrodella sp. indet.									
Ancyrognathus asymmetricus			x						
Ancyrognathus sp. A									
Ancyrognathus sp. indet.				?					
Icriodus alternatus									
Icriodus sp. indet.									

Icriodus alternatus alternatus Branson and Mehl, 1934
Figure 10: 14

Icriodus alternatus BRANSON AND MEHL, 1934, p. 225–226, pl. 13, figs. 4–6 [fig. 4 = lectotype].
Icriodus alternatus alternatus Branson and Mehl. DRUCE, 1976, p. 105–106, pl. 29, figs. 1a–4b; SANDBERG AND DREESEN, 1984, p. 158–159, pl. 2, figs. 5, 11.

Remarks. Icriodus alternatus alternatus is the most common icriodontid in the Java Formation. Specimens identified as *Icriodus* sp. are likely *I. alternatus alternatus*.

Order Ozarkodinida Dzik, 1976
Family Palmatolepidae Sweet, 1988

Genus _Palmatolepis_ Ulrich and Bassler, 1926

Type species. Palmatolepis perlobata Ulrich and Bassler, 1926

Palmatolepis bogartensis (Stauffer, 1938)
Figure 6: 14, 15

Nothognathella bogartensis STAUFFER, 1938, p. 436, pl. 48, fig. 30 [= holotype].
Palmatolepis bogartensis (Stauffer). KLAPPER AND FOSTER, 1993, p. 17–18, figs. 13.4–13.6, 19.1–19.5, 20.1–20.11

Remarks. Lange (1968) and Klapper and Foster (1993) illustrated a complete apparatus of *P. bogartensis*.

Palmatolepis boogaardi Klapper and Foster, 1993
Figure 6: 16

Palmatolepis boogaardi KLAPPER AND FOSTER, 1993, p. 18–22, figs. 13.3, 14.1–14.10, 17.1, 17.2, 19.13–19.15 [fig. 14.5 = holotype].

Palmatolepis clarki Ziegler, 1962
Figure 10: 4

Palmatolepis marginata clarki ZIEGLER, 1962, p. 62–65, fig. 4, pl. 2, figs. 20–25 [figs. 20–22 = holotype].
Palmatolepis delicatula clarki Ziegler. ZIEGLER in Ziegler (ed.), 1973, p. 269–270, Palmatolepis—pl. 3, fig. 4 [reillustration of holotype].
Palmatolepis clarki Ziegler. ZIEGLER AND SANDBERG, 1990, p. 66, pl. 16, fig. 7; BULTYNCK AND MARTIN, 1995, pl. 7, fig. 11.
Palmatolepis clarki clarki Ziegler. SCHÜLKE, 1995, p. 30–31, pl. 9, figs. 4, 5, 8, 9, 13.
Palmatolepis triangularis Sannemann. ZIEGLER AND SANDBERG, 1990, pl. 16, fig. 8, 9 (only).

Remarks. The holotype of *P. clarki* is a large (gerontic?) form atypical of specimens usually identified at *P. clarki*. Distinctive characteristics of the Pa element of the species are a relatively narrow platform, a well-defined outer lobe with a concave anterior outer margin, and a raised platform edge.

The first occurrence of *Palmatolepis clarki* defines the base of the Middle *triangularis* Zone (Ziegler, 1962; Klapper et al., 1994).

TABLE 3. DISTRIBUTION OF CONODONT SPECIES IN THE HANOVER SHALE MEMBER OF THE JAVA FORMATION AND LOWER DUNKIRK FORMATION FROM EXPOSURES AT PT. GRATIOT (LOCALITY 1), WALNUT CREEK (LOCALITY 2), AND IRISH GULF (LOCALITY 3) (continued)

Locality	Irish Gulf (3)											
Sample Interval	IG-CC	IG-01b	IG-01c	IG-03a	IG-03b	IG-03top	IG-04	IG-06	IG-07	IG-08	IG-09	IG-11
Meters from base of Dunkirk	-30	-4.75	-3.2	-3.05	-2.95	-2.95	-2.75	-2.05	-0.72	-0.65	-0.12	0.0–0.2
Palmatolepis bogartensis	x	x	?	?	x	?						
Palmatolepis boogaardi		x		?	?							
Palmatolepis linguiformis		x										
Palmatolepis winchelli	x	x	x	x	x							
Palmatolepis triangularis					x	x	x	x	x	x	x	
Palmatolepis subperlobata							x	x	x	x	x	
Palmatolepis delicatula delicatula							x	x	x	x	x	
Palmatolepis delicatula platys									x			
Palmatolepis clarki												
Palmatolepis sp. indet.	x	x		x	x		x	x			x	x
Polygnathus decorosus	x	x		x								
Polygnathus webbi		x	x									
Polygnathus sp. indet.			x	x		x	x	x				
Ancyrodella buckeyensis		x										
Ancyrodella curvata (late form)	x			x		x						
Ancyrodella sp. indet.												
Ancyrognathus asymmetricus	?		x									
Ancyrognathus sp. A								x				
Ancyrognathus sp. indet.												
Icriodus alternatus						x			x	?		
Icriodus sp. indet.				x		x						

Palmatolepis delicatula delicatula Branson and Mehl, 1934
Figure 10: 10

Palmatolepis delicatula BRANSON AND MEHL, 1934, p. 237, pl. 18, figs. 4, 10 [fig. 4 = lectotype selected by Ziegler (1973, p. 265), from the Saverton Shale at Sees Creek, Missouri].
Palmatolepis delicatula delicatula Branson and Mehl. ZIEGLER AND SANDBERG, 1990, p. 67, pl. 17, figs. 1–3; BULTYNCK AND MARTIN, 1995, pl. 7, fig. 7; SCHÜLKE, 1995, p. 33, pl.10, figs. 16–19.

Remarks. The lectotype of *Palmatolepis delicatula delicatula* represents a relatively rare form that has a reduced outer platform, where the anterior outer margin diverges from the blade at a relatively small acute angle. Most specimens illustrated as *P. delicatula delicatula* (see synonomy) have much wider outer platforms, characterized by a straight or very slightly concave outer margin and reduced ornamentation. Specimens identified as *P. delicatula delicatula* in the Hanover and Dunkirk shales are imbedded in matrix or highly fragmented and have not been figured.

Palmatolepis delicatula platys Ziegler and Sandberg, 1990
Figure 10: 8

Palmatolepis delicatula platys ZIEGLER AND SANDBERG, 1990, p. 67–68, pl. 17, figs. 4–7 [fig. 4 = holotype].
Palmatolepis delicatula delicatula Branson and Mehl. BULTYNCK AND MARTIN, 1995, pl. 7, fig. 6.
Palmatolepis protorhomboidea Sandberg and Ziegler. BULTYNCK AND MARTIN, 1995, pl. 7, fig. 8?

Remarks. Palmatolepis delicatula platys is distinguished from *P. delicatula delicatula* by the convex outer anterior margin and smooth upper surface; from *P. clarki*, which has a longer posterior platform, raised platform edges and ornamentation on the anterior platform margins; and from *P. protorhomboidea*, which has a straight to convex anterior outer margin, parapet on the inner platform, and nodes on the anterior outer platform.

Palmatolepis hassi Müller and Müller, 1957
Figure 6: 4, 5

Palmatolepis (Manticolepis) hassi MÜLLER AND MÜLLER, 1957, p. 1102–1103, pl. 139, fig. 2, pl. 140, figs. 2–4 [fig. 4 = holotype].
Palmatolepis hassi Müller and Müller. KLAPPER AND FOSTER, 1993, p. 22, figs. 15.1–15.9 [fig. 15.4 = reillustration of holotype], 18.12, 19.16–19.18.

Palmatolepis aff. _Palmatolepis hassi_ Müller and Müller, 1957
Figure 6: 2, 3

Remarks. Specimens assigned to *P.* aff. *P. hassi* are similar to *P. hassi* sensu strictu in general outline of the Pa element but are characterized by a broader outer and inner posterior platform and straighter margin (more gentle curve) of the anterior outer lobe.

Palmatolepis juntianensis Han, 1987
Figure 6: 10, 11

Palmatolepis juntianensis HAN, 1987, p. 186, pl. 1, figs. 15, 16 [fig. 15 = holotype]; ZIEGLER AND SANDBERG, 1990, p. 52, pl. 14, figs. 16, 17.

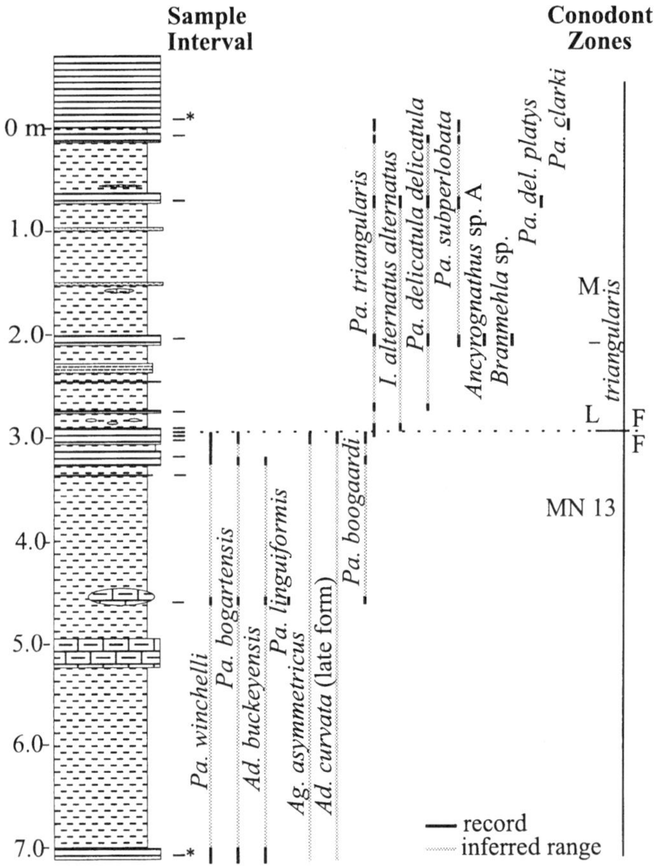

Figure 8. Upper Hanover Shale at Irish Gulf showing lithology and conodont occurrence. (*) denotes sample position and conodont fauna from equivalent strata at Walnut Creek (Locality 2); F/F = Frasnian-Famennian boundary; L = Lower; M = Middle; MN = Montagne Noire conodont zones; *Ad.* = *Ancyrodella; Ag.* = *Ancyrognathus; Pa.* = *Palmatolepis; I.* = *Icridous.* See Figure 7 for lithologic symbols.

Description. Pa element relatively small; anterior carina and free blade straight, posterior carina short, deflected slightly inward, or absent. Outer platform wider than inner platform, with short outer lobe lacking distinctive posterior sinuses, posterior platform very short; platform margins raised and may have small nodes; otherwise the platform is smooth.

***Palmatolepis linguiformis* Müller, 1956**
Figure 6: 13

Palmatolepis (Palmatolepis) linguiformis MÜLLER, 1956, p. 24–25, pl. 7, figs. 1–7.
Palmatolepis linguiformis Müller. ZIEGLER AND SANDBERG, 1990, p. 59–60, pl. 14, figs. 8–10.

***Palmatolepis subperlobata* Branson and Mehl, 1934**
Figure 10: 2, 3, 6, 7, 9

Palmatolepis subperlobata BRANSON AND MEHL, 1934, p. 235, pl. 18, figs. 11, 21 [fig. 11 = lectotype selected by Ziegler (1958, p. 64); from the Saverton Shale, Sees Creek, Missouri]; GLENISTER AND KLAPPER, 1966, p. 822–823, pl. 92, figs. 5, 7 (only) [fig. 6 = P. lobicornis]; SZULCZEWSKI, 1971, p. 40, pl. 13, fig. 12; JI, 1989a, pl. 2, fig. 26, 27 (only); LAZREQ, 1992, pl. 3, fig. 7; METZGER, 1994, p.

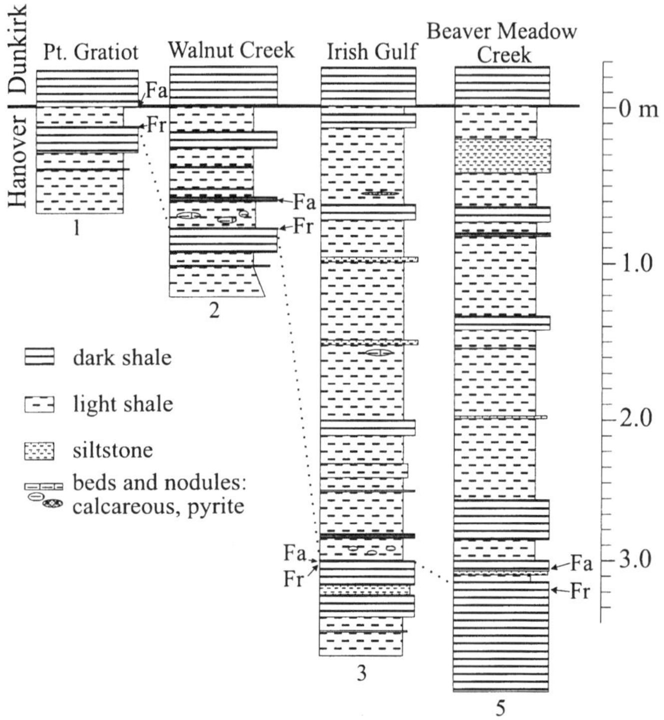

Figure 9. General lithology and position of the Frasnian-Famennian boundary (dotted line) in Hanover Shale. Numbers refer to section locations in Figure 2; Fr is last occurrence of Frasnian conodonts: Fa is first record of Famennian conodonts. Datum = base of Dunkirk Shale of Canadaway Group.

Figure 10. Conodonts from the Hanover Shale Member of the Java Formation and lowest Dunkirk Formation; lectotypes of *Palmatolepis subperlobata* Branson and Mehl and *Palmatolepis delicatula* Branson and Mehl from the Saverton Shale at Sees Creek, Missouri. BMC = Beaver Meadow Creek; IG = Irish Gulf; WCB = Walnut Creek near I-90 bridge. Magnification ×52 for all specimens except 10, which is ×70. 1, 5: *Palmatolepis triangularis* Sannemann, 1955. 1, Pa element with broken outer lobe, NYSM 15652, IG-03b; 5, Pa element with broken anterior inner platform, NYSM 15653, WCB-12. 2, 3, 6, 7, 9: *Palmatolepis subperlobata* Branson and Mehl, 1934. 2, Pa element, small specimen, NYSM 15654, IG-07; 3, Pa element, NYSM 15655, IG-08; 6, Pa element with broken posterior platform, NYSM 15656, IG-07; 7, Pa element, NYSM 15657, IG-09; 9, Pa element, lectotype selected by Ziegler (1958), C330-1, Univ. of Missouri, originally illustrated by Branson and Mehl (1934, pl. 18, fig. 11). 4: *Palmatolepis clarki* Ziegler, 1962, Pa element, NYSM 15658, WCB-12. 8: *Palmatolepis delicatula platys* Ziegler and Sandberg, 1990, Pa element, NYSM 15659, IG-07. 10: *Palmatolepis delicatula delicatula* Branson and Mehl, 1934, Pa element, lectotype selected by Ziegler (1973), C327-1, Univ. of Missouri, originally illustrated in lower view by Branson and Mehl (1934, pl. 18, fig. 4). 11: *Polygnathus decorosus* Stauffer, 1938, Pa element, NYSM 15660, IG-01b. 12: *Polygnathus brevis* Miller and Youngquist, 1947, Pa element, NYSM 15661, BMC-10. 13: *Polygnathus webbi* Stauffer, 1938, Pa element, NYSM 15662, IG-01b. 14: *Icriodus alternatus alternatus* Branson and Mehl, 1934, Pa element, NYSM 15663, IG-07.

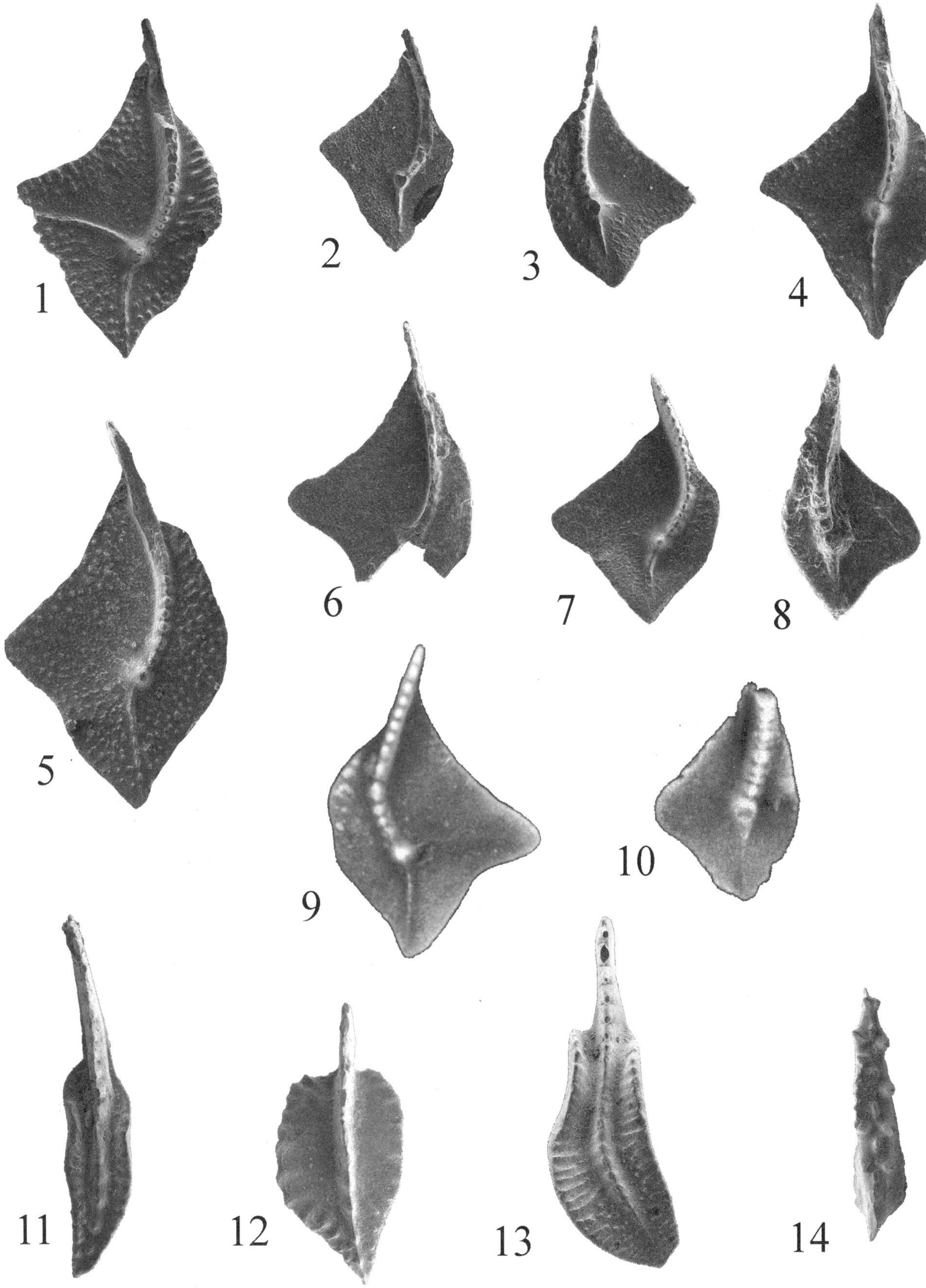
1
2
3
4
5
6
7
8
9
10
11
12
13
14

638–639, figs. 17.2, 17.3, 17.7 17.10 (only) [17.9. = P. lobicornis]; SCHÜLKE, 1995, p. 52–53, pl. 12, figs. 1–4, 7.
?Palmatolepis sp. B JI, 1989a, p. 291, pl. 3, figs. 20–23.
Palmatolepis praetriangularis Ziegler and Sandberg. BULTYNCK AND MARTIN, 1995, pl. 4, figs. 1?, 8, 9? (only).

Description. Pa element roughly triangular; sigmoidal carina, moderately curved anterior and posterior of central node, posterior carina consists of low denticles, is reduced to a low ridge, or absent toward the posterior tip; inner and outer platform are well developed; outer platform originates approximately half way between the tip of blade and origin of inner platform; in some later forms it originates closer to the tip; anterior outer platform margin is concave, outer lobe is rounded, and posterior margin is concave, becoming convex toward the posterior tip; posterior platform rises from the central node and may level to near horizontal; posteriormost tip is deflected upward or downward; ornamentation is smooth, or several small nodes may be developed on the slightly raised anterior inner platform.

Remarks. The variation in platform outline of the Pa element and range of *Palmatolepis subperlobata* are identical to that in *P. triangularis*, differing only in ornamentation. Specimens referred to as *P. subperlobata* n. subsp. A Helms and *P. subperlobata helmsi* Ovnatanova, characterized by long outer lobes and relatively straight posterior outer margins, have been synonymized in *Palmatolepis lobicornis* Schülke. *Palmatolepis subperlobata* first appears in the Lower *triangularis* Zone.

Palmatolepis triangularis Sannemann, 1955

Figure 10: 1, 5

Palmatolepis triangularis SANNEMANN, 1955, p. 327–328, pl. 24, fig. 3 [= holotype];
Palmatolepis triangularis Sannemann. ZIEGLER AND SANDBERG, 1990, p. 64–65, pl. 14, figs. 1–5 only [pl. 16, figs. 8, 9 = P. clarki, fig. 10 = ?]; BULTYNCK AND MARTIN, 1995, pl. 5, figs. 1–9, pl. 6, figs. 1–7; SCHÜLKE, 1995, p. 55-57, pl. 11, figs. 1–3, 5, 6, 9, 10, 14, 15.
Palmatolepis praetriangularis ZIEGLER AND SANDBERG in Sandberg et al., 1988, p. 298–299, pl. 1, figs. 1–4; ZIEGLER AND SANDBERG, 1990, p. 64; BULTYNCK AND MARTIN, 1995, pl. 4, figs. 2–7, 10?, 11 [figs. 1?, 8, 9? = P. subperlobata]; SCHÜLKE, 1995, pl. 11, fig. 10.

Remarks. Palmatolepis praetriangularis is considered here as a junior synonym of *P. triangularis*. The slight downward deflection of the posteriormost tip in specimens identified as *P. praetriangularis* is not considered to be significant enough taxonomically to define a separate species. *Palmatolepis triangularis* with downward and upward deflected posteriormost tips have been described from highest Frasnian and lower Famennian strata in Germany (Schülke, 1995).

Palmatolepis winchelli (Stauffer, 1938)

Figure 6: 6–9

Bryantodus winchelli STAUFFER, 1938, p. 423, pl. 48, fig. 33 [= holotype].
Palmatolepis winchelli (Stauffer). KLAPPER AND FOSTER, 1993, p. 24, 26, 31, figs. 13.1, 13.2, 18.1–18.8, 18.10, 18.11, 19.6-19.12 [fig. 19.12 = reillustration of holotype], 20.12–20.24.

Family Polygnathidae Bassler, 1925

Genus *Ancyrodella* Ulrich and Bassler, 1926

Type species. Ancyrodella nodosa Ulrich and Bassler, 1926

Ancyrodella buckeyensis Stauffer, 1938

Figure 11: 1–9

Ancyrodella buckeyensis STAUFFER, 1938, p. 418, pl. 52, figs., 17, 18, 23, 24 [figs. 23, 24 = lectotype selected by Müller and Müller (1957, p. 1091)]; ?YOUNGQUIST, 1945, p. 356, pl. 54, fig. 11; MÜLLER AND MÜLLER, 1957, p. 1091, pl. 136, figs., 2, 5; ZIEGLER, 1958, p. 40, pl. 11, fig. 7; ETHINGTON, 1965, p. 570, pl. 68, fig. 3; DRUCE, 1976, p. 53–55, pl. 1, fig. 2, pl. 2, fig. 4 [not pl. 1, fig. 4 = ?; nor pl. 2, figs. 1–3, 5 = ?].
?Ancyrodella hamata Ulrich and Bassler. MÜLLER AND MÜLLER, 1957, p. 1091–1092, pl. 136, fig. 4.
?Ancyropenta longidenticulata MÜLLER AND MÜLLER, 1957, p. 1093, pl. 136, fig. 3.
Ancyrodella nodosa Ulrich and Bassler. DRUCE, 1976, p. 61–62, pl. 7, figs. 1, 2?, 3?, pl. 8, figs. 1–3.; JI, 1989a, pl. 1, figs. 27, 28; JI, 1989b, pl. 3, fig. 6; ORCHARD, 1989, pl. 1, fig. 5.

Description. Pa element consists of a high free blade with three to four large laterally compressed denticles, anterior- and posterior-most denticles are lower; anterior carina is low, bearing closely spaced denticles, largest over the basal pit where secondary carinae diverge; posterior carina is gently curved and deflected downward posteriorly; denticles become smaller and more widely spaced near the posterior tip. Secondary outer carina diverges at an acute angle from the main carina, curving strongly anteriorly; wide anterior outer platform with a straight to convex margin parallel to subparallel with the blade; posterior outer platform deflects or curves inward abruptly, terminating near the posterior tip; outer posterior margin is straight to slightly concave or convex. Secondary inner carina diverges at an acute angle from the main carina, extending farther anteriorly than the secondary outer carina; wide anterior inner platform has a convex to straight outer margin and extends farther to the anterior than the outer platform; posterior inner platform is concave, with the constriction forming a shallow to deep sinus anterior of any concavity on the outer platform, narrowing and terminating near the posterior tip. The platform margin is ornamented with robust, moderately spaced transverse ridges (nodes on small specimens); inner platform may bear one or two discrete nodes. Keels completely underlie all carinae; basal pit is diamond shaped with a longer narrow posterior extension and a slightly greater expan-

Figure 11. Conodonts from the Hanover Shale Member of the Java Formation and the lectotype of *Ancyrodella buckeyensis* Stauffer from the Olentangy Shale. BMC = Beaver Meadow Creek; IG = Irish Gulf. 1–9: *Ancyrodella buckeyensis* Stauffer, 1938. 1–3, upper views Pa elements, juveniles, NYSM 15664-NYSM 15666, ×52, BMC-10; 4–6, upper and lower views, Pa elements, intermediate forms, NYSM 15667-15669, ×52, BMC-10; 8, 9, upper and lower views of Pa elements, adult forms, NYSM 15670, NYSM 15671, ×52, BMC-10; 7, Pa element, lectotype selected by Müller and Müller (1957), B4945, Univ. of Minnesota, ×80, locality 120 of Stauffer (1938). 10–14: *Ancyrodella curvata* Branson and Mehl, 1934. 10, 11, Pa elements, juveniles, NYSM 15672, NYSM 15673, ×52, IG-CC; 12, 13, Pa elements, intermediate forms, NYSM 15674, NYSM 15675, ×52, IG-CC; 14, Pa element, adult form, NYSM 15676, ×52, IG-CC. 15: *Ancyrognathus* sp. A, Pa element, NYSM 15677, ×37, IG-06. 16, 17: *Ancyrognathus amana* Müller and Müller, 1957, Pa elements (16 with broken outer lobe), NYSM 15678, ×70, NYSM 15679, ×52, BMC-10. 18: *Ancyrognathus asymmetricus* (Ulrich and Bassler, 1926), Pa element, NYSM 15680, ×52, BMC-10. 19: *Ancyrognathus* aff. *A. altus* Müller and Müller, 1957, Pa element, NYSM 15681, ×52, BMC-10.

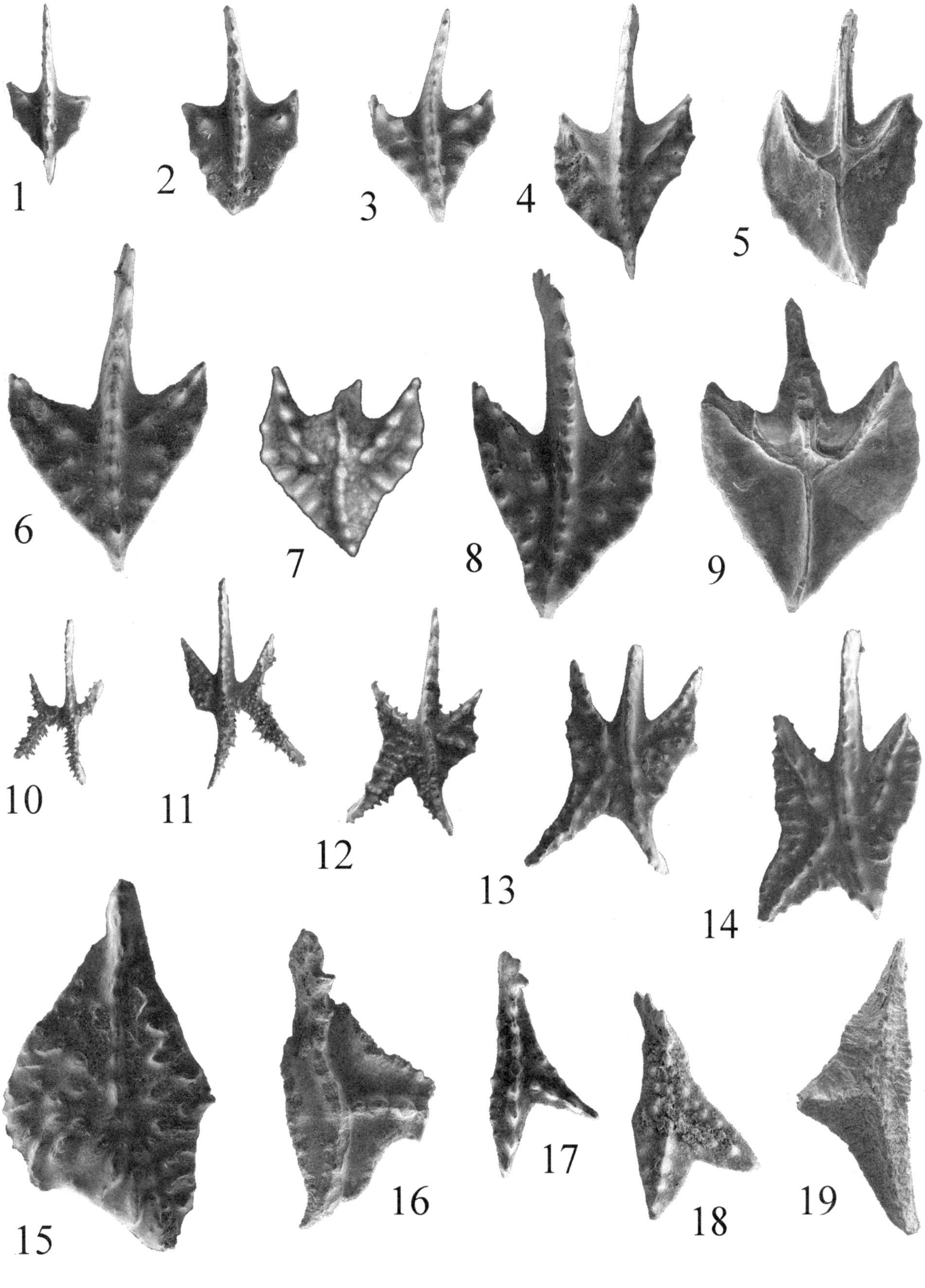
1
2
3
4
5
6
7
8
9
10
11
12
13
14
15
16
17
18
19

sion under the outer secondary carina relative to the inner secondary carina.

Remarks. The lectotype of *Ancyrodella buckeyensis* is a late juvenile form with the free blade missing (Fig. 11: 7). A series of growth forms from the Upper Olentangy Shale and the Hanover Shale clearly indicate the adult form. *Ancyrodella buckeyensis* is differentiated from *A. nodosa* by the termination of the secondary carinae at the main carina and the relative width of the inner and outer posterior platform. The secondary carinae of *A. nodosa* converge with nodes near and parallel to the posterior carina. Also, the outer and inner posterior platforms of *A. nodosa* are approximately the same width, whereas the outer platform of *A. buckeyensis* is noticeably wider. The lectotype of *A. nodosa* is from the lower Rhinestreet Shale at Weyer, in New York State, a stratum that contains a fauna indicative of MN Zone 6 and represents a distinct, older species currently under revision by Klapper and Kirchgasser (personal communication, 1993).

Ancyrodella curvata (Branson and Mehl, 1934) (late form)

Figure 11: 10–14

Ancyrognathus curvata BRANSON AND MEHL, 1934, p. 241, pl. 19, figs. 6, 11 [fig. 6 = lectotype selected by Ziegler (1958, p. 41)].
Ancyrodella curvata (Branson and Mehl). ETHINGTON AND FURNISH, 1962, p. 1261–1262, pl. 172, figs. 12–20; DRUCE, 1976, p. 55–57, pl. 1, figs. 3a–c; KLAPPER, 1989, pl. 3, fig. 20 (figs. 18, 19 = early form); VANDELAER, VANDORMAEL, and BULTYNCK, 1989, p. 329, pl. 3, figs. 1a, b; MATYJA AND NARKIEWICZ, 1992, pl. IV, figs. 12, 16 (not? pl. I, figs. 4–6 = juv. sp. indet.).

Description. Pa element has a secondary anteriorly directed inner carina and lobe; two secondary outer carinae (one directed anterior and one directed posterior), each with a well-developed lobe. Secondary keels underlie the main and all secondary carinae.

Remarks. Feist and Klapper (1985, p. 17; Klapper, 1989, p. 457) differentiated two forms of *Ancyrodella curvata*, an early form that does not have a well-differentiated lobe surrounding the posterior-directed secondary outer carina and a late form, including the type, with a well differentiated outer posterior lobe. A fourth secondary carina may also be developed (Fig. 11: 12).

Growth stages of *Ancyrodella curvata* were described and figured (drawings) by Ethington and Furnish (1962) and Lindström (1964, p. 99). Juvenile forms are characterized by well-developed secondary carinae and needle-like denticles on the posterior and secondary carinae and platforms. In intermediate forms the needle-like denticles form the sites of denticles and nodes or are engulfed by the platform as it expands. Juvenile specimens assigned to *A. curvata* by Matyja and Narkiewicz (1992) lack the secondary outer posterior carina that is an early growth feature in specimens from the Hanover Shale.

Genus *Ancyrognathus* Branson and Mehl, 1934

Type species. Ancyrognathus symmetricus Branson and Mehl, 1934

Ancyrognathus aff. *A. altus* Müller and Müller, 1957

Figure 11: 19

Ancyrognathus aff. *Ancyrognathus alta* MÜLLER AND MÜLLER, 1957, p. 1095, pl. 141, fig. 5.
Ancyrognathus aff. *A. altus* Müller and Müller. KLAPPER, 1990, p. 1001, figs. 2.7–2.9.

Ancyrognathus amana Müller and Müller, 1957

Figure 11: 16, 17

Ancyrognathus amana MÜLLER AND MÜLLER, 1957, p. 1095, pl. 138, fig. 5 [= holotype]; KLAPPER, 1990, p. 1001–1003, figs. 2.1–2.6.

Ancyrognathus asymmetricus (Ulrich and Bassler, 1926)

Figure 11: 18

Palmatolepis asymmetrica ULRICH AND BASSLER, 1926, p. 50, pl. 7, fig. 18 [= holotype].
Ancyrognathus asymmetricus (Ulrich and Bassler). KLAPPER, 1990, p. 1003–1005, figs. 4.1–4.16, 5.9–5.16, 6.7, 6.10, 7.10–7.13.

Ancyrognathus sp. A

Figure 11: 15

Description. Pa element is teardrop-shaped in outline; free blade very short or absent, consisting of low compressed denticles; fixed blade and anterior carina are short or poorly developed; posterior carina is cryptic; secondary posterior-directed outer carina is defined by aligned widely separated denticles, surrounded by the outer lobe, outer platform subequal in width to the inner platform, originating from approximately the same position; margin is irregular, but overall convex, with a shallow anterior sinus; posterior platform between the main and secondary carinae bears a shallow sinus; inner platform is convex, except for a shallow anterior sinus and in some specimens a bulge approximately midway from the anterior end. Ornamentation consists of coarse nodes and short transverse ridges normal to the platform margin.

Remarks. Ancyrognathus sp. A is known from the Middle *triangularis* Zone in the Woodford Shale (Over, 1990), upper Upper Olentangy Shale (author's collections), and a single specimen from the Hanover Shale, also in the Middle *triangularis* Zone. *Ancyrognathus* sp. A has a morphology similar to Famennian ancyrognathids with and without an outer secondary carina and lobe (e.g. *A. bifurcatus* [Ulrich and Bassler] and *A. sinelaminus* [Branson and Mehl], respectively). It is most closely allied with *A. bifurcatus*, which has a distinct outer posterior secondary carina and lobe.

Genus *Polygnathus* Hinde, 1879

Type species. Polygnathus dubius Hinde, 1879

Polygnathus brevis Miller and Youngquist, 1947

Figure 10: 12

Polygnathus brevis MILLER AND YOUNGQUIST, 1947, p. 514, pl. 74, fig. 9 [= holotype]; KLAPPER *in* Ziegler, 1973, p. 341-342, Polygnathus—Plate 1, fig. 3; SAVAGE, 1992, p. 286, figs. 4.1, 4.2, 5.26–5.28.

Polygnathus decorosus Stauffer, 1938

Figure 10: 11

Polygnathus decorosus STAUFFER, 1938, p. 438, pl. 53, figs. 5, 6, 10, 15, 16 (not figs. 1, 20, 30 = Po. sp. indet., nor fig. 11 = *P. timorensis*); KLAPPER AND LANE, 1985, p. 935, fig. 18.7.

Polygnathus samueli Klapper and Lane, 1985

Figure 6: 1

Polygnathus samueli KLAPPER AND LANE, 1985, p. 943–944, figs. 17.13–17.18 [figs. 17.16, 17.17 = holotype]; SAVAGE, 1992, p. 287, fig. 2.13.

Remarks. Specimen 1 in Figure 6 was abraded during preparation, removing the ornamentation, but the outline and overall morphology are distinctive.

Polygnathus webbi Stauffer, 1938

Figure 10.13

Polygnathus webbi STAUFFER, 1938, p. 439, pl. 53, figs. 25, 26, 28, 29 [figs. 28, 29 = lectotype selected by Wittekindt (1966)]; KLAPPER AND LANE, 1985, p. 944–945, fig. 16.18.

ACKNOWLEDGMENTS

Funding for this study was provided by the Donors to the Petroleum Research Fund of the American Chemical Society (Grant 25751-B8), the Geneseo Foundation, and a Geneseo President's Summer Fellowship. G. Baird, C. Brett, W. Kirchgasser, G. Klapper, G. Kloc, R. Pretzer, and C. Ver Straeten are thanked for their open discussions on various topics and assistance in preparation and identifications; G. Klapper provided photographs of several type specimens. Fieldwork and sample processing was ably assisted by C. Conaway, L-J. Davignon, T. Hopkins, K. Kanhalangsy, D. Katz, C. Matzek, M. Rhodes, and J. Wichtowski; thanks are extended to W. Baker, E. Bartz, G. Boldy, R. deRossa, H. Lawson, G. Perry, J. Zalinski, and Byrncliff Resort for access to exposures on their property. J. Boger, G. Klapper, and M. Orchard provided critical reviews of the manuscript.

APPENDIX 1: LOCALITY INFORMATION AND UNIVERSAL TRANSVERSE MERCATOR (UTM) REFERENCE COORDINATES

Locality 1. Point Gratiot at Dunkirk Beach: Dunkirk Quadrangle, Hanover-Dunkirk contact is located on shoreline north of NYS Rt. 5, west of breakwater and east of the lighthouse park, UTM 17TPT3540(E)0560(N).

Locality 2. Walnut Creek: Silver Creek Quadrangle, Pipe Creek Member is exposed in stream bed north of US 20 bridge in Village of Silver Creek, UTM 17TPT5035(E)1098(N); Hanover-Dunkirk contact is exposed in stream bed and bank south of US 20 bridge, under I-90 bridge, UTM 17TPT5013(E)0810(N).

Locality 3. Irish Gulf: Hamburg and Colden quadrangles, Pipe Creek Member is exposed in stream bed under railroad bridge east of Boston State Road, UTM 17TPT8420(E)2630(N); Hanover-Dunkirk contact is exposed upstream near top of the first major waterfall, UTM 17TPT8488(E)2724(N).

Locality 4. Pipe Creek: Colden Quadrangle, Pipe Creek Member is exposed in stream bed east of NYS Rt. 240, UTM 17TPT8965(E)2900(N); Hanover and Dunkirk are exposed upstream.

Locality 5. Beaver Meadow Creek: Strykersville Quadrangle, Pipe Creek Member is exposed in stream bed east of NYS Rt. 78, above Angel Falls, UTM 17TQT1020(E)2725(N); Hanover-Dunkirk contact is exposed upstream, approximately UTM 17TQT1170(E)2695(N).

Locality 6. Arkport Dam: Arkport Quadrangle, Pipe Creek Member and Hanover/Wiscoy are exposed in spillway and north side of dam site, UTM 18TTB7665(E)9720(N).

REFERENCES CITED

Algeo, T. J., Berner, R. A., Maynard, J. B., and Scheckler, S. E., 1995, Late Devonian oceanic anoxic events and biotic crises: "Rooted" in the evolution of vascular land plants?: GSA Today, v. 5, p. 45, 64–66.

Baird, G. C., and Brett, C. E., 1986, Erosion of an aerobic seafloor: Significance of reworked pyrite deposits from the Devonian of New York State: Palaeogeography, Palaeoclimatology, Palaeoecology, v. 57, p. 157–193.

Baird, G. C., and Lash, G. G., 1990, Devonian strata and paleoenvironments:Chautauqua County and adjacent northern Erie County, in Lash, G. G., ed., Field Trip Guidebook, New York State Geological Association, 62nd Annual Meeting, Trip A, 46 p.

Bassler, R. S., 1925, Classification and stratigraphic use of conodonts: Geological Society of America Bulletin, v. 36, p. 218–220.

Becker, R. T., 1993, Anoxia, eustatic changes, and Upper Devonian to lowermost Carboniferous global ammonoid diversity, in House, M. R., ed., The Ammonoidea. environment, ecology, and evolutionary change: Systematic Association Special Volume 47, p. 115–163.

Becker, R. T., Feist, R., Flajs, G., House, M. R., and Klapper, G., 1989, Frasnian-Famennian extinction events in the Devonian at Coumiac, southern France: Comptes Rendus de l'Académie des Sciences de Paris, v. 309, Série 2, p. 259–266.

Brand, U., 1989, Global climatic changes during the Devonian-Mississippian: Stable isotope biogeochemistry of brachiopods: Palaeogeography, Palaeoclimatology, Palaeoecology (Global and Planetary Change Section), v. 75, p. 311–329.

Branson, E. B., 1938, Stratigraphy and paleontology of the Lower Mississippian of Missouri. Part 1: University of Missouri Studies, v. 13, 208 p.

Branson, E. B., and Mehl, M. G., 1934, Conodonts from the Grassy Creek shale of Missouri: University of Missouri Studies, v. 8, p. 171–259. [imprint 1933]

Branson, E. B., amd Mehl, M. G., 1938, The conodont genus *Icriodus* and its stratigraphic distribution: Journal of Paleontology, V. 12, p. 156–166.

Bryant, W. L., 1921, The Genessee conodonts with descriptions of new species: Bulletin of the Buffalo Society of Natural Science, v. 13, 59 p.

Bultynck, P., and Martin, F., 1995, Assessment of an old stratotype: The Frasnian/Famennian boundary at Senzeilles, southern Belgium: Bulletin de l'institute Royal des Sciences Naturelles de Belgique, v. 65, p. 5–34.

Chadwick, G. H., 1923, Chemung stratigraphy in western New York: Geological Society of America Bulletin, v. 34, p. 69.

Chadwick, G. H., 1933, Great Catskill Delta and revision of Late Devonic succession: Pan-American Geology, v. 60, p. 96, 99, 193, 198, 199.

Copper, P., 1986, Frasnian/Famennian mass extinction and cold-water oceans: Geology, v. 14, p. 835–839.

de Witt, W., Jr., 1960, Java Formation of Late Devonian age in western and central New York: American Association of Petroleum Geologists Bulletin, v. 44, p. 1933–1939.

de Witt, W., Jr., and Roen, J. B., 1985, Correlation and geographic extent of some Middle and Upper Devonian and Lower Mississippian black shales in the Appalachian Basin: U.S. Geological Survey Bulletin 1605-A, p. A45–A57.

Druce, E. C., 1976, Conodont biostratigraphy of the Upper Devonian reef complexes of the Canning Basin, Western Australia: Bulletin of the Bureau of Mineral Resources, Geology, and Geophysics, v. 158, 303 p.

Dzik, J., 1976, Remarks on the evolution of Ordovician conodonts: Acta Palaeontologica Polonica, v. 21, p. 395–455.

Ethington, R. L., 1965, Late Devonian and Early Mississippian conodonts from Arizona and New Mexico: Journal of Paleontology, v. 39, p. 566–589.

Ethington, R. L., and Furnish, W. M., 1962, Silurian and Devonian conodonts from Spanish Sahara: Journal of Paleontology, v. 36, p. 1253–1290.

Feist, R., and Klapper, G., 1985, Stratigraphy and conodonts in pelagic sequences across the Middle-Upper Devonian boundary, Montagne Noire, France: Palaeontographica Abt., A., v. 188, p. 1–18.

Geldsetzer, H. H. J., Goodfellow, W. D., McLaren, D. J., and Orchard, M. J., 1987, Sulfur-isotope anomaly associated with the Frasnian-Famennian extinction, Medicine Lake, Alberta, Canada: Geology, v. 15, p. 393–396.

Glenister, B. F., and Klapper, G., 1966, Upper Devonian conodonts from the Canning Basin, Western Australia: Journal of Paleontology, v. 40, p. 777–842.

Goodfellow, W. D., Geldsetzer, H. H. J., McLaren, D. J., Orchard, M. J., and Klapper, G., 1989, The Frasnian-Famennian extinction: Current results and possible causes, in McMillan, N. J., Embry, A. F., and Glass, D. J., eds., Devonian of the world: Calgary, Canadian Society of Petroleum

Geologists Memoir 14, v. 3, p. 19–22. [imprint 1988]

Hall, J., 1843. Geology of New York. Part IV, Comprising the survey of the Fourth Geological District: Albany, Carrol and Cook, 525 p.

Han Y.-J., 1987, Study on Upper Devonian Frasnian/Famennian boundary in Maanshan, Zhongping, Xianzhou, Guangxi: Bulletin Chinese Academy of Science, v. 17, p. 171–194.

Hass, W. H., 1958, Upper Devonian conodonts of New York, Pennsylvania, and interior states: Journal of Paleontology, v. 32, p. 765–769.

Helms, J., 1963, Zur "Phylogenese" und Taxionomie von Palmatolepis (Conodonten, Oberdevon): Geologie, v. 12, p. 449–485.

Hinde, G. J., 1879, On conodonts from the Chazy and Cincinnati group of the Cambro-Silurian and from the Hamilton and Genesee shale divisions of the Devonian in Canada and the United States: Geological Society of London Quarterly Journal, v. 35, p. 351-369.

House, M.R., 1978, Devonian ammonoids from the Appalachian and their bearing on international zonation and correlation: Special Papers in Palaeontology, no. 21, 70 p.

House, M. R., and Kirchgasser, W. T., 1993, Devonian goniatite biostratigraphy and timing of facies movements in the Frasnian of eastern North America, in Hailwood, E. A., and Kidd, R. B., eds., High resolution stratigraphy: Geological Society Special Publication 70, p. 267–292.

Huddle, J. W., assisted by Repetski, J. E., 1981, Conodonts from the Genesse Formation in western New York: U.S. Geological Survey Professional Paper 1032-B, 66 p.

Ji, Q., 1989a, On the Frasnian-Famennian mass extinction event in South China: Courier Forschungsinstitut Senckenberg, v. 117, p. 275–301.

Ji, Q., 1989b, On the Frasnian conodont biostratigraphy in the Guilin Area of Guangxi, South China: Courier Forschungsinstitut Senckenberg, v. 117, p. 303–319.

Joachimski, M. M., and Buggisch, W., 1993, Anoxic events in the late Frasnian—Causes of the Frasnian-Famennian faunal crisis: Geology, v. 21, p. 675–678.

Johnson, J. G., and Klapper, G., 1992, North American Midcontinent Devonian T-R cycles, in Chaplin, J. R., and Barrick, J. E., eds., Special papers in paleontology and stratigraphy: A tribute to Thomas W. Amsden: Oklahoma Geological Survey Bulletin, v. 145, p. 127–135.

Kirchgasser, W. T., 1985, Ammonoid horizons in the Upper Devonian Genesee Formation of New York: Legacy of the Genesee, Portage, and Chemung, in Woodrow, D. L., and Sevon, W. D., eds., The Catskill Delta: Geological Society of America Special Paper 201, p. 225–235.

Kirchgasser, W. T., 1994, Early morphotypes of *Ancyrodella rotundiloba* at the Middle/Upper Devonian boundary, Genesee Formation, west-central New York, in Landing, E., ed., Studies in stratigraphy and paleontology in honor of Donald W. Fisher: New York State Museum Bulletin 481, p. 117–134.

Kirchgasser, W. T., and House, M. R., 1981, Upper Devonian goniatite biostratigraphy, in Oliver, W. A., Jr., and Klapper, G., eds., Devonian Biostratigraphy of New York, part 1: Text: Wahington D.C., International Union of Geological Sciences, Subcommission on Devonian Stratigraphy, p. 39–55. (SUNY Binghamton meeting, 1981)

Kirchgasser, W. T., Over, D. J., and Woodrow, D. L., 1994, Frasnian (Upper Devonian) strata of the Genesee River Valley, western New York State, in Brett, C. E., and Scatterday, J., eds., Field trip guidebook: New York State Geological Association, 66th Annual Meeting, p. 325–358.

Klapper, G., 1981, Review of New York Devonian conodont biostratigraphy, in Oliver, W. A., Jr., and Klapper, G., eds., Devonian biostratigraphy of New York. Part 1, Text: Washington, D.C., International Union of Geological Sciences, Subcommission on Devonian Stratigraphy, p. 57–66. (SUNY Binghamton meeting, 1981)

Klapper, G., 1989, The Montagne Noire Frasnian (Upper Devonian) conodont successions, in McMillan, N. J., Embry, A. F., and Glass, D. J., eds., Devonian of the world: Calgary, Canadian Society of Petroleum Geologists Memoir 14, v. 3, p. 449–468. [imprint 1988]

Klapper, G., 1990, Frasnian species of the Late Devonian conodont genus *Ancyrognathus*: Journal of Paleontology, v. 64, p. 998–1025.

Klapper, G., and Foster, C. T., Jr., 1993, Shape analysis of Frasnian species of the Late Devonian conodonts Genus *Palmatolepis*: The Paleontological Society Memoir 32, p. 1–35.

Klapper, G., and Lane, H. R., 1985, Upper Devonian (Frasnian) conodonts of the *Polygnathus* biofacies N.W.T., Canada: Journal of Paleontology, v. 59, p. 904–951.

Klapper, G., and eight others, 1971, North American Devonian conodont biostratigraphy, *in* Sweet, W. C., and Bergström, S. M., eds., Symposium on conodont biostratigraphy: Geological Society of America Memoir 127, p. 285–316.

Klapper, G., Feist, R., Becker, R. T., and House, H. R., 1994, Definition of the Frasnian/Famennian stage boundary: Episodes, v. 4, p. 433–441. [imprint 1993]

Klapper, G., Kirchgasser, W. T., and Baesemann, J. F., 1995, Graphic correlation of a Frasnian (Upper Devonian) composite standard, *in* Mann, K. O. and Lane, H. R., eds., Graphic correlation: SEPM (Society for Sedimentary Geology) Special Publication 53, p. 177–184.

Kralick, J. A., 1994, The conodont genus *Ancyrodella* in the Middle Genesee Formation (Lower Upper Devonian, Frasnian), in western New York: Journal of Paleontology, v. 68, p. 1384–1395.

Lange, F. G., 1968, Conodonten-Gruppenfunde aus Kalken des tieferen Oberdevon: Geologica et Palaeontologica, v. 2, p. 37–57.

Lazreq, N., 1992, The Upper Devonian of M'rirt (Morocco): Courier Forschungsinstitut Senckenberg, v. 154, p. 107–124.

Lindström, M., 1964, Conodonts: Amsterdam, Elsevier, 196 p.

Matyja, H., and Narkiewicz, M., 1992, Conodont biofacies succession near the Frasnian-Famennian Boundary: Some Polish examples: Courier Forschungsinstitut Senckenberg v. 154, p. 125–147.

McGhee, G. R., Jr., 1989, The Frasnian-Famennian extinction event, in Donovan, S. K., ed., Mass extinctions: Processes and evidence: New York, Columbia University Press, p. 133–151.

McGhee, G. R., Jr., 1996, The Late Devonian mass extinction: The Frasnian/Famennian crisis: New York, Columbia University Press, 303 p.

McGhee, G. R., Jr., Gilmore, J. S., Orth, C. J., and Olsen, E., 1984, No geochemical evidence for an asteroidal impact at late Devonian mass extinction horizon: Nature, v. 308, p. 629–631.

McLaren, D. J., 1970, Presidential address: Time, life and boundaries: Journal of Paleontology, v. 44, p. 801–815.

McLaren, D. J., 1989, Detection and significance of mass killings, in McMillan, N.J., Embry, A. F., and Glass, D. J., eds., Devonian of the world: Calgary, Canadian Society of Petroleum Geologists Memoir 14, v. 3, p. 1–8. [imprint 1988]

Metzger, R. A., 1994, Multielement reconstructions of *Palmatolepis* and *Polygnathus* (Upper Devonian), Famennian, from the Canning Basin, Australia, and Bactrain Mountain, Nevada: Journal of Paleontology, v. 68, p. 617–647.

Miller, A. K., and Youngquist, W., 1947, Conodonts from the type section of the Sweetland Creek shale in Iowa: Journal of Paleontology, v. 21, p. 501–517.

Müller, K. J., 1956, Zur Kenntnis der Conodonten-Fauna des europäischen Devons. 1: Die Gattung *Palmatolepis*: Abhandlungen der Senckenbergischen Naturforschenden Gesellschaft, 494, 70 p.

Müller, K. J., and Müller, E. M., 1957, Early Upper Devonian (Independence) conodonts from Iowa, Part 1: Journal of Paleontology, v. 31, p. 1069–1108.

Oliver, W. A., Jr., de Witt, W., Jr., Dennison, J. W., Hoskin, D. M., and Huddle, J. W., 1968, Devonian of the Appalachian Basin, United States, in Oswald, D.H., ed., International Symposium on the Devonian System, Volume 1: Calgary, Alberta Society of Petroleum Geologists, p. 1001–1040.

Oliver, W. A., Jr., de Witt, W., Jr., Dennison, J. W., Hoskin, D. M., and Huddle, J. W., 1969, Correlation of Devonian rocks in the Appalachian Basin: U.S. Geological Survey Oil and Gas Investigations, Chart OC-64.

Orchard, M. J., 1989, Conodonts from the Frasnian-Famennian boundary interval in western Canada, in McMillan, N. J., Embry, A. F., and Glass, D. J., eds., Devonian of the world: Calgary, Canadian Society of Petroleum Geologists Memoir 14, v. 3, p. 35–52. [imprint 1988]

Over, D. J., 1990, Conodont Biostratigraphy of the Woodford Shale (Late Devonian–Early Carboniferous) in the Arbuckle Mountains, south-central Oklahoma [Ph.D. thesis]: Lubbock, Texas Tech University, 173 p.

Over, D. J., Conaway, C. A., Katz, D. J., Goodfellow, W. D., and Grégoire, D. C., 1997, Platinum group element (PGE) enrichments and possible chondritic Ru:Ir across the Frasnian-Famennian boundary, western New York: Palaeogeography, Palaeoclimatology, Palaeoecology (in press).

Ovnatanova, N. S., 1976, Novye pozdnedevonskie konodonty Russkoy Platformy (New Late Devonian conodonts of the Russian Platform): Paleontologicheskiy Zhurnal, 1976, no. 2, p. 106–115.

Pepper, J. F., and de Witt, W., Jr., 1950, Stratigraphy of the Upper Devonian Wiscoy Sandstone and equivalent Hanover shale in western and central New York: U.S. Geological Survey, Oil and Gas Investigations, Chart OC 37.

Pepper, J. F., de Witt, W., Jr., and Colton, G. W, 1956, Stratigraphy of the West Falls Formation of Late Devonian age in western and central New York: U.S. Geological Survey, Oil and Gas Investigations, Chart OC 55.

Rickard, L. V., 1964, Correlation of the Devonian rocks in New York State: New York State Museum Map and Chart Series, no. 4.

Rickard, L. V., 1975, Correlation of the Silurian and Devonian rocks in New York State: New York State Museum, Map and Chart Series, no. 24.

Sandberg, C. A., and Dreesen, R., 1984, Late Devonian icriodontid biofacies models and alternate shallow-water conodont zonation, in Clark, D. L., ed., Conodont biofacies and provincialism: Geological Society of America Special Paper 196, p. 143–178.

Sandberg, C. A., and Ziegler, W., 1973, Refinement of standard Upper Devonian conodont zonation based on sections in Nevada and West Germany: Geologica et Palaeontologica, v. 7, p. 97–122.

Sandberg, C. A., Ziegler, W., Dreesen, R., and Butler, J. L., 1988, Part 3: Late Frasnian mass extinction: Conodont event stratigraphy, global changes, and possible causes: Courier Forschungsinstitut Senckenberg, v. 102, p. 263–307.

Sannemann, D., 1955, Beitrag zur Unterliederung des Oberdevons nach Conodonten: Neues Jahrbuch für Geologie und Paläontologie, Abhandlungen, v. 100, p. 324–331.

Savage, N. M., 1992, Late Devonian (Frasnian and Famennian) conodonts from the Wadleigh Limestone, southeastern Alaska: Journal of Paleontology, v. 66, p. 277–292.

Schindler, E., 1990, Die Kellwasser-Krise (hohe Frasne-Stufe, Ober-Devon): Göttinger Arbeiten zur Geologie und Paläontologie, no. 46, 115 p.

Schindler, E., 1993, Event-stratigraphic markers within the Kellwasser Crisis near the Frasnian/Famennian boundary (Upper Devonian) in Germany: Palaeogeography, Palaeoclimatology, and Palaeoecology, v. 104, p. 115–125.

Schülke, I., 1995, Evolutive Prozesse bei Palmatolepis in der frühen Famenne-Stufe (Conodonta, Ober-Devon): Göttinger Arbeiten zur Geologie und Paläontologie, no. 67, 108 p.

Sorauf, J. E., and Pedder, A. E. H., 1986, Late Devonian rugose corals and the Frasnian-Famennian crisis: Canadian Journal of Earth Sciences, v. 23, p. 1265–1287.

Stauffer, C. R., 1938, Conodonts of the Olentangy Shale: Journal of Paleontology, v. 12, p. 411–443.

Sutton, R. G., 1963, Correlation of Upper Devonian strata in south-central New York, in Shepps, V. C., ed., Symposium of Middle and Upper Devonian stratigraphy of Pennsylvania and adjacent states: Pennsylvania Geological Survey, General Geology Reports G-39, p. 87–101.

Sweet, W. C., 1988, The Conodonta: Morphology, taxonomy, paleoecology, and evolutionary history of a long-extinct animal phylum: Oxford, Clarendon Press, 212 p.

Szulczewski, M., 1971, Upper Devonian conodonts, stratigraphy and facial development in the Holy Cross Mountains: Acta Geologica Polonica, v. 21, p. 1–129.

Tesmer, I. H., 1963, Geology of Chautauqua County. Part 1: Stratigrapy and paleontology (Upper Devonian): New York State Museum Bulletin 391, 65 p.

Tesmer, I. H., 1968, Upper Devonian stratigraphy and paleontology of southwestern New York State, in Oswald, D. H., ed., International Symposium on the Devonian System, Calgary 1967: Alberta Society of Petroleum Geologists, v. 1, p. 259–269.

Thompson, J. B., and Newton, C. R., 1989, Late Devonian mass extinction: Episodic climatic cooling or warming? in McMillan, N. J., Embry, A. F., and Glass, D. J., eds., Devonian of the world: Calgary, Canadian Society of Petroleum Geologists Memoir 14, v. 3, p. 29–34. [imprint 1988]

Ulrich, E. O., and Bassler, R. S., 1926, A classification of the toothlike fossils, conodonts, with description of American Devonian and Mississippian specimens: Proceedings of the U.S. National Museum, v. 68, art. 12, 63 p.

Vandelaer, E., Vandormael, C., and Bultynck, P., 1989, Biofacies and refinement of conodont succession in the lower Frasnian (Upper Devonian) of the type areas (Franes-Nismes, Belgium): Courier Forschungsinstitut Senckenberg, v. 117, p. 321–352.

Wang, K., Orth, C. J., Attrep, M., Chatterton, B. D. E., Hou, H., and Geldsetzer, H. H. J., 1991, Geochemical evidence for a catastrophic biotic event at the Frasnian/Famennian boundary in south China: Geology, v. 19, p. 776–779.

Wilde, P., and Berry, W. B. N., 1984, Destabilization of the oceanic density structure and its significance to marine "extinction" events: Palaeogeography, Palaeoclimatology, Palaeoecology, v. 48, p. 143–162.

Wittekindt, H. P., 1966, Zur Conodonten Chronologie des Mittledevons: Fortschritte in der Geologie von Rheinland und Westfalen, v. 9, p. 621–646. [imprint 1965]

Woodrow, D. L., Dennison, J. M., Ettensohn, F. R., Sevon, W. T., and Kirchgasser, W. T., 1989, Middle and Upper Devonian stratigraphy and paleogeography of the central and southern Appalachians and eastern midcontinent, U.S.A., in McMillan, N. J., Embry, A. F., and Glass, D. J., eds., Devonian of the world: Calgary, Canadian Society of Petroleum Geologists Memoir 14, v. 3, p. 277–301. [imprint 1988]

Youngquist, W., 1945, Upper Devonian conodonts from the Independence Shale(?) of Iowa: Journal of Paleontology, v. 19, p. 335–367.

Ziegler, W., 1958, Conodontenfeinstratigraphische untersuchungen an der Grenze Mitteldevon/Oberdevon und in der Adorfstuffe: Notizblatt des Hessischen Landesamtes für Bodenforschung, v. 87, p. 7–77.

Ziegler, W., 1962, Taxionomie und Phylogenie oberdevonischer Conodonten und ihre stratigraphische Bedeutung: Abhandlungen des Hessischen Landesamtes für Bodenforschung, no. 38, 166 p.

Ziegler, W., ed., 1973, Catalogue of Conodonts, Volume 1: Stuttgart, E. Schweizerbart'sche Verlagsbuchhandlung, 504 p.

Ziegler, W., ed., 1975, Catalogue of Conodonts, Volume 2: Stuttgart, E. Schweizerbart'sche Verlagsbuchhandlung, 404 p.

Ziegler, W., and Sandberg, C. A., 1990, The Late Devonian standard conodont zonation: Courier Forschungsinstitut Senckenberg, v. 121, 115 p.

MANUSCRIPT ACCEPTED BY THE SOCIETY FEBRUARY 27, 1997

Geological Society of America
Special Paper 321
1997

Use of vertebrates to solve biostratigraphic problems: Examples from the Lower and Middle Devonian of western North America

David K. Elliott and Heidemarie G. Johnson
Department of Geology, Northern Arizona University, Flagstaff, Arizona 86011-4099

ABSTRACT

Because of the absence of invertebrate faunas, previous correlations of Lower and Middle Devonian nonmarine sedimentary rocks in the western United States have resulted in various age assignments. Although vertebrate fossils are present in these strata, they have not been widely used for biostratigraphic correlation in North America. Recent work indicates that, despite problems of endemism of these faunas during the Early Devonian, vertebrate fossils do provide a tentative basis for correlation. The lower part of the Grassy Flat Member of the Water Canyon Formation and the Beartooth Butte Formation at Beartooth Butte are shown to be Emsian in age, whereas the Beartooth Butte Formation in the Bighorn Mountains is late Lochkovian to early Pragian. An endemic fauna restricted to the Sevy Dolomite and Lippincott Member of the Lost Burro Formation is also considered to be Emsian. The Middle Devonian fauna has a more cosmopolitan distribution; similar assemblages at disparate localities in North America provide a framework for the use of the fauna in placing age constraints on stratigraphic units. For the first time pteraspidids are unequivocally identified from Middle Devonian strata. Together with a characteristic fauna of arthrodires and antiarchs, they are used to correlate the Yahatinda Formation of Alberta with the Spring Mountain channel of the Lemhi Range, Idaho, and the upper part of the Grassy Flat Member of the Water Canyon Formation, all of which can now be dated as late Givetian.

INTRODUCTION

Johnson et al. (1989) provided an overview of the distribution of lithofacies across the western United States during the Early and Middle Devonian, at the end of which they noted the current status of several problematic formations. Chief among these was the Water Canyon Formation, the members of which had been assigned very different ages by the same authors on various occasions. Similar problems were evident with other Early Devonian formations, including the Lippincott Member of the Lost Burro Formation in California and the Beartooth Butte Formation of Wyoming and Montana. These formations are all nonmarine fish- and plant-bearing sedimentary rocks with a scattered epicratonic distribution; they are connected with transgressive and regressive events that took place during the Devonian (Johnson et al., 1985; Johnson and Sandberg, 1989). The formations lack age-diagnostic marine invertebrates used in generating the standard zonal scheme for marine sediments of the continental shelf (Johnson et al., 1980; Johnson and Murphy, 1984; Johnson, 1986) and are frequently separated by unconformities from the under- and overlying strata.

Vertebrate faunas have been known from these formations for some time (Branson and Mehl, 1931; Bryant, 1932, 1933, 1934, 1935; Denison, 1952, 1953, 1966, 1967, 1968a, 1968b, 1970; Elliott, 1994; Ilyes and Elliott, 1994; Elliott and Ilyes, 1996a), but assessing their value for dating and correlating non-

Elliott, D. K., and Johnson, H. G., 1997, Use of vertebrates to solve biostratigraphic problems: Examples from the Lower and Middle Devonian of western North America, *in* Klapper, G., Murphy, M. A., and Talent, J. A., eds., Paleozoic Sequence Stratigraphy, Biostratigraphy, and Biogeography: Studies in Honor of J. Granville ("Jess") Johnson: Boulder, Colorado, Geological Society of America Special Paper 321.

marine sediments has scarcely been broached in western North America (Elliott and Ilyes, 1996b). Vertebrates have generally been utilized only in Paleozoic biostratigraphic correlation outside North America (Janvier and Blieck, 1993; Blieck et al., 1987, 1989; Donovan et al., 1974; Jarvik, 1961; Wang, 1993; Long, 1993), although attempts have been made to establish correlations between Lower Devonian vertebrate-bearing successions in the Canadian Arctic and those of western Europe (Dineley and Loeffler, 1976; Elliott, 1984). An outline of Lower Devonian vertebrate biostratigraphy in the western United States has recently been published (Elliott and Ilyes, 1996b). Much of this work has used macrovertebrates that occur in transitional environments and thus, with spores, provide a valuable link between terrestrial sequences and the established marine invertebrate or microfossil zones. One problem is the endemism of many of the groups of macrovertebrates (Young, 1993), which restricts the range of correlation schemes based on them. However, the recent upsurge of studies on microvertebrates, chiefly thelodonts and acanthodians, which may have very wide geographic and environmental distributions, may provide a means of linking local correlation schemes (Turner, 1993).

A first attempt at developing a Lower Devonian vertebrate-based correlation scheme in the western United States was published by Elliott and Ilyes (1996b). In the present chapter we include additional information to correct and extend the Lower Devonian scheme and also provide an initial outline of vertebrate correlations in the Middle Devonian.

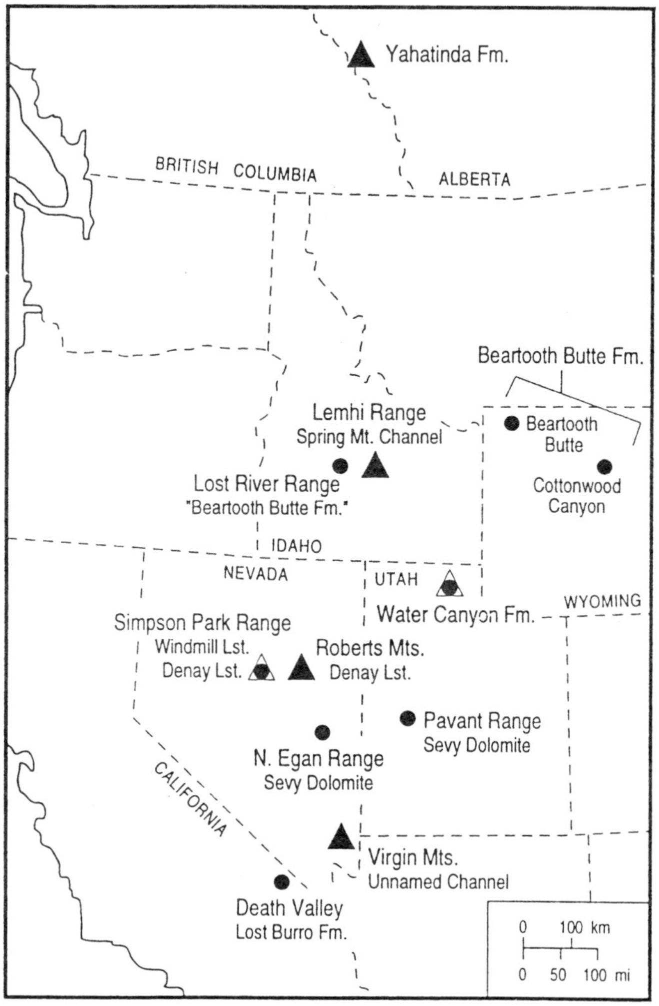

● Lower Devonian
▲ Middle Devonian
Lower and Middle Devonian

Figure 1. Vertebrate localities referred to in the text.

LOWER DEVONIAN FORMATIONS

The Lost Burro Formation (Fig. 1) of Death Valley, California, as originally defined (McAllister, 1952), encompassed all three Devonian series and consisted mainly of limestones and dolomites. The lower part was separated as the Lippincott Member (McAllister, 1955) owing to its distinctive sandy lithology; the lower part of the member has been informally separated as the "cherty argillaceous unit" (Johnson et al., 1989). There is a sharp contrast between this unit and the rest of the Lippincott Member above, which may reflect the presence of a regional unconformity. A late Emsian age *(inversus-serotinus* Zones) has been ascribed to this unit on the basis of age-diagnostic conodonts.

In the Trail Canyon area of the Panamint Range, Death Valley, a large channel cut into the Silurian and Devonian Hidden Valley Dolomite appears to be laterally continuous with the "cherty argillaceous unit" (Elliott and Ilyes, 1996a; Ilyes, 1993). It contains a typical Lower Devonian vertebrate fauna composed of an actinolepid arthrodire, an osteostracan, and heterostracans, including pteraspidids and cyathaspidids (Fig. 2). Unfortunately the heterostracans are mostly new, which limits their biostratigraphic value. However, one of the pteraspidids, *Blieckaspis priscillae* Elliott and Ilyes (1996a), is also found in the lower part of the Grassy Flat Member of the Water Canyon Formation in northern Utah (Fig. 2), suggesting a correlation with that member. The cyathaspidids show close similarity to those found in the Sevy Dolomite of Nevada and Utah (Fig. 2), indicating a correlation with that formation. Although thelodonts are now known from this horizon, they belong to a new genus and species of nikoliviid and are not yet sufficiently well known to be age diagnostic (J. Evans, personal communication, 1995).

The Sevy Dolomite is a widespread carbonate unit of Silurian to Early Devonian age exposed in east-central Nevada and west-central Utah. It is part of the thick carbonate sequence deposited within the Cordilleran region during the early Paleozoic (Osmond, 1954, 1962). The occurrence of chert in the lowermost part of the upper Sandy Member and the abundant sandy and silty layers throughout suggest it is a lithological correlative of the "cherty argillaceous unit" (Johnson et al., 1989). Vertebrate faunas are now known from two localities within the Sandy Member (Figs. 1 and 2). The assemblage found in the northern Egan Range, Nevada, consists of an actinolepid arthrodire, an osteostracan, pteraspidids, cyathaspidids, and acanthodians (Ilyes and

	Lost Burro Formation Lippincott Member Death Valley	Sevy Dolomite Sandy Member		Water Canyon Formation Lower Grassy Flat Member Bear River Range	Beartooth Butte Formation	
		Pavant Range	Northern Egan Range		Beartooth Butte	Cottonwood Canyon
Osteostraci	*Cephalaspis* sp.	*Cephalaspis* sp.	*Cephalaspis* sp.	*Cephalaspis? wyomingensis* *Cephalaspis utahensis* *Cephalaspis brevirostris*	*Cephalaspis wyomingensis*	*Cephalaspis* sp.
Heterostraci *Pteraspididae*	*Blieckaspis priscillae* *Panamintaspis snowi*	New genus and species A & B	*Tuberculaspis elyensis* *Lamiaspis longiripa* *Pirumaspis lancasteri*	*Oreaspis dunklei* *Oreaspis williamsi* *Oreaspis? tenuistriata* *Oreaspis* sp. *Blieckaspis priscillae* *Clydonaspis faberensis* *Pirumaspis* sp. *Cosmaspis* sp.	*Oreaspis ampla* *Protaspis bucheri* *Protaspis dorfi* *Cyrtaspidichthys ovata* *Cyrtaspidichthys sculpta*	*Protaspis mcgrewi* *Protaspis brevispina* *Cosmaspis transversa* *Lampraspis tuberculata*
Cardipeltidae				*Cardipeltis wallacii*	*Cardipeltis wallacii*	*Cardipeltis richardsoni* *Cardipeltis bryanti*
Cyathaspididae	New genus and species B New genus and species C	New genus and species B New genus and species C	*Ctenaspis* sp. New genus and species A New genus and species B New genus and species C	*Allocryptaspis utahensis* *Allocryptaspis* n. sp.	*Allocryptaspis flabelliformis* *Allocryptaspis ellipticus*	
Thelodonti	*Nikoliviids*	*Nikoliviids*		*Nikoliviids* *Canonia?* sp.		
Placodermi *Actinolepidae*	New genus and species	New genus and species	New genus and species	*Simblaspis cachensis* *Aethaspis major* *Aethaspis utahensis* *Bryantolepis* sp.	*Anarthraspis montana* *Anarthraspis chamberlini* *Bryantolepis brachycephala* *Bryantolepis cristata*	*Simblaspis* sp. *Aethaspis* sp.
Acanthodii	*Onchus* sp.	*Cheiracanthoides* sp.	*Onchus* sp.	*Onchus* sp.	*Onchus* sp.	
Dipnoi				*Uranolophus* sp. *Dipterus* sp.	*Uranolophus wyomingensis*	*Uranolophus wyomingensis*

generic similarity - - -   specific similarity ....

Figure 2. Vertebrate faunal lists for the main Pragian and Emsian horizons in the western United States. Adapted from Elliott and Ilyes (1996b).

Elliott, 1994) and is again typically Early Devonian. The cyathaspidids are similar to those in the Death Valley fauna; the connection is now reinforced by the recognition of similar nikoliviid thelodonts in both occurrences (J. Evans, personal communication, 1995). The second locality is in the Pavant Range, west-central Utah (Fig. 1); it contains a fauna that is similar in overall aspect but can be closely correlated only on the presence of several cyathaspidids (Fig. 2). The thelodonts from this locality are abundant and represent a form similar to *Nikolivia elongata*, occurring with a new species of *Nikolivia* (Elliott et al., 1998). However, they are insufficiently diagnostic to give a precise age.

The Water Canyon Formation (Fig. 1) was named by Williams (1948) for a sequence of carbonates lying between the Silurian Laketown Dolomite and the Jefferson Formation in northern Utah. Subsequently it was divided into a lower Card Member and an upper Grassy Flat Member (Williams and Taylor, 1964). An extensive vertebrate fauna has been described from this formation (see references in Elliott and Ilyes, 1996b), but some confusion concerning the position of the boundary between the members, as well as the horizons at which the vertebrates were collected, has caused problems with dating and correlation. An Emsian age has been proposed for the Card Member and a Givetian age for the Grassy Flat Member in several publications (Elliott, 1994; Elliott and Ilyes, 1996a, b; Ilyes and Elliott, 1994). However, it is now clear that Elliott and Ilyes had incorrectly identified the boundary between the two members. The remapping shows that the vertebrate faunas come from the Grassy Flat Member only and that these indicate an Emsian age for the fauna occurring in the lower part, whereas the scarce fauna from the upper part indicates an upper Middle Devonian age (see below). Although the lower vertebrate fauna is extensive (Figs. 2, 3), it shows little similarity to those in the Sevy Dolomite and the Lippincott Member of the Lost Burro Formation, sharing only one pteraspidid with each (Fig. 2). Nikoliviids are again present. Furthermore, the occurrence of the thelodont *Canonia?* sp. (D. Goujet, personal communication, 1995) may indicate that part of the Grassy Flat Member may be older than Emsian, as these thelodonts range from middle Lochkovian to early Pragian in the Canadian Arctic and are found in the *delta* Zone in central Nevada (Turner and Murphy, 1988). The Card Member contains no vertebrates and has been referred to the Lochkovian (Johnson et al., 1989).

In the Lost River Range near Challis, central Idaho, a channel deposit containing vertebrates was correlated with the Beartooth Butte Formation of Wyoming (Johnson and Sandberg, 1977; Johnson et al., 1991) though lithologies are different in the two areas. The Idaho deposits consist of intertidal to subtidal

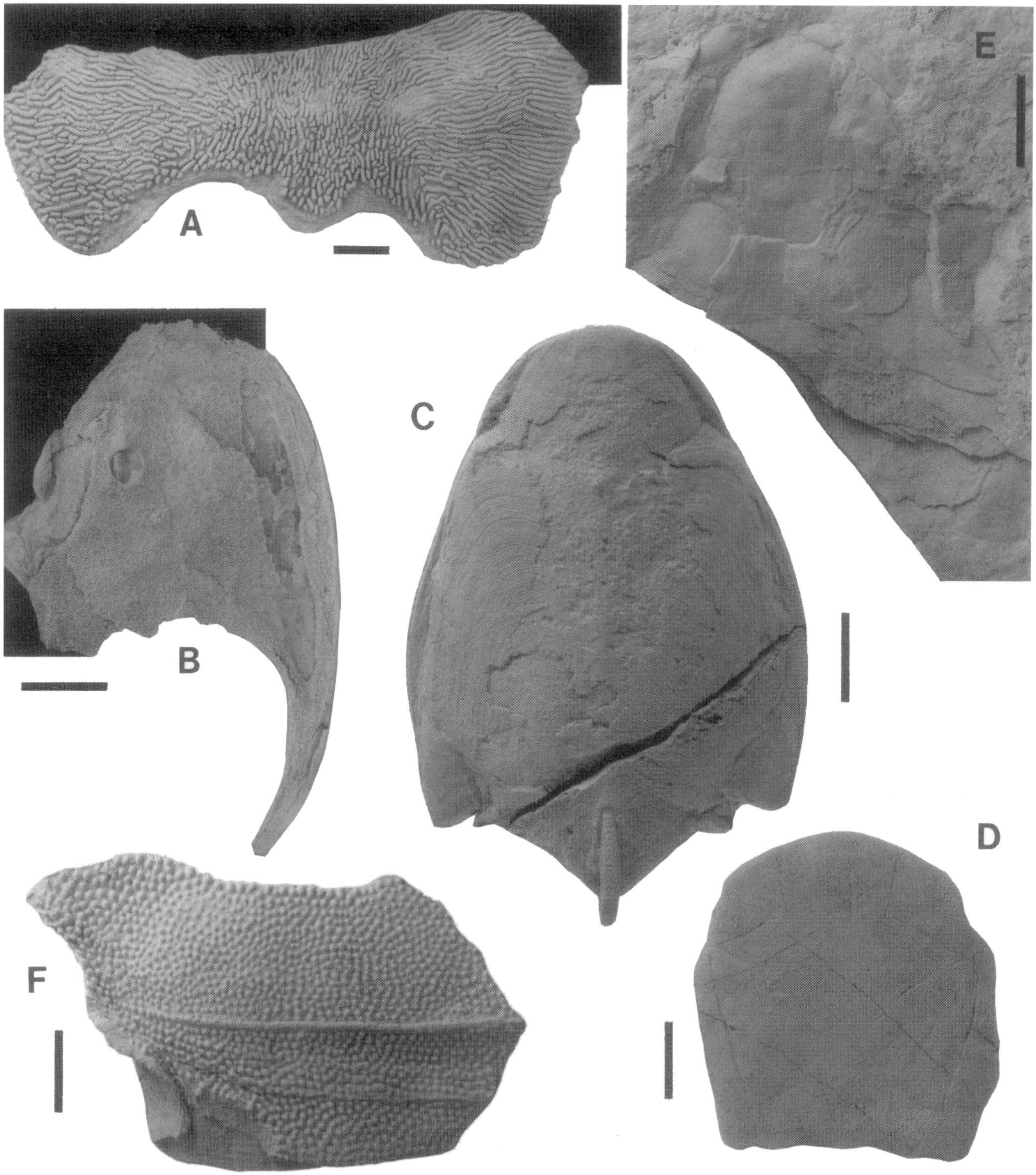

Figure 3. Representative Devonian vertebrates from the western United States. Lower Devonian: A, *Cardipeltis* sp., posterior part of the dorsal shield from the lower Grassy Flat Member of the Water Canyon Formation; B, *Cephalaspis utahensis*, headshield, from the lower Grassy Flat Member; C, *Tuberculaspis elyensis*, head shield, from the Sevy Dolomite; D, *Bryantolepis* sp., skull, from the lower Grassy Flat Member of the Water Canyon Formation. Middle Devonian: E, Osteolepid skull, from the upper Grassy Flat Member of the Water Canyon Formation; F, *Asterolepis* sp., anterior dorsolateral, from the upper Grassy Flat Member. Scale bars are 1 cm.

sandstones and siltstones (Dehler, 1995), which have yielded an actinolepid arthrodire and two new pteraspidids; these provide no help with dating. The microvertebrates, however, include *Nikolivia elongata* and *Turinia pagei* together with tesserate heterostracan elements similar to *Aporemaspis pholidata* and *Lepidaspis serrata* (Elliott et al., 1998). *Nikolivia elongata* was first described from the Lochkovian of the Baltic (Karatajute-Talimaa, 1978) and has since been identified from rocks of Lochkovian age in northern France (Goujet and Blieck, 1979) and Belgium (Blieck and Goujet, 1991). Similar forms have also been reported from the Lochkovian of the Canadian Arctic (Vieth, 1980), the Mackenzie Mountains (Dineley and Loeffler, 1976), and the Windmill Limestone of Nevada (Turner and Murphy, 1988), where they occur with *delta* Zone conodonts. *Turinia pagei* is a Lochkovian indicator in western Europe (Turner, 1973); it has also been reported from Australia (Turner et al., 1981), Spitsbergen (Goujet and Blieck, 1977; Blieck and Goujet, 1978; Karatajute-Talimaa, 1978), and northern France (Goujet and Blieck, 1979). The tesserate heterostracans are known from the Lower Devonian of Arctic Canada (Elliott and Loeffler, 1989) and the Delorme Formation of the Mackenzie Mountains (Dineley and Loeffler, 1976). Similar forms were illustrated by Turner and Murphy (1988) from the Windmill Limestone of Nevada, where they occur in the *delta* Zone. The microvertebrates, therefore, strongly suggest a Lochkovian age, particularly because of similarity to the microvertebrate fauna of the Windmill Limestone (Elliott et al., 1998).

The Beartooth Butte Formation (Fig. 1) was named (Dorf, 1934) for estuarine channel-fill deposits of reddish limestone and limestone conglomerate on the east and west faces of Beartooth Butte, northern Wyoming. At the type locality the channel cuts into the Ordovician Bighorn Dolomite and is overlain by the Jefferson Dolomite. Sandberg (1961) reported similar localities from northern Wyoming and southwestern Montana; the most important is the Cottonwood Canyon channel in the Bighorn Mountains of northern Wyoming, from whence a typical Lower Devonian vertebrate fauna has been described by Bryant and Denison (see references in Elliott and Ilyes, 1996b). Although material was described from the two main localities, it was considered that both were contemporaneous and Pragian in age (Johnson et al., 1989). However, spore determinations indicate that the Beartooth Butte locality is mid-late Emsian in age, whereas the Cottonwood Canyon locality is late Lochkovian to early Pragian (McGregor, in Elliott and Ilyes, 1996b; Tanner, 1983). The vertebrate faunas from the two localities show little in common (Fig. 2) beyond the presence of *Protaspis* and *Cardipeltis*, although the species differ. Elliott and Ilyes (1996b) used the apparent similarity between the Water Canyon and Beartooth Butte type section faunas to suggest they were both Emsian. New information on the arthrodires from Cottonwood Canyon (D. Goujet, personal communication, 1995) indicates that both *Simblaspis* and *Aethaspis* are present, providing a connection at the generic level with the lower part of the Grassy Flat Member of the Water Canyon Formation and weakening the previous correlation; they are nevertheless endemic genera and almost nothing is known of their stratigraphic range. Because the same two species of vertebrates occur in both the lower part of the Grassy Flat Member of the Water Canyon Formation and in the Beartooth Butte Formation at Beartooth Butte (Fig. 2), whereas the similarity between the Water Canyon Formation and the Cottonwood Canyon section is only at the generic level, the original correlation is retained for the time being.

Problems thus remain with the Lower Devonian correlations: poor age constraint on the fossils, similar but not identical species occurring at various localities, and pronounced endemism in the western North American faunas. Moreover, the more northerly faunas, found in the Beartooth Butte and Water Canyon Formations and characterized by *Cardipeltis* and pteraspidids of the *Protaspis* type, contrast with the more southerly cyathaspidid fauna characterizing the Sevy and Lost Burro Formations (Fig. 2). It is suggested that these problems may be reduced eventually by development of a thelodont zonal scheme linked to the conodont standard.

MIDDLE DEVONIAN FORMATIONS

In western North America, vertebrates are locally abundant in channel fills in Middle Devonian formations, thus paralleling the Early Devonian situation. Conodont and spore determinations from some of the localities enable some precision to be placed on the scheme presented here, although knowledge of the range of some of the species is lacking, which may result in the need for future revision. The discussion in the present chapter is limited to Givetian fossil assemblages, although the presence of an Eifelian fauna is noted in Figure 4.

The Yahatinda Formation in Alberta (Fig. 1) was named (Aitken, 1966) for a series of continental to marine-littoral beds overlying irregular topography developed on a profound regional unconformity. It exhibits distinct facies laterally and vertically. The "nonchannel" facies is characteristic of the thinner, more laterally persistent parts of the formation; it consists of breccias and conglomerates overlain by dolomites and siltstones. The "channel" facies occurs in deeper portions of the more pronounced channels; it consists mostly of purple-red to pink dolomite and dolosiltite. Plant and fish material has been collected from the dolomitic siltstones, and although plant and spore determinations have been published (Greggs et al., 1962; McGregor, 1963, 1964), only passing mention has been made of fish. The latter (Fig. 4) include the pteraspidid *"Psephaspis,"* a large tuberculated form previously reported from the Lemhi Range and the Water Canyon Formation, the antiarch placoderm *Asterolepis*, the arthrodire *Holonema*, and a sarcopterygian that may be *Eusthenopteron*. Because of precise palynological dating, the fauna is important; spore and macroplant determinations enable chronologic alignment to be made with similar vertebrate faunas to the south.

The Yahatinda Formation is unconformably overlain by the Flume Formation or the Flume Member of the Cairn Formation.

	Mountain Springs Fm. Virgin Mts. Nevada	Denay Lst. Red Hill Nevada	Denay Lst. Roberts Mts. Nevada	Upper Grassy Flat Mbr. Water Canyon Fm. N. Utah	Spring Mt. Channel Lemhi Range Idaho	Yahatinda Fm. Alberta Canada
Heterostraci						
Pteraspididae				*Psephaspis williamsi* "*Psephaspis*" n.sp.	*Psephaspis idahoensis* "*Psephaspis*" n. sp.	"*Psephaspis*" 2 spp.
Placodermi						
Arthrodira	*Holonema*		*Holonema* Pachyosteina Dinichthyidae *Coccosteus*		*Holonema haiti*	*Holonema* Arthrodira indet.
Antiarcha	*Asterolepis*	*Asterolepis*	*Asterolepis*	*Asterolepis*	*Asterolepis*	*Asterolepis*
Ptyctodontida			*Ptyctodus* *Rhynchodus*			
Petalichthyida			*Macropetalichthys*			
Osteichthyes						
Actinopterygii		*Cheirolepis*	*Cheirolepis*			
Sarcopterygii						
Osteolepiformes	Osteolepiform	*Eusthenopteron*		Osteolepiform		*?Eusthenopteron*
Porolepiformes		*?Glyptolepis*				
Onychodontida			*Onychodus*			
Dipnoi		dipnoan		dipnoan		
Acanthodii		*Persacanthus*	*Persacanthus*			
Chondrichthyes						
Holocephali			*Synthetodus?*		Holocephali?	
incertae sedis			*Petalodontida*			

Figure 4. Vertebrate faunal lists for Middle Devonian horizons. Note that the Eifelian faunas are not discussed in the text.

As the transgressive Flume carbonates have been dated as latest Givetian (*norrisi* conodont Zone) to earliest Frasnian (Lower *asymmetrica* Zone; Geldsetzer and Mallamo, 1991; Uyeno, 1991), it is reasonable to assume that the Yahatinda Formation cannot be younger than latest Givetian. Geldsetzer and Mallamo (1991) concluded on paleogeographic grounds that the Yahatinda channels are remnants of the peritidal belt represented by the Gilwood Member of the Watt Mountain Formation around the Peace River Arch to the north. They inferred a date of late middle and late *varcus* Zone or mid- to early late Givetian for the Yahatinda Formation.

Plant macrofossils were described by Greggs et al. (1962) and figured by McGregor (1963). Scheckler (1978) described *Archaeopteris macilesta*, *Archaeopteris* sp., and *Callixylon* sp. from the same section. Although *Archaeopteris* has generally been reported from strata that are no older than Frasnian, not all the datings are securely based or related to a specified Givetian-Frasnian boundary level (C. McGregor, personal communication, 1992); some "early" Frasnian *Archaeopteris* occurrences may be late Givetian. Spores were described from the formation and have recently been reassessed (McGregor, 1963, 1964, and personal communication, 1992). One of the most abundant spores, *Aneurospora greggsii*, ranges from late Givetian through the Famennian in Canada, although it is most abundant in the late Givetian to very early Frasnian of the Queen Elizabeth Islands. Associated spores are consistent with such an age.

In summary, paleontological and stratigraphic evidence points to a late Givetian age for the Yahatinda Formation. In the spore zonal scheme it correlates with the *obtivus-triangulatus* assemblage Zone, whereas in the conodont zonal scheme it probably falls within the range circumscribed by the Upper *varcus* Subzone to *norrisi* Zone (C. McGregor, personal communication, 1992).

A channel fill in the Lemhi Range, central Idaho (Fig. 1), is referred to here as the Spring Mountain Channel. This large (700-m-wide) channel is cut into the Ordovician Fish Haven Dolomite and Silurian Laketown Dolomite. A vertebrate fauna from this locality (Denison (1968b) includes *Psephaspis idahoensis*, *Holonema haiti*, and *Asterolepis*; our collections confirm this faunal list and indicate that a second species of tuberculate pteraspidid is present (Fig. 4). Because the pteraspidid *Psephaspis* was considered to be restricted to the Lower Devonian and the age range of the placoderms was poorly constrained, Johnson and Sandberg (1977) proposed a lower Devonian (Pragian) age for these strata and included them in the Beartooth Butte Formation. It is now clear from the Yahatinda Formation that pteraspidids range to the top of the Middle Devonian. *Holonema* is a widespread genus but ranges from the Givetian to the Frasnian in the United States, whereas *Asterolepis* is confined

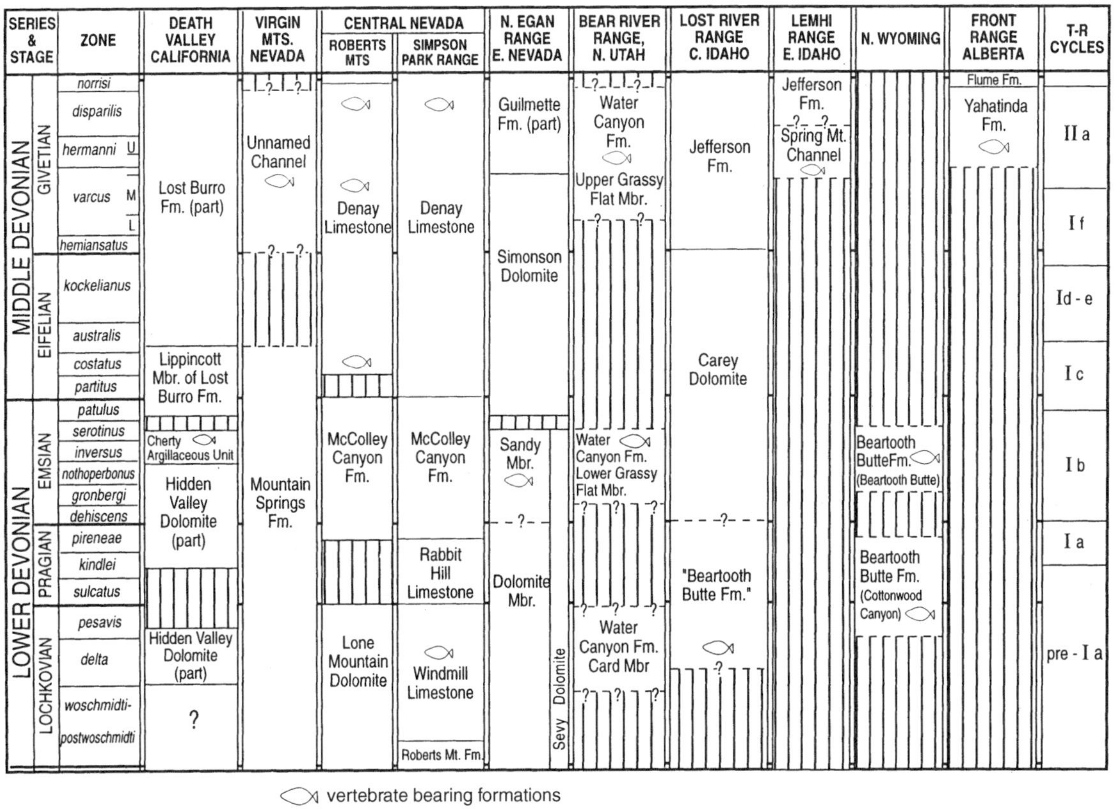

Figure 5. Proposed stratigraphic correlation of vertebrate-bearing formations in the Lower and Middle Devonian of western North America. Note that the vertical lines on the figure indicate periods of time not represented by sediments.

to the Middle Devonian, has a worldwide distribution, and has been reported from the upper Givetian of Nevada (Murphy et al., 1976). Based on similarities between the Yahatinda Formation and Spring Mountain Channel faunas and improving knowledge of the ranges of some of the vertebrates, we propose a Late Givetian age for the Spring Mountain Channel (Fig. 5).

The Grassy Flat Member of the Water Canyon Formation in northern Utah has been correlated with the Lower Devonian Beartooth Butte Formation in the past (Poole et al., 1977). Revisions of the faunal lists (Elliott and Ilyes, 1996b) and new collections indicate a fauna in the upper part of the member that is similar to that of the Spring Mountain Channel and the Yahatinda Formation (Fig. 4). Ørvig (1961) described *Psephaspis williamsi* from the upper Grassy Flat Member and noted the presence of euarthrodires, antiarchs, and sarcopterygians in his collection. An *Asterolepis* anterior dorsolateral plate collected by C. A. Sandberg from the upper Grassy Flat Member at Logan Canyon (Fig. 3F) verifies the presence of antiarchs. Our discovery of articulated osteolepids (Fig. 3E) in the same area shows that "crossopterygians" are also present, and as osteolepids originate in the Middle Devonian, they accord with the Givetian age we propose for the upper part of the Grassy Flat Member. At present, however, the fauna from this member is sparse; further collecting is needed for increased precision.

The Red Hill and Roberts Mountains localities in central Nevada (Fig. 1) are both in the marine Denay Limestone. The Roberts Mountains localities have yielded various vertebrates (Fig. 4) including *Holonema* and *Asterolepis*; the latter also occurs with *Eusthenopteron* in the Red Hill locality. In the southern Roberts Mountains, *Holonema* occurs at the top of the middle Denay Limestone, associated with conodonts of the Middle *varcus* Subzone (Morgan, 1980). *Asterolepis* has been found and described from the northern Roberts Mountains and Red Hill in the Simpson Park Range to the northwest (Murphy et al., 1976; Morgan, 1980; Gregory et al., 1977) but not specifically assigned. Conodonts of the *disparilis* Zone are present (Ziegler et al., 1976) 2.4 m below the fish bed at Red Hill. The occurrence of placoderms in both marine and freshwater sediments indicates their potential value for establishing correlations of nonmarine occurrences with the standard conodont succession.

The Mountain Springs Formation in southeastern Nevada (Fig. 1), resting unconformably on the Cambrian Nopah Formation, has been assigned to the Lower Ordovician to Middle Devonian (Rowlands et al., 1990). The conodonts *Icriodus* sp., *Polygnathus* sp., *Hindeodella* sp., *Spathognathodus* sp., and *Tri-*

chonodella sp. in the Mountain Springs section indicate an age no older than late Early Devonian or possibly early Middle Devonian (Miller and Zilinsky, 1981). An isolated channel fill near the top of the formation in the Virgin Mountains contains a fragmentary vertebrate fauna, including *Holonema*, *Asterolepis* and "crossopterygians" (H. G. J., unpublished data); we place this horizon tentatively in the Givetian (Fig. 5).

Although these correlations (Fig. 5) are in some cases rather tentative we think that the vertebrates can be used effectively for correlation in the Middle Devonian.

CONCLUSION

Because of their tendency to be endemic and to be environmentally restricted, vertebrates pose problems for use as biostratigraphic indicators. However, they can be useful locally in situations in which other biostratigraphic indicators are not available. This study shows that the extensive Lower Devonian vertebrate faunas provide the basis for the development of a scheme for correlating nonmarine strata. There are prospects for tying this to the conodont zonal scheme by continued study of microvertebrates, particularly thelodonts. It is also shown that northern and southern endemic vertebrate faunas are present in the Lower Devonian in the western United States, although the reasons for this differentiation are not currently understood.

Middle Devonian faunas are more cosmopolitan. Pteraspidids, represented by *Psephaspis* and other tuberculate species, continued to the uppermost Middle Devonian, and together with a characteristic fauna of placoderms and sarcopterygians, they can be used to correlate nonmarine vertebrate-bearing strata from Alberta to Nevada. The presence of antiarchs in both marine and nonmarine strata enables a link to be made with the standard zonal scheme.

ACKNOWLEDGMENTS

G. Klapper, C. A. Sandberg, D. Goujet, C. McGregor, R. Reed, and J. Evans gave valuable assistance, M. Wilson and A. Blieck provided helpful comments on the manuscript, and R. O'Reilly drafted four of the figures.

REFERENCES CITED

Aitken, J. D., 1966, Sub-Fairholme Devonian rocks of the eastern Front Ranges, southern Rocky Mountains, Alberta: Geological Survey of Canada Paper 64-33, 88 p.

Blieck, A., and Goujet, D., 1978, A propos de nouveaux materiel de thélodontes (Vertébrés Agnathes) d'Iran et de Thaïlande: Aperçu sur la repartition géographique et stratigraphique des agnathes des "regions gondwaniennes" au Paléozoique moyen: Annales de la Société Géologique du Nord, v. 97, p. 363–372. [imprint 1977]

Blieck, A., and Goujet, D., 1991, Les Vertébrés du Dévonien inférieur d'Arville et de Nonceveux (Belgique): Annales de la Société Géologique du Nord, 2e ser.,v. 1, p. 67–78.

Blieck, A., Goujet, D., and Janvier, P., 1987, The vertebrate stratigraphy of the Lower Devonian (Red Bay Group and Wood Bay Formation) of Spitsbergen: Modern Geology, v. 11, p. 197–217.

Blieck, A., Mark-Kurik, E., and Marss, T., 1989, Biostratigraphical correlations between Siluro-Devonian invertebrate-dominated and vertebrate-dominated sequences: The East Baltic example, in McMillan, N. J., Embry, A. F., and Glass, D. J., eds., Devonian of the world: Calgary, Canadian Society of Petroleum Geologists Memoir 14, v. 3, p. 579–587 [imprint 1988].

Branson, E. B., and Mehl, M. G., 1931, Fishes of the Jefferson Formation of Utah: Journal of Geology, v. 39, p. 509–531.

Bryant, W. L., 1932, Lower Devonian fishes of Beartooth Butte, Wyoming: American Philosophical Society Proceedings, v. 71, p. 225–254.

Bryant, W. L., 1933, The fish fauna of Beartooth Butte, Wyoming; Part I: American Philosophical Society Proceedings, v. 72, p. 285–314.

Bryant, W. L., 1934, The fish fauna of Beartooth Butte, Wyoming; Parts II and III: American Philosophical Society Proceedings, v. 73, p.127–162.

Bryant, W. L., 1935, *Cryptaspis* and other Lower Devonian fossil fishes from Beartooth Butte Formation, Wyoming: Proceedings of the American Philosophical Society, v. 75, p. 111–128.

Dehler, C. M., 1995, Paleoenvironmental and stratigraphic analysis of Lower Devonian Beartooth Butte Formation and associated strata, central-eastern Idaho [M.S. thesis]: Flagstaff, Northern Arizona University, 155 p.

Denison, R. H., 1952, Early Devonian fishes from Utah; Part I. Osteostraci: Fieldiana: Geology, v. 11, p. 265–287.

Denison, R. H., 1953, Early Devonian fishes from Utah; Part II. Heterostraci: Fieldiana: Geology, v. 11, p. 291–355.

Denison, R. H., 1966, *Cardipeltis*, an Early Devonian agnathan of the Order Heterostraci: Fieldiana: Geology, v. 16, p. 89–116.

Denison, R. H., 1967, A new *Protaspis* from the Devonian of Utah, with notes on the classification of the Pteraspididae: Journal of the Linnean Society of London (Zoology), v. 47, p. 31–37.

Denison, R. H., 1968a, Early Devonian lungfishes from Wyoming, Utah, and Idaho: Fieldiana: Geology, v. 17, p. 353–413.

Denison, R. H., 1968b, Middle Devonian fishes from the Lemhi Range of Idaho: Fieldiana: Geology, v. 16, p. 269–288.

Denison, R. H., 1970, A revised classificiation of Pteraspididae with description of new forms from Wyoming: Fieldiana: Geology, v. 20, p.1–41.

Dineley, D. L., and Loeffler, E. J., 1976, Ostracoderm faunas of the Delorme and associated Siluro-Devonian formations, Northwest Territories, Canada: Special Papers in Palaeontology 18, 214 p.

Donovan, R. N., Foster, R. J., and Westoll, T. S., 1974, A stratigraphical revision of the Old Red Sandstone of north-eastern Caithness: Transactions of the Royal Society of Edinburgh, v. 69, p. 167–201.

Dorf, E., 1934, Lower Devonian flora from Beartooth Butte, Wyoming: Geological Society of America Bulletin, v. 45, p. 425–440.

Elliott, D. K., 1984, Siluro-Devonian fish biostratigraphy of the Canadian Arctic Islands: Proceedings of the Linnean Society of New South Wales, v. 107, p. 197–209.

Elliott, D. K., 1994, A new pteraspidid (Agnatha: Heterostraci) from the Lower Devonian Water Canyon Formation of Utah: Journal of Paleontology, v. 68, p. 176–179.

Elliott, D. K., and Ilyes, R. R., 1996a, New Early Devonian pteraspidids (Agnatha, Heterostraci) from Death Valley National Monument, southeastern California: Journal of Paleontology, v. 70, p. 152–161.

Elliott, D. K., and Ilyes, R. R., 1996b, Lower Devonian vertebrate biostratigraphy of the western United States: Modern Geology, v. 20, p. 253–262.

Elliott, D. K., and Loeffler, E. J., 1989, A new agnathan from the Lower Devonian of Arctic Canada, and a review of the tessellated heterostracans: Palaeontology, v. 32, p. 883-891.

Elliott, D. K., Dehler, C. M., and Evans, J., 1998, Toward the development of a Lower Devonian microvertebrate marine-nonmarine biostratigraphic zonation of the western United States: Bulletin de la Société Géologique de France.

Geldsetzer, H. H. J., and Mallamo, M. P., 1991, The Devonian of the southern Rocky Mountains: the Canmore area, in Smith, P.L., ed., A field guide to the paleontology of southwestern Canada: The First Canadian Paleontology Conference, University of British Columbia; Vancouver, p. 83-102.

Goujet, D., and Blieck, A., 1977, La faune de vertébrés de l'horizon "Vogti" (Groupe de Red Bay, Spitsberg). Comparaison avec les autres ichthyologiques du Dévonien inférieur européen: Academy des Sciences, Comptes Rendus Hebdomadaires des Sciences, Serie D, Sciences Naturelles, Paris, v. 284, p. 1513–1515.

Goujet, D., and Blieck, A., 1979, Les vertébrés de l'Assise des Schistes et Grés de Pernes (Dévonien du nord de la France): Annales de la Société Géologique du Nord, v. 98, p. 263-278. [imprint 1978]

Greggs, R. G., McGregor, D. C., and Rouse, G. E., 1962, Devonian plants from the type section of the Ghost River Formation of western Alberta: Science, v. 35, p. 930–931.

Gregory, J. T., Morgan, T. G., and Reed, J. W., 1977, Devonian fishes in central Nevada, in Murphy, M. A., Berry, W. B. N., and Sandberg, C. A., eds., Western North America: Devonian: University of California, Riverside Campus Museum Contribution 4, p. 112–120.

Ilyes, R.R., 1993, New Early Devonian pteraspids (Agnatha, Heterostraci) from Nevada and California [Ph.D. thesis]: Flagstaff, Northern Arizona University, 156 p.

Ilyes, R. R., and Elliott, D. K., 1994, New Early Devonian pteraspidids (Agnatha, Heterostraci) from east-central Nevada: Journal of Paleontology, v. 68, p. 878–892.

Janvier, P. and Blieck, A., 1993, The Silurian-Devonian agnathan biostratigraphy of the Old Red Continent, in Long, J. A., ed., Palaeozoic vertebrate biostratigraphy and biogeography: London, Belhaven Press, p. 67–86.

Jarvik, E., 1961, Devonian Vertebrates, in Raasch, G.O., ed., Geology of the Arctic: Toronto, University of Toronto Press, v.1, p. 197-200.

Johnson, J.G., 1986, Revision of Lower Devonian (Emsian) brachiopod biostratigraphy and biogeography, central Nevada: Journal of Paleontology, v.60, p. 825-844.

Johnson, J.G., and Murphy, M.A., 1984, Time-rock model for Siluro-Devonian continental shelf, western United States. Geological Society of America Bulletin, v. 95, p. 1349-1359.

Johnson, J. G., and Sandberg, C. A., 1977, Lower and Middle Devonian continental-shelf rocks of the western United States, in Murphy, M. A., Berry, W. B. N., and Sandberg, C. A., eds., Western North America: Devonian: University of California, Riverside Campus Museum Contribution 4, p. 121–143.

Johnson, J. G., and Sandberg, C. A., 1989, Devonian eustatic events in the western United States and their biostratigraphic responses, in McMillan, N. J., Embry, D. J., and Glass, eds., Devonian of the World: Calgary, Canadian Society of Petroleum Geologists Memoir 14, v. 3, p. 171–178. [imprint 1988]

Johnson, J. G., Klapper, G., and Trojan, W. R., 1980, Brachiopod and conodont successions in the Devonian of the northern Antelope Range, central Nevada: Geologica et Palaeontologica, v. 14, p. 77–116.

Johnson, J. G., Klapper, G., and Sandberg, C. A., 1985, Devonian eustatic fluctuations in Euramerica: Geological Society of America Bulletin, v. 96, p. 567–587.

Johnson, J. G., Sandberg, C. A., and Poole, F. G., 1989, Early and Middle Devonian paleogeography of the western United States, in McMillan, N. J., Embry, A. F., and Glass, D. J., eds., Devonian of the world: Calgary, Canadian Society of Petroleum Geologists Memoir 14, v. 1, p. 161–182. [imprint 1988]

Johnson, J. G., Sandberg, C. A., and Poole, F. G., 1991, Devonian lithofacies of western United States, in Cooper, J. D., and Stevens, C. H., eds., Paleozoic paleogeography of the western United States-II: Society of Economic Paleontologists and Mineralogists, Pacific Section, v. 67, p. 83–105.

Karatajute-Talimaa, V. N., 1978, Telodonti silura i devona SSSR i Shpitsbergena: Vilnius, Mosklas, 334 p. [In Russian].

Long, J. A., 1993, Morphological characteristics of Palaeozoic vertebrates used in Biostratigraphy, in Long, J. A., ed., Palaeozoic vertebrate biostratigraphy and biogeography: London, Belhaven Press, p. 3–24.

McAllister, J. F., 1952, Rocks and structure of the Quartz Spring area, northern Panamint Range, California: California Division of Mines Special Report 25, 38 p.

McAllister, J. F., 1955, Geology of the mineral deposits in the Ubehebe Peak Quadrangle, Inyo County, California: California Division of Mines Special Report 42, 63 p.

McGregor, D. C., 1963, Palaeobotanical evidence for the age of basal Devonian strata at Ghost River, Alberta: Bulletin of Canadian Petroleum Geology, v. 11, p. 299–303.

McGregor, D. C., 1964, Devonian miospores from the Ghost River Formation, Alberta: Geological Survey of Canada Bulletin, v. 109, 31 p.

Miller, R. H., and Zilinsky, G. A., 1981, Lower Ordovician through Lower Devonian cratonic margin rocks of the southern Great Basin: Geological Society of America Bulletin, v. 92, p. 255–261.

Morgan, T. G., 1980, The Middle Devonian fish faunas of central Nevada [Ph.D. thesis]: Berkeley, University of California, 250 p.

Murphy, M. A., Morgan, T. G., and Dineley, D. L., 1976, *Astrolepis* sp. from the Upper Devonian of central Nevada: Journal of Paleontology, v. 50, p. 467–471.

Ørvig, T., 1961, Notes on some early representatives of the Drepanaspida (Pteraspidomorphi, Heterostraci): Arkiv för Zoologi, v. 12, p. 515–535.

Osmond, J. C., 1954, Dolomites in the Silurian and Devonian of east-central Nevada: American Association of Petroleum Geologists Bulletin, v.38, p. 1911–1956.

Osmond, J. C., 1962, Stratigraphy of Devonian Sevy Dolomite in Utah and Nevada: American Association of Petroleum Geologists Bulletin, v. 46, p. 2033–2056.

Poole, F. G., Sandberg, C. A., and Boucot, A. J., 1977, Silurian and Devonian Paleogeography of the western United States, in Stewart, J. H., Stevens, C. H., and Fritsche, A. E., eds., Paleozoic paleogeography of the western United States, Pacific Coast Paleogeography Symposium 1: Los Angeles, Society of Economic Paleontologists and Mineralogists, Pacific Section, p. 39–65.

Rowlands, S. M., Parolini, J. R., Eschner, E., and McAllister, A. J., 1990, Sedimentologic and stratigraphic constraints on the Neogene translation and rotation of the Frenchman Mountain structural block, Clark County, Nevada: Geological Society of America Memoir 176, p. 99–122.

Sandberg, C. A., 1961, Widespread Beartooth Butte Formation of Early Devonian age in Montana and Wyoming and its paleogeographic significance: American Association of Petroleum Geologists Bulletin, v. 45, p. 1301–1309.

Scheckler, S. A., 1978, Ontogeny of progymnosperms. II: Shoots of Upper Devonian Archaeopteridales: Canadian Journal of Botany, v. 56, p. 3136–3170.

Tanner, W., 1983, A fossil flora from the Beartooth Butte Formation of northern Wyoming [Ph.D. thesis]: Carbondale, Southern Illinois University, 208 p.

Turner, S., 1973. Siluro-Devonian thelodonts from the Welsh borderland: Quarterly Journal of the Geological Society of London, v. 129, p. 557–581.

Turner, S., 1993, Palaeozoic microvertebrate biostratigraphy of eastern Gondwana, in Long, J. A., ed., Palaeozoic vertebrate biostratigraphy and biogeography: London, Belhaven Press, p. 174–207.

Turner, S., and Murphy, M. A., 1988, Early Devonian vertebrate microfossils from the Simpson Park Range, Eureka County, Nevada: Journal of Paleontology, v. 62, p. 959–964.

Turner, S., Jones, P. J., and Draper, J. J., 1981, Early Devonian thelodonts (Agnatha) from the Toko Syncline, western Queensland and a review of other Australian discoveries: Bureau of Mineral Resources Journal of Australian Geology and Geophysics, v. 6, p. 51–69.

Uyeno, T. T., 1991, Pre-Famennian Devonian conodont biostratigrapy of selected intervals in the eastern Canadian Cordillera, in Orchard, M.J., and McCracken, A.D., eds., Ordovician to Triassic conodont paleontology of the Canadian Cordillera: Geological Survey of Canada Bulletin 417, p. 129–161.

Vieth, J., 1980, Thelodontier-, Acanthodier- und Elasmobranchier-Schuppen aus dem Unter-Devon der Kanadischen Arktis (Agnatha, Pisces): Göttinger Arbeiten zur Geologie und Paläontologie, v. 23, 69 p.

Wang, S.-T., 1993, Vertebrate biostratigraphy of the Middle Palaeozoic of

China, in Long, J. A., ed., Palaeozoic vertebrate biostratigraphy and biogeography: London, Belhaven Press, p. 252–276.

Williams, J. S., 1948, Geology of the Paleozoic rocks, Logan Quadrangle, Utah: Geological Society of America Bulletin, v. 59, p. 1121–1164.

Williams, J. S., and Taylor, M. E., 1964, The Lower Devonian Water Canyon Formation of northern Utah: Contributions to Geology of the University of Wyoming, v. 3, p. 38–53.

Young, G. C., 1993, Middle Palaeozoic macrovertebrate biostratigraphy of eastern Gondwana, in Long, J. A., ed., Palaeozoic vertebrate biostratigraphy and biogeography: London, Belhaven Press, p. 208–251.

Ziegler, W., Klapper, G., and Johnson, J. G., 1976, Redefinition and subdivision of the varcus Zone (conodonts, Middle-?Upper Devonian) in Europe and North America: Geologica et Palaeontologica, v. 10, p. 109–140.

MANUSCRIPT ACCEPTED BY THE SOCIETY FEBRUARY 27, 1997

Printed in U.S.A.

Geological Society of America
Special Paper 321
1997

Famennian-Tournaisian conodonts and Devonian–Early Carboniferous transgressions and regressions in northeastern Australia

Ruth Mawson and John A. Talent
Centre for Ecostratigraphy and Palaeobiology, School of Earth Sciences, Macquarie University, N.S.W. 2109, Australia

ABSTRACT

Conodont data are presented from stratigraphic sections and spot samples from the Keelbottom Group (Famennian-Tournaisian) of the Burdekin Basin and the Venetia Formation (early Tournaisian) of the Clarke River Basin of the Townsville hinterland of northeastern Australia. Though much of the Devonian–Early Carboniferous sedimentary sequences of this region is nonmarine, the marine intervals document at least eight transgression-regression (T-R) cycles in the Burdekin Basin and adjacent areas of onlap during the late Emsian–Tournaisian. Three (?four) of these are suggested to be reflections of global transgressions rather than regional tectonically driven events: in the Early *marginifera* and *expansa* zones (Famennian), in the *sandbergi*–Early *crenulata* interval, and seemingly also early in the *sulcata* Zone (Tournaisian, but perhaps commencing late in the Late *praesulcata* Zone). The eighth transgression event, in the upper Hardwick Formation, is not chronologically constrained; it too may have been eustatically driven. The *sandbergi*–Early *crenulata* transgressive interval is clearly expressed in the Venetia Formation of the Clarke River Basin. The major marine event expressed as the Teddy Mountain Formation in the Broken River Province is assumed to align with transgression events 7 and 8, but available data are not compelling.

T-R events additional to the above may be deciphered from the pre-Emsian stratigraphy of the Broken River Province: an early Lochkovian regression event (following the upper limestone "member" of the Jack Formation), plus two more T-R cycles inferred from an unnamed post-Jack unit with allochthonous carbonates and from the three members of the Shield Creek Formation (late Lochkovian–early Pragian). These pre-Emsian events do not coincide with Euramerican T-R events; they are accordingly assumed to have been tectonically driven.

A major early Frasnian regression, inferred to have been tectonically driven (?Tabberabberan Orogeny), is reflected in the Mytton Formation–Bulgeri Formation unconformity and associated dramatic change from shallow marine to fluvial environments in the Broken River Province. This event is presumed to have been contemporaneous with the change from marine Cultivation Gully Formation to nonmarine conditions low in the Vanneck Formation of the Burdekin Basin.

The Early *marginifera* Zone conodont faunas of the Myrtlevale Formation, dominated by species of *Polygnathus* associated with minor palmatolepids, seemingly

Mawson, R., and Talent, J. A., 1997, Famennian-Tournaisian conodonts and Devonian–Early Carboniferous transgressions and regressions in northeastern Australia, *in* Klapper, G., Murphy, M. A., and Talent, J. A., eds., Paleozoic Sequence Stratigraphy, Biostratigraphy, and Biogeography: Studies in Honor of J. Granville ("Jess") Johnson: Boulder, Colorado, Geological Society of America Special Paper 321.

represented the deepest Late Devonian–Tournaisian environments of the region, perhaps "outer shelf" environments. Carbonates from the suggested shallowest marine environments for the late Famennian–Tournaisian interval, exemplified by the Teddy Mountain Formation of the Broken River Province, have produced only bispathodid conodonts. More diverse late Famennian (*expansa* Zone) and Tournaisian faunas include *Polygnathus* and/or siphonodellid conodonts; these are inferred to have been deeper water than contemporaneous faunas (specifically Tournaisian) from the Broken River Province. None of the conodont-bearing Late Devonian–Tournaisian units of the Townsville hinterland had relatively pure palmatolepid or siphonodellid faunas that one might associate with truly pelagic contexts.

The conodont faunas of the Keelbottom Group (Famennian-Tournaisian) and the Venetia Formation (early Tournaisian) are documented. As a consequence of the Famennian correlations having been tied preferentially to palmatolepids, extensions of ranges are suggested for *Ancyrognathus* and for several species of icriodontids and *Polygnathus*. Four new species are proposed, one icriodontid and three polygnathids: *Icriodus scalariformis*, *Polygnathus costulifera*, *P. hieroglyphica*, and *P. wyatti*.

INTRODUCTION

Late Devonian as well as late Givetian and earliest Carboniferous sedimentary and volcanic sequences are widespread in eastern Australia (Fig. 1), but in contrast with the Canning Basin of northwestern Australia, there is a dearth of refined chronologic control. This is due largely to the predominance of nonmarine clastic sequences. This dearth of chronologic precision has inhibited precise dating of vertebrate faunas (De Pomeroy, 1995) and has impacted profoundly on interregional stratigraphic alignments. It should be noted nevertheless that conodont data have been presented for a few Late Devonian–Tournaisian marine horizons in eastern Australia: from the late Frasnian at Trotter Creek south of Mount Morgan in central Queensland (Druce, 1970b), from the late Famennian Mandowa Mudstone of northern New South Wales (Philip and Jackson, 1971; as "Luton Beds"), from Frasnian strata at Ettrema in southeastern New South Wales (Pickett, 1972), from spot samples from Famennian and Tournaisian horizons in the Townsville hinterland (Pickett, 1981), from allochthonous blocks in the Mostyn Vale Formation at Keepit, northern New South Wales (Wright et al., 1990), and from Tournaisian sequences in northern New South Wales and the Rockhampton region of Queensland (Mory and Crane, 1982), focused on the zonally pivotal genus *Siphonodella*. Minor contributions concern Late Devonian conodonts from cherts in the Hastings Block of northeastern New South Wales (Ishiga and Leitch, 1988), a Famennian fauna from the Baldwin Formation in the southern Tamworth Belt (Naka et al., 1988) and ?early Frasnian conodonts from the Broken River region of northeastern Queensland (Mawson and Talent, 1989; Sloan et al., 1995).

The present investigation came about because:

1. Devonian-Tournaisian stratigraphies recently elucidated for the Townsville hinterland (Figs. 1 and 2) of northeastern Australia display dramatic changes in sedimentation patterns through time (Mawson and Talent, 1989; Withnall, 1993; Withnall and Lang, 1993; Draper and Lang, 1994b; Sloan et al., 1995). It is tempting to align these interregionally, but to do so does not answer the inevitable question: how much of these patterns was tectonically driven and how much represented responses to global transgression/regression (T-R) events. We have elsewhere demonstrated that in the Broken River region a major regression occurred very early in the Frasnian (Mawson and Talent, 1989; Sloan et al., 1995) and that in the Burdekin Basin a minor deepening event in the late Givetian Cultivation Gully Formation was followed by change to nonmarine conditions basinwide (low in the Vannek Formation) (Talent and Mawson, 1994).

2. Previous work had demonstrated that for at least part of Famennian time, during accumulation of the Myrtlevale Formation, marine conditions had extended across what was formerly known as the "Star Basin," successor basin to the Burdekin Basin *sensu stricto* (Wyatt, 1968, 1973; Wyatt et al., 1968, 1970; McKellar, 1970; Wyatt and Jell, 1980; Pickett, 1981). Sampling we undertook at Mount Podge (Figs. 2, 3) in a previously unrecognized sequence of Myrtlevale Formation produced a surprising abundance and diversity of *Polygnathus* occurring in association with palmatolepids. The faunas include at least four new species, three of them polygnathids, as well as many forms not previously reported from Australia. Because of the dearth of well-documented Famennian sequences in which polygnathids had been recorded and tightly tied to the palmatolepid-based zonal scheme, the Mount Podge successions were deemed worthy of closer investigation. Other sections in the Townsville hinterland, potentially informative as regards chronology of Late Devonian–Tournaisian transgression/regression (T-R) events, were investigated, namely:

a. The Cultivation Gully Formation, Fanning River (Fig. 2).

b. The Myrtlevale, Lollypop, and Hardwick formations at Mount Podge (Fig. 3).

c. The Hardwick Formation at Corner Creek, a tiny tributary of the Star River (Fig. 2).

d. The Myrtlevale and Hardwick formations, including the

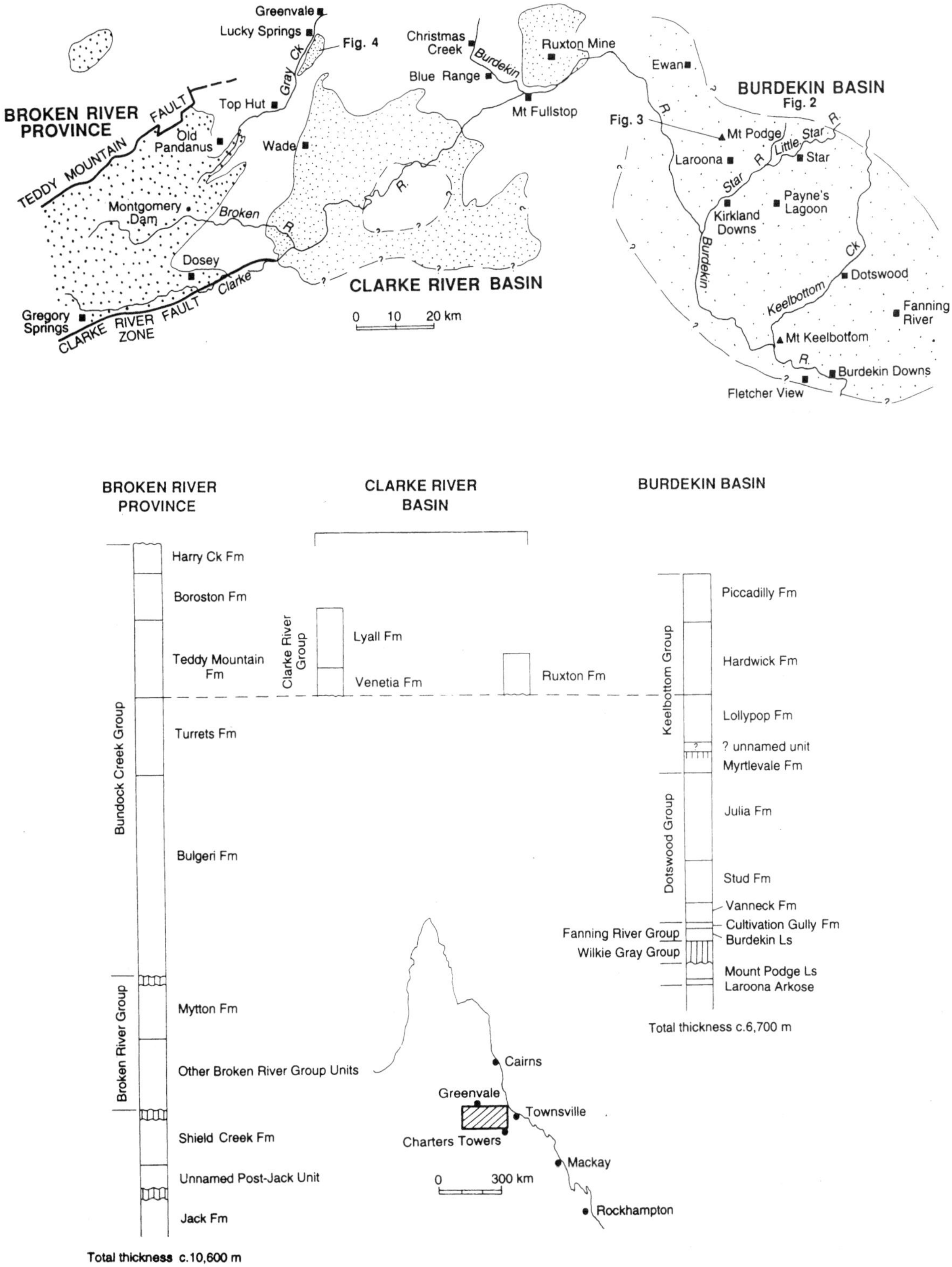

Figure 1. Late Devonian–Early Carboniferous sedimentary basins of the Townsville hinterland, northeastern Australia. Thicknesses of units indicated on the skeletal stratigraphic columns are not maximal but, in general, are based on quoted thicknesses on type sections (Withnall and Lang, 1993; Draper and Lang, 1994b). The alignment assumes a single transgressive event was responsible for initiation of the sedimentation (earliest Carboniferous, but possibly late in the Late *praesulcata* Zone) that gave rise to the Hardwick, Ruxton, Venetia, and Teddy Mountain Formations.

Saint Michael Member at Mount Saint Michael, and various localities on Dotswood Station (Fig. 2).

e. The Venetia Formation (Tournaisian) in the watershed of Big Stockyard Creek on Lucky Downs Station, south of Greenvale (Fig. 4).

f. The Teddy Mountain Formation of the Broken River Province.

Numerous spot samples from carbonates from various units of the Burdekin Basin and Broken River Province were leached in pursuit of improved data; many of these were supplied by Geological Survey of Queensland personnel mapping in the Townsville hinterland.

Our aims may be summarized as follows: (1) To seek increased chronologic precision for the transgressions and regressions, marine to nonmarine and vice versa, for the Famennian-Tournaisian of the Burdekin Basin and to consider these in the broader context of T-R events for the Devonian–Early Carboniferous of northeastern Australia; and (2) to determine to what extent these transgressions may have aligned with the T-R pattern presented by Johnson et al. (1985) for Euramerica.

STRATIGRAPHY

Burdekin Basin

Despite numerous brief notes (e.g. Leichhardt, 1847; Jack, 1879; see Draper and Lang, 1994b) on the stratigraphy of the region now referred to as the Burdekin Basin, little stratigraphic detail became available until the 1960s and 1970s (Wyatt et al., 1968, 1970; Wyatt, 1968, 1973, 1978; Wyatt and Jell, 1980). Mapping since the Wyatt era has resulted in modifications to stratigraphic nomenclature, with considerable attention being given to the changing pattern of sedimentary environments (Draper and Lang, 1994b).

The Devonian-Tournaisian interval is represented by four sequences in the Townsville hinterland (Fig. 5): the Wilkie Gray, Fanning River, Dotswood, and Keelbottom groups, the last of these spanning the Devonian-Carboniferous boundary. The first of these, the Wilkie Gray Group (Laroona Arkose plus Mount Podge Limestone), restricted in outcrop to the vicinity of Mount Podge, was formerly incorporated within the Fanning River Group (Zhen et al., 1993; Draper and Lang, 1994b; Cook, 1995). Because of a substantial diastem equating approximately with all of Eifelian time, expressed as an irregular erosional surface between the Mount Podge Limestone and the overlying Burdekin Limestone, Talent and Mawson (1994) excised the Wilkie Gray Group (= Laroona Arkose plus Mount Podge Limestone) from the Fanning River Group; they pointed out that the incursion connected with these units was from the north and that inception of the Burdekin Basin did not take place until subsequently, close to the Eifelian-Givetian boundary.

Wilkie Gray Group (new name). This group name is proposed to embrace two formations, the Laroona Arkose and Mount Podge Limestone (Fig. 3), excised by Talent and Mawson (1994) from the Fanning River Group and demonstrated by them to have accumulated in the late Emsian (largely if not entirely *serotinus* Zone), several conodont zones prior to inception of the Burdekin Basin *sensu stricto*. This latter event did not take place until close to the Eifelian-Givetian boundary. The substantial diastem between the Wilkie Gray and Fanning River groups is expressed as an irregular erosional surface between the Mount Podge Limestone and the overlying Burdekin Limestone on our section WMP at Mount Podge, just below our sample 274.5, at 96 m above the base of the Mount Podge Limestone (Talent and Mawson, 1994, fig. 2). Excision of the Laroona Arkose and Mount Podge Limestone from the Fanning River Group represents a return to Wyatt's three-fold conception of the Fanning River Group (for literature see below, "Fanning River Group").

The very youngest horizons of the Mount Podge Limestone yielding conodonts produced *Polygnathus serotinus* and *Pandorinellina exigua exigua*. Though the former may extend into the *costatus* Zone, the latter is known to extend to low in the *patulus* Zone and seemingly not higher. The regressive event indicated on Figure 5 has therefore been indicated as about the middle of the *patulus* Zone, but the top of the Mount Podge Limestone, and regressive event R1, could well have been within the *serotinus* Zone. Descriptions of the constituent units of the Wilkie Gray Group have been presented elsewhere (Talent and Mawson, 1994; Draper and Lang, 1994b; Zhen, 1995); its conodonts and rugose corals have been documented (Talent and Mawson, 1994; Zhen, 1995). The name is derived from the county name for the region including Mount Podge, Wilkie Gray County, where the constituent units of the Wilkie Gray Group crop out (Fig. 3).

Fanning River Group. This unit (Wyatt et al., 1968, 1970; Wyatt, 1973, 1978; Wyatt and Jell, 1967, 1980; Levingston, 1981; Zhen, 1991; Zhen et al., 1993; Talent and Mawson, 1994; Cook et al., 1994; Cook, 1995) occurs in numerous outcrop-tracts over an area of 50 by 110 km. Wyatt et al. (1970) viewed it as embracing three stratigraphic units: the Big Bend Arkose, Burdekin Limestone (or Formation) and Cultivation Gully Formation.

As noted above, two additional units, the Laroona Arkose (or Formation) and overlying Mount Podge Limestone, subsequently included within the Fanning River Group were excluded by us (Talent and Mawson, 1994) because of a substantial diastem expressed as an irregular erosional surface between the Mount Podge Limestone and the overlying Burdekin Limestone at Mount Podge.

Conodonts from the Burdekin Limestone (Talent and Maw-

Figure 2 (on facing page). Outcrop-tracts of late Emsian–Tournaisian rock units in the Townsville hinterland, northern Queensland (derived from Draper and Lang, 1994b); only marine or partly marine units are indicated. Not shown is the mosaic of pre-Devonian metamorphics and intrusives, the Late Devonian–Early Carboniferous nonmarine sedimentary units (Stud, Julia, and Piccadilly formations), and Carboniferous to Permian intrusives and extrusives, including the post-Tournaisian sedimentary-cum-extrusive Glenrock Group. For details about Mount Podge, see Figure 3.

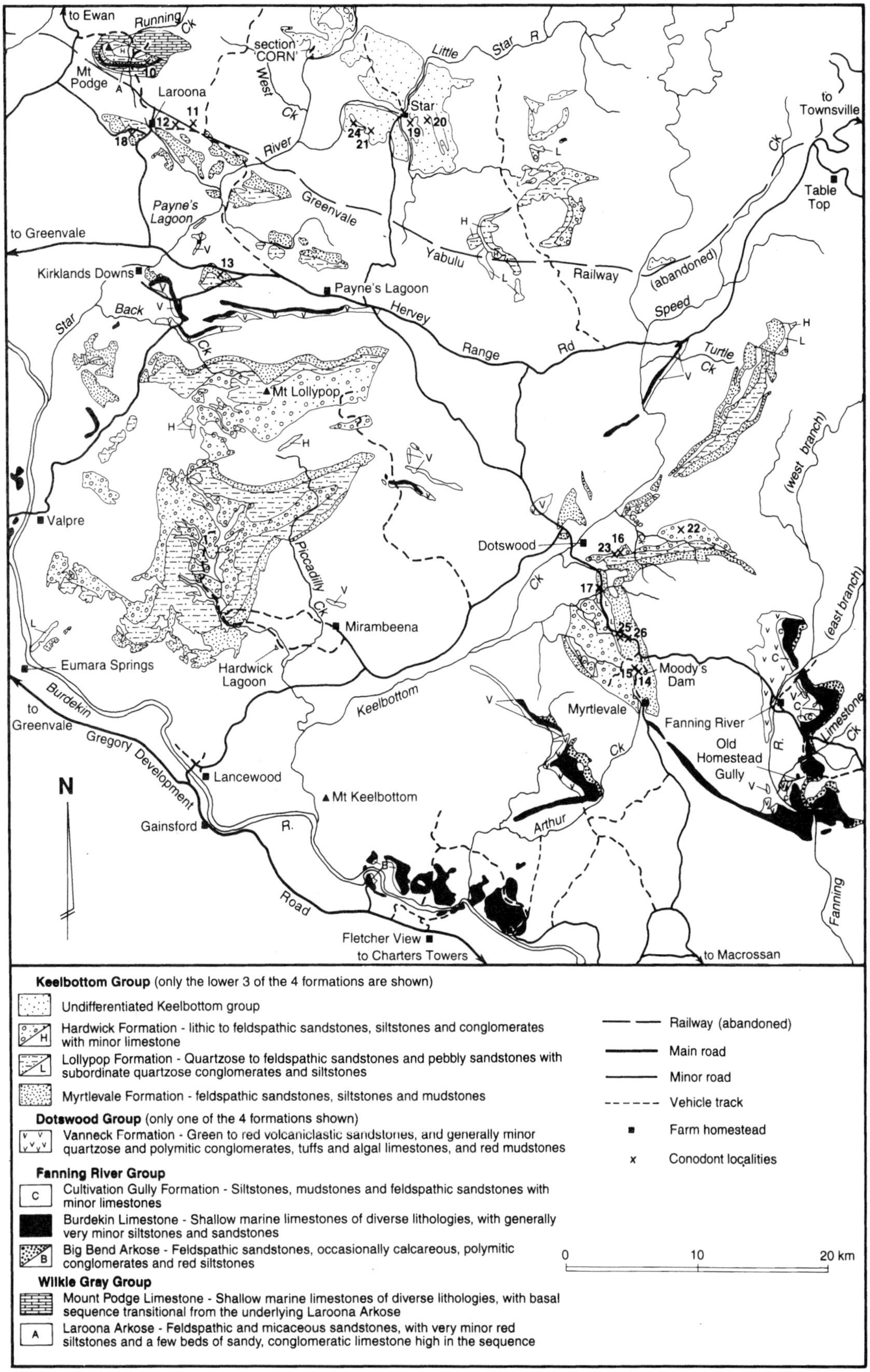
to Ewan
Running Ck
section 'CORN'
Mt Podge
Laroona
West Ck
River
Little Star R
to Townsville
Star
Greenvale
Payne's Lagoon
to Greenvale
Yabulu
Railway
(abandoned)
Table Top
Kirklands Downs
Payne's Lagoon
Back
Star
Ck
Hervey
Range
Rd
Speed
Turtle Ck
Mt Lollypop
(west branch)
Valpre
Dotswood
Piccadilly Ck
Mirambeena
Ck
(east branch)
Eumara Springs
Hardwick Lagoon
Moody's Dam
to Greenvale
Burdekin
Keelbottom
Myrtlevale
Fanning River
Limestone Ck
Gregory Development
Old Homestead Gully
R.
N
Lancewood
Mt Keelbottom
Arthur
Ck
Gainsford
R.
Road
Fanning
Fletcher View
to Charters Towers
to Macrossan
Keelbottom Group (only the lower 3 of the 4 formations are shown)
Undifferentiated Keelbottom group
H Hardwick Formation - lithic to feldspathic sandstones, siltstones and conglomerates with minor limestone
L Lollypop Formation - Quartzose to feldspathic sandstones and pebbly sandstones with subordinate quartzose conglomerates and siltstones
Myrtlevale Formation - feldspathic sandstones, siltstones and mudstones
Dotswood Group (only one of the 4 formations shown)
V Vanneck Formation - Green to red volcaniclastic sandstones, and generally minor quartzose and polymitic conglomerates, tuffs and algal limestones, and red mudstones
Fanning River Group
C Cultivation Gully Formation - Siltstones, mudstones and feldspathic sandstones with minor limestones
Burdekin Limestone - Shallow marine limestones of diverse lithologies, with generally very minor siltstones and sandstones
B Big Bend Arkose - Feldspathic sandstones, occasionally calcareous, polymitic conglomerates and red siltstones
Wilkie Gray Group
Mount Podge Limestone - Shallow marine limestones of diverse lithologies, with basal sequence transitional from the underlying Laroona Arkose
A Laroona Arkose - Feldspathic and micaceous sandstones, with very minor red siltstones and a few beds of sandy, conglomeratic limestone high in the sequence
Railway (abandoned)
Main road
Minor road
Vehicle track
Farm homestead
x Conodont localities
0 10 20 km

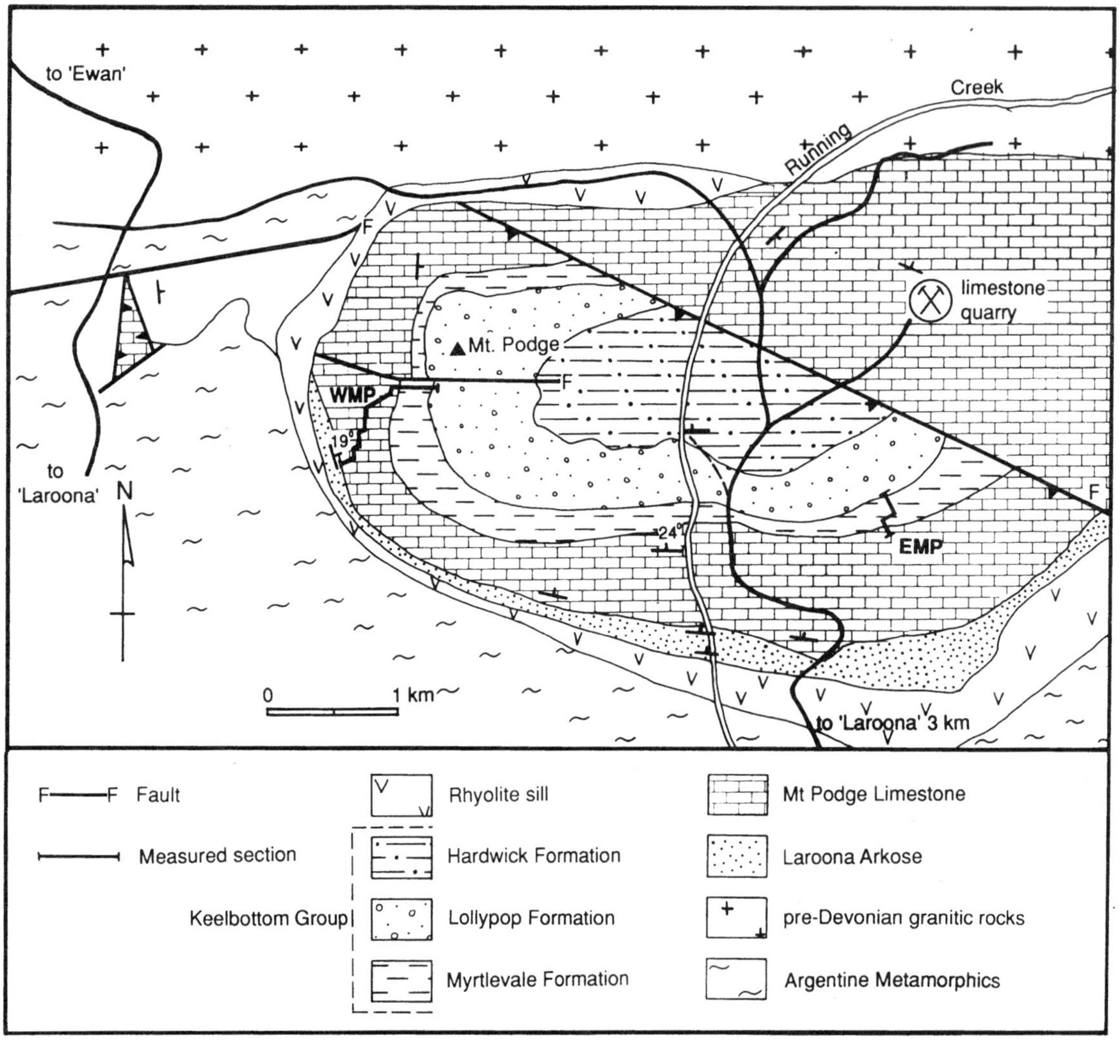

Figure 3. Mount Podge area showing location of stratigraphic sections. Geological boundaries by Geological Survey of Queensland. Numerous rhyolite dykes and sills have been omitted.

son, 1994), though not abundant, have been sufficient to demonstrate an age-range from about *ensensis* Zone (but conceivably a little older) to Late *varcus* Subzone, with maximum flooding having occurred during the interval spanned by the *hemiansatus* Zone and Early *varcus* Subzone.

The stratigraphically highest unit of the Fanning River Group *sensu stricto*, the Cultivation Gully Formation, is gradational from the underlying Burdekin Limestone. It consists of two informal members: a lower sequence of sometimes highly fossiliferous mudstones and siltstones with subordinate calcareous, feldspathic sandstones, limestones, and arenaceous limestones that passes gradually by decrease of arenites and limestones into an upper sequence of generally unfossiliferous siltstones, arenites and shales. The faunas of the "lower member" are diverse (Zhen, 1991; Draper and Lang, 1994, p. 18), being conspicuously dominated by atrypidine and reticulariid brachiopods. A modicum of conodont data from the "lower member" has shown it to be broadly late Givetian in age (Talent and Mawson, 1994). The "lower member" has been interpreted as a low-energy environment, possibly "an inner shelf/nearshore bay" (Draper and Lang, 1994b); this is in accord with the low disarticulation index of the rich brachiopod faunas from that interval. Though reported as 78 to 90 m thick in its type area, 2 km northeast of Fanning River homestead (Draper and Lang, 1994b, p. 18), the Cultivation Gully Formation appears to be appreciably thicker, ca. 135 m, in the vicinity of Old Homestead Gully ca. 6 km south of Fanning River homestead (cf. Draper et al., 1993a). Conodont data from the Cultivation Gully Formation indicate that the Burdekin Basin underwent eversion late in the Givetian, terminating no earlier than the Late *varcus* Subzone, but conceivably as much as two or three conodont zones later (Talent and Mawson, 1994).

Dotswood Group. This unit (Wyatt et al., 1968, 1970; Wyatt, 1973, 1978; Wyatt and Jell, 1980; Lang et al., 1990a; Draper and Lang, 1994b) occurs as several outcrop-tracts in an area of 60 by 70 km between the Fanning and Burdekin rivers; these have been hypothesized to be remnants of a formerly more extensive

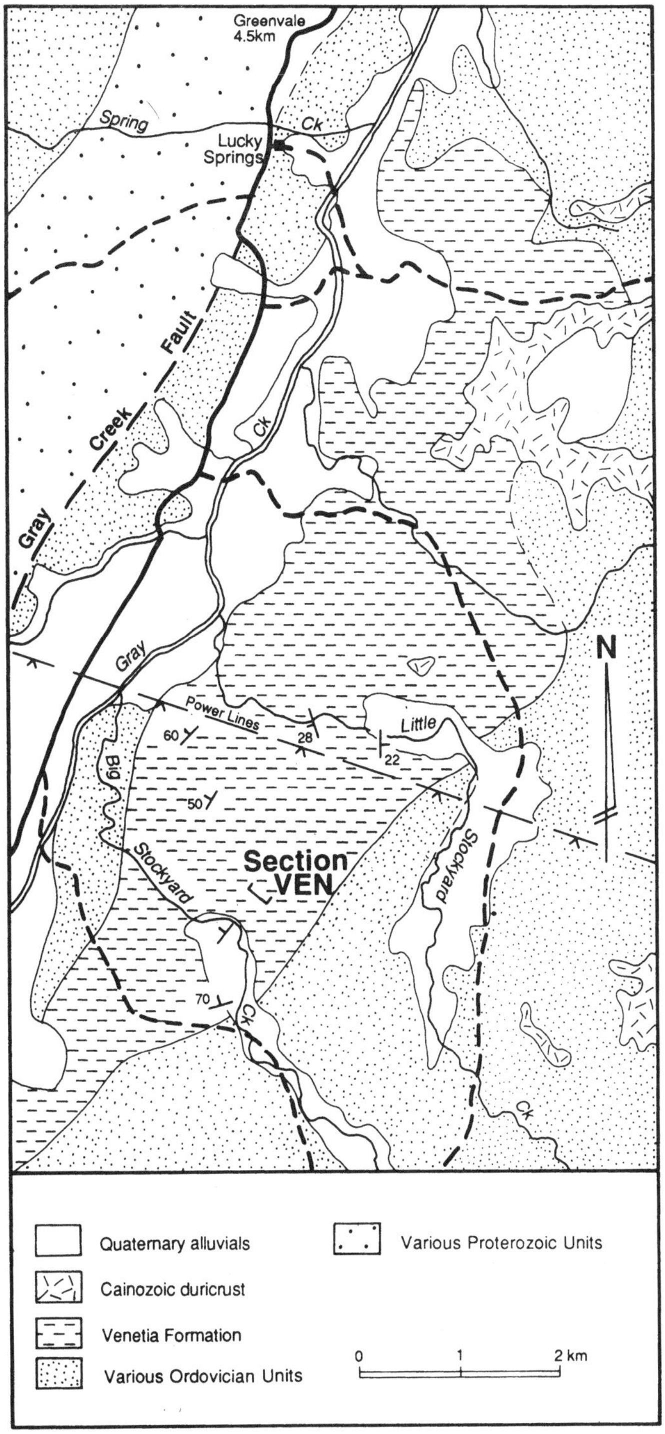

Figure 4. Geology in the vicinity of Big Stockyard Creek, "Lucky Downs," south of Greenvale, based on Withnall et al. (1992).

intermontane basin. The Dotswood Group is now construed (Lang et al., 1994) as consisting of four sedimentary units aggregating up to 2,550 m of arenites, siltstones, polymictic conglomerates with subordinate tuffs, cherts, and rare oolitic and oncolitic carbonates. Relationships to the underlying Fanning River Group range from gradational, as in Old Homestead Gully, to unconformable; the Dotswood Group onlaps northward onto pre–Middle Devonian (including Precambrian) plutonic and metamorphic rocks. Mudcracks, redbeds, and calcrete nodules have been reported (Lang et al., 1994). It should be noted that Lang et al. (1990a, p. 633) redefined the lower boundary of the Vanneck Formation (basal unit of the Dotswood Group) to coincide with the incoming of tuffaceous sediments; this has resulted in incorporation into the basal Dotswood Group of horizons formerly referred to the Cultivation Gully Formation. This redefinition has introduced an unequivocally marine interval with a diverse but undescribed late Givetian–early Frasnian fauna into the base of the Dotswood Formation. Acid-leaching of Dotswood Group carbonates failed to produce conodonts. A broad Frasnian age is, however, indicated (Lang et al., 1994).

Keelbottom Group. This unit (Figs. 2, 3, 5), the principal focus of this report, conformably overlies the Frasnian Dotswood Group (Wyatt et al., 1970; Wyatt, 1978; Wyatt and Jell, 1980; Lang et al., 1990b; Draper and Lang, 1994b). It is construed as consisting of four formations: a basal sequence of generally poorly outcropping fine-grained siliciclastics with generally rare limestones, the Myrtlevale Formation, followed by the boldly outcropping coarse clastics of the mainly nonmarine Lollypop Formation; overlain in turn by the comparatively recessive and predominantly marine Hardwick Formation, and, finally, by the generally coarse clastics of the Piccadilly Formation. Two other stratigraphic names were formerly used for large tracts of Keelbottom Group strata lacking the sort of marker horizons necessary for making satisfactory stratigraphic alignments with typical Keelbottom Group sequences: the "Game Hill Beds" (Wyatt et al., 1968, 1970) and the "Star Beds" (Jack, 1879; Wyatt et al., 1970; Wyatt, 1973; Wyatt and Jell, 1980).

The Keelbottom Group overlaps the Dotswood Group and outcrops over a much greater area; where the two occur in sedimentary contact they are conformable. The Keelbottom Group is unconformable on the Burdekin Limestone at Mount Podge (Fig. 3); in other areas in the north it rests unconformably on pre-Devonian igneous and metamorphic rocks, for instance, east and southeast of Star homestead where it rests on ?Proterozoic Argentine Metamorphics (Wyatt and Jell, 1980, fig. 15).

The Myrtlevale Formation. This formation, the basal unit of the Keelbottom Group (Wyatt et al., 1970; Wyatt and Jell, 1980; Langet al., 1990b; Draper et al., 1994), is a transgressive unit with a lower sequence of typically poorly fossiliferous coarse sandstones with quartz pebble conglomerates and pebbly quartzose sandstones at its base. This unit passes upward gradationally into fossiliferous micaceous mudstones and siltstones with micaceous calcareous sandstones, and sandy and silty limestones; the last are especially well developed at Mount Podge (Fig. 6). The upper part of the formation consists typically of fine-grained sandstones and siltstones with minor mudstones (but see next section). Thicknesses of the Myrtlevale Formation range from a maximum of ca. 460 m, down to 320 to 340 m in the type area near Myrtlevale homestead (Draper et al., 1994), and only

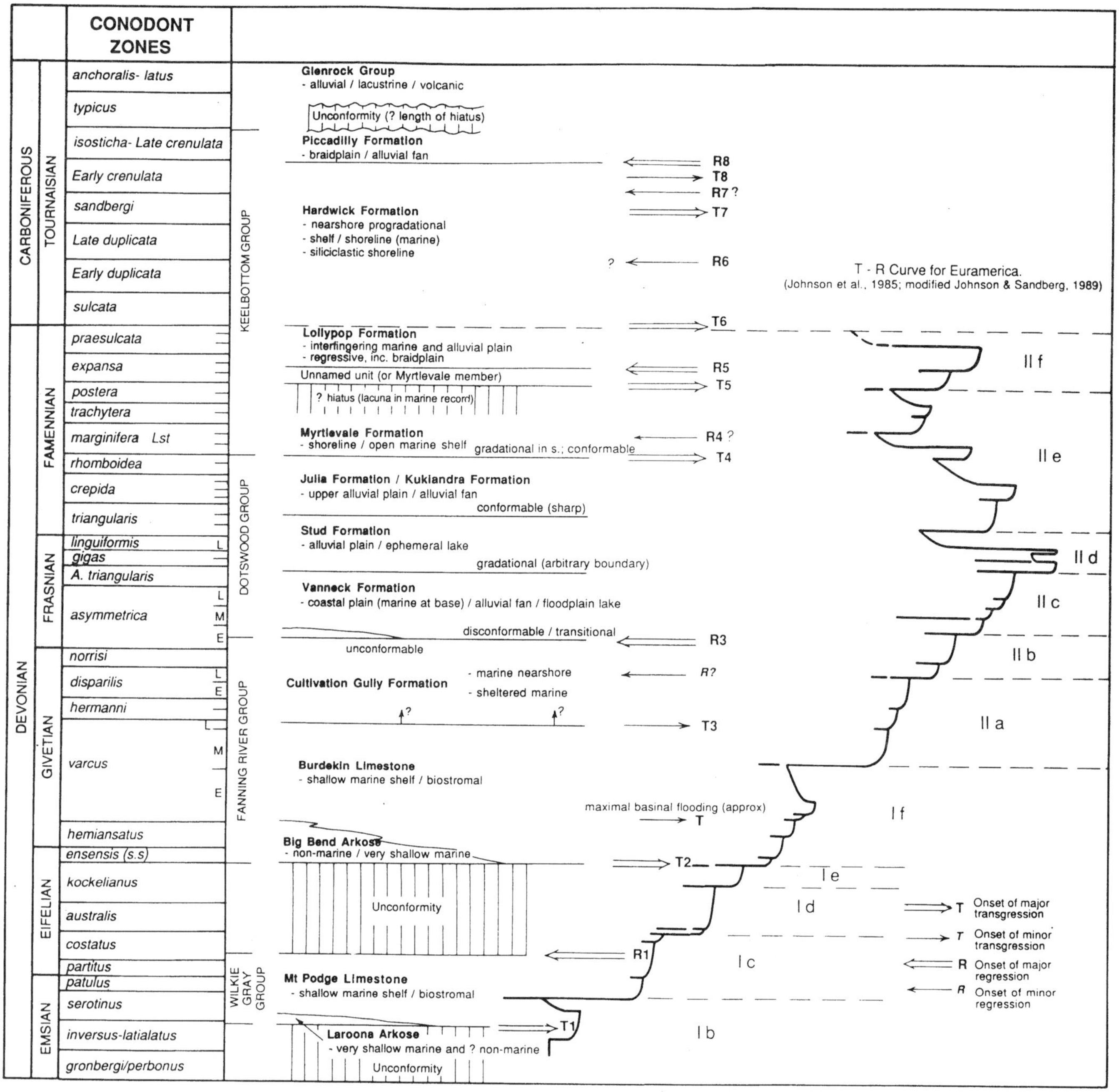

Figure 5. Devonian–Early Carboniferous stratigraphic units of the Townsville hinterland showing relationships, inferred depositional systems (based on Draper and Lang, 1994b), and transgression/regression events. Chronologic alignments are from Talent and Mawson (1994), and data herein. For ease of comparison with the Johnson et al. (1985, modified in Johnson and Sandberg, 1989) T-R curve, the zonal scale used by them has been retained but with minor nomenclatorial modifications. New schemes involving profound changes for the Frasnian have been proposed by Klapper (1989) and Ziegler and Sandberg (1990).

50.5 and 28 m respectively on our two sampled sections at Mount Podge (Figs. 3, 7; Tables 1, 2).

The macrofauna of the Myrtlevale Formation is diverse, consisting principally of chonetidine, productidine, and spiriferidine brachiopods (including *Cyrtospirifer*); occasional gastropods and cephalopods; and about a dozen genera of bivalves, together with rugose corals occurring rarely in the limestones. Apart from conodonts from four spot localities indicating the *marginifera* interval (Pickett, 1981), and productidine brachiopods (McKellar, 1970), the Myrtlevale faunas had not previ-

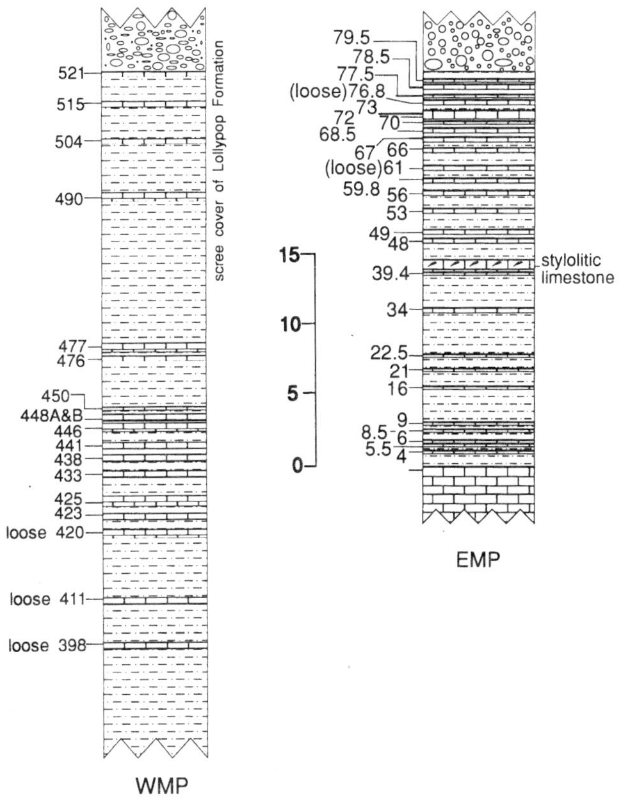

Figure 6. Sampled sections through the EMP and WMP at Mount Podge (Fig. 3). Horizons sampled for conodonts are indicated by a number corresponding to the distance along a metric tape. Dashed shading represents siltstones and fine sandstones; the brick pattern represents carbonate intervals. The overlying Lollypop Formation is indicated by conglomerate pattern.

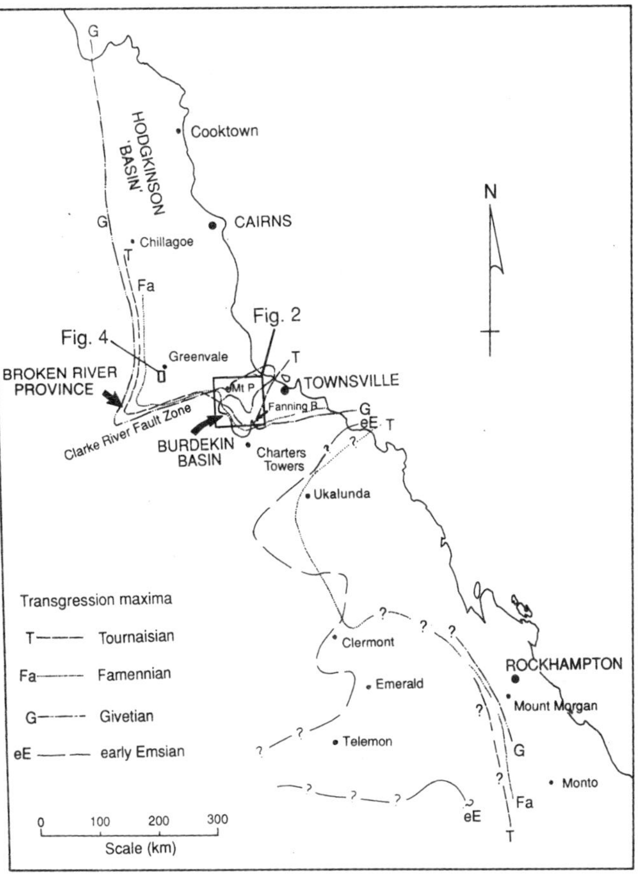

Figure 7. A selection of Devonian–Early Carboniferous transgression maxima (approximate) in northeastern Australia on a nonmobilist base. Note offset, assumed to have been post-depositional and connected with post-Tournaisian movement along the Clarke River Fault Zone. Note also the widespread early Emsian transgression(s) (eE) into central Queensland but absence of such from the Townsville hinterland.

ously been documented. It appears to be the most fossiliferous Famennian sequence in eastern Australia and the most productive to date as regards conodonts; a few of our samples produced close to 100 conodonts per kg.

Unnamed formation (or upper member of Myrtlevale Formation). This formation, the upper part of the relatively recessive interval, currently shown as Myrtlevale Formation on maps of the Burdekin Basin (Draper and Lang, 1994b), is a "package" consisting typically of fine-grained sandstones and siltstones with minor mudstones. This interval includes very minor conodont-bearing limestones at four localities: 17 and 18 just under the Myrtlevale-Lollypop boundary and 11 and 12 in fault-slices in the Argentine Metamorphics, east of Laroona (Fig. 3). These localities have produced conodont faunas indicative of the late Famennian *expansa* Zone.

Conodont faunas from these localities (refer to Table 3; discussed below) are some four zones younger (i.e., without Late and Latest *marginifera, trachytera,* and *postera* zones) than the widespread Early *marginifera* Zone faunas typical of the Myrtlevale Formation. We interpret these occurrences as indicators of a discrete late Famennian transgression, T5 (Fig. 8), presumed to have been from the north. We are reluctant to regard these faunas as being from the Myrtlevale Formation *sensu stricto*, preferring to interpret them as belonging to either a presently undefined upper member of the Myrtlevale Formation or even a post-Myrtlevale–pre-Lollypop formation whose boundary with the Myrtlevale Formation *sensu stricto* has yet to be defined.

Preservation of the *expansa* Zone sequence at localities 11 and 12 as fault-slivers in Proterozoic basement, without an expected complement of older Devonian units (Lassie Creek, Fanning River, and Dotswood groups), is believed by us to be evidence for rapid removal of pre-*expansa* Zone Devonian units from various fault-blocks in the vicinity of Laroona, concurrent with major fault movements during the interval late Emsian to mid-Famennian (pre-*expansa* Zone). Some of these events may have taken place during the interval post-*marginifera* Zone–pre-*expansa* Zone.

The Lollypop Formation. The Lollipop Formation (Wyatt et al., 1970; Draper et al., 1994) is a unit of well-bedded (0.6 to 1.2 m) and cross-bedded sandstones and conglomerates with maxi-

TABLE 1. DISTRIBUTION OF CONODONTS IN PART OF SECTION WMP PASSING THROUGH THE MYRTLEVALE FORMATION

		Meters Above Base of WMP Section																	
		167.5	171	175.5	176.7	177.5	179.6	180.8	181.5	182.9	183.6	183.6	184.1	187.8	188.5	198.9	203.1	206.2	207.9
		Sample																	
Conodont Taxa		WMP 398 L	WMP 411 L	WMP 420 L	WMP 423	WMP 425	WMP 433	WMP 438	WMP 441	WMP 446	WMP 448A	WMP 448B	WMP 450	WMP 476	WMP 477	WMP 490	WMP 504	WMP 515	WMP 521
I. alternatus		1		1													4		1
P. karadjalis		2	17	24	23	19	1				46	13	1		4	1	789	61	310
P. planirostratus		2	2	2		1		1			1	2	1				7	1	21
P. pseudobrevilaminus		4	8	4	4	3		1			3	1		1			5	6	15
P. subnormalis		1	1		1	3	1	2				4	2				46	3	2
Pal. quadrantinodosa quadrantinodosa		1				1						1					3		1
P. lauriformis			2	3	1	3		3			7	8	3	1			92		14
Mehlina strigosa			1									2					3		1
P. bouckaerti				2			2	1									1	1	2
P. costulifera				3		1	4	6			8	4					88	1	27
P. hieroglyphica				1	1	1					8	6	1	1	2		165	12	74
Icriodus sp. A				1									1				1		1
Alternognathus pseudostrigosus						1											4	1	3
Pal. quadrantinodosa inflexa						2		1									4		1
Pal. glabra pectinata							1	3	1		4	5	3		1		12	1	21
Pal. minuta minuta							1												
P. wyatti								1			1		1				2	3	14
P. diversus								1									1		1
Pal. quadrantinodosa inflexoidea								1				1					5	1	4
Apathognathus varians varians								1			1								1
Apparatus A	Pa																1		3
Apparatus A	Pb							2						1			1		24
Apparatus A	M																1		6
Apparatus A	Sa											1	1				1		4
Apparatus A	Sb							1			2		1				1		12
Apparatus A	Sc																		2
Polylophodonta confluens											1						2		1
Pal. marginifera marginifera													3				27	2	13
P. asilbekensis																	1		
Ancyrognathus cf. *irregularis*																	2		1
Pal. falcata																	1		2
P. cf. *communis communis*																	1		
P. szulczewskii																	1	3	5
I. chojnicensis																		1	1
P. nodocostatus																		1	
P. pennatuloideus																		1	1
Polygnathus sp. A																			1
Polygnathus sp. C																			1
Pal. marginifera duplicata																			5
I. scalariformis																			3
Unassigned elements	Pa																1	1	7
	Pb		3		4	1	1	5		1	8	3	1		2		98	2	18
	M	1	1	2	1		1	2			4	2		1	1		21	1	8
	Sa		1														20	1	6
	Sb		1	1	1	1		3			2	3					40	4	22
	Sc		2	4			2	8	2		10	6	1		1	2	80	11	34
	Sd																1		

mum thickness of about 820 m; it is 460 to 475 m at the type locality and about 305 m at Mount Saint Michael near Dotswood. Draper et al. (1994) report three lithotopes. The lowest, a predominantly sandstone sequence, is reported as interfingering with the Myrtlevale Formation and passing upward into coarse, pebbly sandstones and matrix-supported conglomerates, with these in turn passing upward into a further sequence of cross-bedded and graded sublabile sandstones and pebbly sandstones. The Lollypop Formation tends to be topographically prominent; floaters from it, especially from its conglomerates, often obscure the recessive

TABLE 2. DISTRIBUTION OF CONODONTS IN PART OF SECTION EMP PASSING THROUGH THE MYRTLEVALE FORMATION

Conodont taxa		EMP 4	EMP 5.5	EMP 6	EMP 8.5-9	EMP 16	EMP 21	EMP 22.5	EMP 34	EMP 39.4	EMP 48	EMP 49	EMP 53	EMP 56	EMP 61 L	EMP 66	EMP 67	EMP 68.5	EMP 70	EMP 72	EMP 73	EMP 76.8 L	EMP 77.5	EMP 78.5	EMP 79.5
Meters above base of EMP section		1.4	1.9	2	2.9-3	5.4	7.2	7.7	11.6	13.8	16.4	16.75	18.1	19.1	20.9	22.6	22.9	23.4	23.9	24.6	24.9	26.2	26.5	26.8	27.2
P. bouckaerti		1						1	3					1											
P. pseudobrevilaminus		1		3			3	1	3				1												
I. alternatus				1		1		1												2					1
I. chojnicensis				1																1					
P. costulifera				1	1	1		2	5		1	2	2												
P. hieroglyphica				2	1	1		2	10			2	1							4		1	2		2
P. karadjalis				9	12	12	22	6	6		1	4	16		2	1				12		2	4		
P. lauriformis				1					1		1	3	5						1	4				1	
Alternognathus pseudostrigosus				1					3			1		1	2					3	1		1		1
Pal. minuta minuta				1																					
Pal. quadrantinodosa inflexoidea				1																					
P. wyatti					1										2					1	1				
P. nodocostatus						1					1														
P. planirostratus							2		1				3												
P. subnormalis									2			1			1					4					
Pal. marginifera marginifera									2		1		2							2				3	
Mehlina strigosa											1				1								1		
Pal. glabra pectinata												1								2			1	2	1
Apparatus A	Pb												8		2					3	1			1	
Apparatus A	M												2					1		2					
Apparatus A	Sa												1							2	1				
Apparatus A	Sb												2					1		2				1	
Apparatus A	Sc												1			1				1					
P. quadrantinodosa quadrantinodosa													1												
P. q. inflexa													1												
Pal. marginifera duplicata														1				1		3				1	
Apathognathus varians varians																		1		2					
P. diversus																				1					
P. communis communis																				1					
Polylophodonta confluens																				1					
Unassigned elements	Pa	1		1			1		2																
	Pb			7	3	3	2	2	4	1		1	2					1						2	
	M			2	1	1							1			1	2				1			1	
	Sa			1	1	1								1				2							
	Sb			3	2	2				1											1				
	Sc	1	1	8	3	2	1	2	2	1	1	3	4					2			2	2	2	5	2
	Sd																		1	1					

Myrtlevale and Hardwick formations occurring downslope from ridges of Lollypop Formation. Though formerly regarded as unfossiliferous (Wyatt et al., 1970), the Lollypop Formation has produced lycopods including *Leptophloeum* (Draper et al., 1994). Along Running Creek in the vicinity of Mount Podge, it passes gradationally into the Hardwick Formation. The transitional interval has occasional prominent beds of fossiliferous calcareous sandstones (some verging on limestone), a few of them characterized by great abundance of *Cyrtospirifer*.

The Hardwick Formation. This formation is a relatively recessive unit consisting of feldspathic sandstones, subordinate micaceous mudstones and calcareous sandstones, minor limestones (including oolitic limestones), and rare pebbly sandstones. It is thickest (ca. 1,320 m) in the Kirkland Downs–Payne's Lagoon area and up to 820 m south of Mount Saint Michael, where a prominent calcareous interval has been referred to as the Saint Michael Member (Wyatt, 1973; Wyatt and Jell, 1980; Draperet al., 1994). The Saint Michael Member consists of calcareous sandstones, arenaceous limestones and occasionally oolitic calcarenites with shale and mudstone interbeds, generally 0.6 to 1.2 m but up to 3 m in thickness, passing upward into 1.2 to 1.8 m of shales with rare limestone nodules or beds of lime-

TABLE 3. DISTRIBUTION OF CONODONTS IN SPOT SAMPLES FROM THE MYRTLEVALE AND HARDWICK FORMATIONS

Conodont taxa	Sample 10	13	14	15	11	12	17	18	23	16	20L	20U	21L	21U	22	CORN 1	CORN 4	CORN 5	VEN c. 210
I. alternatus	2																		
P. bouckaerti	1																		
P. costulifera	1																		
P. hieroglyphica	1																		
P. karajalis	10																		
P. planirostratus	2																		
P. subnormalis (= *P. webbi* Pickett)	3	311		3															
Pal. glabra pectinata		156																	
Pal. quadrantinodosa inflexoidea		3																	
Pol. confluens		4	1																
Ap. varians varians			1		4	1	1												
I. cornutus			4																
M. strigosa			11				3	3											
P. pennatuloideus			1																
Pal. ex. gr. *quadrantinodosa*			2																
P. nodocostatus				1															
Pal. marginifera marginifera				1															
Polygnathus sp.					2			2											
Bi. aculeatus aculeatus					7	1				7	1							3	1
Bi. aculeatus anteposicornis					2										1				
Bi. stabilis					18	3			1						1			2	
P. communis communis						1	2						1						
Ps. primus						1						1							
I. scalariformis							3	2											
P. homoirregularis							2	2											
P. semicostatus morphotype 1							4												
Pal. perlobata postera							3												
I. cf. *symmetricus*								1											
P. experplexus								3											
P. inornatus								2						1			1	2	
I. cf. *constrictus*									2										
P. vogesi									5										
Bi. spinulicostatus										1		1		3					
Si. duplicata										3									
Si. sulcata										2									
"Apatognathid" sp.											2	1							2
P. lectus											2								
Mehlina sp.											1								
H. crassidentatus												1	1		1			2	1
Ps. multistriatus												1							1
Bi. aculeatus plumulus														1					1
Si. obsoleta														1		1		13	

stone. It outcrops along the southern flank of Mount Saint Michael for ca. 12.8 km. It was from this interval (Wyatt's locality C414) that Pickett (1981) reported a conodont assemblage for which he inferred a broad Tournaisian age, somewhere in the interval Early *duplicata* to Early *crenulata* zones, and from which Wyatt (1973) reported the productidine brachiopods *Avonia* and *Linoproductus*. This horizon is regarded by us as referable to the *sandbergi* Zone. The upper Hardwick Formation, above the Saint Michael Member, consists of interbedded shales and feldspathic sandstones (Wyatt et al., 1970).

Rare calcareous concretions ansd even thin impure limestone beds occur occasionally elsewhere in the Hardwick Formation, for example stratigraphically beneath the Saint Michael Member in outcrops in the gully heads of One Mile Creek immediately south of Mount Saint Michael. Carbonates also occur low in the Hardwick Formation associated with richly fossiliferous shales and siltstones, for instance, along Running Creek and in the amphitheater on the northern flank of Mount Podge. A short stratigraphic interval but with several carbonate horizons and a few exceptionally richly fossiliferous siltstones is well exposed at Corner Creek (Fig. 2).

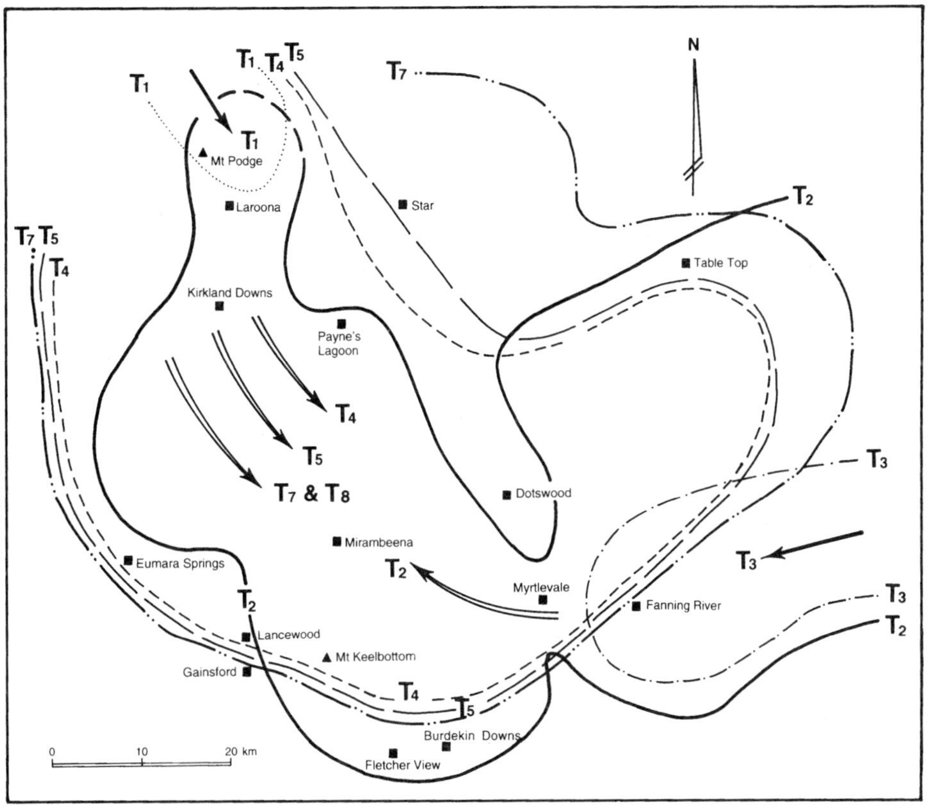

Figure 8. Cartoon showing stages in the paleogeographic evolution of the Townsville hinterland, northeastern Australia, during Emsian-Tournaisian times. Arrows indicate inferred directions of incursion. Note: 1. The late Emsian transgression event T1 from the north penetrated only a very short distance into the region that, close to the Eifelian–Givetian boundary, developed into the Burdekin Basin. 2. Trangression T2 extends into an inferred barred, relatively hypersaline basin west of the Dotswood High (Talent and Mawson, 1994, Text-fig. 1), and sedimentation is restricted during late Givetian times (Cultivation Gully Formation) to regions east of it. 3. The major Famennian-Tournaisian marine incursions T4–T8 (compare Fig. 5) were all seemingly from the north. 4. Not represented are a broadly Frasnian–early Famennian interlude (Dotswood Group) of nonmarine intermontane-basin sedimentation, nor are hypothetical boundaries suggested for the chronologically poorly constrained transgression events T6 and T8. 5. Transgression T7 (*sandbergi* Zone, but possibly extending into the Early *crenulata* Zone) was represented in other areas of Late Devonian–Early Carboniferous sedimentation west and west-northwest of the Burdekin Basin: in the Clarke River Basin (Venetia and Ruxton Formations) and Broken River Province (Venetia Formation). 6. Note the tendency of the heart area of the basin to persist as the primary locus of sedimentary accumulation from the Givetian through into the Early Carboniferous.

The Piccadilly Formation. This formation consists typically of 365 to 460 m (but up to 850 m) of generally thick-bedded quartzose and sublabile sandstones with 1 to 10% of pebbles and cobbles, gradationally overlying the Hardwick Formation (Wyatt et al., 1970; Draper et al., 1994). Poorly preserved plant material is occasionally encountered; the unit appears to be otherwise unfossiliferous.

Broken River Province and Clarke River Basin

Though thick, largely clastic, Late Devonian–Early Carboniferous sequences have long been known from the Broken River Province (Fig. 1), their stratigraphy had not been closely investigated until recent years (Wyatt and Jell, 1980; Lang, 1986, 1993; Draper et al., 1993b). Knowledge of their faunas and floras and of how these might align with the international stages and conodont zones has remained rudimentary. The stratigraphy and sedimentology of the Late Devonian–Early Carboniferous Clarke River and Bundock Creek groups, outcropping extensively in the watersheds of the Clarke and Broken rivers west and west-northwest of the Burdekin Basin (Fig. 1), have been detailed recently (Lang, 1993; Draper et al., 1993b). Though primarily fluviatile, these very thick units (ca. 6,000 m, for instance, for the Bundock Creek Group; Fig. 1) had one very prominent marine interval with carbonates and brachiopod-mollusk faunas. This interval is expressed as the Teddy Mountain Formation (Bundock Creek

Group), as the Ruxton Formation (Clarke River Group), and by a finer-grained recessive interval of the Venetia Formation (also in the Clarke River Group). Both the Clarke River Group and the Bundock Creek Group exhibit similarities in sedimentary succession to the Late Devonian–Early Carboniferous sedimentary sequences (Dotswood and Keelbottom groups) of the Burdekin Basin. They may even have been parts of the same predominantly clastic wedge.

Bundock Creek Group. The Givetian-Frasnian boundary has been argued to occur high in the highly fossiliferous, very shallow marine Stanley Limestone Member of the Mytton Formation (Mawson and Talent, 1989; Sloan et al., 1995). Marine conditions, reflected in diverse brachiopod-mollusk faunas, persisted for a few tens of meters into Mytton Formation clastics above the Stanley Member, but fluviatile conditions with sporadic marine influence then prevailed through the lowest two units of the Bundock Creek Group, the Bulgeri and Turrets formations. Though the Bulgeri Formation rests with low-angle unconformity on the Mytton Formation, it has been assumed that this diastem does not represent a very significant time-break (Lang, 1993). There is no record of "normal" marine sedimentation during accumulation of the Bulgeri and Turrets formations, though sedimentary evidence and very rare occurrences of generally very poorly preserved fossils have been interpreted (Lang, 1993) as sporadic incursions of estuarine conditions; taxa reported from these occurrences have yet to be documented.

More persistent marine conditions returned with the inception of the less coarsely clastic Teddy Mountain Formation in which limestone occurrences and horizons with brachiopod-mollusk faunas are not infrequent. Lang (1986, p. 133–163) reported a sparse bispathodid fauna from sandy calcarenites of the Teddy Mountain Formation exposed in a gully 1.2 km north-northwest of Montgomery Dam. Though the conodont fauna is not chronologically constraining within the interval *expansa* Zone–early Tournaisian, a latest Famennian age was suggested (Lang, 1986). Because resampling of these limestones and sampling of other carbonate horizons in the Teddy Mountain Formation has so far proved disappointing, we limit discussion to a more informative section: in the Venetia Formation of the Clarke River Group.

Venetia Formation. The lower unit of the Clarke River Group in the Clarke River–Broken River area, the Venetia Formation, consists predominantly of coarse to very coarse micaceous lithofeldspathic sandstones with minor pebbly sandstones and conglomerates. The designated type section of the Venetia Formation, located some 35 km south-southeast of our sampled section VEN, has been interpreted as lacking a marine interval; a marine interval is, however, represented in exposures of the Venetia Formation along Gill Creek, 40 km east-southeast of our VEN section. The thickness of the marine interval in the latter area is quoted as "unlikely to exceed 20 m" (Draper et al., 1993b, p. 207) but on our section VEN is more substantial, at least 230 m. The same marine interval is thicker in the Blue Range, ca. 60 km east of our VEN section, where most of a 750-m sandstone-mudstone sequence, referred to as the Ruxton Formation, has been interpreted as marine (Withnall, 1993; Draper et al., 1993b). Conodonts so far obtained from the Ruxton Formation are not constraining relative to positioning of the Devonian-Carboniferous boundary (B. F. Fordham, quoted in Draper et al., 1993b, p. 210).

The marine interval on our section VEN, in the Big Stockyard Creek area, ca. 12 km south of Greenvale (Fig. 4), is conspicuously finer grained than typical Venetia Formation. It comprises siltstones, shales, calcareous siltstones, flaggy sandstones, and occasional fossiliferous calcareous rudites, limestones, and arenaceous limestones.

SAMPLING

Approximately 1 metric ton of samples (5 to 10 kg each) was obtained from measured stratigraphic sections and from spot-sampling other localities. Samples that produced no conodonts after leaching 1.5 to 2 kg were discarded; samples proved productive at this stage were then further leached, generally to completion. Only the productive localities are indicated on the accompanying maps (Figs. 2–4).

ABBREVIATIONS

The following abbreviations are used for generic names of conodonts in the text and on tables: *Alt.* = *Alternognathus*; *Anc.* = *Ancyrognathus*; *Ap.* = *Apatognathus*; *Bi.* = *Bispathodus*; *H.* = *Hindeodus*; *I.* = *Icriodus*; *M.* = *Mehlina*; *P.* = *Polygnathus*; *Po.* = *Polylophodonta*; *Pal.* = *Palmatolepis*; *Ps.* = *Pseudopolygnathus*; *Si.* = *Siphonodella*. Abbreviations used for stratigraphic sections are WMP = western flank of Mount Podge; EMP = south flank of "East Mount Podge"; CORN = section down Corner Creek (tributary of the Star River); VEN = section through portion of Venetia Formation on south side of the divide between Big and Little Stockyard Creek, "Lucky Springs," approximately 12 km south of Greenvale (Fig. 4). Spot localities sampled are shown on Figs. 2 and 3. It should be noted that wherever the word "zone" is used in the following text, the current zonal scheme, based on conodonts, is implied.

IMPLICATIONS OF CONODONT DATA

Chronology

Cultivation Gully Formation. Though we have extensively sampled the Cultivation Gully Formation since recently suggesting a late Givetian age for it (Talent and Mawson, 1994), no additional taxa have been found that would provide a better age estimate. Conodonts from the "lower member" of the Cultivation Gully Formation have a conspicuously greater abundance of *Polygnathus* than any of our sections through the underlying Burdekin Limestone (cf. Talent and Mawson, 1994). This preponderance of polygnathid biofacies in the lower Cultivation Gully Member accords with a deepening event (Fig. 5; Fig. 8; Table 3), and finds support in the remarkably low disarticulation index for the rich brachiopod faunas of this interval. Thus, there was not only a swift diminution in carbonate sedimentation low in the Cultivation Gully Formation but also, at least locally, a significant deepening event, prior to the recession expressed by the "upper member" of the Cultivation Gully Member and basal Vanneck Formation. The pattern does not tie in with the late Givetian-early Frasnian T-R events inferred by Johnson et al. (1985) and Johnson and Klapper (1992); it may therefore have been tectonically controlled.

***Myrtlevale Formation* (sensu stricto).** Diverse conodont faunas (Tables 1–3) indicate that the Myrtlevale Formation *sensu stricto* is referable, broadly if not entirely, to the Early *marginifera* Zone. The conodont faunas are unusual in that they are intermediate between the palmatolepid (pelagic, open ocean) and polygnathid (shallower, but not nearshore) biofacies. The conodont faunas of the Myrtlevale Formation provide a useful spike for aligning less appreciated polygnathid faunas with the standard palmatolepid biozones. There is some deficiency in high precision information regarding the age ranges for polygnathids of the Famennian; finding a rich polygnathid fauna together with several species of palmatolepids might be of some help in tying polygnathid faunas to the palmatolepid-based biozonation.

Important publications documenting faunas of this age include: Vorontsova and Kuz′min (1984) and Kuz′min (1990) from central Kazakhstan; Khalymbadzha et al. (1991) from southern Kazakhstan; Kononova (1979, mostly dealing with palmatolepids) and Khalymbadzha and Mingaleev (1994) from the southern Urals; Matyja (1972) from Pomerania; Matyja (1975, 1987), Matyja and Zbikowska (1974), Szulczewski (1971), and Wolska (1967) from Poland; Helms (1959, 1961) from Thuringia, Germany; Dreesen and Dusar (1974) from Belgium; Metzger (1994) from the Bactrian Mountains of Nevada; Perri and Spalletta (1990) from the Carnic Alps of northeast Italy; Ji and Ziegler (1993) and Bai (1994) from south China; and Glenister and Klapper (1966), and Metzger (1994) from northwestern Australia. Most of the above emphasized palmatolepids in preference to polygnathids.

Of the platform elements in the fauna from limestones in the Myrtlevale Formation at Mount Podge, 90% are species of *Polygnathus*, indicative of a Polygnathid Biofacies normally seen as representing an outer-shelf origin. Pelagic or offshore palmatolepids make up 7% of the fauna, and the remaining 3% consists of a sprinkling of icriodontids, alternognathids, apatognathids, polylophodontids, mehlinans, and ancyrognathids. Because of the palmatolepid content of the fauna, it is suggested that the environment of deposition might, in this case, be in water slightly deeper than what would be expected for a Polygnathid Biofacies.

Traditionally, palmatolepids have been basic to the zonation of the Late Devonian. Their apparent ubiquitous distribution can be explained by the unimpeded seaways during the greatest part of this period of time; the relatively cosmopolitan aspect of most Late Devonian conodont faunas is widely recognized.

Considering the accepted age ranges (e.g., Ziegler and Sandberg, 1984) for the palmatolepids present in the fauna, albeit making up only 7% of the fauna, the sequence of limestones in the Myrtlevale Formation at Mount Podge lies within the Early *marginifera* Zone because forms such as *Pal. quadrantinodosa quadrantinodosa*, *Pal. glabra inflexa*, and *Pal. g. inflexoidea* range spasmodically from bottom to top of both sections at Mount Podge, WMP, and EMP. *Pal. q. quadrantinodosa* has been reported to range as low as the uppermost part of the Late *rhomboidea* Zone (Ziegler in Ziegler, 1977), but we emphasize the occurrence of *P. karadjalis* in the lowest sample from our WMP section. It is a form not known to occur below the base of the Early *marginifera* zone (Vorontsova and Kuz′min, 1984; Vorontsova, 1993, Kuz′min, personal communication, 1995). Accordingly, we conclude that no horizons older than Early *marginifera* Zone are represented. If this is accepted, the Mount Podge sequence presents evidence for some emendments to accepted ranges of a few taxa:

1. The genus *Ancyrognathus* was not known to occur in horizons younger than Late *crepida* Zone; at Mount Podge it occurs high in the Early *marginifera* Zone.

2. An icriodontid, *I. alternatus*, appears in several horizons in both the EMP and WMP sections at Mount Podge. *I. alternatus sensu stricto* has not previously been reported from horizons younger than Latest *crepida* Zone (e.g., Ziegler and Sandberg, 1984).

3. At the top of the Mount Podge section, in samples WMP 515 and WMP 521, an enigmatic icriodontid, apparently a transitional form between *I. iowaensis* and *I. costatus costatus*, occurs. It is described herein as a new species, *I. scalariformis*, and is documented as occurring in both the Early *marginifera* Zone and the *expansa* Zone.

Faunas of this age have been recorded from elsewhere in Australia: from the Ningbing Limestone and Button Beds of the Bonaparte Gulf Basin (Druce, 1970a), and from the Canning Basin (Glenister and Klapper, 1966; Druce, 1976; Metzger, 1994) of northwestern Australia .

Unnamed formation (or upper member of Myrtlevale Formation). The presence of *P. experplexus*, *P. homoirregularis*, and *P. semicostatus* Morphotype 1 at localities 17 and 18 restricts their age to an interval between Early *expansa* Zone and early in the Late *expansa* Zone (cf. Sandberg and Ziegler, 1979). Faunas from both localities also include an icriodontid identified herein as a new species, *I. scalariformis*. Constraints occasioned by the first appearance of subspecies of *Bi. aculeatus* and the last appearance of *Ap. varians varians* [*sensu* Nicoll, 1980] indicate that the faunas dominated by bispathodids from localities 11 (TMR49B) and 12 (TMR47) fall within the interval Middle *expansa* Zone to Late *expansa* Zone.

Elsewhere in Australia, a fauna of a similar age has been recovered from an impure limestone lens, formerly thought to be low in the Luton Formation of northern New South Wales (Philip and Jackson, 1971) but now known (Gould, 1975) to be from the underlying Mandowa Mudstone.

Lollypop Formation. Calcareous arenites and arenaceous limestones with abundant *Cyrtospirifer* and chonetidine brachiopods and/or shell debris occur high in the Lollypop Formation or transitional from Lollypop Formation to Hardwick Formation; these are well displayed along Running Creek immediately east of Mount Podge. No conodonts were obtained from these calcareous beds, so we still lack precise chronologic underpinning for this very prominent transgressive event at or about the base of the Hardwick Formation (Fig. 5, event T6). It could correlate with the onset of the worldwide transgression near the base of the Tournaisian.

Hardwick Formation. As noted above, the age of the base of the Hardwick Formation is not well constrained by conodont data from the Running Creek–Mount Podge area or by spot samples from elsewhere in the region. However, because *Cyrtospirifer* was excessively abundant and cosmopolitan in the Famennian and is generally believed to have extended only briefly into the Tournaisian, this transgressive event, T6, is hypothesized as equating with the major global transgression initiated at or very close to the Devonian-Carboniferous boundary.

Some support for this comes from one of the limestones low in the Hardwick Formation (stratigraphically beneath the Saint Michael Member) outcropping in the head of a gully tributary to One Mile Creek on the south flank of Mount Saint Michael, specifically at locality 23. This horizon produced a conodont fauna obviously predating the widespread *sandbergi*–Early *crenulata* interval (see below). Some constraint comes from the presence of *P. vogesi*, a form known to range from Late *expansa* zone to *sulcata* Zone (e.g., Barskov et al., 1991; Ji and Ziegler, 1993). An icriodontid identified as *I.* cf. *constrictus* would be in accord with a latest Famennian age, but because of the uncertainty connected with specific attribution of this form, we do not attach high chronologic significance to this occurrence. We nevertheless suspect that this fauna is connected with the earliest Tournaisian transgression event. An elegant study (Girard, 1994) of conodont biofacies through the global stratotype for the Devonian-Carboniferous boundary in the Montagne Noire of southern France has demonstrated that this event commenced incipiently in the late *praesulcata* Zone, just below the the Devonian-Carboniferous boundary.

Conodonts from nine localities, including three horizons on our CORN section, and two horizons, localities 20 and 21, are regarded as referable to the *sandbergi*–Early *crenulata* interval on the basis of the presence in the assemblage of forms such as *Si. obsoleta* and *H. crassidentatus*. The former indicates the age can be no older than *sandbergi* Zone; the latter indicates an age not younger than Early *crenulata* Zone. The above localities include the Saint Michael Member of the Hardwick Formation. This unit was previously sampled by Pickett (1981) at Wyatt's locality C414 (= locality 16 herein) on the south flank of Mount Saint Michael; allocation to the Early *duplicata* Zone had been suggested because of the presence of *Si. sulcata* and *Si. duplicata* (Pickett, 1981). Both of these species range into the *sandbergi* Zone (Klapper, in Ziegler, 1973; Sweet, 1988). Because locality 22 in the Hardwick Formation in the same area has produced a fauna referable to the *sandbergi*-Early *crenulata* interval, we suggest C414 (= locality 16 herein) might be best regarded as *sandbergi* Zone on the basis of presently available data.

Australian conodont faunas of this age include those from the Bonaparte Gulf Basin, northwestern Australia (Druce, 1970a, 1976); Wirrabilla, northern New South Wales (Mory and Crane, 1982); a spot sample from the Mandowa Mudstone, near Barraba, New South Wales (Mory and Crane, 1982); and Old Cannindah, Cannindah Creek, and Malchi Creek in central Queensland (Mory and Crane, 1982).

No conodont data are available bearing on the age of the regressive event, represented by siliciclastics, that terminated the marine Hardwick Formation sedimentary episode, nor is there paleontologic evidence bearing on the ages of lithologic units postdating the Hardwick Formation: the Piccadilly Formation (uppermost unit of the Keelbottom Group) and the Glenrock Group, a predominantly volcanic sequence unconformably overlying the Keelbottom Group. Nevertheless, the Glenrock Group is thought, from regional stratigraphic relationships, to be Visean–?early Namurian (Lang et al., 1990b).

Venetia Formation. All but six samples out of 24 calcareous rudites, limestones, and arenaceous limestones from near the top of a relatively recessive 230-m interval of Venetia Formation in the Big Stockyard Creek area ca. 12 km south of Greenvale (section VEN, Fig. 4) proved barren of conodonts; only one of the 24 produced a chronologically interesting fauna. The sparse conodont fauna included *Hindeodus crassidentatus*, *Pseudopolygnathus multistriatus*, and *Bispathodus aculeatus plumulus*. The first of these suggests a *sandbergi*–Early *crenulata* age; the last, however, is normally not known to occur in horizons younger than *sandbergi* Zone (Sandberg, 1979). The inferred *sandbergi* Zone allocation for at least the highest horizons of our VEN section suggests to us that the *sandbergi*–Early *crenulata* transgression in the Burdekin Basin should probably be viewed as having taken place in the *sandbergi* Zone.

Inferred transgression-regression pattern

Burdekin Basin. The Townsville hinterland experienced at least seven major transgressions during the interval late Emsian–Tournaisian (Figs. 5, 7, and 8). The first of these, T1, expressed as the Wilkie Gray Group (Laroona Arkose and Mount Podge Limestone), occurred in the late Emsian and was approximately coeval with a widespread transgressive event in the Broken River region dated as having been primarily *serotinus* Zone but initiated very late in the *inversus* Zone (Mawson and Talent, 1989; Talent and Yolkin, 1987; Talent, 1989b; Sloan et al., 1995). It was clearly an incursion from the north (cf. Fig. 8) into a very small area of what subsequently became the Burdekin Basin *sensu stricto*.

The second of these transgressive events (Figs. 5 and 8,

Table 2), expressed by the transgression at the base of the Fanning River Group (*sensu* Wyatt et al., 1970; Wyatt and Jell, 1980; Talent and Mawson, 1994), is documented as having taken place very late in the Eifelian during the pre-*hemiansatus* portion of the *ensensis* Zone (or conceivably slightly earlier), that is just prior to the Eifelian-Givetian boundary (Talent and Mawson, 1994, p. 68). This transgression may have been partly if not entirely eustatically driven, and may have coincided with the transgression at the base of the Hare Indian Formation of western Canada, the Woodpecker Limestone of central Nevada, and the Oderhausen Limestone of Germany. Johnson et al. (1985) used this Euramerica-wide transgressive event to define the base of their T-R cycle If.

Accumulation of the Burdekin Limestone continued through the Givetian until at least some time into the Late *varcus* Subzone. About 50 m of Burdekin Limestone occurs above the entry of *P. latifossatus*, the entry of which is taken to be an approximate indicator of the base of the Late *varcus* Subzone. Maximum flooding, suggested by the thin occurrence of Burdekin Limestone at Kirkland Downs, was argued (Talent and Mawson, 1994, p. 68, section KD) to fall within the *hemiansatus* Zone/Subzone and to have extended into the Early *varcus* Subzone.

The minor deepening event represented by the clastics and subordinate calcareous "lower member" of the Cultivation Gully Formation (uppermost Fanning River Group) and the major regression very low in the Vanneck Formation (Figs. 5 and 8, events T3 and R3) are not tightly constrained chronologically. On the basis of data presented earlier (Talent and Mawson, 1994) and summarized above, this minor deepening event is inferred to have occurred during the late Givetian, presumably later than the *varcus* Zone.

We lack precise chronologic control on the regression near the base of the Vanneck Formation. We suggest it was probably very early in the Frasnian, contemporaneous with similar events in the nearby Broken River region where the Stanley Limestone Member (of the Mytton Formation) is overlain by an interval of fossiliferous brachiopod-bearing sandstones and siltstones (uppermost Mytton Formation or basal Bulgeri Formation; Talent and Brock, unpublished data), and these in turn by the Bulgeri Formation. In this latter area, the principal chronologic control comes from occurrence of the exclusively post-Givetian *Icriodus symmetricus* at the top of the Stanley Limestone Member; the same species is also known from within the Mytton Formation in the Broken River area, south of Storm Dam at SD 205 (Mawson and Talent, 1989, fig. 5; cf. Sloan et al., 1995, p. 29-30).

There seems not to have been any sharply defined transgressive or regressive events in Euramerica that would correspond to events T3 and R3 in the Burdekin Basin and Broken River area, that is, during the late Givetian *hermanni, disparilis,* and *norrisi* zones. Marine sediments from this particular time interval seem to have been very restricted areally in eastern Australia and difficult to discriminate; this contrasts with the earlier two-thirds of the Givetian. We therefore suggest that the T-R events inferred for the uppermost Burdekin Limestone, Cultivation Gully Formation, and basal Vanneck Formation were largely if not entirely tectonically driven.

According to Draper et al. (1994), the sedimentary history of the Keelbottom Group suggests four significant transgression-regression couples: The Myrtlevale Formation being witness to the first of these; the other three being displayed in the Hardwick Formation. The pattern indicated by our data (Figs. 5 and 8, T5–T7 and R5–R7) diverges somewhat from their model. The major transgressive event T4 at the base of the Myrtlevale Formation extended marine conditions over a large area of the Burdekin Basin, even beyond areas in which the largely if not entirely nonmarine Dotswood Group had accumulated (Wyatt et al., 1970; Wyatt, 1973; Wyatt and Jell, 1980; Draper et al., 1994). Conodont data presented here indicate that T4 occurred in the Early *marginifera* Zone, probably at or very close to its base. It was therefore very close to, if not coeval with, the Early *marginifera* Zone transgressive event responsible for enlargement of the Pilot Basin in Nevada and Utah (Johnson and Sandberg, 1989) and reported widely, for instance, from northern Africa (e.g. Wendt and Belka, 1991). Because of this alignment with transgression events elsewhere, the Early *marginifera* event in the Townsville hinterland may have been, in essence, eustatically rather than tectonically driven. Our data help underpin the global importance of this event.

Major regressive events seem to have been a feature of the Late and Latest *marginifera* zones in Euramerica (Johnson et al., 1985). These events may have had counterparts in the Townsville hinterland, but in the absence of conodont data for the interval Late *marginifera* Zone to Late *postera* Zone, such events cannot be discriminated. Four localities, however, provide evidence for an *expansa* Zone transgression, T5, apparently from the north, that we believe correlates with the Euramerican transgression at the base of transgression IIf of Johnson et al. (1985). Like the Early *marginifera* transgressive event, this event too is strongly displayed in northwest Africa (Wendt and Belka, 1991).

At Mount Podge, the abrupt change from carbonates and fine siliciclastics of the Myrtlevale Formation to coarse siliciclastics including pebbly sandstones and conglomerates (occasionally with angular clasts) of the Lollypop Formation, without an interval referable to the *expansa* Zone, is in accord with existence of a significant disconformity or even a very low angle unconformity. The latter interpretation appeals to us because, viewed regionally, the Lollypop Formation appears to rest on horizons of various ages, some referable to the Early *marginifera* Zone, others to the *expansa* zones; elsewhere, notably in the vicinity of Game Hill, the Lollypop Formation rests directly on Argentine Metamorphics.

The age of blanketing the Townsville hinterland by the coarse Lollypop Formation clastic wedge is not tightly constrained by conodont data, apart from the *expansa* Zone age of limestones immediately beneath it and *sandbergi* Zone well above it in the Hardwick Formation. Onset of the next major transgressive event, T6, evidenced by calcareous sandstones with abundant *Cyrtospirifer* and occasional chonetids occurring high in the Lollypop Formation and transitional to the Hardwick Formation, well displayed along Running Creek, was suggested (see earlier) as a possible analog of the major global transgression initiated at or just before the Devonian-Carboniferous boundary (Ross and Ross, 1988). This event was preceded by a major regressive event that had its inception mid-way through the *praesulcata* Zone (cf. Johnson et al., 1985).

Evidence for transgressive event T7 comes from the limestones of our section CORN at Corner Creek, and from the Saint Michael Member of the Hardwick Formation developed in the vicinity of Dotswood. This tendency to carbonate deposition, an inferred slackening of clastic input, and the presence of a modicum of siphondellid conodonts in the limestones (?somewhat more pelagic, cf. Girard, 1994) is thought to be consistent with an interval of basin-deepening, though, as Draper and Lang (1994a, p. 135) have pointed out, there was local shoaling of the carbonates, specifically in the Mount Saint Michael area. As noted earlier, these limestones have produced conodonts indicative of the *sandbergi*–Early *crenulata* interval. Evidence from our VEN section (see earlier) is in accord with this transgressive events having had its inception in the *sandbergi* Zone.

Draper and Lang (1994a, p. 135) inferred a "third and final Hardwick Formation transgression [that] resulted in the deposition of nearshore and shelfal sedimentary rocks across the area"; they referred the Corner Creek limestones (our section CORN) to this interlude. Our data have led us and J. J. Draper (personal communication, 1995) to believe that the Corner Creek sequence antedated this third Hardwick Formation transgression, T-8 (Fig. 5). We have presently no data bearing on the timing of this event.

The siliciclastic sequence of the upper Hardwick Formation has been interpreted as a regressive sequence reflecting shoreline environments; the nonmarine clastics of the succeeding Piccadilly Formation have been interpreted as braidplain and braidplain delta environments (Fig. 5; Draper and Lang, 1994a).

Viewed together, the Myrtlevale and Hardwick formations exhibit overall finer grain size of sediments, greater frequency of carbonates, and, at least in the case of the Myrtlevale Formation, increased conodont diversity northwestward. This is interpreted as reflecting somewhat greater water depths and increasingly pelagic environments in that direction and perhaps transgression having come from the north or northwest.

Broken River Province and Clarke River Basin. T-R events, additional to those inferred from sedimentary and biostratigraphic patterns for the Burdekin Basin, may be deciphered from the pre-Emsian stratigraphy of the Broken River region. These consist of:

1. A regression event, following the upper limestone "member" of the Jack Formation and inferred to be late *woschmidti* Zone (Simpson *in* Mawson et al., 1989; Sloan et al., 1995) or soon thereafter.

2. A T-R couple represented by the unnamed post-Jack unit with allochthonous carbonates (interpreted as transgressive; Sloan et al., 1995, p. 27, 39), followed by the lower Shield Creek Formation arenites (regressive).

3. A further T-R couple represented by the Arch Creek and Martins Well limestone members of the Shield Creek Formation (transgressive; late in the *pesavis* Zone; Martins Well Transgression of Talent, 1989b) and by the upper Shield Creek Formation arenites (regressive; late *sulcatus* Zone; Mawson et al., 1989).

These Lochkovian-Pragian events seem not to coincide with Euramerican T-R events inferred to have been eustatic; the first of these, interestingly, appears to coincide with the onset of the Bowning orogenic events in southeastern Australia (cf. Talent, 1989a).

If the interval sampled on our VEN section is indeed the only marine incursion into the Clarke River Basin, it might indicate that the pre-*sandbergi* T-R events are not represented in the basin. Such pre-*sandbergi* Zone events, indicated by sparse and poorly preserved fossils as well as the sedimentology, nevertheless occur in the Bulgeri and Turrets formations of the Broken River area, west of the Clarke River Basin. Caution is nevertheless necessary. The critical sample from our VEN section is from near the top of a very considerable marine interval. Pre-*sandbergi* horizons could be represented low in the VEN section; such horizons could also be represented in the even thicker marine section in the Ruxton area of the Clarke River Basin where B. F. Fordham's sampling (Draper et al., 1993b, p. 210) failed to produce chronologically constraining conodonts.

SYSTEMATIC PALEONTOLOGY

As the majority of conodonts in this study have been documented in the Catalogue of Conodonts (Ziegler, 1973, 1975, 1977, 1981, 1991) or described fully in earlier papers (e.g., Dreesen and Dusar, 1974), formal descriptions are restricted to new taxa, with the exception of a few diagnoses for species proposed in rare, not readily accessible, publications. Documentation of other species is limited to illustrations and, where appropriate, discussion. Ziegler and Sandberg's (1984, 1990) revised zonal scheme for the Famennian is used. In some instances, for clarity, square brackets are used to indicate approximate equivalence in terms of the previous zonal scheme (Ziegler, 1962a, 1971), for example, Early *expansa* Zone [= former Upper *styriacus* Zone]. Type and figured specimens are housed in the paleontological collections of the Queensland Museum. Numbers of specimens, precise horizon, and other locality data for each sample can be obtained from Figures 2 and 3 and Tables 1-3.

Order CONODONTOPHORIDA Eichenberg, 1930
Family ICRIODONTIDAE Müller and Müller, 1957

Genus *Icriodus* Branson and Mehl, 1938

Type species. Icriodus expansus Branson and Mehl, 1938.

Icriodus alternatus **Branson and Mehl, 1934a**
Figure 9: 3–4

Icriodus alternatus BRANSON AND MEHL, 1934a, p. 225–226, Pl. 13, Figs. 4–6; SZULCZEWSKI, 1971, p. 21, Pl. 7, Fig. 2; METZGER, 1989, p. 512, Fig. 13.8; WRIGHT ET AL., 1990, Fig. 9C–F.
Icriodus alternatus alternatus Branson and Mehl. SANDBERG AND DREESEN, 1984, p. 158–159, Pl. 2, Figs. 5, 11 (synonymy to 1984); MATYJA, 1993, Pl. 22, Figs. 2–3; Pl. 25, Figs. 1–3; BALINSKI, 1995, Pl. 19, Figs. 1, 2.

Remarks. Specimens of *I. alternatus* from Mount Podge sequence are distinguished from *I. cornutus* Sannemann by lack of a prominent posterior cusp and, in lateral view, absence of posterior arching (Ziegler, 1975, p. 101), diagnostic features of the latter.

Icriodus chojnicensis Matyja, 1972
Figure 9: 2, 5–7

Icriodus chojnicensis MATYJA, 1972, p. 745, Pl. 4, Figs. 1–4.
"Icriodus" chojnicensis Matyja. SANDBERG AND DREESEN, 1984, p. 163–164, Pl. 2, Figs. 12–14 (synonymy to 1984); JOHNSTON AND CHATTERTON, 1991, Pl. 3, Figs. 1–4; MATYJA, 1993, Pl. 27, Figs. 1–4; Pl. 30, Figs. 6–7.
"Icriodus" chojnicensis Matyja. MATYJA, 1987, p. 375, Pl. 22.2, Figs. 1–3.

Remarks. Small, slightly curved icriodontid specimens with a discrete central row of denticles, slightly offset from the lateral rows, and with a prominent posterior cusp were described by Matyja (1972) as a new species, *I. chojnicensis*. Subsequently, faunas of larger specimens were recovered from the Dinant and Vesdre basins, Belgium (Dreesen and Houlleberghs, 1980). Sandberg and Dreesen (1984), in arguing heterogeneity of the species, recognized three morphotypes: Morphotype 1—small with medium-sized cusp, all denticles rounded and the medial row of denticles showing only a slight indication of fusion; Morphotype 2 —larger than Morphotype 1, with well-developed posterior cusp, and round medial denticles tending to fuse to form a partial ridge; Morphotype 3—larger than Morphotype 1, with well-developed posterior cusp and a long, narrow, fused medial ridge with only traces of laterally compressed medial denticles. At Mount Podge, Morphotype 1 of Sandberg and Dreesen (1984) is represented by a specimen with discrete medial denticles and a small posterior cusp, illustrated herein (Figure 9: 5–7). Because of the tendency for the medial denticles to fuse, the specimen illustrated in Figure 9: 2 herein, is identified as *I. chojnicensis* Morphotype 2 *sensu* Sandberg and Dreesen (1984).

Icriodus cf. *constrictus* Thomas, 1949
Figure 10: 6

[cf.] *Icriodus constrictus* THOMAS, 1949, p. 416, Pl. 1, Fig. 25.

Remarks. A single icriodontid specimen is compared to *I. constrictus* on the basis of its overall shape and the anastomosing pattern of lateral ridges. It differs from the holotype from the Maple Hill Shale, Iowa, in the branching of lateral ridges on the platform being less ordered and in having the central ridge forming a herringbone pattern. The specimen of *I. constrictus* illustrated by Klapper (1966, pl. 6, fig. 18) shows a tendency for the medial row of denticles to become diagonally elongate, but they have not coalesced as in the Mount Podge specimen. Variation within this species is difficult to assess as so few have been recorded in the literature. Specimens identified by Anderson (1966) as *I. constrictus* differ from the holotype in having an anterior cusp. As only a few specimens of *I. constrictus* have been described and/or illustrated, it is not possible to identify the Mount Podge specimen with certainty.

Icriodus scalariformis n. sp
Figures 9: 1, 10: 3–5

Derivation of name. Scalaris (Lat.) = pertaining to a ladder, in reference to the ladderlike appearance of the elongate lateral denticles merging with the medial denticles.

Holotype. UQY 7815, the specimen illustrated in Figure 10: 4–5 is from locality 18, a limestone from the basal Hardwick Formation.

Diagnosis. A species of *Icriodus* in which representative Pa elements are generally fusiform in shape with lateral nodes extended transversely to form ridges that meet at the base of the indistinct medial nodes.

Description. The platform of *Icriodus scalariformis* is generally fusiform, with early forms tending to reach maximum width posterior of center. The strongly ridged denticle rows extend across the platform, creating, in upper view, a notched or scalloped platform margin. The indistinct medial denticles are aligned just above the meeting point of the lateral ridges. Posteriorly, the last two ridges or extended denticles are less clearly developed; the terminal denticle forms a modest cusp. The large, open basal cavity is rounded posteriorly and tapers to a point anteriorly.

Remarks. Because of lack of diversity in the late Famennian of icriodontids, *I. scalariformis* is described despite being represented by only seven specimens. It is thought that this species might form a link between older icriodontids having ridges connecting lateral and medial denticles, for example, *I. iowaensis* Youngquist and Peterson, and younger forms including the subspecies of *I. costatus* (Thomas). In the Mount Podge area, the oldest representative of *I. scalariformis* occurs in horizons of Early *marginifera* age. As the limestone in the WMP and EMP sections sampled in this area is no older than this zone, it is possible that *I. scalariformis* might extend into older horizons. According to Sandberg and Dreesen (1984), the upper limit of *I. iowaensis* is the top of the Early *rhomboidea* Zone, leaving a gap of just one conodont zone between the last occurrence of *I. iowaensis* and the first occurrence of *I. scalariformis*. The other occurrences of *I. scalariformis* in the Mount Podge area are in horizons of Early or Middle *expansa* age, thus overlapping in time with subspecies of *I. costatus* that collectively range from Latest *marginifera* Zone possibly to Early or Middle *praesulcata* Zone. Icriodontids are not known for certain from horizons postdating the Hangenberg extinction event during the Middle *praesulcata* Zone.

Specimens illustrated by Dreesen et al. (1976, pl. 9, figs. 1, 4; pl. 11, figs. 1, 3) and referred to *"I." costatus darbyensis* Klapper by Sandberg and Dreesen (1984), share with *I. scalariformis* the characteristic of development on the platform of transverse ridges joining with and/or replacing the lateral nodes. The posterior termination and the ridged platform of the Belgian specimens, however, are clearly differentiated: in *"I." costatus darbyensis* the posterior termination appears to be an extension of the platform, whereas in *I. scalariformis* the ridged platform tapers more gradually. Subspecies of *I. iowaensis* Youngquist and Peterson share with *I. scalariformis* the feature of conspicuous ridges connecting lateral denticles. They have, however, more widely spaced ridges, a much more strongly developed vertical row of elongate medial denticles, and, in most cases, a broader platform and wider, more expanded basal cavity than the new species.

Icriodus cf. *symmetricus* Branson and Mehl, 1934a
Figure 10: 1–2

[cf.] *Icriodus symmetricus* BRANSON AND MEHL, 1934a, p. 226, Pl. 13, Figs. 2–6; SANDBERG AND DREESEN, 1984, p. 157, Pl. 1, Figs. 2–6; MAWSON AND TALENT, 1989, Pl. 2, Figs. 4, 5; JI, 1989, p. 318, Pl. 4, Figs. 9–17; WRIGHT ET AL., 1990, Figure 9 J–L; UYENO, 1991, Pl. 5, Fig. 19; JI AND ZIEGLER, 1993, p. 57–58, Pl. 5, Figs. 11–13; MATYJA, 1993, Pl. 18, Figs.18–19, Pl. 19, Fig. 7.

Remarks. In the original description of *I. symmetricus*, Branson and Mehl (1934a) emphasized the lateral compression of the median row of denticles that "more or less connected antero-posteriorly into a straight, sharply nodose median crest that is somewhat higher than the lateral rows." The single specimen from locality 18 is compared to *I. symmetricus* because of this feature, but the relationship is only slight as the specimen from locality 18 shows no expansion of the basal cavity posteriorly; several specimens would be necessary before a firm attribution could be made. If the form is indeed *I. symmetricus*, this would extend the known range of the species to high in the *expansa* Zone. This is most unlikely, as there do not appear to be any forms closely comparable to *I. symmetricus* in the lower half of the Famennian.

There are no documented pre-Frasnian occurrences of *I. symmetricus*. Dr. R. Thomas Becker has kindly drawn our attention to undocumented occurrences from the late Givetian of Germany and Morocco. *I. symmetricus* has been reported from the Tannenkopf Slate by Leuteritz (*in* von Kamp, 1972, table 5, sample 716) as occurring with *P. varcus*, *P. decorosus*, and *I. obliquimarginatus*. It has been reported as well from the Brilon Reef area of eastern Sauerland by Stritzke (1990, Sections 8 and 9) from two horizons, respectively late in the *hermanni* Zone and most probably from the *disparilis* Zone. In Morocco, Ebert (1993, tables 5 and 6) has reported *I. symmetricus* from the Early *hermanni* Zone in Bed 33b at Bou Tchrafine and from Late *disparilis* Zone in Bed 13 at Hamar Laghdad.

Icriodus sp. A
Figure 9: 8–9

Remarks. Four icriodontids reminiscent of species commonly found in the Middle and Late Devonian have been found in the Myrtlevale Formation at Mount Podge. The base of the spindle is almost semicircular in outline, tapering rapidly to a slender anterior termination. Denticles are discrete, with few (commonly three or four) lateral pairs and twice as many (up to seven or eight) in the medial row, half of which occupy the slender anterior portion of the spindle. There is a posterior extension of the medial row of denticles beyond the spindle.

Family PRIONIODINIDAE Bassler, 1925
Genus *Apatognathus* Branson and Mehl, 1934a

Type species. Apatognathus varians Branson and Mehl, 1934a

Remarks. Because of lack of "cyclic alternation of diagonally offset denticles" (Clark et al., 1981), many species formerly referred to *Apatognathus* have been subsequently rejected. The genus appears to be restricted to the Late Devonian (Klapper, 1966; Nicoll and Druce, 1979; Nicoll, 1980; Stone, 1991; Over, 1992).

Apatognathus varians varians Branson and Mehl, 1934a
Figures 9: 13–14, 11: 22

Apatognathus varians BRANSON AND MEHL, 1934a, p. 201, Pl. 17, Figs. 1–3; GLENISTER AND KLAPPER, 1966, p. 803, Pl. 96, Figs. 14–16 (synonymy to 1966); BOOGAARD AND SCHERMERHORN, 1975, p. 7, Pl. 1, Fig. 2; NICOLL, 1980, p. 138–143, Figs. 3–5; WANG AND YIN, 1988, p. 114, Pl. 33, Fig. 9 (synonymy to 1988); OVER, 1992, p. 308, Figure 6.5.
Apatognathus varians klapperi Druce. JOHNSTON AND CHATTERTON, 1991, Pl. 2, Fig. 7.

Remarks. The specimens of *Ap. v. varians* closely resemble material from northwestern Australia described by Nicoll (1980). The Burdekin Basin occurrences are from horizons referred to the Early *marginifera* Zone (Figure 9: 13, 14) and *expansa* Zone (Figure 11: 22).

Family SPATHOGNATHODONTIDAE Hass, 1959

Genus *Bispathodus* Müller, 1962

Type species. Spathodus spinulicostatus Branson, 1934

Bispathodus stabilis (Branson and Mehl, 1934a)
Figure 11: 10, 11, 13

Spathodus stabilis BRANSON AND MEHL, 1934a, p. 188, Pl. 17, Fig. 20.
Bispathodus stabilis (Branson and Mehl). ZIEGLER ET AL., 1974, p. 103–104, Pl. 3, Figs. 1–3 (synonymy to 1974); SANDBERG AND ZIEGLER, 1979, p. 181–182, Pl. 2, Fig. 12; GAGIEV ET AL., 1987, Pl. 28, Fig. 3; WANG AND YIN, 1988, p. 116, Pl. 24, Fig. 3; METZGER, 1989, p. 516, Figs. 14.10, 14.14, 14.16, 14.20; JOHNSTON AND CHATTERTON, 1991, Pl. 2, Figs. 1, 2; OVER, 1992, p. 307–308, Figs. 6.1–6.16, 6.21, 6.26, 6.28; MATYJA, 1993, Pl. 34, Fig. 2; Pl. 35, Fig. 6; PERRET, 1993, Pl. C 2, Fig. 18.

Remarks. Specimens from the Mount Podge area closely resemble *Bi. stabilis* Morphotype 1 of Ziegler et al. (1974, pl. 3, figs. 1 and 3) wherein the basal cavity is relatively small and slightly asymmetrical and does not reach the posterior tip of the blade. Over (1992) recognized and illustrated all six elements of the apparatus. Ziegler et al. (1974) gave the range of this species as from Latest *marginifera* Zone [formerly Uppermost *marginifera* Zone] to *isosticha*–Late *crenulata* Zone [formerly Upper *crenulata* Zone].

Bispathodus aculeatus aculeatus (Branson and Mehl, 1934a)
Figure 11: 18–21

Spathodus aculeatus BRANSON AND MEHL, 1934a, p. 186, Pl. 17, Figs. 11, 14.
Bispathodus aculeatus aculeatus (Branson and Mehl). ZIEGLER ET AL., 1974, p. 101, Pl. 1, Fig. 5, Pl. 2, Figs. 1–8 (synonymy to 1974); NICOLL AND DRUCE, 1979, p. 21, Pl. 22, Figs. 6, 7; WANG AND ZIEGLER, 1983, Pl. 7, Fig. 3; WANG AND YIN, 1984, Pl. 3, Fig. 22; GAGIEV ET AL., 1987, Pl. 28, Fig. 4; WANG AND YIN, 1988, p.115, Pl. 24, Figs. 4, 8, 9; JI ET AL., *in* HOU ET AL., 1985, p. 100–101, Pl. 39, Figs. 19–22; NEMIROVSKAYA ET AL., 1993, Pl. 4, Figs. 4–6, 9, 12; MATYJA, 1993, Pl. 94, Fig. 7; PERRET, 1993, p. 276, Pl. C 2, Figs. 22, 23.

Remarks. Specimens of *Bi. aculeatus* in which any side denticle is positioned on or posterior to the basal cavity have been assigned to this long-ranging subspecies.

Bispathodus aculeatus anteposicornis (Scott, 1961)
Figure 11: 15–17

Spathognathodus anteposicornis SCOTT, 1961, Text-fig. 2 H-K.
Bispathodus aculeatus anteposicornis (Scott). ZIEGLER ET AL., 1974, p. 101, Pl. 1, Fig. 11; Pl. 2, Fig. 9; Pl. 3, Fig. 25 (synonymy to 1974); BARSKOV ET AL., 1984, Pl. 2, Fig. 1; GAGIEV ET AL., 1987, Pl. 28, Figs. 5, 6; WANG AND YIN, 1988, p. 115, Pl. 24, Figs. 5, 6; METZGER, 1989, p. 516, Fig. 14.22; MATYJA, 1993, Pl. 36, Fig. 2.

Remarks. As pointed out by Ziegler et al. (1974), this subspecies is rare but very useful biostratigraphically. In the specimens from locality 11, the large denticle on the right-hand side of the blade is found adjacent to the anterior margin of the basal cavity.

Figure 9 (on facing page). Numbers 1 ×110; 2–4, 8, 9, 13 ×90; 14 ×75; 5–7, 10–12 ×60; 15–22 ×45. 1: *Icriodus scalariformis* n. sp. Upper view of UQY 7716, WMP 521. 2, 5–7: *Icriodus chojnicensis* Matyja. 2, upper view of UQY 7717, WMP 515; 5–7, lateral, upper, and lower views of UQY 7718, EMP 6. 3, 4: *Icriodus alternatus* Branson and Mehl. Lower and upper views of UQY 7719, EMP 79.5. 8, 9: *Icriodus* sp. A. Lower and upper views of UQY 7720, WMP 420. 10–12: *Ancyrognathus* cf. *irregularis* Branson and Mehl. 10, 11, upper and lower views of UQY 7721, WMP 504; 12, upper view of UQY 7722, WMP 504. 13, 14: *Apatognathus varians varians* Branson and Mehl. 13, upper view of UQY 7723, EMP 72; 14, upper view of UQY 7724, EMP 68.5. 15, 16: *Mehlina strigosa* (Branson and Mehl). Lateral and upper views of UQY 7725, WMP 504. 17: *Palmatolepis quadrantinodosa quadrantinodosa* Branson and Mehl. Upper view of UQY 7726, WMP 448B. 18, 19: *Palmatolepis quadrantinodosa inflexoidea* Ziegler. Upper and lower views of UQY 7727, WMP 448B. 20: *Palmatolepis quadrantinodosa inflexa* Müller. Upper view of UQY 7728, WMP 438. 21, 22: *Palmatolepis glabra pectinata* Ziegler. 21, lower view of UQY 7729, WMP 448B; 22, upper view of UQY 7730, WMP 504.

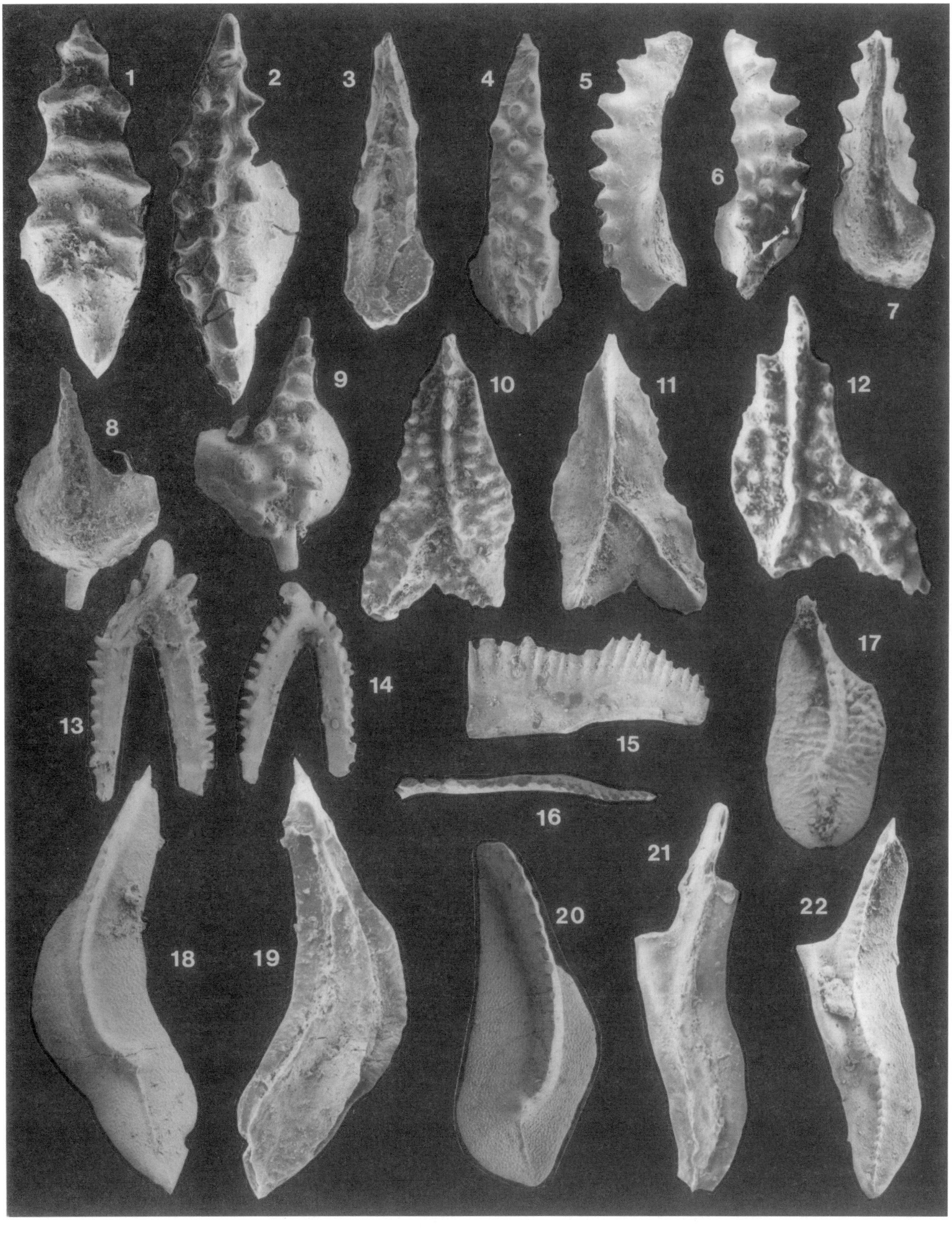
1
2
3
4
5
6
7
8
9
10
11
12
13
14
15
16
17
18
19
20
21
22

Bispathodus aculeatus plumulus (Rhodes, Austin, and Druce, 1969)

Spathognathodus plumulus plumulus RHODES ET AL., 1969, Pl. 1, Figs. 1, 2, 5, 6.
Bispathodus aculeatus plumulus (Rhodes et al.). ZIEGLER ET AL., 1974, p. 101-102, Pl. 2, Figs. 10, 11; Pl. 3, Fig. 24 (synonymy to 1974); NICOLL AND DRUCE, 1979, p. 21–22, Pl. 24, Figs. 1–6 (synonymy to 1979); MORY AND CRANE, 1982, Fig. 8A–C; BARSKOV ET AL., 1984, Pl. 2, Figs. 2–6.
Pandorinellina plumula (Rhodes et al.). MATYJA, 1993, Pl. 36, Fig. 5.

Remarks. A few specimens referred to the distinctive *Bispathodus aculeatus plumulus* were recovered from near the top of the recessive marine interval in the Venetia Formation at VEN ca. 210. This species has been documented from the Canning and Bonaparte Gulf basins of northwestern Australia (Glenister and Klapper, 1966; Druce, 1970a; Druce, 1976; Nicoll and Druce, 1979). Ziegler et al. (1974) give the range of this subspecies of *Bispathodus aculeatus* as from Middle *expansa* Zone [= upper part of the Lower *costatus* Zone] through to the *sulcata* Zone. *Bi. a. plumulus* seems not to have been unequivocally recorded from the *sandbergi* Zone, so this report from the Venetia Formation may be the youngest for this subspecies. However, Sandberg (1979, p. 100) listed *Bi. aculeatus*, without specification of subspecies, among important conodonts found in the *sandbergi* Zone.

Bispathodus spinulicostatus (E. R. Branson, 1934)

Spathodus spinulicostatus E. R. BRANSON, 1934, p. 305, Pl. 25, Fig. 19.
Bispathodus spinulicostatus E. R. Branson. NICOLL AND DRUCE, 1979, p. 22, Pl. 26, Figs. 1–15 (synonymy to 1979); GAGIEV ET AL., 1987, Pl. 28, Fig. 7.

Remarks. Specimens of *Bispathodus spinulicostatus* from localities 16, 20U, and 21U are similar to that described by Ziegler et al. (1974), who give the range for this species as from Middle *expansa* Zone [upper part of the former Lower *costatus* Zone] to ?*crenulata* Zone.

Genus _Mehlina_ Youngquist, 1945

Type species. Mehlina irregularis Youngquist, 1945 (= *Mehlina gradata* Youngquist, 1945).

Mehlina strigosa (Branson and Mehl, 1934a)
Figure 9: 15–16

Spathodus strigosus BRANSON AND MEHL, 1934a, p. 187, Pl. 17, Fig. 17.
Mehlina strigosa (Branson and Mehl). MATYJA, 1987, Pl. 22.6, Figs. 2, 3; METZGER, 1989, p. 517, Figs. 14.13, 14.15 (synonymy to 1989); MATYJA, 1993, Pl. 35, Fig. 2; Pl. 36, Fig. 2.
Spathognathodus strigosus (Branson and Mehl). WANG AND YIN, 1988, p. 145, Pl. 32, Figs. 5–7 (synonymy to 1988).

Remarks. A small number of specimens of *M. strigosa* in the Mount Podge collections are confidently assigned to this species because of similarity to the holotype (Branson and Mehl, 1934a).

Genus _Pseudopolygnathus_ Branson and Mehl, 1934b

Type species. Pseudopolygnathus prima Branson and Mehl, 1934b

Pseudopolygnathus multistriatus Mehl and Thomas, 1947
Figure 11: 12

Pseudopolygnathus multistriatus MEHL AND THOMAS, 1947, p. 16–17, Pl. 1, Fig. 36; NICOLL AND DRUCE, 1979, p. 30, Pl. 19, Figs. 1–6; KLAPPER (*in* Ziegler), 1981, p. 395–396, *Pseudopolygnathus*-Pl. 4, Figs. 2–4 (synonymy to 1981); WANG AND YIN, 1984, Pl. 2, Figs. 14–15; VARKER AND SEVASTOPULO, 1985, Pl. 5.2, Figs. 2, 7, 8; NEMIROVSKAYA ET AL., 1993, Pl. 4, Fig. 16; PERRET, 1993, p. 278–279, Pl. C 4, Figs. 20, 22; PICKETT, 1994, Fig. 2.13–31.

Remarks. Although the elongate nodes are not always well preserved in the specimen from near the top of the recessive marine interval in the Venetia Formation at VEN ca. 210, the nodose carina and configuration of the basal pit allow positive identification. Elsewhere in Australia, *Ps. multistriatus* has been recorded from the Canning Basin, northwestern Australia (Nicoll and Druce, 1979).

Family POLYGNATHIDAE Bassler, 1925

Genus _Ancyrognathus_ Branson and Mehl, 1934a

Type species. Ancyrognathus symmetricus Branson and Mehl, 1934a

Ancyrognathus cf. _irregularis_ Branson and Mehl, 1934a
Figure 9: 10–12

[cf.]*Ancyrognathus irregularis* BRANSON AND MEHL, 1934a, p. 242, Pl. 19, Figs. 4, 16; KLAPPER, 1990, p. 1014–1015, Figs. 3.10, 10.14, 10.15 (synonymy to 1990).

Remarks. Specimens of *Anc. irregularis* illustrated from Dixie, Missouri (Branson and Mehl, 1934a, e.g. pl. 19, the upper surface illustrated in fig. 4, and the lower surface of a second specimen illustrated in fig. 16) resemble the specimens from the Myrtlevale Formation at Mount Podge, especially in respect to alignment of the nodes and the acute angle at the junction of the carinae. However, as the Queensland specimens are proportionally longer and narrower than the type material, the lobe on the outer side less well developed, the sinus at the posterior platform margin deeper, and the bifurcate pit not as well developed, they cannot be regarded as being truly conspecific with *Anc. irregularis*. The possibility of the Mount Podge specimens representing a new species cannot be ruled out, but an insufficient number of specimens are available for a clear understanding of morphologic variability. On the other hand, the Mount Podge specimens are smaller than those from the Holts Summit Sandstone reillustrated by Klapper (1990, figs. 3.10, 10.14, 10.15)—0.9 mm compared with 1.35 and 1.75 mm—suggesting that the differences might be related to the maturity of the specimens. Klapper (1990) noted that the ancyrognathids from Branson and Mehl's (1934a) collection are from a mixed fauna from the Holts Summit Sandstone, the youngest elements of which can be dated as Early *marginifera* Zone. It appears that although the genus *Ancyrognathus* is generally thought of as persisting to horizons as young as Latest *crepida* Zone [formerly Upper *crepida* Zone] (e.g., Ziegler *in* Ziegler, 1973), the present report not only extends the range of the genus to the Early *marginifera* Zone but raises the possibility that the type specimens of *Anc. irregularis* from the Holts Summit Sandstone are not reworked and thus could be as young as Early *marginifera* Zone.

Figure 10 (on facing page). Numbers 1–5, 9, 10, 14–16 ×90; 11–13 ×75; 6–8 ×60. 1, 2: *Icriodus* cf. *symmetricus* Branson and Mehl. Lower and upper views of UQY 7813, Locality 18. 3–5: *Icriodus scalariformis* n. sp. 3, upper view of UQY 7814, Locality 17; 4, 5, upper and lower views of UQY 7815, Locality 18. 6: *Icriodus* cf. *constrictus* Thomas. Upper view of UQY 7816, Locality 23. 7, 8: *Polygnathus* cf. *inornatus* E. R. Branson. Upper and lower views of UQY 7817, Locality 18. 9, 10: *Polygnathus communis communis* Branson and Mehl. Upper and lower views of UQY 7818, Locality 19. 11–13: *Polygnathus lectus* Kononova. 11, upper view of UQY 7819, Locality 20; 12, 13, upper and lower views of UQY 7820, Locality 20. 14: *Polygnathus semicostatus* Branson and Mehl, Morphotype 2 Sandberg and Ziegler. Upper view of UQY 7821, Locality 17. 15, 16: *Polygnathus vogesi* Ziegler. Lower and upper views of UQY 7822, Locality 23.

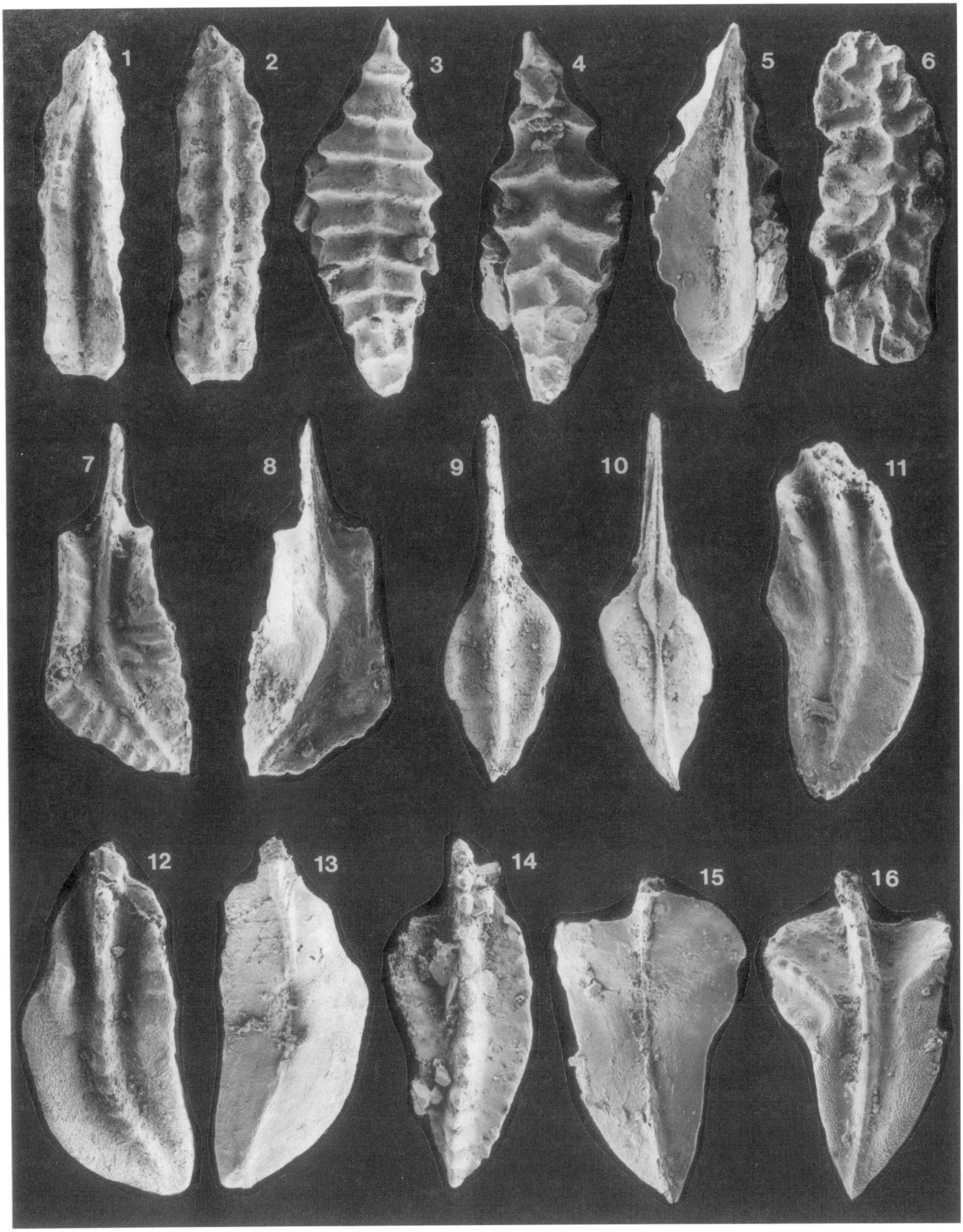
1
2
3
4
5
6
7
8
9
10
11
12
13
14
15
16

Genus *Polygnathus* Hinde, 1879

Type species. Polygnathus dubius Hinde, 1879

Polygnathus asilbekensis Khalymbadzha, Shinkaryov and Gatovskiy, 1991
Figure 12: 23

Polygnathus asilbekensis KHALYMBADZHA ET AL., 1991, p. 58, Pl. 7, Figs. 3–5; BARSKOV ET AL., 1991, p. 104, Pl. 30, Figs. 11–13.

Remarks. One specimen from high in the West Mount Podge section (WMP/504) closely resembles *P. asilbekensis* described from the Famennian of southern Kazakhstan (Khalymbadzha et al., 1991) in having a narrow, only slightly curved platform with upturned margins and a low, smooth carina that extends to the pointed posterior extremity. The platform is smooth. The specimens from Kazakhstan are from horizons dated as *marginifera* Zone.

Polygnathus bouckaerti Dreesen and Dusar, 1974
Figure 12: 12–16

Polygnathus bouckaerti DREESEN AND DUSAR, 1974, p. 11–12, Pl. 1, Figs. 1–7; Pl. 2, Figs. 1–12, Text-fig. 8; BARSKOV ET AL., 1991, p. 74, Pl. 20, Figs. 6–9; MATYJA, 1993, Pl. 30, Fig. 9; Pl. 31, Fig. 1.
Polygnathus bouckerti [sic] Dreesen and Dusar. BALINSKI, 1995, Pl. 21, Figs. 6, 11, 14.
Polygnathus pennatuloidea Holmes. FILIPOVIC ET AL., 1975, Pl. 24, Fig. 22; PICKETT, 1981, Fig. 2.16.

Amended Diagnosis. A species of *Polygnathus* in which representative Pa elements have an elongate or oval platform generally ornamented by nodes that may fuse, forming irregular, transverse ridges in the posterior portion of the platform.

Remarks. Examination of collections from the Famennian of Belgium illustrated by Dreesen and Dusar (1974) indicates that formation of low ridges on the posterior of the platform by fusion of nodes is a feature not consistently present on specimens in the Belgian collection; this can be seen in specimens illustrated by them (Dreesen and Dusar, 1974, pl. 2, figs. 1–12). The original diagnosis has therefore been slightly amended.

Specimens from Mount Podge (Tables 1 and 2) are very similar to Dreesen and Dusar's (1974) description and illustrations of *P. bouckaerti*. The range of variation in the two faunas is also comparable: Some specimens show the formation of low, irregular transverse ridges on the platform posterior to the fading out of the carina (cf. Fig. 12: 13–14 herein with Dreesen and Dusar, 1974, pl. 2, fig. 1) but others do not (cf. Fig. 12: 15 herein with Dreesen and Dusar, 1974, pl. 2, fig. 6). A feature also shared by the latter two specimens is development of incipient costae along the outer anterior margin of the platform.

The specimen from northwest Serbia illustrated by Filipovic et al. (1975, pl. 24, fig. 22) as *P. pennatuloideus* Holmes is referred to *P. bouckaerti* because the shape of the platform and the parallel nature of the nodose rostral ridges of the former are features closely matching those of the latter.

Dreesen and Dusar (1974), Barskov et al. (1991), and Vorontsova (1993) give the range of *P. bouckaerti* as from the base of the *rhomboidea* Zone to the top of the Early *marginifera* Zone.

Polygnathus communis communis Branson and Mehl, 1934b
Figures 10: 9–10; 11: 7

Polygnathus communis communis BRANSON AND MEHL, 1934b, p. 293, Pl. 24, Figs. 1–4; GAGIEV ET AL., 1987, Pl. 26, Figs. 1, 2; SCHÖNLAUB et al., 1988, Pl. 3, Figs. 8–9; WANG AND YIN, 1988, p. 125–126, Pl. 25, Figs. 1–17; Pl. 26, Figs. 1–17 (synonymy to 1988); METZGER, 1989, p. 518, Figs. 7, 9 (synonymy to 1989); PERRI AND SPALLETTA, 1991, p. 70, Pl. 5, Figs. 1–2; JOHNSTON AND CHATTERTON, 1991, Pl. 2, Figs. 11, 12; MATYJA, 1993, Pl. 36, Fig. 1; PERRET, 1993, p. 293, 295, Pl. C 6, Figs. 36, 40, 41, 43; PICKETT, 1994, Fig. 2.1–2; BAI, 1994, p. 176, Pl. 26, Fig. 4; Pl. 29, Fig. 1; BALINSKI, 1995, Pl. 20, Fig. 4.
Neopolygnathus communis (Branson and Mehl). BARSKOV ET AL., 1991, p. 115, Pl. 31, Figs. 6–9.

Remarks. Specimens of *P. c. communis* from Mount Podge conform with the holotype in having a rounded, basinlike platform with generally smooth, upturned margins and, on the underside, a depression behind the basal pit. On the basis of the configuration of the pit and lower surface of *Polygnathus* species commonly known as the *communis* Group, Vorontsova (*in* Barskov et al., 1991) erected a new genus *Neopolygnathus* to separate them from *Polygnathus* in a more restricted sense. Acceptance of the new genus would seem to depend on demonstration that the nonplatform elements are at variance with those of *Polygnathus*.

Polygnathus cf. *communis communis* Branson and Mehl, 1934b
Figure 13: 2, 3

Remarks. A single specimen of *Polygnathus* from high in the WMP section at WMP504 is close to *Polygnathus communis communis*, but the distinctive, short marginal ridges on the platform preclude unequivocal assignment. The upper surface of the Mount Podge specimen closely resembles a specimen of *P. depressus* Metzger (1989, fig. 14.6), but the Mount Podge specimen lacks the deep circular depression immediately posterior to the pit, a diagnostic characteristic of Metzger's species.

Polygnathus costulifera n. sp.
Figure 14: 13–17

?*Polygnathus szulczewskii* MATYJA, 1975, Fig. 1.7.

Derivation of name. Csostula (Latin), diminutive of *costa*, a rib, and *fero* (Lat.), to carry or bear, in reference to the small ribs or ridges ornamenting the platform.

Holotype. UQY 7776, the specimen illustrated in Figure 14: 14–15;

Figure 11 (on facing page). Numbers 13, 14 ×90; 1, 3, 18, 19, 21 ×75; 2, 5–7, 11, 15, 17, 20, 22 ×60; 8, 10, 12, 16 ×45; 4, 9 ×30. 1–4: *Siphonodella obsoleta* Hass. 1, upper view of UQY 7823, Locality CORN 5; 2, upper view of UQY 7824, Locality CORN 5; 3, lower view of UQY 7825, Locality CORN 5; 4, upper view of UQY 7826, Locality CORN 1. 5, 6: *Polygnathus* cf. *inornatus* E. R. Branson. 5, upper view of left-hand form, UQY 7827, Locality CORN 4; 6, upper view of right-hand form, UQY 7828, Locality CORN 5. 7: *Polygnathus communis communis* Branson and Mehl. Upper oblique view of UQY 7829, Locality 12. 8: *Polygnathus experplexus* Sandberg and Ziegler. Upper view of UQY 7830, Locality 18. 9: *Polygnathus homoirregularis* Ziegler. Upper view of UQY 7831, Locality 18. 10, 11, 13: *Bispathodus stabilis* (Branson and Mehl). 10, upper view of UQY 7832, Locality 11; 11, lateral view of UQY 7833, Locality 11; 13, upper view of UQY 7835, Locality CORN 5. 12: *Pseudopolygnathus multistriatus* Mehl and Thomas. Upper view of UQY 7834, Locality POW 3. 14: *Hindeodus crassidentatus* (Branson and Mehl). Lateral view of UQY 7836, Locality, POW 3. 15–17: *Bispathodus aculeatus anteposicornis* (Scott). 15, lateral view of UQY 7837, Locality 11; 16, upper view of UQY 7838, Locality 11; 17, upper view of UQY 7839, Locality 11. 18–21: *Bispathodus aculeatus aculeatus* (Branson and Mehl). 18, upper view of UQY 7840, Locality CORN 5; 19, upper view of UQY 7841, Locality CORN 5; 20, upper view of UQY 7842, Locality CORN 5; 21, upper view of UQY 7843, Locality POW 3. 22: *Apatognathus varians varians* Branson and Mehl. Upper view of UQY 7844, Locality 11.

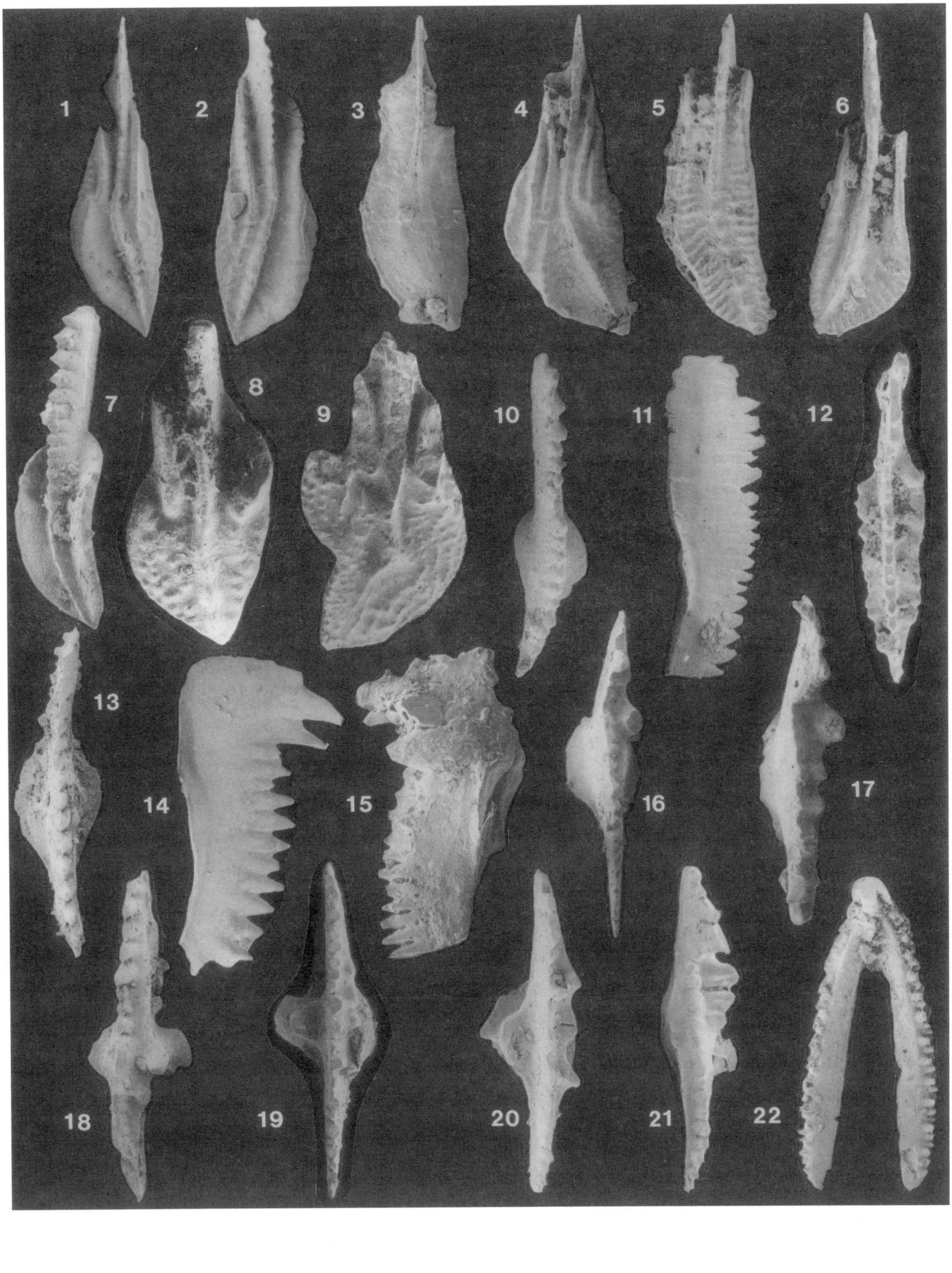
1
2
3
4
5
6
7
8
9
10
11
12
13
14
15
16
17
18
19
20
21
22

it is from EMP 48, 16.4 m stratigraphically above the base of the EMP section through the Myrtlevale Formation at East Mount Podge.

Diagnosis. A species of *Polygnathus* in which representative Pa elements have an asymmetrical, elongate, lanceolate platform, with a strong constriction along the left-hand margin of the anterior platform giving rise to a deeper adcarinal groove than on the right-hand side. The carina, straight anteriorly and gently incurved posteriorly, is clearly developed for three-quarters the length of the platform. Platform ornament consists primarily of transverse ridges.

Description. The asymmetrical, slightly arched platform is elongate, with the length of the platform being greater than twice the maximum width of the unit. The outer anterior margin of the platform meets the blade farther anteriorly than the other (inner anterior) margin. In left-hand forms, a strong constriction in the anterior third of the outer platform margin forms a deep adcarinal groove. In right-hand forms, by contrast, the constriction on the inner anterior margin is more pronounced than that on the higher, outer margin. Adcarinal grooves are confined to the anterior half of the platform. The carina is straight and clearly developed for from half to three-quarters the length of the platform and is gently incurved posteriorly, with the angle formed at the point of curvature being 20 to 25°.

Distinct transverse ridges are generally developed along the platform margins; on the tongue of the platform they may become discontinuous and irregular, with ridges interspersed with nodes close to the margin. The free blade is of moderate length, approximately one-third the length of the complete unit. A minuscule, lenticular basal pit is situated in the anterior third of the platform; a narrow groove runs anteriorly from it onto the free blade; a keel runs to the posterior extremity of the platform.

Remarks. P. costulifera, although similar to *P. planirostratus* in some respects, can be separated from it because of having the following characteristics: an elongate platform (width to length ratio being 1:2+ compared to 2:3 for *P. planirostratus*), a strongly expressed constriction in the anterior platform margin, less curvature of the carina (the angle formed at the point of curvature being 20 to 25° compared to 40°+ in *P. planirostratus*), and only a slight degree of arching of the platform compared to *P. planirostratus*. Although both *P. costulifera* and *P. szulczewskii* are elongate, the former is proportionally a much narrower form, and the lateral margins of the platform are almost parallel compared to those of mature specimens of *P. szulczewskii* that taper toward the anterior from the point of maximum width of the platform. The degree of curvature of the carina is much greater in *P. szulczewskii* (the angle formed at the point of curvature being 30 to 40° compared to 20 to 25° in *P. costulifera*), and in *P. costulifera* the carina is straight for from half to three-quarters the length of the platform compared to only half or less than half in *P. szulczewskii*. In mature specimens of *P. szulczewskii*, the adcarinal grooves are quite short compared to those in correspondingly mature specimens of *P. costulifera*.

Although left-hand specimens of *P. aequalis* have an upturned anterior outer margin, they do not have a marked constriction as is the case in *P. costulifera*. The Frasnian species *P. imparilis* Klapper and Lane (1985) is easily distinguished from *P. costulifera* by its greater curvature of the platform and the numerous fine ribs ornamenting most of the platform. From a fauna from the Lvov Depression, Drygant (1986) described a *crepida* Zone fauna including *P. volhynicus*, which differs from *P. costulifera* in having a carina continuing right to the posterior extremity of the platform, resulting in longer adcarinal grooves, and in having a less well developed marginal constriction. At Mount Podge, *P. costulifera* occurs in the Myrtlevale Formation, in the Early *marginifera* Zone.

Polygnathus diversus Helms, 1959

Figure 12: 19–21

Polygnathus diversa HELMS, 1959, p. 650–651, Pl. 5, Figs. 5–8.

Polygnathus diversus Helms. DREESEN AND DUSAR, 1974, Pl. 3, Fig. 28; AUSTIN ET AL., 1985, Pl. 4.7, Figs. 7–8; OLIVIERI, 1985, Pl. 7, Figs. 2–5.

Remarks. Olivieri (1985) records *P. diversus* from the Late *marginifera* Zone in southwestern Sardinia. Faunas from Central Kazakhstan include specimens of the species from the Late and Latest *marginifera* zones. Sweet (1988) shows the range of *P. diversus* extending from the base of the Early *marginifera* Zone; Barskov et al. (1991) give a generalized *marginifera* age for the species. A small number of specimens of *P. diversus* occur 72 m above the base of the EMP section and 40 m above the base of the WMP section. It should be noted that though the Mount Podge occurrences are from the Early *marginifera* Zone, they are not from early in the zone.

Polygnathus experplexus Sandberg and Ziegler, 1979

Figure 11: 8

Polygnathus experplexus SANDBERG AND ZIEGLER, 1979, p. 185–186, Pl. 4, Figs. 2–6 (synonymy to 1979); PERRI AND SPALLETTA, 1991, p. 70, Pl. 5, Fig. 3; JI AND ZIEGLER, 1993, p. 79, Text-fig. 20, Fig. 7; MATYJA, 1993, Pl. 34, Figs. 8–10.

Polygnathus perplexus Thomas. METZGER, 1989, Figs. 15.13, 14, 25.

Remarks. It has been suggested by Metzger (1989) that *P. perplexus* and *P. experplexus* should be regarded as synonyms. Although confusion might arise in distinguishing right-hand specimens of *P. perplexus* and *P. experplexus*, as suggested by Sandberg and Ziegler (1979), left-hand forms are easily distinguished, the former lacking the strongly divergent rostral ridges and the expanded platform posterior to the rostral ridges. *P. perplexus* is known from older horizons (Early *trachytera* Zone) than *P. experplexus* (first appearing in the Early *expansa* Zone).

Polygnathus hieroglyphica n. sp.

Figure 14: 1–10

Polygnathus sp. C MATYJA AND ZBIKOWSKA, 1974, p. 684, 692, Pl. 7, Fig. 7.

Polygnathus orientalis GAGIEV ET AL., 1987, Pl. 26, Figs. 11a–b only [non Fig. 9 = *P. szulczewskii*; Figs. 10, 12 = *P. planirostratus*]; BARSKOV ET AL., 1991, p. 60, Pl. 14, Fig. 15 [non Fig. 16 = *P. planirostratus*].

Derivation of name. From the Greek, *hieroglyphikos*, meaning holy writing or carving (commonly referring to Egyptian writing), in reference to the ornament of the platform.

Holotype. UQY 7772, the specimen illustrated in Figure 14: 8–10

Figure 12 (on facing page). Numbers 3, 4, 19, 21 ×90; 1, 2, 5, 6 ×75; 7–9, 11, 20, 23 ×60; 10, 12–15, 17, 18, 22 ×45; 16 ×30. 1–11: *Polygnathus wyatti* n. sp. 1, 2, lower and upper views of UQY 7782, WMP 521; 3, 4, upper and lower views of UQY 7783, WMP 521; 5, 6, upper and lower views of UQY 7784, WMP 521; 7, 8, upper and lower views of UQY 7785, WMP 438; 9, upper view of UQY 7786, WMP 515; 10, 11, upper and lower views of holotype UQY 7787, WMP 504. 12–16: *Polygnathus bouckaerti* Dreesen and Dusar. 12, 13 lower and upper views of UQY 7788, EMP 34; 14, upper view of UQY 7789, EMP34; 15, upper view of UQY 7790, WMP 438; 16, upper view of UQY 7791, EMP 4. 17, 18: *Alternognathus pseudostrigosus* (Dreesen and Dusar). Upper and lateral views of UQY 7792, EMP 72. 19–21, *Polygnathus diversus* Helms. 19, 21: upper and lateral views of UQY 7793, WMP 438; 20, lateral view of UQY 7794, WMP 504. 22: *Polygnathus planirostratus* Dreesen and Dusar. Upper view of UQY 7795, WMP 438. 23: *Polygnathus asilbekensis* Khalymbadzha, Shinkaryov and Gatovskiy. Upper view of UQY 7796, WMP 504.

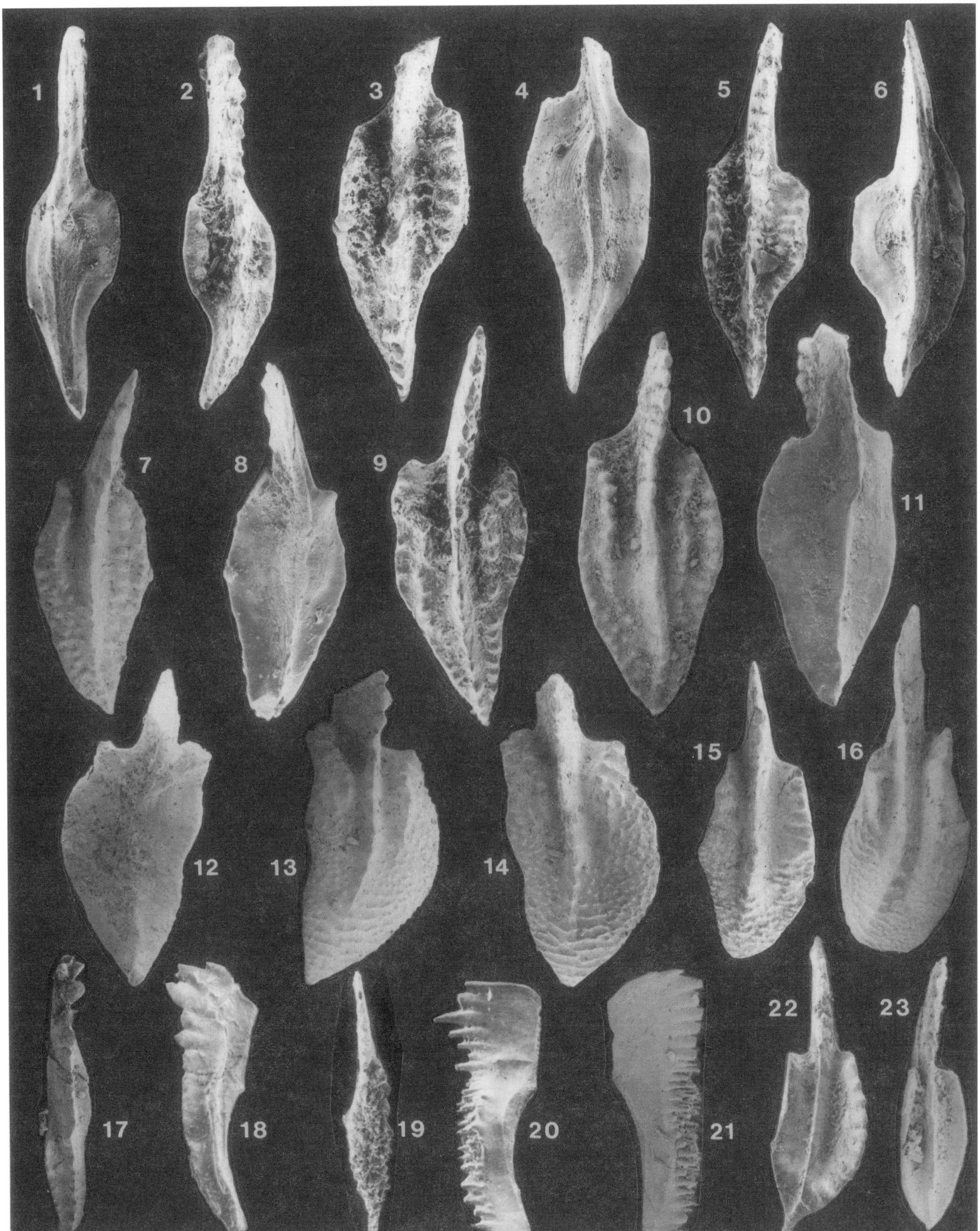
1
2
3
4
5
6
7
8
9
10
11
12
13
14
15
16
17
18
19
20
21
22
23

from sample WMP 504, high in the Myrtlevale Formation, 203.1 m above the base of section WMP at Mount Podge.

Diagnosis. A species of *Polygnathus* in which representative Pa elements having a relatively broad, lanceolate platform constricted symmetrically, anteriorly resulting in short, deep, angular adcarinal grooves and upwardly flexed platform margins posterior to the constriction. Ornament varying from irregular, weak, discontinuous, or anastomosing ridges to a pattern of short ridges and nodes, at times resembling hieroglyphs.

Description. The arched platform is lanceolate, moderately wide at the anterior margin, widest at midlength and constricted anteriorly, resulting in short, deep adcarinal grooves in the anterior quarter of the platform that "V" toward the carina; the posterior part of the constriction is flexed upward. The anterior quarter of the platform is either smooth or bears a few faint, parallel ridges. The free blade, composed of coalescing denticles, occupies about a third of the length of the entire unit; it continues as a rapidly diminishing carina to a point level with the lateral flexure; from there only an indistinct trace can be followed to the rounded posterior extremity. Ornament on the posterior three-quarters of the platform consists of an irregular pattern of weak, discontinuous ridges anastomosing at the margins, breaking into very short segments and nodes. In the vicinity of the upturned flexed margins, the pattern becomes greatly disturbed. The minuscule basal pit is located on the lower surface at the point of greatest curvature, generally from midway to a third of the way anteriorly along the platform. From the basal pit, a keel runs anteriorly to the free blade and posteriorly to the platform's extremity.

Remarks. The flexures in the anterior part of the platform of *P. hieroglyphica* can be compared to those in *P. nodoundatus* Helms (see *P. bicavata in* Ziegler, 1962a, pl. 10, especially fig. 7), but the platform of the latter is far wider and more asymmetrical, and the ornament is almost exclusively nodose. It is clear that *P. bouckaerti* Dreesen and Dusar (1974) belongs to the "*nodocostata Gruppe*" *sensu* Helms (1961); the authors designated two illustrated specimens (Dreesen and Dusar, 1974, pl. 2, figs. 9, 10) as being transitional to *P. nodocostatus*. *P. hieroglyphica* differs from *P. bouckaerti* in being proportionally wider anteriorly and in having distinct, paired constrictions in the anterior margins of the platform, resulting in definite flexures in the platform margin, and in having far less nodose ornament. A common feature for all species of the *nodocostata* group *sensu* Helms (1961) is the narrow anterior margin of the platform at its junction with the free blade, a feature that separates all of them from *P. hieroglyphica*.

Ornament along the anterior upturned margins is reminiscent of adult specimens of *P. brevis* Miller and Youngquist (cf. Szulczewski, 1972, pl. 2, figs. 1–4) in which the nodes are said to be "fused into hieroglyphic ridges." The specimens from Mount Podge differ from *P. brevis* in having a paired constriction and flexure close to the anterior margin of the platform, resulting in short, angular, adcarinal grooves and generally smooth upturned margins compared with the proportionally long, parallel adcarinal grooves and clearly ornamented anterior margins that typify *P. brevis*.

Although incomplete, the specimen figured by Matyja and Zbikowska (1974, pl. 7, fig. 7) from a core in the Lublin Basin is referred here to *P. hieroglyphica*. Gagiev et al. (1987, pl. 26, fig. 11a–b) include in the illustrations of specimens they describe from Magadan as *P. orientalis* (synonymized herein) a specimen that has many features in common with *P. hieroglyphica*.

Polygnathus homoirregularis Ziegler, 1971
Figure 11: 9

Palmatolepis? irregularis THOMAS, 1949, Pl. 2, Fig. 27 [non *Polygnathus irregularis* Cooper, 1939]
Polygnathus homoirregularis ZIEGLER, 1971, p. 269-270; ZIEGLER in ZIEGLER, 1973, p. 365–366, *Polygnathus*-Pl. 3, Figs. 4, 9, 10 (synonymy to 1973); BARSKOV ET AL., 1991, p. 80–81, Pl. 22. Figs. 16–18.

Remarks. Several specimens recovered from localities 17 and 18 are identified as *P. homirregularis* because of the distinctive bend in the carina, the rostral ridges, and shape and ornament of the platform. Ji and Ziegler (1993) give the range of *P. homoirregularis* as extending from within the Late *trachytera* Zone to the end of the Middle *expansa* Zone. Sweet (1988), incidentally, gave a more limited range: from the base of the *postera* Zone to the top of the Early *expansa* Zone.

Polygnathus cf. *inornatus* E.R. Branson, 1934
Figures 10.7-8; 11.5-6

[cf.]*Polygnathus inornata* E. R. BRANSON, 1934, Pl. 25, Figs. 8, 26.
Polygnathus lobatus lobatus Branson and Mehl. RHODES ET AL., 1969, p. 191–192, Pl. 9, Figs. 5–8.
[cf.]*Polygnathus inornatus* E.R. Branson. KLAPPER (*in* ZIEGLER, 1975, p. 294–297, *Polygnathus*-Pl. 4, Figs. 1, 2, 4 (synonymy to 1975); GAGIEV ET AL., 1987, Pl. 27, Figs. 1–3; WANG AND YIN, 1988, p. 126, Pl. 27, Fig. 10; Pl. 31, Figs. 15, 16; PERRET, 1993, p. 298, Pl. C 7, Figs. 5–11, 15, 16, 24, 25; BAI, 1994, p. 178, Pl. 26, Figs. 11, 12.

Remarks. Specimens from Corner Creek identified as *P.* cf. *inornatus* differ from the type material in being less arrowlike in outline. They more closely resemble specimens from England identified as *P. lobatus lobatus* (Rhodes et al., 1969, pl. 9, figs. 5–7) and subsequently included with *P. inornatus* (Klapper *in* Ziegler, 1975, p. 295). In accord with Klapper (*in* Ziegler, 1975, p. 294) they are, therefore, regarded as a left and right of "an asymmetrical curvature pair (Symmetry Class IIIa of Lane, 1968)." *P. inornatus* is long ranging, known to occur from the late Famennian (*expansa* Zone) to almost the Tournasian-Viséan boundary (latest *anchoralis–latus* Zone).

Polygnathus karadjalis Vorontsova and Kuz′min, 1984
Figure 15: 15–24

Polygnathus karadjalis VORONTSOVA AND KUZ′MIN, 1984, p. 61–62, Pl. 1, Figs. 1–5; BARSKOV ET AL., 1991, p. 23, Pl. 4, Figs. 1, 2.

Original Diagnosis (translated from Russian, Vorontsova and Kuz′min, 1984). Conodonts with a "boxlike" platform; anterior lateral edges are raised a little and form deep troughs. The platform surface bears transverse costae. The pitted basal cavity is located near the central part of the platform.

Amended Diagnosis. Representative Pa elements of *P. karadjalis* have a "serotiniform" (i.e., like *P. serotinus*) platform, and can be distinguished by the arrow-shaped structure developed on the undersurface of the platform anterior of the centrally situated basal pit.

Remarks. In the upper view, *P. karadjalis* resembles a slightly wide version of *P. serotinus* but with very deep adcarinal troughs and with anterior margins of the platform widening outward anteriorly and termi-

Figure 13 (on facing page). Numbers 4, 14 ×90; 5, 15 ×75; 1–3, 10, 11, 13 ×60; 7–9, 16–19 ×45; 6, 12 ×30. 1, 5: *Polygnathus pennatuloideus* Holmes. 1, upper view of UQY 7797, WMP 515; 5, upper view of fragment of UQY 7800 showing ornamentation at higher magnification, WMP 515. 2, 3: *Polygnathus* cf. *communis communis*. Lower and upper views of UQY 7798, WMP 504; 4: *Polygnathus nodocostatus* Branson and Mehl. Upper view of broken specimen, UQY 7799, WMP 515. 6–8: *Polylophodonta confluens* (Ulrich and Bassler). Upper oblique, upper, and lower views of UQY 7801, EMP 72. 9: *Polygnathus* sp. A. Upper view of UQY 7802, example of Duplicatio Pathology *sensu* Weddige (1990), WMP 521. 10–19: Apparatus A. All lateral views. 10, ?Pa element, UQY 7803, WMP 504; 11, Sb element, UQY 7804, WMP 504; 12, Sa element, UQY 7805, WMP 504; 13, ?Sb element, UQY 7806, WMP 504; 14, ?Sb element, UQY 7807, WMP 504; 15, Sc element, UQY 7808, WMP 521; 16, ?Pa element, UQY 7809, WMP 521; 17, Pb element, UQY 7810, WMP 521; 18, Pb element, UQY 7811, WMP 504; 19, M element, UQY 7812, WMP 521.

1
2
3
4
5
6
7
8
9
10
11
12
13
14
15
16
17
18
19

nating at right angles to the blade. There is strong downward flexing of the blade. In lower view, from the deepest part of the curved surface anteriorward, the base of the free blade is exposed. The lower platform, reflecting the base of the adcarinal troughs of the upper surface, folds inward tightly against the exposed blade, forming a raised arrowlike structure that is slightly higher than the outer side of the unit. A small basal pit is located slightly posterior to the "arrow tip"; at times it is so tiny that it cannot be easily detected.

Comparisons could be drawn between the upper surface of *P. karadjalis* and *P. serotinus* Telford, or various subspecies of *P. linguiformis*; however, the arrow-shaped structure developed anteriorly on the lower surface of *P. karadjalis* clearly separates it from any of them.

Specimens of *P. karadjalis* from central Kazakhstan (Vorontsova and Kuz′min, 1984) occur in association with *Pal. minuta minuta* Branson and Mehl, *Pal. falcata* (Helms), *Pal. glabra distorta* Branson and Mehl, *P. granulosus* Branson and Mehl, and *I. cornutus chojnicensis* Matyja as well as with two forms Vorontsova and Kuz′min proposed as new species: *P. subnormalis* and *P. argutus*, the latter now synonymized with the former (Kuz′min, 1990). Vorontsova and Kuz′min (1984) considered *P. karadjalis* to occur in limestones referrable to the "*Polmatolepis marginifera–Scaphignathus velifer* zones"; Barskov et al. (1991, table 1, p. 175) show the range to be from Late *marginifera* Zone to the end of the Early *trachytera* Zone, whereas Vorontsova (1993, p. 85) suggested a local *Polygnathus karadjalis* Zone equating with the interval from base of the Early *marginifera* Zone to top of the Late *trachytera* Zone. This range has been confirmed recently (A.V. Kuz′min, personal communication, 1995). To date, *P. karadjalis* has been reported only from central Kazakhstan and northern Queensland.

Polygnathus lauriformis Dreesen and Dusar, 1974
Figure 14: 18–22

Polygnathus lauriformis DREESEN AND DUSAR, 1974, p. 16, Pl. 1, Figs. 8–11; Pl. 3, Figs. 1–12, Text-fig. 11; PERRI AND SPALLETTA, 1990, p. 65, Pl. 5, Figs. 6a–b; MATYJA, 1993, Pl. 29, Fig. 4; BALINSKI, 1995, Pl. 21, Fig. 5.
Polygnathus sp. FILIPOVIC ET AL., 1975, Pl. 24, Figs. 23, 24.

Original Description (Dreesen and Dusar, 1974). A species of *Polygnathus* with a narrow, symmetrical lancet-shaped platform; margins ornamented with coarse, regular transverse ridges, not extending to the high, denticulate or nodose carina that reaches the posterior end of the platform.

Remarks. Material from both sections at Mount Podge, EMP and WMP, conforms with the diagnostic features of the species, namely the narrow, lancet-shaped platform, coarse ornament, and high, robust carina extending to the posterior tip of the platform. The tiny, elongate basal pit is situated on the widened keel, a little anterior of center of the platform. On small specimens the platform is ornamented with regular ridges, and the carina appears denticulate. On mature specimens the ridges become less clearly defined, giving way to nodose ornament with the nodes tending to align parallel to the carina. The change in ornament is accompanied by a slight thickening of the platform margin and a thickening and smoothing of the carina.

Specimens illustrated by Dreesen and Dusar (1974) from Belgium and France appear to be predominantly young forms; adult forms illustrated on their pl. 1, figs. 8, 9, 10, and pl. 3, figs. 11, 12, closely resemble specimens recovered from Mount Podge. Material described from southern Kazakhstan as *P. guttiformis* (Khalymbadzha et al., 1991) resembles *P. lauriformis* in some respects but differs in having a less symmetrical platform and less robust ornament, even in small specimens.

Polygnathus lectus Kononova, 1981
Figure 10: 11–13

Polygnathus communis lectus KONONOVA, in BUSHMINA AND KONONOVA, 1981, p. 96, Pl. 20, Figs. 2, 4–7.
Neopolygnathus lectus (Kononova). BARSKOV ET AL., 1991, p. 116–117, Pl. 31, Figs. 10–12.

Remarks. Originally considered to be a subspecies of *P. communis*, the taxon was given species status in 1991 (Kononova *in* Barskov et al., 1991). The range given for the species is from *praesulcata* Zone to *?duplicata* Zone. Material from undifferentiated Keelbottom Group at locality 20 east of Star homestead is associated with faunas of the *sandbergi*–Early *crenulata* interval.

Polygnathus nodocostatus Branson and Mehl, 1934a
Figure 13: 4

Polygnathus nodocostata BRANSON AND MEHL, 1934a, p. 246–247, Pl. 20, Figs. 9–13; Pl. 21, Fig. 15; HELMS, 1959, Pl. 1, Figs. 17, 21, 23; Pl. 2, Figs. 16–20, 22.
Polygnathus nodocostatus Branson and Mehl. JOHNSTON AND CHATTERTON, 1991, Pl. 2, Fig. 16; JI AND ZIEGLER, 1993, p. 81–82, Pl. 34, Figs. 13–15; Text-fig. 15 (synonymy to 1993); BAI, 1994, p. 181, Pl. 23, Fig. 6.
Polygnathus nodocostatus nodocostatus Branson and Mehl. BALINSKI, 1995, Pl. 21, Figs 1, 2.

Remarks. A few specimens are referred to *P. nodocostatus* because of the arrangement of the rows of fused nodes paralleling the carina. Ji and Ziegler (1993) give the range of *P. nodocostatus* as being from within the Early *crepida* Zone into the Early *expansa* Zone.

Polygnathus pennatuloideus Holmes, 1928
Figure 13: 1, 5

Polygnathus pennatuloidea HOLMES, 1928, p. 32-33, Pl. 11, Fig. 14; GLENISTER AND KLAPPER, 1966, p. 830, Pl. 94, Figs. 11, 12 (synonymy to 1966); JI AND ZIEGLER, 1993, p. 83, Pl. 34, Fig. 12 (?juvenile form).
[non] *Polygnathus pennatuloideus* Holmes. JI AND ZIEGLER, 1993, Text-fig. 20, Fig. 10.
Polygnathus triphyllatus (Ziegler). LONG, 1987, p. 510, Figure 3-A, B, E.

Remarks. This form from near the top of the Myrtlevale Formation on section WMP can be recognized by its asymmetrical platform, with the inner platform being widest anteriorly, and distinctive, nodose ("knotty") ornament tending to become disturbed toward the anterior of the platform. The margin of the outer platform is curved more evenly than for *P. nodoundatus*. Whereas Ji and Ziegler (1993) suggested that *P. pennatuloideus* first makes its appearance around the middle of Late *marginifera* Zone, Barskov et al. (1991) and Sweet (1988) give the range as Early *marginifera* Zone into Late *marginifera* Zone. This accords with our report from horizons regarded as Early *marginifera*

Figure 14 (on facing page). Numbers 18, 21, 22 ×90; 11 ×75; 12, 14, 15, 19, 20 ×60; 3, 7, 13, 16 ×45; 1, 2, 4–6, 8–10, 13, 17 ×30. 1–10: *Polygnathus hieroglyphica* n. sp. 1, upper view of UQY 7765, WMP 504; 2, upper oblique view of UQY 7766, WMP 504; 3, upper view of UQY 7767, EMP 79.5; 4, upper view of UQY 7768, WMP 504; 5, upper view of UQY 7769, WMP 504; 6, upper view of UQY 7770, WMP 504; 7, upper view of UQY 7771, WMP 504; 8–10, upper oblique, lower, and upper view of holotype UQY 7772, WMP 504. 11, 12: *Polygnathus szulczewskii* Matyjia. 11, upper view of UQY 7773, WMP 515; 12, upper view of UQY 7774, WMP 504. 13–17: *Polygnathus costulifera* n. sp. 13, upper view of UQY 7775, WMP 504; 14, 15, lower and upper views of holotype UQY 7776, EMP 48; 16, upper view of UQY 7777, WMP 448B; 17, lower view of UQY 7778, WMP 504. 18–22: *Polygnathus lauriformis* Dreesen and Dusar. 18, upper view of UQY 7779, EMP 49; 19, 20, lower and upper views of UQY 7780, EMP 34; 21, 22, upper and upper oblique views of UQY 7781, WMP 521.

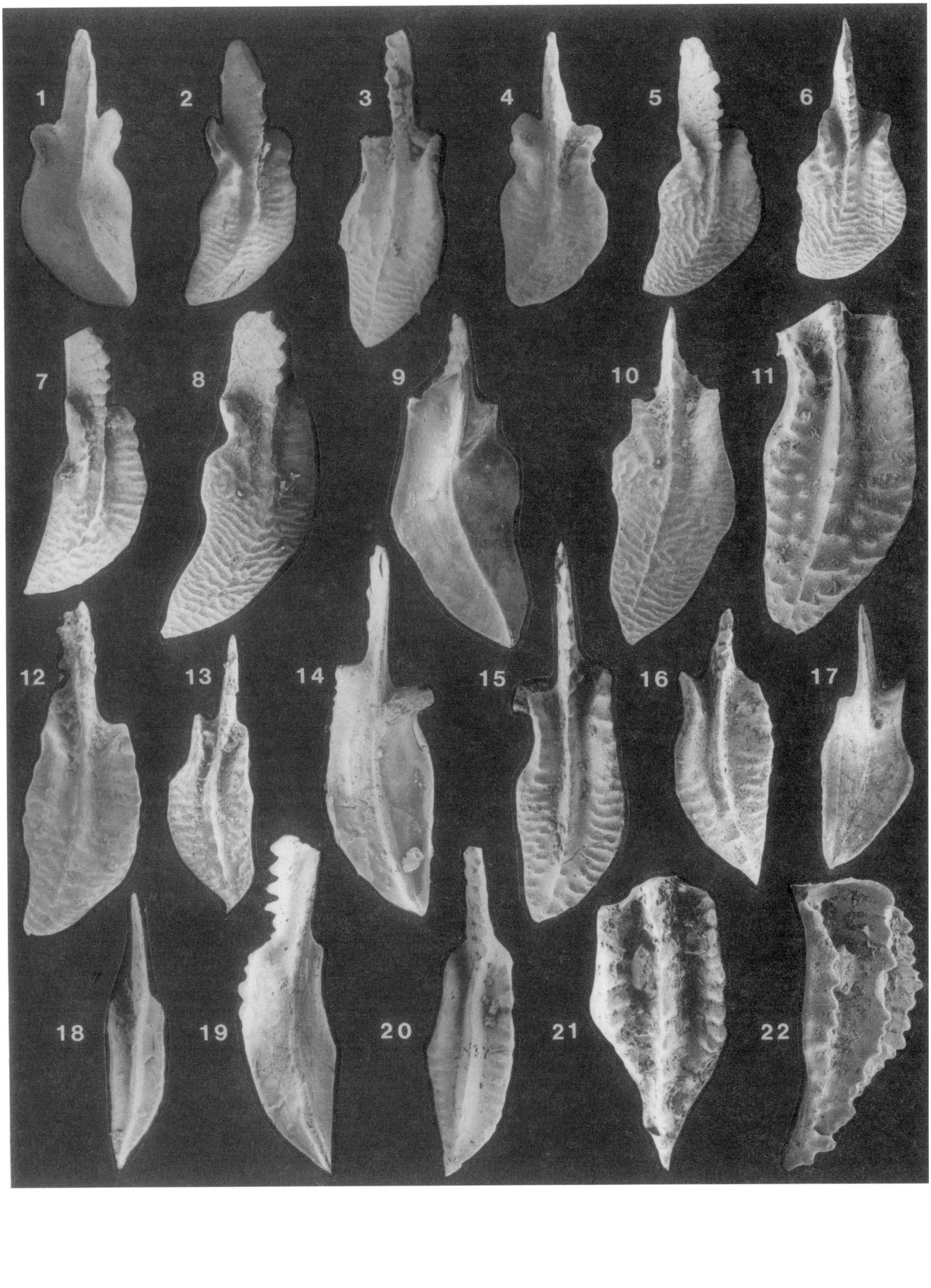
1
2
3
4
5
6
7
8
9
10
11
12
13
14
15
16
17
18
19
20
21
22

Zone, though the Mount Podge occurrences could be viewed as late in the zone.

***Polygnathus planirostratus* Dreesen and Dusar, 1974**
Figure 12: 22

Polygnathus sp. ZIEGLER, 1962a, Pl. 10, Fig. 22.
Polygnathus planirostratus DREESEN AND DUSAR, 1974, p. 18, Text-Fig. 12, Pl. 1, Fig. 14 [non Fig. 15 = *P. szulczewskii*], Pl. 2, Figs. 13–15, 17–18, ?20 [non Figs. 16, 19 = *P. szulczewskii*]; BARSKOV ET AL., 1991, p. 62, Pl. 14, Figs. 13, 14.
Polygnathus orientalis GAGIEV ET AL., 1987, p. 95–96, Pl. 26, Figs. 10, 12 [non Fig. 9 = *P. szulczewskii*; Fig. 11 = *P. hieroglyphica*]; BARSKOV ET AL., 1991, p. 60, Pl. 14, Fig 16 [non Fig. 15 = *P. hieroglyphica*].

Original Diagnosis. "A species of *Polygnathus* with upturned platform margins in the anterior part of the unit, producing adcarinal grooves. The posterior part is rather flat and broad with smooth, short ridges or sparse nodes (forma 1), or bearing widely spaced cross ridges (forma 2)."

Amended Diagnosis. A species of *Polygnathus* with upturned, near parallel platform margins that in the anterior part of the unit produce adcarinal grooves. Carina, straight anteriorly and incurved posteriorly, is clearly developed for half to three-quarters the length of the platform. The posterior platform is rather flat and broad, with ornament of ridges or a combination of aligned short ridges and nodes.

Remarks. Two forms of *P. planirostratus* were defined by Dreesen and Dusar (1974, p. 18). The holotype illustrated in their text-fig. 12 and on their pl. 2, fig. 17, is clearly an example of their "forma 1" with the posterior part of the platform "smooth with isolated nodes or short irregular ridges near the margins, sometimes connected with the nodes of the carina." Dreesen and Dusar's "forma 2," exemplified in their text by the specimen figured by Ziegler (1962a, pl. 10, fig. 22) and illustrated in their paper (Dreesen and Dusar, 1974, pl. 2, figs. 16, 19), presents some problems. Dreesen and Dusar (1974, p. 18) differentiated "forma 2" by the ornament of the platform, but rather than complying with their definition of "forma 2" in "bearing widely spaced cross ridges" (on the posterior platform), the specimens Dreesen and Dusar illustrate (1974, pl. 2, figs. 16, 19) conform more closely with *P. szulczewskii*, a species described by Matyja (1975) from borehole Minkowice 1 in the central part of the Lublin Basin, southeast Poland. The specimen illustrated by Ziegler (1962a, pl. 10, fig. 22) which Dreesen and Dusar (1974, p. 18) refer to in their discussion as an example of their "forma 2," fits within their definition of "forma 1" and is, indeed, very similar to the specimen they illustrate (Dreesen and Dusar, 1974, pl. 2, fig. 13) as an example of "forma 1."

Features that separate *P. szulczewskii* from *P. planirostratus* include the following: The former has a proportionally longer platform (width to length ratio 1:2 compared to 2:3 for *P. planirostratus*), the free blade in the former generally is longer, the lateral margins in *P. szulczewskii* narrow significantly toward the anterior whereas in *P. planirostratus* the lateral margin are close to parallel in the anterior half of the platform, and *P. szulczewskii* shows a slight but definite constriction of the anterior part of the platform. Although not of high taxonomic significance, the platform of *P. szulczewskii* is ornamented anteriorly by weakly expressed ribs giving way posteriorly to a rather irregular pattern of interrupted transverse ribs, nodes, and tubercles, whereas the ornament of *P. planirostratus* consists of a more ordered pattern of ridges or interrupted ridges and isolated nodes that tend to be aligned.

P. planirostratus is represented by 47 specimens from both sections at Mount Podge. It is ornamented by short, irregular ridges or nodes at the margins. Some of our specimens, however, appear to have a more attenuated tongue than the Belgian material. Compare, for example, specimens illustrated by Dreesen and Dusar (1974, pl. 2, figs. 17–18) with Figure 12: 22 herein; the latter, however, may be an end member of the species.

Gagiev et al. (1987), in documenting faunas from Magadan, proposed a new species, *P. orientalis*; the published illustrations indicate that at least some of the specimens included in this species fall within *P. planirostratus* as presently understood.

Dreesen and Dusar (1974, p. 18) state that *P. planirostratus* ranges from the base of the Early *rhomboidea* Zone into the Early *trachytera* Zone [= former Middle *velifer* Zone], with no material being obtained from horizons of Late, or Latest, *marginifera* zones.

***Polygnathus pseudobrevilaminus* Vorontsova, 1993**
Figure 15: 1–8

Polygnathus pseudobrevilaminus VORONTSOVA, 1993, p. 74, Pl. 8, Figs. 21–26.

Diagnosis. A species of *Polygnathus* in which representative Pa elements have an asymmetrical, lanceolate platform with upturned margins of subequal height giving rise to well-developed adcarinal grooves for three-quarters the length of the platform. The carina is generally straight and clearly developed for the length of the platform; platform ornament consists primarily of transverse ridges or, more rarely, irregular nodes.

Remarks. The rather abundant material from Mount Podge shows greater variation in ornament of the platform than the specimens of *P. pseudobrevilaminus* illustrated by Vorontsova (1993, pl. 8, figs. 21–26), ranging from distinct transverse ridges meeting the smooth adcarinal grooves or shorter, less distinct ridges developed only on the margins of the platform (resulting in wide, smooth adcarinal grooves), to the development of tiny nodes along the posterior margins, especially of the tongue. An example of the last is the specimen from EMP/21 (Fig. 15: 6) showing an almost smooth platform with nodes or tiny ridges developed only on the posterior part of the platform and extending along the extreme lateral margins. Such intraspecific variation in ornament is not uncommon. The Frasnian species, *P. aequalis* Klapper and Lane (1985, figs. 16.7–16.14), exhibits similar variability in ornament on the platform. *P. aequalis* is, however, more linguloid in outline and shows a slight but definite constriction of the anterior platform. Right-hand forms of the latter are significantly wider than left-hand forms; this is not the case in *P. pseudobrevilaminus*. *P. pseudobrevilaminus* was thought to range from the Late *marginifera* Zone to the *postera* Zone (Vorontsova, 1993, table 1), but it is present in relative abundance in the Early *marginifera* Zone at Mount Podge.

The left-hand specimens of *P. padovanii* Perri and Spalletta (1990, pl. 6, figs. 1, 2, 5a) from the Carnic Alps, Italy, are similar in many respects to *P. pseudobrevilaminus*, but no specimens of the latter having denticulate anterior margins of the platform occur among the Mount Podge material, and none were illustrated by Vorontsova (1993). The holotype of *P. padovanii* (Perri and Spalletta, 1990, pl. 6, figs. 3a–d) appears to be widest in the anterior part of the platform, and the carina begins to curve anterior of mid-length. By contrast, *P. pseudobrevilaminus* is widest posterior to the mid-point of the platform, coinciding with the curve in the carina.

Figure 15 (on facing page). Numbers 3–5, 10, 13, 14, 17 ×75; 1, 2, 6–8, 11, 12, 18–23 ×60; 9, 15, 16, 24 ×45. 1–8, *Polygnathus pseudobrevilaminus* Vorontsova. 1–2, lower and upper views of UQY 7750, EMP 34; 3–5, upper, oblique upper, and lower views of UQY 7751, EMP 53; 6, upper view of UQY 7752, EMP 21; 7, upper view of UQY 7753, EMP 21; 8, upper view of UQY 7754, EMP 21. 9–14, *Polygnathus subnormalis* Vorontsova and Kuz′min. 9, upper view of UQY 7755, WMP 448B; 10, upper view of UQY 7756, EMP 34; 11, 12, lower and upper views of UQY 7757, WMP 438; 13, 14, lower and upper views of UQY 7758, WMP 411. 15–24, *Polygnathus karadjalis* Vorontsova and Kuz′min. 15, 16, upper and lower views of UQY 7759, WMP 504; 17, lower view of UQY 7760, EMP 34; 18, 19, lower and upper views of UQY 7761, WMP 504; 20, 21, upper and lower views of UQY 7762, WMP 504; 22, 23, lower and upper views of UQY 7763, EMP 21; 24, lower view of UQY 7764, EMP 16.

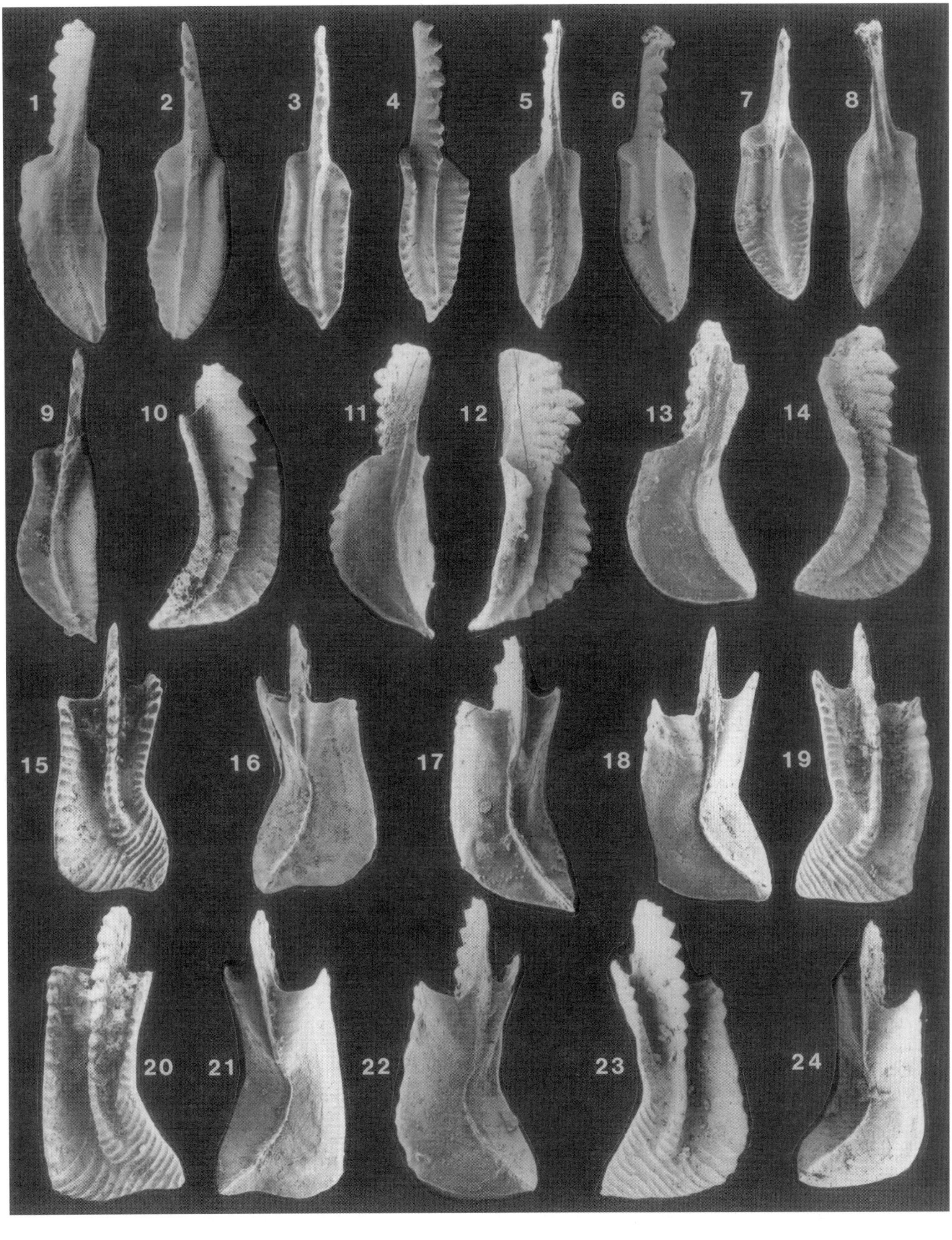
1
2
3
4
5
6
7
8
9
10
11
12
13
14
15
16
17
18
19
20
21
22
23
24

A Givetian subspecies proposed by Drygant (1986), *P. timorensis semitimorensis*, is very similar to *P. pseudobrevilaminus*. The deep adcarinal grooves and upturned platform margins of the latter, however, continue almost to the posterior tip of the platform, and, moreover, Drygant's subspecies has less well defined ornament.

The inference that the range of *P. pseudobrevilaminus* in central Kazakhstan is from Late *marginifera* Zone to somewhere in the *postera* Zone can be gleaned from Vorontsova (1993, p. 74 and Table 1). The lower limit derives from positioning of Vorontsova's informal *sulcifer* horizon, shown by her as commencing in the Late *marginifera* [former Upper *marginifera* Zone]. Because *P. pseudobrevilaminus* is present in the basal beds of both the WMP and EMP sections at Mount Podge, we infer the range to probably extend downward to the vicinity of the base of the Early *marginifera* Zone.

Polygnathus semicostatus Branson and Mehl, Morphotype 2 Sandberg and Ziegler, 1979
Figure 10: 14

Polygnathus semicostata BRANSON AND MEHL, 1934a, p. 247–248, Pl. 21, Figs. 1, 2.
Polygnathus semicostatus Branson and Mehl, Morphotype 2. SANDBERG AND ZIEGLER, 1979, p. 187–188, Pl. 5, Figs. 8–10.

Remarks. Dreesen and Orchard (1974) discriminated several morphologic trends in *P. semicostatus* from horizons ranging in age from Middle *crepida* Zone to Late *trachytera* Zone. Following their lead, Sandberg and Ziegler (1979) identified three morphotypes of this species in younger horizons, that is, from Early *postera* Zone to Late *expansa* Zone. One specimen from the Mount Podge area (Fig. 10: 14) closely resembles a specimen from the Leatham Formation, Wasatch Range, Utah, illustrated by Sandberg and Ziegler (1979, pl. 5, fig. 9).

Polygnathus subnormalis Vorontsova and Kuz′min, 1984
Figure 15: 9–14

Polygnathus subnormalis VORONTSOVA AND KUZ′MIN, 1984, p. 62, Pl. 1, Figs. 6–10; KUZ′MIN, 1990, p. 70–71, Pl. 4, Figs. 8–10 (synonymy to 1990); BARSKOV ET AL., 1991, p. 34, 35, Pl. 4, Figs. 7, 8; METZGER, 1994, p. 640, 644, Figs. 19.10–19.22.
Polygnathus argutus VORONTSOVA AND KUZ′MIN, 1984, p. 62–63, Pl. 1, Figs. 11–13; BARSKOV ET AL., 1991, p. 14, 15, Pl. 3, Figs. 9, 10.

Original Diagnosis. Conodonts with lanceolate platform with raised lateral margins attached asymmetrically to the free blade. Platform surface covered by transverse costae. Basal cavity lenticular, extending as a slit in the free blade.

Remarks. Vorontsova and Kuz′min (1984), in differentiating two new species of *Polygnathus*, *P. argutus* and *P. subnormalis*, pointed out that the latter has a lanceolate, narrow platform with raised lateral margins for its entire length, contrasting with the linguiloid platform of the former in which adcarinal grooves are developed only anteriorly. Realizing that both species occurred in approximately the same numbers in each horizon sampled and that *P. subnormalis* occurred only as a right-hand form whereas *P. argutus* was always a left-hand form, Kuz′min (1990) synonymized the latter with the former, arguing them to be an asymmetric pair *sensu* Klapper (1971a, p. 66). Some specimens from the Mount Podge collection assigned to *P. subnormalis* tend to have less pronounced and less closely spaced transverse ridges along the lateral margins of the platform, but this could be an artifact of preservation. Barskov et al. (1991) give the range for *P. subnormalis* as *rhomboidea* Zone to *trachytera* Zone.

Polygnathus szulczewskii Matyja, 1975
Figure 14: 11, 12

Polyganthus planirostratus DREESEN AND DUSAR, 1974, Pl. 1, Fig. 15; Pl. 2, Figs. 16, 19; VORONTSOVA, 1993, Pl. 9, Figs. 12–15.
Polygnathus szulczewskii MATYJA, 1975, p. 785–786, Figs. 1–4, ?5, 6, 8–10, [non Fig. 7 = ?*P. planirostratus*]; MATYJA AND ZBIKOWSKA, 1974, p. 682–683, Pl. 7, Figs. 2–6; GAGIEV, 1979, p. 45, 46, Pl. 14, Figs. 6, 7 [non Figs. 8, 9]; BARSKOV ET AL., 1991, p. 65, Pl. 13, Figs. 16–18; MATYJA, 1993, Pl. 29, Fig. 3; Pl. 30, Fig. 10; Pl. 32, Fig. 2, non Pl. 34, Fig. 7; BALINSKI, 1995, Pl. 22, Figs. 1–2.
Polygnathus orientalis GAGIEV ET AL., 1987, p. 95–96, Pl. 26, Figs. 9a–b [non Fig. 10, 12 = *P. planirostratus*; Fig. 11 = *P. hieroglyphica*].

Original Diagnosis (Matyja, 1975). A *Polygnathus* with elongated, arched, laterally bent platform. In its anterior part, highly raised margins form rather deep grooves on both sides of the carina. Posterior part of the platform wider than the anterior one and sharply terminated. The platform in its anterior part is either devoid of any sculpture or is ornamented with weak ribs; in its posterior part it is covered with transverse ribs interrupted in their middle parts. Ribs of internal part of the platform shorter than those covering its external part. Long, free blade with high teeth. Carina massive, low, does not reach the posterior end of platform. On the aboral side of the platform keel and long, narrow basal cavity placed in the anterior part.

Remarks. Although Matyja's paper appeared in the Bulletin de L'Académie Polonaise des Sciences, vol. 22, no. 11 for 1974, the date of distribution noted on reprints is 1975. Two specimens from Belgium referred by Dreesen and Dusar (1974, pl. 2, figs. 16 and 19) to their new species, *P. planirostratus*, a form characterized by having a broad, relatively short platform and more regular ornament, are here referred to *P. szulczewskii*. Refer to the Remarks section of *P. planirostratus* and *P. costulifera* for discussion of the differences between these species and *P. szulczewskii.*

Matyja's (1975) specimens were recovered from strata spanning the *crepida* to *marginifera* zones. Gagiev (1979, p. 45–46, pl. 14, figs. 6, 7) has documented small specimens of this species from the Omolon Massif in far northeast Russia.

Polygnathus vogesi Ziegler, 1962a
Figure 10: 15–16

Polygnathus vogesi ZIEGLER, 1962a, p. 94–95, Pl. 11, Figs. 5–7; PHILIP AND JACKSON, 1971, p. 71, Pl. 5, Figs. 8, 19–22; WANG AND WANG, 1978, p. 78, Pl. 7, Figs. 13–16; KONONOVA, 1979, Pl. 2, Fig. 10; WANG AND ZIEGLER, 1983, Pl. 8, Fig. 5; WANG AND YIN, 1984, Pl. 3, Fig. 21; XIONG, J., *in* HOU ET AL., 1985, p. 120, Pl. 34, Figs. 7–6; WANG AND YIN, 1988, p. 129, Pl. 21, Figs. 3, 4; JI ET AL., 1989, p. 90, Pl. 19, Figs. 6a–9b; WANG, 1989, p. 123, Pl. 38, Figs. 16, 17; PERRI AND SPALLETTA, 1991, p. 73, Pl. 7, Figs. 4, 5; JI AND ZIEGLER, 1993, p. 85, Pl. 34, Figs. 1–5; NEMIROVSKAYA ET AL., 1993, Pl. 3, Figs. 22, 23; KORN ET AL., 1994, Pl. 4, Fig. 10; Pl. 6, Fig. 7.

Remarks. This species is easily recognized on the basis of its triangular platform with a pair of nodose secondary carinae converging with the carina in the anterior half of the platform. Philip and Jackson (1971) recorded this species from the "Luton Formation" (actually Mandowa Mudstone, see Gould, 1975) in northeastern New South Wales. In the Lali section, South China, Ji and Ziegler (1993) report *P. vogesi* as occurring from the Early *expansa* Zone into the Middle *praesulcata* Zone. Other workers have reported *P. vogesi* as ranging into even younger horizons: Barskov et al. (1991) to as young as *sulcatus* Zone and Nemirovskaya et al. (1993) into the *duplicata* Zone.

Polygnathus wyatti n. sp.
Figure 12: 1–11

Derivation of name. In honor of Don Wyatt, who did much to unravel the stratigraphy of the Townsville hinterland.

Holotype. UQY 7787, the specimen illustrated in Figure 12: 10–11, is from the Myrtlevale Formation from sample WMP 504, 203.1 m above the base of section WMP at Mount Podge.

Diagnosis. A species of *Polygnathus* in which representative Pa elements have an asymmetrical, narrow, lanceolate platform with the outer margin more regularly curved than the inner margin; in the posterior half of the platform, the outer margin is convex and the inner margin is concave. The carina curves gently for the full length of the platform; the platform ornament consists of marginal nodes or short ridges with a row of nodes developed, paralleling the carina.

Description. The outer platform of juvenile Pa elements of *P. wyatti* is gently curved from the posterior extremity for approximately three-quarters the length of the platform; a convex, lobelike enlargement is developed on the anterior platform, occupying approximately the middle third of the specimen. In more mature specimens, the lobe becomes less distinct, but even in adult specimens the outer margin of the posterior half of the platform is convex, and the corresponding inner margin becomes only slightly concave. In juvenile forms the platform ornament consists of marginal nodes tending to elongate into short ridges. In some specimens (e.g., Fig. 12: 2) the nodes on the inner margin of the platform are positioned slightly in from the extreme edge of the platform. The marginal nodes along the platform tend to stand upright, giving an impression that the narrow platform is upturned. In adult forms the marginal nodes are retained, and a second row of nodes, parallel to the carina, is developed on the anterior half of the platform. The free blade occupies approximately a third the length of the unit. The slightly curved carina, extending to the posterior extremity, is clearly denticulate in young specimens; the denticles tend to fuse with maturity. The lower surface has a conspicuous keel rising immediately anterior to the elongate basal pit; the latter lies in an elongate depression tapering along the crest of the keel for about a quarter the length of the unit.

Remarks. Adult specimens of *P. wyatti* have some resemblance to *P. nodocostatus* Branson and Mehl, but in the former the posterior half of the inner lateral margin is always noticeably incurved, and on the lower surface, immediately posterior of the basal pit, the keel is conspicuously raised (cf. Fig. 12: 11 herein). The diverse *P. nodocostatus* Group includes several chronologically useful index forms; though these have various forms, the outline of the platform, with few exceptions, is elongately oval. The medially constricted platform of *P. triphyllatus* (Ziegler) sets it aside from the rest of the group; the kitelike outline of *P. styriacus* is also unusual. Helms (1961, fig. 17) illustrated an unnamed third exception that he sees occurring as an intermediate between *P. nodocostatus* and *P. granulosus* Branson and Mehl. The platform of the specimen he illustrated is triangular in outline and ornamented by two rows of nodes in the posterior half of the platform. The anterior half of the outer platform is convex, and the corresponding inner platform is slightly convex. If *P. wyatti* is indeed the intermediate alluded to by Helms, then one would expect its oldest occurrence to fall in the Early *marginifera* Zone or thereabouts.

Specimens of *P. brevilaminus* Branson and Mehl (1934a, pl. 21, figs. 3–6) resemble juvenile specimens of *P. wyatti* in respect of their overall size and platform development. The platform in *P. wyatti*, however, is decidedly more asymmetrical, the free blade is never more than half the length of the unit, and the elongate basal pit is positioned more anteriorly than in *P. brevilaminus*. The outline of the platform in specimens illustrated as *P. brevilaminus* by Barskov et al. (1991, pl. 26, figs. 9–11) more closely resembles the Mount Podge specimens, but the length of the free blade and the configuration of the basal pit differ considerably.

Specimens described from Kazakhstan as *P. atasuensis* (Kuz′min, 1992), a Famennian species, are similar in shape to *P. wyatti* but differ from it in having high, vertical platform margins in the anterior half of the platform, with the posterior lateral margins being slightly raised and the ornament of short marginal ribs forming a more or less denticulate edge along the margins. The carina of *P. wyatti* is less curved than in *P. atasuensis* and tends to fade posteriorly instead of consisting of well-developed denticles that reach the posterior tip.

The shape of *P. wyatti* also mimics the outline of the Frasnian species of *Polygnathus*, *P. lodinensis* Pölsler (1969). The latter, however, has a larger and more rounded basal pit and lacks the rows of nodes developed on the anterior platform parallel to the carina.

Polygnathus sp. A
Figure 13: 9

Remarks: This single specimen of *Polygnathus* displays bifurcation of the posterior quarter of the carina. It illustrates a Duplicatio Pathology *sensu* Weddige (1990). Weddige believes that such abnormalities can, at times, be of taxonomic significance. Szulczewski (1971, pl. 18, fig. 4) illustrates a specimen of *P. dubius* Hinde showing similar bifurcation of the carina and keel.

Genus _Polylophodonta_ Branson and Mehl, 1934a

Type species. Polylophodonta gyratilineata Holmes, 1928

Remarks. For discussion of the genus and selection of the type species, see Glenister and Klapper (1966, p. 831–833).

Polylophodonta confluens (Ulrich and Bassler, 1926)
Figure 13: 6–8

Polygnathus confluens ULRICH AND BASSLER, 1926, p. 46–47, Pl. 7, Figs. 14, 15.
Polylophodonta confluens (Ulrich and Bassler). GLENISTER AND KLAPPER, 1966, p. 831–832, Pl. 94, Figs. 10, 11 (synonymy to 1966); WOLSKA, 1967, p. 418, Pl. 17, Fig. 7; DREESEN AND DUSAR, 1974, Pl. 4, Fig. 21; JOHNSTON AND CHATTERTON, 1991, Pl. 2, Fig. 20; BALINSKI, 1995, Pl. 22, Figs. 14, 15.
Polylophodonta gyratilineata Ulrich and Bassler. PICKETT, 1981, Fig. 2, 20, 21.
Polylophodonta cf. *confluens* (Ulrich and Bassler). VORONTSOVA AND KUZ′MIN, 1984, Pl. 1, Fig. 22.
Polylophodonta linguiformis Branson and Mehl. GAGIEV ET AL., 1987, Pl. 28, Fig. 10; PERRI AND SPALLETTA, 1990, p. 67, Pl. 7, Fig. 5a–b.

Remarks. Specimens of *Po. confluens* are ornamented by rows of nodes or short ridges parallel to the carina, swinging around anteriorly to become slightly curved transverse rows in the posterior part of the platform. Branson and Mehl (1934a, p. 258, pl. 20, fig. 2) referred to the specimen they illustrated from Missouri as a "perfect specimen." The specimen from Mount Podge illustrated herein (Fig. 14: 6–8) rivals this northern hemisphere specimen as regards perfection. It has more regular, even margins than the specimen elegantly hand illustrated by Ulrich and Bassler (1926, pl. 7, fig. 14) and later photographically illustrated by Huddle (1968, pl. 17, fig. 1). Only five specimens have been recovered, all from the Mount Podge sections. This rarity accords with the reports of rare occurrences, of one or two specimens at a time, scattered throughout the literature. Perhaps lack of a strengthening carina may have made specimens of *Polylophodonta* particularly vulnerable to breakage.

Family PALMATOLEPIDAE Sweet, 1988

Genus _Palmatolepis_ Ulrich and Bassler, 1926

Type species. Palmatolepis perlobata Ulrich and Bassler, 1926

Palmatolepis falcata (Helms, 1959)
Figure 16: 7, 21

Nothognathella(?) *falcata* HELMS, 1959, p. 645, Pl. 5, Figs. 27 a, b.
Palmatolepis glabra lepta Ziegler and Huddle. JOHNSTON AND

CHATTERTON, 1991, Pl. 1, Fig. 2; MATYJA, 1993, Pl. 30, Fig. 4. *Palmatolepis falcata* (Helms). METZGER, 1994, p. 626–629, Figs. 10.21–10.29 (synonymy to 1994).

Remarks. Boogaard and Kuhry (1979) demonstrated statistically the association of *Palmatolepis* and "*Nothognathella*" (see their diagram 2, p. 13 and diagram 3, p. 14) thus confirming earlier beliefs (e.g., Klapper and Philip, 1972) that palmatolepid and nothognathellan elements occurred together in the same apparatus. Helms (1959) demonstrated that "*Nothognathella*" (?) *falcata* cooccurred in the same Kleinknotiger Kalk sample as *Pal. glabra elongata* (= Pal. g. lepta, see Ziegler, 1977); the latter was the only palmatolepid of *glabra* affinities present in the sample. Boogaard and Kuhry (1979), in examining faunas from Holts Summit, found a similar, exclusive association. Metzger (1994) recovered 292 Pa elements, 23 Pb elements, 4 Sa elements, 39 Sc elements, and one ?Sb element of this apparatus from faunas from the WAPET C section at McWhae Ridge in the Virgin Hills Formation of the Canning Basin of northwestern Australia and from the West Range Limestone at Bactrian Mountain, Nevada.

Specimens from high in our WMP section at Mount Podge appear identical to those illustrated by Helms (1959, pl. 5, fig. 25), Boogaard and Kuhry (1979, fig. 21), and Glenister and Klapper (1966, pl. 95, fig. 1).

Palmatolepis glabra pectinata Ziegler, 1962b
Figures 9: 21–22, 16: 1–6, 12

Palmatolepis glabra pectinata ZIEGLER, 1962b, p. 398–399, Pl. 2, Figs. 3–5; ZIEGLER (in ZIEGLER), 1977, p. 305–307, *Palmatolepis*-Pl. 6, Figs. 7–11 (synonymy to 1977); PICKETT, 1981, Fig. 2.23–26; LONG, 1987, p. 505, Fig. 2–A, B, F, G; JOHNSTON AND CHATTERTON, 1991, Pl. 1, Fig. 5; MATYJA, 1993, Pl. 27, Fig. 9; PHUONG AND WEYANT, 1994, Pl. 1, Figs. 1, 5, 6, 8; BAI, 1994, p. 168, Pl. 13, Figs. 7–10, 16; BALINSKI, 1995, Pl. 19, Fig. 9.

Remarks. This is the most abundant species of *Palmatolepis* (208 specimens) represented in the Myrtlevale Formation. Although exhibiting considerable variation, the population as a whole differs somewhat from the holotype from Sessacher near Oberscheld in the Dill Syncline, Germany (Ziegler, 1962b, pl. 2, fig. 3a, 3b), by being slightly more elongate and a little less strongly sigmoidal; specimens from Mount Podge appear a little narrower than the holotype (reillustrated in Ziegler, 1962a, pl. 4, fig. 16). In the latter, the parapet appears denticulate and is situated midway between the carina and the inner margin of the platform; specimens obtained in the present study commonly have a smooth parapet coinciding with the anterior inner margin. According to Glenister and Klapper (1966, p. 814), denticulation of the parapet "appears sporadically in both *Pal. glabra pectinata* and the nominate subspecies." Compared to material illustrated from the Bonaparte Gulf and Canning basins, northwestern Australia (Druce, 1970a; 1976), the Mount Podge material is generally more elongate and slender, having a proportionally shorter parapet that at times runs at an angle toward the carina rather than parallel to it. Some specimens (Fig. 16: 5) show a tendency to develop a "thornlike" projection of the inner platform; this feature and the nonparallel parapet suggest close relationship to *Pal. g. acuta* Helms (Ziegler *in* Ziegler, 1973, p. 293–295).

Palmatolepis marginifera marginifera Helms, 1959
Figure 16: 9, 10, 13, 17, 18

Palmatolepis quadrantinodosa marginifera HELMS, 1959, p. 649, Pl. 5, Figs. 22, 23; ZIEGLER (in ZIEGLER), 1977, p. 325–330, *Palmatolepis*-Pl. 7, Figs. 17–18; Pl. 8, Fig. 2 (synonymy to 1977 and discussion of the designated holotype); PICKETT, 1981, Fig. 2.2; MORY AND CRANE, 1982, Fig. 9A; OLIVIERI, 1985, p. 298, Pl. 6, Figs. 11–12; LONG, 1987, Fig. 2–E; GAGIEV ET AL., 1987, Pl. 28, Figs. 12, 16, 18; PERRI AND SPALLETTA, 1990, p. 61, Pl. 2, Figs. 3a–4b; JOHNSTON AND CHATTERTON, 1991, Pl. 2, Figs. 21, 25; MATYJA, 1993, Pl. 27, Fig. 5; PHUONG AND WEYANT, 1994, Pl. 1, Figs. 2–4; BAI, 1994, p. 170, Pl. 14, Figs. ?1, 2, 4.

Remarks. Pal. m. marginifera is represented in the Mount Podge fauna by 58 specimens, varying in form. Some are very close to the specimen designated as holotype (Ziegler, 1962b, pl. 2, figs. 7a–c), wherein the parapet extends posteriorly past the central node, forming a shallow, but well-defined gutter between it and the posterior extension of the carina. Other specimens resemble a form illustrated by Sandberg and Ziegler (1973), wherein the parapet is somewhat shorter. Such variation also occurred among specimens from the Canning Basin (Glenister and Klapper, 1966—compare pl. 91, figs. 7, 11, 12, and 13 with pl. 91, figs. 9, 10, and 14). On some specimens from Mount Podge (e.g., Fig. 16: 13, 17), the parapet is denticulate. *Pal. m. marginifera* occurs 11.6 m above the base of the EMP section (EMP/34) and 184.1 m above the base of the WMP section (WMP/450) in the Myrtlevale Formation at Mount Podge.

Palmatolepis marginifera duplicata Sandberg and Ziegler 1973
Figure 16: 19–20

Palmatolepis marginifera duplicata SANDBERG AND ZIEGLER, 1973, p. 105, Pl. 3, Fig. 15–19, 21–25; Pl. 5, Fig. 15; ZIEGLER (*in* ZIEGLER), 1977, p. 331–332, *Palmatolepis*-Pl. 8, Figs. 6–9 (synonymy to 1977); GAGIEV ET AL., 1987, Pl. 28, Figs. 14, 17; JOHNSTON AND CHATTERTON, 1991, Pl. 2, Fig. 26; [non] BAI, 1994, p. 170, Pl. 14, Fig. 1 (= ?*Pal. marginifera marginifera*)

Remarks. The subspecies *Pal. m. duplicata* was erected to accommodate forms of *Pal. marginifera* that developed a second parapet or bulge on the outer platform. The holotype selected (Sandberg and Ziegler, 1973, pl. 3, fig. 15) shows, paralleling the carina, a significant parapet developed on the outer platform. From examination of the Mount Podge faunas and specimens illustrated by Sandberg and Ziegler (1973) and Ziegler (1962a), it appears that the form developing this clearly defined parapet should be considered to be a separate morphotype from those wherein the subspecific feature of a bulge on outer side of the platform is developed. *Pal. m. duplicata* morphotype 1 is, therefore, characterized by a distinct, high parapet extending from the anterior margin of the outer platform to at least, but usually beyond, the central node; two adcarinal grooves are clearly developed (e.g. Sandberg and Ziegler, 1973, pl. 3, figs. 15–19, 25). Morphotype 2 is characterized by a bulge on the outer margin of the platform, paralleling the carina developed (e.g., Sandberg and Ziegler, 1973, pl. 3, figs. 22–24; pl. 5, fig. 15; Ziegler, 1962a, pl. 7, fig. 8). Whether or not the two morphs have

Figure 16 (on facing page). Numbers 8, 11, 13 ×90; 9, 10, 16, 19, 21 ×75; 3, 5, 6, 12, 14, 15, 20, 22 ×60; 1, 2, 4, 17, 18 ×45; 7 ×30. 1–6, 12: *Palmatolepis glabra pectinata* Ziegler. 1, 2, lower and upper views of UQY 7731, EMP 77.5; 3, upper view of UQY 7732, WMP 448B; 4, lower view of UQY 7733, WMP 450; 5, upper view of UQY 7734, WMP 448B; 6, lower view of UQY 7735, WMP 448B; 12, upper view of UQY 7736, EMP 49. 7, 21: *Palmatolepis falcata* (Helms). 7, upper view of UQY 7737, WMP 504; 21, upper view of UQY 7738, WMP 521. 8, 11: *Palmatolepis minuta minuta* Branson and Mehl. 8, upper view of UQY 7739, WMP 433; 11, upper view of UQY 7740, EMP 6. 9, 10, 13, 17, 18: *Palmatolepis marginifera marginifera* Helms. 9, 10, lower and upper views of UQY 7741, EMP 78.5; 13, upper view of UQY 7742, EMP 48; 17, upper view 7743, EMP 34; 18, upper view of UQY 7744, WPM 515. 14, 15: *Palmatolepis quadrantinodosa quadrantinodosa* Branson and Mehl. Lower and upper views of UQY 7745, EMP 53. 16: *Palmatolepis quadrantinodosa inflexa* Müller. Upper view of UQY 7746, EMP 53; 19–20: *Palmatolepis marginifera duplicata* Sandberg and Ziegler. 19, upper view of UQY 7747, EMP 72; 20, upper view of UQY 7748, EMP 72. 22: *Palmatolepis quadrantinodosa inflexoidea* Ziegler. Upper view of UQY 7749, WMP 515.

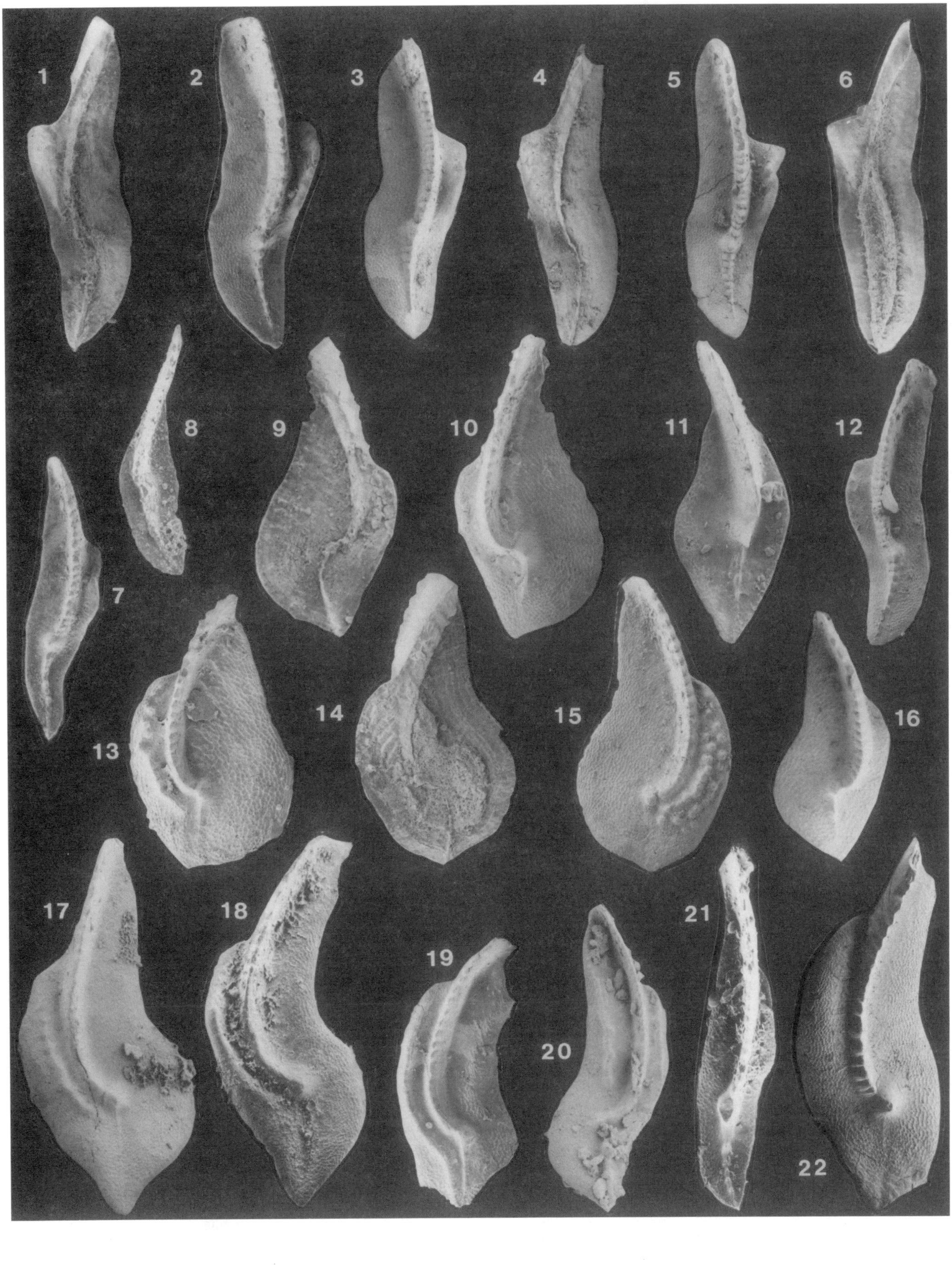
1
2
3
4
5
6
7
8
9
10
11
12
13
14
15
16
17
18
19
20
21
22

biostratigraphic significance has yet to be tested. At Mount Podge, *Pal. m. duplicata* appears first in the Myrtlevale Formation in EMP/56, 7.5 m above the incoming of *Pal. m. marginifera*. In the WMP section *Pal. m. duplicata* was encountered only in the highest horizon, WMP521, 23.8 m above the first occurrence of *P. m. marginifera*.

Palmatolepis minuta minuta Branson and Mehl, 1934a
Figure 16: 8, 11

Palmatolepis minuta BRANSON AND MEHL, 1934a, p. 236–237, Pl. 18, Figs. 1, 6, 7; ZIEGLER (in ZIEGLER), 1977, p. 335–338, *Palmatolepis*-Pl. 9, Figs. 1–5 (synonymy to 1977); LONG, 1987, p. 507, Fig. 2-H–J, N, P; PERRI AND SPALLETTA, 1990, p. 62, Pl. 3, Figs. 1a–b, 7a–b; PERRI AND SPALLETTA, 1991, p. 64, 66; MATYJA, 1993, Pl. 23, Fig. 5; METZGER 1994, p. 632, Figs. 8.1–8.15 (synonymy of apparatus elements); SCHÜLKE, 1995, Pl. 10, Figs. 5, 7–9, 13, 21 (synonymy to 1995).
Palmatolepis (Tripodellus) minuta minuta Branson and Mehl. BOOGAARD AND KUHRY, 1979, p. 42, Fig. 13

Remarks. This long-ranging species (base of the Upper *triangularis* Zone to Upper *trachytera* Zone (= former Upper *velifer* Zone) was obtained from only two horizons, both from the Myrtlevale Formation at Mount Podge, in samples WMP433 and EMP6. Metzger (1994) recognized all elements of the apparatus in material from the Virgin Hills Formation in the WAPET B and WAPET C sections at McWhae Ridge, northwestern Australia, and from the West Range Limestone at Bactrian Mountain, Nevada. He recovered 6100 Pa elements, 1265 Pb elements, 1,048 M elements, 277 Sa elements, 714 Sb elements, and 1,295 Sc elements; we obtained only two elements from Mount Podge!

Palmatolepis perlobata postera Ziegler, 1960

Palmatolepis rugosa postera ZIEGLER, 1960, p. 39–40, Pl. 2, Figs. 10, 11.
Palmatolepis perlobata postera Ziegler. ZIEGLER (in ZIEGLER), 1977, p. 359–360, *Palmatolepis*-Pl. 9, Figs. 14–15 (synonymy to 1977); SANDBERG AND ZIEGLER, 1979, p. 180, Pl. 2, Figs. 1–4; JI AND ZIEGLER, 1993, Text-fig. 15; BAI, 1994, p. 171–172, Pl. 14, Figs. 6, 7.
Palmatolepis perlobata perlobata Ziegler. JI AND ZIEGLER, 1993, p. 67, Fig. 10.

Remarks. The absence or near absence of a lobe on the margin of the outer platform sets this subspecies apart from others. *Pal. p. postera* has been found in faunas associated with *P. homoirregularis* Ziegler and *P. experplexus* Sandberg and Ziegler (Table 3).

Palmatolepis quadrantinodosa quadrantinodosa
Branson and Mehl, 1934a
Figures 9: 17; 16: 14, 15

Palmatolepis quadrantinodosa BRANSON AND MEHL, 1934a, p. 235, Pl. 18, Figs. 3, 17, 20.
Palmatolepis (Palmatolepis) inflexa (MÜLLER), 1956, Pl. 10, Figs. 7, 10.
Palmatolepis quadrantinodosa quadrantinodosa Branson and Mehl. SANDBERG AND ZIEGLER, 1973, Pl. 3, Figs. 27–30; ZIEGLER (*in* ZIEGLER), 1977, p. 371–372, *Palmatolepis*-Pl. 8, Figs. 10–16 (synonymy to 1977); MATYJA, 1993, Pl. 27, Figs. 12–13.

Remarks. The specimens from Mount Podge are very close to two specimens from Thuringia illustrated by Müller (1956, pl. 10, figs. 7, 10) as *Pal. (Pal.) inflexa* n. sp. and referred to *Pal. q. quadrantinodosa* by Ziegler (1977). The two Thuringian specimens differ from the holotype of *Pal. (Pal.) inflexa* Müller (1956, pl. 10, figs. 5a, 5b) by having a more gently curved carina; a broader, more U-shaped posterior tongue; and a more ordered pattern of nodes on the inner platform. On the basis of these differences, Ziegler (1977, p. 371) assigned these specimens to *Pal. q. quadrantinodosa*. The coarse ornament on the outer side of the platform shown herein in Figure 9: 17 is unlike typical specimens of *Pal. q. quadrantinodosa* in which the outer surface is covered by a shagreen rather than nodose surface, but as it closely resembles Müller's 1956, pl. 10, fig. 7, included by Ziegler (1977) in *Pal. q. quadrantinodosa*, it is classified thus.

Sandberg and Ziegler (1973) considered Branson and Mehl's (1934a) material to include two morphotypes: one evolving from *Pal. stoppeli* Sandberg and Ziegler by developing nodes on the ramp (e.g., Sandberg and Ziegler, 1973, pl. 3, figs. 27–28) and the other from *Pal. q. inflexa* by developing nodes on the bulge (e.g., Sandberg and Ziegler, 1973, pl. 3, figs. 29–30). In the two Thuringian specimens (Müller, 1956, pl. 10, figs. 7 and 10) and the specimens from Mount Podge (Fig. 9: 17), the carina extends from the central node to the margin of the broadly tapering tongue of the unit. These features are present in Sandberg and Ziegler's (1973) specimens illustrated on their pl. 3, figs. 29 and 30. These forms are therefore considered to have been derived from *Pal. q. inflexa* even though at Mount Podge a specimen transitional between *Pal. q. inflexa* and *Pal. q. inflexoidea* was obtained from sample WMP438 (Table 1), below the first occurrence of *Pal. q. quadrantinodosa*.

In Australia *Pal. q. quadrantinodosa* has been documented (Druce, 1976) from the Virgin Hills Formation at Needle Eye Rocks in the Canning Basin, northwestern Australia.

Palmatolepis quadrantinodosa inflexa Müller, 1956
Figures 9: 20; 16: 16

Palmatolepis (Palmatolepis) inflexa MÜLLER, 1956, p. 30, Pl. 10, Fig. 5; ZIEGLER (*in* ZIEGLER), 1977, p. 377–379, *Palmatolepis*-Pl. 12, Figs. 3–10 (synonymy to 1974); JOHNSTON AND CHATTERTON, 1991, Pl. 2, Fig. 23; MATYJA, 1993, Pl. 27, Fig. 6; BAI, 1994, p. 172, Pl. 13, ?Fig. 17; Pl. 14, Figs. 10–12.

Remarks. Eight specimens from Mount Podge, superficially similar to *Pal. marginifera*, display a flat inner margin; they are therefore referred to *Pal. q. inflexa*. A specimen from WMP/438 (Figure 9: 20) is transitional between *Pal. q. inflexa* and *Pal. q. inflexoidea*; the central node is proportionally closer to the posterior margin than in *Pal. q. inflexoidea*, and a bulge is developed on the anterior inner platform as occurs in *Pal. q. inflexa*.

Palmatolepis quadrantinodosa inflexoidea Ziegler, 1962a
Figure 9: 18–19; 16: 22

Palmatolepis quadrantinodosa inflexoidea ZIEGLER, 1962a, p. 74, Pl. 5, Figs. 14–18; ZIEGLER (*in* ZIEGLER), 1977, p. 383–384, *Palmatolepis*-Pl. 13, Figs. 1–5 (synonymy to 1976); PICKETT, 1981, Fig. 2.22; LONG, 1987, p. 509, Fig. 2–C, D; GAGIEV ET AL., 1987, Pl. 28, Fig. 15; MATYJA, 1993, Pl. 27, Fig. 8.

Remarks. The 13 specimens from WMP and EMP are characterized by a flat, smooth sigmoidal platform lacking either a parapet of the inner margin or a lobe on the inner margin of the platform. The inner platform margin begins about one-third the length of the carina, whereas the outer margin begins at the anterior extremity of the carina. The carina terminates posterior to the central node. The specimens of *Pal. q. inflexoidea* from Mount Podge resemble more closely those from Bactrian Mountain, Nevada (Sandberg and Ziegler, 1973, pl. 4, figs. 1–3), than specimens from Sessacker near Oberscheld, Germany (Ziegler, 1962a, pl. 5, figs. 14–18) because the carina posterior to the central node continues almost to the posterior margin; it is expressed either as a series of nodes or as a threadlike line unlike the specimens in the original illustrations. The first appearance of *Pal. q. inflexoidea* is in sample WMP/438, 179.6 m above the base of the section at West Mount Podge, and in sample EMP/6, 2 m above the base of the seciton at East Mount Podge. Other occurrences of *Pal. q. inflexoidea* in Australia have been reported by Glenister and Klapper (1966, pl. 93, figs. 11, 12) from the WAPET section H1 in the Virgin Hills, Canning Basin, and by Druce (1976, pl. 56, figs. 4, 5) from McWhae Ridge, Canning Basin, northwestern Australia.

Family ELICTOGNATHIDAE Austin and Rhodes, 1981

Genus *Alternognathus* Ziegler and Sandberg, 1984

Type species. Alternognathus regularis Ziegler and Sandberg, 1984

***Alternognathus pseudostrigosus* (Dreesen and Dusar, 1974)**
Figure 12: 17–18

Polygnathus? pseudostrigosus DREESEN AND DUSAR, 1974, p. 20, Pl. 1, Figs. 17–20; Pl. 3 Figs. 16–24, text-Fig. 13.
Pseudopolygnathus sp. FILIPOVIC ET AL., 1975, Pl. 24, Figs. 28, 29.
Alternognathus pseudostrigosus (Dreesen and Dusar). MATYJA, 1987, Pl. 22.2, Fig. 7; MATYJA, 1993, Pl. 29, Figs. 1–3.

Original Diagnosis (Dreesen and Dusar, 1974). A species related to *Polygnathus* with an incipient to weakly developed asymmetrical platform and a relatively large rhomb-shaped inverted basal cavity.

Remarks. Twenty-three Pa elements from the EMP and WMP sections at Mount Podge are assigned to *Alt. pseudostrigosus*. These are characterized by their very weakly developed asymmetrical platform and proportionally large basal cavity. Some specimens from Mount Podge (e.g., Fig. 12: 17), as well as specimens figured by Dreesen and Dusar (1974, pl. 3, fig. 24; text-fig. 13b), display the slight curvature of the blade reminiscent of that in *P. sinuosus* Szulczewski (1971, pl. 20, figs. 2b, 3a) from the late Frasnian from which it may have evolved. *P. gracilis* Klapper and Lane (1985), another Frasnian species, differs from *Alt. pseudostrigosus* by the degree of curvature of the blade and carina. Specimens from southwestern Serbia identified as *Pseudopolygnathus* sp. (Filipovic et al., 1975) are conspecific with material from Belgium; Dreesen and Dusar (1974) note that the large inverted basal cavity is reminiscent of that found in *Pseudopolygnathus*.

In their original description of this species, Dreesen and Dusar (1974) were hesitant to assign it with certainty to *Polygnathus*. Erection of the genus *Alternognathus* (Ziegler and Sandberg, 1984, p. 188), for *Polygnathus*-like species in which the development of the platform appears to alternate from the left- to right-hand side, comfortably accommodated this species. Ziegler and Sandberg (1984) suggested that *Alternognathus* may have been the precursor to the siphonodellid lineage.

Although Sweet (1988, p. 107) recognized Dreesen and Dusar's (1974) species *Alt. pseudostrigosus* to be the oldest known species of the genus, he indicated the range to be from within the Latest *marginifera* Zone to somewhere within the Late *postera* Zone; this is contrary to Dreesen and Dusar, who stated the range to be from the upper part of the Late *rhomboidea* Zone into the Latest *marginifera* Zone. That *Alt. pseudostrigosus* appears in the Early *marginifera* Zone in Borehole Chojnice 2, southeastern Pomerania (Matyja, 1987), is in accord with the range given by Dreesen and Dusar and is in line with its occurrence in the Early *marginifera* Zone at Mount Podge.

Genus *Siphonodella* Branson and Mehl, 1944

Type species. Siphonodella duplicata (Branson and Mehl, 1934b)

***Siphonodella obsoleta* Hass, 1959**
Figure 11: 1–4

Siphonodella obsoleta HASS, 1959, p. 392–393, Pl. 47, Figs. 1, 2; KLAPPER *in* ZIEGLER), 1973, p. 463–464, *Siphonodella*-Pl. 1, Fig. 7 (synonymy to 1973); MORY AND CRANE, 1982, p. 295, Fig. 11 J–M (synonymy to 1982); JI ET AL., 1984, Pl. 1, Figs. 26, 27; JI ET AL. [*in* HOU ET AL.], 1985, p. 139, Pl. 25, Figs. 11–19; Pl. 26, Figs. 1–8; WANG AND YIN, 1988, p. 140, Pl. 19, Fig. 13; Pl. 29, Figs. 18, 19; JI ET AL., 1989, p. 99, Pl. 14, Figs. 1, 2; JI AND ZIEGLER, 1992, Pl. 2, Figs. 5, 6; KÜRSCHNER ET AL., 1993, Pl. 1, Fig. 3; PERRET, 1993, p. 310, Pl. C 1, Figs. 19–21, 23, 24; PICKETT, 1994, Figure 2.7; BAI, 1994, p. 186, Pl. 31, Figs. 13, 14.

Remarks. The 15 specimens of *Siphonodella obsoleta* recovered, 14 from Corner Creek, are of two types: a slender form with two rostral ridges (e.g., Fig. 11: 1–2), similar to specimens from the Chappel Limestone, Texas (Hass, 1959, pl. 47, figs. 1, 2), and a wider form with four rostral ridges (e.g., Fig. 11: 4) more closely resembling a specimen from the Lodgepole Limestone, Montana (Klapper, 1971b, pl. 1, fig. 25). Mory and Crane (1982, fig. 11 J–M) illustrated specimens of *Si. obsoleta* from Old Cannindah, east-central Queensland. *Si. obsoleta* is known to occur from the uppermost part of the Late *duplicata* Zone into the Early *crenulata* Zone (Sandberg et al., 1978).

Family ANCHIGNATHODONTIDAE Clark, 1972

Genus *Hindeodus* Rexroad and Furnish, 1964

Type species. Spathognathodus cristulus Youngquist and Miller, 1949

***Hindeodus crassidentatus* (Branson and Mehl, 1934b)**
Figure 11: 14

Spathodus crassidentatus BRANSON AND MEHL, 1934b, p. 276, Pl. 22, Figs. 17, 18.
Spathognathodus crassidentatus Branson and Mehl. NICOLL AND DRUCE, 1979, p. 31–32, Pl. 23, Figs. 1–5 (synonymy to 1979).

Remarks. Sweet (1988) considered *Hindeodus crassidentatus* to be the oldest known representative of the genus. Although one of the diagnostic features of the species is the two main denticles at the anterior of the blade, the specimen from near the top of the recessive marine interval in the Venetia Formation at VEN ca. 210 has two tiny denticles on the leading edge of the first main denticle, a characteristic noted by others (e.g. Klapper, 1966, p. 23; Nicoll and Druce, 1979, p. 32). According to Sweet (1988), *H. crassidentatus* occurs from the base of the *sandbergi* Zone to the top of the Early *crenulata* Zone.

Family Unknown
Apparatus sp. A
Figure 13: 10–19

Apparatus sp. A. PICKETT, 1981, Figure 2.6–11.

Remarks. In the upper part of the Mount Podge sections, 94 unassigned nonplatform elements have been recovered that may represent elements of one, two, or three apparatuses. As individual numbers are insufficient to be able to recognize intraspecific variation, they have been assigned temporarily to Apparatus sp. A. Pickett (1981, fig. 2.6-11) illustrated similar elements from Don Wyatt's spot localities C410, C411, and C413 (shown herein on Fig. 2 as Localities 13, 14, and 15). Faunas from each of these localities have been dated as Early *marginifera* Zone.

ACKNOWLEDGMENTS

Don Wyatt, who contributed so notably to unraveling the stratigraphy and structure of the Townsville hinterland, introduced us to the geology of the Burdekin Basin in the early 1970s and insisted we sample his Corner Creek sequence. We have enjoyed interaction with many Geological Survey of Queensland colleagues mapping in the Townsville hinterland and also with the owners of the Burdekin Downs, Dotswood, Fanning River, Kirkland Downs, Laroona, Lucky Springs, Payne's Lagoon, and Star cattle stations, all of whom graciously allowed access to their properties for sampling. The Department of Defence gave permission for collecting on Dotswood. Ian Withnall drew our attention to the occurrence of fossiliferous limestones among the Venetia Formation clastics on Lucky Springs. He, Leanne Gunther, Simon Lang, and Tim McLennan, all of the Geological Survey of Queensland, and Debbie Pojolke of James Cook University, Townsville, provided numerous samples of Keelbottom Group limestones encountered in regional mapping in the Townsville hinterland and kept doing so even when most samples were barren or failed to produce conodonts of stratigraphic value. Glenn Brock, Alan Bagnall, Kirill Simakov, Terry Sloan, and the late Ross Talent enthusiastically helped with sampling; the last drew our attention to and sampled the calcareous horizons in the transitional interval from Lollypop Formation to Hardwick Formation. Michael Engelbretsen looked after the acid-leaching and SEM photography; Margaret Anderson carried out the sodium polytungstate separations, picked many of the samples, and, with George Wilson, assisted with plate-making; Glenn Brock and Terry Furey-Greig helped with compilation of the stratigraphic columns and distribution charts. We are especially indebted to Gilbert Klapper and Brian Glenister of the University of Iowa, who generously gave access to their personal libraries during three all-too-brief sojourns by R.M. at the University of Iowa in 1988, 1992, and 1993. The manuscript benefited greatly from Gilbert's tolerance and well-known prodigality with regard to time when tackling questions raised by his colleagues and from an especially meticulous reading of our manuscript. John Pickett, John Draper and Simon Lang also commented helpfully on our manuscript. Bai Shun-liang and Immo Schülke generously provided copies of their recent monographs. Thomas Becker, rooting around in literature on the late Givetian, came up with undocumented reports of *Icriodus symmetricus* from pre-Frasnian horizons in Germany and kindly conveyed these generally overlooked reports to us. We are especially appreciative of funding provided by the Australian Research Council and Macquarie University.

APPENDIX: LOCALITY REGISTER

Grid references are to Australian National 1:100,000 topographic sheets 8059 Ewan, 8158 Dotswood, 8159 Rollingstone, and 7859 Burges. Preliminary editions of 1:100,000 geological maps of the first three of these sheets have been published: Ewan by Withnall et al. (1991), Dotswood by Draper et al. (1993a), and Rollingstone by Gunther et al. (1991). A 1:100,000 geological map of Burges sheet forms part of the Broken River Special map compiled by Withnall and Lang (1992).

Sampled stratigraphic sections

EMP—Section (Fig. 6) extending from high in the Mount Podge Limestone through the Myrtlevale Formation to the base of the Lollypop Formation commencing at grid reference 923646 approx 0.9 km east of the road up the Running Creek valley to a railway ballast quarry (in Mount Podge Limestone).

WMP—Section (Fig. 6) extending from high in the Laroona Arkose through the Mount Podge Limestone and Myrtlevale Formation to the base of the Lollypop Formation on the southwest flank of Mount-Podge commencing at grid reference 887_5651_2.

CORN—Section through the Hardwick Formation along "Corner Creek," a small, unnamed, sharply incised gully joining the Star River from the left at grid reference 052651_5.

VEN—228.5 m of section through the recessive interval of Venetia Formation on the south flank of the divide between Big Stockyard Creek and Little Stockyard Creek, ca. 12 km south of Greenvale (Fig. 4). The section terminated at grid reference 845843, at the uppermost conglomeratic limestone. The one productive horizon out of 25 samples, VEN ca. 210 in an incised gully head, was ca. 15 m south of the section line; it aligned approximately with VEN 210 on the section line.

Spot localities

Numbering of these continues from the locality numbers given in Talent and Mawson (1994).

10. Myrtlevale Formation limestones and fine sandstones exposed in gully at grid reference 926649; collectors R. Mawson, J. A. Talent and A. Bagnall. The top and bottom of the Myrtlevale are not exposed at this locality.

11. Fault sliver of Myrtlevale Formation in Argentine Metamorphics exposed in a gully approximately 200 m north of the Greenvale-Yabulu railway, at grid reference 962_5603_0; collector T. McLennan.

12. Myrtlevale Formation limestone exposed in unnamed gully, at grid reference 949_0603_5; collector T. McLennan.

13. Myrtlevale Formation—two limestone beds, each a few centimeters thick, cropping out among siltstones and mudstones in an erosion gully between the Hervey Range Road and the road to Kirkland Downs at approximate grid reference 982494; collectors R. Mawson and J. A. Talent. Our locality appears from previously published and recent mapping (Gunther et al., 1991) to be at or very close to Wyatt's notably productive locality C413, but the latter was indicated as being approximately 1.5 km south of Payne's Lagoon homestead (Pickett, 1981, Fig. 1, after McKellar, 1970). Our locality, which accords with the pattern of mapping of geological boundaries by Wyatt and others, however, is 4 km south-southeast of an ephemeral lake known as Payne's Lagoon. This is about 8.3 km west of Payne's Lagoon homestead. It should be noted that our samples have proved much less productive than Wyatt's C413 (Pickett, 1981, Table 1); we therefore suspect they are not precisely the same horizon in the Myrtlevale Formation.

14. Limestone low in the Myrtlevale Formation limestone at D. Wyatt's spot locality C411 (in Pickett, 1981) at grid reference 299_0202_5.

15. Limestone from the Myrtlevale Formation at D. Wyatt's spot locality C410 (in Pickett, 1981) at grid reference 293_1205_1. This is stratigraphically higher in the Myrtlevale Formation than Wyatt's locality C411.

16. Limestones, generally arenaceous, from D. Wyatt's spot locality C414 (in Pickett, 1981), from the Saint Michael Member of the Hardwick Formation exposed in the sharply incised gully-head (tributary to the left branch of Gap Creek) beside the spur running southward from Mount Saint Michael and forming the watershed between Gap Creek and One Mile Creek at approximate grid reference 282288; collectors R. Mawson, J. A. Talent et al.

17. Limestone at the very top of the Myrtlevale Formation (as presently interpreted) at grid reference 268263; collector L. Gunther.

18. Limestone at the very top of the Myrtlevale Formation (as presently interpreted) at grid reference 915596; collector L. Gunther.

19. Limestone from undifferentiated Keelbottom Group south of Star Homestead at grid reference 124602; collector L. Gunther.

20. Limestones from a dam east of Star Homestead at grid reference 136605; collector L. Gunther. Two limestone horizons at this locality are shown as lower (L) and upper (U) on Table 3.

21. Limestones from undifferentiated Keelbottom Group south of Little Star River at grid reference 096598; collector L. Gunther. Two limestone horizons at this locality are shown as lower (L) and upper (U) on Table 3.

22. Limestone in the Hardwick Formation at grid reference 329306; collector L. Gunther.

23. Limestone from near the base of the Hardwick Formation in a gully tributary to One Mile Creek at 279_3286_3; collector L. Gunther.

24. Limestone from undifferentiated Keelbottom Group east of Little Star River at grid reference 083602; collector L. Gunther.

25. Small exposure in Topsy Creek near Topsy Well, at approximate grid reference 282227; collectors A. Bagnall, R. Mawson, T. Sloan, and J. A. Talent.

26. On the Dotswood–Fanning River road at approximate grid reference 287_0226_5; collectors A. Bagnall, R. Mawson, T. Sloan, and J. A. Talent.

REFERENCES CITED

Anderson, W. I., 1966, Upper Devonian conodonts and the Devonian-Mississippian boundary of north-central Iowa: Journal of Paleontology, v. 40, p. 395–415.

Austin, R. L., Orchard, M. J. and Stewart, I. J., 1985, Conodonts of the Devonian System from Great Britain, *in* Higgins, A. C., and Austin, R. L., eds., A stratigraphical index of conodonts: Chichester, British Micropalaeontological Society, Ellis Horwood Limited, p. 93–166.

Bai, S. L., 1994, Systematic paleontology. Part 1: Conodonts and miscellaneous; *in* Bai, S. L., Bai, Z. Q., Ma, X. P., Wang, D. R., and Sun, Y. L, eds., Devonian events and biostratigraphy of South China: Beijing, Peking University Press, p. 161-195.

Balinski, A., 1995, Brachiopods and conodonts biostratigraphy of the Famennian from the Debnik Anticline, southern Poland: Palaeontologia Polonica, no. 54, p. 1–88, 22 plates.

Barskov, I. S., Kononova, L. I, and Migdisova, A. V., 1984, Konodonty nizhneturneiskikh otlozheniy Podmoskovnogo basseyna (Conodonts of the Lower Tournasian beds of the Moscow Basin), *in* Menner, V. V, ed., Paleontologicheskaya kharakteristika stratotipicheskikh i opornykh razrezov karbona: Moskovskoy sineklizy: Moscow, Moskovskiy gosudarstvennyy universitet, p. 3–31, pls. 1, 2.

Barskov, I. S., Vorontsova, T. N., Kononova, L. I, and Kuz′min, A. V., 1991, Opredelitel′ konodontov devona i nizhnego karbona (Index conodonts of the Devonian and Lower Carboniferous), Moscow, Moskovskiy gosudarstvennyy universitet, 183 p.

Boogaard, M. van den, and Kuhry, B., 1979, Statistical reconstruction of the *Palmatolepis* apparatus (Late Devonian conodontophorids) at the generic, subgeneric, and specific level: Scripta Geologica, v. 49, 57 p.

Boogaard, M. van den, and Schermerhorn, L. J. G, 1975, Conodont faunas from Portugal and southwestern Spain. Part 2: A Famennian conodont fauna at Cabezas del Pasto: Scripta Geologica, v. 28, p. 1–36, 17 plates.

Branson, E. B., and Mehl, M. G., 1934a, Conodonts from the Grassy Creek Shale of Missouri: University of Missouri Studies, v. 8, p. 171–259, 9 plates.

Branson, E. B., and Mehl, M. G., 1934b, Conodonts from the Bushberg Sandstone and equivalent formations of Missouri: University of Missouri Studies 8, p. 265–299, 335–338 plates. 22-24.

Branson, E. R., 1934, Conodonts from the Hannibal formation of Missouri: University of Missouri Studies 8, 301–334. 338–343, pls. 25–28.

Bushmina, L. S., and Kononova, L. I., 1981, Mikrofauna i biostratigrafiya pogranichnykh sloev devona i karbona (Microfauna and biostratigraphy of the Devonian and Carboniferous boundary beds): Akademiya Nauk SSSR, Sibirskoe otdelenie, Institut geologii i geofiziki, trudy, v. 459, p. 1–121, 20 pls.

Clark, D. L., and 10 others, 1981, Conodonta, *in* Robinson, R. A., ed., Treatise on invertebrate paleontology, Part W, Miscellanea, Supplement 2: Boulder, Colorado, Geological Society of America (and University of Kansas Press), 202 p.

Cook, A. G., 1995, Sedimentology and depositional environments of the Middle Devonian Big Bend Arkose and Burdekin Formation, Fanning River Group, Burdekin Subprovince, north Queensland, Australia: Memoirs of the Queensland Museum, v. 38, p. 53–91.

Cook, A. G., Zhen, Y. Y, Lang, S. C., and Draper, J. J., 1994, Fanning River Group: Geological Survey of Queensland Record 1994/9, p. 10–26.

Cooper, C. L., 1939, Conodonts from a Bushberg-Hannibal horizon in Oklahoma: Journal of Paleontology, v. 13, p. 379–422, plates. 39–47.

De Pomeroy, A. M., 1995, Australian Devonian fish biostratigraphy in relation to conodont zonation. Courier Forschungsinstitut Senckenberg, v. 182, p. 475–486.

Draper, J. J., and Lang, S. C., 1994a, Geological evolution of the Burdekin Basin: Geological Ssurvey of Queensland Record 1994/9, p. 127–143.

Draper, J. J., and Lang, S. C., eds., 1994b, Geology of the Devonian to Carboniferous Burdekin Basin: Geological Survey of Queensland Record 1994/9, 199 p.

Draper, J. J., Crouch, S. B, and 13 others, 1993a, 1:100,000 geological map 8158 Dotswood, preliminary edition: Brisbane, Queensland Department of Mines and Energy.

Draper, J. J., Scott, M., and Withnall, I. W., 1993, Clarke River Basin and associated rocks: Queensland Geology, v. 4, p. 203–219.

Draper, J. J., Gunther, I. M., and Lang, S. C., 1994, Keelbottom Group: Geological Survey of Queensland Record 1994/9, p. 61–88.

Dreesen, R., and Dusar, M., 1974, Refinement of conodont-biozonation in the Famenne-type area: Service géologique de Belgique Publication 13, 36 p., 7 plates.

Dreesen, R., and Houllebergh, E., 1980, Evolutionary trends of Famennian icriodids in the Dinant and Vesdre Basins (conodonts, Belgian Upper Devonian): Annales de la Société Géologique de Belgique, v. 103, p. 111–141, 7 plates.

Dreesen, R., and Orchard, M., 1974, Intraspecific morphological variation within *Polygnathus semicostatus* Branson and Mehl: Service géologique de Belgique Publication 21, 8 p.

Dreesen, R., Dusar, M., and Groessens, E., 1976, Biostratigraphy of the Yves-Gomezee Road Section (uppermost Famennian): Service géologique de Belgique Professional Paper 1976, no. 6, 20 p., 13 plates.

Druce, E. C., 1970a, Devonian and Carboniferous conodonts from the Bonaparte Gulf Basin, northern Australia: Australia, Bureau of Mineral Resources, Geology and Geophysics Bulletin 98, 242 p., 43 plates.

Druce, E.C., 1970b, Frasnian conodonts from Mount Morgan, Queensland. Australia, Bureau of Mineral Resources, Geology and Geophysics Bulletin 108, p. 75–90, pls. 12–14.

Druce, E. C., 1976, Conodont biostratigraphy of the Upper Devonian reef complexes of the Canning Basin, Western Australia: Australia, Bureau of Mineral Resources, Geology and Geophysics Bulletin 158, 303 p., 97 plates.

Drygant, D. M., 1986, Novye konodonty roda *Polygnathus* Hinde, 1879 iz srednego i verkhnego devona L′vovskogo progiba (New conodonts of the genus *Polygnathus* Hinde, 1879, from the Middle and Upper Devonian of the L′vov Depression): Paleontologicheskiy sbornik (L′vovskiy gosudarstvennyy universitet), no. 23, p. 47–52, 2 plates.

Ebert, J., 1993, Globale events im Grenz-Bereich Mittel-/Ober Devon: Göttinger Arbeiten zur Geologie und Paläontologie, no. 59, 106 p., 1 plate.

Filipovic, I., Pajic, V., and Stojanovic-Kuzenko, S., 1975, Biostratigrafija Devona Severozapadne Srbije (Biostratigraphy of the Devonian in north-west Serbia): Rasprave Zavoda za Geoloska i Geofizicka Istrazivanja (Memoires du Service Géologique et Géofisique), Beograd, v. 12, 91 p., 31 plates.

Gagiev, M. H., 1979, Conodonts from the Devonian-Carboniferous boundary deposits of the Omolon Massif [in Russian with new taxa diagnosed in English], *in* Biostratigraphy and fauna of the Devonian-Carboniferous boundary deposits: Field Excursion Guidebook for Excursion 9, 14th Pacific Science Congress (Khabarovsk, USSR, August 1979): Magadan, The Far East Science Centre of the Soviet Academy of Science, 104 p., 18 plates.

Gagiev, M. K., Kononova, L. I., and Pazukhin, V. N., 1987, Conodonta: *in* Maslov, V. I., ed., Fauna i biostratigrafiya pogranichnykh otlozheniy devona i karbona Berchogura (Mugodzhary) (The fauna and biostratigraphy of Devonian and Carboniferous boundary deposits in the Berchogur area of the Mugodozhary Region): Moscow, Nauka Press, p. 91–97, plates 25–28.

Girard, C., 1994, Conodont biofacies and event stratigraphy across the D/C boundary in the stratotype area (Montagne Noire, France): Courier Forschungsinstitut Senckenberg, v. 168, p. 299–309.

Glenister, B. F., and Klapper, G., 1966, Upper Devonian conodonts from the Canning Basin, Western Australia: Journal of Paleontology, v. 40, p. 777–842, plates 85–96.

Gould, R. E., 1975, The succession of Australian pre-Tertiary megafossil flora: The Botanical Review, v. 41, no. 4, p. 453–483.

Gunther, M. C., Withnall, I. W., McLennan, T. P. T., Gunther, L., Lang, S. C., Macansh, J. D., and Grimes, K. G., 1991, 1;100,000 Geological map 8159 Rollingstone, preliminary edition: Brisbane, Queensland Department of Resource Industries.

Hass, W. H., 1959, Conodonts from the Chappel Limestone of Texas: U.S. Geological Survey Professional Paper 294-J, p. 365–399, plates 46–50.

Helms, J., 1959, Conodonten aus dem Saalfelder Oberdevon (Thüringen): Geologie, v. 8, p. 634–677, 6 plates.

Helms, J., 1961, Die "nodocostata Gruppe" der Gattung *Polygnathus*, oberdevonische Conodonten: Geologie, v. 10, p. 674–711, 4 plates.

Holmes, G. B., 1928, A bibliography of the conodonts with descriptions of Early Mississippian species: U.S. Natural Museum Proceedings, v. 72, 38 p., 11 plates.

Hou, H. F, and eight others, 1985, Muhua sections of Devonian-Carboniferous boundary beds: Beijing, Geological Publishing House (in Chinese with English summary), 226 p., 46 plates.

Huddle, J. W., 1968, Redescription of Upper Devonian conodont genera and species proposed by Ulrich and Bassler in 1926: U.S. Geological Survey Professional Paper 578, 55 p., 17 plates.

Ishiga, H., and Leitch, E. C., 1988, Stratigraphy of the western part of the Hastings Block, New England Fold Belt, *in* Preliminary report on the geology of the New England Fold Belt. Australia, no. 1: Tokushima, Tokushima University, p. 33–45.

Jack, R. L., 1879, Report on the geology and mineral resources of the district between the Charters Towers Goldfield and the coast: Geological Survey of Queensland Report 1, 28 p.

Ji, Q., 1989, On the Frasnian conodont biostratigraphy in the Guilin area of Guangxi, south China: Courier Forschungsinstitut Senckenberg, v. 117, p. 303–319, 4 plates.

Ji, Q., and Ziegler, W., 1992, Phylogeny, speciation and zonation of Siphonodella of shallow water facies (Conodonta, Early Carboniferous): Courier Forschungsinstitut Senckenberg, v. 154, p. 223–251, 4 plates.

Ji, Q. and Ziegler, W., 1993, The Lali section: An excellent reference section for Upper Devonian in south China: Courier Forschungsinstitut Senckenberg, v. 157, 183 p., 45 plates.

Ji, Q., Hou, H. F, Wu, X. H, and Xiong, J. F., 1984, The discovery of conodont *Siphonodella praesulcata* Zone and *S. sulcata* Zone in China, and their stratigraphic value (in Chinese with English abstract): Acta Geologica Sinica, v. 58. no. 2, p. 106–113, 2 plates.

Ji, Q. and nine others, 1989, The Dapoushang section: An excellent section for the Devonian-Carboniferous boundary stratotype in China: Beijing, Science Press, 165 p., 43 plates.

Johnson, J. G., and Klapper, G., 1992, North American Midcontinent Devonian T-R cycles, *in* Chaplin, J. R., and Barrick, J. E., eds., Special papers in paleontology and stratigraphy: A tribute to Thomas W. Amsden: Oklahoma Geological Survey Bulletin 145, p. 127–135.

Johnson, G. G., and Sandberg, C. A., 1989, Devonian eustatic events in the western United States and their biostratigraphic responses, *in* McMillan, N. J., Embry, A. F., and Glass, D. J., eds., Devonian of the world: Calgary, Canadian Society of Petroleum Geologists Memoir 14, v. 3, p. 171–178. [imprint 1988]

Johnson, J. G., Klapper, G., and Sandberg, C. A., 1985, Devonian eustatic fluctuations in Euramerica: Geological Society of America Bulletin, v. 96, p. 567–587.

Johnston, D. I. and Chatterton, B. D. E., 1991, Famennian conodont biostratigraphy of the Palliser Formation, Rocky Mountains, Alberta and British Columbia, *in* Orchard, M. J., and McCracken, A. D., Ordovician to Triassic conodont paleontology of the Canadian Cordillera: Geological Survey of Canada Bulletin 417, p. 163–183, 3 plates.

Kamp, H. von, 1972, Geologische Karte von Nordrhein-Westfalen 1:25000, Erläuterungen zu Blatt 4611 Hohenlimburg: Krefeld, Geologische Landesamt Nordhrein-Westfalen, p. 182.

Khalymbadzha, V. G., and Mingaleev, N. M., 1994, Novye famenskie polignatidnye konodonty iz zilairskoy serii vostochnogo sklona yuzhnogo urala (New Famennian polygnathid conodonts for the Zilar series of the eastern slope of the southern Urals): Byulleten′ Moskovskogo Obshchestva Ispytateley Prirody otdel geologicheskiy, v. 69, no. 3, p. 81–98.

Khalymbadzha, V. G., Shinkaryov, G. Ye, and Gatovskiy, Yu. A., 1991, Novyye famenskiye polignatidy (konodonty) Yuzhnogo Kazakhstana (New Famennian polygnathids [conodonts] from southern Kazakhstan): Paleontologicheskiy Zhurnal, v. 1991, no. 2, p. 56–64. English version in Paleontological Journal, v. 25, no. 2, p. 69–81.

Klapper, G., 1966, Upper Devonian and Lower Mississippian conodont zones in Montana, Wyoming and South Dakota: University of Kansas Paleontological Contributions 3, 43 p., 6 plates.

Klapper, G., 1971a, Sequence within the conodont genus *Polygnathus* in the New York lower Middle Devonian: Geologica et Palaeontologica, v. 5, p. 59–79, 3 plates.

Klapper, G., 1971b, *Patrognathus* and *Siphonodella* (Conodonta) from the Kinderhookian (Lower Mississippian) of western Kansas and southwestern Nebraska: Kansas Geological Survey Bulletin 202, part 3, 14 p., 2 plates.

Klapper, G., 1989, The Montagne Noire Frasnian (Upper Devonian) conodont succession, *in* McMillan, N. J., Embry, A. F., and Glass, D. J., eds., Devonian of the world: Calgary, Canadian Society of Petroleum Geologists Memoir 14, v. 3, p. 449–468. [imprint 1988]

Klapper, G., 1990, Frasnian species of the Late Devonian conodont genus *Ancyrognathus*: Journal of Paleontology, v. 64, p. 998–1025.

Klapper, G., and Lane, R. H., 1985, Upper Devonian (Frasnian) conodonts of the Polygnathus biofacies, N.W.T., Canada: Journal of Paleontology, v. 59, p. 904–951.

Klapper, G., and Philip, G., 1972, Familial classification of reconstructed Devonian conodont apparatuses, *in* Lindström, M., and Ziegler, W., Symposium on conodont taxonomy: Geologica et Palaeontologica, special volume 1, p. 97–114, 4 plates.

Kononova, L. I., 1979, Upper Frasnian, Famennian and Tournaisian conodonts of the Sikaza River section (southern Urals): Service géologique de Belgique Professional Paper 1979/5, no. 161, p. 74–86, 2 plates.

Korn, D., Clausen C.-D., Belka, Z., Leuteritz, K., Luppold, F. W., Feist, R, and Weyer, D., 1994, Die Devon/Karbon-Grenze bei Drewer (Rheinsches Schiefergebirge): Geologie und Paläontologie in Westfalen no. 29, p. 97–147, 7 plates.

Kürschner, W., Becker, R. T, Buhl, D., and Veizer, J., 1993, Strontium isotopes in conodonts: Devonian-Carboniferous transition, the northern Rhenish Slate Mountain, Germany: Annales de la société géologique de Belgique, v. 115, p. 595–621, 4 plates.

Kuz′min, A. V., 1990, Asimmetricheskiye pary platformennykh elementov u nekotorykh predstaviteley roda *Polygnathus* (konodonty) (Asymmetrical paired platform elements and some specimens of the genus *Polygnathus* [conodonts]): Paleontologicheskiy Zhurnal, v. 1990, no. 4, p. 66–74, plate 4. English version in Paleontology Journal, v. 25, no. 4, p. 62–70.

Kuz′min, A. V., 1992, Ispol′zovanie konodontov dlya raschleneniya famenskikh otlozheniy Atasuyskogo rudnogo rayona (Tsentral′nyy Kazakhstan) (Use of conodonts for subdivision of the Famennian deposits of the Atasu ore-bearing regoin [central Kazakhstan]): Rossiyskaya Akademiya Nauk, Izvestiya, Seriya geologicheskaya, v. 1992, no. 8, p. 30–40, 2 plates.

Lang, S. C., 1986, Devonian-Carboniferous stratigraphy of the southeastern Bundock Basin, Broken River area, north Queensland: Geological Survey of Queensland Record 1986/5, v. 1, 213 p.; v. 2, 64 p., 25 figures.

Lang, S. C., 1993, Stratigraphy and sedimentology of the Bundock Creek Group: Queensland Geology, v. 4, p. 129–174.

Lang, S. C., Gunther, L. M., and Rich, N., 1990a, The Devonian-Carboniferous intracratonic Burdekin Basin, north Queensland. II: Volcanoclastic continental facies of the Dotswood Group: Parkville, Australasian Institute of Mining and Metallurgy, Pacific Rim Congress 90, v. 3, p. 631–640.

Lang, S. C., Jell, J. S., and Draper, J. J., 1990, Depositional evolution of the Devonian-Carboniferous intracratonic Burdekin Basin, north Queensland: Parkville, Australasian Institute of Mining and Metallurgy, Pacific Rim Congress 90, v. 3, p. 791–800.

Lang, S. C., Draper, J. J., and Gunther, L. M., 1994, Dotswood Group: Geological Survey of Queensland Record 1994/9, p. 34–60.

Leichhardt, L., 1847, Journal of an overland expedition in Australia from Moreton Bay to Port Essington, a distance of upward of 3,000 miles, during the years, 1844–1845: London, T. and W. Boone, xx + 544 p., 3 maps.

Levingston, K. R., 1981, Geological evolution and economic geology of the Burdekin River region, Queensland: Australia, Bureau of Mineral Resources, Geology and Geophysics Bulletin 208, 48 p., 1 map.

Long, J. A., 1987, Upper Devonian conodonts associated with a large placoderm fish skull from the Canning Basin, Western Australia: Western Australian Museum Records, v. 13, p. 501–513.

Matyja, H., 1972, Biostratigraphy of the Upper Devonian from the borehole Chojnice 2 (western Pomerania): Acta Geologica Polonica, v. 22, 735–750, 4 plates.

Matyja, H., 1975., A new conodont species from the Famennian of Poland: Bulletin de L'Académie Polonaise des Sciences, Série des sciences biologiques Cl. II, v. 22, no. 11: p. 785–787, 1 plate. [imprint 1974]

Matyja, H., 1987, Conodont biofacies in the Famennian Stage of Pomerania, northwestern Poland, *in* Austin, R. L., ed., Conodonts: Investigative techniques and applications: Chichester, British Micropalaeontological Society, Ellis Horwood Limited, p. 363–381.

Matyja, H., 1993, Upper Devonian of western Pomerania: Acta Geologica Polonica, v. 43, p. 27–94, 36 pls.

Matyja, H., and Zbikowska, B., 1974, Stratigrafia dewonu górnego profilu wiercenia Minkowice 1 (basen lubelski) (Stratigraphy of the Upper Devonian from the borehole Minkowice 1 [Lublin Basin]): Acta Geologica Polonica, v. 24, p. 663–694, 8 plates.

Mawson, R., and Talent, J.A., 1989, Late Emsian–Givetian conodont stratigraphy and biofacies—carbonate slope and offshore shoal to lagoon and nearshore carbonate ramp—Broken River, north Queensland, Australia: Courier Forschungsinstitut Senckenberg, v. 117, p. 205–259, 8 plates.

Mawson R., and 13 others, A.G., 1989, Conodont data in relation to resolution of stage and zonal boundaries for the Devonian of Australia, *in* McMillan, N. J., Embry, A. F., and Glass, D. J., eds., Devonian of the world: Calgary, Canadian Society of Petroleum Geologists Memoir 14, v. 3, p. 485–527. [imprint 1988]

McKellar, R. G., 1970, The Devonian productoid brachiopod faunas of Queensland: Geological Survey of Queensland Publication 342 (Palaeontology Paper 18), 40 p., 12 plates.

Mehl, M. G. and Thomas, L. A, 1947, Conodonts from the Fern Glen of Missouri: Denison University Bulletin, Journal of the Scientific Laboratories, v. 40, p. 3–19, 1 plate.

Metzger, R. A., 1989, Upper Devonian (Frasnian-Famennian) conodont biostratigraphy in the subsurface of north-central Iowa and southeastern Nebraska: Journal of Paleontology, v. 63, p. 503–524.

Metzger, R. A., 1994, Multielement reconstructions of *Palmatolepis* and *Polygnathus* (Upper Devonian, Famennian) from the Canning Basin, Australia, and Bactrian Mountain, Nevada: Journal of Paleontology, v. 68, p. 617–647.

Mory, A. J., and Crane, D. T., 1982, Early Carboniferous *Siphonodella* (Conodonta) faunas from eastern Australia: Alcheringa, v. 6, p. 275–303.

Müller, K. J., 1956, Zur Kenntnis der Conodonten-Fauna des europäischen Devons. 1: Die Gattung *Palmatolepis*: Senckenbergische Naturforschende Gesellschaft Abhandlungen, v. 494, 70 p., 11 plates.

Naka, T., Kuwahara, K., Ishiga, H., and Leitch, E. C., 1988, Age and significance of Late Devonian conodonts and radiolarians from the Tamworth Belt, southern New England Fold Belt, eastern Australia: *in* Preliminary report on the geology of the New England Fold Belt, Australia, no. 1: Tokushima, Tokushima University, p. 73–76.

Nemirovskaya, T. I., Chermnykh, V. A., Kononova, L. I. and Pazukhin, V. N., 1993, Conodonts of the Devonian-Carboniferous boundary section, Kozhim, Polar Urals, Russia: Annales de la Société Géologique de Belgique, v. 115, no. 2, p. 629–647.

Nicoll, R. S., 1980, The multielement genus *Apatognathus* from the Late Devonian of the Canning Basin, Western Australia: Alcheringa, v. 4, p. 133–152.

Nicoll, R. S., and Druce, E. C., 1979, Conodonts from the Fairfield Group, Canning Basin, Western Australia: Australia, Bureau of Mineral Resources, Geology and Geophysics Bulletin 190, 134 p., 28 plates.

Olivieri, R., 1985, Middle and Late Devonian conodonts from southwestern Sardinia: Bollettino della Società Paleontologica Italiana, v. 23, p. 269–310, 9 plates.

Over, D. J., 1992, Conodonts and the Devonian-Carboniferous boundary in the upper Woodford Shale. Arbuckle Mountains, south-central Oklahoma: Journal of Paleontology, v. 66, p. 293–311.

Perret, M-F., 1993, Recherches micropaleontologiques et biostratigraphiques (Conodontes-Foraminifères) dans le Carbonifère Pyrénéen: Strata, Série 2, v. 21, p. 1–597, 24 plates.

Perri, M. C., and Spalletta, C., 1990, Famennian conodonts from climenid pelagic limestone, Carnic Alps, Italy: Palaeontographia Italica, v. 77, p. 55–83, 7 plates.

Perri, M. C., and Spalletta, C., 1991, Famennian conodonts from Cava Cantoniera and Malpasso sections, Carnic Alps, Italy: Bollettino della Società Paleontologica Italiana. v. 30, p. 47–78, 9 plates.

Philip, G. M. and Jackson, J. H., 1971, Late Devonian conodonts from the Luton Formation, northern New South Wales: Proceedings of the Linnaean Society of New South Wales, v. 96, p. 66–75, 2 plates.

Phuong, T. H., and Weyant, M., 1994, Discovery of Famennian conodonts from the Quyet Thang Mountain, north of the town of Thanh Hoa, Vietnam: Courier Forschungsinstitut Senckenberg. v. 168, 131–137, 1 plate.

Pickett, J. W., 1972, Late Devonian (Frasnian) conodonts from Ettrema, New South Wales: Journal and Proceedings of the Royal Society of New South Wales, v. 105, p. 31–37, 3 plates.

Pickett, J. W., 1981, Late Devonian and Early Carboniferous conodonts from the Myrtlevale Beds and Hardwick Formation, Star Basin: Queensland Government Mining Journal, v. 82, p. 24–27, 1 plate.

Pickett, J. W., 1994, Tournaisian corals and conodonts from the Slaughterhouse Creek area, New South Wales, Australia: Courier Forschungsinstitut Senckenberg, v. 172, p. 51–67.

Pölsler, P., 1969, Conodonten aus dem Devon der Karnischen Alpen (Findenigkofel, Österreich): Jahrbuch der geologischen Bundesansalt im Wien, v. 112, p. 399–440, 6 plates.

Rhodes, F. H. T., Austin, R. L., and Druce, E. C., 1969, British Avonian (Carboniferous) conodont faunas and their value in local and intercontinental correlation: Bulletin of the British Museum of Natural History, Supplement 5, 313 p., 31 plates.

Ross, C. A., and Ross, J. R. P., 1988, Late Palaeozoic transgressive-regressive deposition, *in* Wilgus, C. A., et al., eds., Sea-level changes, an integrated approach. Society of Economic Paleontologists and Mineralogists Spécial Publication 42, p. 227–247.

Sandberg, C. A., 1979, Devonian and Lower Mississippian conodont zonations of the Great Basin and Rocky Mountains, *in* Sandberg, C. A., and Clark, D. L., eds., Conodont biostratigraphy of the Great Basin and Rocky Mountains: Brigham Young University Geological Studies, v. 26, p. 87–106.

Sandberg, C. A., and Dreesen, R., 1984, Late Devonian icriodontid biofacies models and alternate shallow-water conodont zonation, *in* Clark, D. L., ed., Conodont biofacies and provincialism: Geological Society of America Special Paper 196, p. 143–178, 4 plates.

Sandberg, C. A., and Ziegler, W., 1973, Refinement of standard Upper Devonian conodont zonation based on sections in Nevada and West Germany: Geologica et Palaeontologica, v. 7, p. 97–122, 5 plates.

Sandberg, C. A., and Ziegler, W., 1979, Taxonomy and biofacies of important conodonts of Late Devonian styriacus-Zone, United States and Germany: Geologica et Palaeontologica, v. 13, p. 173–212, 7 plates.

Sandberg, C. A., Ziegler, W., Leuteritz, K. and Brill, S. M., 1978, Phylogeny, speciation, and zonation of *Siphonodella* (Conodonta, Upper Devonian and Lower Carboniferous): Newsletters on Stratigraphy, v. 7, no. 2, p. 102–120.

Schönlaub, H. P., Feist, R., and Korn, D., 1988, The Devonian-Carboniferous boundary at the section "Grüne Schneid" (Carnic Alps, Austria): A preliminary report: Courier Forschungsinstitut Senckenberg, v. 100, p. 149–167, 4 plates.

Schülke, I., 1995, Evolutive Prozesse bei *Palmatolepis* in der frühen Famenne-Stufe (Conodonta, Ober-Devon): Göttinger Arbeiten zur Geologie und Pälontologie, no. 67, 108 p., 14 plates.

Scott, A. J., 1961, Three new conodonts from the Louisiana Limestone (Upper Devonian) of western Illinois: Journal of Paleontology, v. 35, p. 1223–1227.

Sloan, T. R., and 10 others, 1995, Conodont data from Silurian–Middle Devonian carbonate fans, debris flows, allochthonous blocks and adjacent autochthonous platform margins: Broken River and Camel Creek areas, north Queensland, Australia: Courier Forschungsinstitut Senckenberg, v. 182, p. 1–77, 15 plates.

Stone, J. J., 1991, Arundian (Lower Carboniferous) conodonts from South Wales: Special Papers in Palaeontology 46, 63 p., 5 pls.

Stritzke, R., 1990, Die Karbonatsedimentation im Briloner Vorriffbereich: Geologische Jahrbuch, v. 95, p. 253–315.

Sweet, W. C., 1988, The Conodonta: Morphology, taxonomy, paleoecology and evolutionary history of a long-extinct animal phylum: Oxford, Clarendon Press, 212 p.

Szulczewski, M., 1971, Upper Devonian conodonts, stratigraphy and facial development in the Holy Cross Mts: Acta Geologica Polonica, v. 21, p. 1–129, 34 plates.

Szulczewski, M., 1972, Interspecific relation of some Upper Devonian polygnathid conodonts: Acta Geologica Polonica, v. 22, p. 449–458, 2 plates.

Talent, J. A., 1989a, Dating diastrophism in the Devonian of eastern Australia: biostratigraphic and biogeographic constraints, *in* McMillan, N. J., Embry, A. F., and Glass, D. J., eds., Devonian of the world: Calgary, Canadian Society of Petroleum Geologists Memoir 14, v. 2, p. 313–320. [imprint 1988]

Talent, J. A., 1989b, Transgression-regression pattern for the Silurian and Devonian of Australia, *in* Le Maitre, R. W., ed., Pathways in geology—essays in honour of Edwin Sherbon Hills: Carlton, Australia, Blackwells, p. 201–219.

Talent, J. A., and Mawson, R., 1994, Conodonts in relation to the age and environmental framework of the Burdekin Basin (mid-Devonian), northeastern Australia: Courier Forschungsinstitut Senckenberg, v. 168, p. 61–81.

Talent, J. A., and Yolkin, E. A., 1987, Transgression-regression patterns for the Devonian of Australia and southern West Siberia: Courier Forschungsinstitut Senckenberg v. 92, p. 235–249.

Thomas, L. A., 1949, Devonian-Mississippian formations of southeast Iowa: Geological Society of America Bulletin, v. 60, p. 403–437, 4 plates.

Ulrich, E. O., and Bassler, R. S., 1926, A classification of the tooth-like fossils, conodonts, with descriptions of American Devonian and Mississippian species: Proceedings of the U.S. National Museum, v. 68, article 12, 63 p., 11 plates.

Uyeno, T. T., 1991, Pre-Famennian Devonian conodont biostratigraphy of selected intervals in the eastern Canadian Cordillera, *in* Orchard, M. J., and McCracken, A. D., Ordovician to Triassic conodont paleontology of the Canadian Cordillera: Geological Survey of Canada Bulletin 417, p. 129–161, 5 plates.

Varker, W. J., and Sevastopulo, G. D., 1985, The Carboniferous System. Part I: Conodonts of the Dinantian Subsystem from Great Britain and Ireland, *in* Austin, R. L, and Higgins, A. C., eds., A stratigraphical index of conodonts: Chichester, The British Micropalaeontological Society, Ellis Horwood Limited, p. 167–209.

Vorontsova, T. N., 1993, Rod *Polygnathus sensu lato* (konodonty): Filogeniya i sistematika (The genus *Polygnathus sensu lato* [conodonts]: Phylogeny and systematics): Paleontologicheskiy Zhurnal, v. 1993, no. 3, p. 66–78, 2 plates.

Vorontsova, T. N., and Kuz′min, A., 1984, Rasprostranenie novykh vidov konodontov roda Polygnathus v famenskikh otlozheniyakh Tsentral'nogo Kazakhstana (Distribution of new species of conodonts of the genus *Polygnathus* from Famennian beds of central Kazakhstan): Izvestiya Akademiya Nauk SSSR, geologicheskaya seriya, v. 1984, no. 10, p. 58–64, 1 plate.

Wang, C. Y., 1989, Devonian conodonts in Guangxi: Memoirs of Nanjing Institute of Geology and Palaeontology, Academia Sinica 25, p. 1–152, 43 plates. (in Chinese with English summary).

Wang, C. Y., and Wang, Z. H., 1978, Upper Devonian and Lower Carboniferous conodonts from southern Guizhou: Memoirs of Nanjing Institute of Geology and Palaeontology, Academia Sinica 11, p. 51–91, 8 plates. (in Chinese with English summary).

Wang, C. Y, and Yin, B.-A., 1984, Conodont zonations of early Lower Carboniferous and Devonian-Carboniferous boundary in pelagic facies, south China: Acta Palaeontologica Sinica, v. 23, no. 2, p. 224–238, 3 plates.

Wang, C. Y, and Yin, B.-A., 1988, Conodonts, *in* Yu Chang-Min, ed., Devonian-Carboniferous boundary in Nanbiancun, Guilin, China—aspects and records: Beijing, Science Press, p. 105–148, plates 13–33.

Wang, C. Y., and Ziegler, W., 1983, Devonian conodont biostratigraphy of Guangxi, South China, and the correlation with Europe: Geologica et Palaeontologica, v. 17, p. 75–107, 8 plates.

Weddige, K., 1990, Pathological conodonts: Courier Forschungsinstitut Senckenberg, v. 118, p. 563–589, 4 plates.

Wendt, J., and Belka, Z., 1991, Age and depositional environment of Upper Devonian (Early Frasnian to Early Famennian) black shales and limestones (Kellwasser facies) in the eastern Anti-Atlas, Morocco: Facies, v. 25, p. 51–90, plates 15–21.

Withnall, I. W., 1993, Geological map, Camel Creek Special 1:250,000: Brisbane, Queensland Department of Minerals and Energy.

Withnall, I. W., and Lang, S. C., eds, 1993, Geology of the Broken River Province, north Queensland: Queensland Geology, v. 4, 289 p.

Withnall, I. W., and eight others, 1991, 1:100,000 geological map 8059 Ewan, preliminary edition: Brisbane, Queensland Department of Resource Industries.

Withnall, I. W., Lang, S. C., and 12 others, 1992, Geological map, Broken River Special 1;100,000: Brisbane, Queensland Department of Minerals and Energy.

Wolska, Z., 1967, Gorno-dewonskie konodonty z poludniowo-zachodniego regionu Gor Swietokrzyskich (Upper Devonian conodonts from the southwest region of the Holy Cross Mountains, Poland): Acta Palaeontologica Polonica, v. 12, p. 363–456, 19 plates.

Wright, A. J., Pickett, J. W., Sewell, D., Roberts, J., and Jenkins, T. B. H., 1990, Corals and conodonts from the Late Devonian Mostyn Vale Formation, Keepit, New South Wales: Association of Australasian Palaeontologists Memoir 10, p. 211–254.

Wyatt, D. H., 1968, 1:250,000 geological series, explanatory notes—Townsville, Queensland, Sheet SE/55-14: Canberra, Australian Government Publishing Service, 32 p.

Wyatt, D. H., 1973, The stratigraphy and palaeogeography of the Fanning River-Burdekin Downs-Star area [M.Sc. thesis]: Townsville, Australia, James Cook University, 180 p.

Wyatt, D. H., 1978, Fanning River–Dotswood–Charters Towers area, north Queensland, *in* Rubenach, M., ed., Australian Geological Convention 3rd., Excursions handbook, (Townsville, August 1978): Brisbane, Geological Society of Australia, Queensland Division, p. 69–83.

Wyatt, D. H., and Jell, J. S., 1967, Devonian of the Townsville hinterland, Queensland, Australia, *in* Oswald, D. H., ed., International symposium on the Devonian System, Volume 2: Calgary, Alberta Society of Petroleum Geologists p. 99–105.

Wyatt, D. H., and Jell, J. S., 1980, Devonian and Carboniferous stratigraphy of the northern Tasman orogenic zone in the Townsville hinterland, north Queensland, *in* Henderson, R. A., and Stephenson, P. J., eds., The geology and geophysics of northeastern Australia: Brisbane, Queensland Division of the Geological Society of Australia, p. 201–228.

Wyatt, D. H., Dearne, D. W., Paine, A. G. L., Harding, R. R., and Clarke, D. E., 1968, Australia 1:250,000 geological series, Townsville, sheet SE 55-14: Canberra, Bureau of Mineral Resources.

Wyatt, D. H., Paine, A. G. L., Clarke, D. E., and Harding, R. R., 1970, Geology of the Townsville 1:250,000 sheet area, Queensland: Report of the Bureau of Mineral Resources, Geology and Geophysics, v. 127, 85 p., map.

Zhen, Y. Y., 1991, The biostratigraphy and the rugose corals of the Fanning River Group, north Queensland [Ph.D. thesis]: Brisbane, University of Queensland, 394 p.

Zhen, Y. Y., 1995, Late Emsian rugose corals of the Mount Podge area, Burdekin Basin, north Queensland: Alcheringa, v. 19, p. 193–234.

Zhen, Y. Y., Lang, S. C., and Jell, J. S., 1993, A new biostratigraphic framework and lithostratigraphic nomenclature for the Devonian Fanning River Group, Burdekin Basin, north Queensland: Queensland Government Mining Journal, v. 94, no. 1105, p. 7–14.

Ziegler, W., 1960, Paläontologischer Anhang, *in* Kronberg, P., Pilger, A., Scherp, A., and Ziegler, W., Spuren altvariscischer Bewegungen im nordöstlichen Teil des Rheinischen Schiefergebirges, *in* Teichmüller, R., and Paproth, E., eds., Das Karbon der subvariscischen Saumsenke, Ein Symposium. Teil 1: Der Kulm und die flözleere Fazies des Namurs: Fortschritte in der Geologie von Rheinland und Westfalen, v. 3, p. 35–43, 7 plates.

Ziegler, W., 1962a, Taxionomie und Phylogenie Oberdevonischer Conodonten und ihre stratigraphische Bedeutung: Abhandlungen des Hessischen Landesamtes für Bodenforschung no. 38, 166 p. 14 plates.

Ziegler, W., 1962b, Die Conodonten aus den Geröllen des Zechsteinkonglomerates von Rossenray (südwestlich Rheinberg/Niederrhein): Fortschritte in der Geologie von Rheinland und Westfalen, v. 6, p. 391–406, 4 plates.

Ziegler, W., 1971, Conodont stratigraphy of the European Devonian, *in* Sweet, W. C, and Bergström, S. M., eds., Symposium on conodont biostratigraphy: Geological Society of America Memoir 127, p. 227–284, 6 charts.

Ziegler, W., ed., 1973, Catalogue of conodonts, Volume 1: Stuttgart, E. Schweizerbart'sche Verlagsbuchhandlung, 504 p., 27 plates.

Ziegler, W., ed., 1975, Catalogue of conodonts, Volume 2: Stuttgart, E. Schweizerbart'sche Verlagsbuchhandlung, 404 p., 25 plates.

Ziegler, W., ed., 1977, Catalogue of conodonts, Volume 3: Stuttgart, E. Schweizerbart'sche Verlagsbuchhandlung, 574 p., 39 plates.

Ziegler, W., ed., 1981, Catalogue of conodonts, Volume 4: Stuttgart, E. Schweizerbart'sche Verlagsbuchhandlung, 445 p., 41 plates.

Ziegler, W., ed., 1991, Catalogue of conodonts, Volume 5: Stuttgart, E. Schweizerbart'sche Verlagsbuchhandlung, 212 p., 13 plates.

Ziegler, W., and Sandberg, C. A., 1984, *Palmatolepis*-based revision of upper part of standard Late Devonian conodont zonation, *in* Clark, D. L., ed., Conodont biofacies and provincialism: Geological Society of America Special Paper 196, p. 179–194, 2 plates.

Ziegler, W., and Sandberg, C. A., 1990, The Late Devonian standard conodont zonation: Courier Forschungsinstitut Senckenberg, v. 121, 115 p., 17 plates.

Ziegler, W., Sandberg, C. A., and Austin, R. L., 1974, Revision of *Bispathodus* group (Conodonta) in the Upper Devonian and Lower Carboniferous: Geologica et Palaeontologica, v. 8, p. 97–112, 3 plates.

Manuscript Accepted by the Society February 27, 1997

Printed in U.S.A.

Geological Society of America
Special Paper 321
1997

Biogeography of Early Devonian Chitinozoa

Theresa Winchester-Seeto
Centre for Ecostratigraphy and Palaeobiology, School of Earth Sciences, Macquarie University, N.S.W. 2109, Australia

ABSTRACT

The relationship between Early Devonian chitinozoan assemblages from Australia and those of the rest of the world is investigated using simple statistical procedures including the number of taxa in common and the Dice Similarity Coefficient. A moderate degree of generic provincialism is established with 32% of Early Devonian genera being absent from Australian strata, namely *Armoricochitina*, *Cingulochitina*, *Linochitina*, *Margachitina*, *Urnochitina*, and *Urochitina*. In contrast to pre-Devonian times, there is no evidence of provincialism based on paleolatitude. The genera *Ancyrochitina*, *Angochitina*, *Bursachitina*, *Calpichitina*, *Eisenackitina*, *Fungochitina*, *Gotlandochitina*, *Hoegisphaera*, *Sphaerochitina*, and possibly *Muscochitina* and *Bulbochitina* are found to be relatively cosmopolitan in distribution.

INTRODUCTION

An important feature of biostratigraphic studies of Early Devonian chitinozoans of Australia has been the potential for intercontinental correlation. This has been based largely on index species used in biozonations from southwest Europe being present in Australian strata (e.g., Winchester-Seeto, 1993a, 1993b). A simple comparison of Early Devonian chitinozoan faunas from Australia and other continents highlights the wide geographic spread of some important species and of the majority of genera. Nonetheless there are some significant disparities. For instance, six genera common in coeval faunas on other continents have not been detected in Australian strata. Furthermore, the Emsian fauna of the Taravale Formation near Buchan, Victoria (Winchester-Seeto, 1996), has a unique collection of species with no known relationship to contemporaneous assemblages.

There are few biogeographical studies of Chitinozoa, particularly for the Early Devonian. Laufeld (1979) postulated a moderate level of provincialty for Early Devonian genera, falling between the fairly high provincialism of the Ordovician and the relative cosmopolitanism of the Silurian. In contrast, Paris (1981) observed that Early Devonian genera from southwest Europe and North Africa were very similar and that chitinozoan genera from Brazil were well known in Europe. He also noted that the species across the three regions were similar and concluded that, for these regions, there was no real evidence of provincialism at that time.

Studies of older assemblages indicate lack of agreement between chitinozoan workers about the degree of provincialism and possible causes. Cramer (1971) and Cramer and Diez (1972) suggested that Silurian chitinozoans displayed some indications of provincialism related to paleolatitude and observed that Silurian phytoplankton assemblages also displayed this dependence, but to a greater degree. The relationship of phytoplankton to paleolatitude outlined by them has been challenged by Colbath (1990), who contended that more recent plate reconstructions did not support their observations.

Achab (1991) and Achab et al. (1992), in recent studies of chitinozoan provinciality in the Ordovician and Early Silurian, examined statistical data from species of several regions and concluded that eastern Canadian faunas are distinct from those of the north Gondwana margin, low-latitude Laurentia, and Australian faunas. They also suggest paleogeographic distribution to have been linked to paleolatitude.

Paris (1990), in a survey of Ordovician chitinozoan biozones from northern Gondwana, observed that no major differences could be found in the composition of assemblages "between major paleogeographic domains" (p.183). Thus, in contrast to the benthic macrofauna, there are many cosmopolitan species of Chitinozoa.

Winchester-Seeto, T., 1997, Biogeography of Early Devonian Chitinozoa, *in* Klapper, G., Murphy, M. A., and Talent, J. A., eds., Paleozoic Sequence Stratigraphy, Biostratigraphy, and Biogeography: Studies in Honor of J. Granville ("Jess") Johnson: Boulder, Colorado, Geological Society of America Special Paper 321.

Paris noted, however, that five genera were restricted to northern Gondwana, a high-latitude region during the Ordovician.

The biogeographic distribution of chitinozoans has been used to test models of the paleogeographic evolution of Europe in the Paleozoic (Ordovician to Early Devonian). These tests were based on calculated coefficients of similarity between chitinozoan faunas from various regions in Europe. Areas showing a high degree of similarity were taken to indicate geographic proximity, whereas areas sharing a low similarity coefficient were regarded as more distant. Paris (1993) regarded major differences in faunal composition found in the above analysis to primarily reflect latitudinal differences as well as the influence of major ocean currents.

The small number of studies on Early Devonian chitinozoan distribution reflects the paucity of taxonomic studies and the geographically limited area of investigation. The present investigation is a preliminary survey, based on very limited data. The aims of this investigation were (1) to obtain a simple statistical comparison of Early Devonian assemblages from Australia with those from different paleocontinents, both at the specific and generic level; (2) to probe the question of provinciality of chitinozoans in the Early Devonian; and (3) to investigate the paleogeographic distribution of the six genera that are found in coeval assemblages elsewhere but are missing from Australian sequences: *Armoricochitina* Paris, *Cingulochitina* Paris, *Linochitina* Eisenack, *Margachitina* Eisenack, *Urnochitina* Paris, and *Urochitina* Taugourdeau and Jekhowsky.

METHODS

Previous Studies

In a review of biogeographic methods, Newton (1990) outlined two major approaches. The first, cladistic biogeography, is based on construction of an area cladogram, which is then used to interpret geographic relationships. This approach is not really suitable for chitinozoans because it requires phylogenetic cladograms to have been erected for the taxa being considered. Furthermore, according to Newton, the procedure is most suitable for endemic faunas and does not cope well with widespread and overlapping taxa, features often found in chitinozoan distribution.

The more common procedure for extinct groups, or for examining fossil distributions, is phenetic biogeography, wherein similarity coefficients are calculated for whole-fauna comparisons. Many statistical measures of similarity have been proposed; all have advantages and disadvantages. For examples and reviews of the statistical tests, see Campbell and Valentine (1977), Newton (1990), and Shi (1993).

Two studies of chitinozoan biogeography for the Ordovician and Silurian (Achab, 1991; Achab et al., 1992) made use of two simple procedures, among other, more sophisticated methods. These were a listing of "species in common," portrayed on a triangular diagram, and calculation of the Dice (also called the Sørenson) Coefficient of Similarity.

The "species in common" method has the advantage of simplicity in displaying data. Selection of an appropriate measure of similarity, however, becomes quite complicated because no coefficient is perfect. The Dice Coefficient has been favorably reviewed as a relatively reliable indicator of similarity (e.g., Cheetham and Hazel, 1969, Newton, 1990; Shi, 1993) and has been selected for this study to allow a modicum of comparison with the work of Achab et al. (1992). Another version of the Dice Coefficient is a coefficient of similarity used by Clark and Hatleberg (1983) and by Paris (1993). The "species in common" and Dice Coefficients have been used for species and generic comparisons, concentrating on comparison between Australian faunas and those from other regions of the world. These statistical procedures allow the degree of similarity and dissimilarity between regions of the world to be inspected and a general summation of provinciality or cosmopolitanism of species and genera to be inferred (Figs. 1, 2). The known distribution of genera missing from Australian strata has been plotted on two Scotese and McKerrow (1990) plate reconstructions, one for the late Lochkovian–Pragian and one for the Emsian (Figs. 3a, 3b). This has involved "time-averaging" of the Lochkovian and Pragian data.

Selection of data

Talent et al. (1986) set out three principles of data selection that allow some degree of confidence about the calculated similarity or dissimilarity of faunas. The first precept is that comparison should be made between faunas of the same age; second, the taxonomy must be consistent, preferably the work of the same paleontologist or team of workers. Third, the data should be drawn from the same sedimentary environments, to avoid complications of comparing vastly different ecological conditions.

Attempts have been made to follow these logical and important guidelines, but the limited amount of chitinozoan data available would be severely depleted if we were to adhere to all these principles. The faunas selected have come from all three Early Devonian stages—Lochkovian, Pragian, and Emsian. Achab et al. (1992) used relatively large time intervals to increase the number of taxa available for study, thus giving a more reliable statistical basis and avoiding problems arising from needing very precise stratigraphic alignments (although blurring stratigraphically connected results).

The problem of successfully integrating the taxonomic works of many authors is apparent. It is particularly difficult with chitinozoans because it is virtually impossible to reconcile taxonomic work from authors using light microscopy with that from workers whose observations are based on scanning electron microscope (SEM) photography (Paris, 1992; Winchester-Seeto 1993c). For the purposes of this biogeographic study, only work documented by electroscan micrographs has been included. This does not quite meet the requirements set out by Talent et al. (1986) but does allow some degree of comparison. Each of the studies selected has been scrutinized for comparison in biostratigraphic investigations for Australian faunas, though some revi-

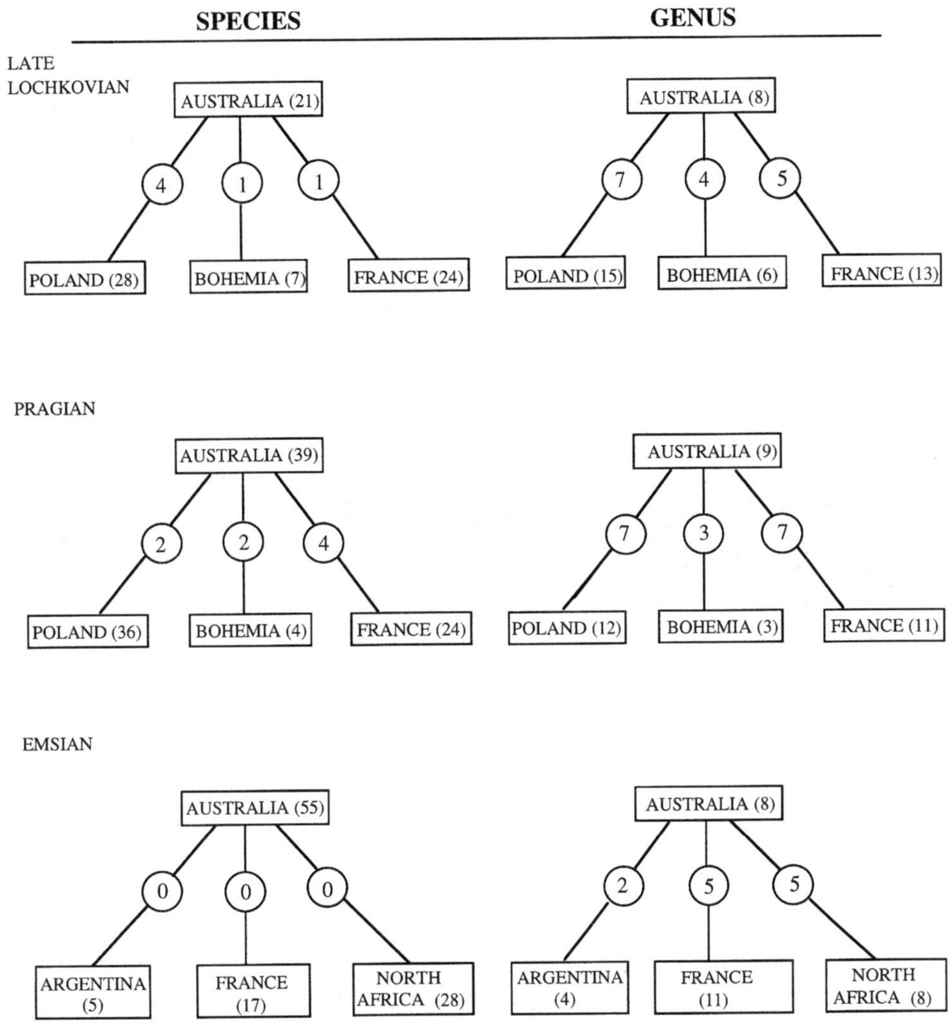

Figure 1. Number of taxa common between Australia and other regions listed for species and genus for each different stage. Numbers in parentheses refer to the number of species or genera present at each location. Numbers inside the circles are the number of common taxa.

SPECIES

LATE LOCHKOVIAN

	Australia
France	0.04
Poland	0.2
Bohemia	0.07

PRAGIAN

	Australia
France	0.13
Poland	0.06
Bohemia	0.09

EMSIAN

	Australia
France	0
North Africa	0
Argentina	0

GENUS

LATE LOCHKOVIAN

	Australia	France	Poland
France	0.48		
Poland	0.64	0.79	
Bohemia	0.57	0.57	0.64

PRAGIAN

	Australia	France	Poland
France	0.7		
Poland	0.66	0.75	
Bohemia	0.5	0.4	0.4

EMSIAN

	Australia	France	North Africa
France	0.55		
North Africa	0.66	0.75	
Argentina	0.3	0.46	0.5

Figure 2. Matrices showing the Dice Similarity Coefficient calculated for regions during the Early Devonian.

sions have proved necessary (Winchester-Seeto, 1993a, 1993b; 1996). Using only studies with SEM micrographs has the benefit of limiting the database to the work of only four different authors or teams, possibly increasing taxonomic consistency. No attempt has been made to differentiate between studies according to sedimentary environment. In most studies, however, chitinozoans were recovered from shallow-marine, low-energy environments.

The Lochkovian data for Australia are represented by the lower segment of the Garra Formation (Winchester-Seeto, 1993a); for France the data are compiled from sections of the Formation de St. Céneré (Paris, 1976, 1981) and Formation de l'Armorique (Paris, 1980, 1981); other faunal lists are for southeastern Poland (Wrona, 1980) and the Barrandian area of the Czech Republic (Chlupác et al., 1985). All of these studies also cover parts of the Pragian; appropriate segments of the rangecharts were examined and data extracted. Additional material for the Pragian in Australia came from the Shield Creek Formation of northern Queensland and the Coopers Creek Limestone of Victoria (Winchester-Seeto, 1993b); the lower segment of Formations de Bois-Roux and de la Foulerie yielded information on Pragian chitinozoans of France (Paris, 1981). Emsian data for Australia were extracted from studies of the Taravale Formation (Winchester-Seeto, 1996); the upper segment of the Formation de la Foulerie and Marettes (Paris, 1981) provided additional information for France; North African data were based on studies by Boumendjel et al. (1988) from cores in the Algerian Sahara; cores from Argentina, documented by Volkheimer et al. (1986), yielded sparse but important data for the Emsian of South America.

All of the sections from Australia, France, and Bohemia used in this study have been independently dated with other microfossils, notably conodonts. Studies by Boumendjel et al. (1988) from Algeria were calibrated against miospore distribution.

The chitinozoan studies from Argentina are less rigorously dated; correlations suggested by Volkheimer et al. (1986) with the Paraná Basin were based on chitinozoans. The age of formations in the Paraná Basin has recently been reassessed by Grahn (1992); his conclusions have been used in this study. For instance, the lower part of the Puesto El Tigre, equivalent to Formation Furnas, has been interpreted as mid-Lochkovian to late Pragian; the upper part, equivalent to the Ponta Grossa Formation, is interpreted as ranging from late Pragian to middle Eifelian.

Some reinterpretation of the position of the Lochkovian-Pragian boundary has been undertaken for Polish core samples by Wrona (1980). Paris (1981) points out that the chitinozoan species *Urochitina simplex* Taugourdeau and Jekhowsky is a good marker species for the late Lochkovian. Thus the presence of this species in the lower Ciepelovian suggests that the boundary between the Bostovian and Ciepelovian does not precisely coincide with the Lochkovian-Pragian boundary (F. Paris, personal communication, 1993). This reinterpretation, where *Urochitina simplex* indicates the strata are still within the Lochkovian, has been used for compiling the species and generic lists used in this study.

Comparison of Early Devonian chitinozoan taxa can be accomplished with a reasonable degree of confidence between Australian assemblages and those from other regions of the world. First-hand knowledge of the Australian faunas gives a solid basis from which to compare SEM micrographs and descriptions from elsewhere. The same assurance cannot be given when comparing the taxonomic work of authors working in other parts of the world, especially at the species level. Lack of intimate knowledge of a fauna from a remote region often leads to a reluctance to synonymize taxa, leading to an apparently lower similarity. This is less of a problem when considering chitinozoan genera because definitions are clearer and differences are more obvious than at the species level. Thus this preliminary survey documents only comparison between Australian chitinozoan assemblages and those of other regions at the species level.

Achab (1991) and Achab et al. (1992) compared Ordovician and Silurian chitinozoan assemblages according to three criteria: number of common index-species used in regional zonations, number of common species in synthetic stratigraphic range charts for different regions compiled by the authors, and number of common species when comparing entire regional faunas. The second of these, concentrating on comparing the occurrence of species in specific stratigraphic settings, was employed in this study. The dispersed nature of the European crustal blocks during the Early Devonian compared with the present day configuration (Paris, 1993; Robardet et al., 1993), limits possibilities for making meaningful lists of "whole regional faunas."

RESULTS

Figure 1 displays the results of examining the number of taxa in common between Australia and other regions at both the specific and generic level for each of the different stages. The number in parentheses denotes the total number of species or genera for each region. The Dice Similarity Coefficients are listed in Figure 2, with a value of 1 indicating perfect correlation and zero indicating no similarity. Figures 3a and 3b show the distribution of the six genera not recovered from Australian strata. For information about the species and genera considered common to two regions, see the appendix.

It is difficult to discern biogeographic trends at the species level. It is clear that there are not very many species in common between Australia and the rest of the world. There is an abrupt increase in dissimilarity between Australia and all other regions in the Emsian, with no evidence of Emsian species in Australia being found in any of the other studies included in this investigation or in any other known study.

Consideration at a higher level of classification shows a

Figure 3. a, Distribution of late Lochkovian and Pragian chitinozoan genera not found in Australian strata (map after Scotese and McKerrow, 1990). Note: the presence of *Linochitina* and *Cingulochitina* in Bohemia is based on unpublished data from F. Paris (personal communication, 1995). b, Distribution of Emsian chitinozoan genera not found in Australian strata (map after Scotese and McKerrow, 1990).

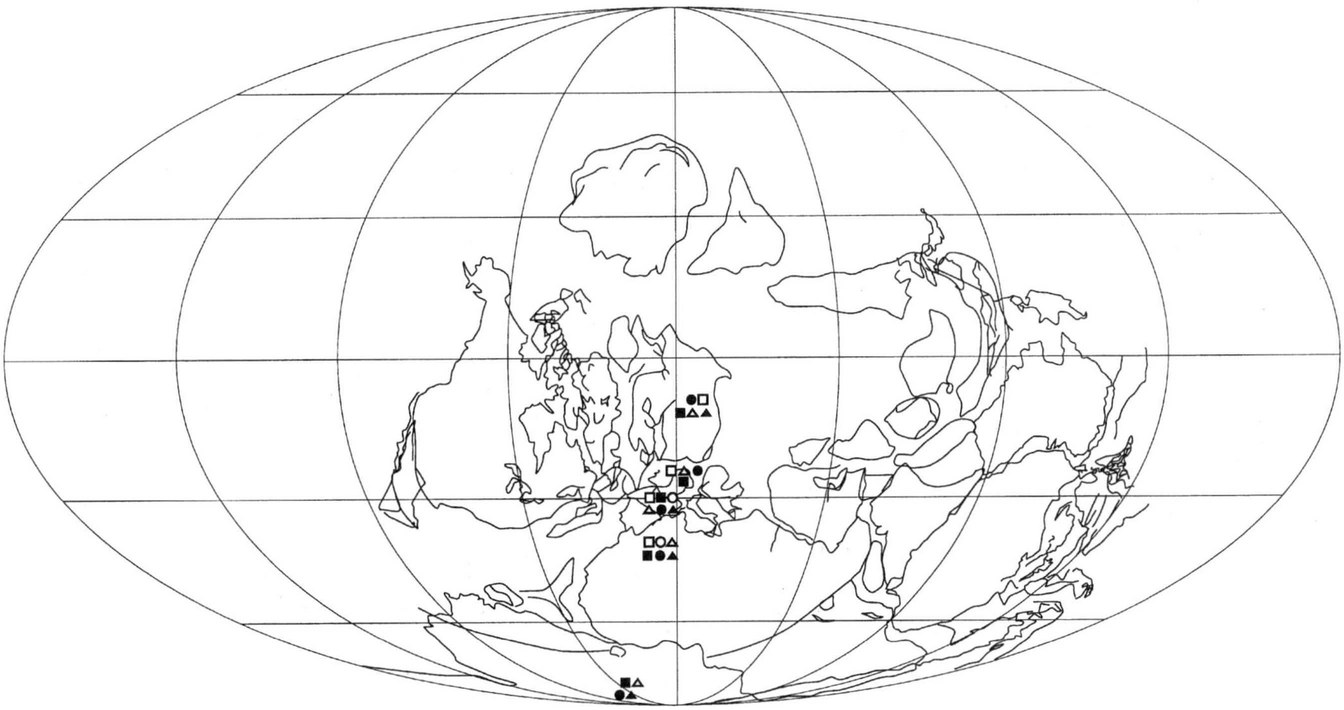

a. Late Lochkovian and Pragian

□ Linochitina ○ Armoricochitina △ Urnochitina
■ Margachitina ● Cingulochitina ▲ Urochitina

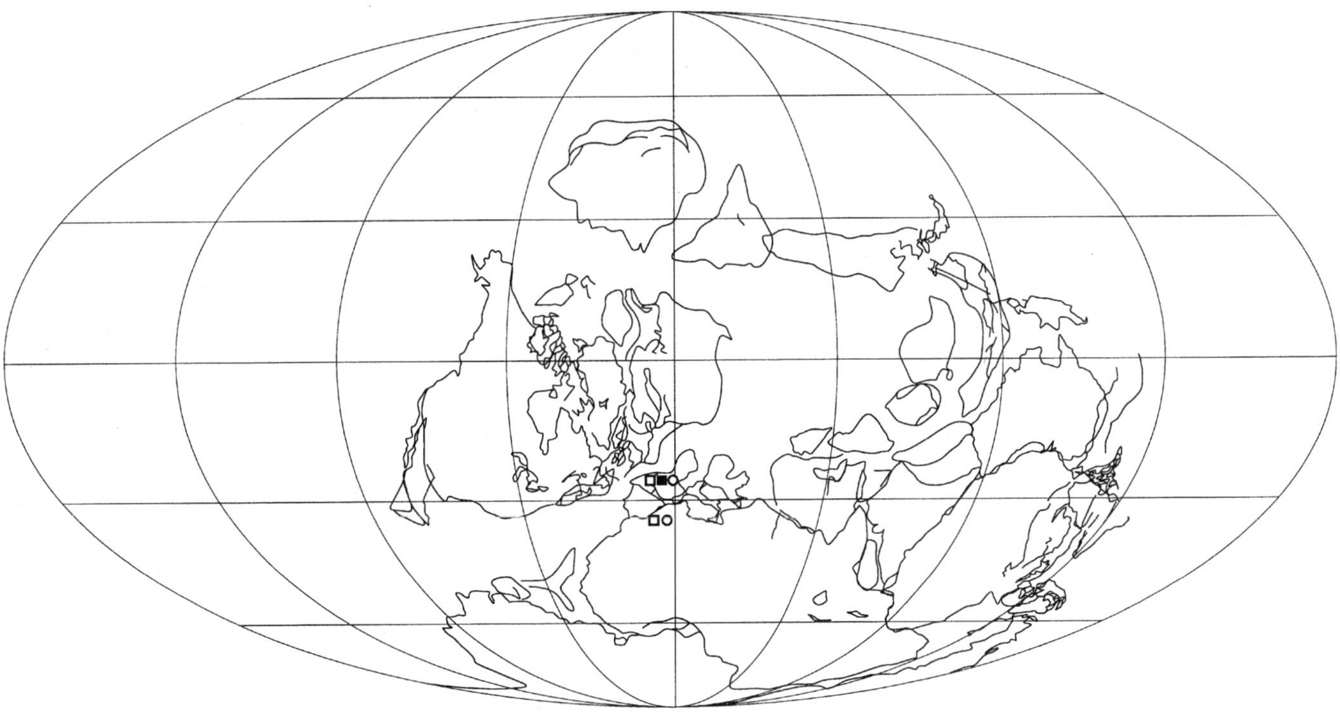

b. Late Early Devonian (Emsian)

□ Linochitina ○ Armoricochitina ■ Margachitina

greater number of genera in common; this is confirmed by the higher similarity coefficients. The results for both Argentina and Bohemia may have been artificially depressed by the relatively smaller number of species found in studies of these sites; the published account for Bohemia, for instance, concerns only a few meters around the Lochkovian-Pragian boundary.

The highest similarity in the Emsian is observed between northern France and North Africa, probably reflecting their close geographic proximity. Northern France and Poland also have a high similarity coefficient for the Lochkovian and Pragian, unexpectedly higher than that between northern France and Bohemia, given the supposedly close proximity between them in the Early Devonian. This anomaly may be the result of a smaller number of species from Bohemia (see above) or the longer time interval from which the French data have been obtained, but may also reflect the absence in Bohemia of two genera: *Armoricochitina* and *Urochitina*, noted by Paris (*in* Chlupác et al., 1985). Australia is shown to have a large number of genera in common with all other parts of the world, although the degree of similarity with Argentina is less than with other regions, possibly owing to the limited number of horizons investigated in the latter. No time-related trend can be determined from these data.

Although the differences are small, Australian faunas tend to show closer similarity to those from Poland (in the Lochkovian and Pragian) and to North Africa (in the Emsian) than to those from France. The reasons for this are unknown but may reflect the larger number of studies drawn upon for this survey from the Pragian of both France and Australia. No latitudinal pattern is apparent in the distribution of Early Devonian chitinozoan genera.

Figures 3a and 3b allow simple inspection of the distribution of six genera whose occurrence has not so far been observed in Australia: *Armoricochitina*, *Cingulochitina*, *Linochitina*, *Margachitina*, *Urnochitina*, and *Urochitina*. For the purposes of this biogeographic survey, paleogeographic reconstructions for the Early Devonian proposed by Scotese and McKerrow (1990) have been uncritically used. This plot shows concentration of these "truant" genera in areas believed to have been in close proximity during the Early Devonian. Australia is believed to have been geographically distant from those particular areas at the time.

The cluster of these six genera—concentrated in southwest and central Europe, North Africa, and, to a lesser extent, South America—may, to some extent, reflect the larger number of chitinozoan studies carried out in these areas. However, the absence of these genera from Australia appears to be real and supports the contention that they were provincial.

DISCUSSION

There is very little similarity at the species level between chitinozoan assemblages from Australia and those from the rest of the world during the Early Devonian. The similarity indices calculated for comparison of the regions are generally much lower than those listed by Achab et al. (1992); this accords with the more cosmopolitan nature of the distribution in the Silurian. The data used by Achab et al. (1992), however, were grouped into very broad time intervals (Early Ordovician, Middle Ordovician, etc.), and this may account for the relatively high similarity indices. Data from Australia were not included in the calculations for the Late Ordovician and Silurian; the apparent increase in the similarity may be an artifact of the method (G. Van Grootel, personal communication, 1995).

The similarity indices are also much lower than those calculated by Paris (1993) for four sets of data between the Ordovician and Early Devonian (early Lochkovian), using a similar coefficient. The regions used by Paris (1993) were very close, both geographically and latitudinally, albeit on two sides of a paleocean; this may explain their close faunal similarity.

At the generic level, there is much higher similarity between all the regions, with many cosmopolitan genera. The pattern of high similarity indices is loosely related to proximity of these regions in the Early Devonian, based on the Scotese and McKerrow (1990) reconstructions. Thus France, Poland, Bohemia, and North Africa were quite close together, with a slightly greater distance to Argentina. Australia, on the other hand, belonged to the same paleocontinent as Argentina (Gondwana), but the two regions were separated by vast oceans and areas of emerged land.

There is no evidence of any change in provinciality at the generic level, but it is clear that the species composition of Australian assemblages is very different in the Emsian from other coeval faunas—more so than those for the Lochkovian and Pragian. These results are based, admittedly, on only one Australian study; there may be additional, ecological reasons for the unique nature of the Taravale Formation faunas (Winchester-Seeto, 1996).

Laufeld (1979) regarded the following genera as being cosmopolitan in the Early Devonian: *Ancyrochitina* Eisenack, *Angochitina* Eisenack, *Eisenackitina* Jansonius, *Bursachitina* Taugourdeau, and *Sphaerochitina* Eisenack; this has been confirmed by studies of Australian faunas. In addition, *Hoegisphaera* Staplin, *Calpichitina* Wilson and Hedlund, *Gotlandochitina* Laufeld, *Fungochitina* Taugourdeau, and possibly *Muscochitina* Paris and *Bulbochitina* Paris can be added to the list of cosmopolitan genera.

Of the six genera absent from Australian assemblages, *Urochitina* was specified by Laufeld (1979) as being restricted in distribution; this is reinforced by the present analysis. Paris (1990) noted that in the Ordovician, *Armoricochitina* was restricted to the north Gondwana area; endemism for this genus appears to have persisted into the Early Devonian.

Though they range across thousands of kilometers from Baltica to South America, *Armoricochitina*, *Cingulochitina*, *Linochitina*, *Margachitina*, *Urnochitina*, and *Urochitina* were provincial in distribution.

Figure 4 shows 19 genera occurring in various parts of the Early Devonian. Of these, six—32% of Early Devonian genera—are provincial.

Examination of the phylogenetic relationship of the provincial genera highlights an interesting pattern. Paris (1981; personal communication, 1993) proposed a phylogenetic tree

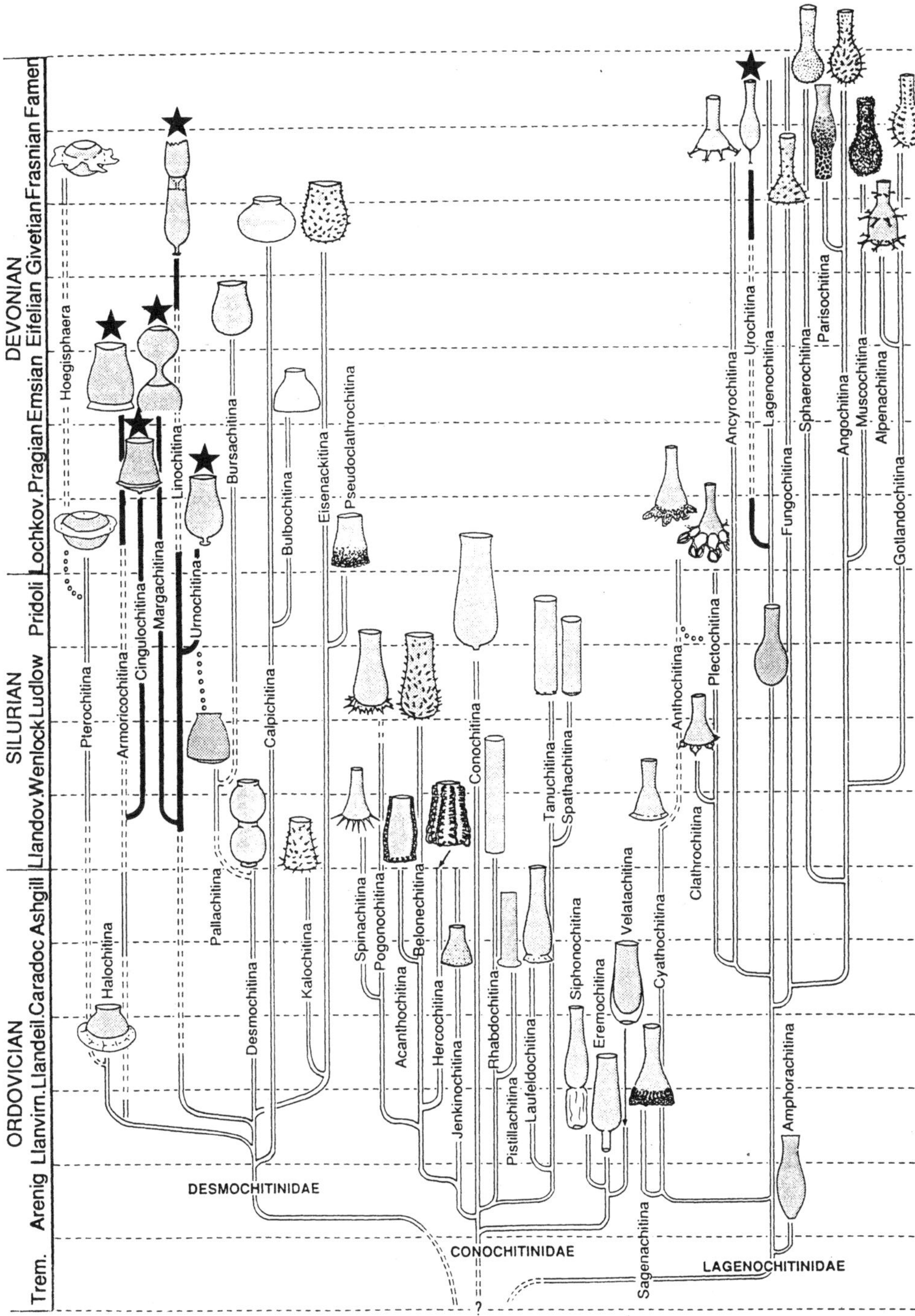

Figure 4. Phylogenetic chart showing relationship among chitinozoan genera. Stars indicate Early Devonian genera not found in Australian strata (chart reproduced and modified with permission from F. Paris, personal communication, 1993).

showing suggested generic interrelationships (Fig. 4). This tree shows *Armoricochitina* and *Cingulochitina* as closely related, with *Cingulochitina* stemming from the *Armoricochitina* branch. Similarly, *Margachitina* and *Urnochitina* are shown as branching off *Linochitina*. The apparent close phylogenetic links between these genera may help explain their restricted pattern of distribution in the Early Devonian. Common evolutionary ancestors may have originated in a restricted area, perhaps in north Gondwana, with subsequent dispersal having been impeded.

The absence of *Urochitina* from Australian strata is more difficult to explain, especially if the phylogenetic position of the genus is as portrayed in Figure 4. In this phylogenetic reconstruction, *Urochitina* appears within Lagenochitinidae, based on the presence of a well-differentiated neck; however, *Urochitina* is unusual because it is the only genus of this family with a peduncle. An intriguing alternative would be to place *Urochitina* within Desmochitinidae and subfamily Margachitininae, associated with three other genera having a similar linking structure (copula or peduncle), that is, *Margachitina*, *Urnochitina*, and *Linochitina*, most probably originating from *Linochitina* (G. Van Grootel, personal communication, 1995). The absence of all four genera from Early Devonian sequences in Australia lends some weight to this suggestion.

Cramer (1971), Cramer and Diez (1972), Achab (1991), and Achab et al. (1992) have suggested that chitinozoans from the Ordovician and Silurian display provincialism related to paleolatitude. The present survey does not support this contention for the Early Devonian. The paleolatitude of Australia and that of northwestern Gondwana, where most provincial genera occur, was very close in the Early Devonian.

If we accept the supposition that the majority of chitinozoans were planktonic in habit or were at some stage of their life cycle, then ocean surface currents must have played an important part in their dispersal. Assuming that the continental reconstructions by Scotese and McKerrow (1990) are correct and that the same principles that govern oceanic circulation today also operated in the past, then it would be reasonable to suggest an anticlockwise oceanic gyre in the Southern Hemisphere. Thus Australia's east coast would have been under the influence of a warm current descending southward from the equator. This is similar to the pattern proposed by Bergström (1990, p. 118) for the Silurian and by Heckel and Witzke (1979) for the Middle Devonian. Eastern Australia would therefore have been effectively isolated from what is now southwest and central Europe (Bohemia and Poland) and North Africa. Australia would also have been separated by a very wide, deep ocean from South America; biogeographic isolation would have been reinforced by the pattern of oceanic circulation. Although this may help explain why many chitinozoan genera are absent from Australia, it does not explain the presence of the more cosmopolitan taxa.

Rickards et al. (1990, p. 141) reviewed the distribution of modern zooplankton and related some aspects to combinations of limiting factors of various oceanic water masses—specifically, temperature, density, turbidity, and water chemistry. It has long been known that water masses are created with unique sets of characteristics and that these are retained despite movement over thousands of kilometers. Some modern zooplankton species are adapted to specific water mass conditions and have a wide geographic distribution predominantly reflecting the path of a particular water mass. Other species can be identified whose distribution is based on response to a particular requirement, for example, depth, latitude, and so on. It may therefore be a mistake to seek a simple, universal answer for endemism of specific species or genera.

CONCLUSIONS

The data used in this survey are quite sparse. Nonetheless there are some salient conclusions:

1. There is a high degree of dissimilarity at the species level between Australia and other regions of the world. This pattern is found throughout the Early Devonian but is more obvious in the Emsian.

2. There is a moderate degree of endemism of genera in the Early Devonian, with evidence of distribution of some genera restricted to the north Gondwana region and what is now central Europe.

3. The genera *Ancyrochitina*, *Angochitina*, *Bursachitina*, *Calpichitina*, *Eisenackitina*, *Fungochitina*, *Gotlandochitina*, *Hoegisphaera*, *Sphaerochitina*, and possibly *Muscochitina* and *Bulbochitina* are found to be relatively cosmopolitan.

4. Of the chitinozoan genera presently known to have existed in the Early Devonian, 32% have not been found in Australian strata.

5. *Armoricochitina*, *Cingulochitina*, *Linochitina*, *Margachitina*, *Urnochitina*, and *Urochitina* are provincial in distribution; all are absent from Australia.

6. Based on worldwide distribution of the genus, *Urochitina* may be more closely related to the subfamily Margachitininae than previously thought.

7. No evidence of provincialism based on paleolatitude is apparent for the Early Devonian. The distribution of genera appears more likely to have been connected to palaeoceanic currents.

ACKNOWLEDGMENTS

The author deeply appreciates the advice and assistance given by Ruth Mawson and John Talent. The encouragement and generous support of Florentin Paris is gratefully acknowledged. Constructive criticism and thoughtful suggestions given by Geert Van Grootel greatly improved the manuscript. Drafting assistance by Judy Davis is much appreciated.

REFERENCES CITED

Achab, A., 1991, Biogeography of Ordovician Chitinozoa, *in* Barnes, C. R., and Williams, S. H., eds., Advances in Ordovician geology: Geological Survey of Canada Paper 90-9, p. 135–142.

Achab, A., Bertrand, R., and Van Grootel, G., 1992, Chitinozoan contribution to the Ordovician and Lower Silurian paleobiogeography: Journal of Geology, v. 100, p. 621–629.

Bergström, S. M., 1990, Relations between conodont provincialism and the changing palaeogeography during the Early Palaeozoic, *in* McKerrow, W. S., and Scotese, C. R., eds., Palaeozoic palaeogeography and biogeography: Geological Society of London Memoir 12, p. 105–121.

Boumendjel, K., Loboziak, S., Paris, F., Steemens, P. and Streel, M., 1988, Biostratigraphie des miospores et des chitinozoaires du Silurien supérieur et du Dévonien dans le Bassin D'Illizi (S.E. Sahara, Algérien): Géobios, v. 21, p. 329–357.

APPENDIX. SPECIES AND GENERA CONSIDERED TO BE "IN COMMON" BETWEEN VARIOUS REGIONS STUDIED.

Region	Species/Genera in Common
Late Lochkovian	
Australia vs. France	*Eisenackitina oviformis* (Eisenack); *Angochitina* Eisenack, *Bursachitina* Taugourdeau, *Eisenackitina* Jansonius, *Gotlandochitina* Laufeld, *Hoegisphaera* Staplin.
Australia vs. Poland	*H.* cf. *glabra* Staplin, *C. velata* (Wrona), *Angochitina* sp. A (= *Sphaerochitina sphaerocephala* Eisenack), *Bursachitina (Amplichitina) cupellata* (Wrona): *Angochitina, Bursachitina, Calpichitina* Wilson and Hedlund, *Eisenackitina, Gotlandochitina, Hoegisphaera, Sphaerochitina* Eisenack.
Australia vs. Bohemia	*B. (A) cupellata* (Wrona); *Bursachitina, Calpichitina, Eisenackitina, Gotlandochitina.*
France vs. Poland	*Ancyrochitina* Eisenack, *Angochitina, Bursachitina, Cingulochitina* Paris, *Eisenackitina, Gotlandochitina, Hoegisphaera, Linochitina* Eisenack, *Margachitina* Eisenack, *Urnochitina* Paris, *Urochitina* Taugourdeau and Jekhowsky.
France vs. Bohemia	*Ancyrochitina, Bursachitina, Eisenackitina, Gotlandochitina, Margachitina, Urnochitina.*
Bohemia vs. Poland	*Ancyrochitina, Bursachitina, Calpichitina, Eisenackitina, Gotlandochitina, Margachitina, Urnochitina.*
Pragian	
Australia vs. France	*Angochitina comosa* Taugourdeau and Jekhowsky, *A. caeciliae* Paris, *A.* cf. *A. callawayensis* Urban and Kline, *Bulbochitina bulbosa* Paris, *Angochitina, Bursachitina, Bulbochitina, Calpichitina, Eisenackitina, Gotlandochitina, Hoegisphaera.*
Australia vs. Poland	*A. cactula* Winchester-Seeto (= *A.* cf. *A. crassispina* Eisenack), *C. velata* (Wrona), *Ancyrochitina, Angochitina, Bursachitina, Calpichitina, Eisenackitina, Gotlandochitina, Sphaerochitina.*
Australia vs. Bohemia	*A. comosa* Taugourdeau and Jekhowsky, *Gotlandochitina phillipotti* (Paris)?; *Angochitina, Bursachitina, Gotlandochitina.*
France vs. Poland	*Ancyrochitina, Angochitina, Bursachitina, Cingulochitina, Eisenackitina, Gotlandochitina, Hoegisphaera, Linochitina, Margachitina.*
France vs. Bohemia	*Angochitina, Bursachitina, Gotlandochitina.*
Poland vs. Bohemia	*Angochitina, Bursachitina, Gotlandochitina.*
Emsian	
Australia vs. France	*Ancyrochitina, Angochitina, Bursachitina, Fungochitina, Gotlandochitina.*
Australia vs. North Africa	*Ancyrochitina, Angochitina, Bursachitina, Fungochitina, Gotlandochitina.*
Australia vs. Argentina	*Ancyrochitina, Angochitina.*
France vs. North Africa	*Ancyrochitina, Angochitina, Armoricochitina, Bursachitina, Fungochitina, Gotlandochitina, Hoegisphaera, Linochitina.*
France vs. Argentina	*Ancyrochitina, Hoegisphaera.*
North Africa vs. Argentina	*Ancyrochitina, Angochitina, Hoegisphaera.*

Campbell, C. A., and Valentine, J. W., 1977, Comparability of modern and ancient marine faunal provinces: Paleobiology, v. 3, p. 49–57.

Cheetham, A. H., and Hazel, J. E., 1969, Binary (presence-absence) similarity coefficients: Journal of Paleontology, v. 43, p. 1131–1136.

Chlupác, I., Lukes, P., Paris, F., and Schönlaub, H. P., 1985, The Lochkovian-Pragian boundary in the Lower Devonian of the Barrandian Area (Czechoslovakia): Jahrbuch der Geologischen Bundesanstalt, v. 128, p. 9–41.

Clark, D. L., and Hatleberg, E. W., 1983, Palaeoenvironmental factors and the distribution of conodonts in the Lower Triassic of Svalbard and Nepal: Fossils and Strata, v. 15, p. 171–175.

Colbath, G. K., 1990, Palaeobiogeography of Middle Palaeozoic organic-walled phytoplankton, *in* McKerrow, W. S., and Scotese, C. R., eds., Palaeozoic palaeogeography and biogeography: Geological Society of London Memoir 12, p. 207–213.

Cramer, F. H., 1971, Distribution of selected Silurian acritarchs: Revista Español di Micropaleontologia, Numero extraordinario, p. 1–203.

Cramer, F. H., and Diez, M. d. C. R., 1972, North American Silurian palynofacies and their spatial arrangement: Acritarchs: Palaeontographica Abteilung B, v. 138, p. 107–180.

Grahn, Y., 1992, Revision of Silurian and Devonian strata of Brazil: Palynology, v. 16, p. 35–61.

Heckel, P. H., and Witzke, B. J., 1979, Devonian world palaeogeography determined from distribution of carbonates and related lithic palaeoclimatic indicators, *in* House, M. R., Scrutton, C. T., and Bassett, M. G., eds., The Devonian System: Special Papers in Palaeontology, v. 23, p. 99–123.

Laufeld, S., 1979, Biogeography of Ordovician, Silurian and Devonian chitinozoans, *in* Gray, J., and Boucot, A. J., eds., Historical biogeography, plate tectonics, and the changing environment: Corvallis, Oregon State University Press, p. 75–90.

Newton, C. R., 1990, Palaeobiogeography, *in* Briggs, D. G., and Crowther, P. R., eds., Palaeobiology: Cambridge, Blackwell Scientific; p. 452–460

Paris, F., 1976, Les Chitinozoaires, *in* Lardeux, H., ed., Les schistes et calcaires éodévoniens de Saint-Céneré (Massif armoricain, France), Sédimentologie, Paléontologie, Stratigraphie: Mémoires de la Société Géologique et Minéralogique de Bretagne, v. 19, p. 93–133.

Paris, F., 1980, Les Schistes et calcaires de l'Armorique: Mémoires de la Société Géologique et Minéralogique de Bretagne, v. 23, p. 111–128.

Paris, F., 1981, Les Chitinozoaires dans le Paléozoïque du sud-ouest de l'Europe: Mémoires de la Société Géologique et Minéralogique de Bretagne, v. 26, p. 1–412.

Paris, F., 1990, The Ordovician chitinozoan biozones of the northern Gondwana Domain: Review of Palaeobotany and Palynology, v. 66, p. 181–209.

Paris, F., 1992, Application of chitinozoans in long-distance Ordovician correlations, *in* Webby, B. D., and Laurie, J. R., eds., Global perspectives on Ordovician geology: Rotterdam, Balkema, p. 23–35.

Paris, F., 1993, Évolution paléogéographique de l'Europe au Paléozoïque inférieur : le test des Chitinozoaires: Comptes Rendus de l'Académie des Sciences, Paris Série 2, v. 316, p. 273–280.

Rickards, R. B., Rigby, S. and Harris, J. H., 1990. Graptoloid biogeography: Recent progress, future hopes, *in* McKerrow, W. S., and Scotese, C. R., eds., Palaeozoic palaeogeography and biogeography: Geological Society of London Memoir 12, p. 139–145.

Robardet, M., and 13 others, 1993, Paléogéographie de l'Europe occidentale de l'Ordovicien au Dévonien: Bulletin de la Société géologique de France, v. 164, p. 683–695.

Scotese, C. R. and McKerrow, S. F., 1990, Revised world maps and introduction, *in* McKerrow, W. S., and Scotese, C. R., eds., Palaeozoic palaeogeography and biogeography: Geological Society of London Memoir 12, p. 1–21.

Shi, G., 1993, A comparative study of 39 binary similarity coefficients: Memoirs of the Association of Australasian Palaeontologists, v. 15, p. 329–341.

Talent, J. A., Gratsianova, R. T., and Yolkin, E. A., 1986, Prototethys: Fact or phantom? Palaeobiogeography in relation to the crustal mosaic for the Asia-Australia hemisphere in Devonian–Early Carboniferous times, *in* McKenzie, K. G., ed., Shallow Tethys: Rotterdam, Balkema, p. 87–111.

Volkheimer, W., Melendi, D. L., and Salas, A., 1986, Devonian chitinozoans from northwestern Argentina: Neues Jahrbuch für Geologie und Paläontologie, Abhandlungen, v. 173, p. 229–251.

Winchester-Seeto, T., 1993a, Expanded biostratigraphic perspectives for Devonian chitinozoans: Memoirs of the Association of Australasian Palaeontologists, v. 15, p. 249–254.

Winchester-Seeto, T., 1993b, Chitinozoa from the Early Devonian (Lochkovian-Pragian) Garra Limestone, central New South Wales, Australia: Journal of Paleontology, v. 67, p. 738–758.

Winchester-Seeto, T., 1993c, Chitinozoan assemblages from the Pragian of eastern Australia: Proceedings of the Royal Society of Victoria, v. 105, p. 85–112.

Winchester-Seeto, T., 1996, Emsian Chitinozoa from the Buchan area, southeastern Australia: Acta Palaeontolgica Polonica, v. 41 p. 149–230.

Wrona, R., 1980, Upper Silurian-Lower Devonian chitinozoa from the subsurface of southeastern Poland: Palaeontologia Polonica, v. 41, p. 103–165.

MANUSCRIPT ACCEPTED BY THE SOCIETY FEBRUARY 27, 1997

Printed in U.S.A.

Geological Society of America
Special Paper 321
1997

Phylogeny and biogeography of Tecnocyrtina *(Brachiopoda-Spriferinida) in the Devonian (Givetian-Frasnian) of North America*

Jed Day
Department of Geography-Geology, Illinois State University, Normal, Illinois 61790-4400

ABSTRACT

Taxonomy and biostratigraphic and geographic distribution of North American species of the medially plicate cyrtinid genus *Tecnocyrtina* demonstrate that this distinctive brachiopod lineage is late Givetian to early Frasnian. The stratigraphic distribution appears to have been connected with sea-level rises that initiated deposition of four subdivisions of Devonian eustatic Transgressive-Regressive (T-R) cycles IIa and IIb of Johnson and others and of Johnson and Klapper.

Two major divisions of the North American *Tecnocyrtina* lineage are designated as the *missouriensis* and *billingsi* groups. The *missouriensis* group consists of six forms, including the oldest-known tecnocyrtinids; these apparently originated in central Canada. The *Cyrtina-Tecnocyrtina* phyletic transition occurred after the start of Devonian eustatic T-R cycle IIa, or T-R cycle IIa-1 of Day and others, probably coinciding with the upper part of the Middle *varcus* Subzone or lower part of the undifferentiated *hermanni* Zone.

Renewed transgression related to the sea level rise of T-R cycle IIa-2 in the late Givetian resulted in migration of *Tecnocyrtina* and allopatric evolution of three geographic species and three subspecies of the *missouriensis* group in six North American shelf settings. The oldest species of the *billingsi* group evolved phyletically in the Great Basin from a subspecies of the *missouriensis* group during the later part of T-R cycle IIa-2 in the late Givetian. The regression terminating T-R cycle IIa in North America and the widespread emergence of carbonate platforms coincided with extinction of nearly all species of the *missouriensis* group, except in the U.S. midcontinent.

The *billingsi* group underwent radiation as a consequence of migration from the Great Basin into western Canadian shelves, and in shelf habitats from Poland to Germany; this coincided with the sea-level rise that initiated T-R cycle IIb during the latest Givetian *norrisi* Zone. The last element of the *missouriensis* group evolved phyletically from *T. johnsoni* n. sp. at this time in the U.S. midcontinent. Extinction of the last known elements of both *Tecnocyrtina* groups in central and western North America occurred in the upper part of T-R cycle IIb in beds with conodonts of Frasnian Montagne Noire Zone 4 of Klapper.

INTRODUCTION

Middle-Upper Devonian strata in North America and western Europe yield at least seven species of a distinctive group of medially plicate cyrtinid brachiopods assigned to *Tecnocyrtina* Johnson and Norris (1972). Because of the widespread occurrence, distinctive morphology, and restricted stratigraphic ranges of tecnocyrtinid brachiopods, they have proven to be useful tools

Day, J., 1997, Phylogeny and biogeography of *Tecnocyrtina* (Brachiopoda-Spriferinida) in the Devonian (Givetian-Frasnian) of North America, *in* Klapper, G., Murphy, M. A., and Talent, J. A., eds., Paleozoic Sequence Stratigraphy, Biostratigraphy, and Biogeography: Studies in Honor of J. Granville ("Jess") Johnson: Boulder, Colorado, Geological Society of America Special Paper 321.

for correlation of late Givetian and early Frasnian strata in central and western North America. All tecnocyrtinids known from North America are restricted to strata deposited during Devonian eustatic Transgressive-Regressive (T-R) cycles IIa and IIb of Johnson et al. (1985), Johnson and Klapper (1992), or subdivisions thereof (Day et al., 1994, 1996).

Reliable range-data for known forms allow accurate analysis of changes in the temporal and spatial distribution of *Tecnocyrtina,* indicating that the timing of major migrations and most phylogenetic events in the main North American *Tecnocyrtina* species-groups coincided with major sea-level rises that initiated late Givetian and early Frasnian T-R cycles. Prior to the extinction of *Tecnocyrtina* in the early Frasnian, the most important event during the history of the lineage was the extinction of most elements of the *missouriensis* group in western and central North America; this coincided with a profound regression that terminated T-R cycle IIa in the late Givetian.

Recent investigations of brachiopod and conodont sequences in the Middle and Upper Devonian of central North America (Day, 1989, 1992, 1996; Day et al., 1991, 1996) and new data outlined in this study establish geographic occurrences and biostratigraphic ranges of three species (one new) of *Tecnocyrtina.* These new data and published information on the geographic and stratigraphic ranges of North American forms provide the basis for analysis of temporal and spatial distribution patterns, and of morphologic trends within two main species groups of the genus (Fig. 1).

The first of two groups that evolved during the Givetian includes seven described forms of the *T. missouriensis* group of Johnson and Trojan (1982, p. 130), designated simply as the *missouriensis* group (Fig. 1). The species of this group are characterized by development of two to five medial plications on well-defined dorsal folds and comparable numbers of plications in the sulcus of their ventral valves. The oldest known, but undescribed, tecnocyrtinid (*T.* sp.) was recently discovered in beds of Member C of the Dawson Bay Formation of the Manitoba Group in southwestern Manitoba by A. W. Norris (personal communication, 1995). This form is tentatively included in the *missouriensis* group (Figs. 1 and 2).

Discovery of this form indicates that the *Cyrtina*-to-*Tecnocyrtina* transition (Johnson and Norris, 1972; Johnson and Trojan, 1982) probably took place in central Canada as early as the Middle *varcus* Subzone. Evidence for the timing of evolution of the first tecnocyrtinids is provided by conodont correlations of the Dawson Bay Formation of Manitoba (Norris et al., 1982; Braun et al., 1989; Day et al., 1996, fig. 2). Deposition of this formation coincided with the lower part of T-R cycle IIa of Johnson et al. (1985) or T-R cycle IIa-1 of Day et al. (1994, 1996). All other species of the *missouriensis* group, including *T. johnsoni*n. sp of this report, occur in younger late Givetian and early Frasnian strata deposited during T-R cycles IIa-2, IIb-1, and IIb-2 (Figs. 1–3).

The second *Tecnocyrtina* lineage is designated as the *billingsi* group; its members are geographically restricted to western North America and western Europe (Figs. 1–3). The *billingsi* group is distinguished by development of more numerous medial and flank plications and reduction or almost total elimination of the dorsal fold and ventral sulcus. The group includes three species representing the "*fissiplicata-billingsi* stage of evolution" (Johnson, in Johnson and Trojan, 1982, p. 123). Two elements of the *billingsi* group occur in western Europe; these are a late Givetian form (*T.* aff. *T. billingsi* of Racki et al., 1985) and a latest Givetian–early Frasnian form, *T. beckmanni*, described by Johnson (1973) from Germany.

The oldest species of the *billingsi* group appears in the Great Basin, where *T. fissiplicata* Johnson and Trojan (1982) apparently evolved from a species of the *missouriensis* group in the upper part of the Upper *disparilis* Subzone (Johnson and Trojan, 1982; Johnson, 1990) during the sea-level highstand phase of T-R cycle IIa-2 (Figs. 1 and 2). Younger species of the *billingsi* group from western Canada (Figs. 1 and 2) and Europe most likely evolved from the Great Basin *T. fissiplicata* stock during migration associated with the sea-level rise and the resultant widespread transgression that initiated T-R cycle IIb in the latest Givetian.

DISTRIBUTION AND BIOSTRATIGRAPHIC RANGE OF *TECNOCYRTINA* IN NORTH AMERICA

Johnson and Trojan (1982) described major aspects of three species sequences of *Tecnocyrtina* developed in Middle-Upper Devonian successions of western Canada, Nevada, and the Iowa Basin of the U.S. midcontinent. The new form from Manitoba, reported by A.W. Norris, demonstrates a fourth species sequence consisting of *T.* sp. and *T. missouriensis missouriensis* (Day, 1992, Day et al., 1996).

Great Basin–Nevada

Two species of *Tecnocyrtina* occur in a well-established late Givetian and early Frasnian sequence in the Antelope Range of central Nevada. Ranges were delimited in a series of studies of associated brachiopods and conodonts (Johnson and Norris, 1972; Johnson, 1977, 1979, 1990; Johnson et al., 1980; Johnson and Trojan, 1982).

The oldest form, *T. missouriensis teleta*, occurs with brachiopods of the Lower *Tecnocyrtina* Community, equivalent to the lower part of Faunal Interval 27 (F.I. 27l) of Johnson (1990), and conodonts of the Upper *disparilis* Subzone. *Tecnocyrtina missouriensis teleta* ranges through F.I. 27 and is joined by its descendant, *T. fissiplicata*, in the Middle *Tecnocyrtina* Community of Johnson (1990) in the upper part of F.I. 27. *T. fissiplicata* ranges into Faunal Interval 28, where it occurs with brachiopods of the Upper *Tecnocyrtina* Community and conodonts of the *norrisi* Zone (Klapper and Johnson, *in* Johnson, 1990, figs. 2 and 49). Johnson (1990, fig. 50) provisionally placed the upper range of *T. fissiplicata* in Faunal Interval 29, coincident with conodonts of Klapper's (1989) Montagne Noire (M. N.) Frasnian Zones 1–4 (Johnson, 1990, fig. 2).

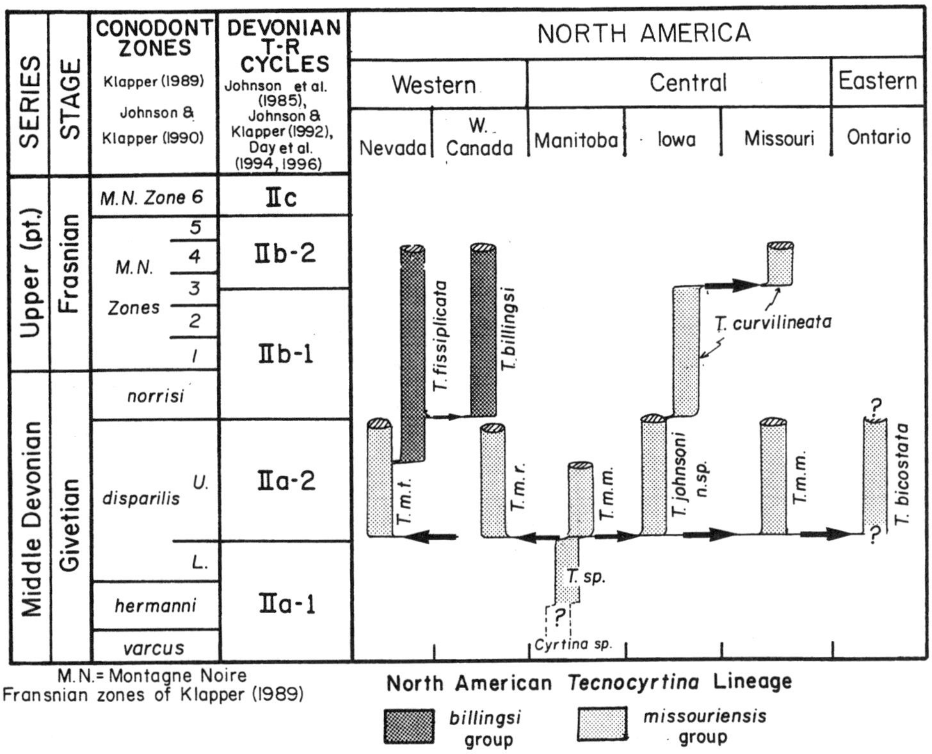

Figure 1. Conodont zones, Devonian eustatic Transgressive-Regressive (T-R) cycles, and spatial and temporal distribution of elements of the *missouriensis* and *billingsi* groups comprising the North American *Tecncocyrtina* lineage. Subspecies of *Tecnocyrtina missouriensis* are *missouriensis* (*T. m. missouriensis*), *raaschi* (*T. m. raaschi*), and *teleta* (*T. m. teleta*). Black horizontal arrows indicate migrations and related phyletic speciation or geographic speciation events. Labeled vertical columns depict known ranges of each taxon. Splits of vertical columns indicate speciation events in known species sequences or species migrations. Note major speciation events in the *missouriensis* group in the interval of the Upper *disparilis* Subzone during Devonian T-R cycle IIa-2 and subsequent speciation event in the western North American *billingsi* group in the *norrisi* Zone during Devonian T-R cycle IIb-1. Subdivisions of Devonian T-R cycles of Johnson et al. (1985) and Johnson and Klapper (1992) are discussed in Day et al. (1994, 1996). Upper Givetian conodont zonation of Johnson and Klapper (*in* Johnson, 1990); Frasnian conodont zonation after Klapper (1989).

Alberta and British Columbia

Two species of *Tecnocyrtina* (*T. missouriensis raaschi* and *T. billingsi*) are found in sequence in late Givetian and early Frasnian rocks in British Columbia (Maurin and Raasch, 1972; Johnson and Trojan, 1982). *T. billingsi* is the most widespread, occurring throughout the Northwest Territories, British Columbia, and Alberta (Norris, *in* Johnson and Norris, 1972; Fig. 3 herein).

The oldest *Tecnocyrtina* species from the region, *T. missouriensis raaschi* Johnson and Trojan (1982), was first identified as *T.* cf. *missouriensis* on specimens from the lower part of the Flume Formation in northeastern British Columbia at the Kakwa Lake North and Wallbridge Mountain sections of Maurin and Raasch (1972, p. 63–64, addendum p. 80). It occurs in Assemblage II of Maurin and Raasch (1972) with brachiopods including *Schizophoria meeki*, *Athyris vittata*, and *Independatrypa* cf. *randalia*. This fauna is similar to that found in the lower Cou Falls Member of the Coralville Formation of Iowa. As currently understood (Johnson and Trojan, 1982; Norris and Uyeno, 1983), the lower Flume strata yielding *T. m. raaschi* and brachiopods of Assemblage II of Maurin and Raasch (1972) are correlated with F.I. 271 of the Great Basin Devonian and the *Tecnocyrtina johnsoni* Zone (defined herein) of the Iowa Devonian brachiopod zonation. The latter corresponds to Iowa Faunal Interval 8 of Day (1996).

At the Wallbridge Mountain section of Maurin and Raasch (1972), *T. m. raaschi* is succeeded by *T. billingsi* in the uppermost beds of their Assemblage III and the lower half of their Assemblage IV in the upper half of the Flume (Norris, in Norris and Uyeno, 1983). Other important elements of the brachiopod fauna in the uppermost part of Assemblage III and Assemblage IV include *Allanella minutilla*, *Eleutherokomma killeri*, *E. jasperensis*, and *Strophodonta* cf. *iowensis*. The lowest occurrences of *T. billingsi* are in association with conodonts (Johnson and Trojan, 1982) now assigned to the *norrisi* Zone (Fig. 1).

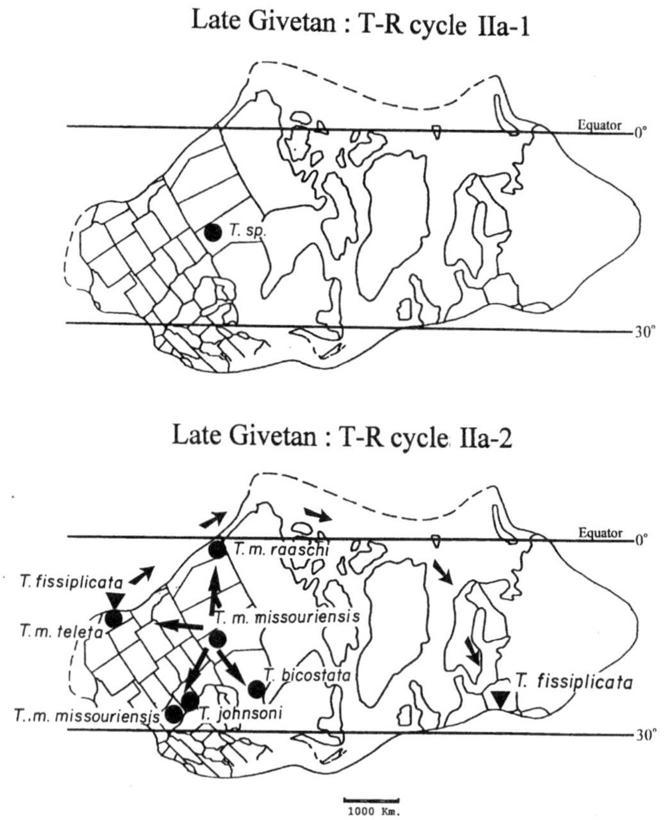

Figure 2. Middle Devonian paleogeographic base maps of Euramerica showing distribution of *Tecnocyrtina* during Devonian T-R cycles IIa-1 (upper map) and IIa-2 (lower map). The upper map shows plot of *Tecnocyrtina* distribution during the interval of the Middle *varcus* Subzone–Lower *disparilis* Subzone. The lower map is a plot of *Tecnocyrtina* distribution in deposits of Devonian T-R cycle IIa-2 of Day et al. (1994, 1996) associated with Upper *disparilis* Subzone/Upper *subterminus* Fauna conodonts. Occurrences of species and subspecies of the *missouriensis* group shown as solid circles, and the oldest element of the *billingsi* group in the Great Basin and Poland is shown as a solid inverted triangle. Inferred migration routes indicated by solid arrows. Middle Devonian paleogeographic base map modified from fig. 1 of Day (1992). Devonian paleolatitudes of Euramerica approximated from North American Devonian geomagnetic paleopole data of Day et al. (1991) and Parse and Plumley (1992).

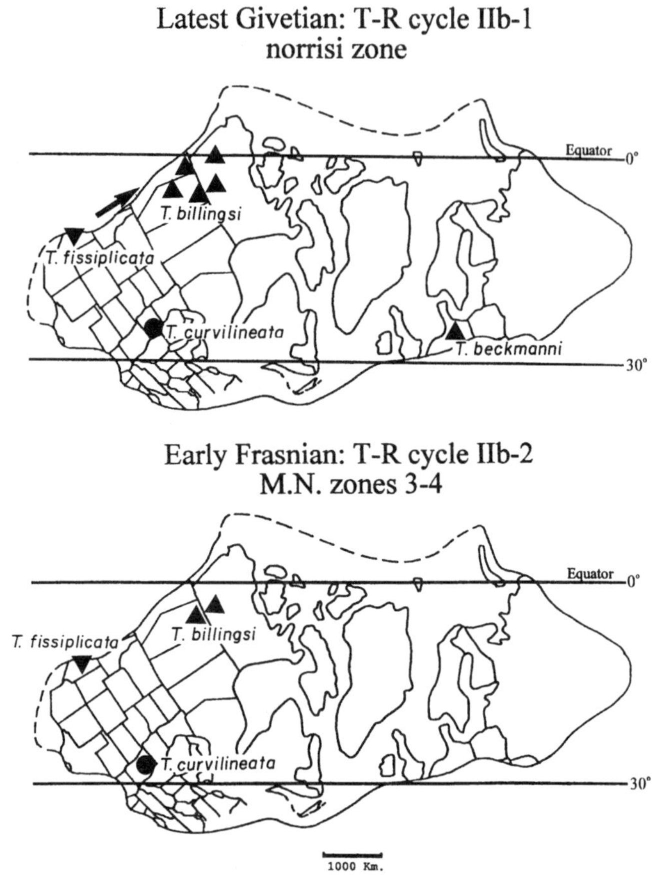

Figure 3. Devonian paleogeographic base maps of Euramerica showing distribution of species of the North American *billingsi* (triangles) and *missouriensis* (circles) groups of the *Tecnocyrtina* lineage. The upper map plots known *Tecnocyrtina* distribution after the initial trangression of Devonian T-R cycle IIb-1 of Day et al. (1994, 1996) and illustrates western North American and European distribution of species of the *billingsi* group. The lower map shows the distribution of species of the *billingsi* and *missouriensis* groups in early Frasnian rocks correlated with the lower part of T-R cycle IIb-2 of Day et al. (1994, 1996) just prior to extinction of *Tecnocyrtina* in the early Frasnian. Source of base map as in Figure 2. Devonian conodont zones as in Figure 1.

In the Waterways Formation of northeastern Alberta, the first occurrence of *T. billingsi* is in the basal Firebag Member, and it ranges into the lower part of the Moberly Member (Norris, in Norris and Uyeno, 1983). In the Gypsum Cliffs area of northeastern Alberta, it first occurs in association with conodonts of the *norrisi* Zone in the Peace Point Member of the Waterways (Norris and Uyeno, 1983). The last occurrence of *T. billingsi* is in the lower part of the Moberly Member associated with conodonts correlated with M. N. Zone 4. The latter correlation is based on Moberly faunas recovered during recent study of the subsurface Waterways conodont sequence by T. T. Uyeno (personal communication, 1995).

Southwestern Manitoba

Two forms of the *missouriensis* group occur in sequence in the late Givetian Manitoba Group in central Canada. These are *Tecnocyrtina* sp. from Member C of the Dawson Bay Formation (A. W. Norris, personal communication, 1995), succeeded by *T. missouriensis missouriensis* in the lower part of the Souris River Formation. Norris, Uyeno, and Day (in Day et al., 1996, fig. 2) correlated Member C of the Dawson Bay with the middle to upper part of the *varcus* Zone and referred the Dawson Bay Formation to T-R cycle IIa-1 of Day et al. (1994, 1996).

The presence of *Tecnocyrtina* in the overlying Souris River Formation was first reported in discussions of correlations of

Manitoba and Iowa Devonian strata by Day (1992, p. 75), in which it was identified as *T. missouriensis* (Swallow, 1860). Day and Norris (in Day et al., 1996, fig. 5) did not assign this form to one of the three subspecies of *T. missouriensis*. All *Tecnocyrtina* material from the Argillaceous limestone beds of the Point Wilkins Member of the Souris River Formation is assigned here to *T. m. missouriensis*.

At both Point Wilkins and Mafeking Quarry (Day et al., 1996, figs. 3–5), *T. m. missouriensis* occurs in association with conodonts of the Upper *subterminus* Fauna. Day et al. (1994, 1996, fig. 2) correlated the argillaceous limestone beds of the Souris River Formation with the Coralville Formation of the Iowa Basin and the Slave Point Formation in northeastern Alberta and southern Northwest Territories, all of which were deposited during T-R cycle IIa-2.

Iowa Basin–U.S. midcontinent

Three species of *Tecnocyrtina* are now known from the late Givetian and early Frasnian age in central North America (Fig. 1): *T. johnsoni* n. sp., *T. missouriensis missouriensis*, and *T. curvilineata* (White, 1862). New brachiopod and conodont data from Missouri and previously undescribed Devonian rocks in southeastern Iowa constrain the ranges of these species.

The Iowa *T. missouriensis* Zone of Day (1992, 1996) is renamed the *T. johnsoni* Zone (Fig. 4). Refinement of the Iowa Devonian brachiopod zonation is made by definition of the *Orthospirifer missouriensis* Zone intervening between the *Strophodonta callawayensis* Zone of Day (1992) and the *Tenticospirifer shellrockensis* Zone of the Shell Rock Formation (Fig. 4). Previously undescribed Devonian rocks in southeastern Iowa (Fig. 5) are designated below as the Buffalo Heights Member of the Lithograph City Formation.

Iowa Basin: Central and eastern Iowa. Two species of the *missouriensis* group occur in carbonate platform rocks of the Cedar Valley Group in eastern and central Iowa (Fig. 4): *T. johnsoni* n. sp. from the Coralville Formation and *T. curvilineata* from overlying beds of the Lithograph City Formation. The former species is restricted to the *T. missouriensis* Zone of older reports by Day (1992, 1994, 1996), (= *T. johnsoni* Zone herein). The latter species first appears in latest Givetian beds included in the lowermost part of the *Allanella allani* Zone and ranges through most of the early Frasnian *Strophodonta callawayensis* Zone.

The form described herein as *T. johnsoni* n. sp. from the Coralville Formation was identified as *T. missouriensis* in earlier studies (Day, 1989, 1992), and later designated as a new subspecies (Day, 1994, 1996; Day et al. 1996). Thus far, *T. johnsoni* n. sp. occurs at one locality in the lower part of the Gizzard Creek Member of the Coralville Formation at the Glory Quarry, southeastern Blackhawk County, central Iowa (Day, 1992, p. 66, fig. 7, units 12–14). There it occurs with the conodonts *Icriodus subterminus* and *Mehlina gradata* (Day, 1992, fig. 7, conodont sample 7C; Rogers, 1990).

The oldest *T. curvilineata* occur in the lowermost beds of the Andalusia Member and correlative beds of the State Quarry Member in eastern and southeastern Iowa, respectively. The State Quarry Member yields conodont faunas of the *norrisi* Zone (Metzger, 1993). Brachiopod faunas in both units are included in the *Allanella allani* Zone (Fig. 4, Table 1). The upper part of the range of *T. curvilineata* (Fig. 1) is based on its occurrence in the lower part of the Snyder Creek Shale of central Missouri (Figs. 4 and 6), where it occurs in brachiopod faunas indicative of the *Strophodonta callawayensis* Zone of Day (1992), now restricted by definition of the *Orthospirifer missouriensis* Zone.

Tecnocyrtina curvilineata was not illustrated nor was the type locality documented by White (1862). Labels with the syntypes in the collections at the University of Michigan indicate that they were collected at "New Buffalo, Iowa." The syntypes and other specimens later illustrated in Hall (1867), Hall and Clarke (1894), and Stainbrook (1943a, Pl. 70, figs. 9–11 only) were all collected from shales now included in the lowermost part of the Andalusia Member of the Lithograph City Formation at the Lafarge Corporation Quarry (= Buffalo Quarry and "New Buffalo" of older reports) near the town of Buffalo in Scott County in southeastern Iowa (Fig. 5, unit 8; Table 1, sample interval 10). The lowermost beds of the Andalusia Member coincide with the "*Strophodonta iowensis* zonule of the *waterlooensis* Zone of Stainbrook (1938, 1943a, 1943b).

In east-central Iowa, *T. curvilineata* occurs in the State Quarry Member of the Lithograph City Formation in outcrops on the east side of Coralville Lake in Johnson County. There it occurs at the Indian Cave locality of Day (1992, figs. 15, 16, 18) and locality 7 of Witzke and Bunker (1994, figs. 15–17). Additional material from the subsurface State Quarry Member was also recovered by Day (1989) from eastern Johnson County in the Mid-America Pipeline Core Number 4 (Day, 1989; 1992, fig. 15).

New stratigraphic unit in the Cedar Valley Group. Newly discovered Devonian rocks in Scott County, Iowa are referred to here as the Buffalo Heights Member, a new member of the Lithograph City Formation. It is named for the community of Buffalo Heights in Scott County, Iowa, one mile (1.6 km) north and west of the type section in the Lafarge Corporation Quarry (Fig. 5). The Buffalo Heights Member consists of calcareous mudstones and shales, ranging from 0 to 4.3 m, (0-14.1 ft.) overlying the pyrite-encrusted disconformity developed on the uppermost beds of the Andalusia Member. The Buffalo Heights brachiopod and conodont faunas are similar to those seen in the uppermost New Bloomfield and Craghead Branch Members of the Snyder Creek Shale in central Missouri (Figs. 4 and 6).

The brachiopod fauna of the Buffalo Heights Member (Table 1) includes *Strophodonta* (S.) *callawayensis* (Swallow, 1860), *Schizophoria athabaskensis* (Warren, 1944), *Eostrophalosia callawayensis* (Swallow, 1860), *Allanella annae* (Swallow, 1860),

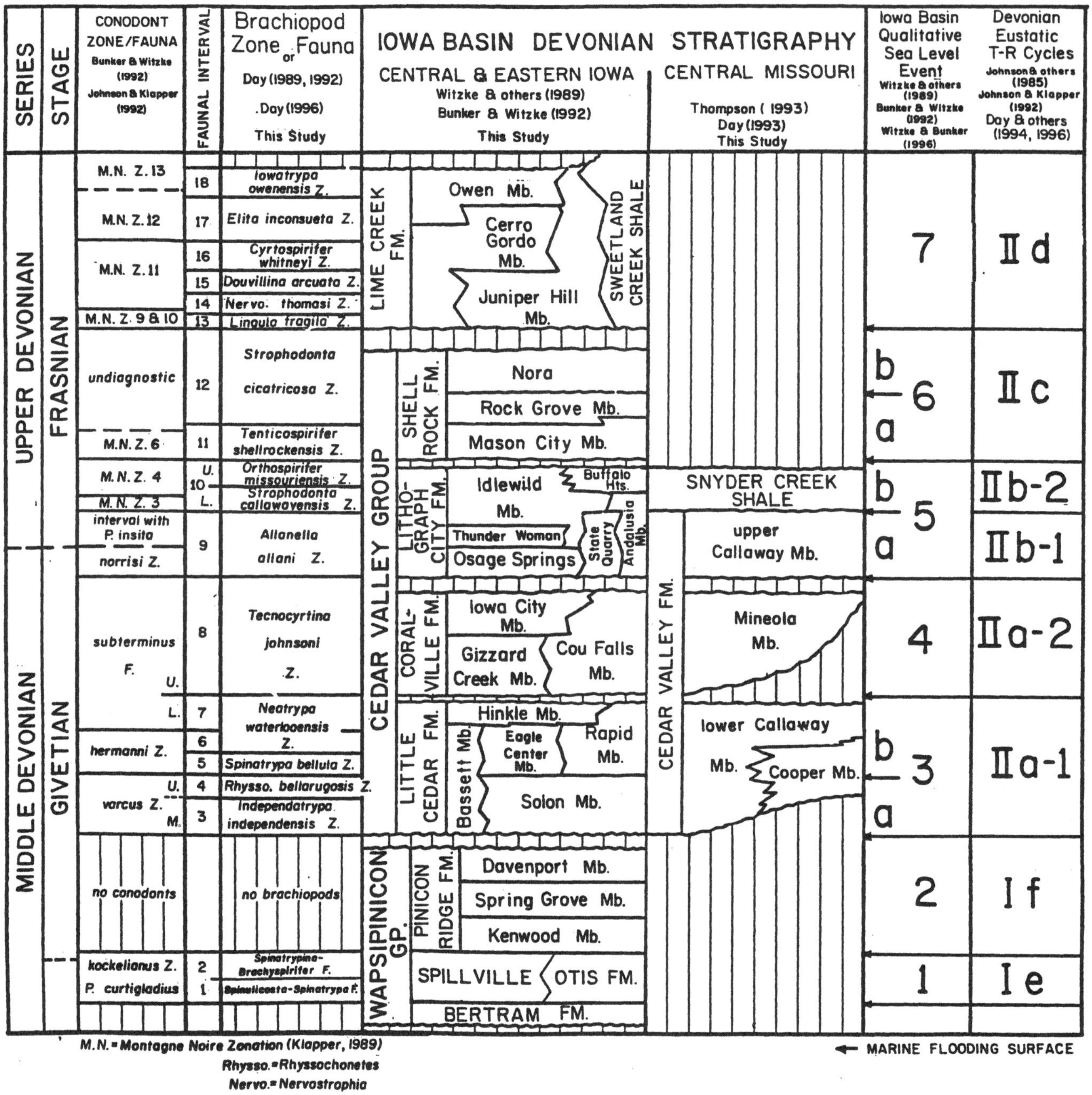

Figure 4. Stratigraphic and biostratigraphic framework for the Middle-Early Upper Devonian (Eifelian-Frasnian) strata of the Iowa Basin, showing relationships between the Iowa Devonian sea–level curve (Witzke et al., 1989; Bunker and Witzke, 1992; Witzke and Bunker, 1996) and Devonian eustatic sea-level curve (Johnson et al., 1985; Day et al., 1994, 1996). Iowa Devonian brachiopod zonation after Day (1989, 1992, 1996) and this study. Devonian composite brachiopod-conodont faunal intervals from Day (1996).

and *Orthospirifer missouriensis* Pitrat, (1975). The lower part of the Buffalo Heights Member (Fig. 5, Table 1, units 17–19) yields all of these, whereas the upper part (Fig. 5, Table 1, units 20 and 21) yields *S. callawayensis* and rare *A. annae*. *Eostrophalosia callawayensis* was previously known only from the New Bloomfield Member of the Snyder Creek Shale in central Missouri (Branson, 1922, 1944). In Iowa, *Orthospirifer missouriensis* in the Buffalo Heights Member is in direct sequence above *O. capax* in the underlying Andalusia Member (Figs. 4 and 5, Table 1); this provides the basis for definition of the *O. missouriensis* Zone

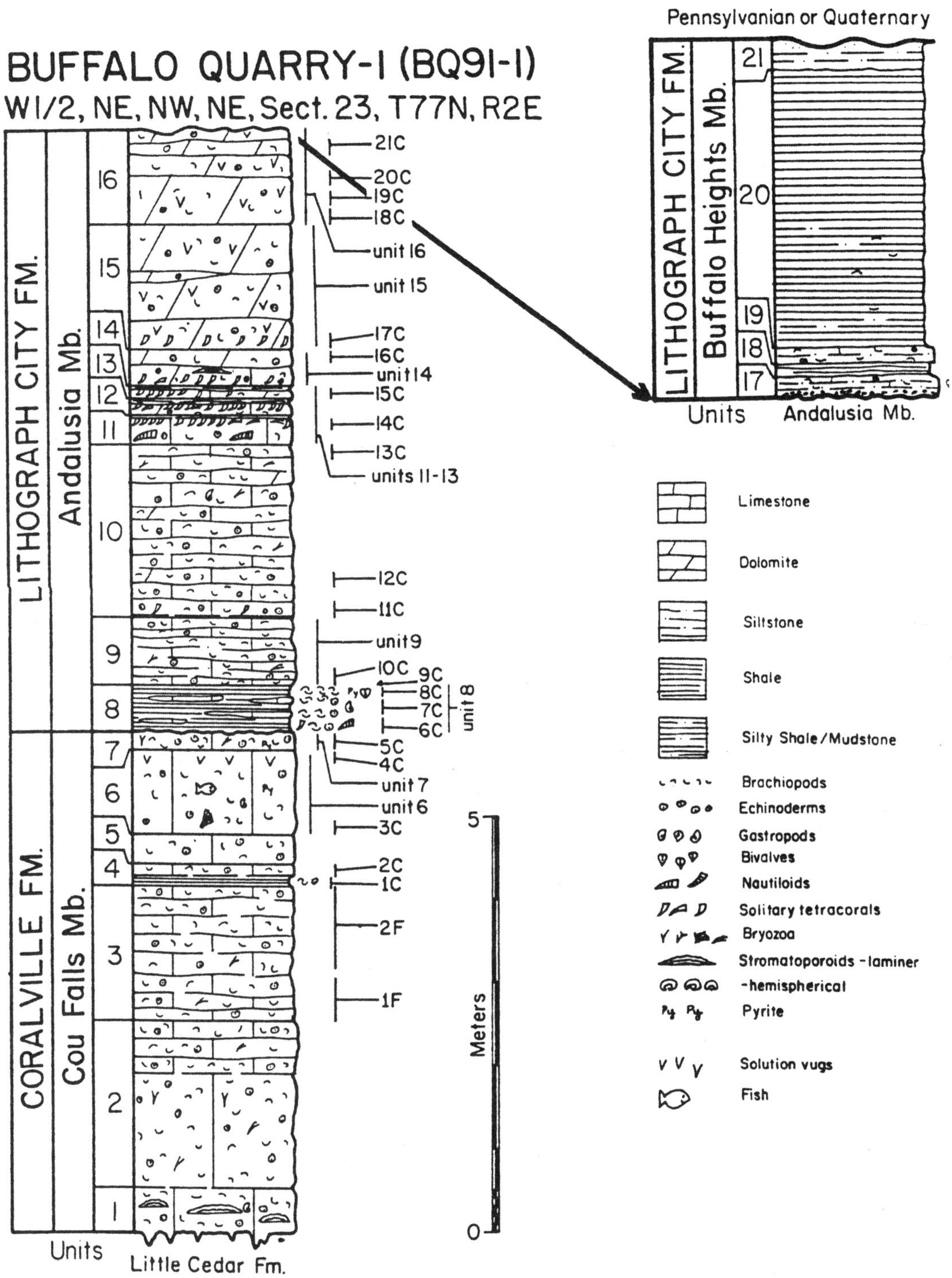

Figure 5. Stratigraphy of the Coralville and Lithograph City Formations of the upper part of the Cedar Valley Group at the Lafarge Quarry (= Buffalo Quarry) in Scott County in southeastern Iowa. Buffalo Heights Member (new) of the Lithograph City is defined in this study. Modified from figure 10 of Day (1992).

(new) in the Iowa Devonian brachiopod sequence and subdivision of Faunal Interval 10 of Day (1996) into lower (*S. callawayensis* Zone) and upper (*O. missouriensis* Zone) intervals (Fig. 4).

Conodonts from the basal 20 cm of the Buffalo Heights Member (Fig. 5, unit 17) include *Ancyrodella africana*, *A.* sp., *Icriodus* cf. *I. symmetricus*, *Mesotaxis* cf. *M. asymmetrica*, *Pandorinellina insita*, and *Polygnathus* sp. A sample from 50 to 80 cm above the base of the member (Fig. 5, lower unit 20) yields *A. africana*, *I.* cf. *I. symmetricus*, and *Polygnathus* sp. These collections are correlated with M. N. Zone 4 of Klapper (1989; Fig. 4 herein).

UNIT	1	2	3	4	5	6	7	8	9	10	11	12	13	14	15	16	17	18	19	20	21	22	23	24	25	26	27	28	29	30	31
Formation	CORALVILLE FM.								LITHOGRAPH CITY FM.																						
Member	Cou Falls Mb.								Andalusia Mb.																			Buffalo Heights Mb.			
ELEVATION	0.00	0.54	2.54	3.17	4.27	4.79	5.62	5.82	6.02	6.58	6.78	7.41	7.81	8.12	9.30	9.50	9.85	10.05	10.20	10.55	10.80	10.98	11.29	11.42	11.62	12.15	13.25	13.30	13.70	14.50	15.25
SAMPLE NUMBER	unit 1	unit 2	1F	2F	2C	3C	4C	5C unit 7	unit 8	10C	unit 9	11C	12C	point sample	13C	unit 11	unit 12	unit 13/15C	unit 14a	unit 14b	unit 14c	unit 15a	unit 15b	unit 15c	unit 15d	18C	21C	unit 17	unit 19	unit 20	
SAMPLE INTERVAL	1	2	3	4	5	6	7	8	9	10	11	12	13	14	15	16	17	18	19	20	21	22	23	24	25	26	27	28	29	30	31
Athyris vittata	●		●	●			●												●			●	●								
Independatrypa randalia	●					●	●	●																							
Pseudoatrypa minor	●	●	●	●	●	●																									
Schizophoria laudoni	●																														
Orthospirifer euruteines		●	●	●																											
Strophodonta (S.) randalia		●		●		●	●																								
S. (S.) parva		●	●				●	●																							
Cyrtina umbonata		●	●																												
Tylothyris subvaricosa		●					●	●																							
Allanella annae									●		●		●			●			●	●	●	●							●	●	
Strophodonta (S.) iowensis									●		●				●																
Eosyringothyris occidentalis									●																						
Floweria n. sp.									●																						
Independatrypa scutiformis									●	●	●	●	●		●							●									
Schizophoria lata									●		●		●																		
Orthospirifer capax										●	●			●	●						●										
Tecnocyrtina curvilineata										●																					
Cyrtina sp. A											●					●	●		●		●										
Strophodonta (S.) callawayensis																●	●	●	●	●	●	●	●	●	●	●	●	●	●	●	●
Schizophoria athabaskensis																●			●									●	●		
Cranaena depressa																	●		●				●								
Tabulophyllum callawayensis									●							●	●	●	●	●		●									
Hadrorhynchia solon																							●								
Paracyclus rowleyi																				●											
Eostrophalosia callawayensis																												●	●		
Orthospirifer missouriensis																												●	●		
BRACHIOPOD ZONE	*Tecnocyrtina johnsoni* Zone								*Allanella allani* Zone							*Strophodonta callawayensis* Zone												*Orthospirifer missouriensis* Z.			

Table 1. Megafossil (mainly brachiopods) sequence developed in the Coralville and Lithograph City formations at the Lafarge Quarry in Scott County, Iowa (sample and unit elevations shown in Fig. 5). *Tecnocyrtina curvilineata* occurs in sample interval 10 in the lower part of the Andalusia Member of the Lithograph City Formation. Modified from fig. 11 of Day (1992).

Revisions to Iowa Basin Devonian brachiopod zonation. Changes to the Iowa Devonian brachiopod zonation (Day, 1989, 1992) include renaming of the *Tecnocyrtina missouriensis* Zone of Day (1992) and definition of the *Orthospirifer missouriensis* Zone. The *T. missouriensis* Zone of Day (1992) is here renamed the *T. johnsoni* Zone. This is because all specimens from the Gizzard Creek Member of the Coralville Formation identified as *T. missouriensis* in Day (1989, 1992) constitute the types for *T. johnsoni* n. sp. Consequently, the base of the *T. johnsoni* Zone coincides exactly with the base of the former *T. missouriensis* Zone as defined at the Glory Quarry in Blackhawk County in central Iowa (Day, 1992, p. 87). The base of this zone also coincides with the oldest conodonts of the Upper *subterminus* Fauna in the Iowa Basin. In central Missouri, the part of Mineola Member of the Cedar Valley Formation that yields *T. m. missouriensis*, associated with conodonts of the Upper *subterminus* Fauna, is correlated with the *T. johnsoni* Zone (Fig. 4).

The *Orthospirifer missouriensis* Zone is defined here by the entry of the nominal species in the lowermost 40 cm of the Buffalo Heights Member of the Lithograph City Formation (= base of unit 17, Fig. 5; Table 1) at the Lafarge Corporation Quarry in Scott County, Iowa. The base of the zone is within strata of M. N. Zone 4, according to associated conodonts in the basal beds of the member.

The *O. missouriensis* Zone is correlated with the upper part of the Snyder Creek Shale (uppermost New Bloomfield, Craghead Branch, and Cow Creek Members) based on the first occurrence of *O. missouriensis* in association with conodonts of Frasnian M. N. Zone 4 in the uppermost beds of the New Bloomfield Member of the Snyder Creek Shale in unit 25 of the section at the Craghead Creek Quarry in eastern Callaway County (Fig. 6, Table 3).

Southern Iowa Basin: central Missouri. Two species of *Tecnocyrtina* occur in the Devonian of central Missouri: *T. missouriensis missouriensis* (Swallow, 1860) from the Mineola Member of the Cedar Valley Formation (Thompson, 1993) and *T. curvilineata* (White, 1862) from the lower Snyder Creek Shale (Figs. 4, 6). The brachiopod sequence in the Cedar Valley Formation and Snyder Creek Shale at three major reference sections (Fig. 6) is shown in Tables 2, 3, and 4. Current correlations

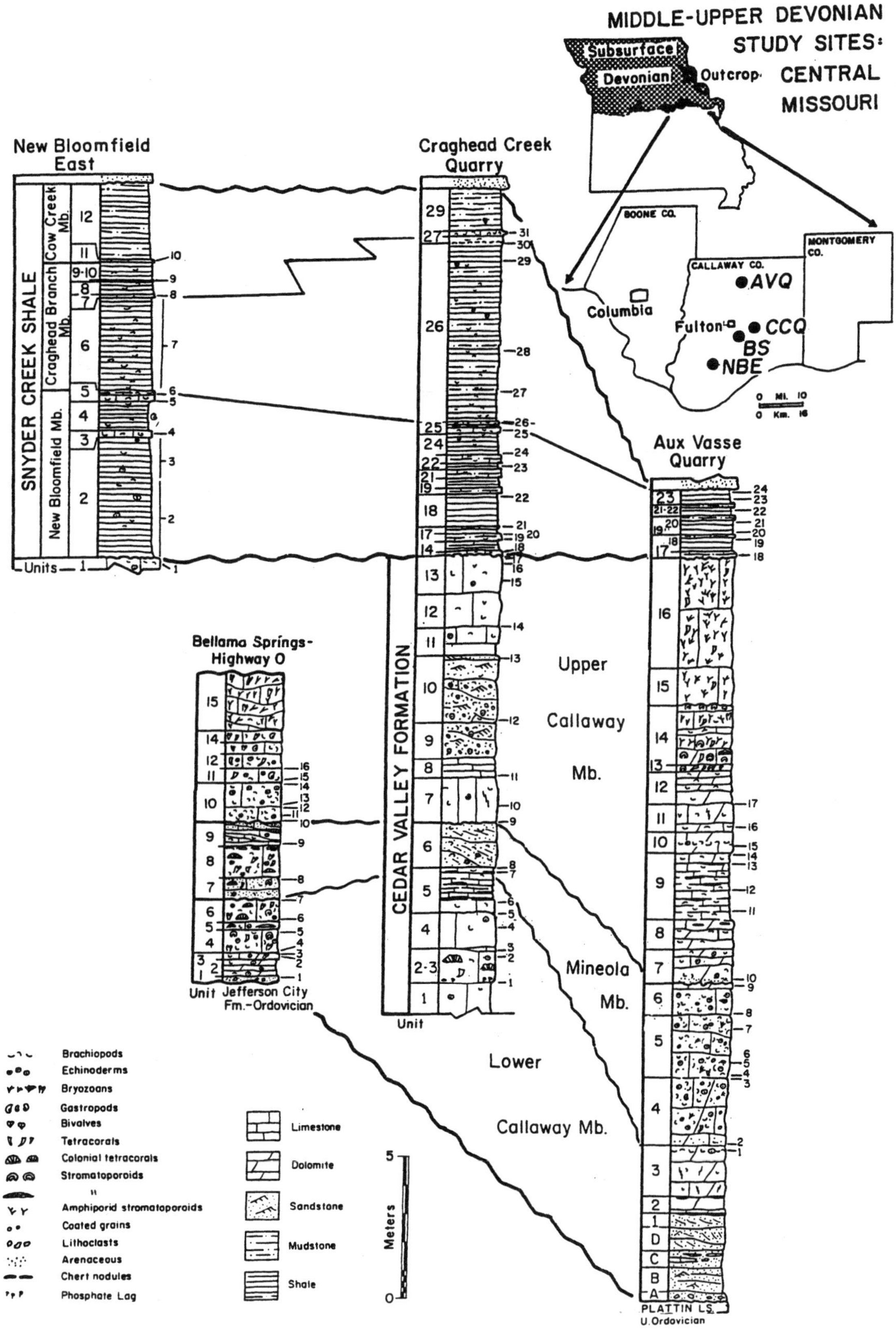

Figure 6. Stratigraphy of the Cedar Valley Formation and Snyder Creek Shale at reference sections of Cedar Valley Group equivalents (Fig. 4) in central Missouri (locations in upper right). Conodont/brachiopod sample numbers and positions are shown to the right of each stratigraphic column. Missouri Devonian stratigraphic nomenclature modified from Thompson (1993).

Aux Vasse Quarry	*Tecnocyrtina johnsoni* Zone								*Allanella allani* Zone							*S. c.* Zone	
	Sample																
BRACHIOPOD TAXA	2	Unit 4	3	5	6	7	8	9	10	11	12	13	15	16	17	18	19
Indenpendatrypa sp.	●																
Pentamerella nitida	●	●	●														
Tecnocyrtina missouriensis			●		●	●		●									
Cyrtina triquetra			●		●												
Schizophoria laudoni		●	●	●	●		●										
Eosyringothyris sp.			●														
Tylothyris subvaricosa		●	●	●	●			●									
Cranaena iowensis		●	●		●			●									
Petrocrania familica			●		●			●									
Athyris fultonensis			●	●	●	●											
Psuedoatrypa			●	●				●									
Strophodonta mineolaensis						●		●									
Allanella annae									●							●	
Independatrypa cf. *scutiformis*										●	●	●	●	●	●		
Strophodonta iowensis										●		●	●	●	●		
Strophodonta callawayensis																●	
S. boonensis sp.																●	
Productella sp.																●	
Eostrophalosia callawayensis																●	
Spinatrypa n.sp.																	●

Table 2. Brachiopod sequence in the Cedar Valley Formation and Snyder Creek Shale at the Aux Vasse Quarry in Callaway County, Missouri (Fig. 6, locality AVQ). All sample positions shown in Figure 6. Brachiopod zonal intervals in upper row, as in Figure 4.

of Devonian rocks of central Missouri and Iowa are shown in Figure 4.

Specimens designated as the lectotypes of *T. m. missouriensis* by Johnson and Trojan (1982, p. 130, Pl. 8, figs. 27, 28, 31, 32, 36) were collected from Greger's (1909) type section of the Callaway Limestone near Bellama Springs, in eastern Callaway County, Missouri. The stratigraphy at the Callaway type section is shown in Figure 6 (Bellama Springs–Highway O locality, map locality BS). This locality corresponds to Schumacher's (1972) locality 13 and Thompson's (1993, p. 111, figs. 74–76) Crow's Fork section. My sampling of the Mineola Member at Bellama Springs section (Fig. 6) did not yield any *Tecnocyrtina* material.

Dozens of specimens of *T. m. missouriensis* were obtained from the Mineola Member at the Aux Vasse Quarry (Fig. 6, units 4–6; Table 2, samples 2–9). The range of phenotypic variation displayed by these new specimens indicates that specimens of *Tecnocyrtina* illustrated by Branson (1922, Pl. 16, 19) and a single specimen illustrated by Stainbrook (1943a, Pl. 70, figs. 6–8 only) are consistent with *T. m. missouriensis*. In both of these studies the material came from unspecified localities in the Callaway Limestone of Callaway County. In my samples from units 4 through 6 of the Mineola Member at the Aux Vasse Quarry (Fig. 6, map locality AVQ), *T. m. missouriensis* and other associated brachiopods (Table 2) occur with *Icriodus subterminus* and *Mehlina gradata* in samples 2 and 3 (Fig. 6). This conodont fauna correlates with the Upper *subterminus* Fauna (Fig. 4). In Iowa, the brachiopods *Tecnocyrtina johnsoni* n. sp., *Cranaena iowensis*, and *Pentamerella nitida* occur with the same conodonts in the Coralville Formation (Day, 1989, 1992).

At the Aux Vasse Quarry (Fig. 6), the Mineola is overlain by rocks of the Upper Callaway Member that yield conodonts of the *insita* Fauna (Fig. 6, sample interval 10) and brachiopods of the *Allanella allani* Zone (Table 2). I did not find *Tecnocyrtina* in the Upper Callaway Member.

Day (1996, table 3) listed "*T. missouriensis curvilineata*" from the New Bloomfield Member of the Snyder Creek Shale. This was based on five specimens (SUI 84313–84317), assigned here to *T. curvilineata*, from the lower Snyder Creek Shale at the New Bloomfield East section (Fig. 6, map locality NBE). This indicates that the upper range of this form in the Iowa Basin is in the *Strophodonta callawayensis* Zone (Fig. 4). Conodonts collected in association with brachiopods of the *S. callawayensis* Zone in the lower Snyder Creek at the Craghead Creek Quarry (Fig. 6, map locality CCQ; Table 3) were correlated with the upper part of M. N. Zone 3 and part of Zone 4 by Day (1996). At this locality (Fig. 6) the base of M. N. Zone 4, indicated by the first occurrence of *Ancyrodella africana* in sample 21 (Day, 1996), occurs at 93 cm above the base of the Snyder Creek Shale.

Craghead Creek Quarry

	A. allani Zone				*Strophodonta callawayensis* Z.					*Orthospirifer' missouriensis* Z.				
	Sample													
BRACHIOPOD TAXA	10	15	16	17	18	19	20	21	25	26	28	29	30	31
Psuedoatrypa missouriensis	●			●										
Allanella annae		●	●				●			●			●	●
Schizophoria athabaskensis					●	●		●	●	●			●	●
Strophodonta callawayensis					●	●		●						
S. linfexa					●	●		●						
S. boonensis					●	●								
Seratrypa callawayensis					●	●	●			●				
Cranaena infrequens						●								●
Strophodonta navalis						●				●				
S. cymbiformis								●						
S. equicostata									●	●				
Eostrophalosia callawayensis									●	●	●			
Hadrorhynchia solon									●	●				
Orthospirifer missouriensis										●				●
Cyrtina sp. A										●				●
Ripidiorhynchus gregeri										●				
Strophodonta n. sp.												●		●
Seratrypa snyderensis													●	●

Table 3. Early Frasnian brachiopod sequence in the Cedar Valley Formation and Snyder Creek Shale at the Craghead Creek Quarry in Callaway County, Missouri (Fig. 6, locality CCQ). All sample positions are shown in Figure 6. Brachiopod zonal intervals in upper row, as in Figure 4.

Eastern North America

Tecnocyrtina bicostata was described from the uppermost beds of the Williams Island Formation in the Moose River Basin of the Hudson Bay Lowlands of Ontario by Norris (in Norris et al., 1992). This species has two to three very weakly developed plicae on the dorsal fold and ventral sulcus; these features are characteristic of the *missouriensis* group (Figs. 1, 2). *T. bicostata* occurs with an erosionally admixed brachiopod and conodont fauna, with conodonts ranging in age from *disparilis* Zone (undifferentiated) through M. N. Zone 3 (Uyeno, in Norris et al., 1992). The conodonts suggest a maximum age for *T. bicostata* as no older than the *disparilis* Zone; that is, presumably no older than other central North American forms of the *missouriensis* group (Fig. 1). The *missouriensis* "stage of evolution" of this form is consistent with a Givetian age, though an early Frasnian age is nevertheless possible.

New Bloomfield East

	Strophodonta callawayensis Zone					*Orthospirifer missouriensis* Zone			
	Sample								
BRACHIOPOD TAXA	2	3	4	5	6	7	8	9	10
Psuedoatrypa missouriensis	●	●	●		●	●	●	●	●
Strophodonta navalis	●			●				●	
S. callawayensis		●		●				●	
S. boonensis				●		●			
Schizophoria athabaskensis				●		●			
Allanella annae				●	●	●	●	●	●
Cyrtina sp. A					●				
Athyris sp.					●		●	●	
Floweria n. sp.						●			
Productella sp.						●		●	
Seratrypa snyderensis						●		●	
Eostrophalosia callawayensis						●			●
Cranaena infrequens								●	

Table 4. Early Frasnian age brachiopod sequence in the Snyder Creek Shale at the New Bloomfield East section in Callaway County, Missouri (Fig. 6, locality NBE). Sample positions shown in Figure 6.

TECNOCYRTINA EVOLUTION IN NORTH AMERICA

The foregoing analysis of stratigraphic occurrences of *Tecnocyrtina* indicates that the timing of major biotic changes in the *missouriensis* and *billingsi* groups coincided with major sea-level changes that initiated Devonian T-R cycles IIa-1, IIa-2, and IIb-1. The oldest-known tecnocyrtinids appear to be members of the *missouriensis* group that evolved from a species of multiplicate *Cyrtina* during the sea-level highstand or stillstand phase of Devonian T-R cycle IIa-1; the *Cyrtina-Tecnocyrtina* transition appears to have taken place in carbonate platform habitats in southwestern Manitoba (Figs. 1, 2).

Two important evolutionary events occurred shortly after the appearance of the oldest *Tecnocyrtina* during T-R cycle IIa-2 (late Givetian). The first was a major speciation event referred to here as the *missouriensis* radiation, followed by evolution of the oldest species of the *billingsi* group from a population of *T. missouriensis* in the Great Basin and its migration from the Great Basin to Poland. These developments were followed by extinction of most elements of the *missouriensis* group at the close of T-R cycle IIa and subsequent radiation of the *billingsi* group during the early stages of T-R cycle IIb during the latest Givetian.

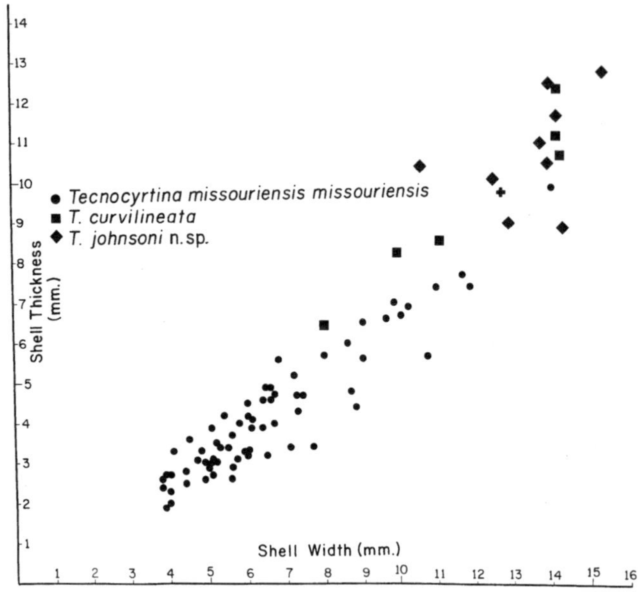

Figure 7. Comparisons of shell width and thickness for measured specimens of *Tecnocyrtina missouriensis missouriensis* from Missouri and Manitoba (plotted as a cross); *T. johnsoni* n. sp. from eastern Iowa; and *T. curvilineata* from eastern Iowa and central Missouri. All measured specimens are housed in the type collections at the University of Iowa Paleontology Repository (= SUI), Iowa City, Iowa.

The missouriensis *radiation*

The *Tecnocyrtina* lineage experienced a major radiation in the late Givetian (Figs. 1, 2). This is marked by the sudden appearance of three species and three subspecies of the *missouriensis* group (Figs. 1, 2). This radiation was probably a consequence of migration of *Tecnocyrtina* from central Canada and rapid allopatric speciation of isolated populations that colonized newly available shelf habitats in carbonate platforms in the Great Basin, the U.S. midcontinent, and western and eastern Canada during the sea-level rise that began in the upper part of the *disparilis* Zone (Figs. 1, 2).

Johnson and Trojan (1982, p. 130, fig. 2) recognized the *T. missouriensis* stage of evolution on the basis of occurrences of three correlative subspecies of *T. missouriensis* in three successions in North America. These subspecies were thought to have evolved independently from an older geographically isolated ancestral cyrtinid stock. I prefer an allopatric model for the radiation of the *missouriensis* group, with *T. missouriensis missouriensis* arriving in northern Missouri by migration from central Canada and remaining isolated in inner shelf shoal or tidal channel habitats along the southern margin of the Iowa Basin. There it is found in abundance in channel or shoal skeletal-peloidal grainstones and packstones of the Mineola Member of the Cedar Valley Formation (Figs. 4 and 6).

The coeval, but larger species, *T. johnsoni* n. sp. (Figs. 7 and 8,) most likely evolved from larval populations of the *T. m. missouriensis* stock that successfully colonized and adapted to different habitats developed in the central part of the carbonate platform in the main Iowa Basin. In central Iowa, dolomites and dolomitic limestones of the Gizzard Creek Member of the Coralville Formation yielding *T. johnsoni* n. sp. accumulated on an inner platform stromatoporoid-coral thicket.

The missouriensis *group extinctions*

Most species of the *missouriensis* group became extinct in the upper part of T-R cycle IIa-2; this coincides broadly with the major sea-level lowstand and profound regression that terminated T-R cycle IIa-2 deposition (= latest *disparilis* Zone lowstand or regression of Day et al. [1994, 1996]; and Johnson et al., [1996]). The only place that members of the *missouriensis* group survived was in the Iowa Basin, where *T. curvilineata* occurs in beds deposited immediately after the extinction interval during Devonian T-R cycle IIb (Figs. 1, 3).

The terminal T-R cycle IIa extinction was not limited to tecnocyrtinids. North American brachiopod faunas as a whole underwent major restructuring and reduction in diversity at that time (Johnson, 1970, 1990; McLaren, 1970; Day, 1996). Other major fossil groups affected during this event include colonial rugose corals, trilobites, and bryozoans (Day, 1996).

Origin of the billingsi *group*

The oldest species of the *billingsi* group, *T. fissiplicata*, first appeared in beds correlated with the upper part of T-R cycle IIa-2 in the Great Basin (Figs. 1, 2). This occurred shortly after the *missouriensis* radiation. According to Johnson and Trojan (1982), *T. fissiplicata* evolved from *T. missouriensis teleta* late in the Upper *disparilis* Subzone or in the upper part of Great Basin Devonian Faunal Interval 27 (Johnson and Trojan, 1982; Johnson, 1990). Based on material described and illustrated by Johnson and Trojan (1982), this occurred through an increase in the number of plications on the dorsal fold, from two to three in *T. m. teleta* (Johnson and Trojan, 1982, Pl. 7, figs. 6–25) to as many as seven in *T. fissiplicata* (Johnson and Trojan, 1982, Pl. 7, figs.

26–39, Pl. 8, figs. 1–26); an increase in the number of flank plications; and development of grooved plications in some specimens of the latter species.

The billingsi *radiation*

The main speciation event affecting the western North American and western European *T. billingsi* group coincided with the initial transgression of T-R cycle IIb (Figs. 1, 3). Johnson and Trojan (1982, fig. 2), speculated that *T. billingsi* possibly evolved from *T. curvilineata* by migration of members of the latter species from the U.S. midcontinent into western Canada; this was based in part on comparisons of two specimens of *T. curvilineata* with *T. billingsi*. External shell features of the syntypes of *T. curvilineata* and additional specimens from new collections are typical of the *missouriensis* group and lead me to view it as not ancestral to species seen in T-R cycle IIb strata in western North America. For discussion of divergent morphologies of *T. billingsi* and *T. curvilineata*, see the Systematic Paleontology section for the latter species.

It is suggested here that *T. fissiplicata* is ancestral to *T. billingsi* and *T. beckmanni* as well. These probably evolved rapidly during migration of elements of the *fissiplicata* stock from the Great Basin into western Canada and from carbonate platforms in Poland into Germany, respectively (Figs. 1 and 3). Evolution would have been by continuation of trends involved in the evolution of *T. fissiplicata* from its presumed ancestor, *T. missouriensis* teleta. Evolution of *T. billingsi* from *T. fissiplicata* is viewed as having involved increase in the number of inserted medial and flank plications, an increase in maximum adult size (up to 16 mm in width), and reduction of the fold, sulcus, and size of flank plications leading to the essentially gentle uniplicate anterior commissure of *T. billingsi.*

Additional support for a western North American origin for *T. billingsi* comes from comparisons of T-R cycle IIb brachiopod faunas. Recent studies by Day (1992, 1996) and Day et al. (1996) suggest that many widespread species, especially orthids, atrypids, stropheodontids, rhynchonellids, and productids, migrated from western and central Canada into the Iowa Basin during the *Allanella allani* and *Strophodonta callawayensis* zones of T-R cycle IIb. That migrations in the other direction were generally not favored is attributed to current circulation patterns developed in the cratonic seaway at that time.

SYSTEMATIC PALEONTOLOGY

Order SPIRIFERINIDA Ivanova, 1972
Suborder CYRTINIDINA Carter and Johnson, 1994
(in Carter et al., 1994)
Family CYRTINIDAE Frederiks, 1911
Genus *Tecnocyrtina* Johnson and Norris, 1972

Type Species. Cyrtina billingsi MEEK, 1868, p. 97, Pl. 14, figs. 6a–6d.

Discussion. Johnson and Norris (1972) proposed *Tecnocyrtina* as a cyrtiniform genus with multiplicate flanks and development of medial plications. Three distinctive forms were known at that time from late Givetian–early Frasnian strata of the U.S. midcontinent and western North America. By 1982 the number of described forms of *Tecnocyrtina* had increased to six as a consequence of studies by Johnson (1973) and Johnson and Trojan (1982). In the latter study, six forms were described from three geographic species-sequences of *Tecnocyrtina* developed in late Givetian and early Frasnian beds in the Great Basin, western Canada, and the U.S. midcontinent; ranges of the species were tied in to the existing Devonian conodont zonation, but ranges of the U.S. midcontinent forms were poorly known. Two additional forms, one of them *T. johnsoni* n. sp., and new occurrences of *Tecnocyrtina* have been reported from Givetian and Frasnian rocks in central and eastern North America by Day (1989, 1992, 1996), Norris (in Norris et al., 1992), and Day et al. (1996). An additional western European occurrence of a form identified as *T.* aff. *T. billingsi* from late Givetian beds in Poland was reported by Racki et al. (1985). This form is here tentatively assigned to *T. fissiplicata* (Fig. 2), based on the presence of 4 to 5 medial plications on its well-developed dorsal fold (Racki et al., 1985, Pl. X, fig. 3b).

Currently seven species and three subspecies of *Tecnocyrtina* are known from Middle-Upper Devonian rocks in North America and Europe. These include the type species *T. billingsi* (Meek, 1868), *T. fissiplicata* Johnson and Trojan 1982; *T. beckmanni* Johnson, 1973; *T. missouriensis missouriensis* (Swallow, 1860), *T. m. raaschi* Johnson and Trojan, 1982; *T. m. teleta* Johnson and Trojan, 1982; *T. bicostata* Norris (in Norris et al., 1992); *T. curvilineata* (White, 1862); and *T. johnsoni* n. sp. The first three forms comprise the western European and western North American *billingsi* group; the latter six species and subspecies make up the *missouriensis* group. The geographic and stratigraphic distribution of all North American forms is summarized in Figures 1 through 3.

Revisions relating to the systematic position of the punctate spiriferids recently introduced by Carter et al. (1994) are adopted here. Large numbers of newly available specimens of *T. missouriensis missouriensis* warrant discussion of ontogenetic changes and phenotypic variation in this highly variable form. A revised synonymy and brief discussion of *T. curvilineata* is given below, although this form in not redescribed or illustrated in this study.

***Tecnocyrtina johnsoni* n. sp.**
Figure 8: 1–13

Diagnosis. A large species of *Tecnocyrtina* with an inflated ventral valve, shell width exceeding thickness, ventral valve inflated and thicker than other species of the genus, wide (up to one half total shell width), with distinct sulcus and fold, and with distinct, 0 to 3 variably developed ventral median plications and up to 3 medial plications on the low rounded dorsal fold.

Exterior. Adult growth stage large for genus (width exceeds 15 mm); shell inflated ventribiconvex to hemipyramidal, with maximum shell width greater than thickness (Fig. 7) and located just anterior of hinge; ventral interarea apsacline, gently curved or arched; deltidium lacking owing to shell damage or nonpreservation; delthyrium generally open; cardinal angles rounded, obtuse; plications simple, up to 11 on each flank, rounded in cross-sectional profile, separated by narrower simple grooves or interspaces with rounded floors, generally added by insertion along posterolateral margin of shell; plications in the ventral sulcus, sometimes tending toward obsolescence anteriorly; dorsal fold with three plications. Sulcus originating at beak, broad and U-shaped, well defined, with anterior margin of shell in the sulcus extended to form short rounded tongue. Dorsal fold well developed in large specimens, less defined to subdued in smaller growth stages. Concentric ornament consisting of fine closely spaced well-defined growth lines, with prominent irregularly spaced disturbance or growth ridges produced during slower intervals of shell growth.

Figure 8. *Tecnocyrtina johnsoni* n. sp. All views magnified at ×3 unless otherwise indicated. 1, posterior view of ventral valve, paratype SUI 81337, ×2. 2, side view of ventral valve, paratype SUI 81337. 3, internal view of ventral valve, SUI paratype 81337, ×3.5. 4 and 5, ventral and anterior views of SUI paratype 81337. 6, dorsal view of isolated brachial valve, paratype SUI 51785. 7, upper view of ventral valve of holotype SUI 65978. 8, anterior view of holotype SUI 65978. 9, posterior view of holotype SUI 65978. 10, side view of holotype SUI 65978. 11, posterior view of paratype SUI 51786. 12, view of dorsal valve of paratype SUI 51786. 13, anterior view of paratype SUI 51786. All from the Gizzard Creek Member of the Coralville Formation at the Glory Quarry, from unit 14 (Day, 1992, fig. 7).

Interior. Hinge of dorsal valve unknown; preserved silicified filaments visible in one specimen, indicating spiralia directed ventrolaterally, with as many as six volutions. Tichorhinum present, elevated on ventral median septum and located in posterior chamber created by fusion of dental lamellae and median septum. Posterior chamber extending anteriorly from inner surface of posterior of the ventral valve, occupying one quarter of valve length. Median septum sandwiched by fused dental plates, extending anteriorly for three quarters of length of valve. Teeth simple, extending dorsally from hinge, projecting from posterior where bases of dental plates arise from shell of interarea.

Shell Substance. Presumably endopunctate; all specimens silicified.

Comparison. The inflated ventral valve, large maximum adult size (Figs. 7 and 8), relatively broad sulcus, and variably developed plicae (generally two, up to three or absent) in the ventral sulcus readily distinguishes *T. johnsoni* from closely related *T. m. missouriensis* and its descendant *T. curvilineatea*. Other forms of the *missouriensis* group are discussed by Johnson and Trojan (1982) and Norris (in Norris et al., 1992).

Holotype. SUI 65978 (Figs. 8: 7–10).

Measurements. Plots of shell width versus thickness are shown in Figure 7.

Occurrence. All figured (Fig. 8) and measured (Fig. 7) specimens are from the Iowa Basin Devonian *Tecnocyrtina johnsoni* Zone, from the lower part of the Gizzard Creek Member of the Coralville Formation, unit 14 at the Glory Quarry (Day, 1992, fig. 7, p. 66) in Blackhawk County, Iowa.

Etymology. In honor of Dr. John Granville Johnson, formerly of Oregon State University.

Tecnocyrtina missouriensis missouriensis (Swallow, 1860)

Cyrtia missouriensis SWALLOW, 1860, p. 647.
Cyrtina missouriensis (Swallow). BRANSON, 1922, p. 108, Pl. 16, fig. 9, Pl. 19, figs. 1–4; STAINBROOK, 1943a, p. 447, Pl. 70, figs. 6–8 only; FRAUNFELTER, 1974, p. 68–69, Pl. 16, fig. 7.
Tecnocyrtina missouriensis (Swallow). JOHNSON and NORRIS, 1972, p. 566, Pl. 1, figs. 5–18.
Tecnocyrtina missouriensis missouriensis (Swallow). JOHNSON *in* Johnson and Trojan, 1982, p. 130–131, Pl. 8, figs. 27, 28, 31, 32, 36.

Discussion. The shell features of specimen (SUI 81335; indicated with a cross in Fig. 7) from the Souris River Formation of southern Manitoba permit inclusion of that form in *Tecnocyrtina missouriensis missouriensis*. Seventy specimens of *T. m. missouriensis* (Fig. 7) from the Mineola Member of the Cedar Valley Limestone of central Missouri display distinctive and systematic changes in morphological features as shell size increases (shell width serves as the proxy feature for discussion of growth stages herein). One observed trend is that the average number of medial plications (three) inserted on the dorsal fold occurred on most shells by the time they achieved an average width of 7 mm. Generally the number of medial plications remains fixed at three, but specimens with as many as five are present. An average number of four medial plications is observed on the cohort of largest specimens with widths greater than 10 mm.

Another distinctive ontogenetic trend involves systematic changes in inclination of the ventral interarea. In the early stages of shell ontogeny, juveniles (less than 5 mm) are characterized by development of pronounced procline ventral interareas. This change during subsequent shell growth resulted in development of shells with nearly catacline to rounded gently inclined apsacline interareas below recurved beaks, as seen in larger shells (usually greater than 7 mm). As in other cyrtinids, changes in the inclination and degree of distortion (lateral twisting or torsion) of the ventral interareas and beaks are highly variable in larger specimens.

Measurements. Plots of shell width versus thickness of all measured specimens are shown in Figure 7.

Material. Specimens labeled SUI 84276-84310 came from the Mineola Member of the Cedar Valley Formation at the Aux Vasse Quarry (Fig. 6, units 5 and 6), northern Callaway County, Missouri. SUI specimen 81335 is from the lower bed (Day et al., 1996, fig. 5, unit 3) of the Argillaceous limestone beds of the Point Wilkins Member of the Souris River Formation at Point Wilkins, west shore of Dawson Bay (Day et al., 1996, fig. 3).

Tecnocyrtina curvilineata (White, 1862)

Cyrtia curvilineata WHITE, 1862, p. 25
Cyrtina curvilineata [?] (White). HALL, 1867, p. 270–271, Pl. 44, figs. 53–55.
Cyrtina curvilineata (White). HALL and CLARKE, 1894, p. 46, Pl. 28, figs. 11 and 2.
Cyrtina missouriensis (Swallow). STAINBROOK, 1943b, p. 447–448, Pl. 70, figs. 9–11 (only).

Discussion. Johnson and Trojan (1982, p. 130) distinguished *Tecnocyrtina curvilineata* from *T. missouriensis* on the basis of features displayed by its syntypes. Additional well-preserved specimens from the State Quarry Member of the Lithograph City Formation and the lower Snyder Creek Shale clearly display features typical of forms of the *T. missouriensis* group.

Larger specimens of *T. curvilineata* now available (up to 14.7 mm in width) have flank plications ranging from seven to 11 in number versus as many as 17 to 19 on specimens of *T. billingsi* of similar size (Johnson and Trojan, 1982, Pl. 8, figs. 33, 34). Plications on the dorsal fold of *T. curvilineata* range from three to five on larger specimens, versus five to 11 on the folds of *T. billingsi* illustrated by Johnson and Norris (1972), Johnson and Trojan (1982), and Norris and Uyeno (1983).

Tecnocyrtina curvilineata is a rare species found in latest Givetian and early Frasnian strata in Iowa and Missouri only. Available specimens indicate that it achieved a large size for the genus, comparable in size to its presumed ancestor *T. johnsoni* (Fig. 7). In both species shell width is greater than thickness, and both developed similar numbers of flank and medial plications. Large specimens of the two species can readily be distinguished by development of a strongly recurved apsacline interarea on shells of *T. curvilineata*; ventral valves of *T. johnsoni* have gently apsacline to catacline interareas.

ACKNOWLEDGMENTS

I gratefully acknowledge the Donors of the Petroleum Research Fund of the American Chemical Society for supporting my study of Middle and Upper Devonian successions in Iowa, Missouri, and Manitoba in 1991 and 1992. Julia Golden, University of Iowa Paleontology Repository, provided much curatorial support. Dan Fisher and Greg Gunnel of the University of Michigan Museum of Paleontology loaned type specimens of *Tecnocyrtina curvilineata*; Willy Norris of the Geological Survey of Canada freely shared critical information on occurrences of undescribed *Tecnocyrtina* from the Northwest Territories and Manitoba; Remy Gourvennec, University of Bretagne Occidentale, Brest, France, and John A. Talent, Macquarie University, Australia, commented constructively on the manuscript. I am especially grateful to Kathy and Jamie Day for exceptional patience during preparation of the manuscript.

This paper is dedicated to Jess Johnson, my former undergraduate adviser, mentor, and friend.

REFERENCES CITED

Branson, E. B., 1922, The Devonian of Missouri: Missouri Bureau of Geology and Mines, v. 17, Second Series, 279 p.

Branson, E. B., 1944, Geology of Missouri: University of Missouri Studies, v. 19, no. 3, 535 p.

Braun, W. K., Norris, A. W., and Uyeno, T. T., 1989, Late Givetian to early Frasnian biostratigraphy of western Canada: The Slave Point-Waterways boundary and related events, *in* McMillan, N. J., Embry, A. F., and Glass, D. J., eds., Devonian of the world: Calgary, Canadian Society of Petroleum Geologists Memoir 14, v. 3, p. 93–112.

Bunker, B. J., and Witzke, B. J., 1992, An upper Middle through lower Upper Devonian lithostratigraphic and conodont biostratigraphic framework of the midcontinent Carbonate Shelf area, Iowa, *in* Day, J. and Bunker, B. J., eds., The stratigraphy, paleontology, depositional and diagenetic history of the Middle-Upper Devonian Cedar Valley Group of central and eastern Iowa: Iowa Department of Natural Resources, Geological Survey Bureau Guidebook Series 16, p. 3–25.

Carter, J. L., Johnson, J. G., Gourvennec, R., and Hou, H.-F.,1994, A revised classification of the spiriferid brachiopods: Annals of the Carnegie Museum, v. 63, p. 327–374.

Day, J., 1989, The brachiopod succession of the late Givetian-Frasnian of Iowa, *in* McMillan, N. J., Embry, A. F., and Glass, D. J., eds., Devonian of the world: Calgary, Canadian Society of Petroleum Geologists Memoir 14, v. 3, p. 303-326.

Day, J., 1992, Middle-Upper Devonian (late Givetian–early Frasnian) brachiopod sequence in the Cedar Valley Group of central and eastern Iowa, *in* Day, J. and Bunker, B. J., eds., The stratigraphy, paleontology, depositional and diagenetic history of the Middle-Upper Devonian Cedar Valley Group of central and eastern Iowa: Iowa Department of Natural Resources, Geological Survey Bureau Guidebook Series 16, p. 53–105.

Day, J., 1993, Revised lithostratigraphy and biostratigraphy of the Middle-Upper Devonian strata of central Missouri, southern part of Iowa Basin: Geological Society of America Abstracts with Programs, v. 25, no. 3, p. 16.

Day, J., 1994, The late Givetian and early Frasnian (Middle-Upper Devonian) brachiopod faunas of southeastern Iowa and northwestern Illinois, *in* Bunker, B. J., ed., Paleozoic stratigraphy of the Quad-Cities region of east-central Iowa, northwestern Illinois: Geological Society of Iowa Guidebook 59, p. 65–84.

Day, J., 1996, Faunal signatures of Middle-Upper Devonian depositional sequences and sea level fluctuations in the Iowa Basin: U.S. midcontinent, *in* Witzke. B. J., Ludvigson, G. A., and Day, J., eds., Paleozoic sequence stratigraphy: Views from the North American Craton: Geological Society of America Special Paper 306, p. 277–300.

Day, J., Plumely, P., and Parse, D., 1991, Correlation of the Givetian-Frasnian age strata of Iowa and Manitoba and preliminary Late Middle Devonian geomagnetic paleopole determination from the Iowa cratonic Devonian: Geological Society of America Abstracts with Programs, v. 23, no. 3, p. 10.

Day, J., Uyeno, T. T., Norris, A. W., Witzke, B. J., and Bunker, B. J., 1994, Middle-Upper Devonian relative sea level histories of central and western North American interior basins: Geological Society of America Abstracts with Programs, v. 26, no. 5, p. 12.

Day, J., Uyeno, T. T., Norris, A. W., Witzke, B. J., and Bunker, B. J., 1996, Middle-Upper Devonian relative sea level histories of central and western North American interior basins, in Bunker, B. J., Ludvigson, G. A., and Day, J., eds., Paleozoic sequence stratigraphy: Views from the North American Craton: Geological Society of America Special Paper 306, p. 259–275.

Fraunfelter, G. H., 1974, Invertebrate megafauna of the Middle Devonian of Missouri: University Museum, Southern Illinois University at Carbondale, Southern Illinois Studies, Research Records 13, p. 43–53.

Frederiks, G. N., 1911, Zametka o nekotorykh′ verkne-paleozoyskikh′ okamenelostyakh′ iz′ okrestnostey gor. Krasnoufimska: Prilozhenie k′ protokolam′ zasedaniy Obschchestva Yestestvoispytateley pri Imperatorskom′ Kazanskom′ Universitete, no. 269, p. 1-12.

Greger, D. K., 1909, The Devonian of central Missouri: American Journal of Science, 4th ser., v. 27, p. 374.

Hall, J., 1867, Descriptions and figures of the fossil Brachiopoda of the Upper Helderberg, Hamilton, Portage, and Chemung groups: Geological Survey of the State of New York, Palaeontology, v. 4, part 1, 428 p.

Hall, J., and Clarke, J., 1894, An introduction to the study of the genera of Palaeozoic Brachiopoda: Geological Survey of the state of New York, Palaeontology, v. 8, part 2, 394 p.

Ivanova, E. A., 1972, Osnovye zakonomernosti evolyutsii spiriferid (Brachiopoda): Paleontologicheskii zhurnal, no. 3, p. 28–42. English translation in Paleontological Journal, 1972, no. 3, p. 309–320.

Johnson, J. G., 1970, Taghanic onlap and the end of North American Devonian provinciality: Geological Society of America Bulletin, v. 81, p. 2077–2105.

Johnson, J. G., 1973, *Tecnocyrtina beckmanni*, a new species of brachiopod from the lower Upper Devonian of Germany: Seckenbergiana Lethaea, v. 53, p. 517–520.

Johnson, J. G., 1977, Lower and Middle Devonian faunal intervals in central Nevada, based on brachiopods, *in* Murphy, M. A., Berry, W. B. N., and Sandberg, C. A., eds., Western North America: Devonian: University of California, Riverside Campus Museum Contribution 4, p. 16–32.

Johnson, J. G., 1979, Devonian brachiopod biostratigraphy, *in* House, M. R., Scrutton, C. T., and Bassett, M. G., eds., The Devonian System: Special Papers in Palaeontology 23, p. 291–306.

Johnson, J. G., 1990, Lower and Middle Devonian brachiopod-dominated communities of Nevada, and their position in a biofacies-province-realm model: Journal of Paleontology, v. 64, p. 902–941.

Johnson, J. G., and Klapper, G., 1990, Revisions of Devonian Conodont Zones, *in* Johnson, J. G., 1990, Journal of Paleontology, v. 64, p. 934–936.

Johnson, J. G., and Klapper, G., 1992, North American Midcontinent Devonian T-R cycles, *in* Chaplin, J. R., and Barrick, J. E., eds., Special papers in paleontology and stratigraphy: A tribute to Thomas W. Amsden: Oklahoma Geological Survey Bulletin 145, p. 127–135.

Johnson, J. G., and Norris, A. W., 1972, *Tecnocyrtina*, a new genus of Devonian brachiopods: Journal of Paleontology, v. 46, p. 565–572.

Johnson, J. G., and Trojan, W. R., 1982, The *Tecnocyrtina* brachiopod fauna (?Upper Devonian) of central Nevada: Geologica et Palaeontologica, v. 16, p. 119–150.

Johnson, J. G., Klapper, G., and Trojan, W. R., 1980, Brachiopod and conodont successions in the Devonian of the Antelope Range, central Nevada: Geologica et Palaeontologica, v. 14, p. 77–116.

Johnson, J. G., Klapper, G., and Sandberg, C. A., 1985, Devonian eustatic fluctuations in Euramerica: Geological Society of America Bulletin, v. 96, p. 567–587.

Johnson, J. G., Klapper, G., and Elrick, M., 1996, Devonian transgressive-regressive cycles and biostratigraphy, northern Antelope Range, Nevada: Establishment of reference horizons for global cycles: Palaios, v. 11, p. 3–14.

Klapper, G., 1989, The Montagne Noire Frasnian (Upper Devonian) conodont succession, *in* McMillan, N. J., Embry, A. F., and Glass, D. J., eds., Devonian of the world: Calgary, Canadian Society of Petroleum Geologists Memoir 14, v. 3, p. 449–468.

Klapper, G., and Johnson, J. G., 1990, Revisions of Middle Devonian conodont zones, *in* Johnson, J. G., Lower and Middle Devonian brachiopod-dominated communities of Nevada, and their position in a biofacies-province-realm model: Journal of Paleontology, v. 64, p. 934–936.

Maurin, A. F., and Raasch, G. O., 1972, Early Frasnian stratigraphy, Kakwa-Cecilia Lakes, British Columbia, Canada: Compagnie Française des Pétroles, Notes et Mémoires 10, 80 p.

McLaren, D. J., 1970, Time, life, and boundaries: Journal of Paleontology, v. 44, p. 801–815.

Meek, F. B., 1868, Remarks on the geology of the valley of the Mackenzie River, with figures and descriptions of fossils from that region, in the museum of the Smithsonian Institution, chiefly collected by the late Robert Kennicott, Esq.: Transactions of the Chicago Academy of Science, v. 1, p. 61–114.

Metzger, R., 1993, Conodonts from the State Quarry Limestone (Middle Devonian; uppermost Givetian; *norrisi* Zone) in the midcontinent: Geological Society of America Abstracts with Programs, v. 25, no. 6, p. A53.

Norris, A. W., and Uyeno, *T.* T., 1983, Biostratigraphy and paleontology of Middle-Upper Devonian boundary beds, Gypsum Cliffs area, northeastern Alberta: Geological Survey of Canada Bulletin 313, 65 p.

Norris, A. W., Uyeno, *T.* T., and McCabe, H. R., 1982, Devonian rocks of the Lake Winnepegosis–Lake Manitoba outcrop belt, Manitoba: Geological Survey of Canada Memoir 392, 280 p.

Norris, A. W., Uyeno, T. T., Sartenaer, P., and Telford, P. G.,1992, Brachiopod and conodont faunas from the uppermost Williams Island Formation and lower Long Rapids Formation (Middle and Upper Devonian), Moose River Basin, northern Ontario: Geological Survey of Canada Bulletin 434, 133 p.

Parse, D. F., and Plumley, P. W., 1992, A paleomagnetic study of Middle Devonian carbonates from Iowa and Manitoba, in Day, J., and Bunker, B. J., eds., The stratigraphy, paleontology, depositional and diagenetic history of the Middle-Upper Devonian Cedar Valley Group of central and eastern Iowa: Iowa Department of Natural Resources, Geological Survey Bureau, Guidebook Series 16, p. 35–52.

Pitrat, C. A., 1975, *Orthospirifer*, a new genus of Devonian spinocyrtid brachiopods: Journal of Paleontology, v. 49, p. 387–394.

Racki, G., Gluchowski, E., and Mallec, J., 1985, The Givetian to Frasnian succession at Kostomloty in the Holy Cross Mts., and its regional significance: Bulletin of the Polish Academy of Sciences, v. 33, p. 159–171.

Rogers, F. S., 1990, Stratigraphy, depositional history, and conodonts of the Little Cedar and lower Coralville formations of the Cedar Valley Group (Middle Devonian) of Iowa [Ph.D. thesis]: Iowa City, University of Iowa, 97 p.

Schumacher, D., 1972, Conodont biostratigraphy and paleoenvironments of the Middle-Upper Devonian boundary beds, Missouri [Ph.D. thesis]: Columbia, University of Missouri, 107 p.

Stainbrook, M. A., 1938, *Atrypa* and *Stropheodonta* from the Cedar Valley beds of Iowa: Journal of Paleontology, v. 12, p. 229–256.

Stainbrook, M. A., 1943a, Spiriferacea of the Cedar Valley Limestone of Iowa: Journal of Paleontology, v. 17, p. 417–450.

Stainbrook, M. A., 1943b, Strophomenacea of the Cedar Valley Limestone of Iowa: Journal of Paleontology, v. 17, p. 39–59.

Swallow, G. C., 1860, Description of new fossils from the Carboniferous and Devonian of Missouri: Saint Louis Academy of Science Transactions, v. 1, p. 635–660.

Thompson, *T.* L., 1993. Paleozoic succession in Missouri Part 3: Silurian and Devonian Systems: Missouri Department of Natural Resources, Division of Geology and Land Survey Report of Investigations 70, 228 p.

Warren, P. S., 1944, Index brachiopods of the Mackenzie River Devonian: Transactions of the Royal Society of Canada, 3d Ser., v. 38, p. 105–135.

White, C. A., 1862, Description of new species of fossils from the Devonian and Carboniferous rocks of the Mississippi Valley: Boston Society of Natural History Proceedings, v. 9, p. 8–33.

Witzke, B. J., and Bunker, B. J., 1994, Classic geological exposures, old and new, Coralville Lake and spillway: Geological Society of Iowa Guidebook 60, 68 p.

Witzke, B. J., and Bunker, B. J., 1996, Relative sea-level changes during Middle Ordovician through Mississippian deposition in the Iowa area, North American craton, in Witzke, B. J., Ludvigson, G. A., and Day, J., eds., Paleozoic sequence stratigraphy: Views from the North American Craton: Geological Society of America Special Paper 306, p. 307–330.

Witzke, B. J., Bunker, B. J., and Rogers, F. S., 1989, Eifelian through lower Frasnian stratigraphy and deposition in the Iowa area, central midcontinent, U.S.A, in McMillan, N. J., Embry, A. F., and Glass, D. J., eds., Devonian of the world: Calgary, Canadian Society of Petroleum Geologists Memoir 14, v. 1, p. 221–250.

Manuscript Accepted by the Society February 27, 1997

Printed in U.S.A.

Geological Society of America
Special Paper 321
1997

Autochthonous Middle Devonian–Lower Mississippian strata in the southern Hot Creek Range, Nye County, Nevada: Constraints on the placement of the Roberts Mountains thrust

John P. Graham
Geology Department, Colby College, Waterville, Maine 04901

ABSTRACT

The Devonian and Mississippian facies at the southern end of the Hot Creek Range, Nye County, Nevada, represent a nearly continuous depositional record of deep-marine facies, are autochthonous, and require positioning of the Roberts Mountains thrust west of the range. Outer-shelf basin to foreland basin lithotopes correlate along strike with deep-marine facies that extend from Eureka County to northern Nye County.

The autochthonous, outer-shelf basin Denay Limestone in the southern Hot Creek Range is composed of both hemipelagic limestone and allodapic interbeds of pelletal-bioclastic wackestone, packstone, and grainstone. Thinly bedded, Devonian-Mississippian radiolarian chert, deposited in a starved basin, overlies the Denay. Mississippian radiolarians in the upper part of the chert correlate with the *isosticha*–Upper *crenulata* Zone, based on conodonts.

Hemipelagic lime mudstones ranging from *isosticha*–Upper *crenulata* to *typicus* Zone (Kinderhookian-Osagean) are interbedded with the chert beds near the top of the chert sequence. Allodapic limestones within the interval contain reworked Kinderhookian (Early Mississippian) and Famennian (Late Devonian) age conodonts. The Lower Mississippian lithotopes were deposited in a foreland trough developing in front of the Roberts Mountains thrust.

Hydrothermal activity, Tertiary volcanism, and Miocene extension have all affected the Devonian and Mississippian rocks near Warm Springs. West-verging folding of Devonian-Mississippian chert beds is local and related to normal faulting and is not related to the Antler orogeny.

INTRODUCTION

Interpretations of the map trace of the eastern limit of the Roberts Mountains thrust (RMT) have been progressively modified since first proposed in 1958 (Roberts et al., 1958) and are now more or less agreed upon for Eureka County, Nevada. The position of this trace to the south in Nye County, however, is still being debated (Fig. 1). Some authors (Johnson and Sandberg, 1977; Johnson and Pendergast, 1981) extend the leading edge of the thrust system southward from the northern Antelope Range to the Monitor Range, west of the Hot Creek Range in northern Nye County, but they do not attempt a more southerly projection. Others (Poole, 1974; Stewart and Poole, 1974; Poole and Sandberg, 1977, 1991, 1992; Poole et al., 1977; Stewart, 1980; Coles, 1991) extend the toe of the thrust into the Warm Springs area at the southern end of the Hot Creek Range. The thrust is then projected southwestward toward southern Esmeralda County (Fig. 1). Data presented here suggest that the allochtho-

Graham, J. P., 1997, Autochthonous Middle Devonian–Lower Mississippian strata in the southern Hot Creek Range, Nye County, Nevada: Constraints on the placement of the Roberts Mountains thrust, *in* Klapper, G., Murphy, M. A., and Talent, J. A., eds., Paleozoic Sequence Stratigraphy, Biostratigraphy, and Biogeography: Studies in Honor of J. Granville ("Jess") Johnson: Boulder, Colorado, Geological Society of America Special Paper 321.

nous plate of the RMT was not emplaced in the southern Hot Creek Range.

These conflicting points of view depend on the interpretation of the Middle Devonian–Mississippian section in the southern Hot Creek Range (Fig. 2). If the Middle Devonian Denay Limestone is allochthonous and overlain depositionally by Mississippian rocks equivalent to the Pinecone sequence, then the eastern limit of the RMT lies to the east of the Hot Creek Range. However, if the Middle Devonian–Mississippian rocks are both autochthonous, the eastern limit of the RMT is to the west of the southern Hot Creek Range.

West of Warm Springs, Nevada (Fig. 3), Devonian limestones and the Devonian-Mississippian chert sequence have been interpreted as allochthonous; Mississippian limestones equivalent to the Tripon Pass Limestone (Fig. 2) have been interpreted as overlap deposits and their association as evidence for early Kinderhookian emplacement of the RMT in this part of Nevada (Stewart, 1980; Kleinhampl and Ziony, 1985; Poole and Sandberg, 1991, 1992). The rocks equivalent to the Pinecone sequence on Figure 2 were assigned by Stewart and Poole (1974) to a Devonian "shale and chert facies" reflecting deep-marine conditions westward of the outer carbonate shelf edge. This interpretation was reemphasized by Stewart (1980) in a paragraph wherein he also referred to the Devonian rocks near Warm Springs as "the easternmost outliers of the Roberts Mountains allochthon." In 1979, a road log prepared for the Basin and Range Symposium mentioned a Kinderhookian klippe west of Warm Springs resting on Middle Devonian argillite and chert (French et al., 1979). Preliminary comments by Poole and Sandberg (1991, 1992) have alluded to the possibility of the allochthon in the Warm Springs area. All of these interpretations include the Denay Limestone and Devonian–Mississippian chert sequence (Fig. 2) within the Roberts Mountains allochthon.

However, Graham (1982) showed that the offshore regional depositional patterns at Warm Springs (Fig. 3) represent a normal succession of facies that do not support the emplacement of the RMT into the southern Hot Creek Range. Rather, he suggested that the thrust was emplaced west of the southern Hot Creek Range, supporting the interpretation of Johnson and Sandberg (1977) and Johnson and Pendergast (1981).

The following lithologic descriptions use the limestone classification scheme of Dunham (1962).

DEVONIAN DEPOSITIONAL ENVIRONMENTS

Middle Devonian to possibly Upper Devonian limestones in the southern Hot Creek Range (Fig. 3) correlate with the outer-

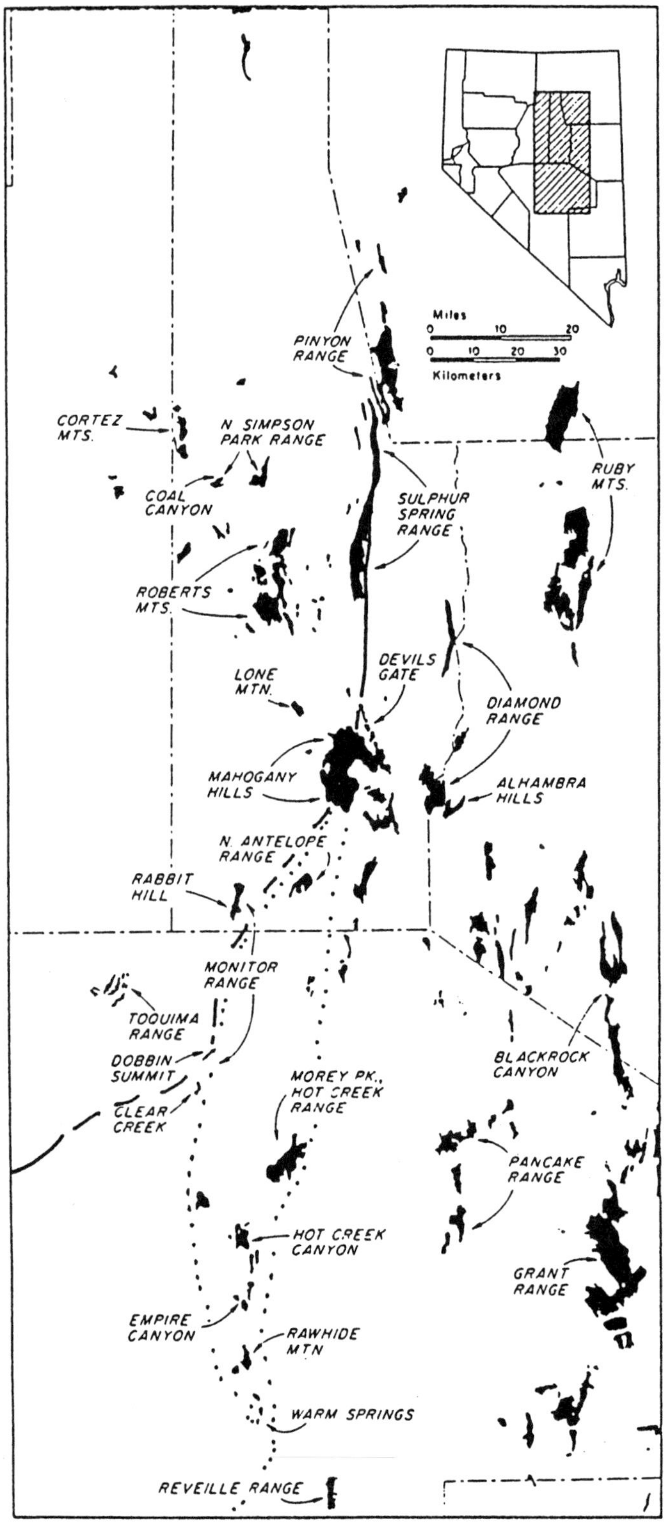

Figure 1. Index map of central Nevada with Devonian outcrops shown in black. Inset in upper right shows area covered. Devonian outcrops in the southern Hot Creek Range include Empire Canyon, Rawhide Mountain, and Warm Springs. The dashed line represents the interpretation supported by this report wherein the toe of the Roberts Mountains thrust did not advance into the southern Hot Creek Range. The dotted lines represent an alternative interpretation wherein the southern Hot Creek Range is included in the allochthon. Some authors project the Roberts Mountains thrust through the southern Fish Creek Range and along the eastern side of the Hot Creek Range, whereas others project the Roberts Mountains thrust southeastward from the Monitor Range to the southern Hot Creek Range.

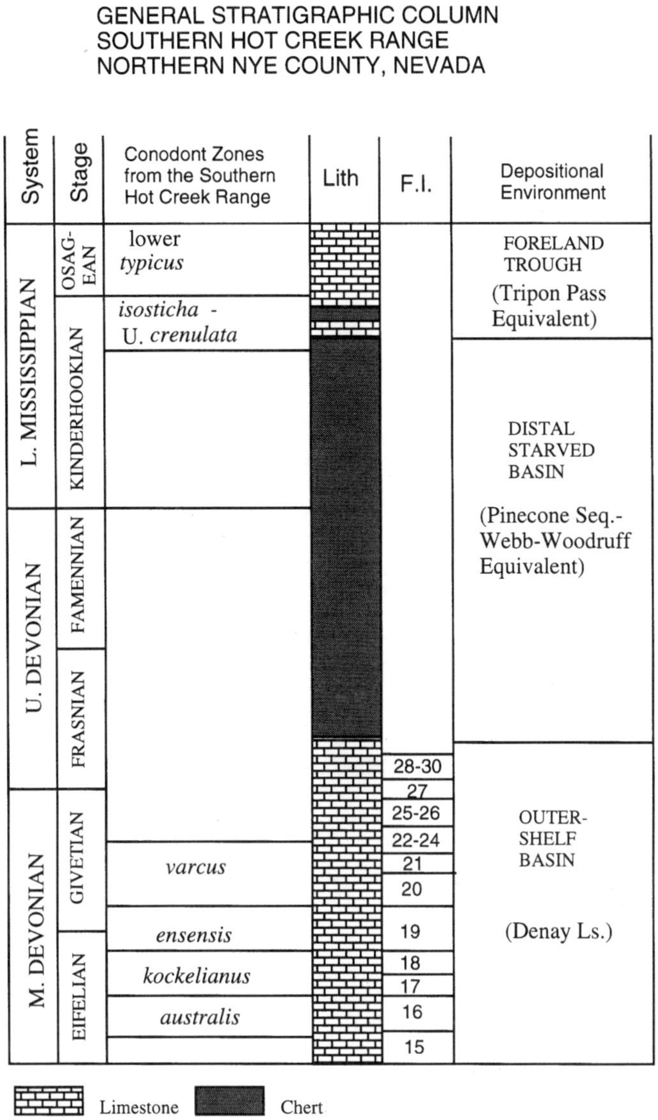

Figure 2. Generalized stratigraphic column, southern Hot Creek Range, northern Nye County, Nevada. Conodont zones determined by Klapper in Graham (1982). F.I.: Faunal intervals based on the correlation of both brachiopods and conodonts (Johnson et al., 1980).

shelf basin lithotopes of the Denay Limestone (Graham, 1982). Regionally, these Middle Devonian facies correspond with a belt of laminated limestone, lime mudstone, and allodapic limestone trending north-south through central Nevada (Murphy, 1977; Johnson et al., 1985; Johnson et al., 1991). From northern Eureka County to Hot Creek Canyon (Fig. 1), autochthonous Middle Devonian Denay Limestone overlies autochthonous Lower Devonian rocks.

Furthermore, autochthonous lower Denay limestones are found at Rawhide Mountain, 13 km (8 mi) north of Warm Springs, and in Empire Canyon, 15 km (9 mi) north of Rawhide Mountain (Fig. 1). At Empire Canyon (Figs. 4, 5), dark gray to black mudstones contain *kockelianus* Zone conodonts. These mudstones are overlain by Upper Middle *varcus* to Upper *varcus* Subzone wackestones and mudstones containing disarticulated brachiopods, crinoids, and corals (Fig. 5; Graham, 1982). Thinly bedded mudstone overlies the fossiliferous wackestone, and the measured section is capped by an interval of interbedded mudstone and wackestone with intraclasts containing bryozoa, solitary coral, stromatoporoid fragments, and dolomite.

The fine-grained Denay limestones are dark and evenly bedded with planar or slightly undulating contacts and contain millimeter-thick laminae. Radiolarians, sponge spicules, conodonts, and conodont pearls are common in these limestones. Locally, dacryoconarids on bedding surfaces are abundant. Relatively small brachiopod shells and spines, crinoid debris, and ostracode valves are minor components of these mudstones, as are disseminated silt particles.

The age and lithology of the Denay at Rawhide Mountain and Empire Canyon are similar to the Denay limestones found in isolated outcrops at the southern terminus of the Hot Creek Range (Figs. 1, 3). Conodonts and brachiopods collected from limestones at location WS-4 in the Warm Springs area (Fig. 3) indicate the Middle Devonian, in the range from the *ensensis* Zone to the Middle *varcus* Subzone (Graham, 1982). *Subrensselandia* and *Stringocephalus*, ranging from upper Interval 19 to Interval 21 in the Givetian (Fig. 2), were collected from a crinoidal-brachiopod wackestone interbedded with gray mudstones near the contact with the overlying chert (Graham, 1982). *Subrensselandia* disappears in Nevada after Interval 21, and *Stringocephalus* enters the record in Interval 19 (Johnson, 1977; Johnson et al., 1980).

The recessive-weathering Denay limestone varies from a light gray, very thinly bedded tentaculitid mudstone to thin to massive interbeds of mudstone and medium-gray, coralline-crinoid-brachiopod wackestone. Petrographically, the mudstone contains minute dolomite crystals floating in an argillaceous micrite. The wackestone, heavily veined with calcite, is a pelletal, bioclastic allodapic limestone containing broken and disarticulated, silicified fossil fragments.

Thin, widely spaced allodapic interbeds of pelletal-bioclastic wackestones, packstones, and grainstones are present at all three localities. Brachiopod valves commonly are disarticulated and broken. Fragmented crinoid ossicles are mixed with bryozoan debris. The fossil debris is aligned subparallel to bedding, and the coarser-grained fossiliferous, pelletal grainstone grades to pelletal wackestone.

The depositional fabric of the Denay, the cosmopolitan distribution pattern of conodont species identified from these localities (Graham, 1982; Klapper and Johnson, 1980), the presence of radiolarians, and the pelagic mode of life for dacryoconarids indicate hemipelagic deposition below storm or fair-weather wave base. Commonly, the mudstones have a fetid odor on a fresh surface, suggesting the possible deposition within an anaerobic, reducing environment. Interbedded with these hemipelagic mudstones are allodapic wackestones shed into the Denay basin.

During this time, the Denay basin may have been sheltered from circulating, normal marine waters by a carbonate barrier or series of carbonate buildups located west and northwest of the

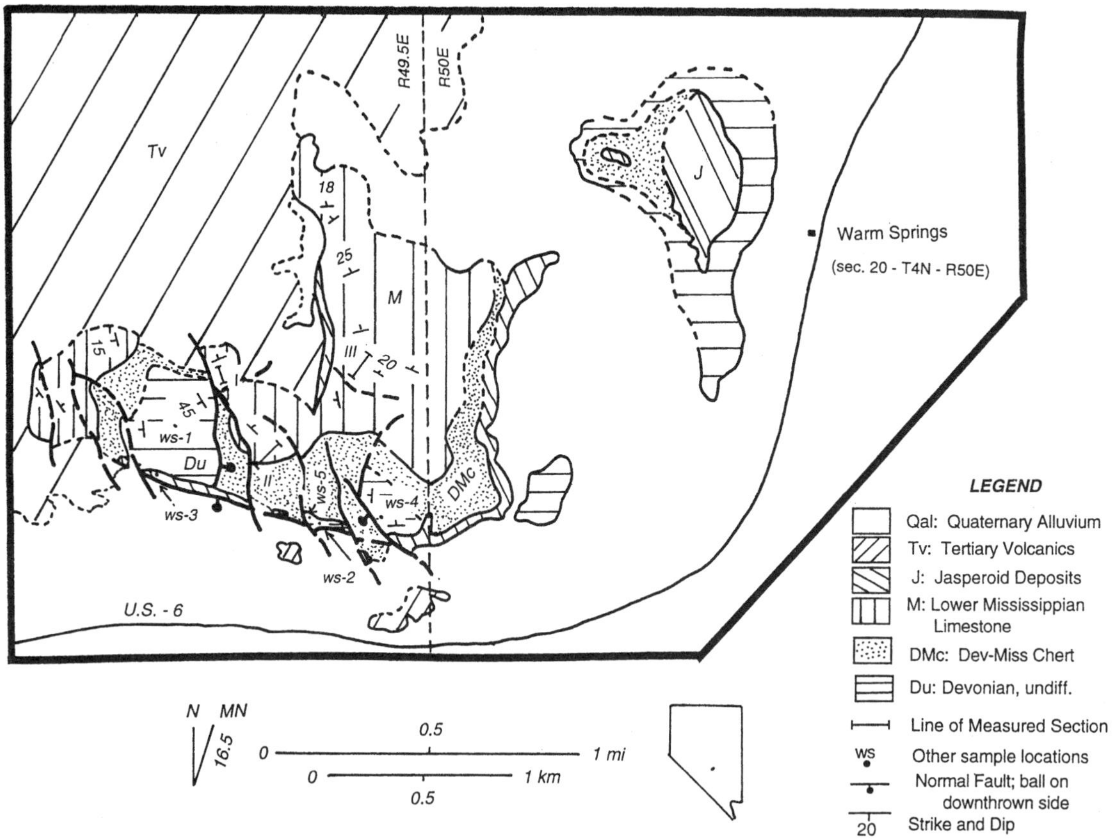

Figure 3. Geologic map of the Warm Springs area in the southern Hot Creek Range, Nye County, Nevada. Modified from Graham (1982).

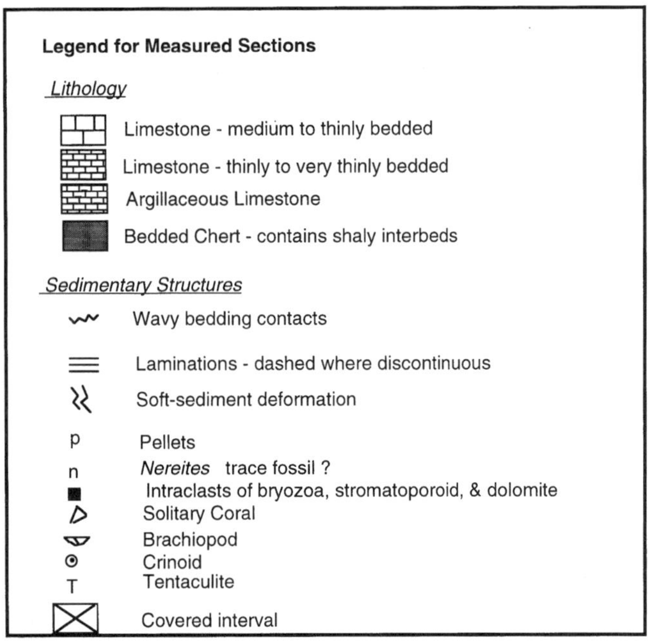

Figure 4. Legend for measured sections EC-II (Fig. 5) and WS-II (Fig. 7).

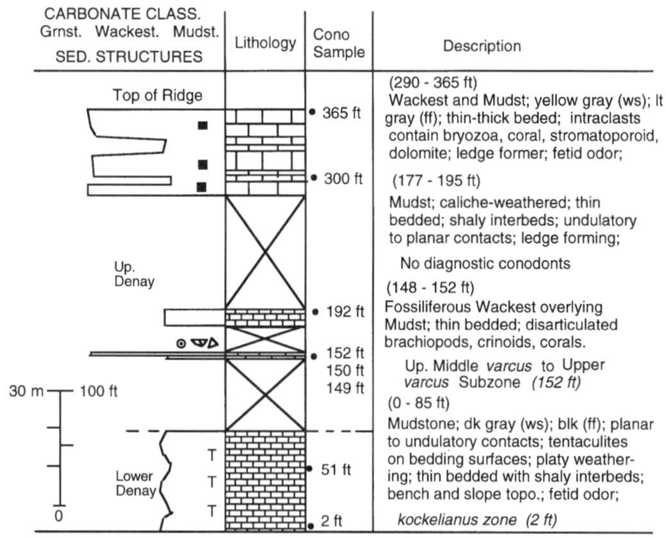

Figure 5. Measured section EC-II of Devonian Denay Limestone at Empire Canyon, southern Hot Creek Range. Section is located approximately 1.6 km (1.0 mi) south 55° east from Cine Mountain and 3 km (2 mi) west of the eastern boundary of T6N, R49E, northern Nye County, Nevada. Empire Canyon is located on Figure 1. See Figure 4 for key to symbols. Modified from Graham (1982).

southern Hot Creek Range in the general area of the antecedent Toiyabe Ridge (Murphy, 1977). The Eifelian- and Givetian-aged deep-water limestones (Fig. 2) also correspond to the transgressive-regressive cycles I-d to II-a of Johnson et al. (1985). The Denay at Warm Springs represents a deep marine lithotope similar in age and lithology to the autochthonous Denay Limestone, which can be traced from Eureka County southward to the Reveille Range (Fig. 1; Ekren et al., 1973; Johnson et al., 1980; Johnson, personal communication, 1992). It is reasonable to interpret the Denay in the Warm Springs area as compatible with other Denay limestones along depositional strike and therefore autochthonous.

DEVONIAN-MISSISSIPPIAN CHERT

A normal succession of deep marine lithotopes continues with the deposition of the Devonian-Mississippian chert, considered to be equivalent to the Pinecone sequence, above the Denay Limestone (Fig. 2). The distal starved-basin deposits of the Pinecone sequence overlie the outer-shelf basin Denay Limestone and are subjacent to Tripon Pass equivalent limestones deposited in a foreland trough formed in front of the Roberts Mountains thrust. The stratigraphic position of the chert is not unusual for a developing foreland basin.

The chert beds (Fig. 6) contain very thin interbeds of shale and are interbedded with Mississippian limestones in the first 21 m (68 ft) of measured section WS-II (Figs. 3, 7; Graham, 1982). Thin sections from samples 1.2 m (4 ft) above and 4.2 m (14 ft) below the base of WS-II disclose siliceous microspicules and radiolarians up to 0.2-mm (.008 in) diameter. Although extensive on the southern end of the Hot Creek Range, north of Highway 6, the chert generally weathers recessively beneath the more resistant Tripon Pass equivalent limestone that caps the peaks.

Extensive Cenozoic faulting at the south end of the Hot Creek Range prevents an accurate measurement of the relatively thick section of well-indurated, thinly bedded, maroon to black, organic-rich radiolarian chert (Fig. 6). Six meters (20 ft) of similar black, well-bedded chert followed by 34 m (110 ft) of interbedded shales, sandstones, and chert were measured by Wise (1976) south of Dobbin Summit in the Monitor Range (Fig. 1). Coles (1991) described dark-gray to black chert in his informal "Pinecone sequence," a pelagic, hemipelagic, and mass-gravity flow deposit in the Toquima Range, the first mountain range west of the Monitor Range. Smith and Ketner (1968) estimated a thickness of several hundred meters for the Woodruff Formation, an Early to Late Devonian formation composed primarily of siliceous mudstone and black chert, in the Carlin-Piñon range area, Elko County, and Desborough et al. (1987) estimated as much as 100 m (330 ft) for the upper Woodruff in the southern Fish Creek Range, northern Nye County. Regardless of its thickness at Warm Springs, the chert is certainly not a depositional anomaly.

Figure 6. Radiolarian chert beds approximately 4.3 m (14 ft) below the base of WS-II. Chert beds are in undulating to planar contact and are relatively undeformed compared to those in Figure 8. Rock hammer for scale.

The age of the Warm Springs chert is uncertain and is considered possibly to span the Upper Devonian–Mississippian boundary. Poole and Sandberg (1992) proposed an early to late Famennian age for the upper part of a siltstone and chert interval near Warm Springs, dated by conodonts on bedding surfaces. Radiolarians collected by the author from the bedded chert between Mississippian and Devonian limestones were not definitive. Conodonts collected from an interval of interbedded chert and limestone, above the bedded chert, are no older than *isosticha*–Upper *crenulata* Zone (Figs. 2, 7).

The relative bathymetry, especially as it changed through the Upper Devonian–Mississippian, remains speculative. Gutschick et al. (1980) and Sandberg et al. (1983) postulated a paleobathymetry of 100 to 200 m (330 to 660 ft) for the foreland trough based on sedimentary structures, and Poole (1974, fig. 25) implies a depth of approximately 1,200 m (3,960 ft) for the trough in the Early Mississippian. Coles (1991) also suggests bathyal depths for the Antler foreland basin. The carbonate compensation depth (CCD) has deepened from about 2,000 m (6,600 ft) to 4,500 m (14,850 ft) since the advent of calcareous nannoplankton and microplankton in the Jurassic (Berger and Winterer, 1974; Garrison and Fischer, 1978). Whether as the result of a rapidly subsiding foreland trough or from physical parameters such as rates of carbonate supply and dissolution, bottom circulation, vertical mixing, sea-level rise, or temperature effects, the Warm Springs chert is interpreted by the author to be in continuum with the deepening facies of the underlying Devonian limestones.

Rather than being part of an allochthonous Slaven Chert sequence as suggested by Poole and Sandberg (1992), the chert at Warm Springs is more nearly synonymous with the Pinecone sequence of the Toquima Range. Like the Pinecone sequence, the Warm Springs chert represents deep-water, starved foreland basin sedimentation. Distal-starved lithotopes resulting from Frasnian transgressions would not be uncommon in places like the Monitor Range and southern Hot Creek Range during the Late Devonian (Johnson et al., 1985; Johnson, personal communication, 1992).

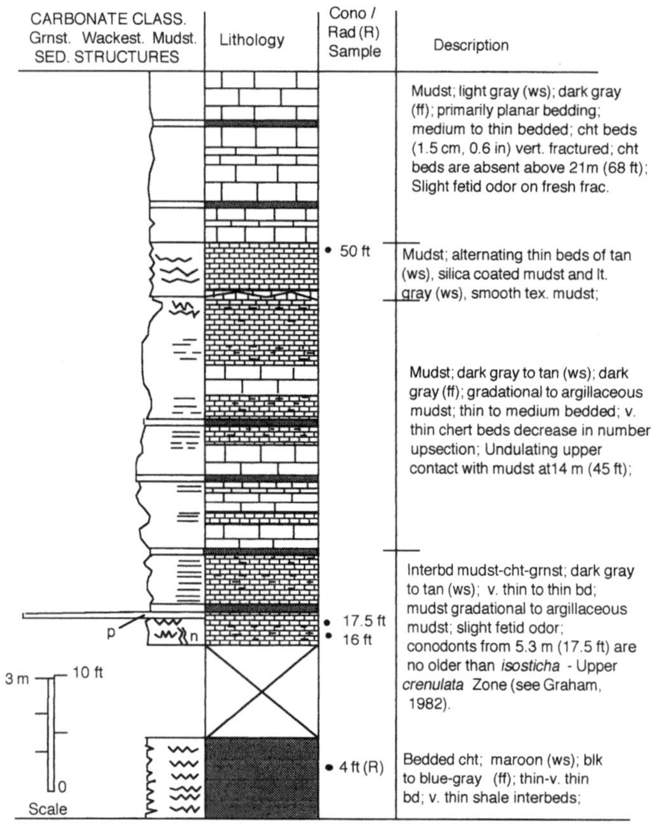

Figure 7. Basal 20 m (65 ft) of measured section II, Warm Springs area, southern Hot Creek Range, northern Nye County, Nevada. Section is located on Figure 3. See Figure 4 for key to symbols. Modified from Graham (1982).

LOWER MISSISSIPPIAN LIMESTONES AT WARM SPRINGS

Both turbidite and hemipelagic deposits are represented by the Mississippian limestones at Warm Springs. The Denay Limestone, the Devonian-Mississippian chert sequence, and the overlying Mississippian limestones are genetically related and do not have any major structure separating them.

The first limestone bed above the bedded chert in measured section WS-II contains meandering indigenous grazing trails of the *Nereites* community (Fig. 7). This ichnofauna implies a deep-water, bathyal marine environment for these Tripon Pass equivalent limestones (Fig. 2). Aside from minor laminations and undulatory bedding contacts in section WS-II (Fig. 7), sedimentary structures are scarce. Identifiable macrofossils are also sparse. Very thin to thin-bedded mudstone is gradational to argillaceous mudstone. Medium-bedded mudstone increases upsection (Fig. 7). When cracked, the dark-gray, dense, argillaceous mudstone emits a fetid odor.

These deep-water, Lower Mississippian limestones have been documented all along the Antler foreland flysch trough from the Monitor Range in northern Nye County (Wise, 1976) to Wells, Nevada, in Elko County (Johnson and Visconti, 1992). The Tripon Pass limestone turbidite beds are considered by Poole (1974), Poole and Sandberg (1977, 1991), Sandberg et al. (1983), Poole and Claypoole (1984), and Johnson and Visconti (1992) to be eastern-derived, basin-fill deposits derived from the tilting Joana bank along the western margin of a regional submarine rise. Goebel (1991) hypothesized a forebulge migration as a mechanism for relative sea-level lowering and subsequent emergence of the carbonate platform east of the foreland basin.

Late Kinderhookian conodonts are found in the Mississippian limestones near Warm Springs (Fig. 3). *Gnathodus punctatus*, *G. delicatus*, *Pseudopolygnathus marginatus*, *Ps. triangulus triangulus*, *Siphonodella cooperi,* and *S. duplicata* were among the conodonts collected from sample WSI-1, 0.3 m (1 ft) above the contact of the limestone with the bedded chert (Graham, 1982; Fig. 2). *Gnathodus punctatus* first occurs in the *isosticha*–Upper *crenulata* Zone, the base of which is defined by the first appearance of *G. delicatus* (Sandberg, 1979). Although *Pseudopolygnathus marginatus* and *Ps. triangulus triangulus* have been found in the *sandbergi–duplicata* Zone in the Midcontinent (Collinson et al., 1971), they first appear in the next younger Lower *crenulata* Zone in the Great Basin and Rocky Mountain regions (Sandberg, 1979). As *G. punctatus* and *G. delicatus* both range across the Kinderhookian-Osagean boundary (Sandberg et al., 1978; Sandberg, 1979), the accuracy of a stage assignment is questionable unless certain *Siphonodella* species can be shown to be indigenous (Sandberg, 1979; Johnson and Pendergast, 1981). One sample from the Mississippian limestones (WS-II-216B, Graham, 1982, p. 117) has *Gnathodus cuneiformis* and is thus Osagean.

Poole and Sandberg, in a 1992 abstract, date the base of the Mississippian sequence as late Kinderhookian (Lower *crenulata* Zone), but details of the biostratigraphy have not been documented.

A mixed conodont fauna and fining upward sequences of silty, pelletal grainstones to spicular-calcisphere mudstones are used to recognize limestone turbidites at Warm Springs. Several stratigraphic levels yielded *Gnathodus punctatus, Siphonodella isosticha,* and other conodonts of possible *isosticha*–Upper *crenulata* Zone age mixed with reworked Kinderhookian *Siphonodella* species and Late Devonian *Palmatolepis* species (Graham, 1982). Similar collections have been used to interpret limestone turbidites in the Camp Creek sequence or its equivalent at Swales Mountain (Johnson and Pendergast, 1981), the central Monitor Range (Wise, 1976), the northern Antelope Range (Trojan, 1978), and near Morey Peak in the northern Hot Creek Range (Potter, 1975).

The diverse conodont fauna present in the allodapic Warm Springs collections (Graham, 1982) reiterates the problems concerning the zonal determination that were outlined by Dombrowski and Klapper (*in* Johnson and Pendergast, 1981) and that remain unresolved. The *sandergi-duplicata* Zone, a Kinderhookian conodont zone separated from the *isosticha*–Upper *crenulata* Zone by the Lower *crenulata* Zone, may be represented at Warm Springs by *Siphonodella cooperi* and *S. duplicata*, but these species are known to range higher. The *Siphonodella* species,

S. obsoleta, S. cf. *S. isosticha, S. quadruplicata,* and *S. lobata,* first occur below the Lower *crenulata* Zone but range into the *isosticha*–Upper *crenulata* Zone (Sandberg et al., 1978). *Siphonodella crenulata* and *S. isosticha*, restricted to the interval of the Lower *crenulata* to *isosticha*–Upper *crenulata* Zones, are also found 0.3 m (1 ft) above the base of section WSI. If these *Siphonodella* species could be shown to be indigenous, then the rocks at Warm Springs could be assigned more accurately to the Kinderhookian. Based on the fauna listed in Graham (1982) and the above uncertainty, the Mississippian limestones at Warm Springs are interpreted to range across the Kinderhookian-Osagean boundary, from the *isosticha*–Upper *crenulata* Zone to the *typicus* Zone (Fig. 2).

Figure 8. West-verging, folded chert beds near the normal fault west of section II (Fig. 3). Chert beds can be traced to the east where the folding diminishes and bedding contacts become undulatory (Fig. 6). Rock hammer for scale.

DEFORMATION

Locally, chert beds are folded, but the deformation is not related to Roberts Mountains thrusting. Chert beds in the Warm Springs area have been deformed into tight and similar, west-verging folds near normal faults (Fig. 8). The chert beds in Figure 6 are 4.3 m (14 ft) below the base of WS-II, just east of the folded chert beds in Figure 8, and illustrate both the nondeformed nature and regular bedding of the chert beds not associated with local faulting. In general, folds are not abundant within the thrust plates of the Roberts Mountains allochthon (Stewart, 1980), but where present, the geometry suggests an east to east-southeast direction of tectonic movement, not westward (Stewart, 1980; Murphy et al., 1984). The deformed chert beds (Fig. 8) relax away from the faults, and the isoclinal folding becomes simply undulating bedding contacts (Fig. 6). The westward vergence, internal bedding integrity, and transition from tight folds into gently undulating bedding within a relatively short distance suggest that the deformation is local and not related to regional tectonic thrusting.

The timing of the deformation of Middle and Upper Devonian rocks at Warm Springs is problematic. The springs are the result of hot waters rising along the range-front faults, which were produced by Miocene extensional tectonics. The Hot Creek Range is a west-dipping block. The Paleozoic beds in the southern Hot Creek Range dip from 19° to as much as 70° near north-south–trending normal faults. The mapped normal faults (Graham, 1982; Fig. 3) may be related to Miocene extension. Tertiary uplift generated gravity slide blocks of brecciated autochthonous lower Paleozoic rocks on the east side of the southern Hot Creek Range south of Hot Creek Canyon (Kleinhampl and Ziony, 1985).

The north-south–trending faults at Warm Springs may also be genetically related to the major east-west, left-lateral lineament postulated to exist between the southern Hot Creek Range and the Kawich Range to the south (Kleinhampl and Ziony, 1985). The down-to-the-south fault is marked by jasperoid deposits, and the fault trends subparallel to the postulated lineament (Graham, 1982). Normal faults are consistently at an oblique angle to the lineament, a geometry common to strike-slip movement.

Prior to Basin-and-Range extension in the Miocene, the area was buried by Tertiary volcanics. Documented caldera complexes exist to the east in the Hot Creek Valley–Pancake Range area, and an inferred volcanic center has been mapped in the Kawich Range, south of the study area (Stewart, 1980; Kleinhampl and Ziony, 1985). The welded tuff of the northern Kawich Range has been intruded by a large, irregular body of quartz latite porphyry (Kleinhampl and Ziony, 1985). Mining activity has defined a northwest-trending structural feature, which parallels the west-northwest–trending terminal fault in the southern Hot Creek Range as mapped by Graham (1982 and this chapter). Again, this feature may be related to the northwest-striking left-lateral shear zone between the Hot Creek Range and the Reveille Range (Kleinhampl and Ziony, 1985).

Immediately west of Warm Springs (Fig. 3), an isolated block of fractured and brecciated dolomitic limestone crops out. Fossil recovery from this isolated block was not successful, but brachiopods from sample WS-4-82 (Fig. 2), a wackestone in a similar lithostratigraphic position southwest of Warm Springs, indicate a Givetian age for these dolomitic limestones (Graham, 1982) (upper part of Interval 19 to Interval 21).

The dolomitic limestone west of Warm Springs is overlain by chert beds and is capped by a massive, dark brown, aphanitic jasperoid (Fig. 3). Jasperoid deposits form prominent, blocky cliffs stretching along the southern, western, and eastern flanks of the range as well as forming a formidable wall south of Rawhide Mountain.

The isolated outcrops of Denay limestone are fractured by Mesozoic and Cenozoic tectonic activity but are not extensively folded. Localized folding within the poorly exposed chert cannot be traced on a regional basis and is not present in Empire Canyon or Hot Creek Canyon. This analysis of folding, combined with the complex structural history of the area, does not support allochthonous emplacement of the Middle and Upper Devonian rocks at Warm Springs.

CONCLUSION

The succession of Upper Devonian–Mississippian facies at Warm Springs is not allochthonous but represents a nearly continuous record of in situ deposition. Outer-shelf basin lithotopes in the Devonian Denay Limestone are overlain by deep marine, organic-rich, bedded cherts deposited in a starved basin. The upper part of the chert unit is interbedded with overlying Lower Mississippian limestones deposited in a foreland trough developing in front of the RMT. Conodonts collected from the measured sections indicate a Kinderhookian-Osagean age for these limestone deposits. The facies represent deep-water sedimentation typical of basal Mississippian units described by Poole and Sandberg (1977, 1991) and Johnson and Pendergast (1981).

Tongues of limestone turbidites containing Mississippian conodonts and reworked Devonian conodonts are interbedded with the hemipelagic limestones as part of the basin-fill sequence. The reworked Devonian fauna may have originated to the east of Warm Springs where Upper Devonian rocks were subaerially exposed during a sea-level fall (Johnson et al., 1985, 1986, 1991). Elsewhere, inverted clast stratigraphy is reported in the Tripon Pass Limestone.

Paleozoic deformation, if a principal feature in the southern Hot Creek Range, has been overprinted by Mesozoic and Cenozoic tectonics. Faulting may have been caused by pluton emplacement, Cenozoic extensional tectonics, or left-lateral shear. The chert, rhythmically bedded below planar bedded Mississippian limestone, is only locally folded. Within the folds, the chert beds are coherent. The transition from chert to Lower Mississippian limestones in the southern Hot Creek Range is the regionally conformable offshore contact represented by the regional Upper Devonian–Mississippian unconformity to the east.

Johnson and Pendergast (1981; dashed line in Fig. 1) swept the trace of the RMT from the western edge of the northern Antelope Range southwestward to Dobbin Summit. Such a reconstruction represented a significant shift from previous interpretations of RMT geometry. The lack of thrust-related rocks at Warm Springs and the succession of Devonian to Mississippian deep-water lithotopes support their interpretation. The toe of the RMT appears to continue in a southwestward direction from Dobbin Summit in the Monitor Range and is located west of the southern Hot Creek Range.

ACKNOWLEDGMENTS

This chapter is a compilation of field studies and discussions with Jess Johnson since 1980. He did not get a chance to read the present manuscript. To his students, Jess was a brilliant, compassionate, and caring mentor who made the study of geology a lively, active science. He was also responsible for identifying the brachiopods from the Warm Springs collections. G. Klapper identified the conodonts from the Warm Springs collections listed in Graham (1982). E. Moores, M. A. Murphy, P. J. Noble, and F. G. Poole provided useful reviews.

REFERENCES CITED

Berger, W. H., and Winterer, E. L., 1974, Plate stratigraphy and the fluctuating carbonate line, *in* Hsü, K. J., and Jenkyns, H. C., eds., Pelagic sediments: On land and under the sea: International Association of Sedimentologists Special Publication 1, p. 11–48.

Coles, K. S., 1991, Depositional provinces in Nevada during the Famennian, *in* Cooper, J. D., and Stevens, C. H., eds., Paleozoic paleogeography of the western United States—II: Society of Economic Paleontologists and Mineralogists, Pacific Section, v. 67, p. 747–766.

Collinson, C., Rexroad, C. B., and Thompson, T. L., 1971, Conodont zonation of the North American Mississippian, *in* Sweet, W. C., and Bergström, S. M., eds., Symposium on conodont biostratigraphy: Geological Society of America Memoir 127, p. 353–394.

Desborough, G. A., Poole, F. G., Hose, R. K., and Green, G. N., 1987, Metalliferous oil shale in the Upper Devonian Gibellini facies of the Woodruff Formation in the southern Fish Creek Range, Nevada, *in* Bush, A. L., ed., Contributions to mineral resources research, 1984: U.S. Geological Survey Bulletin 1694-H, p. 91–104.

Dunham, R. J., 1962, Classification of carbonate rocks according to depositional texture, *in* Ham, W. E., ed., Classification of carbonate rocks—A Symposium: American Association of Petroleum Geologists Memoir 1, p. 108–121.

Ekren, E. B., Rogers, C. L., and Dixon, G. L., 1973, Geologic and bouger gravity map of the Reveille quadrangle, Nye County, Nevada: U.S. Geological Survey Map I-806, scale 1:48,000.

French, D., Kleinhampl, F., Jensen, M. L., and Tschanz, C. M., 1979, Road log—Ely to Las Vegas, *in* Newman, G. W., and Goode, H. D., eds., 1979 Basin and Range Symposium: Denver, Colorado, Rocky Mountain Association of Geologists and Utah Geological Association, p. 651–652.

Garrison, R. E., and Fischer, A. G., 1978, Deep-water limestones and radiolites of the Alpine Jurassic, *in* Cys, J. M., and Mazzullo, S. J., eds., Sedimentary processes: Depositional processes in ancient carbonates: Society of Economic Paleontologists and Mineralogists Reprint Series 7, p. 192–228.

Goebel, K. A., 1991, Paleogeographic setting of Late Devonian to Early Mississippian transition from passive to collisional margin, Antler foreland, Eastern Nevada and Western Utah, *in* Cooper, J. D., and Stevens, C. H., eds., Paleozoic paleogeography of the western United States—II: Society of Economic Paleontologists and Mineralogists, Pacific Section, v. 67, p. 401–418.

Graham, J. P., 1982, Devonian and Mississippian stratigraphy of the southern Hot Creek Range, Nye County, Nevada [M.S. thesis]: Corvallis, Oregon State University, 125 p.

Gutschick, R. C., Sandberg, C. A., and Sando, W. J., 1980, Mississippian shelf margin and carbonate platform from Montana to Nevada, *in* Fouch, T. D., and Magathan, E. R., eds., Paleozoic paleogeography of the west-central United States: Society of Economic Paleontologists and Mineralogists, Rocky Mountain Section, Rocky Mountain Paleogeography Symposium 1, Denver, Colorado, p. 111–128.

Johnson, J. G., 1977, Lower and Middle Devonian faunal intervals in central Nevada, based on brachiopods, *in* Murphy, M. A., Berry, W. B. N., and Sandberg, C. A., eds., Western North America: Devonian: University of California, Riverside Campus Museum Contribution 4, p. 16–32.

Johnson, J. G., and Pendergast, A., 1981, Timing and mode of emplacement of the Roberts Mountains allochthon, Antler orogeny: Geological Society of America Bulletin, v. 92, p. 648–658.

Johnson, J. G., and Sandberg, C. A., 1977, Lower and Middle Devonian continental-shelf rocks of the western United States, *in* Murphy, M. A., Berry, W. B. N., and Sandberg, C. A., eds., Western North America: Devonian: University of California, Riverside Campus Museum Contribution 4, p. 121–143.

Johnson, J. G., and Visconti, R., 1992, Roberts Mountains thrust relationships in a critical area, northern Sulphur Spring Range, Nevada: Geological Society of America Bulletin, v. 104, p. 1208–1220.

Johnson, J. G., Klapper, G., and Trojan, W. R., 1980, Brachiopod and conodont successions in the Devonian of the northern Antelope Range, central Nevada: Geologica et Palaeontologica, v. 14, p. 77–116.

Johnson, J. G., Klapper, G., and Sandberg, C. A., 1985, Devonian eustatic fluctuations in Euramerica: Geological Society of America Bulletin, v. 96, p. 567–587.

Johnson, J. G., Klapper, G., and Sandberg, C. A., 1986, Late Devonian eustatic cycles around margin of Old Red Continent: Annales de la Société géologique de Belgique, v. 109, p. 141–147.

Johnson, J. G., Sandberg, C. A., and Poole, F. G., 1991, Devonian lithofacies of western United States, *in* Cooper, J. D., and Stevens, C. H., eds., Paleozoic paleogeography of the western United States—II: Society of Economic Paleontologists and Mineralogists, Pacific Section, v. 67, p. 83–106.

Klapper, G., and Johnson, J. G., 1980, Endemism and dispersal of Devonian conodonts: Journal of Paleontology, v. 54, p. 400–455.

Kleinhampl, F. J., and Ziony, J. I., 1985, Geology of northern Nye County, Nevada: Reno, Nevada Bureau of Mines and Geology Bulletin 99A, 172 p.

Murphy, M. A., 1977, Middle Devonian rocks of central Nevada, *in* Murphy, M. A., Berry, W. B. N., and Sandberg, C. A., eds., Western North America: Devonian: University of California, Riverside Campus Museum Contribution 4, p. 190–199.

Murphy, M. A., Power, J. D., and Johnson, J. G., 1984, Evidence for Late Devonian movement within the Roberts Mountains allochthon, Roberts Mountains, Nevada: Geology, v. 12, p. 20–23.

Poole, F. G., 1974, Flysch deposits of Antler foreland basin, western United States, *in* Dickinson, W. R., ed., Tectonics and sedimentation: Society of Economic Paleontologists and Mineralogists Special Publication 22, p. 58–82.

Poole, F. G., and Claypoole, G. E., 1984, Petroleum source-rock potential and crude-oil correlation in the Great Basin, *in* Woodward, J., Meisner, F. F., and Clayton, J. L., eds., Hydrocarbon source rocks of the greater Rocky Mountain Region: Denver, Colorado, Rocky Mountain Association of Geologists, p. 179–229.

Poole, F. G., and Sandberg, C. A., 1977, Mississippian paleogeography and tectonics of the western United States, *in* Stewart, J. H., Stevens, C. H., and Fritsche, A. E., eds., Paleozoic paleogeography of the western United States: Society of Economic Paleontologists and Mineralogists, Pacific Section, Pacific Coast Paleogeography Symposium 1, Bakersfield, California, p. 67–85.

Poole, F. G., and Sandberg, C. A., 1991, Mississippian paleogeography and conodont biostratigraphy of the western United States, *in* Cooper, J. D., and Stevens, C. H., eds., Paleozoic paleogeography of the western United States—II: Society of Economic Paleontologists and Mineralogists, Pacific Section, v. 67, p. 107–136.

Poole, F. G., and Sandberg, C. A., 1992, Roberts Mountains allochthon and overlapping Lower Mississippian syntectonic foredeep deposits in the southern Hot Creek Range, central Nevada: Geological Society of America Abstracts with Programs, v. 24, no. 6, p. 57.

Poole, F. G., Sandberg, C. A., and Boucot, A. J., 1977, Silurian and Devonian paleogeography of the western United States—I, *in* Stewart, J. H., Stevens, C. H., and Fritsche, A. E., eds., Paleozoic paleogeography of the western United States: Society of Economic Paleontologists and Mineralogists, Pacific Section, Pacific Coast Paleogeography Symposium 1, Bakersfield, California, p. 39–66.

Potter, E. C., 1975, Paleozoic stratigraphy of the northern Hot Creek Range, Nye County, Nevada [M.S. thesis]: Corvallis, Oregon State University, 129 p.

Roberts, R. J., Hotz, P. E., Gilluly, J., and Ferguson, H. G., 1958, Paleozoic rocks of north-central Nevada: American Association of Petroleum Geologists Bulletin, v. 42, p. 2813–2857.

Sandberg, C. A., 1979, Devonian and Lower Mississippian conodont zonation of the Great Basin and Rocky Mountains, *in* Sandberg, C. A., and Clark, D. L., Conodont biostratigraphy of the Great Basin and Rocky Mountains: Brigham Young University Geology Studies, v. 26, pt. 3, p. 87–106.

Sandberg, C. A., Zeigler, W., Leuteritz, K., and Brill, S. M., 1978, Phylogeny, speciation, and zonation of *Siphonodella* (Conodonta, Upper Devonian and Lower Carboniferous): Newsletters on Stratigraphy, v. 7, p. 102–120.

Sandberg, C. A., Gutschick, R. C., Johnson, J. G., Poole, F. G., and Sando, W. J., 1983, Middle Devonian to Late Mississippian geologic history of the Overthrust belt region, western United States: Denver, Rocky Mountain Association of Geologists, Geologic studies of the Cordilleran thrust belt, v. 2, p. 691–719.

Smith, J. F., Jr., and Ketner, K. B., 1968, Devonian and Mississippian rocks and the date of the Roberts Mountains thrust in the Carlin-Piñon Range area, Nevada: U.S. Geological Survey Bulletin 1251-I, p. I1–I18.

Stewart, J. H., 1980, Geology of Nevada: Nevada Bureau of Mines and Geology Special Publication 4, 136 p.

Stewart, J. H., and Poole, F. G, 1974, Lower Paleozoic and uppermost Precambrian Cordilleran miogeocline, Great Basin, western United States, *in* Dickinson, W. R., ed., Tectonics and sedimentation: Society of Economic Paleontologists and Mineralogists Special Publication 22, p. 28–57.

Trojan, W. R., 1978, Devonian stratigraphy and depositional environments of the northern Antelope Range, Eureka County, Nevada [M.S. thesis]: Corvallis, Oregon State University, 134 p.

Wise, D. C., 1976, Paleozoic geology of the Dobbin Summit–Clear Creek area, Monitor Range, Nye County, Nevada [M.S. thesis]: Corvallis, Oregon State University, 137 p.

Manuscript Accepted by the Society February 27, 1997

Printed in U.S.A.

Geological Society of America
Special Paper 321
1997

European Province Late Silurian brachiopods from the Ciudad Victoria area, Tamaulipas, northeastern Mexico

A. J. Boucot and Robert B. Blodgett
Department of Zoology, Oregon State University, Corvallis, Oregon 97331
John H. Stewart
U.S. Geological Survey, MS 901, 345 Middlefield Road, Menlo Park, California 94025

ABSTRACT

Late Silurian (late Wenlock–early Ludlow, and late Ludlow–early Pridoli) brachiopods from siliciclastic rocks of the Ciudad Victoria area, Tamaulipas, northeastern Mexico, are of European Province rather than North American Province affinities and have close affinities to brachiopods in northern South America. The Silurian sequence is overlain by Early Mississippian (Early Osagean), Pennsylvanian, and Permian strata of North American biogeographic affinities. These relations suggest that the Ciudad Victoria area may originally have been part of South America but by Early Mississippian time have been close enough to North America for faunal interchange. The European Province brachiopods from the Ciudad Victoria area are *Isorthis (Ovalella)*, *Mclearnitesella*, *Baturria*, and *Amphistrophiella (Sulcastrophiella)*. New taxa described are *A. (Sulcastrophiella) stinnesbecki* n. subgen., n. sp., and *Amerista carillobravoi* n. gen., n. sp.

INTRODUCTION

Silurian rocks, first noted in the Ciudad Victoria region (Fig. 1) by Carillo-Bravo (1959, 1961), have been regarded as late Llandovery–Ludlow (Berry and Boucot, 1970). This age assignment was maintained by Harper and Boucot (1978, Pl. 34, Figs. 6–9), who illustrated an undescribed new species of *Amphistrophiella* (*Amsdenostrophiella*) from USNM locality 13005 in the Ciudad Victoria region. In 1993, the authors of this chapter spent several weeks mapping the area, focusing on the fossiliferous Silurian strata. The resultant relatively large fossil collections enable better understanding of the paleoecology and biogeographic affinities of the Silurian faunas.

The Ciudad Victoria Silurian rests unconformably on Precambrian crystalline basement. It has biogeographic affinities with the European rather than the North American Province (Boucot, 1990). This evidence suggests that the Silurian rocks were deposited far from the North American Platform, perhaps in northern South America where the Silurian fauna is similar to that in the Ciudad Victoria area. In contrast, the fauna in the overlying Lower Mississippian (Early Osagean) rocks have strong affinities to North American faunas, in particular to that of the Borden delta (Waverly) of central and eastern Ohio (John L. Carter, personal communication, 1996), and uncertain affinities to poorly known Lower Mississippian faunas, from other parts of Mexico and South America (John L. Carter, personal communication, 1996). These relations suggest that what is now the Ciudad Victoria area may have originally been part of South America but by Mississippian time have been close enough to North America for faunal interchange.

Boucot, A. J., Blodgett, R. B., and Stewart, J. H., 1997, European Province Late Silurian brachiopods from the Ciudad Victoria area, Tamaulipas, northeastern Mexico, *in* Klapper, G., Murphy, M. A., and Talent, J. A., eds., Paleozoic Sequence Stratigraphy, Biostratigraphy, and Biogeography: Studies in Honor of J. Granville ("Jess") Johnson: Boulder, Colorado, Geological Society of America Special Paper 321.

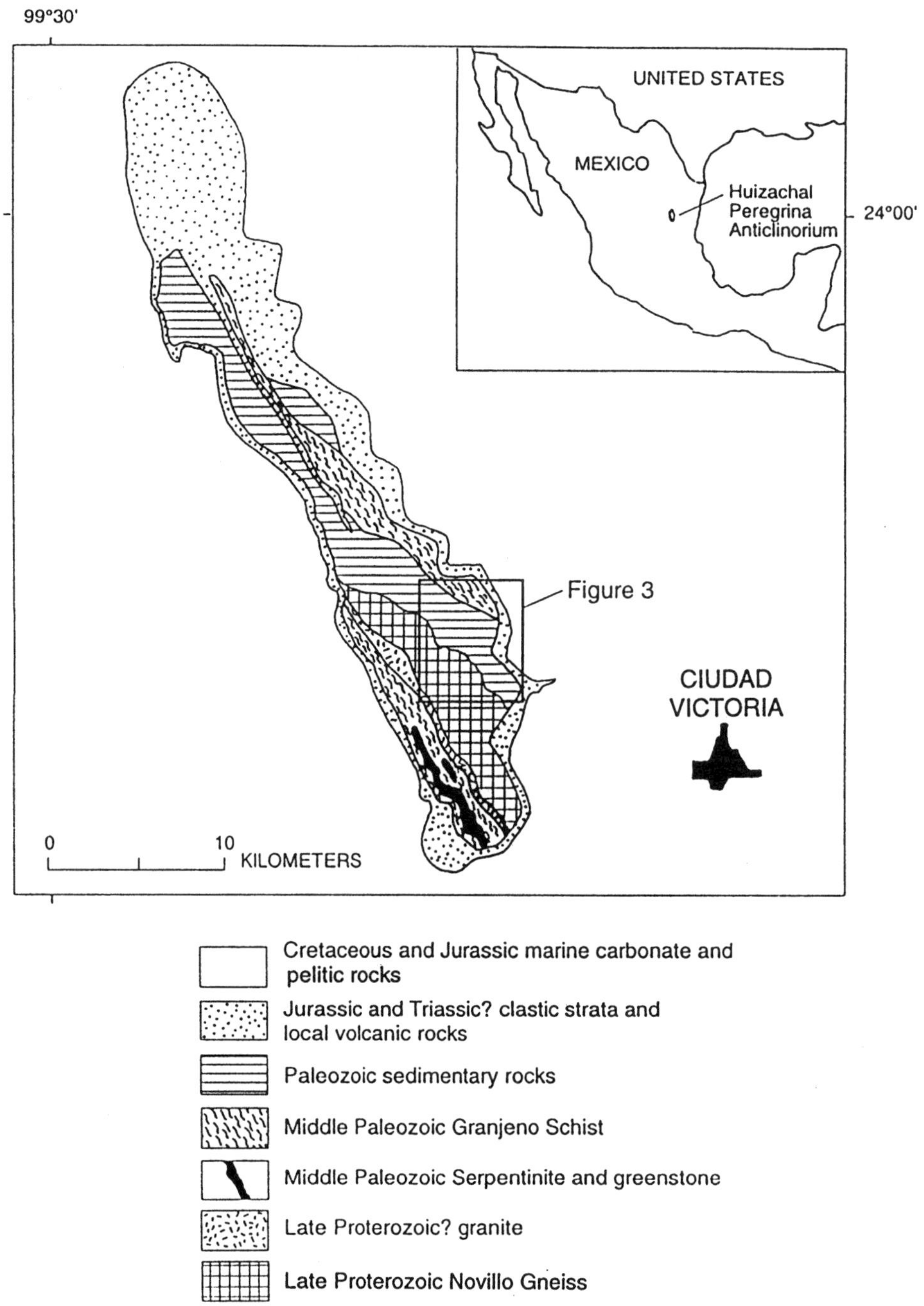

Figure 1. Generalized geologic map of the Huizachal-Peregrina anticlinorium, after Ramirez-Ramirez (1992).

GEOLOGY

The Novillo Gneiss consists of an anorthositic-quartzofeldspathic orthogneissic complex and a paragneissic sequence of quartzofeldspathic gneiss with some layers of marble and graphic schist (Ortega-Gutierrez, 1978). It has been dated by K-Ar methods (Fries et al., 1962; Denison et al., 1971) and by Sm-Nd methods (Ruiz et al., 1988) as between 860 and 930 Ma, by Rb-Sr methods as 1140 ± 80 Ma (Garrison et al., 1980), and by U-Pb zircon methods as 1018 ± 3 Ma (Silver et al., 1994).

In tectonic contact with the Novillo Gneiss, or separated from it by a granite, is the Granjeno Schist. The latter consists of low-grade mica-schist interbedded with abundant greenstone, metachert, and rare carbonate rocks (Carillo-Bravo, 1961; Cserna and Ortega-Gutierrez, 1977, 1978; Garrison et al., 1980; Ramirez-Ramirez, 1992; Ortega-Gutierrez et al., 1993). It includes large bodies of serpentinite with relic textures and structures indicating it was originally a layered ultramafic complex (Ortega-Gutierrez et al., 1993). Seven isotopic ages on the Granjeno Schist (Garrison, 1978; Ramirez-Ramirez, 1992, p. 113) range from 257 to 330 Ma, indicating a mid-Paleozoic metamorphic age for the unit. Pillow basalt, olistolithic blocks (Ramirez-Ramirez, 1992), and metachert suggest that the Granjeno Schist was originally deposited in a deep ocean basin adjacent to a convergent continental margin (Ramirez-Ramirez, 1992; Ortega-Gutierrez et al., 1993).

Relatively unmetamorphosed Paleozoic rocks rest unconformably on the Novillo Gneiss (Carillo-Bravo, 1959, 1961; Murray et al., 1960; Tellez-Giron, 1970; Conklin et al., 1974; Lopez-Ramos, 1981; Gursky and Ramirez-Ramirez, 1986; Gursky and Michalzik, 1989; Ramirez-Ramirez, 1992). These consist of Silurian, Lower Mississippian, Lower and Middle Pennsylvanian, and Permian sedimentary rocks, and a latest Mississippian or earliest Pennsylvanian rhyolite (Fig. 2). Previous reports of Cambrian?, Ordovician?, and Devonian rocks (Carillo-Bravo, 1961) have proven incorrect (Ramirez-Ramirez, 1992).

The Silurian Cañón de Caballeros Formation (Carillo-Bravo, 1961) rests unconformably on the Novillo Gneiss. It is divided into three members: an unnamed basal member, the Santa Ana Limestone member, and an unnamed upper member. The basal member is about 10 m thick and consists of interstratified quartz-rich fine-grained to very coarse-grained sandstone to conglomerate overlain by silicic volcanic arenite. The latter is fine- to coarse-grained and typically composed of 40% finely crystalline to microcrystalline lithic grains, 35% embayed angular to rounded subhedral to euhedral quartz, 15% angular to subangular to anhedral potassium feldspar, sparse siltstone clasts, and about 10% matrix. Most of the finely crystalline to microcrystalline lithic grains are probably of volcanic material. A few of them are definitely rhyolite or felsite. The euhedral to subhedral, unabraded, embayed grains of quartz indicate a nearby silicic volcanic source, apparently from an unconsolidated or relatively unconsolidated tuff, presumably coeval with deposition of the Cañón de Caballeros Formation.

The Santa Ana Limestone member of the Cañón de Caballeros Formation is named for outcrops along the Santa Ana River in Cañón de Caballeros. It is only 2.5 m thick but is a distinctive bioclastic grainstone commonly with abundant coarse grains and granules of polycrystalline and monocrystalline quartz, apparently derived from metamorphic rocks and/or quartz veins like those common in the Novillo Gneiss. The upper member of the Cañón de Caballeros Formation is about 100 m thick and consists of shaley siltstone and very fine-grained sandstone and a few beds of fine- to medium-grained sandstone. The fine- to medium-grained sandstone contains abundant finely crystalline to microcrystalline lithic grains, some of which can be identified as rhyolitic or felsitic rock. Many of the finely crystalline to microcrystalline grains appear to be volcanic material. As discussed below, the Cañón de Caballeros Formation is inferred to be late Wenlock to Pridoli in age. Based on the presence of abundant, unabraded (or relatively unabraded) shallow-water brachiopods, trilobites, and corals, it is considered to be a shallow-water deposit.

The Vicente Guerrero Formation (Carillo-Bravo, 1961) overlies the Cañón de Caballeros Formation without apparent discordance. It consists of a lower member, about 30 m thick, of fine- to medium-grained quartzitic sandstone and minor amounts of conglomerate and an upper member of at least 50 m of siltstone and shale. The sandstone in the lower member consists of quartz and lithic arenite containing finely crystalline to microcrystalline lithic clasts; some are rhyolite felsite and other fine-grained volcanic rocks. Abundant unabraded brachiopods, gastropods, and corals indicate shallow water. Locally abundant brachiopods in the upper member indicate an Early Mississippian (early Osagean, Burlington) age (J. L. Carter, in Stewart et al., 1993).

The El Aserradero Rhyolite is the next younger formation in the Huizachal-Peregrina Canyon area; it is flow-banded and considered to be mainly a lava flow but may be intrusive locally. It was originally considered to be a Devonian novaculite, similar to the Devonian novaculite of west Texas (Flawn et al., 1961), but was shown to be a rhyolite by Gursky and Ramirez-Ramirez (1986) and Ramirez-Ramirez (1992). Field exposures are not adequate for conclusively establishing the age of the El Aserradero Rhyolite relative to the Silurian Cañón de Caballeros Formation or the Mississippian Vicente Guerrero Formation, but on the basis of stratigraphic position and the presence of coarse clasts of rhyolite in the Del Monte Formation, it appears to be pre–Del Monte Formation (Early and Middle Pennsylvanian). This age is confirmed by a lower intercept U-Pb zircon age of 334 ± 39 Ma for the rhyolite (Robert Lopez, written communication).

The Del Monte Formation is about 200 m thick (Carillo-Bravo, 1961) and consists of bioclastic grainstone with abundant fine- to medium-quartz sand. Higher in the formation limey sandstone and sandstone are more abundant; still higher, limey siltstone, silty limestone, and siltstone are dominant. The characteristically thick, graded beds suggest deposition by sand or debris flows in presumably deep water. The Del Monte Formation has been considered to be Early Pennsylvanian in age on the basis of fusulinids, corals, and goniatites (Murray et al., 1960; Carillo-Bravo, 1961), but on the basis of conodonts (Norman Savage, personal communication, 1995) may be as young as Middle Pennsylvanian (Desmoinesian). It rests with apparent unconformity on the El Aserradero Rhyolite and unconformably, locally with high angularity, on the Mississippian Vicente Guerrero Formation and the Silurian Cañón de Caballeros Formation. This relation suggests significant pre-Early to Middle Pennsylvanian deformation in the Huizachal-Peregrina anticlinorium. Such deformation is also suggested by the presence of a basal conglomerate with clasts up to 20 cm.

The youngest Paleozoic formation in the Huizachal-Peregrina anticlinorium is the 1,260-m-thick Lower Permian Guacamaya Formation (Carillo-Bravo, 1961; Gursky and Michalzik, 1989). It consists of graded, very fine to fine-grained sandstone interstratified with siltstone; it also contains fine- to coarse-grained volcaniclastic sandstone and locally coarse conglomerate rich in volcanic detritus. Being characterized mainly by

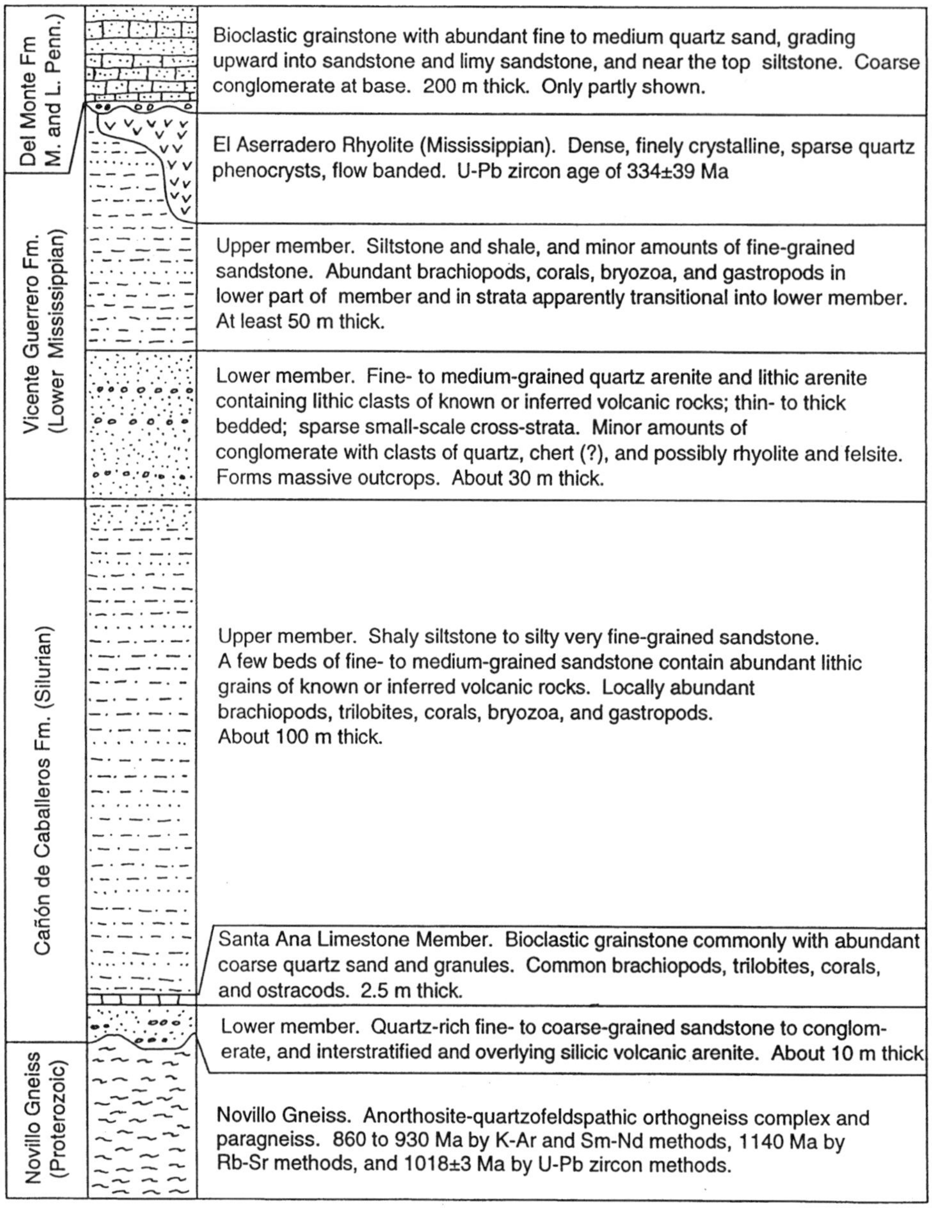

Figure 2. Stratigraphic column of late Proterozoic, Silurian, Lower Mississippian, and Lower and Middle Pennsylvanian rocks in the Huizachal-Peregrina anticlinorium.

fine-grained turbidite layers, it has been interpreted as a deep-water, distal flysch deposit (Gursky and Michalzik, 1989). Local coarse layers may indicate more proximal sources and higher energy environments. The Guacamaya Formation is in fault contact with the Del Monte Formation.

The Novillo Gneiss, Granjeno Schist, and the Paleozoic rocks in the Huizachal-Peregrina anticlinorium are unconformably overlain by folded but conspicuously less deformed Mesozoic rocks. These consist of Lower or Middle Jurassic siltstone, sandstone, and conglomerate (Imlay, 1938, 1943; Mixon et al., 1959; Carillo-Bravo, 1961; Belcher, 1979; Clark and Hopson, 1985; Fastovsky et al., 1987, 1995) and Upper Jurassic and Cretaceous carbonate and evaporitic rocks (Carillo-Bravo, 1961).

PALEONTOLOGY

We have restudied fossils collected by José Carillo-Bravo, R. B. Mixon, and David D. Johnson in the 1950s from the Cañón de Caballeros Formation (USNM locality numbers 12599, 13000, 13001, 13002, 13003, 13004, 13005) and made new collections at 11 localities (JS93-76, 79, 82-84, 86-88, 105, 116, 121) (Fig. 3). Fossils in these collections consist primarily of brachiopods (31 taxa); these are figured herein and, where relevant, discussed. Smaller numbers of trilobites, corals, and other organisms are also present. Norman Savage (personal communication, 1995) obtained a single distacodont conodont from a Cañón de Caballeros Formation limestone sample, but

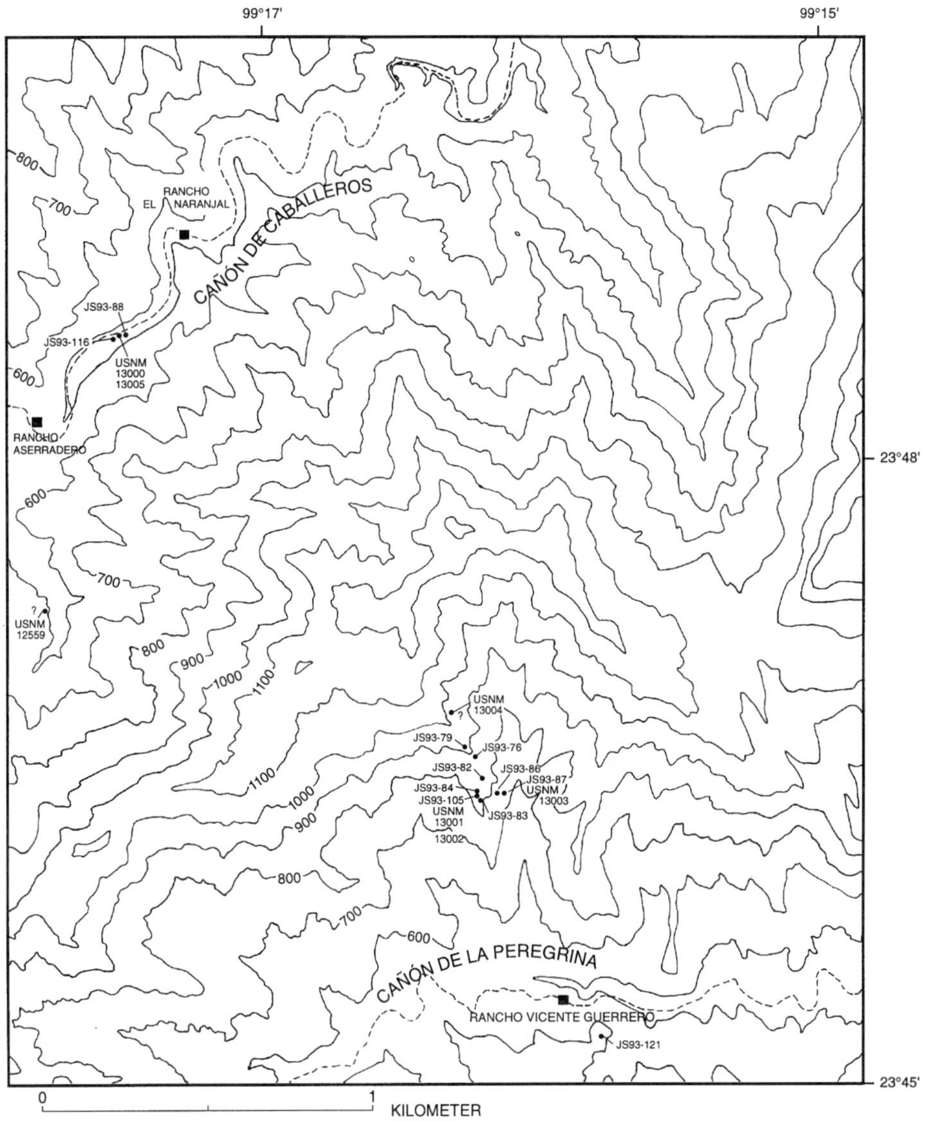

Figure 3. Location of fossil collections. Localities with JS93 prefix are collections made during this study. Localities with USNM (U.S. National Museum) prefix are collections made by other geologists (see list of localities). Exact position of some of the localities with USNM numbers is uncertain and shown with a query. Dashed lines in Cañón de Caballeros and Cañón de la Peregrina are unimproved dirt roads.

the specimen has little chronologic significance (Early Ordovician–Middle Devonian).

AGE

The few fossils obtained from the lower member of the Cañón de Caballeros Formation are suggestive of a Wenlock age. The presence of a costellate stricklandiid at locality JS93-82 (fauna listed in Appendix 2) is consistent with an early to mid-Wenlock age, but the particular form is not known elsewhere.

The limited collections from the Santa Ana Limestone member of the Cañón de Caballeros Formation (Table 1) accord with a late Wenlock–Ludlow age, the presence of *Sphaerirhynchia* in Peregrina Cañón indicating an age probably no older than late Wenlock. However, the occurrence of *Pentamerus* in the Santa Ana Limestone member in the Cañón de Caballeros Formation at locality JS93-88 accords best with a late Llandovery age; this genus is known only from the late Llandovery, except for interior and Appalachian parts of North America where it ranges into the early and middle Wenlock. J. M. Adrain (personal communication, 1995) comments that the presence of common *Stenopareia* in the limestones immediately above those yielding *Pentamerus* indicates an age no younger than Wenlock and more likely Llandovery. It is thus possible that the Santa Ana Limestone member in Cañón de Caballeros to the north might be slightly older than the outcrops

TABLE 1. BRACHIOPOD GENERA AND OTHER FOSSILS FROM THE LATE WENLOCK–LUDLOW SANTA ANA LIMESTONE MEMBER OF THE CAÑÓN DE CABALLEROS FORMATION*

Taxa	Localities							
	13001	13002	13004	76	83	88	105	Stinnesbeck Coll
Orbiculoidea sp						1		
Resserella sp.						6		3
Dalejina sp.							1	
Isorthis sp.								1?
Dicoelosia sp.			x					
Dalmanellids	x		x					
Leptaena "rhomboidalis"	x	x				1		4
Strophoprion sp.		x			1?			
Stropheodontid	x	x	x			3		3
Orthotetacid			x					
Chonetid, coarse ribbed			x					
Pentamerus sp.						5		
Sphaerirhynchia sp.							10	
"Rhynchonella" aff. *stricklandi*	x					1		2
Rhynchonellid					1			
Atrypa "reticularis"	x	x			2	9		17
Meristina sp.	x							
Rostrospiroid			x			1		
Eospirifer sp.	x					1		1
Cyrtia sp.								1
Delthyris sp.							2	
Howellella sp.		x				2?		
Brachiopods indet.	x					11		
Pelmatozoan columnals							x	
Cornulites sp.						1		1
Stenopareia n. sp.						x		
Dalmanitid n. sp.						x		
Dalmanitid							1	
Calymenid						x		
Proetid						x		
Trilobites		x					1	15
Bellerophontid					1			
Rugose corals	x			1	1		3	4
Favositid								1
Pterineoid								2

*x = present but not counted.

in Peregrina Cañón, that is, that the member is diachronous or perhaps that two limestone units are involved. At present, however, we conclude that only one rock unit is involved and that its age needs more investigation.

Two Silurian faunas occur in the upper member of the Cañón de Caballeros Formation. The older fauna (Table 2) includes *Macropleura* and *Sphaerirhynchia*, suggesting a late Wenlock–early Ludlow age; this age is supported by the presence of *Amphistrophiella (Sulcastrophiella) carillobravoi* n. sp. known previously as *Amphistrophia* sp. B (Boucot et al., 1972), from late Wenlock–Ludlow horizons of the Merida Andes. The coarsely plicate *Delthyris* from the younger fauna is a form known from the pre-late Pridoli, whereas *D. (Quadrifarius)* is present in the late Pridoli–Lochkovian.

The presence of *Baturria* in the upper member at JS93-87, float block 3 (Table 3), indicates a late Ludlow–Pridoli age. As mentioned in Boucot (1975), the *Baturria* is more like forms from the Spanish Ludlow–Pridoli than Devonian species. The first collection with *Baturria* (USNM 13003, Ca 1775) was assigned to the Pridoli–Lochkovian largely because the associated form of *Delthyris* was assigned to *D. (Quadrifarius)*. Reconsideration of this identification indicates it was erroneous; the material is merely a large, pre–late Pridoli–Lochkovian type of typical *Delthyris* that can range downwards into the late Ludlow–early Pridoli. Additionally, the presence of a large *Dalejina* in this fauna further contrasts it with late Wenlock–early Ludlow faunas.

TABLE 2. BRACHIOPOD GENERA AND OTHER FOSSILS FROM THE LATE WENLOCK–LUDLOW LOWER PART OF THE UPPER MEMBER OF THE CAÑÓN DE CABALLEROS FORMATION*

	Localities									
Taxa	121	121B	121A	79	86	84	87	13005	12599	13000
Craniops sp.	4		1	1				5		6
Skenidioides sp.								2		
Isorthis sp.			1							x
I. (Ovalella) sp.	17			10	4	1	3	1		
Dalejina sp.	8		3	1	4			6	4	2
Resserella sp.	26	1	11	25	3	2	1	147	9	31
Leptaena sp.	20		2	6	2	1	3	24	4	8
Leptostrophid										11
Mclearnitesella sp.	15		3		4			24	1	
Strophoprion sp.	19		3	1	7	2		79		
A. (Sulcastrophiella) sp.	9	2		4		1		42	x	x
Mesopholidostrophia sp.	7							5		
Stropheodontid	1									
Coolinia sp.	41			8	10	3	4	43	5	
Gypidula sp.				28					1	6
Sphaerirhynchia sp.	8		1	14	1	2	2	13	6	
"Rhynchonella" aff. *stricklandi*					1					
Rhynchonellid	4			3	8		3			
Atrypa sp.	23		4	43	1	3		43	7	13
Meristina sp.	2?	1		7	1?	2		19	2	
Amerista sp.	1	1	1?	3	7	1	1		1?	
Nucleospira sp.	35	2	3	7	1			56		
Cyrtia sp.			1					8		
Eospirifer sp.	6	1	4	12	14	2		67	3	2
Macropleura sp.	2							1?		
Delthyris sp.	50	2	2	69	19	12	4	98	1?	
Howellella sp.										40
Brachiopod indet.	1									
Trilobite indet.				1						
Exallaspis n. sp.	2		1							
Bryozoan							1			
Fenestellids	4		1	4						
Ptilodictyid						1				
Tetracoral				1			1			
Tabulate coral			1							
Gastropod					x					
Poleumita sp.	x			3						
Tropidodiscus sp.	x		1		x					
Euomphalopterus sp.	x				x					

*x = present but not counted.

ECOLOGY

The fauna from the Santa Ana Limestone member at locality JS93-88 represents a rough water *Pentamerus* Community assigned to Benthic Assemblage (B.A.) 3. Dominance of *Atrypa* in Stinnesbeck's collection from the same member suggests a discrete ecological setting, although assignment to Benthic Assemblage 3 is likewise indicated. The latter fauna contrasts with collections from the upper member of the Cañón de Caballeros Formation, the small size of the collection notwithstanding. The abundance of trilobites in Stinnesbeck's collection and in JS93-88 also indicates ecologic contrast when compared with the collections from mudstones. Collections from the upper member of the Cañón de Caballeros Formation also belong to benthic Assemblage 3 but to the *Janius* Community Group, representing high-diversity, level-bottom, moderately turbulent environments.

BIOGEOGRAPHY

Discrimination of European Province from North American Province faunas is based on hundreds of collections and several hundred genera; they represent a broad spectrum of environments and community types. Boucot (1990) listed genera typical of the European and North American provinces; *Baturria*,

TABLE 3. BRACHIOPOD GENERA FROM THE LATE LUDLOW–EARLY PRIDOLI INTERVAL OF THE UPPER MEMBER OF THE CAÑÓN DE CABALLEROS FORMATION*

Taxa	Localities 13003	87 (float block 3)
Isorthis sp.	12	
I. (Ovalella) sp.	x	
Large *Dalejina* sp.	6	1
Baturria mexicana	46	2
Leptaena "rhomboidalis"	11	x
Leptostrophia sp.	8	
Strophoprion cf. *euglypha*	6	
Orthotetacid	6	1
Uncinuloid	27	
Rhynchonellid	9	x
Amerista? (brachial valve)		x
Nucleospira sp.	36	
Large *Delthyris* sp.	16	x

*x = present but not counted.

Mclearnitesella, *Ovalella*, *Plectotreta*, and *Eatoniodes* should be added to the list for the former and *Sulcatina*, *Neumanella*, *Navispira*, and *Onychotreta* to the list for the latter. Although the Silurian faunas from the Merida Andes of Venezuela (Boucot et al., 1972) and from the Ciudad Victoria area of Mexico were obtained from very limited areas and similar environments and consist of genera that are relatively cosmopolitan, extending through both provinces of the Atlantic Region, provincially distinctive elements are present. The late Wenlock–early Ludlow portion of the upper member of the Cañón de Caballeros Formation includes *Mclearnitesella* and abundant *Isorthis (Ovalella)*, both indicative of the European Province. *Amphistrophiella (Sulcastrophiella) carillobravoi* is known elsewhere only from the Merida Andes Silurian of Venezuela, a region also assigned to the European Province. The endemic new genus *Amerista* is not helpful biogeographically. Because no deeper-water, Benthic Assemblage 4 or 5 faunas are present, it is possible that additional endemic European- or Venezuelan-type genera might have been present in contexts no longer available for sampling in this region of northeastern Mexico.

TAXONOMIC NOTES

Although the brachiopods from Ciudad Victoria are not particularly well preserved nor biologically unique, the fauna is biogeographically significant and is therefore illustrated. Formal descriptions are limited to the new genus *Amerista* and the new subgenus *Sulcastrophiella*. Unless specified, all illustrated specimens were collected by the authors in 1993. The arrangement follows the 1965 Treatise on Invertebrate Paleontology (Williams et al., 1965), except for the stropheodontids, for which Harper and Boucot (1978) is followed.

Family SKENIDIIDAE Kozlowski, 1929

Genus *Skenidioides* Schuchert and Cooper, 1931

***Skenidioides* sp.**
Figure 4: 4–5

A few specimens were obtained from the late Wenlock–Ludlow portion of the upper member of the Cañón de Caballeros Formation at one locality; they indicate proximity to the B.A. 4-5 boundary.

Family DALMANELLIDAE Schuchert, 1913

Genus *Isorthis* Kozlowski, 1929

I. (Ovalella) Walmsley and Boucot, 1975

***Isorthis (Ovalella)* sp.**
Figure 4: 6–9

The upper member of the Cañón de Caballeros Formation yielded

Figure 4 (on facing page). Abbreviations accompanying the catalog numbers of figured specimens in Figures 4 through 10 are as follows: IGM - Instituto de Geologia, Universidad Nacional Autonoma de Mexico, Ciudad Universitaria, Mexico, D.F.; USNM - Department of Paleobiology, National Museum of Natural History, Washington, DC. 1-2, *Craniops* sp., impressions of brachial valve interior and exterior, ×4 and ×5, locality JS-93-121, IGM 6857, 6858. 3: *Orbiculoidea* sp., impression of pedicle valve exterior, ×3, locality JS-93-88, IGM 6859. 4–5: *Skenidioides* sp., impression of pedicle valve exterior and interior, ×6, locality USNM 13005, IGM 6860a, b. 6–7: *Isorthis (Ovalella)* sp., impression of pedicle valve interior, brachial valve interior, ×3, locality JS-93-121, IGM 6861, 6862. 8–9: *Isorthis (Ovalella)* sp., impressions of pedicle valve and brachial valve interiors, ×3, locality JS-93-87-1. IGM 6863. 10–11: *Isorthis* sp., impressions of pedicle valve interiors, ×2, locality USNM 13003, IGM 6865, 6866. 12–17: *Dalejina* sp. 1, impression of pedicle valve interior, rubber replica of pedicle valve interior, impression of pedicle valve exterior, rubber replica of pedicle valve exterior, impression of brachial valve interior, rubber replica of brachial valve interior, ×2, ×2, ×2, ×2, ×3, ×3 respectively, from localities JS-93-86, IGM 6867, 6867, 6867, 6867, 6868, 6868 respectively. 18–19: *Dalejina* sp. 2, impressions of brachial valve interior, ×1.5, locality USNM 13003, IGM 6869, 6879. 20–23: *Resserella* sp., three impressions of brachial valve interiors and impression of brachial valve exterior ×2, ×1.5, ×2, ×2 respectively, locality JS-93-121, IGM 6871, 6872, 6873, 6873 respectively. 24–26: *Resserella* sp., impression of pedicle valve exterior, rubber replica of pedicle valve exterior, and impression of pedicle valve interior, all ×2, JS-93-79, IGM 6874a, 6874a, 6874b. 27: *Resserella* sp., impression of pedicle valve interior, ×2, locality USNM 12599, IGM 6875. 28: *Dicoelosia* sp., exterior of pedicle valve, ×3, locality USNM 13004, IGM 6876. 29–32: *Leptaena "rhomboidalis"* (Wilckens), impression of pedicle valve exterior, rubber replica of pedicle valve exterior, impression of pedicle valve exterior, rubber replica of pedicle valve exterior, all × 1.5, locality JS-93-79, IGM 6877, 6877, 6878, 6878 respectively. 33–34: *Leptaena "rhomboidalis"* (Wilckens), impressions of brachial and pedicle valve interiors, ×1.5, locality JS-93-121, IGM 6879, 6880. 35: *Leptaena "rhomboidalis"* (Wilckens), impression of pedicle valve exterior, ×1.5, locality USNM 13005, IGM 6881. 36: *Leptaena "rhomboidalis"* (Wilckens), impression of pedicle valve interior, ×1.5, locality USNM 13003, IGM 6882.

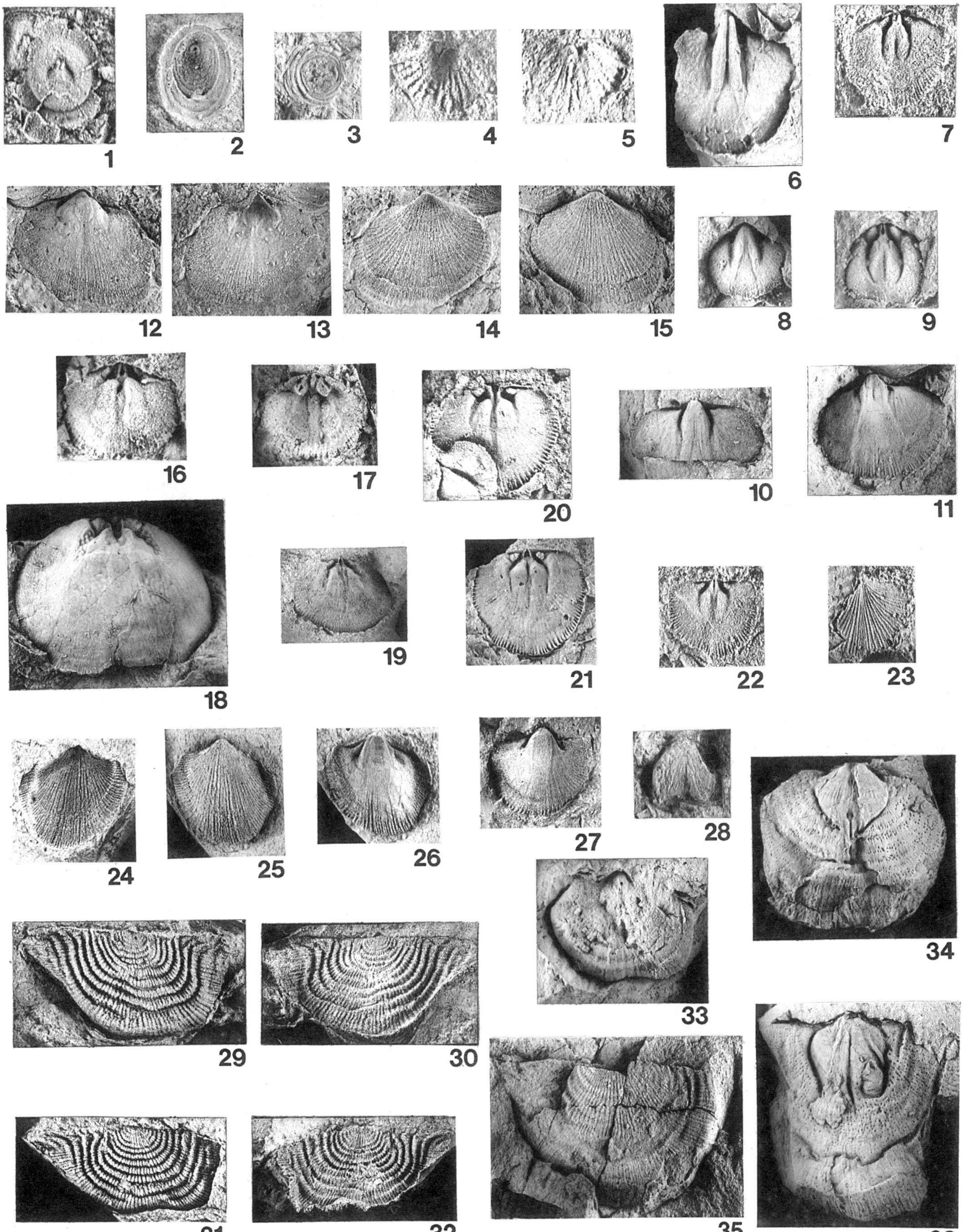
1
2
3
4
5
6
7
12
13
14
15
8
9
16
17
20
10
11
18
19
21
22
23
24
25
26
27
28
34
33
29
30
35
36
31
32

a moderate number of *Isorthis* specimens, having the characteristic pedicle valve muscle field of the European Province subgenus *Ovalella*. Several specimens from JS-93-121A and from USNM locality 13003 could not be identified subgenerically (including those illustrated in Fig. 4:10–11) but probably belong to *Ovalella*.

Genus *Dalejina* Havlí ek, 1956

***Dalejina* sp.**
Figure 4: 12–19

A typically small *Dalejina*, here identified as *D.* sp. 1, occurs in the late Wenlock–Ludlow upper member of the Cañón de Caballeros Formation. These specimens differ considerably in their smaller size from those identified as *D.* sp. 2. from the same unit.

Genus *Baturria* Carls, 1974

***Baturria mexicana* Boucot, 1975**

Baturria mexicana BOUCOT, 1975, p. 350–351, Plate II, Figs. 1–9.

Baturria mexicana occurs in the late Ludlow–Pridoli portion of the upper member of the Cañón de Caballeros Formation. It is most like the Spanish Ludlow-Pridoli *B. simonae* (Boucot, 1975, p. 350–351).

Family DICOELOSIIDAE Cloud, 1948

Genus *Dicoelosia* King, 1850

***Dicoelosia* sp.**
Figure 4: 28

A single, very poorly preserved pedicle valve from the Santa Ana Limestone member belongs to the short-lobed group of *Dicoelosia* and indicates proximity to the B.A. 3-4 boundary.

Family LEPTAENIDAE Hall & Clarke, 1894

Genus *Leptaena* Dalman, 1828

***Leptaena "rhomboidalis"* (Wilckens)**
Figure 4: 29–36

Typical *L. "rhomboidalis"* occurs in the upper member of the Cañón de Caballeros Formation and in the Santa Ana Limestone Member in late Wenlock–Ludlow and late Ludlow–Pridoli horizons. Stinnesbeck's Santa Ana Limestone member locality also yielded this form.

Family LEPTOSTROPHIIDAE Caster, 1939

Genus *Leptostrophia* Hall and Clarke, 1892

***Leptostrophia* sp.**
Figure 5: 1–3

Several specimens from USNM locality 13003 in the late Ludlow–Pridoli part of the upper member of the Cañón de Caballeros Formation are referred to *Leptostrophia* sp. rather than to *Protoleptostrophia*, despite the absence of the critical brachial valve interior. The uniformly costellate, rather than parvicostellate, ornamentation combined with the form of the pedicle valve muscle field agrees with *Leptostrophia*. Moreover, *Protoleptostrophia* is not known from pre-Oriskany (= approximately Pragian) strata.

Leptostrophiid indet.
Figure 5: 4–5

A leptostrophiid pedicle valve from the late Wenlock–Ludlow part of the upper member of the Cañón de Caballeros Formation is parvicostellate and has a leptostrophiid muscle field. Absence of a brachial valve interior precludes generic identification.

Family DOUVILLINIDAE Caster, 1945

Genus *Mclearnites (Mclearnitesella)* Harper and Boucot, 1978

***Mclearnites (Mclearnitesella)* sp.**
Figure 5: 6–9

The evenly costellate external ornament contrasts with the coarser ornament of *Mclearnites (Mclearnites)*; the gentle convexity and form of the pedicle valve muscle field also help in making the generic determination. This form is present at several late Wenlock–Ludlow localities in the Cañón de Caballeros Formation.

Family AMPHISTROPHIIDAE Harper and Boucot, 1978

Genus *Amphistrophiella* Harper and Boucot, 1978

Amphistrophiella (Sulcastrophiella)
Boucot and Blodgett, new subgenus

Type species. A. (Sulcastrophiella) stinnesbecki Boucot and Blodgett, new species.

Diagnosis. Amphistrophiella with a relatively narrow, pronounced brachial valve sulcus and corresponding pedicle valve fold.

Comparison. Both *A.(Amphistrophiella)* and *A.(Amsdenostrophiella)* lack a pronounced brachial valve sulcus and pedicle valve fold, as does *Amphistrophia*.

Amphistrophiella (Sulcastrophiella) stinnesbecki
Boucot and Blodgett, new species
Figures 5: 17–21, 6: 1–2

Figure 5 (on facing page). 1–3: *Leptostrophia* sp., impressions of pedicle valve interior, exterior, and interior respectively, all ×1.5, locality USNM 13003, IGM 6883a, b, 6884. 4–5: Leptostrophiid indet., impressions of pedicle valve interiors, ×1.5, locality USNM 13000, IGM 6885a, b. 6–9: *Mclearnitesella* sp., impressions of pedicle valve interior, pedicle valve exterior, pedicle valve exterior, and pedicle valve interior respectively, ×2, ×1.5, ×2, ×2, locality JS-93-121, IGM 6886, 6887, 6888a, b. 10–12: *Strophoprion* sp., impressions of pedicle valve exterior, pedicle valve interior, and pedicle valve interior respectively, ×1.5, locality JS-93-121, IGM 6889a, b, 6890. 13–16: *Strophoprion* sp., impression of pedicle valve exterior, brachial valve interior, pedicle valve interior, and pedicle valve exterior respectively, ×1.5, locality USNM 13003, IGM 6891, 6891, 6893a, 6893b. 17–18: *Amphistrophiella (Sulcastrophiella) stinnesbecki* n. subgen., n. sp., impression of brachial valve exterior, rubber replica of pedicle valve exterior (note well-developed sulcus), ×2, locality JS-93-84, IGM 6894a, b. 19–21: *Amphistrophiella (Sulcastrophiella) stinnesbecki* n. subgen., n. sp., impression of brachial valve interior (note well-developed sulcus), impression of pedicle valve exterior, and impression of pedicle valve interior (note well-developed fold), ×2, USNM locality 13005. Harper and Boucot, 1978, Pl. 34, Figs. 9, 6a, 6b, USNM 220900, 220896, 220897.

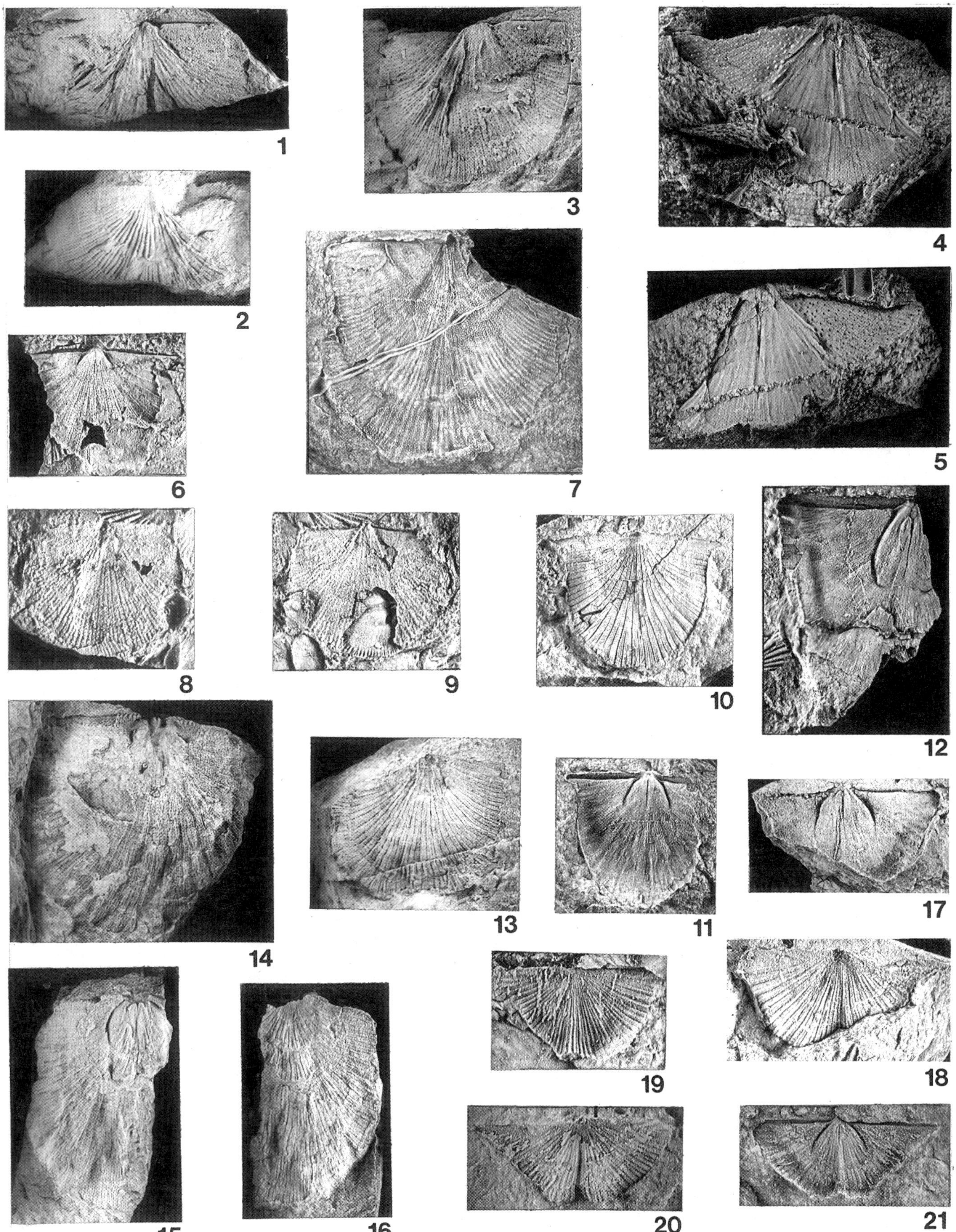

Amphistrophia species B, BOUCOT et al., 1972, p. 627–628, Lam. 18, Figs. 1–11.

Amphistrophiella (Amsdenostrophiella) n. sp. HARPER AND BOUCOT, 1978, Pl. 34, Figs. 6–9.

Exterior. Valve about one and a half times to twice as wide as long, alate; pedicle valve initially gently concave, becoming resupinate and then sharply geniculate; brachial valve nearly flat and then becoming sharply geniculate; pedicle valve bears a relatively high median fold and brachial valve a corresponding sulcus; exterior of both valves parvicostellate; pedicle valve interarea apsacline, flat, narrow.

Pedicle valve interior. Muscle field semielliptical in outline, extending anteriorly from one-half to three-quarters valve length, bounded laterally by curved lateral ridges that are anteriorly open; adductor scars longitudinally elongate, elliptical, confined to posterior part of muscle field; myophragm divides muscle field medially; internal surface endospinose and marked by the impress of the external ornamentation antero-laterally.

Brachial valve interior. Not available, but presumed similar to that of the two other subgenera of *Amphistrophiella* (see Harper and Boucot, 1978).

Occurrence. Localities JS-93-121, 121B, 79, 84, and USNM 13005, 12599 and 13000 in Tamaulipas and from the Ludlow fauna in the Merida Andes (Boucot et al., 1972, identified as Amphistrophia sp. B).

Holotype. Figure 5: 20–21, counterparts USNM 220896, 220897.

Family STRICKLANDIIDAE Schuchert and Cooper, 1931

Stricklandiid indet.
Figure 6: 16–18

The short spondylium and median septum, relatively modest convexity, and outline of the shell support the stricklandiid assignment of a single finely costellate pedicle valve from the lower member of the Cañón de Caballeros Formation. No previously described stricklandiid genus has costellate ornament of the type displayed by this specimen. It probably represents a new genus. In overall morphology it differs, for instance, from *Plicostricklandia* Boucot and Ehlers (1963), but its morphology nevertheless accords with a Wenlock age. It should be noted that several hours of collecting at the locality by several collectors produced only a single specimen. The matrix is a coarse sand with common, well-rounded pebbles; the fossils are primarily fragmented.

Family UNCINULIDAE Rzhonsnitskaya, 1956

Uncinuloid indet.
Figure 7: 9–16

The late Ludlow–Pridoli fauna in the upper member of the Cañón de Caballeros Formation contains a unique, coarsely costate uncinuloid. The pedicle valve muscle field is typically uncinuloid in its large, almost subquadrate to subcircular diductor field, posteriorly enclosing a deeply impressed adductor field. Both valves are moderately convex, with a well-developed fold and pedicle valve sulcus containing a median plication; there are about three lateral plications. The brachial valve interior features a well-developed median septum, sessile cruralium, and relatively large cardinal process. This shell probably belongs to a new genus, but in the absence of more material we are reluctant to erect one.

Family Uncertain

"*Rhynchonella*" aff. *stricklandii* (Sowerby)
Figure 7: 17–23

This spectacularly large rhynchonellid displays a prominent brachial valve median septum supporting a sessile cruralium with, posteriorly, several low ridges for diductor attachment. It has some external similarity to *Rhynchonella stricklandii* (Sowerby, 1839), but little is known about the interior of the latter. In form and ornament it somewhat resembles the North American Province endemic *Sulcatina* but is internally very distinct. It occurs in the upper member of the Cañón de Caballeros Formation in late Wenlock–Ludlow horizons as well as in the Santa Ana Limestone member at one locality.

Rhynchonellids indet.
Figures 7: 24–26, 8: 1–5

Several indeterminate rhynchonellids occur in both the late Wenlock–Ludlow and late Ludlow–Pridoli intervals of the upper member of the Cañón de Caballeros Formation. These shells are of the type that in the Silurian are commonly assigned to *Camarotoechia*, that is, they have a well-developed brachial valve median septum supporting a cruralium. Absence of crenulated dental sockets makes assignment to that genus doubtful.

Family MERISTELLIDAE Waagen, 1883

Subfamily MERISTELLINAE Waagen, 1883

Genus *Meristina* Hall, 1867

***Meristina* sp.**
Figure 8: 11–12

The upper member of the Cañón de Caballeros Formation of late Wenlock–Ludlow age contains *Meristina* pedicle valves. Brachial valves having internal features that may belong to *Amerista* (Figure 8: 21–22) might belong here, but the very small size of the cruralium suggests this might not be so. The morphology of the pedicle valves is typical for the genus.

Subfamily MERISTINAE Hall and Clarke, 1895

Figure 6 (on facing page). 1–2: *Amphistrophiella (Sulcastrophiella) stinnesbecki* n. subgen., n. sp., impressions of pedicle valve interior and pedicle valve exterior, ×2, locality USNM 13005. Note well-developed fold. Harper and Boucot, 1978, Pl. 34, Figs. 7, 8, USNM 220898. 3–7: *Mesopholidostrophia* sp., impressions of pedicle valve interior, pedicle valve interior, pedicle valve exterior, pedicle valve interior, and pedicle valve exterior respectively, ×3, locality JS-93-121, IGM 6895, 6896a, 6896b, 6897a, 6897b. 8–12: *Coolinia* sp., impressions of two pedicle valve interiors and a pedicle valve exterior, rubber replica of pedicle valve exterior, and impression of pedicle valve interior, ×1.5, locality JS-93-121, IGM 6898, 6898, 6899, 6899, 6900. 13: *Coolinia* sp., impression of pedicle valve interior, ×1.5, JS-93-121, IGM 6901. 14–15: orthotetacid, impressions of pedicle valve interior and exterior, ×1.5, locality USNM 13003, IGM 6902a, b. 16–18: Stricklandiid indet., posterior view of pedicle valve interior showing median septum and small, posteriorly situated spondylium; ventral view of impression of pedicle valve interior showing median septum; impression of exterior of pedicle valve showing relatively fine costellae, ×1.5, locality JS-93-82, IGM 6903a, 6903a, 6903b. 19–21: *Pentamerus* sp., anterior, transverse section of pedicle valve showing free margin of spondylium; cross section of pedicle valve showing thick fibrous shell layer and duplex nature of median septum; side view of median septum showing curving growth lines on its median plane, ×1.5, locality JS-93-88, IGM 6904, 6905, 6906. 22: *Gypidula* sp., impression of interior of pedicle valve showing the triangular cross section of the median septum, ×1.5, locality JS-93-79, IGM 6907.

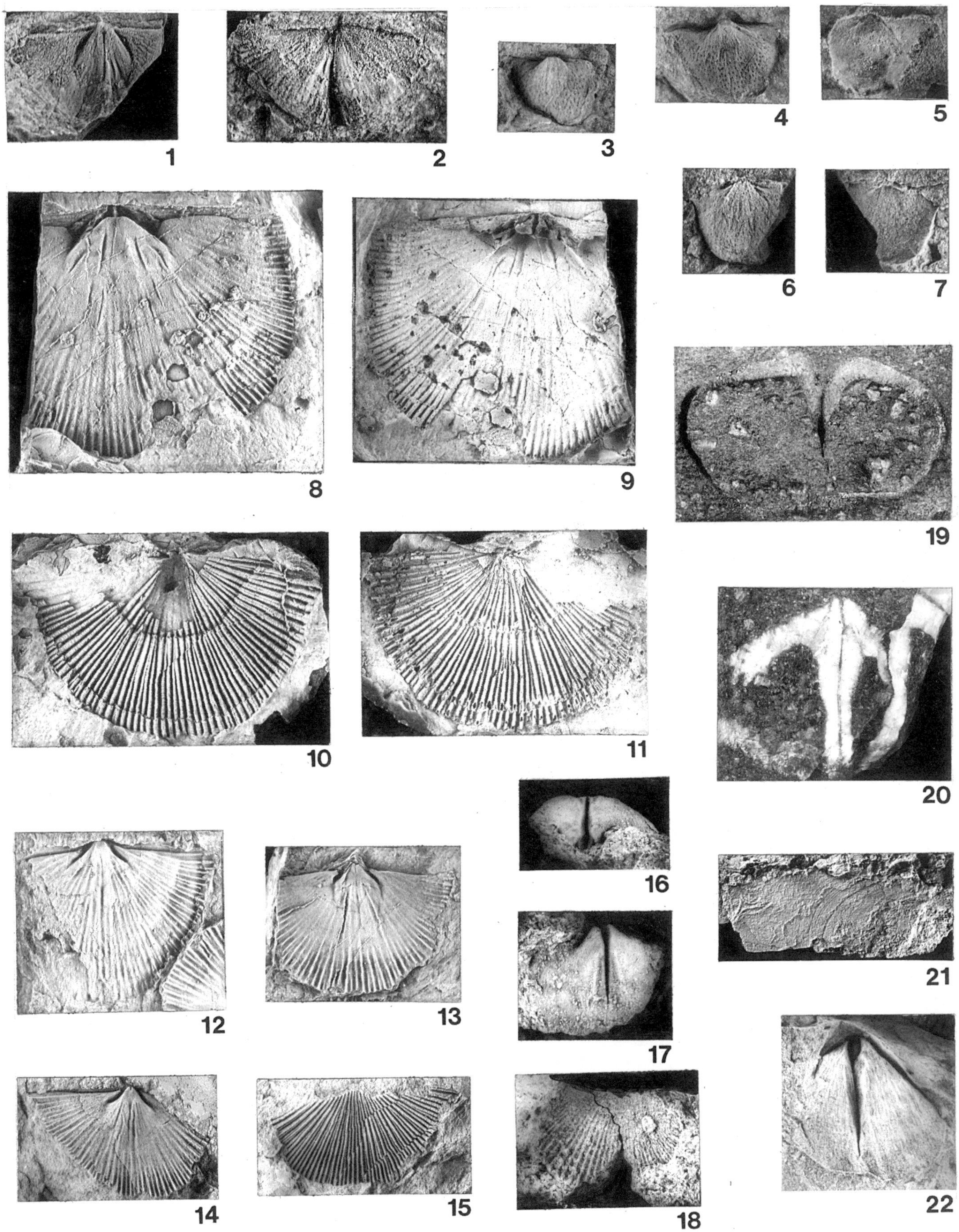
1
2
3
4
5
6
7
8
9
10
11
12
13
14
15
16
17
18
19
20
21
22

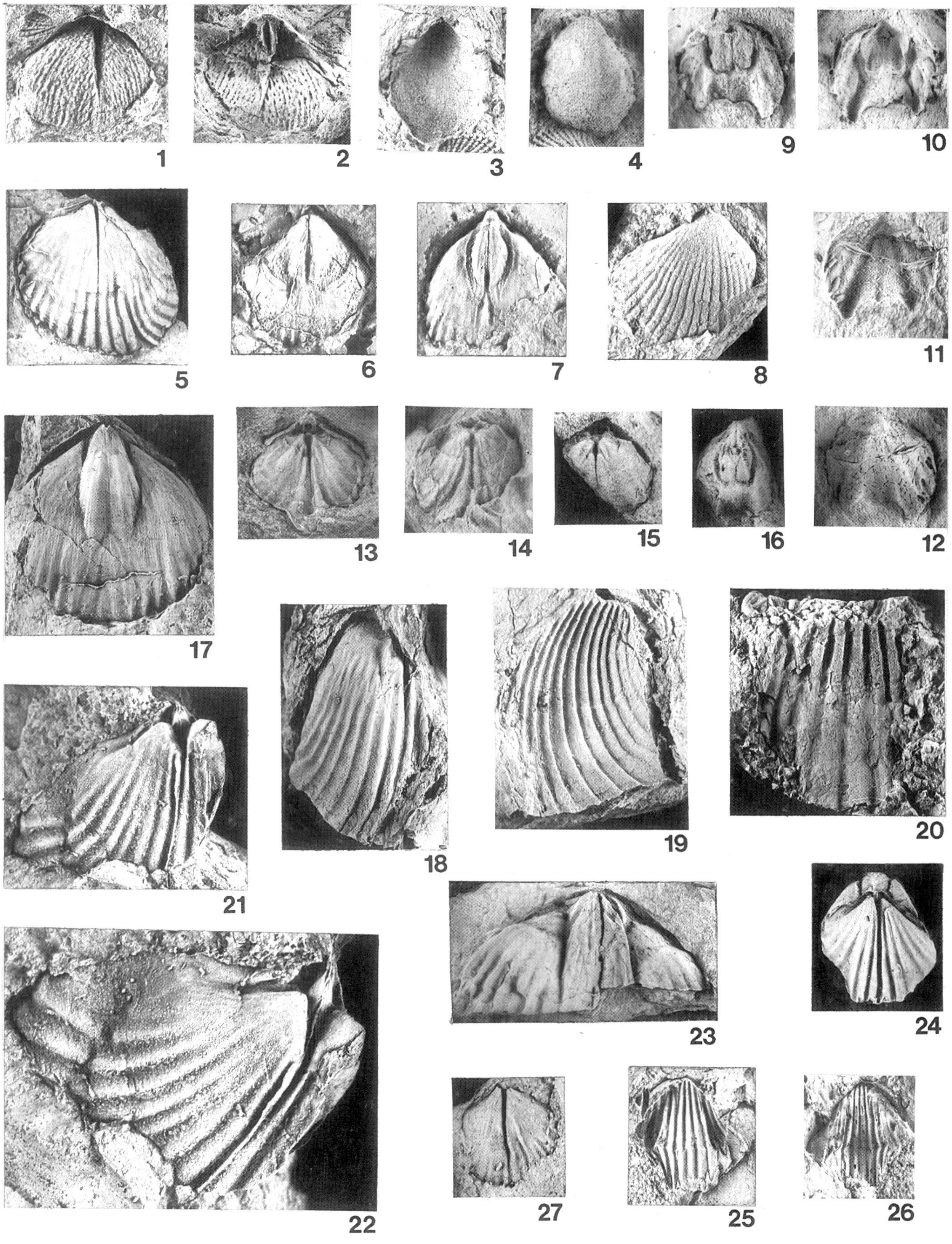
1
2
3
4
9
10
5
6
7
8
11
17
13
14
15
16
12
18
19
20
21
22
23
24
27
25
26

Genus *Amerista* Boucot and Blodgett, new genus

Type species. Amerista carillobravoi Boucot and Blodgett, new species.

Diagnosis. Meristellid pedicle valve interior with the anterior part of the muscle field featuring a central ridge to which the muscles were attached and with, in some specimens, antero-laterally directed grooves on the central ridge.

Comparison. Amerista lacks the shoe lifter process of *Merista* and *Diambonia.*

***Amerista carillobravoi* Boucot and Blodgett, new species**
Figure 8: 13–20 and possibly 8: 21–22

Exterior. Pedicle valve moderately convex; questionably assigned brachial valve also moderately convex; outline subcircular with maximum width near midlength; lateral margins rounded; anterior margin not observed.

Pedicle valve interior. Dental lamellae at anterior of umbonal chamber, short, obsolete; hinge teeth stout; umbonal cavity filled with secondary material; muscle field, anterior of hemispherical umbonal chamber, borne on a central ridge that includes an angle of about 120° and is bounded laterally by deep grooves (sunk into secondary material)—anteriorly it may bear antero-laterally directed ridges.

Brachial valve interior. Owing to absence of specimens with conjoined valves, we are uncertain whether or not the brachial valve (Fig. 8: 21–22) questionably assigned to *Amerista* actually belongs to that genus or to *Meristina*, which also occurs in the fauna. The brachial valve features a narrow median septum, reaching to about mid-length, and supporting a well-developed cardinal plate—formed probably from medially conjunct hinge plates—with a relatively narrow, short, median cruralium-type depression.

Occurrence. Amerista is known in low abundance from late Wenlock–Ludlow beds in the upper member of the Cañón de Caballeros Formation at localities JS-93-79, 84, 86, 87, 121, 121A, 121B, and USNM 12599.

Holotype: Figure 8: 16, IGM 6938.

Family DELTHYRIDIDAE Waagen, 1883

Subfamily DELTHYRIDINAE Phillips, 1841

Genus *Delthyris* Dalman, 1828

***Delthyris* spp.**
Figures 9: 22–26, 10: 1–9

The upper member of the Cañón de Caballeros Formation yields two forms of *Delthyris.* The average-size form is restricted to the beds of late Wenlock–Ludlow age at localities JS-93-79, 84, 86, 87, 121, 121A, 121B, USNM 13005, and questionably USNM 12599, whereas the large-size form occurs in late Ludlow–Pridoli beds at USNM 13003. The latter specimens were misidentified by Boucot (1975) as the late Pridoli and Lochkovian descendant form *Quadrifarius*, which has a median plication in the sulcus of the pedicle valve and a corresponding groove on the fold.

Figure 7 (on facing page). 1–4: *Gypidula* sp., impression of pedicle valve interior, showing median septum and pitted interior; rubber replica of pedicle valve interior showing short spondylium supported by a relatively long median septum; impression of pedicle valve exterior; rubber replica of pedicle valve exterior, ×1.5, locality JS-93-79, IGM 6908, 6908, 6909, 6909. 5: *Sphaerirhynchia* sp., impression of brachial valve interior; note long median septum, ×1.5, locality JS-93-86, IGM 6910. 6: *Sphaerirhynchia* sp., impression of brachial valve interior, ×1.5, locality JS-93-121, IGM 6911. 7–8: *Sphaerirhynchia* sp., impressions of pedicle valve interior and exterior respectively, ×2, locality JS-93-105, IGM 6912, 6913. 9–16: Uncinuloid indet., impression of pedicle valve interior, rubber replica of pedicle valve interior, impression of pedicle valve exterior, rubber replica of pedicle valve exterior, impression of three brachial valve interiors, and impression of pedicle valve interior respectively, ×2, locality USNM 13003, IGM 6914a, 6914a, 6914b, 6914b, 6915, 6916, 6917, 6918. 17–19: *"Rhynchonella"* aff. *stricklandii* (Sowerby), impressions of pedicle valve interior, brachial valve interior, and brachial valve exterior respectively, ×l.5, locality USNM 13000, IGM 6919, 6902a, 6902a. 20: *"Rhynchonella"* aff. *stricklandii* (Sowerby), impression of pedicle valve exterior, ×1.5, locality USNM 13001, IGM 6921. 21–22: *"Rhynchonella"* aff. *stricklandii* (Sowerby), impressions of two brachial valve interiors; note absence of a cruralium, ×1.5, locality JS-93-86, IGM 6922. 23: *"Rhynchonella"* aff. *stricklandii* (Sowerby), impression of pedicle valve interior, ×1.5, locality USNM 13000, IGM 6923. 24–26: Indeterminate rhynchonellid, brachial view of internal impression, impression of pedicle valve exterior, and rubber replica of pedicle valve exterior, ×2, ×1.5, ×1.5 respectively, locality USNM 12599, IGM 6924a, 6924b, 6924b. 27: Indeterminate rhynchonellid, impression of brachial valve interior, ×2, locality USNM 13003, IGM 6925.

ACKNOWLEDGMENTS

The National Geographic Society generously funded our work. Wolfgang Stinnesbeck, Linares, provided specimens from the Santa Ana Limestone member, directed us to the Ciudad Victoria localities, and gave much advice on the area. Norman Savage, University of Oregon, vigorously pursued Paleozoic conodonts in what proved to be a very low yield area. Jonathan M. Adrain, Natural History Museum, London, kindly reported on the trilobites. John L. Carter, Carnegie Museum of Natural History, Pittsburgh, undertook study of the Mississippian brachiopods from the area.

APPENDIX 1. NEW OCCURRENCE OF MAGNESIUM CARBONATES

Pebbles of serpentine, apparently derived from serpentinites associated with the Granjeno Schist, are abundant on the banks of the small stream in Peregrina Canyon. While breaking some of these, Boucot noted one with abundant small millimeter-size specks of a somewhat platey, violet mineral resembling stichtite (a magnesium, chromium carbonate, hydroxide hydrate); associated brownish specks escaped notice at the time. At his behest, Collections Manager Paul W. Pohwat, Division of Mineralogy, National Museum of Natural History, Washington, D.C., had the minerals X-rayed by D. R. Ross, who confirmed the presence of stichtite as well as pyroaurite (a magnesium, iron carbonate hydroxide hydrate) and some sjogrenite (a magnesium, iron carbonate hydroxide hydrate). These are the first occurrences of these three rare minerals in Mexico. The two halves of the pebble are now in the collections of the Division of Mineralogy, National Museum of Natural History.

APPENDIX 2. LOCALITIES

(Ca with a number refers to original field stations of Carillo-Bravo.)

USNM 12599. Upper member of the Cañón de Caballeros Formation. Caballeros Canyon, hillside 1 km south of station Ca 1690, which is in arroyo of Rio Caballeros approximately 0.5 km downstream from Rancho Aserradero. Collector Carrillo-Bravo.
USNM 13000. Upper member of the Cañón de Caballeros Formation. Caballeros Canyon, north bank of Rio Caballeros, 590 m south (upstream) of the Rancho El Naranjal; at the western edge of the fault block of metamorphic rock, Ca 1684. Collector Carillo-Bravo.
USNM 13001. Santa Ana Limestone member of the Cañón de Caballeros Formation. Peregrina Canyon, 1.2 km northwest of Rancho Vicente Guerrero, field no. J-1. Collector David D. Johnson.
USNM 13002. Santa Ana Limestone member of the Cañón de Caballeros Formation, Peregrina Canyon, field no. Ca 1774. Collector Carillo-Bravo.
USNM 13003. Upper member of the Cañón de Caballeros Formation, Peregrina Canyon, field no. Ca 1775. Collector Carillo-Bravo.
USNM 13004. Santa Ana Limestone member of the Cañón de Caballeros Formation, Peregrina Canyon, field no. Ca 1777. Collector Carillo-Bravo.
USNM 13005. Upper member of the Cañón de Caballeros Formation. Caballeros Canyon, north bank of the Rio Caballeros, 590 m south (upstream) of the Rancho El Naranjal; at western edge of fault block of metamorphic rock, field no. Ca 1684. Collector R. B. Mixon (same locality as USNM 13000).
JS93-76. Santa Ana Limestone member of Cañón de Caballeros Formation, near crest of ridge midway up north side of Cañón de la Peregrina, 0.8 km north-northwest of Rancho Vicente Guerrero.
JS93-79. Lower part of upper member of Cañón de Caballeros Formation, near crest of ridge midway up north side of Cañón de la Peregrina, 0.8 km north-northwest of Rancho Vicente Guerrero.
JS93-82. Lower member of Cañón de Caballeros Formation near crest of ridge midway up north side of Cañón de la Peregrina, 0.8 km north-northwest of Rancho Vicente Guerrero. This locality produced the only fauna from the lower member: tabulates (4), fenestellid (1), new plicate pedicle valve of ?stricklandiid (1), Delthyris? sp. (1), Atrypa? "reticularis" (1), unidentified brachiopods (5), ?trilobite fragment (1), rostroconch (1), fenestellid (1), tabulates (4), rugose corals (2).
JS93-83. Santa Ana Limestone member of Cañón de Caballeros Formation, near crest of ridge on north side of Cañón de la Peregrina, 0.7 km north-northwest of Rancho Vicente Guerrero.
JS93-84. Lowermost part of Upper Member of Cañón de Caballeros Formation, near crest of ridge midway up north side of Cañón de la Peregrina, 0.7 km north-northwest of Rancho Vicente Guerrero.
JS-93-86. Middle or upper part of upper member of Cañón de Caballeros Formation on east side of ridge midway up north side of Cañón de la Peregrina, 0.7 km north-northwest of Rancho Vicente Guerrero.
JS93-87. Middle or upper part of upper member of Cañón de Caballeros Formation, on east side of ridge midway up north side of Cañón de la Peregrina, 0.7 km north-northwest of Rancho Vicente Guerrero. Fossils in float material, designated as float blocks 1, 2, or 3.
JS93-88. Santa Ana Limestone member of Cañón de Caballeros Formation, north bank of river in Cañón de Caballeros, 0.4 km south-southwest of Rancho El Naranjal.
JS93-105. Santa Ana Limestone member of Cañón de Caballeros Formation, near crest of ridge on north side of Cañón de la Peregrina, 0.7 km north-northwest of Rancho Vicente Guerrero.
JS93-116. Upper member of Cañón de Caballeros Formation, north bank of river in Cañón de Caballeros, 0.4 km south-southwest of Rancho El Naranjal.
JS93-121. Lower part of upper member of Cañón de Caballeros Formation, south side of Cañón de la Peregrina, 0.2 km southeast of Rancho Vicente Guerrero. JS93-121A is the main collection; JS93-121B is a small supplemental collection from a bed 3 m east of JS93-121A. JS93-121 includes fossils from beds at JS93-121A and from several layers within 3 m to west.
Stinnesbeck collection. Santa Ana Limestone member of Cañón de Caballeros Formation, north side of Cañón de la Peregrina, about 0.7 km north-northwest of Rancho Vicente Guerrero.

Figure 8 (on facing page). 1–3: Indeterminate rhynchonellid, impression of three brachial valve interiors, ×3, locality JS-93-86, IGM 6926, 6927, 6928. 4–5: Indeterminate rhynchonellid, impressions of brachial valve interiors (note cruralium filling), ×3, locality JS-93-87-2, IGM 6929. 6: *Atrypa "reticularis"* (Linné), impression of pedicle valve exterior, ×1.5, locality JS-93-121, IGM 6930. 7–10: *Atrypa "reticularis"* (Linné), impression of pedicle valve interior, impression of brachial valve exterior, rubber replica of brachial valve exterior, and impression of brachial valve interior respectively, ×1.5, locality JS-93-79, IGM 6931, 6932a, 6932a, 6932b. 11–12: *Meristina* sp., impressions of pedicle valve interiors, ×1.5, locality USNM 12599, IGM 6933, 6934. 13-14, *Amerista carillobravoi* Boucot and Blodgett n. gen. n. sp., impressions of pedicle valve interiors, ×2, locality JS-93-86, IGM 6935, 6936. 15: *Amerista carillobravoi* Boucot and Blodgett n. gen. n. sp., impression of pedicle valve interior, ×2, locality USNM 12599, IGM 6937. 16–20: *Amerista carillobravoi* Boucot and Blodgett n. gen. n. sp., impression of pedicle valve interior, rubber replica of pedicle valve interior, impression of pedicle valve exterior, rubber replica of pedicle valve exterior, and impression of pedicle valve interior (note anterior striations in diductor attachment area), ×2, ×2, ×2, ×2, ×3, locality JS-93-79, IGM 6938a, 6938a, 6938b, 6938b, 6939. 21–22: *Amerista?* sp., impression of brachial valve interior (note length of median septum), and posterior view of brachial valve impression (note filling of small cruralium), ×1.5, ×3, locality JS-93-86, IGM 6940.

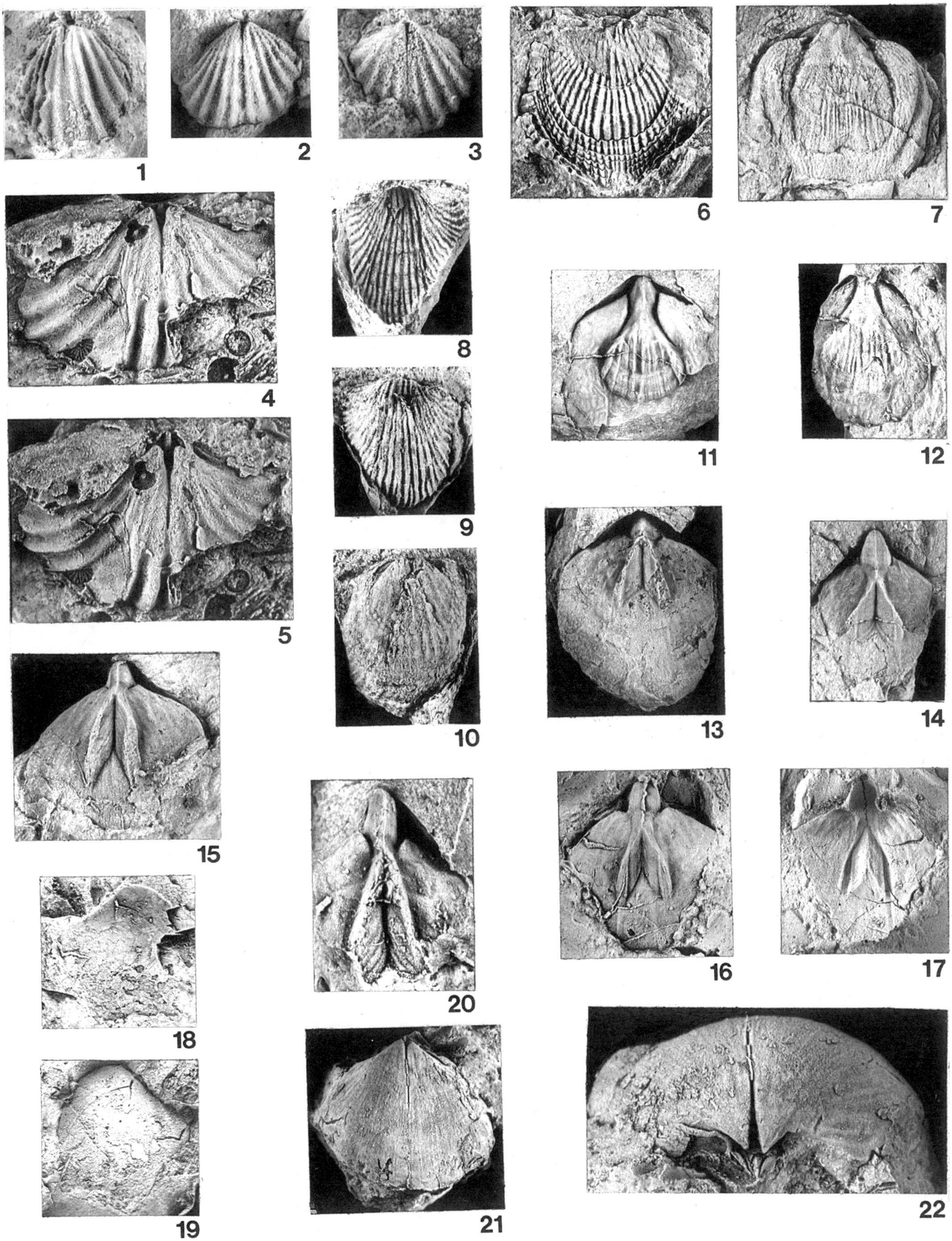

1
2
3
4
5
6
7
8
9
10
11
12
13
14
15
16
17
18
19
20
21
22

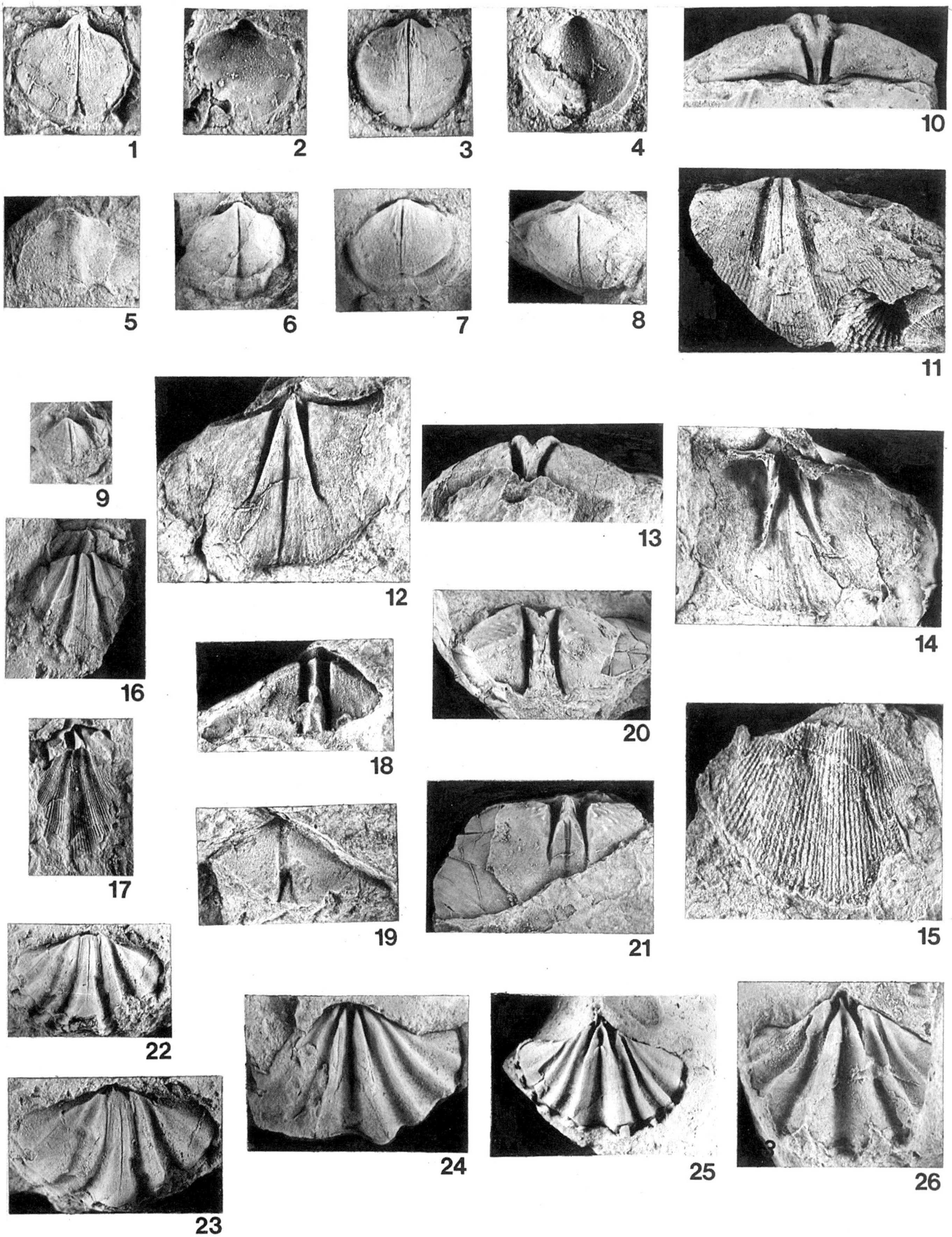
1
2
3
4
10
5
6
7
8
11
9
12
13
14
16
18
20
17
19
21
15
22
24
25
26
23

Figure 9 (on facing page). 1–4: *Nucleospira* sp., impressions of pedicle valve interior, pedicle valve exterior, brachial valve interior, and brachial valve exterior respectively, ×3, locality JS-93-121, IGM 6941a, 6941b, 6942a, 6942b. 5–9: *Nucleospira* sp., impressions of pedicle valve exterior, pedicle valve interior, brachial valve interior, pedicle valve interior, and pedicle valve interior respectively, ×2, locality USNM 13003, IGM 6943, 6944, 6945, 6946, 6947. 10–15: *Eospirifer* sp., posterior view of internal impression of pedicle valve, impressions of two pedicle valve interiors, posterior view of pedicle valve impression, rubber replica of pedicle valve interior, and impression of pedicle valve exterior respectively (note ornamentation), all ×1.5 except the last, which is ×2, locality JS-93-79, IGM 6948, 6948, 6949a, 6949a, 6949a, 6949b. 16–17: *Macropleura* sp., respectively impressions of brachial valve interior and brachial valve exterior (note ornamentation), ×2, locality JS-93-121, IGM 6950a, 6950b. 18–19: *Cyrtia* sp., posterior view of pedicle valve interior impression and posterior view of pedicle valve external impression (note deltidium), ×1.5, locality USNM 13005, IGM 6951a, 6951b. 20–21: *Cyrtia* sp., posterior view of pedicle valve internal impression and impression of pedicle valve interior, ×1.5, locality JS-93-121, IGM 6952. 22–25: *Delthyris* sp., impressions of two brachial valve interiors (note short crural plates) and two pedicle valve interiors (note posterior end of stout median septum), ×2, ×3, ×2, ×2 respectively, locality JS-93-121, IGM 6953, 6953, 6954, 6955. 26: "Medium-size" *Delthyris* sp., impression of pedicle valve interior, ×2, locality JS-93-121, IGM 6956.

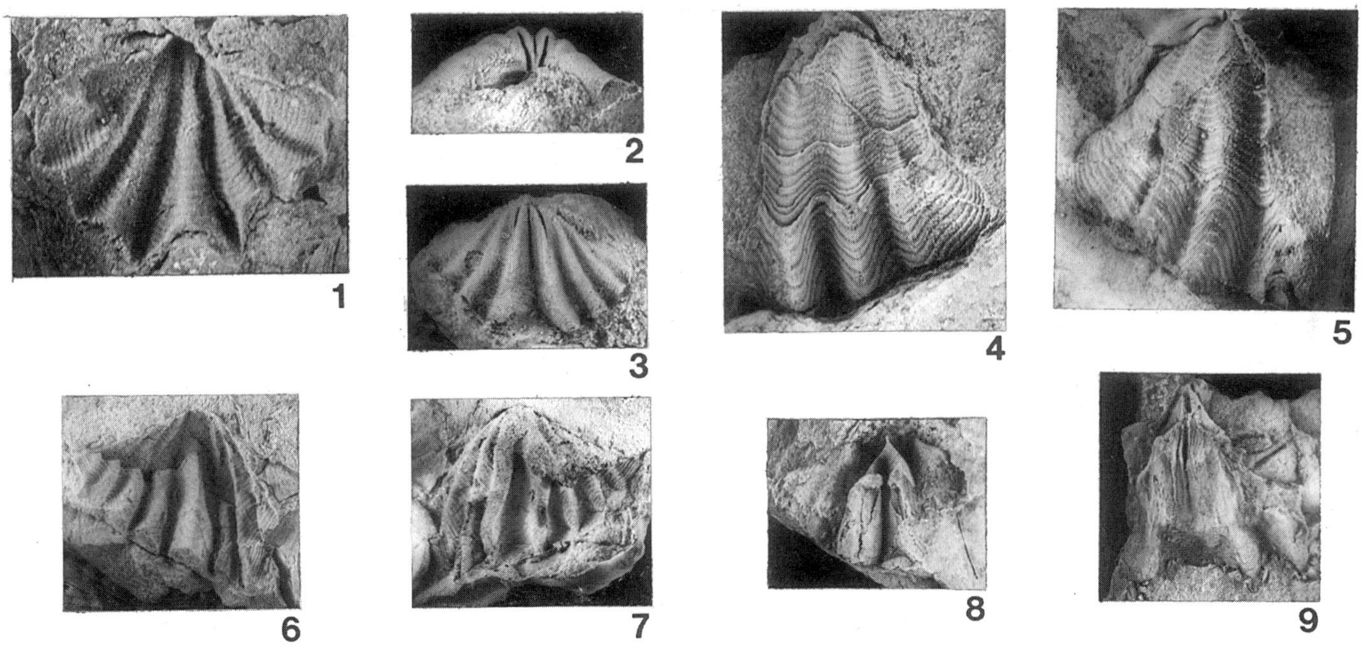

Figure 10. 1–4: "Medium-size" *Delthyris* sp., impressions of pedicle valve exterior (note concentric ornamentation), posterior of pedicle valve (note impression of blade-like median septum), pedicle valve interior, and pedicle valve exterior respectively (note concentric ornamentation and spinule fringe on the last), ×4, ×2, ×2, ×3 respectively, locality USNM 13003, IGM 6957, 6958, 6958, 6959. 5–9: "Large-size" *Delthyris* sp., rubber replica of pedicle valve exterior (note ornament), impression of pedicle valve exterior, rubber replica of pedicle valve exterior, impression of pedicle valve interior (note impression of median septum), and impression of pedicle valve interior respectively, ×3, ×1.5, ×1.5, ×2, ×2, locality USNM 13003, IGM 6959, 6960, 6960, 6961, 6962.

REFERENCES CITED

Belcher, R. C., 1979, Depositional environments, paleomagnetic and tectonic significance of the Huizachal Redbeds (lower Mesozoic), northeastern Mexico [Ph.D. thesis]: Austin, University of Texas, 286 p.

Berry, W. B. N., and Boucot, A. J., 1970, Correlation of the North American Silurian rocks: Geological Society of America Special Paper 102, 289 p.

Boucot, A. J., 1975, Evolution and extinction rate controls: Amsterdam, Elsevier, 427 p.

Boucot, A. J., 1990, Silurian biogeography, in McKerrow, W. S., and Scotese, C. R., eds., Palaeozoic palaeogeography and biogeography: Geological Society of London Memoir 12, p. 191-196.

Boucot, A. J., and Ehlers, G. M., 1963, Two new genera of stricklandid brachiopods: Contributions from the Museum of Paleontology, Michigan University, v. 18, p. 47–66.

Boucot, A. J., Johnson, J. G., and Shagam, R., 1972, Braquipodos del Silurico de los Andes de Merida: Memoria, Cuarto Congreso Geologico Venezolano, Tomo II, Boletin de Geologia, Publicacion Especial 5, p. 585–727.

Carillo-Bravo, J., 1959, Notas sobre el Paleozoico de la Regíón de Ciudad Victoria, Tamps.: Boletin de la Asociacion Mexicana de Geologos Petroleros, v. 11, p. 673–681.

Carillo-Bravo, J., 1961, Geologia del anticlinorio Huizachal-Peregrina al N-W de Ciudad Victoria, Tamps.: Boletin de la Asociacion Mexicana de Geologos Petroleros, v. 13, p. 1–99.

Carls, P., 1974, Die Proschizophoriinae (Brachiopoda; Silurium-Devon) der Östlichen Iberischen Ketten (Spanien): Senckenbergiana lethaea, v. 55, p. 153–227.

Caster, K. E., 1939, A Devonian fauna from Colombia: Bulletins of American Paleontology, v. 24, 218 p.

Caster, K. E., 1945, New names for two homonyms: Journal of Paleontology, v. 19, p. 319.

Clark, J. M., and Hopson, J. A., 1985, Distinctive mammal-like reptile from Mexico and its bearing on the phylogeny of the Tritylodontidae: Nature, v. 315, p. 398–400.

Cloud, P. E., Jr., 1948, Dicaelosia versus Bilobites: Journal of Paleontology, v. 22, p. 373–374.

Conklin, J., Corpstein, P., Fralick, K., Mikels, J., Morales, N., and Rodgers, M., eds., 1974, Geology of Huizachal-Peregrina anticlinorium: Edinburg, Texas, Southwestern Association of Student Geological Societies, Field Conference Guidebook, Pan American Geological Society, 134 p.

Cserna, Z. de, and Ortega-Gutierrez, F., 1977, Aloctono del Paleozoico inferior en la regíón de Ciudad Victoria, Estado de Tamaulipas: Universidad Nacional Autonoma de Mexico, Instituto de Geologia, Revista, v. 1, p. 33–43.

Cserna, Z. de, and Ortega-Gutierrez, F., 1978, Reinterpretation of isotopic age data from the Granjeno Schist, Ciudad Victoria, Tamaulipas y Reinterpretation tectonica del Esquisito Granjeno de Ciudad Victoria, Tamaulipas; Contestacion: Universidad Nacional Autonoma de Mexico, Instituto de Geologia, Revista, v. 2, p. 212–215.

Dalman, J. W., 1828, Uppställning och beskrifning af de i Sverige funne Terebratulifer: Svenska Vetenskaps-academiens, Handlingar, ser. 3, v. 15–16, p. 85–155. [imprint 1827]

Denison, R. E., Burke, W. H., Jr., Hetherington, E. A., Jr., and Otto, J. B., 1971, Basement rock framework of parts of Texas, southern New Mexico and northern Mexico, in Seewald, K., and Sundeen, D., eds., The geologic framework of the Chihuahua Tectonic Belt: Midland, Texas, West Texas Geological Society, p. 3–14.

Fastovsky, D. E., Clark, J. M., and Hopson, J. A., 1987, Preliminary report of a vertebrate fauna from an unusual paleoenvironmental setting, Huizachal Group, Early or Mid-Jurassic, Tamaulipas, Mexico, in Currie, P. J., and Koster, E. H., eds., Fourth Symposium on Mesozoic Terrestrial Ecosystems, Short Papers: Tyrrell Museum of Paleontology Occasional Paper 3, p. 82–87.

Fastovsky, D. E., Clark, J. M., Strater, N. H., Montellano, M., Hernandez, R., and Hopson, J. A., 1995, Depositional environments of a Middle Jurassic terrestrial vertebrate assemblage, Huizachal Canyon, Mexico: Journal of Vertebrate Paleontology, v. 15, p. 561–575.

Flawn, P. T., Goldstein, A., Jr., King, P. B., and Weaver, C. E., 1961, The Ouachita System: University of Texas Publication 6120, 401 p.

Fries, C., Jr., Schmitter, E., Damon, P. E., Livingston, D. E., and Erickson, R., 1962, Edad de las rocas metamorficas en los Cañónes de la Peregrina y de Caballeros, parte centro-occidental de Tamaulipas: Universidad Nacional Autonoma de Mexico, Instituto de Geologia, Boletin, v. 64, part 4, p. 55–69.

Garrison, J. R., Jr., 1978, Reinterpretation of isotopic age data from the Granjeno Schist, Ciudad Victoria, Tamaulipas: Universidad Nacional Autonoma de Mexico, Instituto de Geologia, Revista, v. 2, p. 87–89.

Garrison, J. R., Jr., Ramirez-Ramirez, C., and Long, L. E., 1980, Rb-Sr isotopic study of the ages and provenance of Precambrian granulite and Paleozoic greenschist near Ciudad Victoria, Mexico, in Pelger, R. H., Jr., ed., The origin of the Gulf of Mexico and the early opening of the Central North Atlantic Ocean: Baton Rouge, Louisiana State University, p. 37–49.

Gursky, H.-J., and Michalzik, D., 1989, Lower Permian turbidites in the northern Sierra Madre Oriental: Zentralblatt für Geologie und Paläontologie, 1989, p. 821–838.

Gursky, H.-J., and Ramirez-Ramirez, C., 1986, Notas preliminares sobre el descubrimiento de volcanitas acidas en El Cañón de Caballeros (Nucleo del Anticlinorio Huizachal-Peregrina), Tamaulipas, Mexico, in Barbarin, J. M., and Gursky, H.-J., eds., Aspectos geologicos del nordeste de Mexico: Actas de la Facultad de Ciencias de la Tierra, Universidad Autonoma de Nueva Leon, Linares, v. 1, p. 11–22.

Hall, J., 1867, Descriptions and figures of the fossil Brachiopoda of the Upper Helderberg, Hamilton, Portage, and Chemung groups: Geological Survey of New York, Palaeontology, v. 4, part 1, 428 p.

Hall, J., and Clarke, J. M., 1892, An introduction to the study of the genera of Palaeozoic Brachiopoda: Geological Survey of New York, Palaeontology, v. 8, part 1, 367 p.

Hall, J., and Clarke, J. M, 1894, An introduction to the study of the Brachiopoda: 13th Annual Report of the New York State Geologist for the year 1893, Palaeontology, part 2, p. 751–943.

Hall, J., and Clarke, J. M., 1895, An introduction to the study of the genera of Palaeozoic Brachiopoda: Geological Survey of New York, Palaeontology, v. 8, part 2, p. 319–394.

Harper, C. W., Jr., and Boucot, A. J., 1978, The Stropheodontacea, Part II: Douvillinidae, Telaeoshaleriidae, Amphistrophiidae and Shaleriidae: Palaeontographica, Abteilung A, Bd. 161, p. 119–175.

Havlí ek, V., 1956, Brachiopods of the Braník and Hlubo epy Limestones in the immediate vicinity of Prague: Sborník ústredního ústavu geologického, v. 22, p. 535–650 [in Czech].

Imlay, R. W., 1938, Studies of the Mexican geosyncline: Geological Society of America Bulletin, v. 49, p. 1651–1694.

Imlay, R. W., 1943, Jurassic formations of the Gulf region: American Association of Petroleum Geologists Bulletin, v. 27, p. 1407–1533.

King, W., 1850, A monograph of the Permian fossils of England: Palaeontographical Society Monograph 3, 258 p.

Kozlowski, R., 1929, Les brachiopodes gothlandiens de la Podolie Polonaise: Palaeontologia Polonica, v. 1, 254 p.

Lopez-Ramos, E., 1981, Geologia de Mexico, v. 3: Mexico City, published privately, 446 p.

Mixon, R. B., Murray, G. E., and Diaz, T., 1959, Age and correlation of Huizachal Group (Mesozoic), State of Tamaulipas, Mexico: American Association of Petroleum Geologists Bulletin, v. 43, p. 757–771.

Murray, G. E., Furnish, W. M., and Carillo B., J., 1960, Carboniferous goniatites from Caballeros Canyon, State of Tamaulipas: Journal of Paleontology, v. 34, p. 731–737.

Ortega-Gutierrez, F., 1978, El Gneis Novillo y rocas metamorficas asociadas en los Cañónes del Novillo y de la Peregrina, area de Ciudad Victoria, Tamaulipas: Universidad Nacional Autonoma de Mexico, Instituto de Geologia, Revista, v. 2, p. 19–30.

Ortega-Gutierrez, F., Centeno-Garcia, E., Moran-Zenteno, D. J., and Gomez-Caballeros, A., eds., 1993, Pre-Mesozoic basement of NE Mexico, lower crust and mantle xenoliths of central Mexico, and northern Guerrero terrane; in Guidebook of field trip A: First Circum-Pacific and Circum-Atlantic Terrane Conference; Ciudad Universitaria, Universidad Nacional Autonoma de Mexico, Instituto de Geologia, 96 p.

Phillips, J., 1841, Figures and descriptions of the Palaeozoic fossils of Cornwall, Devon, and West Somerset: London, Longman, 231 p.

Ramirez-Ramirez, C., 1992, Pre-Mesozoic geology of Huizachal-Peregrina anticlinorium, Ciudad Victoria, Tamaulipas, and adjacent parts of eastern Mexico [Ph.D. thesis]: Austin, University of Texas, 317 p.

Ruiz, J., Patchett, P. J., and Ortega-Gutierrez, F., 1988, Proterozoic and Phanerozoic basement terranes of Mexico and Nd isotopic studies: Geological Society of America Bulletin, v. 100, p. 274–281.

Rzhonsnitskaya, M. A., 1956, K systematike rhynkhonellid: Congreso Geologico International, 20th, Ciudad de México, Seccion 7, Paleontología, Taxonomía, y Evolución, p. 107–121.

Schuchert, C., and Cooper, G. A., 1931, Synopsis of the brachiopod genera of the suborders Orthoidea and Pentameroidea, with notes on the Telotremata: American Journal of Science, ser. 5, v. 22, p. 241–251.

Silver, L. T., Anderson, T. H., and Ortega-Gutierrez, F., 1994, The "thousand million year" orogeny in eastern and southern Mexico: Geological Society of America Abstracts with Programs, v. 26, no. 7, p. A48.

Sowerby, J. de C., 1839, in Murchison, R. I., The Silurian System: London, John Murray, 768 p.

Stewart, J. H., Blodgett, R. B., Boucot, A. J., and Carter, J. L., 1993, Middle Paleozoic exotic terrane near Ciudad Victoria, northeastern Mexico, and the southern margin of Paleozoic North America, in Ortega-Gutierrez, F., Coney, P. J., Centeno-Garcia, E., and Gomez-Caballero, A., eds., Proceedings First Circum-Pacific and Circum-Atlantic Terrane Conference, Guanajuato, Mexico: Ciudad Universitaria, Universidad Nacional Autonoma de Mexico, Instituto de Geologia, p. 147–149. Also published in: Ortega-Gutierrez, F., et al., 1993 (see citation above).

Tellez-Giron, C., 1970, Microfacies y microfosiles Paleozoicos del area de Ciudad Victoria, Tamps., N.E. de Mexico: Instituto Mexicano del Petroleo, Serie Monografica 1, Publicacion 70 AG/068, 117 p.

Waagen, W. H., 1883, Salt Range fossils. Part 4 (2): Brachiopoda: Palaeontologica Indica Memoir, series 13, v. 1, p. 391–546.

Walmsley, V. G., and Boucot, A. J., 1975, The phylogeny, taxonomy and biogeography of Silurian and Early to mid-Devonian Isorthinae (Brachiopoda): Palaeontographica, Abteilung A, v. 148, p. 1–108.

Williams, A., and 18 others, 1965, Brachiopoda, in Moore, R. C., ed., Treatise on invertebrate paleontology, V. 1 and 2, Boulder, Colorado Geological Society of America (and University of Kansas Press), 927 p.

Manuscript Accepted by the Society February 27, 1997

Printed in U.S.A.

Geological Society of America
Special Paper 321
1997

Sequence of Devonian thelodont scale assemblages in East Gondwana

Susan Turner
Queensland Museum, P.O. Box 3300, South Brisbane, Queensland 4101, Australia

ABSTRACT

Eight thelodont scale assemblages, predominantly taxa of the genera *Turinia* Powrie, *Jesslepis* Turner, and *Australolepis* Turner and Dring, provide the basis for biostratigraphic subdivision of the latest Pridoli to Early Frasnian, a span of about 40 m.y. The assemblages are: (1) *fuscina*- sp. cf. *polita*, (2) *Turinia australiensis*, (3) *T.* sp. cf. *australiensis*, (4) *Jesslepis johnsoni*, (5) *Turinia gavinyoungi*, (6) *T.* sp. cf. *pagoda*- sp. cf. *hutkensis*, (7) *Turinia antarctica*, and (8) *Australolepis seddoni*. Turiniid thelodonts achieved almost worldwide distribution within the 0 to 60° latitudes during this interval, and their scales occur through a range of depositional settings, making them useful for marine/nonmarine correlation.

INTRODUCTION

The Devonian in East Gondwana (Australia and Antarctica) has considerable thicknesses of shallow marine and nonmarine deposits for which vertebrates provide the primary age control. During the last decade, systematic collecting has provided much new information on Paleozoic vertebrate microfossils (Turner, 1993; Young, 1993; De Pomeroy, 1995). These data are helping to establish correlation from nonmarine to marine sections and allowing integration of microvertebrate and other biostratigraphies. In the 30 years since thelodont scales were first recorded in Australia and the Southern Hemisphere (Jones, *in* Johnstone et al., 1967), many new occurrences, particularly of Turiniidae, have been noted in sediments interpreted to be both marine and nonmarine in origin (Fig. 1; Gross, 1971; Young and Gorter, 1981; Turner et al., 1981; Long et al., 1988). This chapter reviews the biostratigraphy of East Gondwanan thelodonts, puts forward a new subdivision of the latest Pridoli to early Frasnian based on eight assemblages of turiniids (Fig. 2), and where possible relates the assemblages to conodont dating.

Scales of cartilaginous thelodont fishes are often found in shallow-water sedimentary rocks (marine and inferred nonmarine) in Australia and neighboring countries. Sometimes they are the only identifiable and/or biostratigraphically useful remains. Some Devonian microvertebrate assemblages in Australia come from bone beds and are extremely rich; others are known from only a few specimens. Marine limestones typically have low yields of thelodonts; thelodonts seem to be more abundant in sediments from marginal marine environments. Thelodonts appear in East Gondwana by mid-Ludlow (Silurian) time and range to early Frasnian (Late Devonian) time (Fig. 1; Turner, 1993), but only scales of thelodontid type occur (*sensu Thelodus parvidens* histological type of Gross, 1967). Most of the Devonian thelodont scales now known in Gondwana belong to the genus *Turinia* or its close relatives (Family Turiniidae). Scales of *Thelodus* sensu stricto (type taxon of the Thelodonti and family Thelodontidae), however, have been found in the Silurian of Western Australia (Turner, 1993). Scales of Thelodontidae are generally simpler than scales of Turinidae, especially in the crown, and thus are kept in a separate family, but both families share typical orthodentine as the main structure of crown and neck, and both are now known to exhibit complex crown microornament (Turner, 1994; Turner and Blieck, 1995a).

Young (1995) has summarized recent work on Australian microvertebrates and discussed Turner's (1993) provisional scheme of five turiniid thelodont assemblages: the oldest spanning the Siluro-Devonian boundary; the youngest early Late Devonian. Turner (1995a, b) has revised the taxonomy of the

Turner, S., 1997, Sequence of Devonian thelodont scale assemblages in East Gondwana, *in* Klapper, G., Murphy, M. A., and Talent, J. A., eds., Paleozoic Sequence Stratigraphy, Biostratigraphy, and Biogeography: Studies in Honor of J. Granville ("Jess") Johnson: Boulder, Colorado, Geological Society of America Special Paper 321.

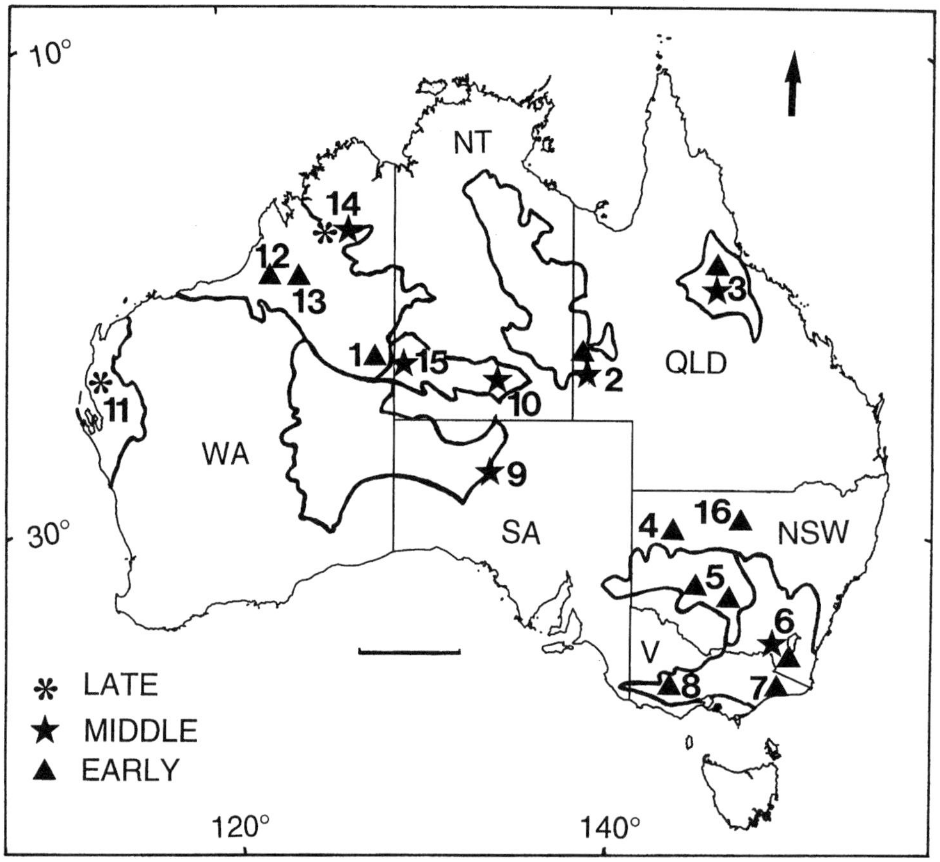

Figure 1. Generalized location of turiniid sites in the Devonian basins of Australia: 1, Wilson's Cliffs (Gross, 1971); 2, Toomba Range, Toko Syncline (Turner et al., 1981); 3, Broken River district; 4, Mt. Jack/Cobar (Turner et al., 1981); 5, Mineral Hill region, Murda Syncline, Trundle, Boor Hill west of Parkes; 6, Hatchery Creek Conglomerate site (Young and Gorter, 1981) and nearby Taemas/Wee Jasper and Tumut Block; 7, Buchan; 8, Grampians (Turner, 1986); 9, Munyurai-1 subsurface in Officer Basin (Long et al., 1988); 10, Ross River (Young et al., 1987); 11, type Gneudna section, Williambury Station in Carnarvon Basin (Turner and Dring, 1981); 12, Gingerah Hill 1 borehole; 13, Pegasus 1 borehole; 14, BC Lawford Range (Turner, 1994) and Horse Springs section; 15, Mt Winter, Cleland Hills (Young and Turner, 1992); 16, Mt Gunderbooka. Of the numbered locations, 1 and 12-14 in Canning Basin; 2 is in Georgina Basin; 3 and 6–8 are in Lachlan fold belt; 4 and 5 are in Darling Basin; 10 and 15 are in Amadeus Basin. States of Australia: NSW—New South Wales; NT—Northern Territory; QLD—Queensland; SA—South Australia; V—Victoria; WA—Western Australia. (Map modified from Palfreyman, 1984; scale bar = 500 km).

thelodonts and reassessed their biostratigraphic and paleobiogeographic significance. Thelodont taxa that may provide the basis for regional zonation are designated. The Australian assemblages (Fig. 1) are compared with those known from elsewhere in Gondwana, especially the Devonian turiniid scale assemblages known from conodont-dated marine sequences in West Yunnan, China, and in Iran. They occur also in nonmarine or marginal marine settings in Antarctica and Bolivia.

Explanations of thelodont biology and methodology of defining thelodont taxa can be found in Karatajute-Talimaa (1978), Märss (1986), and Turner (1991b), hence will be reviewed only briefly here. In the absence of complete fish or articulated squamation, as with conodonts, the method employed is a multielement approach that uses an assemblage of associated scales rather than creating new species on individual scale morphology. An attempt to reconstruct a squamation pattern or "map" is made by recognizing scales from the head, thorax (= combined cephalothorax), and trunk and fin regions by comparison with those morphological series already worked out from articulated specimens. In the usual case, all scales from a single sample are placed in the same species, providing they are morphologically and histologically similar.

HISTORICAL BACKGROUND

When thelodont scales were first found in Australia, it was reasonable to compare or identify them with those then known in the Northern Hemisphere. At that time, turiniid thelodonts

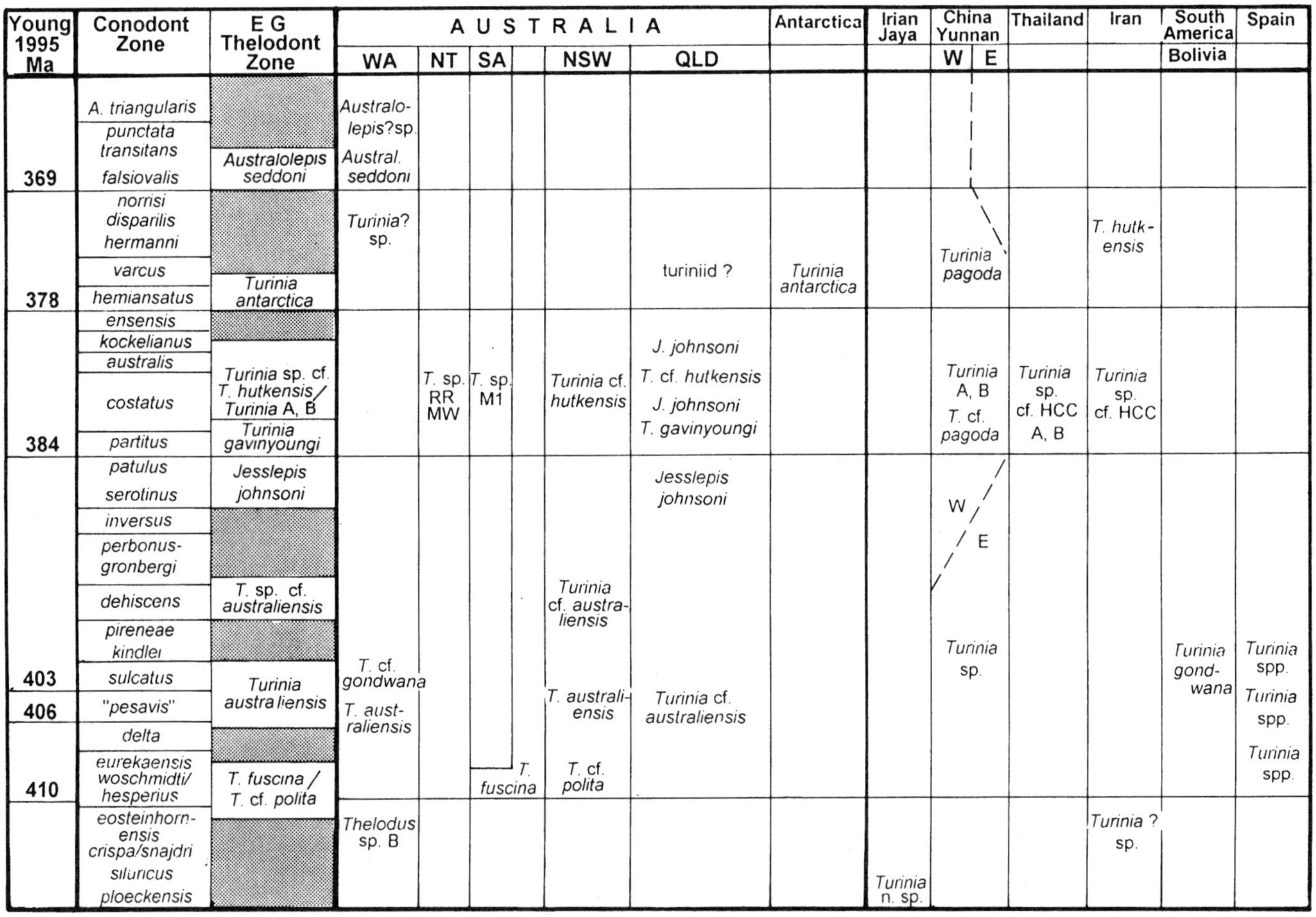

Figure 2. Upper Silurian and Devonian thelodont occurrences in East Gondwana and suggested correlations with Devonian zonation chart for East Gondwana. Dates and conodont zonation from Young (1995). Abbreviations: Ma—million years; EG—East Gondwana; HCC—Hatchery Creek Conglomerate; M1—Munyurai 1 borehole; MW—Mount Winter; RR—Ross River. Other abbreviations explained in text.

were known only from the type species, *Turinia pagei* Powrie, from the Lower Devonian Old Red Sandstone of Scotland (Traquair, 1899) and elsewhere in the Northern Hemisphere and from what were thought until recently to be exclusively nonmarine sediments (Jones, *in* Johnstone et al., 1967; Turner et al., 1981). In 1971, Walter Gross made the first formal identification of thelodonts from Australia when he described a new species, *Turinia australiensis*, based on scales from borehole Wilson Cliffs No. 1, Southeast Canning Basin, Western Australia (Fig. 1, location 1). The scales came from beds containing Permian spores, but Gross argued for a Lower Devonian age because the turiniid scales resembled those of *Turinia pagei* known from "Gedinnian-early Siegenian" in Europe. He regarded the scales as possibly reworked into the Permian. Turiniid scales (*T. pagei* and other taxa) are now known in abundance from marine sequences of Devonian age, and it is believed that these thelodonts lived in shallow marine environments but were capable of entering freshwater river systems (Turner, 1997). Although it now seems that there are no *T. pagei* scales in East Gondwana, comparison of Australian turiniid scales with those of *T. pagei* is still useful because the latter is the only turiniid known from articulated specimens. These are the holotype and two small patches of a few scales each from Scotland and one further partially articulated specimen from the Welsh Borderland (Turner, 1997). *T. pagei* thus illustrates the range of variation of scales within a taxon. The relatively good preservation of the holotype has provided a model for recognition of a wide range of scale morphotypes in the taxon (Gross, 1967). Large thelodonts, such as *Turinia pagei*, are 400 mm long and would have had a potential of 30,000 to 40,000 scales in the exoskeleton (Turner, 1991b). Even if the rate of scale shedding was somewhat slower than in modern sharks, this explains the large accumulations of scales found in Silurian and Devonian bone beds. The wide range of scale size (generally 0.5 to 3.0 mm) and particularly the morphological variation in turiniid scales, exemplified by *T. pagei* (e.g., Karatajute-Talimaa, 1978), needs to be taken into account in establishing a new taxon.

Scales of *T. pagei* predominate in many Lochkovian sites

around the Old Red Sandstone Continent and as far east as the Urals (Talimaa, 1995; Turner, 1997) and are thought to range into the Pragian (Turner, 1973; Talimaa, 1995). *T. pagei* is associated in an assemblage with one or more other turiniids, *T. polita* Karatajute-Talimaa (Lochkovian), *Boreania minima* Karatajute-Talimaa (latest Pridoli–Lochkovian), nikoliviids, and *Gampsolepis insueta* Karatajute-Talimaa. East Gondwana scales have been compared with each of these taxa in the past (Turner et al., 1981; Pickett et al., 1985), and based on these comparisons, they were thought to be indicative of earliest Devonian (Gross, 1971; Turner, 1973; Turner et al., 1981; Pickett et al., 1985). *T. pagei* has been recognized in Early Devonian microvertebrate assemblages in other presumed Gondwana terranes: in northern Spain (D. Goujet, personal communication, 1980); Turner 1984; Wang, 1993) and possibly also in the Nowshera Limestone of Pakistan. However, none of the above taxa, which are used as zone fossils in the Northern Hemisphere, have been confirmed in East Gondwana (Turner, 1993). Subsequent work and discoveries of additional assemblages throughout the Devonian in East Gondwana have brought to light several forms (Turner, 1995b), possibly endemic, that are considered here. This reevaluation of taxa has led to revision of the age of the first discovered turiniid assemblages in Australia. Those from Wilson's Cliffs, Canning Basin (Fig. 1, location 1) and from the Georgina Basin (Fig. 1, location 2), which Gross (1971) and Turner et al. (1981) thought were earliest Devonian, are late Pragian to early Eifelian (see below). Equivalent-aged thelodont scales from Europe come from Spitsbergen, the Baltic states, and western Russia (Karatajute-Talimaa, 1978) and belong to the genus *Amaltheolepis*. The scales of the latter genus are similar in histological structure to those of *Turinia*, but they differ morphologically (Karatajute-Talimaa, 1978). *Amaltheolepis* has not been found in the Southern Hemisphere.

Abbreviations

Abbreviations used in the text and in some of the figure captions are as follows: AGSO—Australian Geological Survey Organization, Canberra; BRJ—Broken River, J. S. Jell sample number; C numbers—John Pickett sample numbers in the collection at GSNSW; CPC—Commonwealth Palaeontological Collection, Canberra; CZ—conodont zone; GR—grid reference; GSNSW—Geological Survey of New South Wales, Micropaleontological Collection, Sydney; GSWA—Geological Survey of Western Australia; M1—Munyurai-1 Borehole; MC—Mount Crean; MMMC—Mineralogical and Mining Museum Collection, Sydney; MUCEP—Macquarie University Centre for Ecostratigraphy and Palaeobiology, Sydney; MW—Mount Winter; NSW—New South Wales; NT—Northern Territory; QD (or QLD)—Queensland; QMF—Queensland Museum, South Brisbane, fossil collection; QML—QD Museum locality number; RR—Ross River; SA—South Australia; SP—Shot Point Sample; UQL—University of Queensland, St. Lucia, Department of Geology locality numbers; V—Victoria; UWA—University of Western Australia, Nedlands, Geology Department; WA—Western Australia; WAM—Western Australia Museum, Perth; WCB—West Canning Basin sample.

MICROVERTEBRATE FAUNAL ASSEMBLAGES

Based on data in Turner (1993) and Young (1995) and the faunal assemblages listed below, I revise the scheme of Young (1995) to include eight Devonian East Gondwana turiniid scale assemblages, some of which occur with conodonts (Fig. 2). The following is a summary of the characteristic elements of each assemblage, the associated conodont fauna (if present), and the age and habitat inferred for each of them. Further discussion of the assemblages is given under stage and/or conodont zone headings in the subsequent section, "Fauna Assemblages."

Fauna 1

The *fuscina*- sp. cf. *polita* assemblage of Turner (1993) and Young (1995) is earliest Devonian, probably *woschmidti* CZ or slightly younger. *Turinia fuscina* is most like the contemporaneous *T. pagei* of the Northern Hemisphere (Turner, 1986). Other simple monolithic and crenulated scales in the fauna are comparable with *T. polita* (Pickett et al., 1985). Both types are typical of the Northern Hemisphere Lochkovian (Talimaa, 1995). There are no supporting conodont data in eastern Australia, but new information from Iran and Spain suggests a range from latest Silurian to earliest Devonian (Wang, 1993). The assemblage is believed to have been nonmarine in Australia (Young, 1995).

Fauna 2

The *Turinia australiensis* assemblage (Figs. 3, 4; Turner, 1993; Young, 1995) appeared during the late Lochkovian–early Pragian highstand in Australia. It is not possible to estimate the full variation expressed in the scales of *T. australiensis* on the basis of the seven syntype scales from the Canning Basin, WA (Turner, 1995a), but sufficient similarity exists between the syntypes and scales from central-western NSW to gauge the squamation. The lower range of the taxon is no older than upper *delta* CZ. The upper range is to at least *sulcatus* CZ in Australia. The

Figure 3. *Turinia australiensis* Gross, 1971. Scanning electron photomicrographs of examples of the range of variation seen in cephalopectoral scales from the Jerula Limestone Member (Gleninga Formation), late Lochkovian–early Pragian of western NSW (J. W. Pickett collection, C numbers, Appendix). Scale bars for A, D, F–H, J, K, M and N equal 500 µm; for B, C, E, I, and L, 200 µm. A–I—oral/head scales and J–N—cephalopectoral scales: A, GSNSW 2362 from C628; B, GSNSW2363 from C628; C, GSNSW 2364 from C626; D, GSNSW 2365 from C626; E, GSNSW 2366 from C625; F, GSNSW 2367 from C626; G, GSNSW 2368 from C626; H, GSNSW 2369 from C626; I, GSNSW 2370 from C 626; J, GSNSW 2371 from C626; K, GSNSW 2372 from C625; L, GSNSW 2373 from C625; M, GSNSW 2374 from C231. N, GSNSW 2375 from C231; GSNSW numbers = Geological Survey of New South Wales, Sydney.

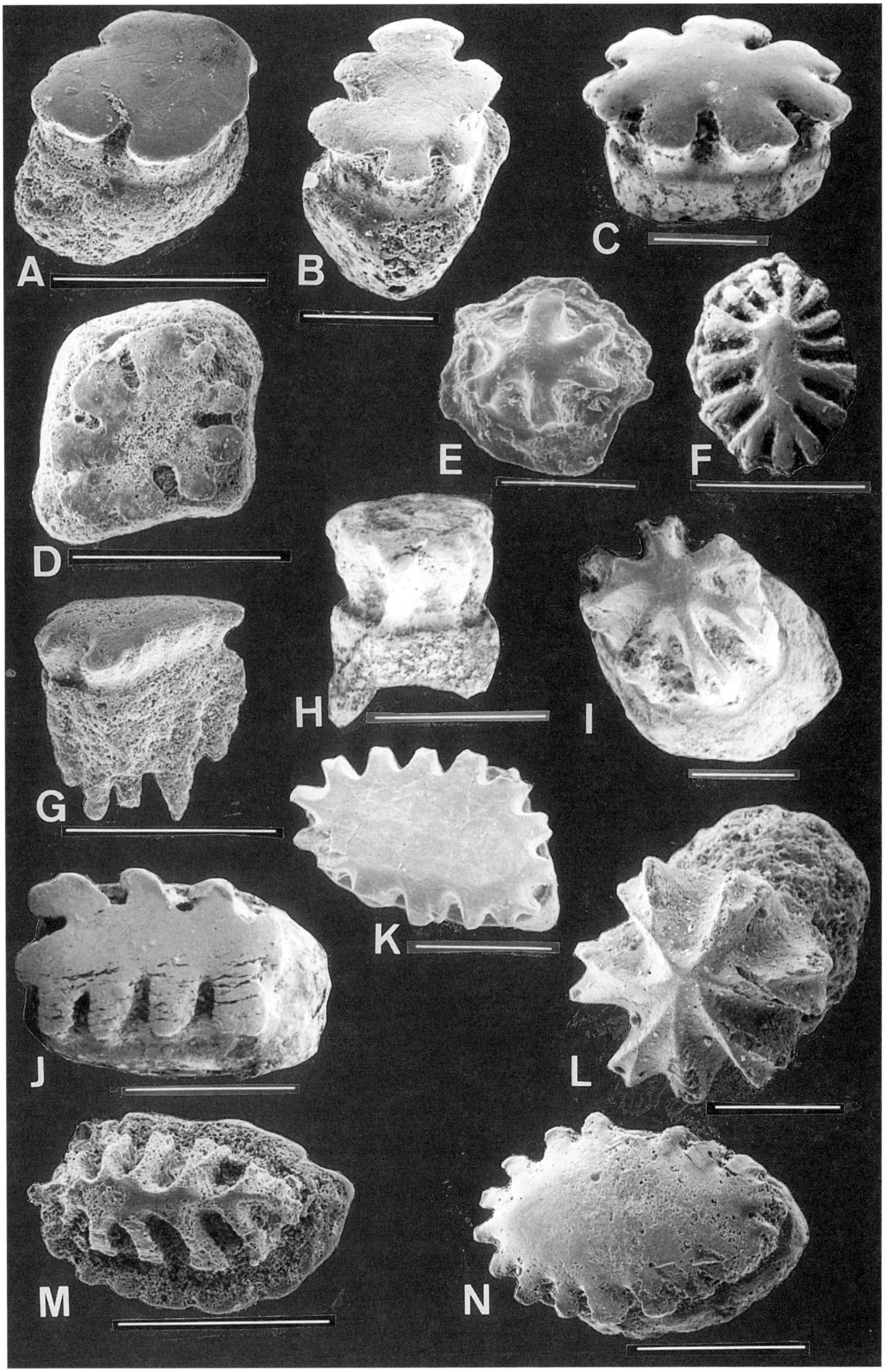
A
B
C
D
E
F
G
H
I
J
K
L
M
N

upper age limit could be early Emsian because of the possibly younger age of the Gleninga Formation (Fig. 5; Sherwin, 1994) and because of similarity to a few scales from Buchan and Wee Jasper of *dehiscens-perbonus* CZ age and to scales in some horizons of the Mulga Downs Formation. *Turinia australiensis* has not been encountered outside of Australia but seems closely related to scales of *Boreania minima* from the late Pridoli to early Lochkovian of the Old Red Sandstone Continent. The habitat is shallow marine in Australia.

Fauna 3

Fauna 3 is the *T.* sp. cf. *australiensis* Assemblage. The Taemas-Buchan microvertebrate fauna (Young, 1995) is now known to yield rare *T.* sp. cf. *australiensis* from the late *pireneae–dehiscens*–age limestones of the Cavan Formation and lower Buchan Group (A. De Pomeroy, personal communication, 1995), which may represent a new species. The habitat is marine.

Fauna 4

Fauna 4 is termed the *Jesslepis johnsoni* assemblage and comprises *J. Johnsoni* Turner (1995b) and *Turinia* spp. (Fig. 6 C–H). The assemblage appears in the late Emsian *serotinus* CZ and ranges into the Eifelian *kockelianus* CZ in the Broken River area, north Queensland (Fig. 1, location 3). The habitat is marine.

Fauna 5

The *Turinia gavinyoungi* Turner (1995b) fauna is found in the Cravens Peak Beds, Queensland. The age and range of this species are uncertain. It could be as old as the *pesavis* CZ if the original geological interpretation of the in situ beds is correct (Turner et al., 1981) or of Eifelian or even early Givetian age on the basis of the associated vertebrates (Young, 1995). Early Eifelian is the preferred top of the range because of similar early Eifelian scales in the Broken River Group (Turner, 1995b). The habitat is marginal marine to nonmarine.

Fauna 6

The *T.* sp. cf. *pagoda*-sp. cf. *hutkensis* Assemblage of Turner (1993) and Young (1995) comprises the Hatchery Creek Conglomerate fauna scales and the central Australian basin occurrences in eastern WA and NT. Similarities to *Turinia* spp. A and B of West Yunnan suggest an early Eifelian age (*costatus* CZ), but the assemblage may range into the Givetian. This assemblage is present also in Thailand and Iran. The habitat is nonmarine to marine.

Fauna 7

The *Turinia antarctica* Assemblage of Young, (1995) is early to middle? Givetian and occurs only in Antarctica (Fig. 7). A form that possibly is related, however, occurs in the late Givetian or early Frasnian of WA (J. A. Long, personal communication, 1996). The habitat is marginal marine to nonmarine.

Fauna 8

The *Australolepis seddoni* Assemblage (Turner, 1993; Young, 1995) is dated as early Frasnian in the Carnarvon Basin (Fig. 1, location 11) and in eastern Iran. *Australolepis seddoni* was described by Turner and Dring (1981) from the early Late Devonian Gneudna Formation, Carnarvon Basin, WA (Figs. 1, location 11; 8; 9, location 23). The thelodont-bearing beds were dated using conodont and palynomorph/acritarch evidence as earliest Frasnian. Scales are abundant from two horizons in the measured section (Turner and Dring, 1981, fig. 2). A possibly closely related form occurs in the Halls Creek district, north Canning Basin of WA in younger Frasnian-age rocks. The habitat is marine.

FAUNAL ASSEMBLAGES

The East Gondwana data sources (occurrences, stratigraphy, comments) for the faunal assemblages and proposed zones based upon them are given below in ascending stratigraphic order, with reference to conodont zone where applicable. Comparison is then made with similar forms from elsewhere in Gondwana to show the potential biostratigraphic utility of the assemblages. Locality information is given in the Appendix.

Late Silurian–Early Devonian (Pragian) assemblages

Although not yet included in the proposed sequence, turiniid and other thelodonts appeared in East Gondwana in the Late Silurian, as indicated by the localities mentioned below and in Turner (1991a, 1993). Thus, there is a possibility of extending thelodont biostratigraphy into the Silurian. Turner et al. (1995) described a probable mid-Ludlow thelodont fauna with turiniid and nikoliviid scales from erratic blocks dated by conodonts from Irian Jaya (Fig. 9, location 50). The turiniid scales resemble some from the upper Pridoli of Greenland. One trunk scale resembles *T. fuscina* from the late Pridoli? or early Lochkovian in Australia (see *woschmidti* CZ

Figure 4. *Turinia australiensis* Gross, 1971. Scanning electron photomicrographs of examples of the range of variation seen in postpectoral trunk scales from the Jerula Limestone Member (Gleninga Formation), late Lochkovian–early Pragian of western NSW (J. W. Pickett collection, C numbers, Appendix). Scale bars for A–F, H, and J–M scale bar equals 500 μm; for G and I, 200 μm. All figures except E, J, and L exhibit fine parallel ribbed microornament. A, GSNSW 2376 from C625; B, GSNSW 2377 from C233; C, GSNSW 2378 from C625; D, GSNSW 2379 from C625; E, GSNSW 2380 from C232; F, GSNSW 2381 from C231; G, GSNSW 2382 from C232; H, GSNSW 2383 from C231; I, GSNSW 2384 from C628; J, GSNSW 2385 from C231; K, GSNSW 2386 from C231, L, 2387 from C232; M, GSNSW 2388 from C232.

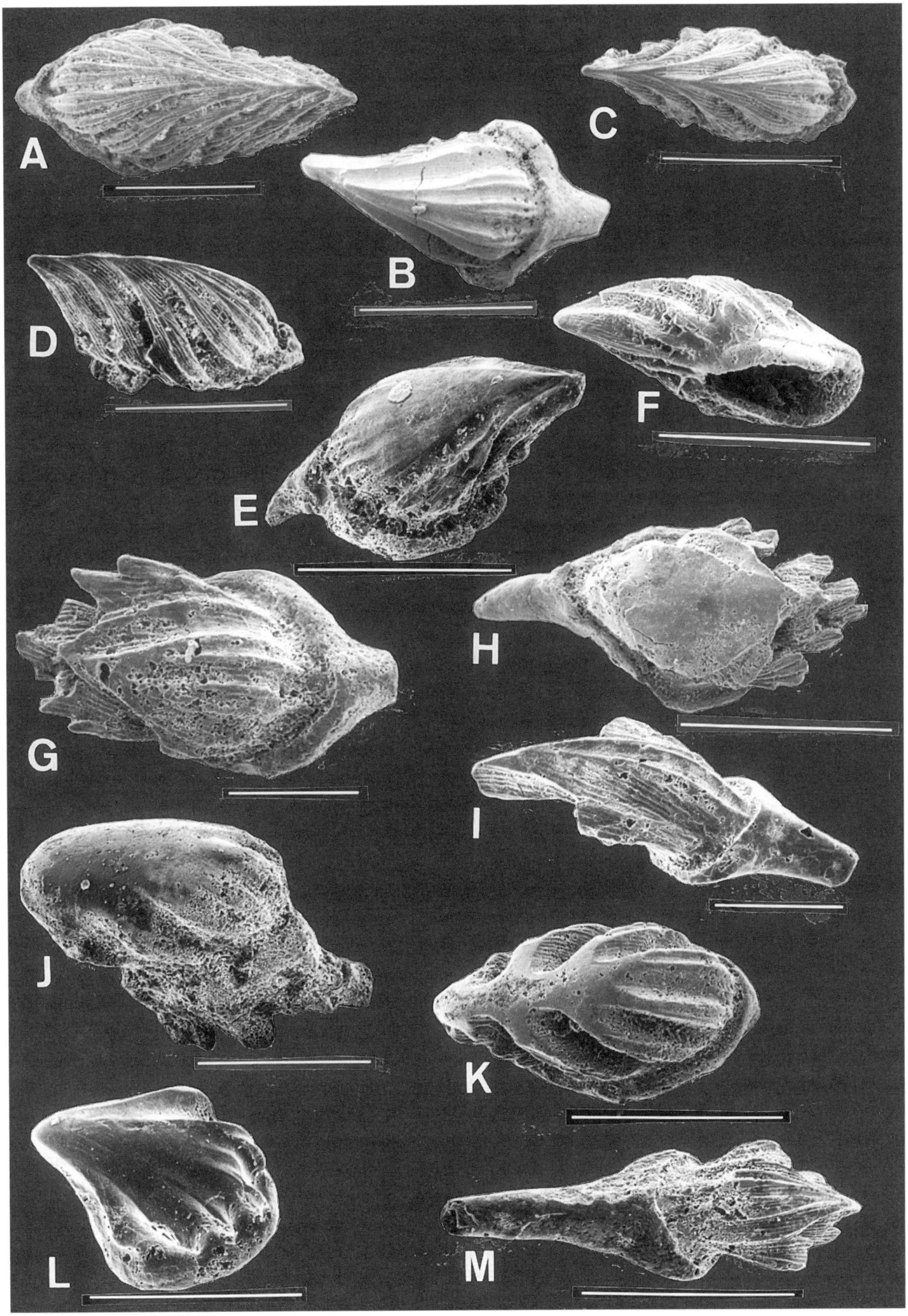
A
B
C
D
E
F
G
H
I
J
K
L
M

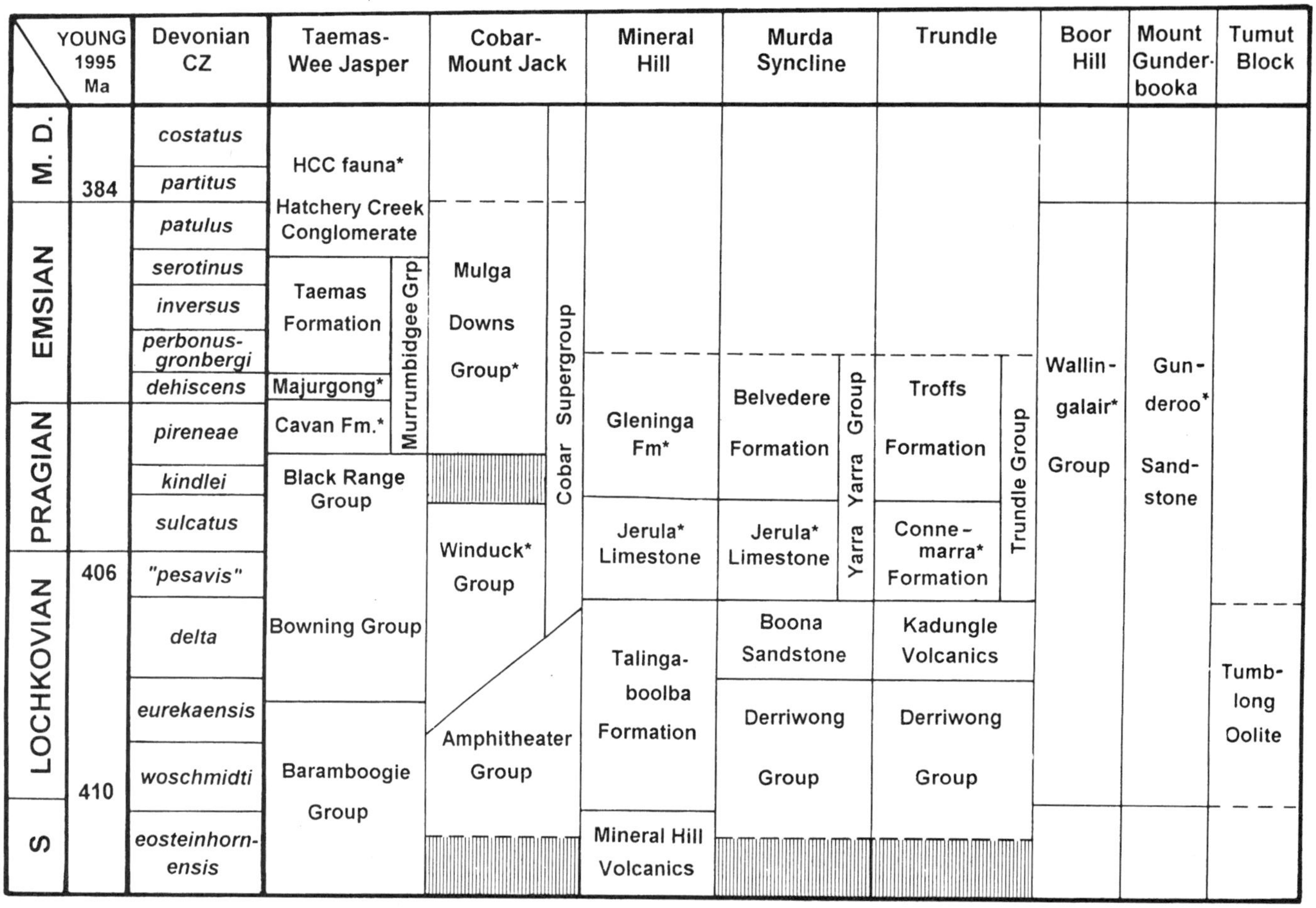

Figure 5. Early to Middle Devonian stratigraphy of south and central western NSW (Fig. 1, locations 4–7). General level of thelodont-bearing beds with known relation to conodont biostratigraphy marked with an asterisk. Stratigraphy based on information in Appendix and from J. W. Pickett, L. Sherwin, and J. A. Talent, personal communications, 1995–1996; Sherwin (1994). M.D.—Middle Devonian, S—Silurian, HCC—Hatchery Creek Conglomerate.

below) and a scale from Spain, which is now generally accepted as a former part of Gondwana (Fig 9, location 9), is identified as "*Canonia*" sp. by R.-H. Wang (1993), and is Lower Devonian (lower "Siegenian"). Young (in Turner and Blieck, 1995b) reported possible Silurian thelodonts from a borehole, Pendock 1 (Carnarvon Basin, WA), and a Silurian microfauna from Iran (G. C. Young, personal communication, 1995) contains scales that resemble *T. fuscina*. Some localities from NSW (e.g., C091, Appendix) with apparent Pridoli conodont dating, however, are suspect (J. Pickett, personal communication, 1996) because thelodont scales and other microvertebrates are like those from late Lochkovian–Pragian samples from the Yarra Yarra Creek and Trundle Groups.

Woschmidti ***conodont zone***. The *woschmidti* CZ ranges a little below the Silurian-Devonian boundary and so begins in the upper Pridoli. The thelodont assemblage found associated with conodonts of this zone is characterized by *Turinia fuscina* Turner and other taxa from the Silverband Member, Grampians Group,

Figure 6. Scanning electron photomicrographs of early Middle Devonian thelodont scales from Broken River Group limestones, Broken River Province, north Queensland, Australia (Fig. 1, location 3). Scale bar in all figures equals 500 μm. C–H from undifferentiated Broken River Group. A, *Turinia* cf. T. *australiensis* Gross, head scale QMF 33640, anterolateral view, from UQL 4736, Martins Well Limestone Member, Shield Creek Formation; B; lateral view of QMF 33640; C, *Jesslepis johnsoni* Turner, head scale QMF 33734, lateral view, from UQL 4704; D, *J. johnsoni* Turner, trunk scale, QMF 33735, lateral view from UQL 4704; E, J. *johnsoni* Turner, head scale, QMF 33736, postero-lateral view, from UQL 4707; F, *J. johnsoni* Turner, trunk? scale QMF 33737, ventral view, from UQL 4707; G, *J. johnsoni* Turner, trunk scale QMF 33738, lateral view, from UQL 4706; H, *J. johnsoni* Turner, trunk scale QMF 33739, lateral view, from UQL 4704; I, *Turinia* sp., head scale QMF 33641, lateral view, from UQL 5950; Chinaman Creek Limestone Member, Broken River Group; J, crown view of QMF 33641; K, *Turinia* sp. aff. *T.* sp. B of West Yunnan (Wang et al., 1986), head scale, QMF 33642, lateral view, from UQL 4706; L, *Turinia* sp. aff. *T.* sp. B, trunk scale QMF 33643, from UQL 5950; M, Postero-lateral view of QMF 33643.

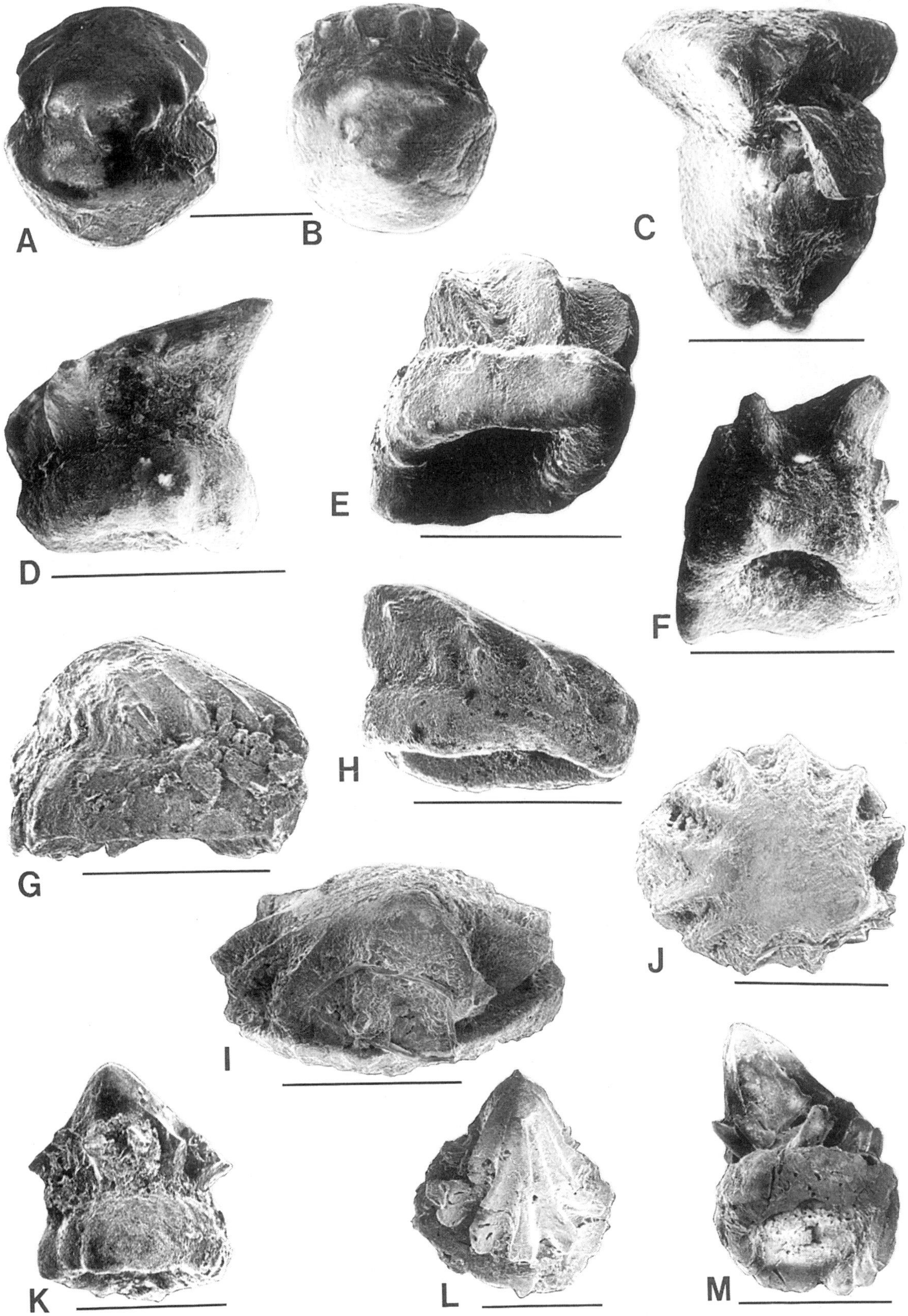
A
B
C
D
E
F
G
H
I
J
K
L
M

of the Grampians of western Victoria (Fig. 1, location 8). Isotopic dates for associated intrusive rocks indicate a youngest age for the Grampians Group of 400 ± 3 Ma (Young, 1995).

The Tumblong Oolite microfauna (sample C800, Pickett et al., 1985; Ross Talent collection at MUCEP) is rich in simple turiniid and nostolepid scales associated with scales of the form genus *Ohiolepis*. Tricuspidate trunk scales from Tumblong resemble those of *T. fuscina* and, along with the *T. polita*–like scales, suggest an early rather than late Lochkovian date (Pickett et al., 1985).

***Upper Lochkovian–Pragian,* "pesavis"-sulcatus *conodont zones*.** The upper Lochkovian *pesavis* Zone of the Cordilleran region of North America has been revised by Valenzuela-Ríos (1994), but the term "*pesavis*" Zone is used here in the sense of upper Lochkovian in accordance with practice in Australia to this point.

Rare limestone outcrops in central-west NSW and in the Mineral Hill and Trundle areas have yielded varying quantities of thelodont scales, referred here to *Turinia australiensis* or *T.* cf. *australiensis* (Fig. 5; C localities, Appendix). Because the range of variation is greater and preservation better than the type material of both taxa (Turner, 1995a), additional examples are illustrated here (Figs. 3, 4). Parallel anterior-posterior microornament on the slanting crown ribs on nearly all trunk scales is present (Fig. 4). However, some are relatively simple and not typical of *T. australiensis*. They are more like *T. polita* in their high sided crowns with simple crown rims (Fig. 3D).

Certain beds within the Trundle Group, especially Connemara Formation, and Yarra Yarra Creek Group (Fig. 5; Gleninga Formation, Jerula Limestone Member, terminology from Sherwin, 1994, 1997, and personal communication, 1995) are characterized by abundant *Turinia australiensis.* Identifiable conodonts all fall within the "*pesavis*"-*sulcatus* zones (Pickett, 1992; Young, 1995). Samples C214–216, C231–234, C287, C600, C624–628, and C661–662 (Appendix) are particularly rich in *T. australiensis* scales (Figs. 3, 4) and could all represent the same horizon (J. Pickett, personal communication, 1995). The Connemara Formation contains a *Howellella jaqueti* brachiopod fauna; some beds yield "*pesavis*" Zone conodonts (Pickett, 1992). According to Pickett (1992), the turiniid-bearing samples from the Jerula Limestone Member contain conodonts of the *sulcatus* CZ. Sherwin (1997) regards at least part of this formation as late Pragian to early Emsian on the evidence of brachiopods, especially abundant *Spinella* assemblages.

Turner (1995a) discussed the age and significance of *Turinia australiensis* based on the material available to Gross (1971). She also published scanning electron microscope (SEM) photomicrographs of the type species in support of Gross's decision to erect a new taxon based on a microornament of minor cross ribbing on the main crown ribs that seems to be a significant taxonomic character (Turner, 1994, discussion below). Turner (1993) referred scales, especially those from shallow marine deposits dated by conodonts from central-western NSW (Fig. 1, location 5) to *T. australiensis,* but the variation exhibited in her sample (Figs. 3, 4) is greater than that shown by the few specimens Gross illustrated.

The syntypes of *T. australiensis* are small scales that include characteristic oral and cephalopectoral scales with a simple crenulated rim, but trunk scales are lacking. The greater variety of *T. australiensis*–type scales seen in the central-western NSW Early Devonian assemblages means that additional species may be present or that the new sample is more representative of the taxon. Based on the conodont-dated samples from western NSW, Turner (1995a) postulated a late Lochkovian–early Pragian range for *T. australiensis*, although the range may extend into Emsian sequences (Young, 1995). Turner (1973) also compared most of the scales from the Toko Syncline in western Queensland (Fig. 1, location 2) with *T. australiensis* and noted a close similarity of the smaller scales with turiniids found in the Lower Devonian ("Gedinnian") of the Welsh Borderland and Podolia. In Turner et al. (1981), some of the Welsh Borderland scales were figured and compared with *T. australiensis*. In that paper, scales from the rich bone beds of the Mulga Downs Group of NSW (Fig. 1, location 4), also presumed to be *T. australiensis*, were illustrated. Turner (1984) identified the Welsh Borderland scales as a new form, *Turinia minima*, which Karatajute-Talimaa (1985) subsequently referred to a new genus, *Boreania*, based on material from Podolia and the Baltic. Some of the Australian turiniid scales, particularly those from the Georgina Basin shot-point samples (Turner et al., 1981, fig. 4), resemble those of *Boreania minima*, which Talimaa has now used as an indicator for highest Pridoli–Lochkovian. The Australian forms are retained in the genus *Turinia*, as they differ from *Boreania* in the presence of a style of microornament that is shared with other species of *Turinia* from East Gondwana (*T. australiensis*, *T. pagoda*, *T. antarctica*, *T.* cf. *pagoda*, and *Australolepis*? sp.). The recognition of microornament as a useful taxonomic character in East Gondwana turiniids, appearing first in *T. australiensis*, may help define phylogenetic lineages.

One of the current problems in Australia for understanding the range and distribution of *T. australiensis* and other turiniids is the age range of Young's (1993) *Wuttagoonaspis* macrovertebrate assemblage. Long et al. (1988) argued for a late Emsian–early

Figure 7. *Turinia antarctica* Turner and Young, 1992, from the Aztec Siltstone, well-preserved material from Mount Crean, Transantarctic Mountains, Antarctica (J. A. Long collection, all but B and D from MC3, Queensland Museum), with up to five side lappets with parallel ridge microornament and showing the tuberculated ridges considered to be the specific character for the taxon. Scale bar in all figures except B and G is 500 μm; B and G—1 mm. C, D, F, H, and I show microornament. A, head scale QMF 33810, crown view; B, head scale QMF 33811 from MC4, lateral view; C, head scale, QMF 33812, posterior view; D, cephalopectoral scale QMF 33813 MC 4, anterolateral view; E, cephalopectoral scale QMF 33814 with outturned ribs at crown-neck interface, lateral view; F, trunk scale QMF 33815, crown view; G, trunk scale QMF 33816 with long basal process, anterodorsal view; H, trunk scale QMF 33817 crown view; I, broken trunk scale QMF 33818, lateral view.

A
B
C
D
E
F
G
H
I

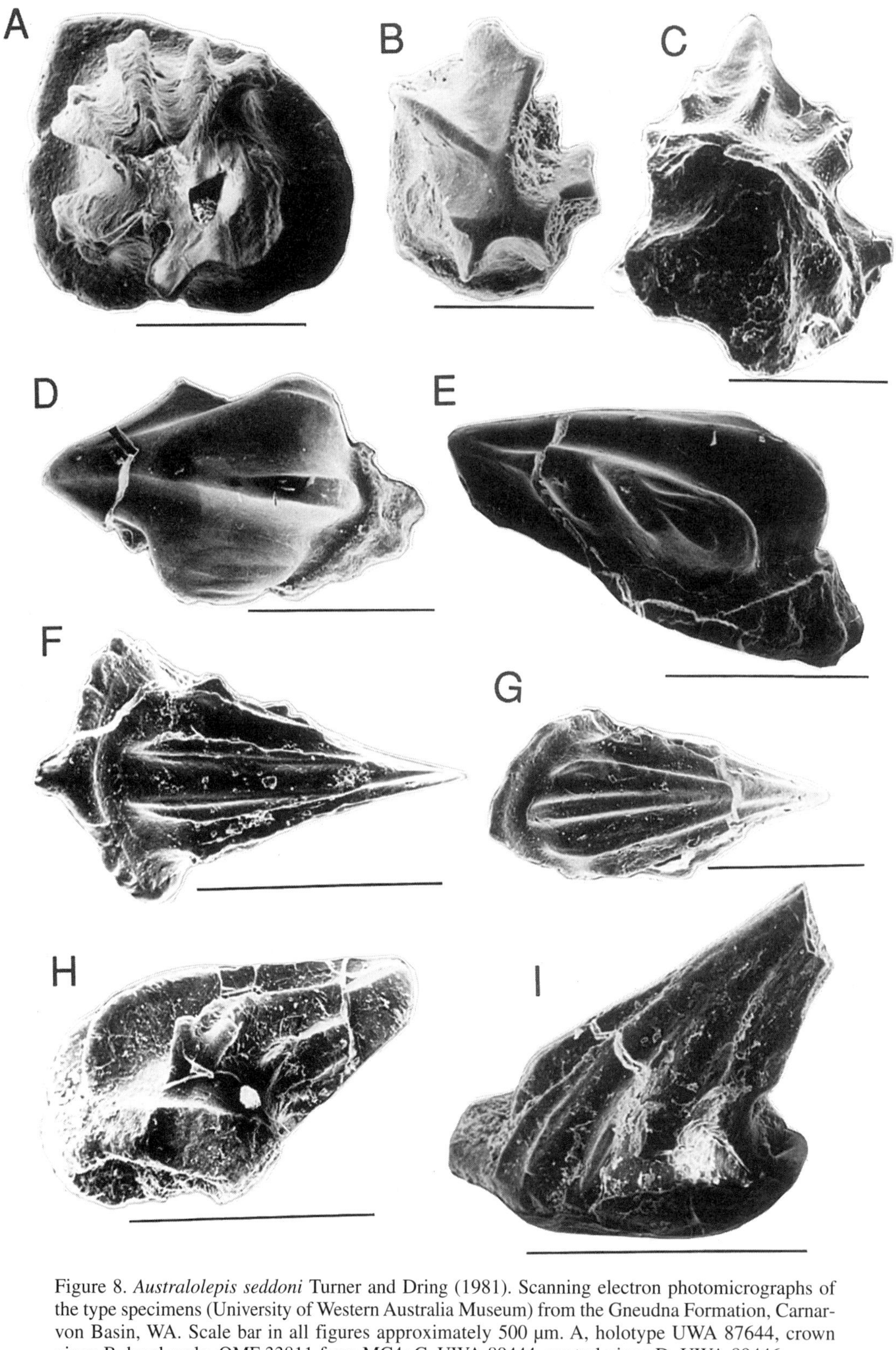

Figure 8. *Australolepis seddoni* Turner and Dring (1981). Scanning electron photomicrographs of the type specimens (University of Western Australia Museum) from the Gneudna Formation, Carnarvon Basin, WA. Scale bar in all figures approximately 500 µm. A, holotype UWA 87644, crown view; B, head scale, QMF 33811 from MC4; C, UWA 89444, ventral view; D, UWA 89446 crown view; E, UWA 89441, lateral view; F, UWA 89438, crown view; G, UWA 88440, crown view; H, UWA 88441, lateral view; I, UWA 89451, lateral view. A and I broken during photography.

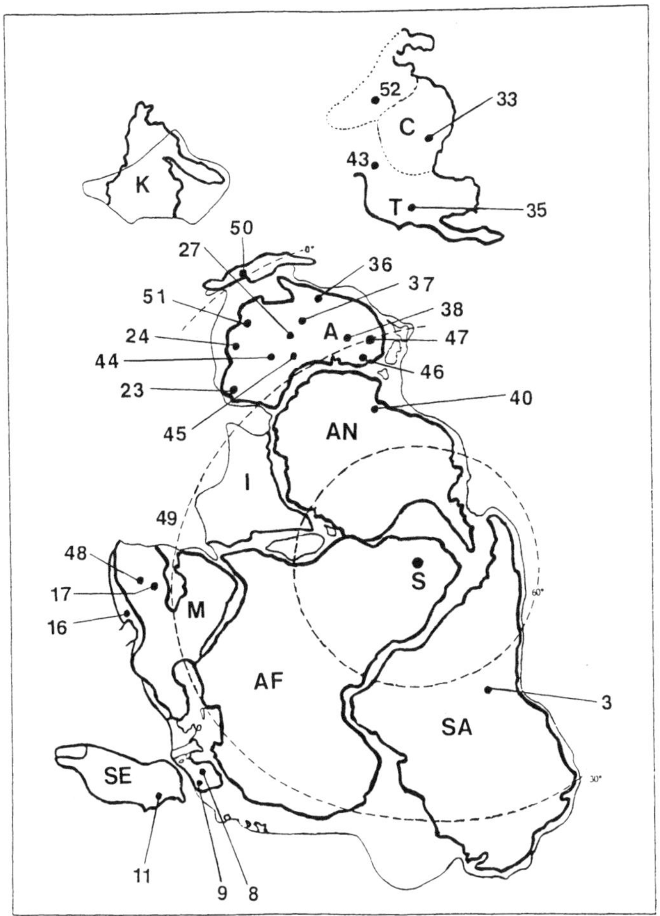

Figure 9. Turiniid distribution in Gondwana. Map modified from Young (1987, fig. 1). Numbered localities between 3 and 40 after Young (1987) where they contain thelodont faunas; additional numbered localities are discussed in the text: 3, *T. gondwana* Turner (in Gagnier et al., 1988), Bolivia; 8, 9, *T. pagei/fuscina* assemblage (Turner, 1984; 1997; Wang 1993), Spain; 11, *T. pagei* assemblage (Turner, 1997), N. France–Belgium; 16, *T.* sp. (Turner and Janvier, 1979), Khush-Yeilagh, Iran; 17, *T. hutkensis* Blieck & Goujet (1979), Hutk, Iran; 23, *A. seddoni* Turner & Dring (1981), Carnarvon Basin, WA; 24, *Turinia* spp. Canning Basin, WA; 27, *Turinia* cf. *pagoda*, (Young et al., 1987), west and east Amadeus Basin, central Australia; 33, *Turinia* spp. (e.g., Wang, 1992), S. China including Yunnan and Guangxi Provinces; 35, *Turinia* sp., Chiang-Mai; 36, *Turinia* spp. + *Jesslepis johnsoni* Turner (1995b), Broken River, north Queensland; 37, *T. australiensis* (Turner et al., 1981), *T. gavin-youngi* Turner (1995b), Toomba Range, western Queensland; 38, *T. australiensis*, Cobar district, western NSW; 40, *T. antarctica* Turner and Young (1992), southern Victoria Land; 43, *T. pagoda* and *T.* spp. A, B Wang et al. (1986) Baoshan, W. Yunnan; 44, *T. australiensis* Gross (1971) Wilson's Cliffs, WA; 45, *T.* cf. *pagoda* and *T.* sp. (Long et al., 1988), eastern Officer Basin, Southern Australia; 46, *T. fuscina* Turner (1986), Grampians, Victoria; 47, *T.* cf. *hutkensis* (Young and Gorter, 1981); *T.* spp. (Pickett et al., 1985), Buchan, Wee Jasper, Tumblong, NSW; 48, *A.* cf. *seddoni*, eastern Iran; 49. *T.* cf. *pagei* (?) Pakistan; 50, *T.* sp. (Turner et al., 1995), Lorentz River, eastern Irian Jaya; 51, *A.?* sp. cf. *A. seddoni* Hall's Creek district, WA; 52, *T.* sp. (Wang 1995b), Qinling Mts., Gansu Province, north China. A = Australia; AF = Africa; AN = Antarctica; C = China; I = India; K = Kolyma; M = Middle East; S = South Pole; SA = South America; SE = southern Europe, T = Thailand.

Eifelian age for the assemblages containing turiniid scales and *Wuttagoonaspis* because it is constrained by occurrences in the Cobar Supergroup and differences from the Hatchery Creek fauna (Fig. 5) and those in central Australia (Munyurai-1, Ross River, Mount Winter; Fig. 1, locations 9, 10, 15 respectively). With new evidence of similar turiniid scales in the *dehiscens-perbonus* CZs from the Taemas-Buchan sections in NSW, and Young's (1995) assessment that the *Wuttagoonaspis* assemblage could be as old as early Emsian, contemporary with the Taemas-Buchan faunas, then the Mulga Downs Group should not be much younger than the Amphitheatre Group (Sherwin, 1995, fig. 3, cf. with Fig. 5)

In some horizons within the Mulga Downs Group (e.g., AMF56264, Mount Jack, Fig. 1, location 4), *T.* sp. cf. *australiensis* is present as an almost continuous layer of scales surrounding *Wuttagoonaspis* and other placoderm plates (e.g., Turner et al., 1981, fig. 16). Thus, Turner et al. (1981) made a tentative correlation of a similar assemblage in the Georgina Basin with the *Wuttagoonaspis* fauna of western NSW. The *Wuttagoonaspis* macrofauna is well represented in the Georgina Basin (Cravens Peak Beds and lower Dulcie Sandstone), where a marine interval of short duration was suggested on the evidence of associated ostracods and eridostracans in a basal limestone unit of the Cravens Peak Beds (Turner et al., 1981). The *Wuttagoonaspis* assemblage has also been found in the upper part of the Mereenie Sandstone of Mount Winter in the Amadeus Basin (Young and Turner, 1992) and other localities in central NT (Young, 1995). This apparent short-lived transgression in central Australia was thought to be at or near the Emsian-Eifelian boundary (Long et al., 1988) but based on the presence of *T. australiensis* or *T.* cf. *australiensis* could have been as early as the "*pesavis*"-*sulcatus* highstand or related to the *pireneae-dehiscens* transgression, as discussed by Talent and Yolkin (1987).

No bone bed has been found in limestone units in the Broken River region of north Queensland (Fig. 1, location 3), but microvertebrates are common in detrital lenses (Turner, 1991a, 1993; De Pomeroy, 1994), even though thelodont scales are scarce. The Lower Devonian Martins Well Limestone Member, Shield Creek Formation, which spans the *pesavis-sulcatus* CZ (Lochkovian-Pragian) boundary (Mawson et al., 1989), has yielded a few scales attributed here to *Turinia australiensis* (Fig. 6 A, B).

T. australiensis from shot-point samples from the Cravens Peak Beds of Georgina Basin, western Queensland, are rich in oral/head scales comparable with the type specimens of the species (Turner et al., 1981, figs. 3, 9, 10A–I). These samples could have come from as low as the "*pesavis*" CZ.

Borehole information from the Canning Basin (Fig. 1, locations 12, 13) shows a wide distribution of *T. australiensis* in WA and central Australia, probably related to the "*pesavis*"-*sulcatus* highstand. At Gingerah Hill No. 1, interval 800.5 to 801.5 m, there are 15 *Turinia* cf. *australiensis* scales and one resembling *T. gondwana* Turner (in Gagnier et al., 1988). This assemblage suggests a Pragian or Emsian age (CPC, G. C. Young stub 89/30). *Turinia* sp. cf. *T. australiensis* and *Onychodus* teeth in supposed

Tandalgoo Formation from Pegasus 1 borehole indicate a maximum age range of late Lochkovian to early Emsian (CPC, G. C. Young stub 93/4).

Based on the distribution data, the range of *T. australiensis* in East Gondwana may extend from Lochkovian (upper "*pesavis*" CZ) to early Emsian (*dehiscens-perbonus* CZs at the youngest). The remaining problem relates to the definition of the taxon because of the apparent great variation of scale types. Most samples contain only cephalothoracic scales and not trunk scales. The bias may be due to winnowing in the high-energy, shallow-water limestones, with the trunk scales tending to be more vulnerable to breakage and wear. However, as seen in some NSW examples (Fig. 4; locality details in Appendix), the trunk scales can be perfectly preserved and show a far greater range of complexity than seen in the type material illustrated by Gross (1971). This complexity may be from either specific differences or infraspecific variation.

Emsian, **dehiscens-perbonus** *conodont zones*

Thelodont scales have been found in the Cavan and Marjurgong Formations, Taemas Group, Taemas/Wee Jasper, NSW (Fig. 1, location 6) and the Buchan Caves Limestone, Buchan Group, Buchan, Victoria (Fig. 1, location 7) in measured, conodont-dated sections (Mawson, 1987; De Pomeroy, personal communication, 1995). The turiniid scales found in marine limestones dated as *dehiscens-perbonus* CZ (SLO and SL samples of Mawson et al., 1992, from the Basin, northeast of Buchan, Section SLO 60) are most like *Turinia australiensis* from Pickett sample C626 (Figs. 3N, 4M). A few head scales also occur in samples SL 32, SLO 40, SLO 60-80 (locality details in Mawson et al., 1992.) and one trunk scale possesses multiple parallel microornament and a complex, partitioned crown. The head/cephalopectoral scales, however, are typically turiniid, but notable are the rounded cleft ribs on the crown rim, which are comparable with those from the Cravens Peak shot-point samples. These scales resemble *T. australiensis* but could be a new species. A single head scale from the top of the Cavan Formation found by A. De Pomeroy (MUCEP sample YWJ117, *dehiscens* CZ) also has deeply cleft ribs similar to scales from the Cravens Peak Beds shot-point samples.

Thelodont scales as natural molds

A. Ritchie has identified fish remains from the Coonardoo Sandstone (Wallingalair Group, late Lochkovian–Pragian), Boor Hill, west of Parkes, central western NSW (Fig. 1, location 5) as plates of *Wuttagoonaspis*? (Sherwin, personal communication, 1993). Associated are molds of *Turinia* sp. scales. Similar preservation is found from the Gundaroo Sandstone at Mt. Gunderbooka, NSW (Figs. 1, location 16, 5; Appendix) and the Merrimerriwa Formation (Mulga Downs Group) near Cobar, NSW (QMF25956-25960; Appendix). These examples are identified as *T.* sp. cf. *T. australiensis.* They may, therefore, be of late Pragian to Emsian age (*pireneae-perbonus* CZs). As the scales are preserved only as natural molds, there is no possibility of determining their histology as an aid to identification. That the Mulga Downs Group scales may belong to a higher Emsian assemblage cannot be ruled out, especially now that scales comparable with *T. australiensis* have been found in *dehiscens-perbonus* CZ samples from Buchan and Wee Jasper.

Late Early to Middle Devonian assemblages

Serotinus-kockelianus ***conodont zones.*** This part of the column is represented in Queensland by a small number of thelodont scales mostly referred to *Jesslepis johnsoni*, named in honor of the late J. G. "Jess" Johnson, to whose memory this volume is dedicated. These rare specimens of late Emsian–Eifelian age that occur in the marine limestones of the Broken River Group (Withnall and Lang, 1993; Sloan et al., 1995) are believed to be endemic (Turner, 1993; 1995b). The type locality for the new taxon (Digger's Creek traverse, UQL4704) is within the upper *serotinus* CZ (R. Mawson, personal communication, 1995) and its range spans the *serotinus* to *costatus* CZs (Turner, 1995b). The scales come from samples in the "Burges Formation" (Withnall and Lang, 1993 = undifferentiated Broken River Group of Sloan et al., 1995), notably from Mawson and Talent's (1989) "SAG" section through the Chinaman Creek Limestone (Jessy Springs outcrop tract). R. Mawson (personal communication, 1995) reports that the SAG horizon predates the *ensensis* CZ. One other scale from the Eifelian, initially assigned to *Turinia* sp. (De Pomeroy, 1995), has been referred to *Jesslepis johnsoni* (Turner, 1995b).

A few scales of *Turinia* (Fig. 6 I, J, L, M) were also found in the Chinaman Creek Limestone (UQL5950) and one other scale of this type (Fig. 6 K) comes from the Diggers Creek locality UQL4706 (Turner, 1995b). These few scales resemble the lower Eifelian scales from Ross River Syncline, Hatchery Creek Conglomerate, and West Yunnan, China and, thus, are potentially helpful for correlating marine and nonmarine strata in East Gondwana.

The basal Mt. Podge Limestone of Mt. Podge, Burdekin region, north Queensland (QML1011), has yielded one cephalopectoral turiniid scale identified as *Jesslepis*? sp. from a block with the pentamerid brachiopod, *Zdimir*. Conodonts from this sample indicate the *serotinus* CZ (Talent and Mawson, 1994).

The Cravens Peak Beds of the Georgina Basin (Fig. 1, location 2) in western Queensland have yielded thelodonts that may represent two assemblages of different ages, according to Turner (1991a, 1993) and Young (1995). Several occurrences in marginal marine and presumed nonmarine sediments in the Georgina Basin, the Amadeus Basin (at Ross River and Mt. Winter), and the Officer Basin (Munyurai-1 borehole) were previously thought to be approximately coeval with the *T. australiensis* microfauna but are now considered to be younger (Turner, 1993; Young et al., 1987; Long et al., 1988; Young and Turner, 1992). These are provisionally placed in the Eifelian, although a younger age cannot be excluded (Young, 1993).

Turner (1995b) classified the advanced turiniid scales that Turner et al. (1981) compared with *T. pagei* as a new species, *T. gavinyoungi*, which may be closely related to the Hatchery Creek Conglomerate scales of NSW. The Queensland scales are better preserved and exhibit multiple ribbing and tiers of lappets, as well as microornament. Most come from the Cravens Peak Beds limestone that yielded teeth of the shark, *Mcmurdodus* (Young, 1995), now regarded as younger than the Cravens Peak shot-point samples (see "*pesavis*"-*sulcatus* CZ section) and are characteristic of the in situ Cravens Peak assemblage. *Mcmurdodus* is a genus known also from the Aztec Siltstone, Antarctica, in presumed younger Givetian horizons (Turner and Young, 1992; Young, 1995). Young (1988a), however, considered that the Cravens Peak Beds and Dulcie Sandstone *Wuttagoonaspis* assemblage could not be as young as the Hatchery Creek Conglomerate fauna because of the absence of bothriolepids.

Costatus-kockelianus? *conodont zones.* Twenty scales from a gray-green sandstone 1900 m above the base of the Hatchery Creek Conglomerate near Wee Jasper, NSW (Fig. 1, location 6), were referred to *Turinia* sp. cf. *T. hutkensis* Blieck and Goujet by Young and Gorter (1981). They compared the scales with those of *Turinia hutkensis* of Hutk, Iran, of supposed late Givetian or earliest Frasnian age (Lelièvre et al., 1993) and also noted that the Hatchery Creek Conglomerate scales are more elongate and thornlike than those of *T. pagei*, *T. polita*, and *T. australiensis*. Additional material exhibits characters shared with other Eifelian turiniid scales: a high smooth-topped or grooved ridge on the crown, often with prominent lateral projections, and a ventral crown rib or neck spurs (Young and Gorter, 1981, fig. 2A3). Because of similarity, especially to *Turinia* spp. A and B of the *costatus* CZ of West Yunnan (Wang et al., 1986), *Turinia* cf. *T. hutkensis* is suggested as an indicator for the *costatus* CZ rather than late Givetian (Young, 1995). No microornament is seen in the Hatchery Creek Conglomerate scales, but they are worn, and preservation may be a factor.

The Hatchery Creek Conglomerate thelodont scales are associated with vascular plant stems, carbonized spores, and macrovertebrate remains (Young, 1993). Young and Gorter (1981) postulated a low-energy lacustrine environment, but it could easily have been a marginal deltaic setting (Young, 1995). The Hatchery Creek fauna includes the bothriolepidid, *Monorolepis verrucosa*. The only other occurrence where turiniids and bothriolepids are associated is in the basal bothriolepid zones of the Aztec Siltstone of Antarctica (Young, 1988b). An early Middle Devonian age is most probable, representing overlap in stratigraphic range of the turiniids with the earliest presumed freshwater bothriolepid antiarchs. A few *Turinia* cf. *pagoda*-type scales (Fig. 6 K–M) are found in a conodont-dated limestone in the Broken River Group (*costatus* just into *kockelianus* CZs, R. Mawson, personal communication, 1995). Otherwise, the closest scale assemblages to those of the Hatchery Creek Conglomerate are from central Australia (Ross River, Munyurai-1 borehole, and Mt. Winter) from poorly constrained stratigraphic settings (Young, 1995).

Turiniid thelodont scales occur in the N'Dahla Member, Ross River Syncline, eastern Amadeus Basin (Fig. 1, location 10; Young et al., 1987, figs. 1, 5). They appear closest to the Hatchery Creek thelodonts of southern NSW but also exhibit similarities to *T. gavinyoungi* to *T.* spp. A and B (both referred to as cf. *pagoda* by Young, 1995), and to undescribed scales from the Cleland Hills at Mt. Winter (Fig. 1, location 15) These assemblages all have scales with projections on the lateral crown ribs.

The Deering Siltstone Member (Parke Siltstone, Pertnjara Group) at Mt. Winter (Fig. 1, location 15) has produced scales from a bone bed in calcareous sandstones less than 20 m above the top of the Mereenie Sandstone (Young et al., 1987; Young, 1988a; Young and Turner, 1992). Although the thelodont scales are worn and rolled as at Wilson's Cliffs, Canning Basin (Fig. 1, location 1), they seem comparable with those from the N'Dahla Member and Hatchery Creek assemblages. Those that show detail (e.g. CPC, G. C. Young stub 94/24) are similar to some scales from Ross River. Young (1988a) stated that the Pertnjara Group is entirely nonmarine, but the implication of widespread occurrences of thelodont scales across the central Australian basins is that the beds may be marginal marine, albeit shallow water.

Long et al. (1988) described turiniid scales from presumed Emsian-Eifelian sandy shale in borehole Munyurai-1, eastern Officer Basin (Fig. 1, location 9). This find significantly increased the age of the basin sequence, then regarded as entirely Permian. The sediments had been identified as Devonian on spores, but the fish assemblage restricts the range to late Emsian and Eifelian, possibly *costatus-kockelianus* CZ.

Givetian. No definite Givetian turiniid assemblage has been found in Australia like the *Turinia antarctica* assemblage from Antarctica and *T. pagoda* from West Yunnan, both thought to be early Givetian (see section "Comparison with Other Gondwana Turiniids"). However, a few scales of *T.* cf. *antarctica* that may be Givetian have been found in the Canning Basin (see below). A single scale, *Australolepis?*, from UQL4454 in the Papilio Formation, Broken River, if confirmed, will be the youngest horizon for thelodonts in north Queensland.

Upper Devonian, Early Frasnian

Carnarvon Basin. The youngest known turiniid scales are from *Australolepis seddoni* from the Gneudna Formation, WA (Fig. 1, location 11; Turner and Dring, 1981). Nikoliviid thelodont scales, identified by Turner et al. (1981, "*Gampsolepis*") and Turner and Dring (1981), have not been confirmed by more recent studies. No more nikoliviid–like scales have been found. It is now certain from a wider range of scales from Gondwana that some scales with nikoliviid characteristics (elliptical anteriorly placed base with a large pulp opening) are actually modified turiniid scales.

Canning Basin. Thelodont scales from the northern part of the Hall Creek district, north Canning Basin (Fig. 1, location 14), come from BC Lawford Range 1 borehole (GSWA, B. Cowan collector: 18° 40′S, 126° 05′ E). Two turiniid scales with caplike crowns and large oval pulp cavities show affinity to *Aus-*

tralolepis. Both have pronounced microornament and crown rib tubercles (Turner, 1994, fig. 2). The scales are comparable with *Turinia antarctica* Turner and Young (1992) and *T. pagoda* (Wang et al., 1986) from the early Givetian of Antarctica and China. Associated are Late Devonian *Acanthodes*-type scales. The geological setting suggests that the sample came from the Pillara Limestone (late Givetian to early Frasnian at youngest).

Sample WCB363/2 from the Virgin Hills Formation in the Horse Spring Range, east of Fitzroy Crossing near Fossil Downs (Fig. 1, location 14, GR sheet 4161: 861.852, R. S. Nicoll, collector; *A. triangularis?* CZ, R. S. Nicoll, personal communication, 1996; Becker et al., 1991) contains broken scales that resemble *Australolepis*. These scales are like those from the Lawford Range with narrow, out-turned ribs. Two scales have crowns with an anterior tripartite section with split ribs like *Australolepis seddoni*. This record currently marks the last occurrence of thelodont scales in Australia (Fig. 2). Too few scales are known from the northern Canning Basin to suggest a zonation for the later Frasnian.

COMPARSION WITH OTHER GONDWANA TURINIIDS

East Gondwana and neighboring terranes

Antarctica. Young (1988b, 1995) noted that thelodonts and bothriolepids occur together only in the Aztec Siltstone of Antarctica and in the Hatchery Creek Conglomerate, NSW. He concluded that the Aztec Siltstone fauna was Givetian. Turner and Young (1992) described *Turinia antarctica* based mostly on material preserved as molds in siltstone from southern Victoria Land. J. A. Long subsequently found better preserved scales from Mount Crean in the Transantarctic Mountains (Fig. 9, location 40), first illustrated here. Long's collection of *T. antarctica* scales shows the advanced nature of the scalloped microsculpture of the ridges of the crown (Fig. 7, A, C, E, F, I), confirming this as a specific characteristic. His collection also shows a microornament typical of East Gondwanan turiniid scales (Fig. 7 C, D, F, H). G. C. Young (personal communication, 1994) confirms the feature in scales he has studied from new Mount Crean samples (V2616 of Young collection).

Concentrations of thelodont scales (Fig. 7) from Mount Crean in the Lashly Range (samples MC 3 and MC 4; Young, 1988b, p. 8, 11, figs. 3, 4) are preserved in presumably coprolitic, calcareous nodules in a black, micritic limestone unit with no marine invertebrates and are associated with black shales. The shales contain a typical Aztec Siltstone fauna from the *Bothriolepis askinae* Zone (Young, 1988b). Turiniids occur in the same beds as the chondrichthyans *Antarctilamna* and *Mcmurdodus* (Young, 1988b). As noted above, another species of *Mcmurdodus* is known from the supposed early Eifelian Cravens Peak Beds with *T. gavinyoungi*, but the East Gondwana *Antarctilamna* seems to be of middle Givetian to early Frasnian age (Young, 1993), although the genus is also recorded in the Lochkovian-Pragian Santa Rosa Formation of Bolivia associated with *T. gondwana* (Gagnier et al., 1988). The bothriolepid-turiniid combination and the similarity of some scales to those in the Hatchery Creek fauna suggest that Young's (1988b) lowest zones of the Aztec Siltstone (*Bothriolepis askinae* and *B. kohni*) could be as old as early Eifelian but no older than the *costatus* CZ. Although Young (1988b) reported the presence of the spore *Geminospora lemurata*, a basal Givetian fossil, in the lower part of the Aztec Siltstone, he has noted that the Aztec Siltstone is not necessarily restricted to the Middle Devonian (Young, 1995). Conditionally, the Aztec thelodonts are here considered early Givetian.

South China. Late Silurian? to Early Devonian scales from the Xitun Formation (Cuifengshan Group) of east Yunnan, China (Fig. 1, location 33), which Wang, (1984) called *Turinia? asiatica* (= *Turinia asiatica* Wang and Klapper, 1987, pl. 1, fig. 6), are closer morphologically to *Thelodus* scales, especially trilobatiform morphs (Turner, 1997). Wang (1995b) has now placed them in a new genus, *Parathelodus*, which he reports is associated with *Turinia* sp. and *Canonia* sp. in the lower Putongguo Formation (*woschmidti* CZ) of west Qinling Mountains, Gansu Province, north China. The scale illustrated by Wang and Klapper (1987, pl. II, fig. 1), however, appears to have neck foramina that are unknown in thelodont scales.

One cephalopectoral scale of *Turinia* sensu stricto has been documented from the early Emsian (*dehiscens* CZ) Yujiang or (Yukiang) Formation of Guangxi (Wang, 1992, fig. 2, pl. 1, fig. 1). It is associated with one scale of Nikoliviidae, genus and species indeterminate (Wang, 1995a).

Shan-Thai terrane, West Yunnan. Long (1993) discussed the similarity of fish material from the Shan-Thai terrane with that of northeastern Australia. *Turinia* spp. A and B come from the early Eifelian (*costatus* CZ) Mulutang and Xibiantang Formations and *Turinia pagoda* from the early Givetian (lower varcus CZ), lower Heyuanzhai Formation of West Yunnan (Wang et al., 1986). The specimens come from conodont-dated sequences (Wang et al., 1986, figs. 2, 3) in a continuous Middle Devonian section in the Baoshan region (Fig. 9, location 43). Unfortunately, the taxa were based on small samples or single scales, and therefore it is difficult to assess the range of scale variation within these species. The few scales of *Turinia* spp. A and B that may belong to the same taxon most closely resemble those of the central Australian Hatchery Creek Conglomerate and *T. hutkensis* of Iran. *Turinia* spp. A and B, which may be closely related to *T. pagoda*, are thus the basis for the *T.* cf. *T. pagoda* designation of scales from the *costatus* CZ (Young, 1995). *Turinia* sp. B most closely resembles scales from Khush-Yeilagh, Iran, and Thailand (see below), which would support an Eifelian age for the fish-bearing horizons of those two areas. *Turinia* sp. A is a head scale and thus difficult to classify. Wang et al. (1986) compared *T. pagoda* with scales of *T. hutkensis* of Iran and *T.* cf. *T. hutkensis* of Young and Gorter (1981). *T. pagoda*–like scales also occur in the central Australian Ross River and Munyurai-1 borehole samples (Long et al., 1988). Turner (1993) and Young (1995) thus proposed a cf. *pagoda*/cf. *hutkensis* assemblage for the Eifelian.

Thailand. A thelodont scale from a sandy limestone (unnamed formation) in Tchop-Ton "Quarry" or road cut, 2 km north of Mae-Saryang in the Chiang-Mai region of northern Thailand (Blieck and Goujet, 1978; Fig. 9, location 35) is here identified as *Turinia* sp. cf. *T.* sp. B of West Yunnan. The scale is broken but most closely resembles scales from West Yunnan (Wang et al., 1986, fig. 4M) and the Hatchery Creek fauna. An early Eifelian date (*costatus* CZ) seems most likely. The associated inarticulate brachiopods and arthropod fragments suggest a near-shore environment.

Western Gondwana

Iran. *Turinia hutkensis* from Hutk in the Kerman region of central Iran (Fig. 9, location 17) came from fine-grained limestone. Blieck and Goujet (1978) interpreted its environment as marine, and by comparison with known turiniid distribution, they suggested an Early Devonian age (possibly Emsian at the youngest). The thelodont-bearing bed lies between undated Paleozoic (or Precambrian) and overlying lower Frasnian limestone (P. Janvier, personal communication, 1992; Blieck and Goujet, 1978, p. 364). Conodonts in the rocks above indicate a latest Givetian age, whereas brachiopods and macrovertebrates (e.g., Moythomasia-type paleoniscoid scales) suggest earliest Frasnian (Lelièvre et al., 1993).

Thelodont scales have been recorded from two horizons in northern and central Iran in the Cimerian terrane (Lelièvre et al., 1993, figs. 7.2F, 11, and 14). Five scales were described by Turner and Janvier (1979), and three more were figured by Hamdi and Janvier (1981). All came from the late Early Devonian or early Middle Devonian Khush-Yeilagh Formation in the eastern Alborz Mountains, northeast Iran (Fig. 9, location 16). Turner and Janvier (1979) compared the scales with those of *Turinia hutkensis* and a Middle Devonian taxon from the Northern Hemisphere, *Skamolepis* Karatajute-Talimaa, and concluded that the sample was probably early Eifelian. The scales are from a carbonaceous layer in sandstones and associated with *Orvikuina*-like actinopterygian scales, charophytes, and conodont elements suggestive of Emsian-Eifelian (Hamdi and Janvier, 1981).

Blieck et al. (1980) showed the position of the thelodont-bearing bed KH 2 and referred to the scales as "*Amaltheolepis*-like." A rich fish horizon, KH 3, above KH 2, contains middle Givetian brachiopods and trilobites (Blieck et al., 1980). The Khush-Yeilagh scales are poorly preserved but exhibit elongate and posteriorly expanding crowns. All have well-ribbed crowns and anteriorly placed, low bases. *T. hutkensis* was distinguished from other species by the double (or greater) ribbing extending all over the crown. This character occurs in Australian scales from Eifelian assemblages (Fig. 1, locations 2, 6, 9, 10). However, all type and topotype *T. hutkensis* scales are very worn, and it is difficult to evaluate their ornament; no microornament has been found on *T. hutkensis* scales.

The presence of early Frasnian thelodonts in Iran is demonstrated by one scale of *Australolepis seddoni* found at Niaz-Kale, near Sardar, Tabas, eastern Iran (Fig. 9, location 48). The specimen comes from sample 101 (Mehdi Yazdi collection), base of Howz-e-Dorah section through the Shishtu Formation, a biostromal limestone dated by conodonts as probably earliest Frasnian (Mehdi Yazdi, personal communication, 1995).

South America. *Turinia gondwana* (Turner in Gagnier et al., 1988, fig. 5) from near Seripona, Bolivia (Fig. 9, location 3), occurs in the late Lochkovian–Pragian Santa Rosa Formation dated by a chitinozoan assemblage (Gagnier et al., 1989). The turiniid scales were found in a rich bone bed that crops out at several localities (Lelièvre et al., 1993, p. 146, fig. 7.2A.3). The environment of deposition is regarded as marginal marine or nonmarine. The taxon has large scales, some with complex three-tiered crowns, split ribbing, and many posterior points, comparable with those from the Middle Devonian elsewhere in Gondwana. The Santa Rosa bone bed recalls those of central-western NSW in composition of the assemblage and relatively good preservation of scales, but the size of the Bolivian scales and complexity of the crowns, along with assessment of the associated vertebrates, suggest a younger age (Lelièvre et al., 1993) and closer relationships to *T.* spp. cf. *pagoda*/cf. *hutkensis*, *T. hutkensis*, and *T. antarctica*.

CONCLUDING REMARKS

Thelodonts are widely distributed in Australia and former Gondwana regions (Figs. 1, 2, 5, 9) in the Lower Devonian. The distribution of the *Turinia australiensis* fauna—present across the Canning Basin (Fig. 1, locations 1, 12, 13), in the eastern Georgina Basin (Fig. 1, location 2), and the Broken River region of north Queensland (Fig. 1, location 3), and in central-western NSW—is consistent with ease of communication between these regions prior to or during the "*pesavis*"-*sulcatus* highstand via a seaway across central Australia. A major transgressive phase also appears to have occurred across central Australia during or after Eifelian times when thelodonts flourished. Long et al. (1988) and Young (1988a) thought the early Eifelian *Wuttagoonaspis* and *Turinia* faunas were contemporaneous. It seems likely, however, that there were at least two, and possibly three, transgressive phases, as indicated by the thelodont occurrences: one or two with *australiensis*-type scales just older than, or contemporary with the *Wuttagoonaspis* fauna and a younger assemblage with scales of *T.* cf. *hutkensis* and *T.* cf. *pagoda*.

A new form, *Jesslepis*, from conodont-dated marine sequences in the Broken River region of north Queensland appears by late Early Devonian.

The Early and Middle Devonian deposits of Bolivia, Hatchery Creek, NSW, and Antarctica have all been interpreted as nonmarine (e.g. Young, 1993) However, the almost worldwide distribution of *Turinia* in the Devonian does not accord with a totally freshwater habitat for its entire life span. It seems more likely, as Blieck (1985) has argued for the Northern Hemisphere, that the Aztec Siltstone and other sediments from Australia and Bolivia were either marginal marine, lacking typical marine invertebrates and conodonts, or that *Turinia* inhabited both freshwater and marine habitats.

One possible late Middle Devonian turiniid resembling the Antarctic, *T. antarctica*, occurs in the northern Canning Basin, WA. Late Devonian (early Frasnian) thelodonts are known only from WA, from one site in the Carnarvon Basin, and a possible occurrence in the northern Canning Basin. The Middle and Upper Devonian turiniids in Australia and Antarctica share more characters in common with those in West Yunnan, Thailand, Iran, and Bolivia, all old Gondwana blocks, than with taxa in the Northern Hemisphere.

Turiniid thelodont distribution throughout Gondwana, presented here on a paleogeographic reconstruction (Fig. 9), suggests that turiniids were capable of dispersal around Gondwanan shorelines anywhere within the 0 to 60° latitudinal span on the globe, particularly within shallow marine and marginal marine, even freshwater, environments. Because turiniids are known to occur in the Early Devonian of western Gondwana, one might predict their occurrence in the Falkland Islands and South Africa where other Devonian fish have been recorded (Long, 1993). However, using Young's (1987) reconstruction, these regions were located above 60° S latitude. Such high latitudes may have been inhospitable for thelodonts. In an earlier assessment of thelodont distribution, Turner and Tarling (1982) tried to take paleoclimatological factors into account and postulated marine currents to aid dispersal of thelodonts around northern Gondwana. The apparent absence of thelodonts from China and South America, noted by them, has now been resolved.

What has emerged from the work of the last decade is that turiniid thelodonts were widespread in Gondwana and that they occur from latest Silurian to Late Devonian, thus diverging from the Northern Hemisphere record. As yet, we have no clear idea whether turiniids originated in the Southern Hemisphere or in Arctic Canada, as Turner (1997) has suggested. Similarities of the Eifelian and Givetian thelodonts of Iran, West Yunnan, and Thailand to those of central Australia support the close relationship of several Middle Eastern and Asian terranes to East Gondwana at least during late Early to Middle Devonian.

There appears to have been relatively rapid turnover of turiniid taxa in Gondwana in the Devonian. These changes may have been related to major regressive events, becoming apparent with subsequent transgressions. The temporal range of the last thelodont taxon, *Australolepis seddoni*, was relatively short, from near the Givetian-Frasnian boundary, dated at about 369 Ma (Young, 1995), probably to the top of the *A. triangularis* CZ at around 366 Ma (Young, 1995) with occurrences in WA and possibly eastern Iran. Because of the paucity of data from long, well-dated stratigraphic sections, many refinements in present knowledge should be expected. However, characteristic morphological features of the scales, such as parallel-ribbed microornament, tuberculations, and split ribs, are being recognized as taxonomically significant and important for tracing lineages within the Gondwana region and related terranes.

ACKNOWLEDGMENTS

Work on microvertebrate material from central western NSW and western QD was initiated in the 1970s through the courtesy of Dr. J. W. Pickett (GSNSW) and Dr. Peter Jones (AGSO). Drs. J. S. Jell (University of Queensland, Bisbane), John A. Long (WAM), and R. S. Nicoll and G. C. Young (AGSO) have allowed me to examine thelodont scales from their collections and given other help. Professors John A. Talent and Ruth Mawson (MUCEP) allowed me to examine microvertebrates from their residues and gave information regarding sections and conodont dating. For unpublished data and discussion on biostratigraphy, I thank A. Cook, A. De Pomeroy (now Basden), S. Hocknull, P. Janvier, J. S. Jell, P. J. Jones, J. A. Long, T. Märss, R. S. Nicoll, J. W. Pickett, P. Playford, A. Ritchie, T. Sharp, L. Sherwin, M. Yazdi, and G. C. Young. Professors David K. Elliott, Michael A. Murphy, and John A. Talent reviewed drafts of the manuscript. Work on Australian microvertebrates has been supported by grants from the Australian Research Council (ARC) and currently an Australian Research Fellowship. Dora Aitken (ARC-funded), Alison Green (Queensland Museum), Diane Phipps (University of Queensland), and Malcolm Willis (University of Newcastle upon Tyne) made the SEM photomicrographs. Queensland Museum photographers provided other photographic support.

APPENDIX—LOCALITY REGISTER

Herein are further data on Australian localities mentioned in this chapter and not previously published. For published data see De Pomeroy, 1994, 1995; Long, 1993; Long et al., 1988; Mawson et al,. 1989, 1992; Pickett, 1980, 1992; Pickett and McClatchie, 1991; Pickett et al., 1985; Sherwin, 1980, 1995, 1997; Sloan et al, 1995; Talent and Mawson, 1994; Turner, 1986, 1991a, 1993, 1995a, b, 1997; Turner and Dring, 1981; Turner et al., 1981; Young 1988a, b, 1993; Young and Gorter 1981; Young and Turner, 1992; Young et al., 1987.

Central western NSW

Comprising Mineral Hill, Murda Syncline, and Trundle districts shown in Fig. 5: John W. Pickett collection, GSNSW. All below have *Turinia australiensis* or *T.* cf. *australiensis* (number of scales in bracket from 3-kg sample if 10 or more). Nearly all have conodonts of *pesavis* and/or *sulcatus* zones (see Pickett, 1992) or contain a comparable microvertebrate fauna.

C007 (= C054) Limestone south of the Troffs Station, Narromine 1:250000 *sulcatus* CZ, GR574.942 (Pickett, 1992).

C008 Limestone Narromine GR518.958 (= C089), Gleninga Formation: head scales basal Pragian *sulcatus* CZ (Pickett and McClatchie, 1991) .

C083 1.5 km west of Gleninga Homestead, Mineral Hill, GR500.976 Nymagee 1:250,000, top limestone Yarra Yarra Creek Group, head scales only.

C086 as C083.

C089 Limestone GR518.958 Narromine 1:250,000, top limestone Yarra Yarra Creek Group, Gleninga Formation.

C091 Limestone GR499.978, Mineral Hill, Gleninga Formation? Nymagee 1:250,000, very rich fish, no conodonts (10).

C092 Limestone on track from Gleninga Homestead in front of

Bogong Woolshed, Mineral Hill, GR500.976 Nymagee 1:250,000, top limestone Yarra Yarra Creek Group (cf. C091, C083, C216).

C214 Limestone near Bogong Woolshed, Mineral Hill, Gindoono 1:100000, GR403.781.5 Nymagee 1:250,000, Yarra Yarra Creek Group: rich very small scales (47).

C215 as C214, GR403.779.5 (very shallow water): rich (82).

C216 as C214, GR398.778.

C231 Limestone GR5267.9467 4 km west of Meloola, 22 km north of Fifield, Gleninga Formation, Yarra Yarra Creek Group, Narromine 1:250,000: (close to C008, C086, C214).

C232 Limestones within Murda north syncline 22 km north of Fifield, GR525.946, Lara Station north of Condobolin, Jerula Limestone Member, Narromine 1:250,000, *sulcatus* CZ Pickett and McClatchie (1991): (56).

C233 as C232.

C234 as C233.

C287 Limestone biomicrite GR57726.94603, Narromine 1:250,000.

C595 Limestone GR5710.9515 3km northeast of Kadungle, Narromine 1:250,000, unnamed beds, (cf. C092/ C287).

C597 Limestone GR5700.9445 2 km north of the Troffs, low in Trundle Beds, Narromine 1:250,000. Two "generations" of scales: one pale yellow, one more gray and worn.

C598 Limestone GR571.938 as C597.

C600 Limestone GR526.946, Kadungle, the Troffs, Meloola, Gleninga Formation, sulcatus CZ (Pickett, 1992): rich v. small (43).

C608 Limestone 3 km west of Meloola Station, GR526.946 Narromine SI/55-3, Condobolin 8332-III, Gleninga Formation, "*pesavis*"-*sulcatus* CZ (Pickett and McClatchie, 1991).

C620 Limestone 32 m above base GR500E 975N T.M. Nymagee 1:250,000 SI/55-3 (Gindoono 8232-1), 1 km north of Bogong Homestead, Mineral Hill Volcanics?: scales cf. C600.

C623 section south of Bogong Woolshed GR500E 976N Mineral Hill column Gleninga Formation, Yarra Yarra Creek Group as C620, "*pesavis*"-early *sulcatus* CZ (Pickett and McClatchie, 1991).

C624 section 8 m from top, limestone south of Bogong Woolshed GR498.8E 977.6N, Mineral Hill column Gleninga Formation, Yarra Yarra Creek Group, rich (39).

C625 as C624, 2 m below youngest beds exposed at top of hill (6 m limestone): abundant thelodont scales (250+), head scales predominate.

C626 as C624, "*pesavis*"-*sulcatus* CZs.

C627 1 m limestone, 8 m below C624: inverts, very rich, striated nikoliviid type? (Turner, 1982) (small, 99).

C628 as C627, 0 m base of section, limestone.

C651 Limestone GR518.958 Gleninga Formation, Yarra Yarra Creek Group, Boona Gap 212m, Narromine 1:250,000, "*pesavis*"-*sulcatus* (Pickett and McClatchie, 1991).

C654 Limestone GR524.944 Gleninga Formation, Yarra Yarra Creek Group, Lara Station, Condobolin, Narromine 1:250,000, *pesavis-sulcatus* (Pickett and McClatchie, 1991): variety of color and preservation.

C655 (H) Limestone 3 km north of Meloola Station, GR627.947 Gleninga Formation, Narromine 1:250,000, Fifield 1:5000.

C657 Northeast Kadungle, Sherwin's (1980) top section 2, top or just above Trundle beds, slightly older than C625 and C627, Narromine 1:250,000.

C659 Limestone float 2.5 km northeast of the Troffs railway station GR571.942, Narromine 1:250,000, Trundle Group.

C661 Limestone GR573.942 tank south of the Troffs, Narromine 1:250,000: T. sp.

C662 Limestone GR575.943 dam at foot of volcanic hill, Troffs Station : rich.

NB: C660–662 = L. Sherwin's measured section 8 (Sherwin, 1980).

C665 Limestone 3 km southeast Troffs, GR572.938: *Turinia* sp.

C667 (H) Limestone east side of road 2.5 km south of Troffs GR571.958: worn thelodonts, *Turinia?*, cf. *Gampsolepis*.

C800 also Tumblong sample J. J. Talent coll. MUCEP: Tumblong Oolite, "*pesavis*"/*sulcatus* or older Tumut Block (details in Pickett et al., 1985) (100+).

Other NSW samples

Mt. Gunderbooka Wrightville Sheet 1:100,000 (unpublished Turner report on T. Sharp collection., University of Technology, Sydney). Gundaroo samples on loan to Queensland Museum, (*Turinia* sp. cf. *australiensis*, *Wuttagoonaspis fletcheri*).

QMF25956–25960 Merrimerriwa Formation (Mulga Downs Group) at road exposure southwest of the escarpment between south Wongin Tank and Yathong Gap, near Cobar, GR 715.745, Wrightville Sheet (*Turinia* sp. cf. *australiensis*, *Wuttagoonaspis fletcheri*) T. Spencer collection.

North Queensland

UQL4736 and UQL4737 J. S. Jell samples BRJ 111B and C from a tributary of Turtle Creek, GR 746.668, Shield Creek Formation ("*pesavis*" CZ).

UQL5950 J. S. Jell sample 133G, Jessey's Springs section, Broken River, Chinaman Creek Limestone, Broken River Group (*costatus* just into *kockelianus* CZs, R. Mawson, personal communication).

UQL4454 J. S. Jell sample BRJ63.4 from a creek bed 1 km south of Storm Dam: GR532.364, in the Papilio Formation, Broken River Group.

QML1011 (A. Cook sample) block from a small tributary of Running Creek, 150 m upstream from confluence, south of Mount Podge, Burdekin region (19° 17′S, 145° 57′E), basal Mount Podge Limestone, *serotinus* CZ age (Talent and Mawson, 1994).

REFERENCES CITED

Becker, R. T., House, M. R., Kirchgasser, W. T., and Playford, P. E., 1991, Sedimentary and faunal changes across the Frasnian/Famennian boundary in the Canning Basin of Western Australia: Historical Biology, v. 5, p. 183–196.

Blieck, A., 1985, Paléoenvironnements des Hétérostracés, vertébrés agnathes ordoviciens et dévoniens: Paris, Bulletin de la Muséum nationale d'histoire naturelle, v. 7, p. 143–155.

Blieck, A., and Goujet, D., 1978, A propos de nouveau matériel de Thélodontes (Vertébrés Agnathes) d'Iran et de Thaïlande: aperçu sur la répartition géographique et stratigraphique des Agnathes des "régions gondwaniennes" au Paléozoique moyen: Lille, Annales de la Société géologique du Nord, v. 97, p. 363–372.

Blieck, A., Golshani, F., Goujet, D., Hamdi, A., Janvier, P., Mark-Kurik, E., and Martin, M., 1980, A new vertebrate locality in the Eifelian of the Khush-Yeialgh Formation, Eastern Alborz, Iran: Montpellier, Palaeovertebrata, v. 9, p. 133–154.

De Pomeroy, A., 1994, Mid-Devonian shark scales from the Broken River, north Queensland, Australia: Memoirs of the Queensland Museum, v. 37, p. 87–114.

De Pomeroy, A., 1995, Australian Devonian fish biostratigraphy in relation to conodont zonation: Courier Forschungsinstitut Senckenberg, v. 182, p. 475–486.

Gagnier, P-Y., Turner, S., Friman, L., Suarez-Riglos, M., and Janvier, P., 1988, The Devonian vertebrate and mollusc fauna from Seripona (Dept. of Chuquisica, Bolivia): Neues Jahrbuch für Geologie und Paläontologie, Abhandlungen 176, p. 269–297.

Gagnier, P-Y., Paris, F., Rachebouef, P., Janvier, P., and Suarez-Riglos, M., 1989, Les Vertébrés de Bolivie: Donnés biostratigraphiques et anatomiques complémentaires: Bulletin de l'Institut Français Études

Andines 1989, v. 18, p. 75–93.

Gross, W., 1967, Über Thelodontier-Schuppen: Palaeontographica, Abteilung A, v. 127, p. 1–67.

Gross, W., 1971, Unterdevonische Thelodontier-und Acanthodier-Schuppen aus Westaustralien: Paläontologische Zeitschrift, v. 45, p. 97–106.

Hamdi, B., and Janvier, P., 1981, Some conodonts and fish remains from the Lower Devonian (lower part of the Khoshyeylaq Formation), North East Shahrud, in Contribution to the palaeontological researches in North and Central Iran: Geological Survey of Iran Report 49, p. 195–212.

Johnstone, M. H., Jones, P. J., Koop, W. J., Roberts, J., Gilbert-Tomlinson, J., Veevers, J. J., and Wells, A. T., 1967, The Devonian of Western and Central Australia, *in* Oswald, D. H., ed., International Symposium on the Devonian System, Volume 1, 1967: Calgary, Alberta Society of Petroleum Geologists, p. 599–612.

Karatajute-Talimaa, V. N., 1978, Telodonti silura i devona S.S.S.R. i Shpitsbergena (Silurian and Devonian thelodonts of the Soviet Union and Spitsbergen): Vilnius, Mosklas, 334 p.

Karatajute-Talimaa, V. N., 1985, Telodonti Podyomnenskoy svity (nijniy devon, zhedin) Severnoy Zemli i ikh korrelyatsionnoye znacheniye (Thelodonts of the Podyomnenskoy Formation [Lower Devonian, Gedinnian] of Severneya Zemlya and their correlative importance): Nauchie Trudi Visshikh Ochebnikh Zavedenii Litovskoi SSR Geologiya, v. 6, p. 50–60 (English summary, p. 60).

Lelièvre, H., Janvier, P., and Blieck, A., 1993, Silurian-Devonian vertebrate biostratigraphy of Western Gondwana and related terranes (South America, Africa, Armorica-Bohemia, Middle East), *in* Long, J. A., ed., Palaeozoic vertebrate biostratigraphy and biogeography: London, Belhaven Press, p. 139–173.

Long, J. A., 1993, Palaeozoic vertebrate biostratigraphy of southeast Asia and Japan, *in* Long, J.A., ed., Palaeozoic vertebrate biostratigraphy and biogeography: London, Belhaven Press, p. 277–289.

Long, J. A., Turner, S., and Young, G. C., 1988, A Devonian fish fauna from subsurface sediments in the eastern Officer Basin, South Australia: Alcheringa, v. 12, p. 61–78.

Märss, T., 1986, Silurian vertebrates of Estonia and West Latvia: Fossilia Baltica, v. 1, p. 1–104 (in Russian).

Mawson, R., 1987, Early Devonian conodont faunas from Buchan and Bindi, Victoria, Australia: Palaeontology, v. 30, p. 251–273.

Mawson, R., and Talent, J. A., 1989, Late Emsian–Givetian stratigraphy and conodont biofacies—carbonate slope and offshore shoal to sheltered lagoon and nearshore carbonate ramp—Broken River, north Queensland, Australia: Courier Forschungsinstitut Senckenberg v. 117, p. 205–259.

Mawson, R., and 13 others, 1989, Conodont data in relation to resolution of stage and zonal boundaries for the Devonian of Australia, *in* McMillan, N. J., Embry, A. F. and Glass, D. J. eds., Devonian of the world: Calgary, Canadian Society of Petroleum Geologists Memoir 14, v. 3, p. 485–527.

Mawson, R., Talent, J. A., Brock, G. A., and Engelbretsen, M. J., 1992, Conodont data in relation to sequences about the Pragian-Emsian boundary (Early Devonian) in southeastern Australia: Proceedings of the Royal Society of Victoria, v. 104, p. 23–56.

Palfreyman, W. D., 1984, Guide to the geology of Australia: Bureau of Mineral Resources Bulletin 181, p. 1–111.

Pickett, J. W., 1980, Conodont assemblages from the Cobar Supergroup (Early Devonian), New South Wales: Alcheringa, v. 4, p. 67–88.

Pickett, J. W., 1992, Review of selected Silurian and Devonian conodont assemblages from the Mineral Hill-Trundle area: Geological Survey of New South Wales, unpublished record 92/1, 8 p.

Pickett, J. W., and McClatchie, L., 1991, Age and relations of stratigraphic units in the Murda Syncline area: Geological Survey of New South Wales, Quarterly Notes, no. 85, p. 9–32.

Pickett, J., Turner, S., and Meyers, B., 1985, The age of oolitic sediments from Tumblong, New South Wales: Geological Survey of New South Wales, Quarterly Notes, no. 58, cover photo + p. 1, 12–14.

Sherwin, L., 1980, Faunal correlation of the Siluro-Devonian units, Mineral Hill–Trundle–Peak Hill area: Geological Survey of New South Wales, Quarterly Notes, no. 39, p. 1–14.

Sherwin, L., 1994, Palaeozoic stratigraphy of the Narromine 1:250,000 sheet area: Geological Survey of New South Wales, Quarterly Notes, no. 96, p. 1–37.

Sherwin, L., 1995, Siluro-Devonian brachiopods from the Amphitheatre and Winduck Groups (Cobar Supergroup), Western New South Wales: Memoirs of the Association of Australasian Palaeontologists, v. 18, p. 61–96.

Sherwin, L., 1997, Geology of the Narromine 1:250,000 sheet SI/55-3: Notes to accompany the geological map: Sydney, Geological Survey of New South Wales 104 p.

Sloan, T. R., and 10 others, 1995, Conodont data from Silurian–Middle Devonian carbonate fans, debris flows, allochthonous blocks and adjacent autochthonous platform margins: Broken River and Camel Creek areas, north Queensland, Australia: Courier Forschungsinstitut Senckenburg, v. 182, p. 1–77.

Talent, J. A. and Mawson, R., 1994, Conodonts in relation to the age and environmental framework of the Burdekin Basin (mid-Devonian), northeastern Australia: Courier Forschungsinstitut Senckenberg, v. 168, p. 61–81.

Talent, J. A., and Yolkin, E. A., 1987, Transgression-regression patterns for the Devonian of Australia and southern West Siberia: Courier Forschungsinstitut Senckenberg, v. 92, p. 235–249.

Talimaa V. N., 1995, Vertebrate complexes in the heterofacial Lower Devonian deposits of Timan-Pechora Province, *in* Turner, S., ed., International Geological Correlation Program 328: Moscow-94 workshop Palaeozoic microvertebrates—State of research in 1995: Socorro, New Mexico, J. Zidek services, Ichthyolith Issues, Special Publication 1, p. 39–42.

Traquair, R. H., 1899, On Thelodus *pagei* Powrie sp. from the Old Red Sandstone of Forfarshire: Transactions of the Royal Society of Edinburgh, v. 39, p. 595–602.

Turner, S., 1973, Siluro-Devonian thelodonts from the Welsh Borderland: Journal of the Geological Society of London, v. 129, p. 557–584.

Turner, S., 1982, Thelodonts and correlation. Contributions to Australian fossil fish biostratigraphy, *in* Rich, P., and Thompson, I., eds., The fossil vertebrate record of Australasia: Melbourne, Monash Offset Printing, p. 128–132.

Turner, S. 1984, Studies of Palaeozoic Thelodonti (Craniata: Agnatha) [Ph.D. thesis]: Newcastle-upon-Tyne, United Kingdom, University of Newcastle-upon-Tyne, 2 vols., 233 p.

Turner, S., 1986, Vertebrate fauna of the Silverband Formation, Grampians, western Victoria: Proceedings of the Royal Society of Victoria, v. 98, p. 53–62.

Turner, S., 1991a, Palaeozoic vertebrate microfossils in Australasia, *in* Vickers-Rich, P., Monaghan, J. M., Baird, R. F., and Rich, T. H., eds., Vertebrate palaeontology of Australasia: Lilydale, Melbourne, Pioneer Design Studios/Monash University Publications Committee, p. 429–464.

Turner, S., 1991b, Monophyly and interrelationship of the Thelodonti, *in* Chang, M-m., Liu Y-h., and Zhang, G-r., eds., Early vertebrates and related problems of evolutionary biology: Symposium on early vertebrates, September 1987: Beijing, Science Press, p. 87–119.

Turner, S., 1993, Palaeozoic microvertebrates of Eastern Gondwana, *in* Long, J. A., ed., Palaeozoic vertebrate biostratigraphy and biogeography: London, Belhaven Press, p. 174–207.

Turner, S., 1994, Thelodont squamation: Ichthyolith Issues, no. 13, p. 12–15.

Turner, S., 1995a, Defining Devonian microvertebrate faunal assemblages or zones for Eastern Gondwana, *in* Turner, S., ed., International Geological Correlation Program 328: Moscow-94 workshop and Palaeozoic microvertebrates—state of research in 1995: Socorro, New Mexico, J. Zidek Services, Ichthyolith Issues, Special Publication 1, p. 43–52.

Turner, S., 1995b, Devonian thelodont scales (Agnatha, Thelodonti) from Queensland: Memoirs of the Queensland Museum, v. 38, p. 677–685.

Turner, S., 1997, Wenlock- Lochkovian (Gedinnian) thelodont assemblages and their possible ecological significance, *in* Boucot, A. J. and Lawson, J., eds., Paleocommunities: A case study from the Silurian and Lower Devonian, International Geological Correlation Programme, Project Ecostratigraphy, Final Report: Cambridge, Cambridge University Press, p. 42–78.

Turner, S., and Blieck, A., 1995a, Conodont-vertebrate comparison and significance of microornament: Abstracts and Programme, First Australian conodont symposium (AUSCOS-1) and the Boucot Symposium, 18–21 July 1995: Sydney, Macquarie University, p. 105.

Turner, S. and Blieck, A., 1995b, IGCP 328 State of play, *in* Turner, S., ed., International Geological Correlation Program 328: Moscow-94 workshop and Palaeozoic microvertebrates—State of research in 1995: Socorro, New Mexico, J. Zidek Services, Ichthyolith Issues, Special Publication 1, p. 5–8.

Turner, S., and Dring, R. S., 1981, Late Devonian thelodonts (Agnatha) from the Gneudna Formation, Carnarvon Basin, Western Australia: Alcheringa, v. 5, p. 39–48

Turner, S., and Janvier, P., 1979, Middle Devonian Thelodonti (Agnatha) from the Khush-Yeilagh Formation, north-east Iran: Geobios, v. 12, p. 889–892.

Turner, S., and Tarling, D., 1982, Thelodont and other agnathan distributions as tests of Lower Palaeozoic continental reconstructions: Palaeogeography, Palaeoclimatology, Palaeoecology, v. 39, p. 295–311.

Turner, S., and Young, G. C., 1992, Thelodont scales from the Middle–Late Devonian Aztec Siltstone, southern Victoria Land, Antarctica: Antarctic Science, v. 4, p. 89–105.

Turner, S., Jones, P. J., and Draper, J. J., 1981, Early Devonian thelodonts (Agnatha) from the Toko Syncline, western Queensland, and a review of other Australian discoveries: Bureau of Mineral Resources Journal of Australian Geology and Geophysics, v. 6, p. 51–69.

Turner, S., Vergoossen, J. M. J., and Young, G. C., 1995, Fish microfossils from Irian Jaya: Memoirs of the Association of Australasian Palaeontologists, v. 18, p. 165–178.

Valenzuela-Ríos, J. I., 1994, The Lower Devonian conodont *Pedavis pesavis* and the *pesavis* Zone: Lethaia, v. 27, p. 199–207.

Wang C.-Y., and Klapper, G., 1987, On the genus *Fungulodus* (Conodonta): Acta Micropalaeontologica Sinica, v. 4, p. 369–374. [Editorial note: Correction to the authorship of this paper is indicated by Wang, C.-Y., in Acta Micropalaeontologica Sinica, 1989, v. 6, p. 371, as Wang was the sole author.]

Wang, N.-Z., 1984, Thelodont, acanthodian and chondrichthyan fossils from the Lower Devonian of southwest China: Proceedings of the Linnean Society of New South Wales, v. 107, p. 419–441.

Wang, N.-Z., 1992, Microremains of agnathans and fishes from Lower Devonian of central Guangxi, with correlation of Lower Devonian between central Guangxi and eastern Yunnan, south China: Acta Palaeontologica Sinica, v. 31, p. 280–303.

Wang N.-Z., 1995a, Silurian and Devonian jawless craniates (Galeaspids, Thelodonti) and their habitats in China, *in* Arsenault, M., Lelièvre, H., and Janvier, P., eds., Études sur les vertébrés inférieurs, 7th Symposium international, Parc de Miguasha, Québec, 1991: Bulletin du Muséum national d'Histoire Naturelle, Section C, 4 ser., v. 17, p. 57–84.

Wang N.-Z., 1995b, Thelodonts from the Cuifengshan Group of East Yunnan, China and its biochronological significance, *in* Lelièvre, H., Wenz, S., Blieck, A., and Cloutier, R., eds., Premiers vertébrés et vertébrés inférieurs, 8th Congress International, Paris, 4-9 septembre 1995: Geobios Mémoire Spécial 19, p. 403–409.

Wang R.-H., 1993, Taxonomie, Palökologie und Biostratigraphie der Mikroichthyolithen aus dem Unterdevon Keltiberiens, Spanien: Courier Forschungsinstitut Senckenberg, v. 161, p. 1–205.

Wang S.-T., Dong Z.-Z., and Turner, S., 1986, Discovery of Middle Devonian Turiniidae (Thelodonti: Agnatha) from western Yunnan, China: Alcheringa, v. 10, p. 315–325.

Withnall, I. W., and Lang, S. C., eds, 1993, Geology of the Broken River Province, North Queensland: Queensland Geology, v. 4, 292 p.

Young, G. C., 1987, Devonian vertebrates of Gondwana, *in* Gondwana six: American Geophysical Union, Geophysics Monograph 41, p. 41–50.

Young, G. C., 1988a, New occurrences of phyllolepid placoderms from the Devonian of central Australia: Bureau of Mineral Resources, Journal of Geology and Geophysics, v. 10, p. 363–376.

Young, G. C., 1988b, Antiarchs (Placoderm fishes) from the Devonian Aztec Siltstone, southern Victoria Land, Antarctica: Palaeontographica Abteilung A, v. 202, p. 1–125.

Young G. C., 1993, Middle Palaeozoic macrovertebrate biostratigraphy of Eastern Gondwana, *in* Long, J. A., ed., Palaeozoic vertebrate biostratigraphy and biogeography: London, Belhaven Press, p. 208–251.

Young, G. C., 1995, Timescales. 4: Devonian. Australian Phanerozoic Timescales, biostratigraphic charts and explanatory notes, Second Series: Australian Geological Survey Organisation, Record 1995/33, 47 p.

Young, G. C., and Gorter, J. D., 1981, A new fish fauna of Middle Devonian age from the Taemas/Wee Jasper region of New South Wales: Bureau of Mineral Resources Bulletin 209, p. 83–147.

Young, G. C., and Turner, S., 1992, IGCP 328: New evidence for correlation of Palaeozoic fish between central Australia and the Canning and Georgina Basins: 11th Australian Geological Congress, 11th, Ballarat, 18–25 January, Geological Society of Australia, Abstracts no. 32, p. 177.

Young, G. C., Turner, S., Owen, M., Nicoll, R. S., Laurie, J. R., and Gorter, J. D., 1987, A new Devonian fish fauna, and revision of post-Ordovician stratigraphy in the Ross River Syncline, Amadeus Basin, central Australia: Bureau of Mineral Resources, Journal of Australian Geology and Geophysics, v. 10, p. 233–242.

Manuscript Accepted by the Society February 27, 1997
This Is a Contribution to IGCP 328: Palaeozoic Microvertebrates

Printed in U.S.A.

Geological Society of America
Special Paper 321
1997

Evolutionary relationships of the Zaphrentidae and Craspedophyllidae (rugose corals, Devonian) in eastern North America

William A. Oliver, Jr.
U.S. Geological Survey (Emeritus), and Department of Paleobiology, U.S. National Museum of Natural History, MRC 137, Smithsonian Institution, Washington, D.C. 20560

ABSTRACT

Several models showing possible evolutionary relationships of the genera of two small families of Early and Middle Devonian rugose corals are analyzed. Cladograms based on parsimonious character transpositions disagree with what is evident in the paleontologic record and are rejected. Those based on the record, using character state polarization where the record is not clear, are more satisfactory even though much less parsimonious. The apparent failure of cladistic methods in this analysis reflects the low number of available morphologic characters and the number of these characters that are convergent. This is a common problem in fossil invertebrates, and it is suggested that the principal value of cladistics in such groups is to force a more rigorous consideration of the characters on which taxa are based.

INTRODUCTION

The Zaphrentidae and Craspedophyllidae are small families of rugose corals that range through the upper Lower and Middle Devonian (Pragian or Emsian through Givetian) in eastern North America (ENA). All of the included genera seem to have originated as ENA endemics, and the known record of most of them is limited to the Eastern Americas Biogeographic Realm (EAR). I have previously analyzed the relationships of the two families and their genera using cladistic methods (Oliver, 1993). The results were questioned because of the small number of morphologic characters and the difficulty of recognizing ancestral and derived states. I manually derived "most parsimonious" cladograms (trees) for each family and for the two together but subjectively modified these for one character (aulos) that the paleontologic record indicates was independently derived several times. J. M. Pandolfi (personal communication, 1991; discussed in Oliver, 1993), using PAUP (Phylogenetic Analysis Using Parsimony), showed that each of my three trees was one of numerous (six to 162) equally parsimonious ones. In spite of this, the exercise had useful results: (1) it satisfied me that previously recognized subfamilies of the Craspedophyllidae (Oliver, 1976) should be abandoned because they were masking some generic relationships, (2) it forced me to examine and reconsider the morphologic criteria on which the families and genera were based, and (3) it indicated the limitations of the methodology in the families studied and in the rugose corals generally.

In this second "cladistic" study I first erect trees based on characters that are considered to be particularly important in the two families. These trees portray evolutionary scenarios that are more or less compatible with the paleontologic and stratigraphic records and worth further consideration. Trees that are more parsimonious are constructed for comparison. Consistency indices (CI) are calculated for each tree, and each tree is discussed. Finally, a tree is proposed, based on the morphologic characters that seem most important, that is most compatible with the known record.

Character states and their ordering are modified from the

Oliver, W. A., Jr., 1997, Evolutionary relationships of the Zaphrentidae and Craspedophyllidae (rugose corals, Devonian) in eastern North America, *in* Klapper, G., Murphy, M. A., and Talent, J. A., eds., Paleozoic Sequence Stratigraphy, Biostratigraphy, and Biogeography: Studies in Honor of J. Granville ("Jess") Johnson: Boulder, Colorado, Geological Society of America Special Paper 321.

earlier analysis (Oliver, 1993), and one taxon is added; these changes are described and discussed in following sections. References to cladistic methods and discussion of procedures are in the earlier paper. The general pattern used in my cladograms is that of Cairns (1984).

TAXA

In Oliver (1993), I analyzed the zaphrentids and craspedophyllids separately as well as together on the assumption that they were distinct, if related, families. The 1993 analysis supported this assumption, but various kinds of relationships can be postulated and these are examined in this chapter by creating and using a single data set for the two families.

Zaphrentidae

Four genera are included in the Zaphrentidae, but five morphologic states in *Heliophyllum* are treated separately as in the 1993 study. A single solitary rugose coral from the Indian Cove Formation, Gaspé, Québec, is considered to be the oldest known zaphrentid (late Pragian or early Emsian). This was discussed in the earlier paper (Oliver, 1993, p. 135) but not included in the analysis; here it is treated as a separate taxon (Indian Cove 1). This gives a total of nine zaphrentid taxa (Table 1).

Craspedophyllidae

Six genera are included in the Craspedophyllidae, but two states in each of two genera are treated separately, making a total of eight taxa (Table 1). These are the same as in the 1993 study.

Outgroup

Selection of an outgroup for the cladistic analysis was discussed in the 1993 paper; the Family Cyathophyllidae and the Genus *Phaulactis* (Lykophyllidae) were considered and the latter selected (Oliver, 1993, p. 133). I have reexamined the question because of the close similarities of some cyathophyllids to some of the genera being analyzed, but the cyathophyllids that are morphologically most similar are too young to have given rise to the EAR families, although a common ancestor might be postulated. I selected *Phaulactis* again because it is stratigraphically older, is ontogenetically similar to the earlier zaphrentids, and makes sense as a tool for ordering several of the character states.

CHARACTERS AND STATES

Characters [1–11] and states (21 total) used in the analysis of the zaphrentid and craspedophyllid genera are listed in Table 2. Polarization or ordering of the states is based on the outgroup, ontogeny, stratigraphic record, or intuition. The ancestral state [0] in each character is that of the outgroup. In all 2-state characters, the other state [1] is (by definition) derived. Multistate characters were ordered as follows:

Character 1. Coloniality follows the commonly accepted "progression" from solitary to phaceloid to cerioid to astreoid; there is ample evidence in the fossil record that this is a common, but not necessary, sequence in the Rugosa. A semicerioid state is

TABLE 1. CHARACTER STATE MATRIX FOR TAXA OF THE ZAPHRENTIDAE AND CRASPEDOPHYLLIDAE WITH *PHAULACTIS* AS THE OUTGROUP*

Character	1	2	3	4	5	6	7	8	9	10	11	12
Phaulactis	0	0	0	0	0	0	0	0	0	0	0	0
Zaphrentidae												
1. Indian Cove 1	0	0	0	1	0	0	0	0	0	0	0	1
2. *Heliophyllum* A (solitary)	0	0	0	3	0	1	0	0	0	0	0	2
3. *H.* B (phaceloid)	1	0	1	3	0	1	0	1	1	0	0	3
4. *H.* C-1 (cerioid)	2	0	1	3	0	1	0	1	1	2	0	2
5. *H.* C-2 (cerioid w. aulos)	2	0	1	3	1	1	0	1	1	2	0	3
6. *H.* D (astreoid)	3	0	1	3	0	1	0	1	1	2	0	2
7. *Zaphrentis*	0	1	0	3	0	0	2	0	0	0	0	4
8. *Aemulophyllum*	0	1	(1)	3	0	1	1	0	0	0	0	2a
9. *Cyathocylindrium*	1	0	1	2	0	1	0	1	1	0	1	2
Craspedophyllidae												
10. *Cylindrophyllum*	1	1	1	2	0	0	1	1	1	1	1	2
11. *Acinophyllum*	1	1	1	2	0	0	2	1	2	1	1	2
12. *Eridophyllum*	1	1	1	2	1	0	1	1	2	1	1	3
13. *Grewgiphyllum*	(2)	1	1	2	1	0	1	1	2	(2)	1	2
14. *Prismatophyllum* A	2	1	1	2	0	1	1	1	2	2	1	2
15. *P.* B (with aulos)	2	1	1	2	1	1	1	1	2	2	1	4
16. *Asterobillingsa* A	3	1	1	2	0	1	1	1	2	2	1	2
17. *A.* B (with aulos)	3	1	1	2	1	1	1	1	2	2	1	2

*The key to characters and states is in Table 2. "Character" 12 is time of first appearance in the geologic record (see Table 2).

considered to be separately derived from phaceloid rather than an intermediate step. Figure 1A illustrates the series used and an alternative that is discussed below.

Character 3. Both the cylindrical and depressed states are derived from the ancestral conical state (Fig. 1B).

Character 4. Carinae states [0] and [1] are ordered in accordance with the stratigraphic sequence. States [2] and [3] appeared at approximately the same time, but given the early appearance of [1], [2] logically precedes [3]. This ordering is changed from Oliver (1993) as a result of recognizing Indian Cove 1 as a separate taxon but is appropriate for an analysis of the combined families in which yard-arm and zigzag carinae are equally important. The tree resulting from the 1993 ordering is discussed as Scenario 1.

Characters 7 and 9. Ordering of dissepimentarium width and offsets is based on the outgroup and the apparent sequence in the stratigraphic record.

Character 10. Support system states could be logically ordered by considering mechanical strength only [0→1→(2)→2], but massive [2] was derived from proximal [0] or processes [1] in most or all cases, and the origin of multidirectional [(2)] is questionable. The ordering used in the analysis and possible alternatives are shown in Figure 1C.

SCENARIOS

Figures 2 through 7 are cladograms based on morphologic characters that were clearly important in the development of the two families. Modifications suggested by the paleontologic or stratigraphic record are discussed and, in some cases, illustrated. Scenario 1 uses the 1993 tree that was "objectively" based on the morphology. Scenarios 2 to 8 are illustrated by new, manually derived trees. The Consistency Index (CI) and its calculation are defined in the explanation of Table 2.

The final scenario (8) is based on my experience in working with the taxa over a period of 35 years (Oliver, 1958, 1976, 1981, 1993, 1997; Oliver and Sorauf, 1989, 1994; Sorauf and Oliver, 1976). It is a subjective evaluation of morphology and the paleontologic record.

Scenario 1

The 1993 analysis (Oliver, 1993, p. 132–136, fig. 2) was based on a cladogram of the taxa listed in Table 1 minus Indian Cove 1. The 1993 tree is not reproduced here, but it is identical to the Scenario 3 tree (below) except for the insertion of Indian Cove 1 between *Phaulactis* (outgroup) and *Heliophyllum* A, in the left side of the tree. In the 1993 analysis there were significant differences from this study in character state ordering for carinae (character 4) and minor changes in support systems (10). With 21 character states and my subjective decision that the aulos (5) developed independently in each of the five taxa in which it occurs, the 1993 tree required 33 steps (character state changes) for a relatively high consistency index (CI=0.64). A different handling of the aulos would have produced the same tree more

TABLE 2. CHARACTERS AND CHARACTER STATES FOR GENERA OF THE ZAPHRENTIDAE AND CRASPEDOPHYLLIDAE*

1. Coloniality (0.57)
   - 0 solitary
   - 1 phaceloid
   - (2) semi-cerioid
   - 2 cerioid
   - 3 astreoid

2. Corallite diameter (0.33)
   - 0 large
   - 1 small

3. Corallite shape (0.67)
   - 0 conical
   - 1 cylindrical
   - (1) depressed

4. Carinae (1.00)
   - 0 none
   - 1 incipient, zigzag
   - 2 predominantly zigzag
   - 3 predominantly yard-arm

5. Aulos (0.20)
   - 0 absent
   - 1 present

6. Calice (0.25)
   - 0 V-shape
   - 1 with broad platform

7. Dissepimentarium (0.40)
   - 0 wide
   - 1 narrow
   - 2 very narrow (1–3 rows)

8. Neanic dilation (0.50)
   - 0 present
   - 1 absent

9. Offsets (0.67)
   - 0 none (solitary)
   - 1 lateral and peripheral
   - 2 lateral only

10. Support system (0.60)
    - 0 proximal attachment only
    - 1 lateral processes
    - (2) multidirectional expansions
    - 2 massive

11. Symmetry (1.00)
    - 0 markedly bilateral
    - 1 biradial

12. First known occurrence
    - 0 pre-Pragian
    - 1 Pragian/early Emsian
    - 2 mid/late Emsian (Schoharie)
    - 2a late Emsian (upper Schoharie)
    - 3 early Eifelian (Edgecliff)
    - 4 middle Eifelian (Moorehouse)

*See text discussion of character polarization or ordering. Numbers in parentheses are the consistency indices (CI) for each character in the Scenario 8 tree (Fig. 7). The CI is defined as the number of character states (excluding 0) for the character divided by the number of times the character changes on the cladogram. A CI of 100 indicates perfect congruence, whereas a low CI suggests that the character is of little value in the analysis. In the same way, a CI can be calculated for each cladogram (see Table 3) to provide a basis for comparing the "efficiency" with which it handles the morphologic changes. "Character" 12 is time of first appearance in the geologic record; this is not used in the cladistic analysis but is referred to in discussions.

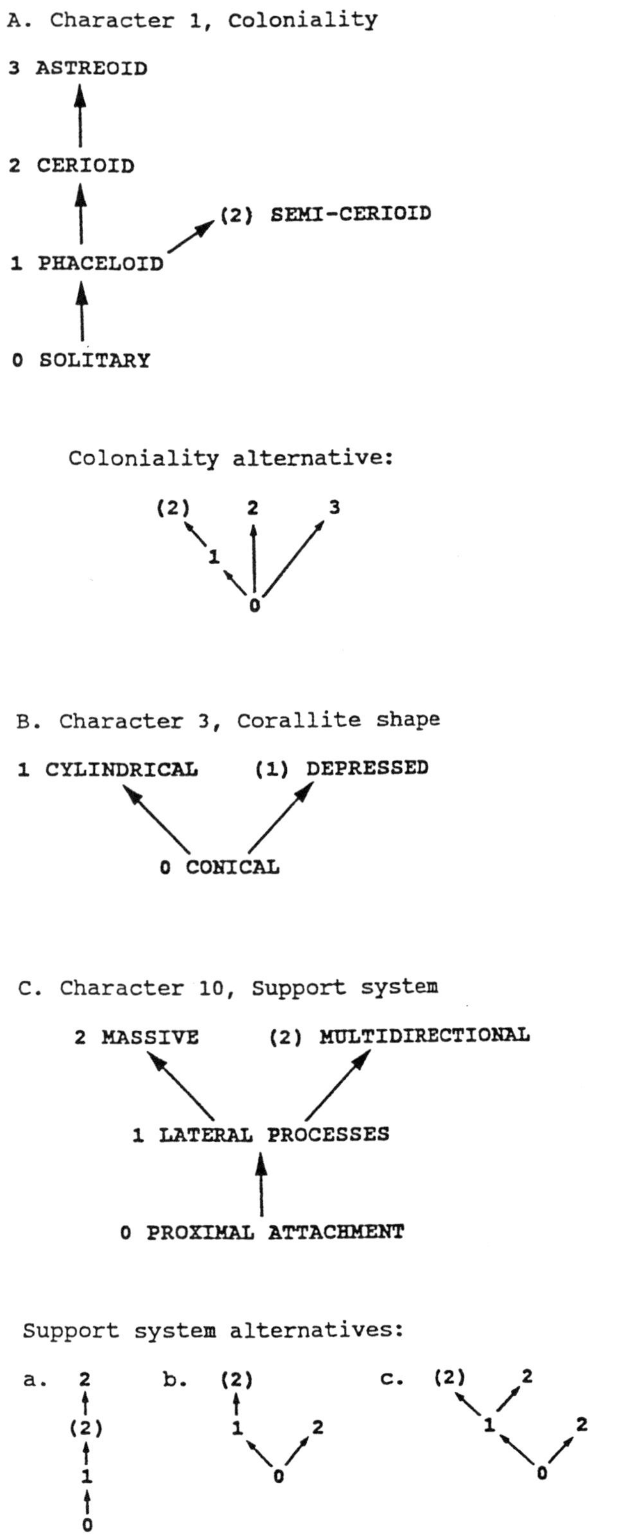

Figure 1. Transformation series for selected multistate characters of Zaphrentidae and Craspedophyllidae. Character number and name are above each series. In the alternative series, numbers rather than names of states are used; the alternatives are discussed in the text. A, Character 1, coloniality. B, Character 3, corallite shape. C, Character 10, support system.

parsimoniously (32 steps; CI=0.66; see Table 3 for comparisons). PAUP analysis by J. M. Pandolfi (discussed in Oliver, 1993) found numerous (162) 32-step trees. Significantly, most or all of the PAUP trees grouped the craspedophyllid genera together, giving support to the morphologic basis for separating the two families as a practical matter while raising questions about monophyly in the families.

Scenario 2

The tree in Figure 2 is based on the assumption that the coloniality states (character 1) developed in the simple progression indicated in Table 2 and Figure 1A; no reversals or convergences are permitted. The tree requires 41 steps (CI=0.51) because of the number of reversals and convergences in other characters that are required to accommodate the coloniality progression. The model is rejected because the morphologic similarities (other than coloniality) of the craspedophyllid genera are too great for them to be separated as they are in this scenario. In addition, it is clear from the paleontologic record of the Rugosa that coloniality developed many times and in many ways, so that there is no reason to give it primacy. All of the other scenarios group the craspedophyllids.

TABLE 3. NUMBER OF CHARACTER STATE CHANGES (STEPS) REQUIRED FOR TREE CONSTRUCTION, AND CONSISTENCY INDICES (CI) FOR EACH SCENARIO DISCUSSED IN TEXT*

Scenario	Steps	CI	Notes
1	33	0.64	1993; Strict character analysis except for independent origin of aulos.
1a	32	0.66	1993; Parsimonious development of aulos.
2	41	0.51	Coloniality sequence strictly followed; aulos as in 1.
3†	36	0.58	Coloniality developed separately in each family; aulos as in 1.
4	35	0.60	As 3 but aulos develops once in each family.
5†	39	0.54	Carinae sequence strictly followed.
6	39	0.54	Craspedophyllids moved; topologically identical to 5.
7†			Independent derivation of solitary genera; adds 2 steps to any of the above.
8†	41	0.51	"Most likely"; combines 5 and 7.

*Scenario 1 is not strictly comparable to the others (see text discussion). Scenarios most compatible with the paleontologic record are marked with a dagger (†).

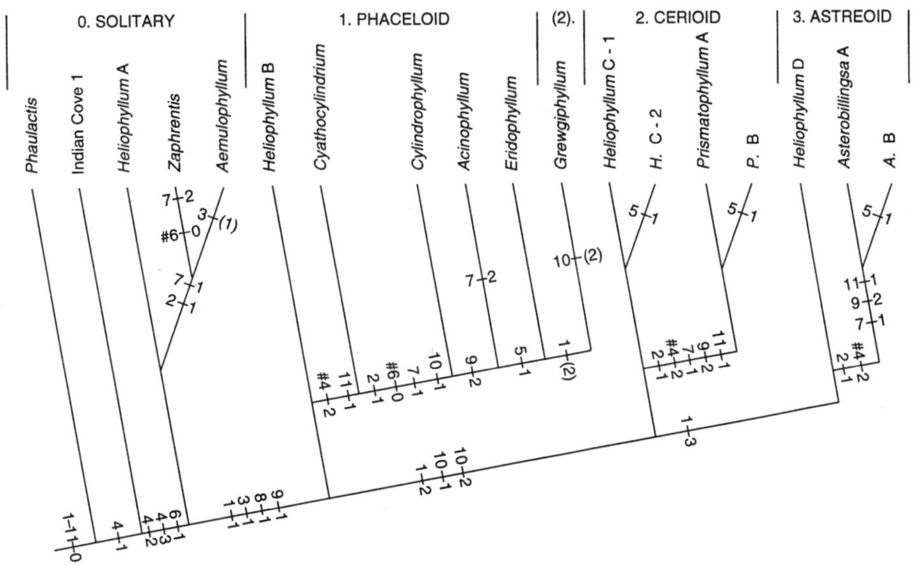

Figure 2. Cladogram of the Zaphrentidae plus Craspedophyllidae based on the simple progression of coloniality character states indicated by the headings (Scenario 2). Each character transformation is marked by a short line with a number at either end; the character number is to the left and the state to the right. Pound signs (#) mark character state reversals. 41 steps; CI = 0.51.

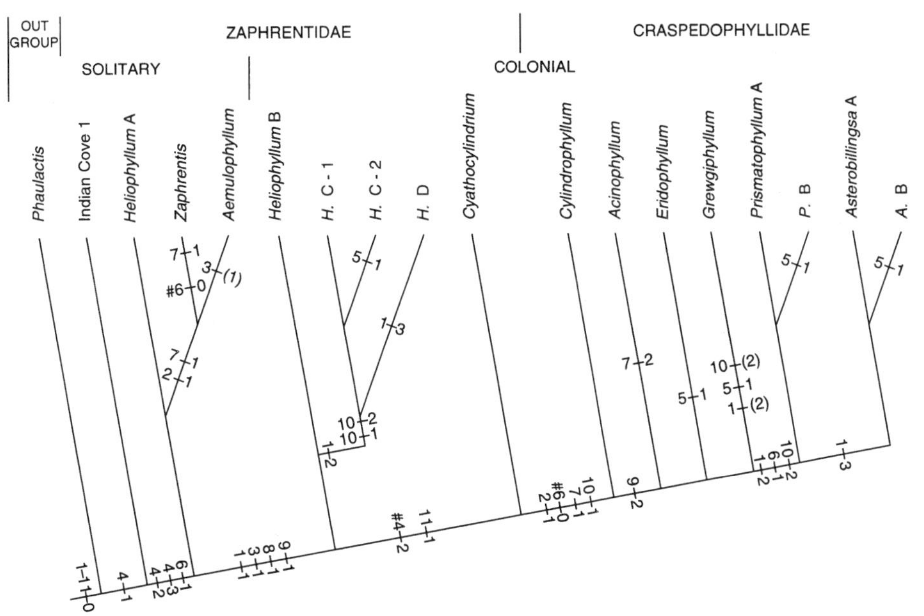

Figure 3. Cladogram of the Zaphrentidae plus Craspedophyllidae with coloniality series developed independently in each family (Scenario 3). See Figure 2 for key. 36 steps; CI = 0.58.

Scenario 3

This scenario (Fig. 3) independently develops each family in the coloniality sequence (character 1) and the craspedophyllids, as a group, from the morphologically most similar zaphrentid, *Cyathocylindrium*. This is the 1993 tree (Scenario 1) with the addition of the taxon Indian Cove 1 and reversal of the carinae sequence (character 4). It requires 36 steps (CI=0.58) and is both more parsimonious than scenario 2 (Table 3) and more in accord with the paleontologic record because it develops the aulos (5) independently in each genus in which it occurs.

Scenario 4

This scenario (Fig. 4) is identical to Scenario 3 (Fig. 3) except that it develops the aulos (character 5) only once in each family group. This saves only one step (35 steps) because it requires two reversals and is thus only marginally more parsimonious (CI = 0.60; Table 3). It is unsatisfactory because the paleontologic record clearly indicates independent development of the aulos in each genus (Oliver, 1981, p. 873–874).

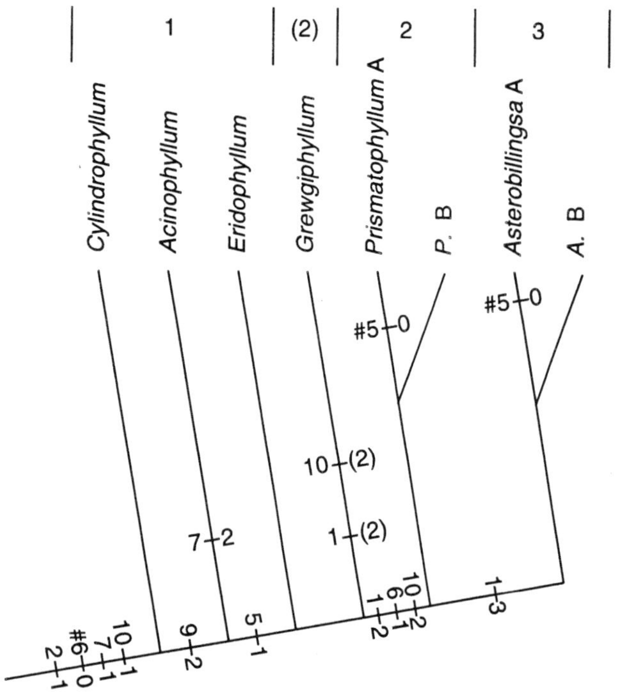

Figure 4. Cladogram in Figure 3 with the aulos (character 5) developing only once in each family (Scenario 4). Only craspedophyllid half of cladogram is shown; numbers at top are coloniality states. See Figure 2 for key. 35 steps; CI = 0.60.

Scenario 5

Carinae (character 4) is a defining character in the two families. If the character state order shown in Table 2 is strictly followed, the tree shown in Figure 5 is the result. It requires 39 steps (CI = 0.54) so is not parsimonious (Table 3), although it is very much in accordance with the paleontologic record. Note further discussion in next section (Scenario 6).

Scenario 6

The tree in Figure 6 illustrates the effect of appearance. It is topologically identical to that of Scenario 5 (Fig. 5) but shows the craspedophyllids between zaphrentids with zigzag carinae (on left) and those with mostly yard-arm carinae (on right). In each tree, Indian Cove 1 has the same two lines of descendants, but Scenario 6 more clearly questions whether either or both families are monophyletic.

Scenario 7

Scenarios 2 to 6 develop the solitary zaphrentids (there are no solitary craspedophyllids) parsimoniously (see the left, "solitary" part of Fig. 2); Scenario 1 differs only in lacking Indian Cove 1. However, the paleontologic record indicates that *Aemulophyllum* and *Zaphrentis* were independently derived from solitary *Heliophyllum* (left side of Fig. 7). This adds two steps to the length of any tree in which independent derivation is used but is more likely to be correct.

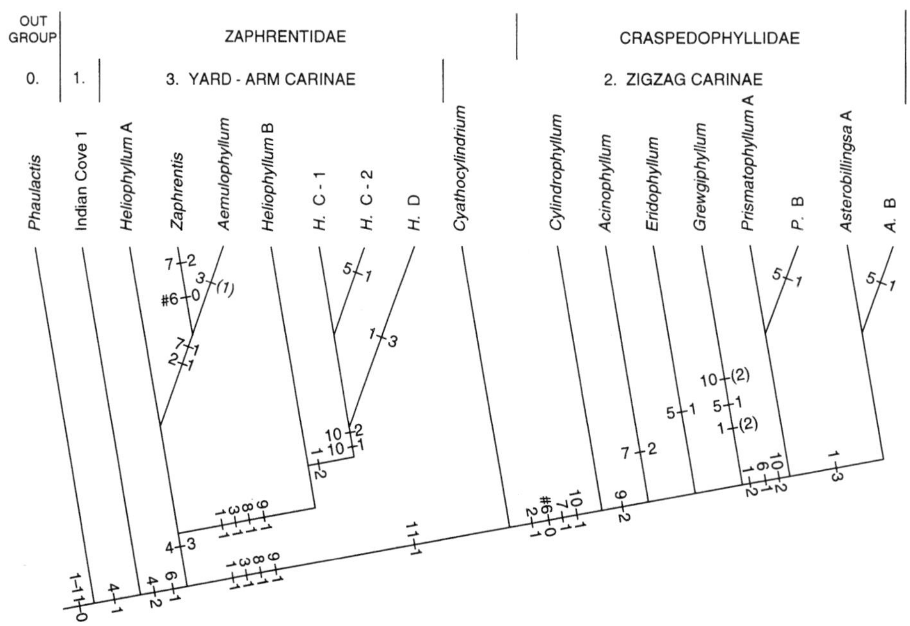

Figure 5. Cladogram of Zaphrentidae plus Craspedophyllidae based on primacy of carinae (character 4), Scenario 5; carinae states are at top. See Figure 2 for key. 39 steps; CI = 0.54.

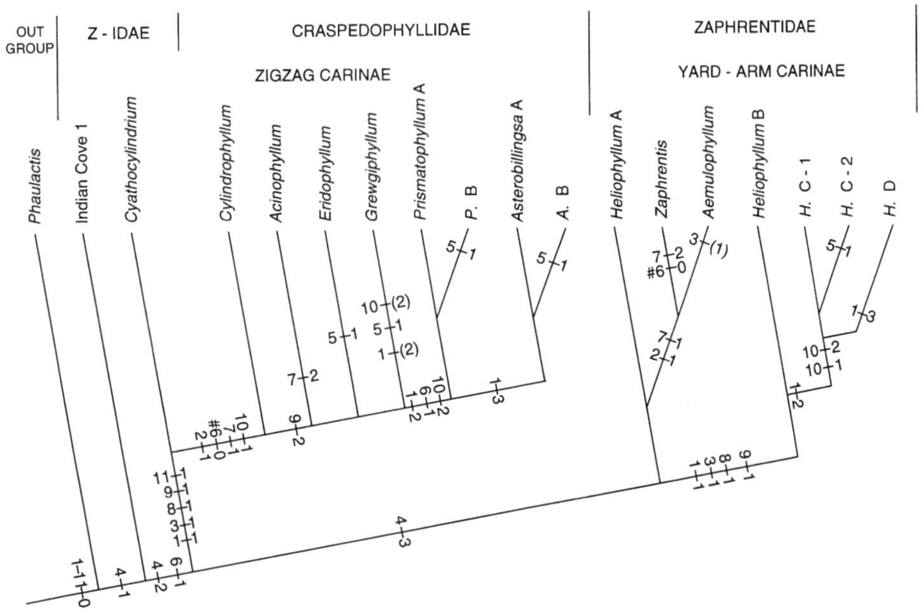

Figure 6. Cladogram of Zaphrentidae plus Craspedophyllidae; topologically identical to cladogram in Figure 5 but with craspedophyllid genera positioned between zigzag zaphrentids and yard-arm zaphrentids (Scenario 6). See Figure 2 for key. 39 steps; CI = 0.54.

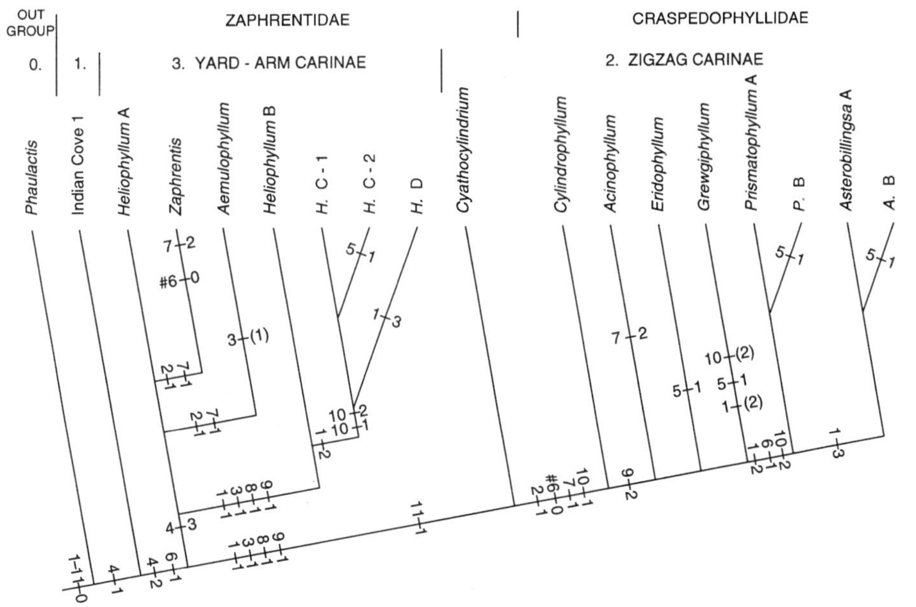

Figure 7. "Most likely" cladogram of Zaphrentidae plus Craspedophyllidae based on paleontologic record and character state sequences (Scenario 8). Carinae states are at top. See Figure 2 for key. 41 steps; CI = 0.51.

Scenario 8

Figure 7 is my interpretation of the genus-level relationships in the Zaphrentidae and Craspedophyllidae. It is partly based on the character state sequences but modified where the paleontologic record seems compelling (points 1 to 4 in following discussion). It is a combination of Scenarios 5 and 7, requires 41 steps (CI = 0.51), and is thus one of the least parsimonious of all the trees discussed (Table 3).

DISCUSSION

The principal value of cladistics in small taxonomic groups with relatively few characters is to force a critical examination of

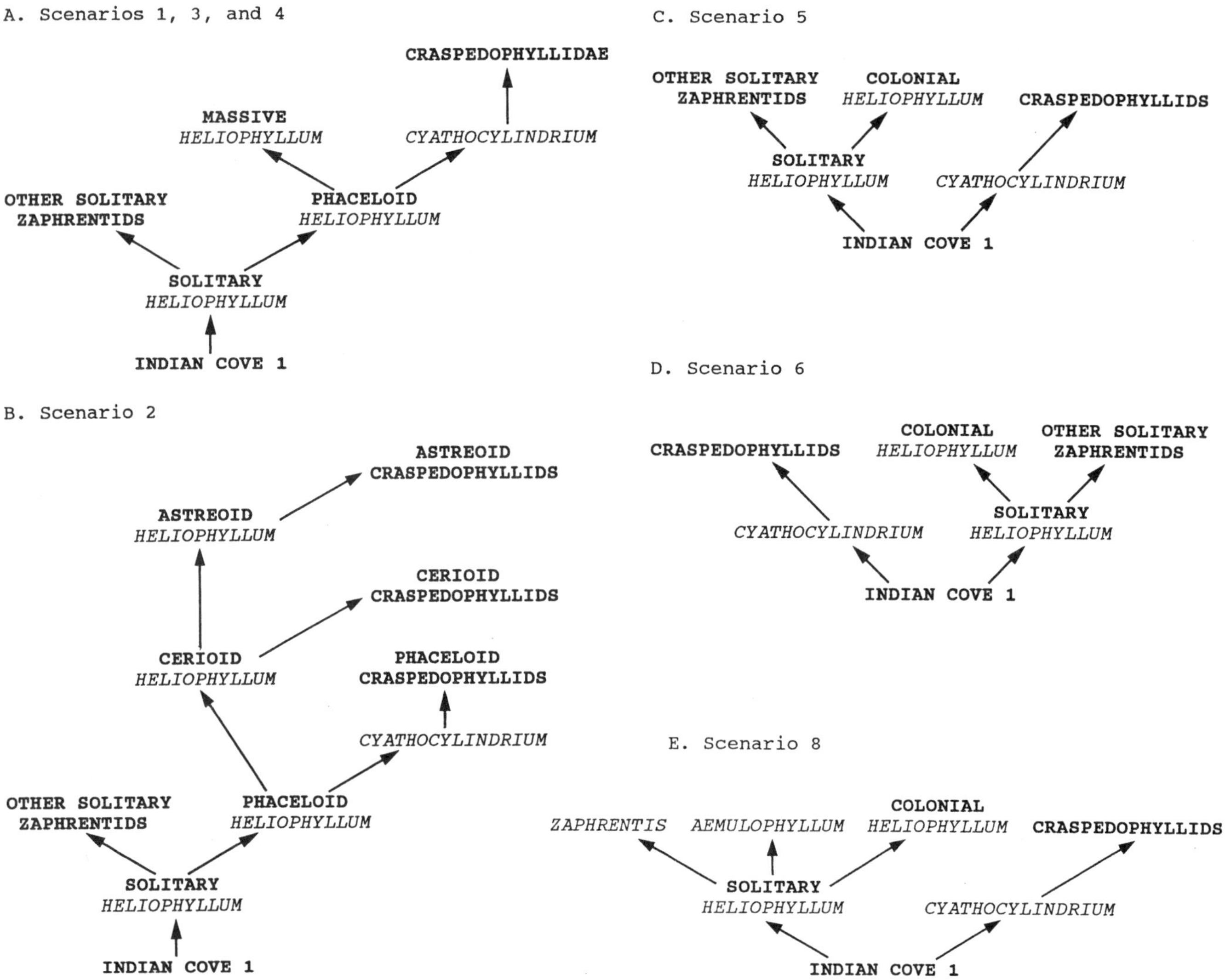

Figure 8. Simplified, conventional trees of the various scenarios. A, Scenarios 1, 3, and 4. B, Scenario 2. C, Scenario 5. D, Scenario 6. E, Scenario 8, combining scenarios 5 and 7.

the available characters and a subjective evaluation of their relative importance. All characters are not equally informative; this imbalance can apparently be ignored in taxa with many characters but must be considered in those with few.

Good stratigraphic and paleontologic records can establish sequence and the existence of homoplasy (see general discussions in Fisher, 1991; Huelsenbeck, 1994; Webb, 1994; Wagner, 1995; and included references). In the two families being considered, the record of parallel or convergent development of four of the characters seems clear: (1) The aulos was independently developed in each genus in which it occurs (Oliver, 1981). Its presence is treated as a generic character only in *Eridophyllum* and *Grewgiphyllum.* (2) Coloniality is an important character, but the character states developed independently in each family (in some cases, in genera and species). The ordering used in the analysis is conventional; there is ample suggestion in the record of the rugose corals that this is a common, but not necessary, sequence. Within the Zaphrentidae, each of the three principal colonial states developed directly from solitary in at least some cases (Fig. 1A, coloniality alternative; Oliver, 1997). (3) Carinae are important and suggest a close relationship between the two families; the sequence of character states given in Table 2 is most compatible with the paleontologic record, but a different sequence was used in the 1993 analysis (Scenario 1) and a near-random development is conceivable. (4) The solitary genera, *Zaphrentis* and *Aemulophyllum*, as well as the colonial *Heliophyllum,* were independently derived from solitary *Heliophyllum.* These four points are used to suggest the "most likely" scenario (Scenario 8 above).

The sequence of first appearances in the stratigraphic record (of the zaphrentid and craspedophyllid taxa; see Table 1, "character 12) is less helpful than the record of homoplasy, but is mostly compatible with Scenario 8. Most of the first known appearances of the colonial taxa are middle to late Emsian and do not help order the coloniality states. The later appearance of *Heliophyllum* B is anomalous, and I suspect that earlier forms will be found. The later appearances of *Heliophyllum* C-2, Eridophyllum, and Prismatophyllum B and the sequence of the solitary taxa fit the cladogram sequence. The first known *Cyathocylindrium* and *Heliophyllum* are roughly the same age, so the record does not support an ordering of the more advanced carinae states.

Recent discussions of the use of cladistics in systematic paleontology and in the study of living corals have made many of the points illustrated here while emphasizing their usefulness where the right kinds of information are available. Forey (1992) and Padian et al. (1994) gave general reviews of paleontologic work and analyses of both positive and negative aspects. Veron (1995) postulated a reticulate phylogeny for scleractinian corals and considered this, and other factors, as making cladistics useful only as an adjunct to other taxonomic methods in coral systematics (p. 24).

CONCLUSIONS

Figure 8 gives simplified, conventional trees for each of the scenarios. Scenarios 1, 3, and 4 (Fig. 8A) are basically the same and much simpler than Scenario 2 (Fig. 8B), but all are rejected for one or another reason as discussed above. The mirror image relationship of Scenarios 5 and 6 is more clearly seen in Figures 8C and D than in the cladograms (Figs. 5, 6). The concluding Scenario 8 (Fig. 8E) is Scenario 5 (or 6), modified to show the independent origins of the younger solitary zaphrentids (Scenario 7). Compilation of the Consistency Indices for the scenarios (Table 3) illustrates the inappropriateness of depending on parsimony when there are few characters and better means than cladistics for evaluating them. All cladistic analyses are subjective to some extent, as has been well argued by Crowson (1991).

If Scenario 8 is approximately correct, the Craspedophyllidae are monophyletic and the Zaphrentidae are paraphyletic. As Valentine (1990, p. 131–133) and Fisher (1991) have pointed out, these terms have limited significance in discussions of relationships of fossil taxa above the species level. Given an ancestor-descendant relationship between two families, the ancestor family is, by definition, paraphyletic. The trees support the separation of the two groups of corals under study, and they are logically recognized as separate families.

ACKNOWLEDGMENTS

I am not aware that Jess Johnson made any cladistic analyses of his brachiopods, and I do not remember ever discussing cladistics with him. However, the application of cladistics to paleontological systematics is just the kind of exercise that would have interested him, and I am pleased to be able to contribute this chapter to a volume in his honor.

Drafts of this chapter were reviewed by A. F. Budd, University of Iowa, Iowa City, and J. Pojeta Jr., U.S. Geological Survey, Washington, D.C. I am particularly indebted to Professor Budd for her in-depth criticisms and suggestions. Figures 2 through 7 were computer generated by K. Moore, Washington, D.C.

REFERENCES CITED

Cairns, S. D., 1984, An application of phylogenetic analysis to the Scleractinia: Family Fungiidae: Palaeontographica Americana, no. 54, p. 49–57.

Crowson, R. A., 1991, A critique of current cladistics: Journal of Natural History, v. 25, p. 3–5.

Fisher, D. C., 1991, Phylogenetic analysis and its application in evolutionary paleobiology, *in* Glinsky, N. L., and Signor, P. W., eds., Analytical paleobiology: Paleontological Society, Short Courses in Paleontology, no. 4, p. 103–122.

Forey, P. L., 1992, Fossils and cladistic analysis, *in* Forey, P. L., Humphries, C. J., Kitching, I. L., Scotland, R. W., Siebert, D. J., and Williams, D. M., Cladistics, a practical course in systematics: Systematics Association Publication 10, p. 124–136.

Huelsenbeck, J. P., 1994, Comparing the stratigraphic record to estimates of phylogeny: Paleobiology, v. 20, p. 470–483.

Oliver, W. A., Jr., 1958, Significance of external form in some Onondagan rugose corals: Journal of Paleontology, v. 32, p. 815–837.

Oliver, W. A., Jr., 1976, Noncystimorph colonial rugose corals of the Onesquethaw and early Cazenovia stages (Lower and lower Middle Devonian) in New York and adjacent areas: U.S. Geological Survey Professional Paper 869, 156 p.

Oliver, W. A., Jr., 1981, The Middle Devonian rugose coral *Prismatophyllum conjunctum* (Davis) and the age of the "Columbus" Limestone at Ingersoll, Ontario: Geological Society of America Bulletin, v. 92, p. 873–877.

Oliver, W. A., Jr., 1993, Origins and relationships of Devonian Rugosa endemic to the Eastern Americas Realm: Courier Forschungsinstitut Senckenberg, v. 164, p. 131–140.

Oliver, W. A., Jr., 1997, Origins and relationships of colonial *Heliophyllum* in the upper Middle Devonian (Givetian) of New York, *in* Proceedings, International Symposium on Fossil Cnidaria 7th, Madrid, September, 1995 (in press).

Oliver, W. A., Jr., and Sorauf, J. E., 1989, *Heliophyllum* Hall and *Charisphyllum* n. gen. (Devonian rugose corals) from Spain: Trabajos de Geologia, Universidad de Oviedo, v. 17, p. 3–17.

Oliver, W. A., Jr., and Sorauf, J. E., 1994, Branching *Heliophyllum* (Devonian rugose corals) from New York and Ohio: Journal of Paleontology, v. 68, p. 1183–1201.

Padian, K., Lindberg, D. R., and Polly, P. D., 1994, Cladistics and the fossil record, the uses of history, *in* Annual Review of Earth and Planetary Sciences, v. 22, p. 63–91.

Sorauf, J. E., and Oliver, W. A., Jr., 1976, Septal carinae and microstructure in Middle Devonian *Heliophyllum* (Rugosa) from New York State: Journal of Paleontology, v. 50, p. 331–343.

Valentine, J. W., 1990, The macroevolution of clade shape, *in* Ross, R.M., and Allmon, W.D., eds., Causes of evolution, a paleontological perspective: Chicago, University of Chicago Press, p. 128–150.

Veron, J. E. N., 1995, Corals in space and time, the biogeography and evolution of the Scleractinia: Sydney, University of New South Wales Press, 321 p.

Wagner, P. J., 1995, Stratigraphic tests of cladistic hypotheses: Paleobiology, v. 21, p. 153–178.

Webb, G. E., 1994, Parallelism, non-biotic data and phylogeny reconstruction in paleobiology: Lethaia, v. 27, p. 185–192.

Manuscript Accepted by the Society February 27, 1997

Printed in U.S.A.

Geological Society of America
Special Paper 321
1997

A global perspective of the Silurian tetracoral Goniophyllum

Anthony J. Wright
School of Geosciences, University of Wollongong, Wollongong, N.S.W. 2522, Australia

ABSTRACT

The pyramidal operculate tetracoral *Goniophyllum* was restricted to low latitude (less than about 30° S), shallow marine environments of Baltica, Laurentia, and Avalonia, unlike other generally cosmopolitan Early Silurian genera. The two named species are *G. pyramidale* (late Llandovery to Wenlock) and *G. fletcheri* (probably late Llandovery to Wenlock, possibly to early Ludlow). Material from several occurrences is described and illustrated, mostly for the first time.

North American representatives of the genus are known from Indiana, Iowa, Maine, Ontario, and Wisconsin. The relatively abundant Iowa material and the unique Wisconsin specimen are assigned to *G. p. pyramidale*. The few Ontario specimens and the two small specimens from Indiana are tentatively assigned to *G. p. primigena*.

One of the few specimens from the early Wenlock Woolhope Limestone, England, described here as *G.* sp. nov.?, is distinguished by its convex cardinal face and lack of a well-developed cardinal fossula. A further specimen from the *Petalocrinus* bed (late Llandovery) at Woolhope is identified as *G. fletcheri*; this is the oldest record for this species.

***Petraia quadrata* M'Coy, 1846 from the late Llandovery Kilbride Formation of Ireland, is refigured and referred with reservation to *G. pyramidale*. *Cyathophyllum tetragona* Quenstedt, 1879 from Gotland is refigured, and a lectotype is designated; this species is considered a junior subjective synonym of *G. pyramidale*.**

***G. p. pyramidale* is selected as type species of both *Rizocalceola* Scalia, 1922 and *Tetracalceola* Scalia, 1922, which thus become objective junior synonyms of *Goniophyllum*. *Protaeropoma* Ting, 1937 is a junior synonym of *Goniophyllum*; its type species species, *P. wedekindi* Ting, 1937, is a subjective synonym of *G. pyramidale*.**

In *G. p. pyramidale*, insertion of septa on the counter side of the alar septa (the prominent septa on the alar sides of the corallites) is demonstrated for the first time. In the calyx, minor septa arise simultaneously late in ontogeny, by "splitting" of major septa. The cardinal operculum is distinguished by its elevated, furrowed median septum that articulated with the depressed, simple distal tip of the cardinal septum; in contrast, the median septum of the counter operculum has a large median posterior notch.

INTRODUCTION

Johannessen (1993) thoroughly documented and reviewed the morphology of the two European species of *Goniophyllum* Milne-Edwards and Haime, 1850: *G. pyramidale* (Hisinger, 1829) and *G. fletcheri* Milne-Edwards and Haime, 1851. The latter, a Llandovery-Wenlock species, is known almost exclusively from England as generally deformed and recrystallized specimens. The only known specimen of *G. fletcheri*, from Gotland, is a nonoperculate but otherwise exquisitely preserved specimen (BMNH 48273: Johannessen, 1993, figs. 17 D–E) in the Lindström collection in the British Museum. A single specimen from the Woolhope area, England, is here identified as a late Llandovery occurrence of this species.

Wright, A. J., 1997, A global perspective of the Silurian tetracoral *Goniophyllum*, *in* Klapper, G., Murphy, M. A., and Talent, J. A., eds., Paleozoic Sequence Stratigraphy, Biostratigraphy, and Biogeography: Studies in Honor of J. Granville ("Jess") Johnson: Boulder, Colorado, Geological Society of America Special Paper 321.

The intent of this chapter is to supplement Johannessen's (1993) study of European (mostly Gotland) occurrences of *Goniophyllum* by documentation and discussion of occurrences of the genus in Ireland, England, Canada, and the United States, as well as by examination of aspects of the morphology of the genus.

MORPHOLOGICAL INTERPRETATIONS

Goniophyllum consisted of a large pyramidal corallite with four opercula, two asymmetrical triangular (alar) ones and two trapezoidal (cardinal and counter) ones. As suggested by Johannessen, diameter is measured along the cardinal-counter plane unless otherwise stated. The identification here of the counter (K) and cardinal (C) septa as seen near the calical lip of the corallite follows Birenheide (1974) and Johannessen (1993). The C septum is commonly subdued and located within a groove (Johannessen, 1993, figs. 6 A, 6 J; Fig. 1: B, I herein), and its articulating tip is depressed well below the level of the calical lip. In contrast, the K septum has a prominently swollen, cleft, elevated distal tip. These protosepta also identify the C (gently concave to gently convex) and K (convex) faces of the *G. pyramidale* corallite and are also vital for interpretation of opercula and varied forms of corallites. In transverse view, the C side is shorter than the K side in *G. pyramidale*.

Assuming the position of the C fossula is confirmed, where are the alar fossulae? Milne-Edwards and Haime (1855, p. 290) considered the "Septal fossulae . . . [as] corresponding to the angles of the visceral chamber." Clearly this cannot be so for all four corners, as the C fossulae flank the C septum. Further, the presence of enlarged septa in a median position on the "alar" edges invites the inference that the alar fossulae were located in that position (Johannessen, 1993). Rarely preserved septa within the calyx (Fig. 1: D) and septal traces on the external faces (Fig. 1: E, F, M, N) indicate that these were indeed the alar septa, with septa arising on the K side of, and abutting against, the alar septa.

Johannessen (1993) indicated that there are up to 140 septa in *G. p. pyramidale*. In specimens I have studied there appear to be about 21 major septa on the K edge, 18 on the C edge, and about 19 on the alar edges, generally with 10 on the K side of the alar septum, indicating about 77 major septa on a mature corallite.

Fossulae and septal insertion

Septal arrangement in *G. pyramidale* can be observed on the normally convex internal cardinal face (Johannessen, 1993, fig. 6 E; Fig. 1: B) and on the external surface of exceptionally finely preserved specimens (e.g., Fig. 4: C). The median C septum is located within the prominent cardinal depression (i.e., between the cardinal fossulae); in mature corallites, the C fossulae are flanked laterally by two prominent convex regions (Fig. 1: A, I). Newly inserted major septa diverge from the C septum low in the calyx but then curve so that, higher in the calyx, they parallel the C septum, owing to cessation of insertion well before the mature corallite size was reached.

Hill and Jell (1969) noted the difficulty of distinguishing minor septa from major septa in the Goniophyllidae; this is certainly so in thin sections. In all too few specimens, major and minor septa are apparent either on the external surface or in the calyx. On the external surface the septa appear as low, rounded ridges separated by fine grooves; along well-preserved calical rims, septa occur as low, short projections (Fig. 1: H, I), minor septa being the less prominent. In well-preserved specimens (e.g., Fig. 1: I), septa of both orders extend as low, rounded to acanthine ridges some distance down into the calyx, minor septa being relatively short. Minor septa arise rather late on all internal faces of the corallite, each minor septum appearing upon "bifurcation" or "splitting" of a major septum (Fig. 1: L). This "bifurcation" has not been discerned on the outer surface of any corallites, and traces of the minor septa are visible only in the distal regions of the corallite. Hill and Jell (1969, p. 456) noted that, in one specimen of *Calceola*, a series of minor (?) septa appeared simultaneously. This mode of abrupt minor septal appearance thus may be characteristic of the Goniophyllidae.

Birenheide (1974) documented the absence of minor septa adjacent to the K septum in *G. pyramidale*; this has been confirmed in this study. Moreover, minor septa appear to be lacking adjacent to the C septum in mature corallites.

In the depression between adjacent septa, an array of fine pits or lateral projections from the septa may be seen, apparently representing multitrabecular septal structure. Only major septa have been observed crossing dissepiments (Fig. 1: J); they occur as rows of acanthine nodes often separated by pits.

Opercula

All known operculate specimens of *G. pyramidale* are from Gotland, except for a single Iowa specimen, which, according to Weller and Davidson (1896), retains an operculum. Operculate specimens of *G. p. pyramidale* from Gotland I have examined are Cn 16057 (Visby, four opercula); two speci-

Figure 1 (on facing page). A–F: *Goniophyllum pyramidale pyramidale* (Hisinger), all × 1.5, specimen number ENSM e-146/3, de Verneuil Collection, Université Claude Bernard. Card with specimens states "Gotland, Coll. Verneuil. 1873," but a later annotation by Dr. P. Semenoff-Tian-Chansky indicates this specimen was probably illustrated by Milne-Edwards and Haime (1851, p. 404, pl. 2, figs. 4, 4a). A, calical view; B, specimen tilted to show cardinal side; C, tilted to show counter side; D, tilted to show alar face; E, counter face; F, alar face showing insertion. G: *G. p. pyramidale* (Hisinger), × 1.5, RM Cn 15905, Visby, Gotland. Three opercula visible. H: *G. p. pyramidale* (Hisinger), MCZ 9396, Gotland, × 1.5; calical view, showing dissepiments and cornulitid in calyx. I–L: *G. p. pyramidale* (Hisinger), MCZ 9399, Gotland. I, calical view × 1.5; J, counter face, showing dissepiments overgrown by major septa, × 10; K, counter face of calyx showing general view of dissepiments, × 3.5; L, enlargement of area of cardinal fossula, showing sudden appearance of minor septa (arrowed), × 8. M, N: Alar faces of two specimens of *G. p. pyramidale* (Hisinger), showing septal insertion in alar fossulae. Counter face to right. M, × 2, ENSM e-146/1; N, × 3, ENSM e-146/2.

A
B
C
D
E
F
G
H
I
J
K
L
M
N

mens from the "Upper Visby Marl," at Lummelunda; Cn 15904 (four opercula) and Cn 15905 (Fig. 1: G, with three opercula); and a single specimen with one operculum (ENSM e-147/1) from the de Verneuil Collection, Université Claude Bernard. Other operculate *Goniophyllum* have been illustrated by Lindström (1882) and Johannessen (1993). Opercula in *G. fletcheri* are much rarer, having been reported only by Salter (1873) and Lindström (1882). (See section "Taxonomy" for explanation of abbreviations.)

Lindström (1871, p. 123) advanced the possibility that the opercula in *Goniophyllum* "overlapped in a sort of spiral way" and asserted that the opercula showed "slightly curved, rarely straight . . . lateral borders of all the valves." Such a spiral arrangement is not supported by any specimen studied by me or illustrated by previous workers. Rather, the tightly fitting opercula (Johannessen, 1993, fig. 7 B; Fig. 1: G herein) show that, because of articulatory structures along the marginal edge of the opercula, these could only rotate (open or close) with that edge as a hinge axis (as in strophic brachiopods). The fit between opercula was close; the longitudinal furrow at each corallite corner and the sharply rounded form of the opercula ensured that their corners were in close contact even in the corner regions. Johannessen claimed that the trapezoidal C operculum of *G. pyramidale* was characterized by a rectangular furrow on the outer surface; additional specimens I have examined have this feature (e.g., Fig. 2: E), accentuated by crushing, but in the K operculum rather than the C septum.

As shown by Johannessen (1993, fig. 7 F, 7 J), the two types of trapezoidal operculum are distinct in that one type has a "socket" excavated into the elevated posterior (i.e., marginal) face of the median septum; this type of opercular median septum must have articulated with the simple, depressed edge of the C septum. The other type has a simple, less elevated median septum (Fig. 2: C, D, F) that must have articulated with the more elevated, cleft K septum.

Opercula illustrated by Johannessen (1993, fig. 7 F, 7 J) as K and C respectively are, therefore, believed to be reversed. His fig. 7 F shows an operculum with a median septum with a large notch on the posteriorly facing edge; I suggest this is a C operculum. His fig. 7 J shows no such feature and is, therefore, a K operculum.

The above assertions are supported by latex molds of the

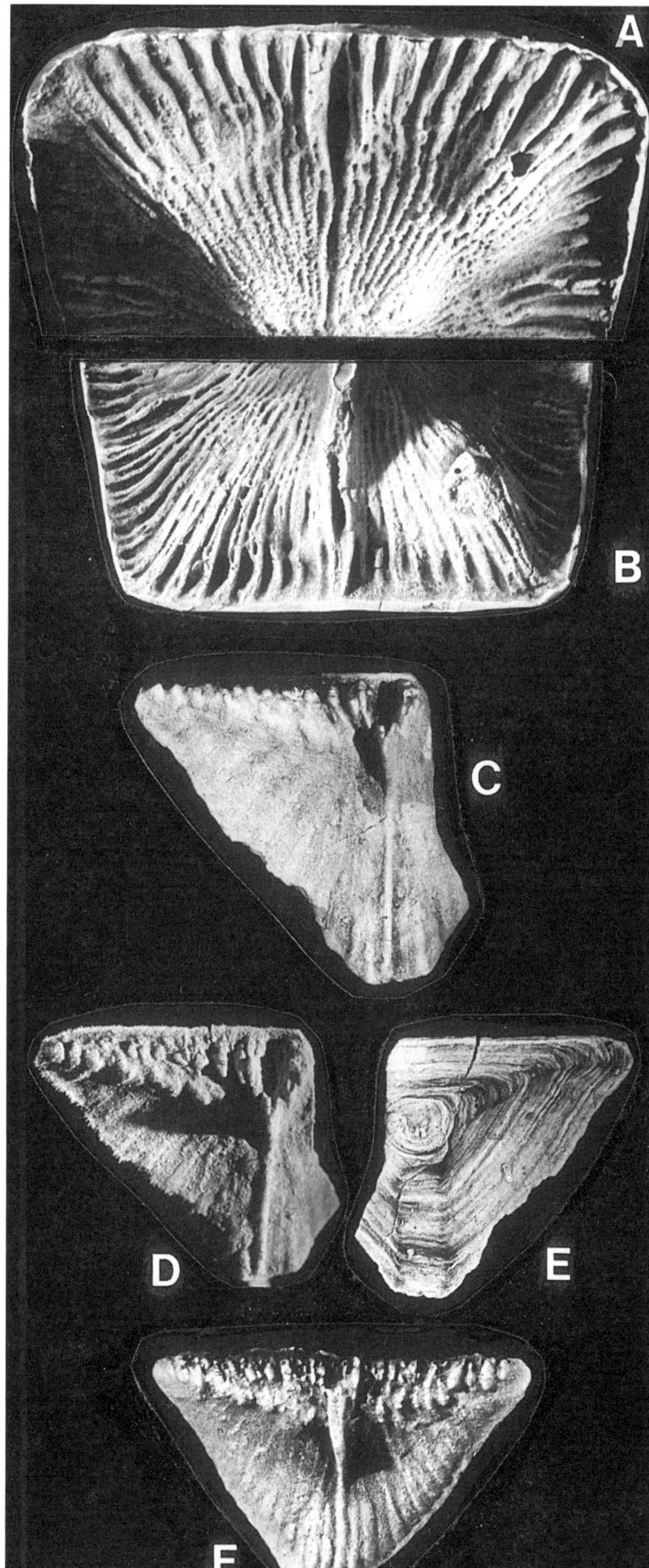

Figure 2. A–F: *Goniophyllum pyramidale pyramidale* (Hisinger), all Gotland. A, rubber mold of counter portion of calyx of BMNH 46391, × 4; note particularly the long depression corresponding to K septum on operculum, suppressed minor septa, and multitrabecular septa. B, rubber mold of cardinal portion of calyx of MCZ 9393, × 3; note the long, furrowed depression corresponding to median cardinal opercular septum. C–E: incomplete counter operculum, RM Cn 15945; C, D, dorsal (× 3.4) and tilted (× 4) views showing spacious, long "socket" as in rubber cast in A; E, external surface showing parallel-sided trench. F, RM Cn 15960, counter operculum showing "large socket" in which K septum articulated, × 4.

calices of suitably preserved corallites, assuming that the impression of the C and K calical edges should reflect the approximate form of the C and K opercula respectively. The relevant portions of two molds are illustrated (Fig. 2: A, F); similarities between the molds and opercula are evident. The C opercular septum is a long, elevated structure with a pronounced median groove (Fig. 2: C, D, F), whereas the K operculum bears a wide median depression with no significant elevation above the inner surface of the operculum.

Lindström (1871, p. 124) asserted that the operculate coral *Araeopoma prismaticum* (Lindström, 1868) shed its opercula at intervals. However, in *G. pyramidale*, growth lines on the opercula unequivocally indicate holoperipheral increments throughout all opercular growth, from the subcircular juvenile stage to the triangular or trapezoidal mature stage (Fig. 2: E). *G. fletcheri* from Dudley exhibits not only considerable rejuvenescence but several generations of branching (Lindström, 1882, p. 56) requiring significant changes in corallite diameter; it is uncertain whether or not preexisting opercula were retained after rejuvenescence. *G. pyramidale* usually shows only weak rejuvenescence (e.g., Cn 15904) varying from normal rugosity to significant changes in diameter. Lindström (1882, pl. 5, fig. 1) and Johannessen (1993, fig. 6 G) illustrated a *G. pyramidale* with a counter operculum attached to the epitheca just below a line of rejuvenescence; this may be original or an operculum from another corallite.

TAXONOMY

Material studied (with acronyms used here) is from the following institutions: Natural History Museum (British Museum of Natural History), London (BMNH); Geological Survey of Canada, Ottawa (GSC); Palaeontological Department, Swedish Museum of Natural History, Stockholm (RM Cn); de Verneuil Collection, Département des Sciences de la Terre, Université Claude Bernard, Villeurbanne (ENSM); Museum of Comparative Zoology, Harvard University, Cambridge (MCZ); National Museum of Ireland, Dublin (NMI); collections of Dr. John Penn, School of Geological Sciences, Kingston University, Kingston-upon-Thames, England (P); Institut und Museum für Geologie und Paläontologie, Universität Tübingen, Tübingen (S); Sedgwick Museum, University of Cambridge, Cambridge (SM); University of Iowa, Iowa City (SUI); National Museum of Natural History (Smithsonian Institution), Washington, D.C. (USNM).

Note that in Figures 1 through 4 the convention is that the counter edge of corallites in calical or transverse view is toward the top of page. Prior to photography all specimens other than Figure 4: G have been blackened with graphite and whitened with ammonium chloride sublimate.

Family CYSTIPHYLLIDAE Milne-Edwards and Haime, 1850
Subfamily GONIOPHYLLINAE Dybowski, 1873

Genus *Goniophyllum* Milne-Edwards and Haime, 1850

Goniophyllum MILNE-EDWARDS AND HAIME, 1850, p. lxix; MILNE-EDWARDS AND HAIME, 1851, p. 404.
Rizocalceola SCALIA, 1922, p. 98.
Tetracalceola SCALIA, 1922, p. 99.
Protaeropoma TING, 1937, p. 413.

Type species. Turbinolia turbinata var. *pyramidalis* Hisinger, 1829; selected by Milne-Edwards and Haime (1850, p. lxix).
Diagnosis. Goniophyllidae with four opercula, including trapezoidal cardinal and counter opercula, and two triangular alar opercula; rootlets low on counter side; corallites rhomboidal to square in cross section, usually with counter side longer than cardinal side and with shallow grooves at edges of counter face; cardinal and counter septa prominent, at the mid-length of walls, with alar septa located at mid-length of alar walls; cardinal fossulae prominent; septa acanthine with multitrabecular microstructure; dissepiments prominent in mature specimens, associated with tabellae.
Remarks. As Johannessen (1993) remarked, this genus was considered a coral by Bromell (1727–1728) and by most subsequent authors, including Lindström. Davidson (1853), however, considered it a brachiopod. The genus contains two species, *G. pyramidale* (Hisinger, 1829) and *G. fletcheri* Milne-Edwards and Haime, 1851; the latter species is neither discussed extensively nor described here, having been fully documented by Johannessen (1993).

Among the more bizarre commentaries on *Goniophyllum* was that by Scalia (1922), as follows. Scalia (1922, p. 98–99) concluded: first, that *Calceola* was a brachiopod; second, that *Rizophyllum* (*sic*) *gotlandicum* was a *Calceola* and that there was no need to create a new genus such as *Rizocalceola* (*sic*); and third, that *G. pyramidale* was apparently formed by the

> coming together of four shells of *Calceola*, perfectly bound together and grown together on the convex part of the ventral valve, as one can easily become aware by the disposition of the relative edge, perfectly similar to those of *Calceola*. . . . The external faces of the pyramid formed by the four calceolidae which comprise the so-called *Goniophyllum*, are the surfaces of the ventral valves; the very strange, supposed operculum, formed by four triangular pieces, is simply constituted by the dorsal valves of each of the four individuals, which even though they are so indissolubly bound together amongst themselves, and most probably, with their respective stomach pouches fused into one common cavity conserved, nonetheless, a certain individual freedom which allowed each of them to collaborate with a certain independence to the well-being of the community. Because of the unique coming together of the four calceolids I propose the institution of the new genus *Tetracalceola*, which in my opinion indicates well its true nature and its exact systematic position.

This imaginative view recalls Davidson's belief that *Calceola* was a brachiopod. As the generic name *Tetracalceola* has remained unused for almost 75 years, it may be discarded as a *nomen oblitum*. Nevertheless, to rectify this situation beyond doubt, I here select *Goniophylum pyramidale pyramidale* (Hisinger) as type species of *Tetracalceola* Scalia. The same author introduced the name *Rizocalceola*, apparently without formally proposing the name; to preempt any confusion about this nominal genus, I select *G. p. pyramidale* (Hisinger) as type species for it. Both *Tetracalceola* and *Rizocalceola* are thus made objective junior synonyms of *Goniophyllum*.

Ting (1937) selected *G. pyramidale primigena* Lindström (Ting, 1937, figs. 1 a–c, 2) as "genotype" of *Goniophyllum* and erected a new genus *Protaeropoma* based on a new species *P. wedekindi* (Ting, 1937, figs. 3a–b). Ting's selection of *G. p. primigena* as genotype (type species) of *Goniophyllum* is invalid as Milne-Edwards and Haime (1850) had already selected *Turbinolia pyramidalis* Hisinger as type species. Ting argued that *P. wedekindi* differed significantly from *G. pyramidale primigena* in having septal cones clearly separated by dissepimental tissue and an apical angle of the order of 90°, but these characters are shared by *G. p. pyramidale*.

Hill and Jell (1969) stated that the range of *Goniophyllum* was "Upper Llandoverian to Ludlovian"; if *G. fletcheri* at Dudley ranges to the top of the Much Wenlock Limestone Formation, the genus may extend into the Ludlow (see Cocks et al., 1992, p. 6, for discussion of the age of this formation).

Goniophyllum pyramidale pyramidale (Hisinger, 1829)
Figures 1: A–M, 2: A–F, 3: A–N, ?3: T–U, 4: K–M, ?4: E

Turbinolia turbinata var. *pyramidalis* HISINGER, 1829, p. 22.
Turbinolia pyramidalis Hisinger. HISINGER, 1831, p. 128, pl. 7, fig. 5; HISINGER, 1837, p. 101, pl. 28, fig. 12.
? *Petraia quadrata* M'COY, 1846, p. 60, pl. IV, fig. 18.
Cyathophyllum tetragonum QUENSTEDT, 1879, p. 407, Pl. 156, figs. 82–85.
Goniophyllum pyramidale mutatio *prima* LINDSTRÖM, 1882, p. 45, pl. 1, fig. 1.
Goniophyllum pyramidale mutatio *secunda*, LINDSTRÖM, 1882, p. 45, pl. 1, fig. 4; pl. 6, fig. 1.
? *Goniophyllum pyramidale* Hisinger. WELLER AND DAVIDSON, 1896, p. 170, pl. VI, figs. 6–8.
Protaeropoma wedekindi TING, 1937, p. 414, figs. 3a–b.
Goniophyllum pyramidale Hisinger. JOHANNESSEN, 1993, p. 130, figs. 6 A–J, figs. 7 A–P, figs. 8 A–E, figs. 9 A–O, figs. 10 A–M, figs. 11 A–H, figs. 12 A–C, figs. 13 D–F.

Material. The lectotype is RM Cn 54963 (see Johannessen, 1993 for details). Material studied herein (all Gotland unless noted) includes: BMNH R 21659, R 46390-1; Cn 15904-5, 15933-50, 15952-53, 15956-65, 16057; ENSM (de Verneuil Collection) e-146/1–3, e-147/1–3; MCZ 9393-9399; NMI 4527 (Ireland: holotype of *Petraia quadrata* M'Coy), 4528, 4590, 13577; SM A 7496, A 7534 (Woolhope), A 7571-3, A 8543 (Woolhope), A 16503 (Woolhope), A 46402 (base of Högklint Series, Irevik, Gotland), A 48264; SUI 47023-31 (plus other specimens), all Iowa material; Tübingen S 407 (lectotype of *Cyathophyllum tetragona* Quenstedt); USNM 42712, 81135, 92119, 112403, 162035-40 (all Gotland material); USNM 42713 (plaster replica, Lamotte, Iowa), 50227 (Monticello, Iowa).

Remarks. Apart from the morphological observations above, no comment is made here on Gotland material of the type species. Johannessen (1993) rationalized the infraspecific taxonomy of *G. pyramidale* proposed by Lindström (1882), retaining the nominate subspecies *G. p. pyramidale* from the type locality of the "Visby Beds" (= Visby Formation of Riding and Watts, 1991) and *G. p. primigena* Lindström, 1882 (formerly *G. p.* forma *primigena*) from the underlying *Arachnophyllum* Marl (not renamed by Riding and Watts, 1991). Two of Lindström's other infraspecific taxa, *G. p.* mutatio *prima* Lindström, 1882 and *G. p.* mutatio *secunda* Lindström, 1882, were regarded by Johannessen as junior synonyms of *G. p. pyramidale*.

Johannessen (1993) discriminated the two subspecies morphologically by the latter's tending to lack or have fewer rootlets, be smaller and more square in cross section, have 60 to 120 septa (compared with 80 to 140 in *G. p. pyramidale*), and exhibit a cardinal fossula in late growth stages. It seems possible that *Goniophyllum* might exhibit a serial change in shape through the sequence at Visby, as indicated by Lindström (1882, fig. on p. 91). However, because specimens illustrated by Johannessen (1993, figs. 14 B, D) as *G. p. primigena* are distinctive, I provisionally accept these two subspecies, *G. p. pyramidale* and *G. p. primigena*, in the sense of Johannessen.

A largely unnoticed species (of *Goniophyllum*) is *Cyathophyllum tetragona* Quenstedt, 1879 (pl. 156, figs. 82–84). The specimen illustrated by Quenstedt as figure 82 is missing. I have, however, studied the other two type specimens. The specimen of Quenstedt's figure 83 (Tübingen S 407) is a poorly preserved *G. p. pyramidale* with an encrustation on the K septum; the specimen illustrated in his figure 84 (like that of his figure 85) is a broken *G. fletcheri*. Thus, in view of the clearly stated derivation of the specimen in his figure 83 from Gotland, the lectotype selected and illustrated here (Fig. 4: K–M) is that of Quenstedt (1879, fig. 83: Tübingen S 407). *C. tetragona* is clearly a junior synonym of *G. p. pyramidale*.

Description. See Johannessen (1993).

Occurrences.

Gotland. On the island of Gotland, with the exception of the unique specimen of *G. fletcheri* from an undetermined Gotland horizon, *Goniophyllum* has been collected from three units, traditionally called (Hede, 1960) the *Arachnophyllum* Marl (lowest), the Lower Visby Beds, and the Upper Visby Beds (highest). The two latter units have been renamed the Ygne Member and the Rövar Lilja Member, respectively, by Riding and Watts (1991), who also modified the name Visby Beds (Upper and Lower Visby marls of Hede, 1960) to Visby Formation. According to Johannessen (1993), *G. p. primigena* is known from the *Arachnophyllum* Marl. *G. p. pyramidale* occurs in the Ygne Member (four specimens) and commonly in the upper Rövar Lilja Member. A specimen of *G. pyramidale* (SMA 46402) in the Sedgwick Museum is recorded as being derived from the Högklint Beds at Irevik. Jeppsson (1983) concluded that the *Pterospathodus amorphognathoides* Zone occurs in the "Lower Visby Beds," with the base of the Wenlock also occurring in that unit. *G. p. pyramidale* thus ranges from the late Llandovery (Telychian) into the Wenlock.

United States. The earliest reference to *Goniophyllum* in the United States is an anonymous, very brief (17 lines) note attributed by Bassler (1915) to Winchell (see [Winchell], 1890), stating that the existence of *G. pyramidale* in America had been known "to some geologists for a number of years." He noted its collection from strata of the "Niagara period" near La Motte, Dubuque County, Iowa, and from other localities near "Monticello, Maquoketa, and a few other places in the same state." Lindström (1882) had mentioned the presence of *Goniophyllum* in Tennessee and informed Weller (Weller and Davidson, 1896, p. 167) that he had a specimen from there; to the best of my knowledge, this Tennessee occurrence has not been confirmed.

A striking, unfortunate aspect of many North American specimens is their coarse silicification: most show strong beekite rings so that, in most specimens, the corallite surface and some skeletal detail are not preserved. Further, there is no clear picture of corallite shape in North American assemblages. For instance, as only a few specimens are known from Iowa, no particular significance can be attached to the small proportion of Iowa specimens with strongly asymmetrical cross-sectional shape. Several specimens display, on or near the K face, rounded tubes, often inferred to be the negative record of rootlets, as in the Canadian material. Some specimens (Fig. 3: B), however, display such holes elsewhere on the coral; these are probably mostly borings.

Iowa. Weller and Davidson (1896) described and illustrated *G. pyramidale* (comparing their material to *G. p.* mutatio *secunda*), *Petalocrinus* n. gen., and *Pentamerus*, from collections from Jones County, Iowa, and noted the presence of *Stricklandia castellana* White among their materials. Bassler (1950) listed Iowa occurrences and others from the Hopkinton Dolomite at Masonville, Iowa, and from the top

Figure 3 (on facing page). A–N, *Goniophyllum pyramidale pyramidale* (Hisinger), Hopkinton Dolomite, Jones County, eastern Iowa; late Llandovery (Telychian). A, B, SUI 47028, × 1.5, in calical and lateral view. C–F: calical views, × 1.5; C, SUI 47025; D, SUI 47030; E, SUI 47029; F, SUI 47026. G, H, SUI 47031, calical and lateral views, both × 2. I, J, SUI 47027, calical and lateral views, × 1.5. K, L, calical and lateral views, SUI 47023, × 1.5. M, N, calical and lateral views, SUI 47024, × 1.5. A, B, D, E, G, H, K, L from "Niagaran coral reef, coll. 21 June, 1917, A. O. Foerste. Florine Farm, Jones County, Iowa." C, F, I, J, M, N from "*Petalocrinus* beds, A. O. Foerste, 1919." O–S: *G. pyramidale ?primigena* Lindström. O–Q: GSC 113387, all × 1.5. Fossil Hill Formation, Manitoulin Island, Ontario; Llandovery. O, calical view; P, apical view showing holes left by weathered rootlets; Q, side view showing parallel-sided corallite. R–S: GSC 20536, both × 2. R, calical view; S, lateral view showing the characteristic parallel sides. T, U: *G.* sp. cf. *G. p. pyramidale* Lindström, USNM 488278, both × 2.5. T, calical view; U, lateral view. 12.9 km southwest of Kewaunee, Wisconsin.

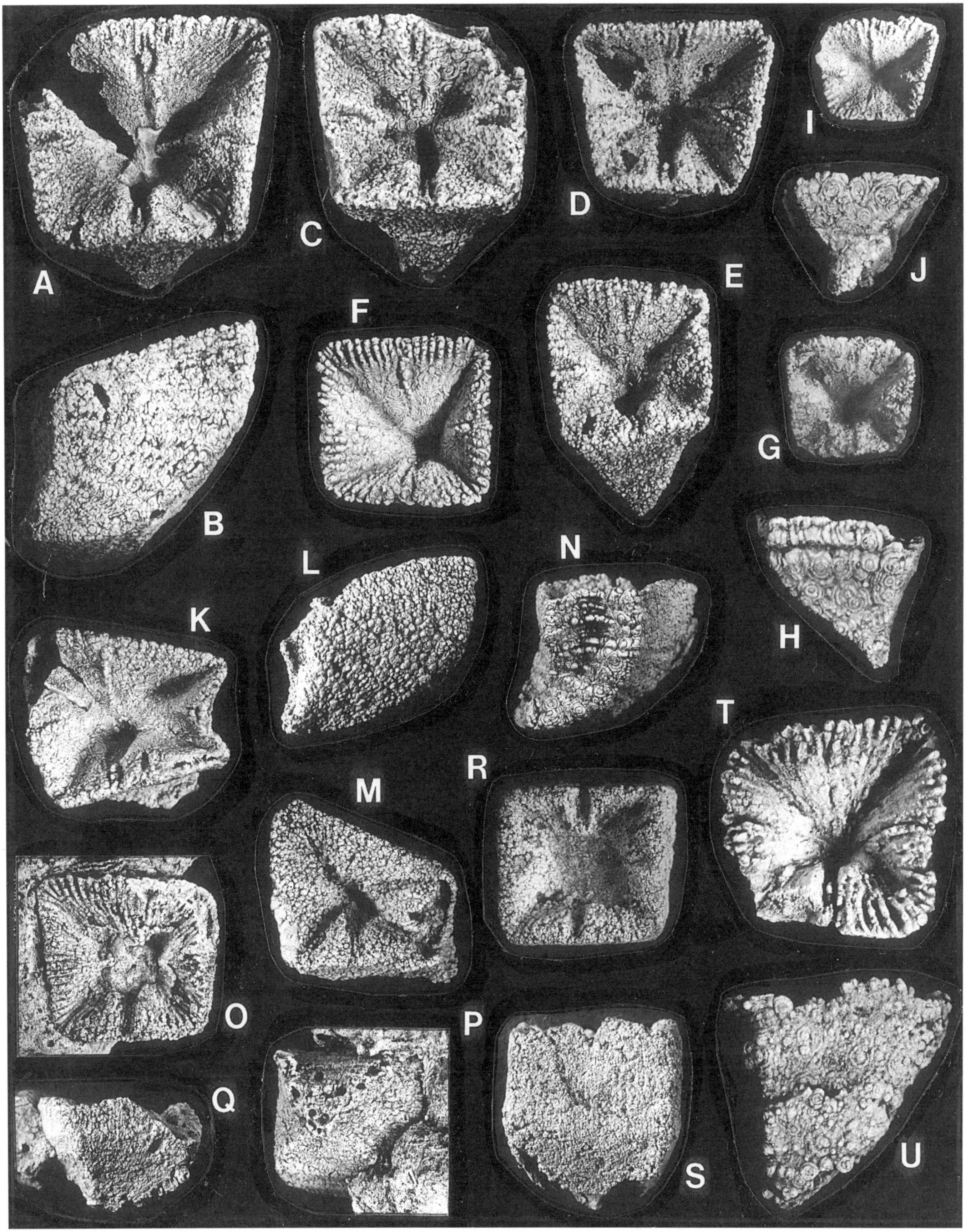
A
B
C
D
E
F
G
H
I
J
K
L
M
N
O
P
Q
R
S
T
U

of the Laurel Formation, St. Paul, Indiana. Foerste (1935, p. 194) reported *Goniophyllum* and *Petalocrinus* from the Laurel of St. Paul, Indiana, as well as the "Hopkinton" of Jones and Dubuque counties, Iowa; the Indiana occurrence is described below as *G. pyramidale ?primigena* Lindström, 1882.

I have studied 24 poorly preserved, silicified specimens from two University of Iowa collections made by A. O. Foerste. Data for these collections indicate that one is from "Niagaran coral reef. A. Florine farm, Jones County, Iowa. 5 1/2 miles SE Monticello. Coll. 6/21/17," and the other is from the "*Petalocrinus* beds, Florine farm, Jones Co., Iowa. Coll. A. O. Foerste, June 28, 1919." The age of this occurrence is late Llandovery (C_5: Johnson, 1980).

These collections are interesting in that about 20% of the corallites is misshapen, having one strongly concave alar side (Fig. 3: K, M). Rootlets are not preserved, although the K surface of one specimen (SUI 47027: Fig. 3: J) bears what may be rootlet bases. A few corallites possess the normal cardinal trough, but the crude silicification makes it impossible to estimate the degree of development of dissepiments; it is possible that this material differs from true *G. p. pyramidale* in lacking abundant dissepiments, in being more heavily calcified, and in having more convex and prominent alar regions. One specimen (SUI 47026: Fig. 3: F) has 8+ septa on the K edge. Juvenile specimens show the prominent transition from the calceoloid form to the pyramidal form at a diameter of about 5 mm (SUI 47031: Fig. 3: H; SUI 47027: Fig. 3: J).

Wisconsin. A new occurrence is represented by a specimen (USNM 488278, Fig. 3: T, U) from Kewaunee county, Wisconsin, collected by Mrs. Selner of Kewaunee and passed on to me by Walter Schilling. Its poor preservation allows counting of about seven septa lateral to K at a diameter of 11 mm, but it closely resembles similar-sized Iowa specimens and is accordingly placed with reservation in *G. p. pyramidale*.

England. One specimen from the Woolhope area (SM A 7534, Woolhope Limestone, 1 m above *Petalocrinus* bed; Fig. 4: E) may belong to this taxon.

Ireland. A Silurian goniophyllid coral from Galway, Ireland, was described by M'Coy (1846) as a new species, *Petraia quadrata*. The illustrated specimen was assigned by Milne-Edwards and Haime (1850–1855) and Bassler (1950) to *G. pyramidale*. As stated by Johannessen (1993, p. 127), this occurrence was interpreted by Lindström (1882) as *G. pyramidale* mutatio *prima*. M'Coy's species was stated to be "Common in the slates of Ardaun and Kilbride, Cong, County Galway" by M'Coy (1846, p. 60). The fossils were collected during geological mapping, under the direction of Sir R. Griffith, from strata termed by Gardiner and Reynolds (1912, 1914) and McKerrow and Campbell (1960) the Finny School Beds, of late Llandovery age (C_{4-5}, McKerrow and Campbell, 1960, table 3). The Kilbride area has been restudied by Piper (1967) and Laird and McKerrow (1970); Piper combined the C_5 Annelid Grit (lower) and the C_6 Finny School Beds into the Kilbride Formation. Harper et al. (1995) indicated that the Finney School Member of the Kilbride Formation probably should be assigned to the *griestoniensis* zone (Cocks et al., 1992), reinforcing the late Telychian age.

Material. NMI F4527, "Kilbride, Cong," holotype of *Petraia quadrata* M'Coy); three previously unfigured paratypes of *Petraia quadrata* M'Coy: NMI F4528, Ardaun, Cong; F4390, Ardaun, Cong; and F13577, Kilbride, Cong.

Description. The only specimen illustrated by M'Coy (1846, pl. 4, fig. 18) is a small calcareous *Goniophyllum* (Fig. 4: H, I) that lacks rootlets (although the distal region of the almost flat K face is broken), reaches a diameter of 20 mm, and has a K face about 22 mm long.

Paratype NMI F 4528 (Fig. 4: J), is from "Ardaun, Cong," and is a well-preserved but incomplete calical mold of a large specimen with about 100 major septa and a clear K septum. The rim of the calyx is about 33 mm in diameter and shows most of three sides, including all of the K side, the two alar sides, and part of the C side; the corners of the corallite are conspicuously rounded. Distal septal edges are clearly acanthine. About 12 major septa occur between the prominent K septum and the corner of the corallite; those adjacent to the K septum are strongly deflected toward it deep in the calyx; minor septa are perhaps represented by slight ridges. Dissepiments are obvious high in the calyx; swollen regions on the K calical surfaces are clearly visible. The K edge of the corallite appears longer than the C edge, as is normal for this genus.

Paratypes NMI F 4390 and 13577 are small calical molds; both are low, broadly flaring corallites, with approximate diameters of only 12 and 14 mm, and show abundant dissepiments.

Remarks. The four known specimens in NMI collections are provisionally considered conspecific. The material clearly belongs to *Goniophyllum*; it is tentatively assigned here to *G. p. pyramidale*.

G. pyramidale primigena Lindström, 1882

Goniophyllum pyramidale mutatio *primigena* LINDSTRÖM, 1882, p. 44, pl. 1, figs. 2–3.

? *Goniophyllum pyramidale* forma *minor* Lindström. NEUMAN, 1982, pl. 2, figs. 5–6; JOHANNESSEN, 1993, p. 135.

Goniophyllum pyramidale primigena Lindström. JOHANNESSEN, 1993, p. 134, figs. 14 A–V, 15 A–S, 16 A–H.

Material. The lectotype selected by Johannessen (1993) is RM Cn 15734.

Remarks. In this small broad subspecies the apical angle may be quite high, with rootlets sometimes missing or confined to the apical portion, and up to 120 septa (Johannessen, 1993). Some specimens assigned to this species (e.g., Johannessen, 1993, fig. 14 B) are distinctive in their parallel-sided form.

G. p. ?primigena Lindström, 1882

Figures 3: O–S, 4: N, O

Goniophyllum pyramidale (Hisinger). BOLTON, 1957, pl. XI, fig. 1; BOLTON, 1966, pl. 7, figs. 1–2, 4.

Material. I have not examined material studied by Johannessen (1993) and previous European workers but have examined the following specimens: GSC 11065, 20536, 20538, 23593, 113387, all from Manitoulin Island, Canada; and USNM 79957 and 162041 from Indiana.

Occurrences.

Canada. Bolton (1957, pl. XI, fig. 1; 1966, pl. 7, figs. 1, 2, 4) illustrated *G. pyramidale* from the Fossil Hill Formation, Manitoulin Island; I have reexamined this and additional material, some of which is illustrated here (Fig. 3: O–S). Confident identification of this Canadian material is precluded by poor preservation. Nevertheless, this silicified material

Figure 4 (on facing page). A–D, *Goniophyllum* sp. nov.?, USNM 99837, all × 2.5. Woolhope Limestone, Wych, Malvern, England; late Llandovery. A, calical view; B, lateral view, showing concave counter face to left; C, convex cardinal face showing septal insertion; D, view of concave counter side with rootlets. E: *G. pyramidale*, SMA 7534a, Woolhope Limestone, 1 m above *Petalocrinus*-band, 0.4 km S of Woolhope Church, Herefordshire, England; late Llandovery. Coll. J. K. St. Joseph, 1935. E, calical view, × 2. F, G: *G. fletcheri*, P 3953, Petaloe Brook, Woolhope; coll. Dr. John Penn. F, side view × 2.5; G, transverse section of mature zone of corallite (see F for location of section), × 8. H, I: *G.* sp. cf. *pyramidale* (Hisinger), both × 1.5. NMI F 4527, calcareous holotype of *Petraia quadrata* M'Coy, "Kilbride, Cong," Ireland; Kilbride Formation, late Llandovery. H, counter surface; I, alar surface. J: *G.* sp. indet., × 4. NMI F 4528, paratype of *Petraia quadrata* M'Coy; latex cast of calical mold. "Ardaun, Cong," Galway, Ireland; Kilbride Formation, late Llandovery. K–M: *G. pyramidale pyramidale* (Hisinger), × 1.5. Lectotype, Tübingen S-407, of *Cyathophyllum tetragona* Quenstedt, 1879, Gotland. K, calical view; L, lateral face; M, counter face. N, O: *G. pyramidale ?primigena* Lindström, USNM 79957, both × 1.5. N, calical view; O, lateral view showing parallel-sided corallite. "Laurel, St. Paul, Indiana."

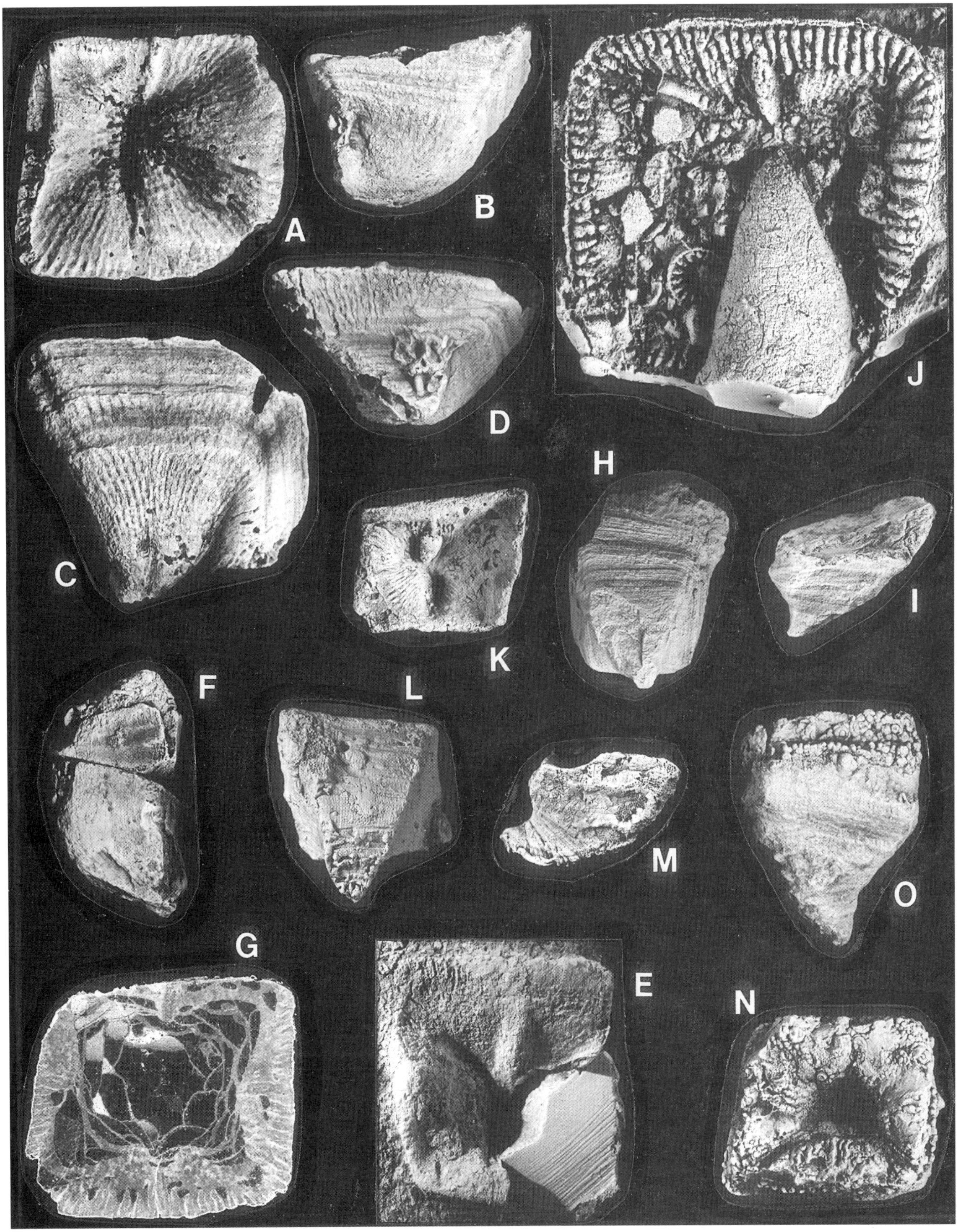
A
B
J
D
H
C
K
I
F
L
M
O
G
E
N

appears to show holes where rootlets once were (Fig. 3: P) and does show a tendency to become asymmetrical in calical view (Fig. 3: O), although not to the extent seen in Iowa material, and parallel-sided in side view (Fig. 3: Q, S). The Ontario material, here assigned tentatively to *G. pyramidale primigena* Lindström, 1882, includes both broadly expanding and more rectangular specimens; the significance of this variation is uncertain.

Indiana. Two small, very poorly preserved, subrectangular corallites (USNM 79957 and 162041) from St. Paul are provisionally assigned to this taxon because one shows the parallel-sided form (Fig. 4: N–O).

Discussion. This was the only one of Lindström's three infraspecific taxa retained by Johannessen (1993); it is the oldest of the Gotland *Goniophyllum* taxa, being from the *Arachnophyllum* Marl. In both Gotland and Norwegian material this subspecies tends to exhibit a parallel-sided mature corallite stage (Johannessen, 1993, figs. 14 B, 16 H).

Dr. J. M. Whittaker kindly sent me a photograph of a *Goniophyllum* observed in outcrop of the upper Llandovery Étage 7c b at Skien, southern Norway. Precise identification is not possible, but Norwegian material from Étage 7c a, Asker area, and Étage 7c g, Oslo region, was recorded as *G. p. primigena* by Johannessen (1993).

Goniophyllum sp. nov.?

Figure 4: A–D

Material. USNM 99837, Woolhope Limestone; specimens doubtfully assigned include SM A 8543 (Woolhope Limestone, Malvern Tunnel Tip), SM A 16503 (base of Woolhope Limestone), P 3952 (Linden). All from the Woolhope district, England.

Discussion. Most English *Goniophyllum* material belongs to *G. fletcheri* Milne-Edwards and Haime, 1851, being derived from Wenlock strata at Dudley in particular and possibly from Wenlock Edge (Bassler, 1950). However, *Goniophyllum* of the *pyramidale* type was first reported from England as *G. pyramidale* by Davidson (1866, p. 283) from the Upper Wenlock Shale. Holl (1866, p. 430) described the occurrence of the latter species from the Malvern district, in the May Hill Sandstone and also in the upper shales "interbedded with the Woolhope Limestone" or "base of the Wenlock Shale" and, perhaps significantly, stressed the similarity of the Malverns form to *Petraia quadrata* M'Coy from Galway. Inadequate material precludes erection of a new taxon; too few specimens are presently available from this region, and their precise stratigraphic source is not always clear. This is unfortunate because, chronologically, they appear to be from the interval during which *G. fletcheri* may be inferred to have evolved from *G. pyramidale*.

Pocock (1930) first recorded the association of *G. pyramidale* and *Petalocrinus* from Woolhope, Herefordshire, in the distinctive *Petalocrinus* bed. Dr. John Penn has informed me (personal communication, 1970) he had observed *Goniophyllum* only in the Woolhope Limestone where it occurs in the band of tabulate corals associated with the *Petalocrinus* bed. Pocock (1930) had noted that the *Petalocrinus* bed occurred in the "Woolhope Shales" at the Wych at Malvern and at May Hill. Pocock recorded *Gladiograptus geinitzianus* var. *angustoides* in the purple transitional shale beds between "typical Upper Llandovery sands and overlying brownish shale containing nodular limestones of Woolhope Limestone types." Some uncertainty stems from changing stratigraphic terminology.

One specimen from the Woolhope Limestone, Wych, Malvern (USNM 99837, Fig. 4: A–D) shows a remarkable concave K side with rootlets and about nine septa lateral to the C septum. This specimen is very distinctive and is almost certainly a new form; it is not named here pending better understanding of this intriguing Woolhope fauna. A second, but poorly preserved, specimen (P 3952) from a quarry at Linden near the Wych cutting, Woolhope, has a similar but not quite so flaring form, but no septa can be counted, and neither rootlets nor the C and K sides can be recognized with certainty.

Goniophyllum fletcheri Milne-Edwards and Haime, 1851

Figure 4: F, G

Goniophyllum fletcheri MILNE-EDWARDS AND HAIME, 1851, p. 405; MILNE-EDWARDS AND HAIME, 1855, p. 290, pl. 68, figs. 3, 3A; LINDSTRÖM, 1882, p. 55, pl. 6, figs. 10–12; JOHANNESSEN, 1993, p. 137, figs. 13 A–C, 17 A–M, 18 A–I.

Material. P 3953, late Llandovery *Petalocrinus* bed (limestone), Petaloe Brook near "Mangerdine," Woolhope; coll. Dr. John Penn.

Remarks. This taxon has been well revised by Johannessen (1993). I have studied only one specimen from Woolhope (Fig. 4: F); it is calcareous but poorly preserved externally. It is strongly parallel-sided as in *G. fletcheri*, has about 16 major septa on the convex K face (identified by the parallel arrangement of septal traces), and has a diameter of about 15 mm at a length of 25 mm; there is no sign of rootlets. If correctly identified, this specimen is the first reported pre-Wenlock *G. fletcheri*. If it is indeed descended from *G. pyramidale*, the transition would seem to have been very rapid.

Johannessen (1993, p. 141) stated that there are up to 70 septa in *G. fletcheri*; his illustrations (figs. 17 d, 18 c) suggest that his figure relates only to major septa. Thus, this Woolhope specimen appears to be a good *G. fletcheri* that arose in the late Llandovery when other morphological novelties (e.g., *G.* sp. nov.?) were developing in the lineage, possibly by what McNamara (1990) called post-displacement paedomorphosis. One specimen (MCZ 9401) is possibly another further compound example of the species.

Goniophyllum species indet.

Occurrence. The most easterly North American occurrence of *Goniophyllum* is a single specimen from northeast Maine from a probable late Llandovery calc-silicate boulder derived from an unnamed unit (Boucot, 1969, p. 46). This specimen, of biogeographic significance only, appears to be lost.

BIOGEOGRAPHY

Pickett (1975) suggested *Goniophyllum* as being potentially biogeographically significant, although he concluded that, in keeping with the brachiopod data, provinciality of Silurian corals was largely weak. Cocks and Fortey (1990) remarked that the "late Llandovery and early Wenlock benthic shelly faunas," rebounding after the end-Ordovician glaciation, were "amongst the most cosmopolitan in the whole geological record," as had been suggested previously (e.g., Boucot and Hurst, 1979). The operculate *Rhizophyllum* was widely distributed, whereas *Goniophyllum* was strongly constrained geographically.

Goniophyllum is recognized here as a late Llandovery to Wenlock "Old World" genus with a very restricted geographic distribution, ranging from western Europe across North America as far as Iowa and Indiana (Fig. 5). In paleocontinental terms, it was restricted to Baltica, Avalonia, and Laurentia. Scotese and McKerrow (1990, fig. 10), Cocks and Fortey (1982, fig. 5), and Copper and Brunton (1991, text-fig. 2) have all indicated that this region lay south of the Equator at about 15 to 30° S latitude, extending across about 45° of longitude.

In England, Gotland, and Iowa, *Goniophyllum* occurs in the late Llandovery associated with the crinoid *Petalocrinus* (Pocock, 1930; Bather, 1898; Weller and Davidson, 1896). This

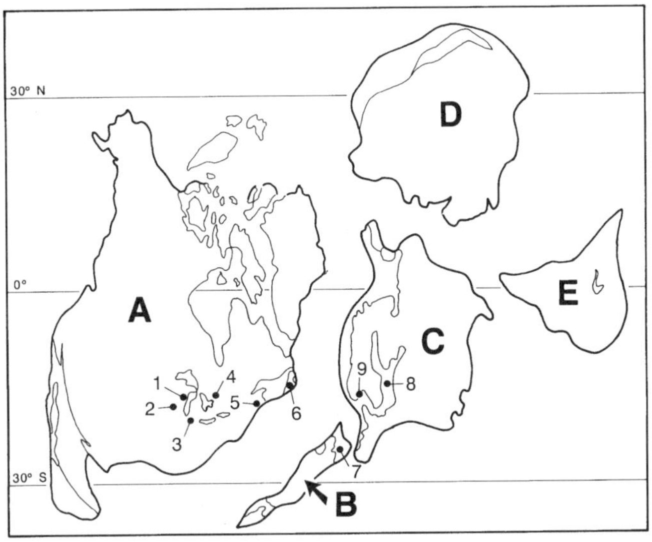

Figure 5. Diagrammatic arrangement for the Llandovery (Lower Silurian) of some continental masses after Copper and Brunton (1991), including Laurentia (A), Avalonia (B), and Baltica (C); also shown are "Siberia" (D) and "Kazakhstania" (E). Areas cited in the text as yielding *Goniophyllum* are: 1, Wisconsin; 2, Iowa; 3, Indiana; 4, Manitoulin Island; 5, Maine; 6, Galway, Ireland; 7, Dudley area, Woolhope area, Tortworth inlier; 8, Gotland; 9, Oslo region.

crinoid occurs elsewhere: in the late Llandovery of the Sandpile Group of British Columbia (Norford, 1962; Witzke et al., 1979, p. 120, p. 125–126); in the late middle Llandovery to early late Llandovery Xiangshuyuan Formation of the Yangtze Platform (Mu et al., 1986, p. 18, fig. 1); and in Nevada, Arctic Canada, Ohio, and Illinois (Witzke et al., 1979). On Gotland, *Petalocrinus* ranges from the Llandovery Ygne Member to possibly Ludlow strata (Bather, 1898); its geographic range was thus greater than that of *Goniophyllum*.

Tasmania

Bigsby (1867, p. 11) reported the genus from west Tasmania; I have been informed by Dr. M. R. Banks that this cannot be substantiated and that the record almost certainly refers to *Rhizophyllum* sp. cf. *enorme* Etheridge, 1903, an Early Devonian form occurring in the Point Hibbs Limestone, western Tasmania (Hill and Jell, 1969, p. 539).

Turkey

Goniophyllum was reported from Llandovery strata at Pendik, Turkey, by Sayar (1962). A photograph kindly sent to me by Dr. C. Sayar shows a mold of a solitary coral described by her (personal communication, February 2, 1970) as "nearly square in section." Further material is required to confirm this identification.

PALEOECOLOGY

Spassky (1967) suggested that tetracorals supported by rootlets grew on the substrate in high-energy zones. This is not supported by the excellent preservation of the largely autochthonous Gotland faunas, nor by the mostly limey and muddy lithofacies. Further, the prolific epifauna (bryozoans, tabulate corals, cornulitids, bacterial filaments?) encrusting both outer and calical surfaces indicates intermittent, and probably low rates of, sedimentation for the Gotland environments colonized by *G. pyramidale*. Because a small proportion of specimens of *G. pyramidale* from Gotland possess *in situ* opercula (recalling the way "petals" of associated *Petalocrinus* are occasionally found *in situ*), the environment is inferred to have been one of low energy. However, there does not appear to have been a discrete *Goniophyllum* biofacies. In terms of Boucot's community model, occurrences of *Goniophyllum* range from BA 4-5 for the "Lower Visby Beds" (Brett et al., 1993) of Gotland to strata in England interpreted as rather shallower and more turbulent (Pocock, 1930; Scoffin, 1971). The *Arachnophyllum* Marl of Gotland probably represents a depth of <60 m (Brett et al., 1993). The occurrence of *Goniophyllum* in Ireland has been assigned to the *Clorinda* community with a suggested depth of 50 to 60 m by Harper et al. (1995).

In a general way, *Goniophyllum* is associated with reefal sediments (Hill et al., 1936; Scoffin, 1971; Riding and Watts, 1991; Copper and Brunton, 1991), but it was not involved in construction of frameworks. Most *Goniophyllum* in Gotland, Indiana, Iowa, and Ontario are found in biostromal beds. Prof. M. Johnson (personal communication, 1980) has informed me that the *Goniophyllum* from Jones County, Iowa, was obtained from a coral-algal horizon overlying a *Pentamerus suberectus* horizon. *Petalocrinus* occurs there in the *Pentameroides* beds (although it has a wide stratigraphic range) and in the lower Bioherm Beds (Johnson, 1980); the Manitoulin Island occurrences are also in a coral-algal community. Professor Johnson has also pointed out that many such fossils in Iowa occur mostly in surface-weathered material known as "geest"; the host stratum of the Iowa material is thus difficult to determine.

In England, *G. fletcheri* occurs abundantly at Dudley but has not been reported from the more reefal sediments (Scoffin, 1971) at Wenlock Edge. In the Woolhope occurrence, the *Petalocrinus* bed of the Upper Haugh Woods Beds is only about 38 mm thick, resting on a bed (50 to 130 mm thick) of large tabulate corals and always occurring about a meter below the transition beds from sandy Llandovery to calcareous Woolhope facies (Pocock, 1930).

EVOLUTION IN *GONIOPHYLLUM*

The broad succession of species in *Goniophyllum* is apparently from *pyramidale* (Visby Formation: late Llandovery to early Wenlock) to *fletcheri* (Much Wenlock Limestone: late Wenlock to early Ludlow), although the late

Llandovery *G. fletcheri* described here from Woolhope complicates this generality.

The oldest occurrences of *Goniophyllum* of the *pyramidale* group are all Late Llandovery from the middle to late Telychian Kilbride Formation of Ireland (*griestoniensis* Zone: Cocks et al., 1992; Harper et al., 1995); the Ygne Member of the Visby Formation (C_6; *Pterospathodus amorphognathoides* Zone: Jeppsson, 1983, fig. 2); the *Petalocrinus* bed in the Woolhope area (upper Telychian: Cocks et al., 1992, fig. 3); the coral-algal bed of the Hopkinton Dolomite (C_5: Johnson, 1980); and the Fossil Hill Formation, Manitoulin Island, Ontario, Canada (C_5 to C_6: Berry and Boucot, 1970, p. 150; Johnson, 1980, and personal communication, 1980). Age data are not available for the Maine (probably late Llandovery, but possibly early Wenlock: Boucot, 1969, p. 46), Indiana, and Wisconsin occurrences. In addition, *G. fletcheri* occurs in the late Llandovery of the Woolhope area (Fig. 4: F) where a mixture of taxa seems to be present, including *G.* sp. nov.? and *G. pyramidale?*

Rhizophyllum is also an early goniophyllid (Hill and Jell, 1969). The earliest *Rhizophyllum* is from the late Llandovery Quarry Creek Limestone, New South Wales (McLean, 1974; Rickards et al., 1995). In Gotland, *R. gotlandicum* Lindström (1882) occurs in the Hemse Beds and the Eke Beds (both early Ludlow: Neuman and Kershaw, 1991); a figure presented by Lindström (1882, p. 91) gives his interpretation of its occurrence. A specimen (SM A 16607) described by Reed (1924) as *R.? tortworthense*, probably (Dr. M. L. K. Curtis, personal communication, 1970) from the Damery Beds (C_5 age) in the Tortworth Inlier, is a generically indeterminate conical calical mold; it should no longer be considered a possible *Rhizophyllum* (see also McLean, 1974). *Rhizophyllum* survived until at least the Early Devonian, late Emsian, and was very widespread (Hill and Jell, 1969, p. 539), whereas the short-lived and geographically constrained *Goniophyllum* may be viewed as an experimental and relatively unsuccessful form.

Previously recognized occurrences of *G. fletcheri* are, as stated, in the late Wenlock to early Ludlow (?) Much Wenlock Limestone or its correlative or perhaps slightly older "Dudley" Limestone at Dudley. The possible *G. fletcheri* reported here from the late Llandovery Woolhope area, as discussed above, is the earliest occurrence of that species.

ACKNOWLEDGMENTS

Numerous colleagues have assisted with material, discussion, and advice on this topic. In particular, I wish to thank Ruth Banger (Earth Sciences library, Cambridge), who provided information about older literature, and Professor Markes Johnson, who provided information about occurrences in the United States. I am very grateful to Dr. John Penn and Walter Schilling for loaning Woolhope material and the Wisconsin specimen respectively. Photography is by Nelson Shupe and the author. The manuscript has benefited from comments from Drs. John Jell, Björn Neuman, and Yong-yi Zhen. This chapter is dedicated to the memory of my good friend and colleague, the late Jess Johnson, whose studies of Silurian and Devonian faunas stand as a benchmark for us all.

REFERENCES CITED

Bassler, R. S., 1915, Bibliographic index of American Ordovician and Silurian fossils: United States National Museum Bulletin 92, v. I, p. 1–718; v. II, p. 719–1521.

Bassler, R. S., 1950, Faunal lists and descriptions of Paleozoic corals: Geological Society of America Memoir 44, 315 p.

Bather, F. A., 1898, *Petalocrinus*, Weller & Davidson: Quarterly Journal of the Geological Society of London, v. 54, p. 401–441.

Berry, W. B. N., and Boucot, A. J., 1970, Correlation of the North American Silurian rocks: Geological Society of America Special Paper 102, 289 p.

Bigsby, J. J., 1867, Thesaurus Siluricus. The flora and fauna of the Silurian Period. With addenda (from recent acquisitions): London, John van Voorst, lii + 214 p. [Not seen]

Birenheide, R., 1974, Zur Herkunft der devonischen cystimorphen Rugosa: Senckenbergiana lethaea, v. 54, p. 453–474.

Bolton, T. E., 1957, Silurian stratigraphy and palaeontology of the Niagara Escarpment: Geological Survey of Canada Memoir 289, 145 p.

Bolton, T. E., 1966, Illustrations of Canadian fossils. Silurian faunas of Ontario: Geological Survey of Canada Paper 66-5, 46 p.

Boucot, A. J., 1969, Geology of the Moose River and Roach River Synclinoria, northwestern Maine: Maine Geological Survey Bulletin 21, p. 1–117.

Boucot, A. J., and Hurst, J. M., 1979, Silurian biogeography: Revision of the North Atlantic Realm in the Llandovery: Rapport Grønlands geologiske undersogelse, v. 91, p. 57–60.

Brett, C. E., Boucot, A. J., and Jones, B., 1993, Absolute depths of Silurian benthic assemblages: Lethaia, v. 26, p. 25–40.

Bromell, M. von, 1727–1728, De corallis fossibilus. Lithographica Svecana: Acta Literaria (et Scientarium) Sveciae publicata, v. 2, 446 p.

Cocks, L. R. M., and Fortey, R. A., 1982, Faunal evidence for faunal separations in the Palaeozoic of Britain: Journal of the Geological Society of London, v. 139, p. 465–478.

Cocks, L. R. M., and Fortey, R. A., 1990, Biogeography of Ordovician and Silurian faunas, *in* McKerrow, W. S., and Scotese, C. R., eds., Palaeozoic palaeogeography and biogeography: Geological Society of London Memoir 12, p. 97–104.

Cocks, L. R. M., Holland, C. H., and Rickards, R. B., 1992, A revised correlation of Silurian rocks in the British Isles: Geological Society of London Special Report 21, 32 p.

Copper, P., and Brunton, F., 1991, A global view of Silurian reefs: Special Paper in Palaeontology 44, p. 225–259.

Davidson, T., 1853, British fossil Brachiopoda. Introduction. On the classification of the Brachiopoda: Palaeontographical Society Monograph, v. 1, pt. 3, p. 41–136.

Davidson, T., 1866, *Goniophyllum pyramidale* in the Wenlock Shale: Geological Magazine, v. 3, p. 283.

Dybowski, W. N., 1873, Monographie der Zoantharia Sclerodermata Rugosa aus der Silurformation Estlands, Nord-Livlands und der Insel Gotland: Archiv für die Naturkunde Liv-, Est- und Kurlands, series 1, v. 5, p. 257–414.

Etheridge, R., Jr., 1903, An unusually large form of *Rhizophyllum* lately discovered in New South Wales: Records of the Geological Survey of New South Wales, v. 7, p. 232–233.

Foerste, A. F., 1935, Relation of Silurian formations in southwestern Ohio, southeastern Indiana, Kentucky and western Illinois: Denison University Scientific Laboratories Bulletin, v. 35, p. 119–205.

Gardiner, C. I., and Reynolds, S. H., 1912, The Ordovician and Silurian rocks of the Kilbride Peninsula (Mayo): Quarterly Journal of the Geological Society (London), v. 68, p. 75–102.

Gardiner, C. I., and Reynolds, S. H., 1914, The Ordovician and Silurian rocks of the Lough Nafooey area (Co. Galway): Quarterly Journal of the Geological Society (London), v. 70, p. 104–118.

Harper, D. A. T., Scrutton, C. T., and Williams, D. M., 1995, Mass mortalities on an Irish seafloor: Journal of the Geological Society of London, v. 152, p. 917–922.

Hede, J. E., 1960, The Silurian of Gotland, *in* Regnéll, G., and Hede, J. E., The Lower Palaeozoic of Scania. The Silurian of Gotland: International Congress, 21st, Guide to excursions numbers A 22 and C 17: Stockholm, Geological Survey of Sweden, p. 44–87.

Hill, D., and Jell, J. S., 1969, On the rugose coral genera *Rhizophyllum* Lindström, *Platyphyllum* Lindström and *Calceola* Lamarck: Neues Jahrbuch für Geologie, Paläontologie und Mineralogie, v. 1969/9, p. 534–551.

Hill, D., Butler, A. J., Oakley, K. P., and Arkell, W. J., 1936, Report of "Coral Reef" meeting at Wenlock Edge, the Dudley district, and the Oxford district: Proceedings of the Geologists' Association, v. 47, p. 130–139.

Hisinger, W., 1829, Esquisse d'un tableau des pétrifactions de Suède, distribuées en ordre systématique (first edition): Stockholm, P. A. Norstedt, 27 p. (This is apparently the first French edition of "Profiler och Tabeller ofver hojdmatningar i Sverige och Norrige," published in Stockholm in 1827. [Not seen]).

Hisinger, W., 1831, Antekningar i Physik och Geognosie under resor uti Sverige och Norrige (Notes in physics and geognosy while traveling in Sweden and Norway), Volume 5: Stockholm, Bernh. M. Bredberg, 174 p. (This series was published in seven parts from 1819 to 1840. [Not seen]).

Hisinger, W., 1837, Lethaea Svecica seu Petrificata Sveciae, iconibus et characteribus illustrata (Swedish rocks and Swedish petrifications, with illustrations of morphological characteristics): Stockholm, P. A. Norstedt and sons, 112 p., with supplement p. 113–124. [Not seen]

Holl, H. B., 1866, *Goniophyllum pyramidale*, His.: Geological Magazine, v. 3, p. 430–431.

Jeppsson, L., 1983, Silurian conodont faunas from Gotland: Fossils and Strata, v. 15, p. 121–144.

Johannessen, W. H., 1993, Species of the Silurian operculate genus *Goniophyllum*: Geologiska Föreningens i Stockholm Förhandlingar, v. 115, p. 119–143.

Johnson, M. E., 1980, Paleoecological structure in Early Silurian platform seas of the North American Mid-Continent: Palaeogeography, Palaeoclimatology, Palaeoecology, v. 30, p. 191–216.

Laird, M., and McKerrow, W. S., 1970, The Wenlock sediments of NW Galway, Ireland: Geological Magazine, v. 107, p. 297–317.

Lindström, G., 1868, Om tvenne nya öfversilurska koraller från Gotland (About two new Upper Silurian corals from Gotland): Kungliga Svenska Vetenskapsakademiens Handlingar, v. 25, p. 419–428.

Lindström, G., 1871, On some operculated corals, Silurian and Recent: Geological Magazine, v. 8, p. 122–126.

Lindström, G., 1882, Om de Palaeozoiska Formationernas Operkelbärande Koraller (About the operculate corals from Palaeozoic formations): Bihang Till Kungliga Svenska Vetenskapsakademiens Handlingar, v. 7, pt. 4, p. 1–112.

McKerrow, W. S., and Campbell, C. J., 1960, The stratigraphy and structure of the Lower Palaeozoic rocks of northwest Galway: Scientific Proceedings of the Royal Dublin Society, series A, v. 1, p. 27–52.

McLean, R. A., 1974, Cystiphyllidae and Goniophyllidae (Rugosa) from the Lower Silurian of New South Wales: Palaeontographica, v. 147A, p. 1–38.

McNamara, K. J., 1990, Heterochrony, *in* Briggs, D. E. G., and Crowther, P. R., eds., Palaeobiology. A synthesis: Oxford, Blackwell, p. 111–119.

M'Coy, F., 1846, A synopsis of the Silurian fossils of Ireland. Collected from the several districts by R. Griffith. The whole being named, and the new species drawn and described, by Frederick M'Coy. (Addenda by J. W. Salter): Dublin, printed at the University Press by M. H. Gill, p. 1–72.

Milne-Edwards, H., and Haime, J., 1850–1855, A monograph of the British fossil corals: Palaeontographical Society of London Monographs, lxxxv + 322 p.

Milne-Edwards, H., and Haime, J., 1851, Monographie des polypiers fossiles des terrains Palaeozoïques, précéde d'un tableau général de la classification des polypes: Archives du Muséum d'Histoire Naturelle (Paris), tome 5, 1–488.

Mu Enzhi, Boucot, A. J., Chen Xu, and Rong Jiayu, 1986, Correlation of the Silurian rocks of China: Geological Society of America Special Paper 202, 80 p.

Neuman, B., 1982, Early Silurian corals of the Oslo region. A preliminary report: Palaeontological Contributions from the University of Oslo 278, p. 33–42.

Neuman, B., and Kershaw, S., 1991, Gotland/Sweden. Silurian reefs and coral-bearing strata, *in* International Symposium on Fossil Cnidaria including Archaeocyatha and Porifera, VI, Excursion A1 Guidebook: Bergen, University of Bergen, 111 p.

Norford, B. N., 1962, The Silurian fauna of the Sandpile Group of northern British Columbia: Geological Survey of Canada Bulletin 78, 51 p.

Pickett, J. W., 1975, Continental reconstructions and the distribution of coral faunas during the Silurian: Journal and Proceedings of the Royal Society of New South Wales, v. 108, p. 147–156.

Piper, D. J. W., 1967, A new interpretation of the Llandovery sequence of north Connemara: Geological Magazine, v. 104, p. 253–267.

Pocock, R. W., 1930, The *Petalocrinus* Limestone horizon at Woolhope (Herefordshire): Quarterly Journal of the Geological Society of London, v. 86, p. 50–62.

Quenstedt, F. A., 1879, Petrefactenkunde Deutschlands 6: Die Röhren- und Sternkorallen: Leipzig, Fues's Verlag, p. 145–624.

Reed, F. R. C., 1924, Some new Ordovician and Silurian fossils: Geological Magazine, v. 61, p. 25–31.

Rickards, R. B., Packham, G. H., Wright, A. J., and Williamson, P. L., 1995, Wenlock and Ludlow (Silurian) graptolite faunas, the Quarry Creek district, New South Wales: Memoirs of the Association of Australasian Palaeontologists, v. 17, 68 p.

Riding, R., and Watts, N. R., 1991, The lower Wenlock reef sequence of Gotland, facies and lithostratigraphy: Geologiska Föreningens i Stockholm Förhandlingar, v. 113, p. 343–372.

Salter, J. W., 1873, Catalogue of the collection of Cambrian and Silurian fossils contained in the Geological Museum of the University of Cambridge: Cambridge, Cambridge University Press, 204 p.

Sayar, C., 1962, New observations on the Palaeozoic sequences of the Bosphorus and adjoining regions, Istanbul, Turkey, *in* Erben, H. K., ed., Proceedings, 2. Internationale Arbeitstagung über die Silur/Devon Grenze und die Stratigraphie von Silur und Devon Bonn-Bruxelles 1960: Stuttgart, E. Schweizerbart'sche Verlagsbuchhandlung, p. 222–223.

Scalia, S., 1922, Può la *Calceola sandalina* Lam. essere ragionevolmente considerata come un polipaio fornito di opercolo? (Can the Calceola sandalina Lam. be reasonably considered as a polypary with an operculum?): Bollettino della Societa Geologica Italiana, v. 41, p. 90–99.

Scoffin, T. P., 1971, The conditions of growth of the Wenlock reefs of Shropshire (England): Sedimentology, v. 17, p. 173–219.

Scotese, C. R., and McKerrow, W. S., 1990, Introduction, *in* McKerrow, W. S., and Scotese, C. R., eds., Palaeozoic palaeogeography and biogeography: Geological Society of London Memoir 12, p. 1–21.

Spassky, N. Ya, 1967, Paleoecology of tetracorals: Palaeontological Zhurnal 1967/2, p. 1–6.

Ting, T. H., 1937, Zur Kenntnis der Gattung "Goniophyllum": Zentralblatt für Mineralogie, Geologie und Paläontologie, series B, v. 10, p. 411–415.

Weller, S., and Davidson, A. D., 1896, *Petalocrinus mirabilis* (n. sp.) and a new American fauna: Journal of Geology, v. 4, p. 166–173.

[Winchell, A. N.], 1890, The occurrence of *Goniophyllum pyramidale* Hisinger, in America: The American Geologist, v. 6, p. 326.

Witzke, B. J., Frest, T. J., and Strimple, H. L., 1979, Biogeography of the Silurian–Lower Devonian echinoderms, *in* Gray, J., and Boucot, A. J., eds., Historical biogeography, plate tectonics and the changing environment: Corvallis, Oregon State University Press, p. 117–129.

MANUSCRIPT ACCEPTED BY THE SOCIETY FEBRUARY 27, 1997

NOTES ADDED IN PROOF

Weyer (1996) has argued that the correct name for this group of operculate tetracorals is Calceolidae King, 1846. The erection of this family in a synoptic table by King (1846) satisfies the requirements of Article 11 (f) of the ICZN (Ride et al., 1985). Thus, the subfamily name used herein should be amended to

Calceolinae King, 1846; relegating Cystiphyllidae Milne-Edwards & Haime, 1850 to the level of a junior synonym of Calceolidae King is not recommended.

ADDITIONAL REFERENCES

King, W., 1846, Remarks on certain Genera belonging to the Class Palliobranchiata: Annals and Magazine of Natural History, v. 18, p. 26–42, 83–94.

Ride, W. D. L., Sabrosky, C. W., Bernardi, G., and Melville, R. V., 1985, International Code of Zoological Nomenclature (third edition): Berkeley and Los Angeles, University of California Press, 338 p.

Weyer, D., 1996, Calceolidae versus Goniophyllidae (Anthozoa, Rugosa; Silur-Devon): Abhandlungen und Berichte für Naturkunde, Magdeburg, v. 19, p. 69–71.

Printed in U.S.A.

Geological Society of America
Special Paper 321
1997

Lochkovian Rugosa of Nevada

Alan E. H. Pedder
Geological Survey of Canada, 3303 33rd Street, N.W., Calgary, Alberta T2L 2A7, Canada
Michael A. Murphy
Geology Department, University of California, Davis, California 95616

ABSTRACT

Eighteen genera and 34 species of Early Devonian corals, largely independently dated by conodonts or graptolites, are described or discussed from the Tuscarora and Roberts Mountains and the Simpson Park and Toquima Ranges. A range chart of their occurrences, calibrated against conodont and graptolite zones and Johnson's Nevada Faunal Intervals, is presented. Three species of Rugosa from the Toquima Range, previously considered to be from Merriam's Lower Silurian coral zone A, are probably Lochkovian. Lochkovian corals from the Vaughn Gulch Limestone and Hidden Valley Dolomite of southeastern California apparently belong to the Nevada biogeographic province, as do corals from member 3 of the Gazelle Formation in the Yreka terrane of northern California. Comparisons are made, using the Otsuka Coefficient, between the Nevada faunas and larger Lochkovian coral faunas of North America, the Ural Mountains, Asia, and Australia. Most results are compatible with the Scotese Lochkovian geographic reconstruction, except for the position of the Altay-Sayan fold belt, which appears to be positioned at too high a latitude on the Scotese reconstruction. *Pararhizophyllum* and *Ecnomiophyllum* are new genera. New species proposed are *Pararhizophyllum parvulum*, *Breviphrentis johnsoni*, *Carlinastraea ampla*, *Aphroidophyllum inceptor*, *Embolophyllum carbonense*, *Toquimaphyllum vulgare*, *T. sentum*, *Ecnomiophyllum merriami*, and *E. fascis*. Thin sections of the holotype of the type species of *Briantia* are figured for the first time.

INTRODUCTION

Previous descriptions and illustrations of Lochkovian rugose corals from Nevada are concentrated almost exclusively in a work by Merriam (1974a) and another by Merriam and McKee (1976). J. W. Huddle (*in* Merriam and McKee, 1976, p. 39) recognized that some of the Lochkovian corals studied by Merriam are associated with Lower Devonian conodonts, and W. B. N. Berry (*in* McKee et al., 1972) established that others were interbedded with *Monograptus uniformis*, the defining graptolite for both the base of the Devonian and the base of the Lochkovian. Nevertheless, Merriam himself seems to have steadfastly held the view that the Lochkovian coral faunas of Nevada belong to the Silurian (zones D and E, Merriam and McKee, 1976, table 1).

Merriam's Nevada Lochkovian corals came from the Tuscarora, Cortez (one specimen only), and Roberts Mountains and the Simpson Park and Toquima Ranges (Fig. 1). With the exception of the Boulder Creek locality in the southern Tuscarora Mountains and the Cortez Mountains locality, which are no longer available because of gold mining operations, we have collected from the same general areas. Our collections include new specimens of all but three of the definite Lochkovian species named by Merriam, resulting in substantial improvement in the knowledge of the morphology, generic affinities, and ranges of several of the previously described species. Other new materials have led to the recognition of two new genera and 15 new species, which are either named or left in open nomenclature, depending on the quality of the available material. The known

Pedder, A. E. H., and Murphy, M. A., 1997, Lochkovian Rugosa of Nevada, *in* Klapper, G., Murphy, M. A., and Talent, J. A., eds., Paleozoic Sequence Stratigraphy, Biostratigraphy, and Biogeography: Studies in Honor of J. Granville ("Jess") Johnson: Boulder, Colorado, Geological Society of America Special Paper 321.

Lochkovian Rugosa of Nevada, and their estimated temporal ranges relative to other faunal zones, are shown in Figure 2. No formal rugose coral biozonation is proposed for the Lochkovian of Nevada, pending clarification of community structures and acquisition of data that can improve our range chart, especially in respect to the middle part of the stage.

The largest part of this work is devoted to systematic descriptions and taxonomy. Discussion of the lithostratigraphy, faunal calibration, and general analysis of the faunas has had to be much abbreviated.

DEVONIAN LITHOSTRATIGRAPHY OF CENTRAL NEVADA

The northern side of the North American continent during Early and Middle Devonian time, presently oriented approximately north-south, was occupied by a broad, shallow shelf, bordered by a deeper complex of outer-shelf basins. Ridges partitioned the outer shelf so that each basin has a unique history of sedimentation that reflects the relative elevation of the ridges with respect to sea level. Relationships of rock units within the basins have been summarized by Johnson and Murphy (1984) and Murphy and Anderson (1991).

CONODONT, GRAPTOLITE AND BRACHIOPOD CALIBRATION

Few of the Lower Devonian coral-bearing beds of Nevada have yielded conodonts more age diagnostic than the corals themselves, and other than species of the order Metriophyllina, it is unlikely that any Lochkovian corals lived in pelagic to hemipelagic graptolitic facies. Nevertheless, because so many of the world's Lower Devonian corals are known from debris-flow and like deposits, conodonts and graptolites have proved the most important fossils for independent dating of Lochkovian corals. Brachiopods are much less important than either conodonts or graptolites for mondial correlation, but untransported Devonian corals can commonly be directly related to benthic brachiopod zonations that are viable within a faunal province.

The most widely applicable zonation for Lower Devonian rocks currently in use was based originally on conodont sequences of western North America (Klapper, 1977). The Lochkovian part was based entirely on taxa with restricted geographic distribution, whose phylogenies are still not known. In a companion chapter in this volume, Valenzuela-Ríos and Murphy propose a new biostratigraphic classification for the Lochkovian. Global conodont zones shown in Figure 2 correspond to those proposed in the companion chapter.

Graptolite faunas in Nevada are known only as high as the Pragian *thomasi* Zone (Jaeger, 1989, Fig. 2). However, they are closely integrated with the conodont and brachiopod scales, and through most of the Lochkovian, the sequence of their occurrences can be classified to conform with the graptolite zonation advocated by Jaeger (1989). The two most important levels in the

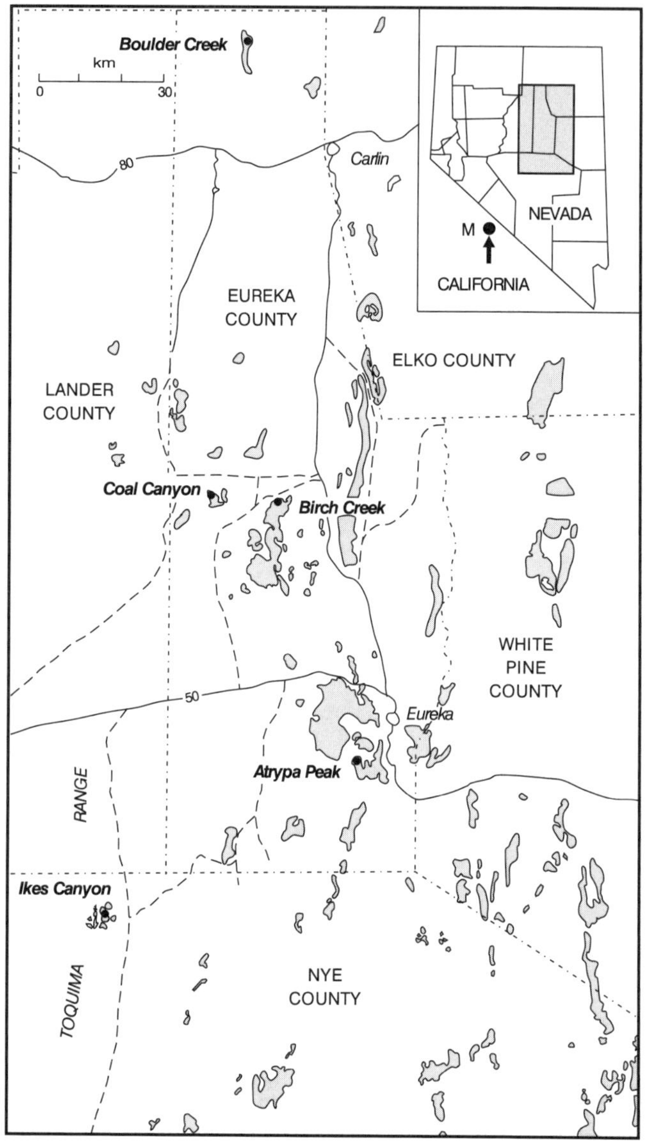

Figure 1. Locations of Lochkovian rugose coral faunas in central Nevada. Atrypa Peak is an Emsian coral locality. Shaded areas are Devonian outcrops. In the right top inset, the letter M indicates the position of the Californian Mazourka Canyon Lochkovian coral fauna relative to Nevada.

zonation are the entries of *Monograptus uniformis* at the base of the Devonian (Berry and Murphy, 1975; Murphy and Berry, 1983) and *M. hercynicus* in the *eleanorae* Zone. Both entries have worldwide correlation significance. Occurrences of *Monograptus birchensis*, which ranges from just below the base of the *uniformis* Zone, in high Pridoli beds (Murphy, 1989a, Fig. 132), to approximately the top of the *uniformis* Zone, have been especially useful in dating some of the Nevada Lochkovian corals.

J. G. Johnson (1977), to whom this volume is dedicated, introduced a concept of faunal intervals based on Devonian brachiopods from Nevada. At first, the intervals were related only to conodont zones, but later they were also integrated with Transgression/Regression cycles. Although more loosely defined than conodont or graptolite zones, Johnson's faunal intervals have been repeatedly tested within central Nevada. After some revision, especially near the base of the Lochkovian, where it is known now that faunal intervals one and two are largely coeval, Johnson's intervals give a close approximation of the conodont zonation. Brachiopod-dominated communities in the Lower and Middle Devonian faunal intervals have been analyzed by Johnson (1990), and the most recent iteration of the succession of intervals is in Johnson et al. (1991).

SOUTHERN TUSCARORA MOUNTAINS

Rugosa from 180 to 240 m above the base of the exposed Roberts Mountains Formation, in the "middle coral-brachiopod zone" of Merriam and McKee (1976, p. 17), on Boulder Creek (Merriam's "Bootstrap Hill" USGS localities M1314, M1315) include *Stylopleura berthiaumi* Merriam, *Stauromatidium* sp. C, and a probable species of *Neomphyma*, recorded as *N.* sp. B. These corals are interbedded with graptolites identified by W. B. N. Berry as *Monograptus uniformis* and *M. angustidens*, indicating levels near the base of the Devonian.

Tryplasma sp. cf. *T. duncanae* Merriam, *Carlinastraea halysitoides* (Etheridge), and unfigured corals listed by Merriam (*in* Merriam and McKee, 1976, p. 17), occur 385 m above base and 70 m below top of the exposed Roberts Mountains Formation at USGS Locality M1313 on Boulder Creek. This fauna is not precisely dated. It was said to occur 200 m above occurrences of *Monograptus uniformis* and *M. angustidens* but is probably not as young as the upper Lochkovian coral faunas.

COAL CANYON, SIMPSON PARK RANGE

A debris flow, 2.1 m above the base of the Windmill Limestone in SP I section (Berry and Murphy, 1975, Fig. 14), west of Coal Canyon Fault, has yielded several favositids and a single large specimen of *Stylopleura nevadensis* Merriam. It also yielded conodonts of the upper part of the lower Lochkovian (493-m level of Klapper, 1977, Fig. 4, p. 39). No rugose coral is known from lower Lochkovian beds east of Coal Canyon Fault.

The thickness of the Windmill Limestone west of Coal Canyon Fault, on "Pine Hill" (Merriam and McKee, 1976, Fig. 2), was given as 120 m by Merriam and about 125 m by Murphy (*in* Johnson and Murphy, 1969, Fig. 2, p. 1278). Merriam's *Carlinastraea* beds are above an occurrence of *Monograptus birchensis*, originally identified as *M. praehercynicus* by W. B. N. Berry (*in* Merriam and McKee, 1976, p. 13), and were said to occur in a zone about 20 m above the middle of the Windmill Limestone. If it is assumed that Merriam's *Carlinastraea* beds are in the order of 60 to 80 m above the base of the Windmill Limestone, they are likely to be of late middle Lochkovian age. The *Carlinastraea* fauna west of Coal Canyon Fault comprises *Tryplasma* sp. cf. *T. duncanae* Merriam, *Stylopleura berthiaumi* Merriam, *Tonkinaria simpsoni* Merriam from USGS Locality M1383, *Carlinastraea*

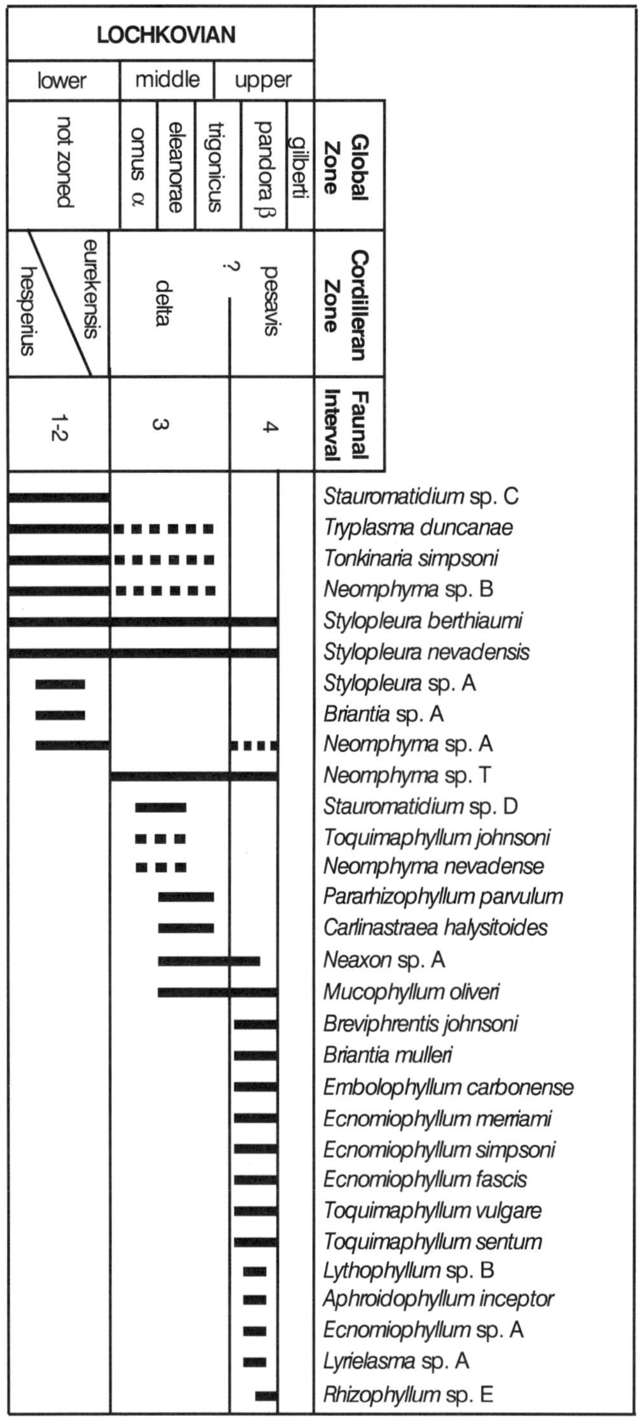

Figure 2. Range chart for Lochkovian rugose coral species in central Nevada. Broken range bars indicate uncertainty concerning either the species' identification or its stratigraphic range. Calibration is based on the graptolite zones of Jaeger (1989), global conodont zones of Valenzuela-Ríos and Murphy (this volume), Cordilleran conodont zones of Klapper (1977), and Faunal Intervals of Johnson et al. (1991).

halysitoides (Etheridge) from USGS Locality M1384, and a possible but indeterminate species of *Breviphrentis*, identified *in* Merriam and McKee (1976, Plate 5, Fig. 25, 26) as *Brachyelasma* sp. c., from USGS Locality M1383. *Stauromatidium* sp. D, from the Windmill Limestone on "Pine Hill," is not the same species as *Stauromatidium* sp. C from the lower Lochkovian of the southern Tuscarora Mountains and is likely to have come from a middle Lochkovian horizon.

East of Coal Canyon Fault, small middle Lochkovian rugose coral faunas have been collected at the 214 ft and 394 ft levels of SP VII section (Berry and Murphy, 1975, Fig. 16). The lower of these includes a fragment of the earliest certain *Breviphrentis* known from the North American craton and a small indeterminate species of *Lyrielasma.* The 394 ft level is in the *trigonicus* Zone and within the range of *Monograptus hercynicus* (Springer and Murphy, 1994, Fig. 2). Rugosa identified from the 394 ft level include *Pararhizophyllum parvulum* gen. and sp. n., *Mucophyllum oliveri* Merriam, and small silicified fragments of *Tryplasma* and *Stylopleura* spp. indet.

The most diverse rugose coral fauna from the Devonian of Nevada comes from Nevada Faunal Interval 4, in debris-flows known as the *Quadrithyris* breccia in the upper Windmill Limestone of Coal Canyon. Few diagnostic conodonts have been recovered from the breccia, but conodonts of the *trigonicus* Zone occur below the breccia and others of the *pandora* beta fauna at the top of the breccia. The main flow, which mixes benthos from more than one community, crops out on both sides of the canyon owing to repetition by the Coal Canyon Fault (Johnson and Murphy, 1969, Fig. 2). The fauna is *Lythophyllum* sp. A, *Rhizophyllum* sp. E, *Tryplasma* sp. indet., *Mucophyllum oliveri* Merriam, *Stylopleura berthiaumi* Merriam, *S. nevadensis* Merriam, *Breviphrentis johnsoni* sp. n., *Briantia mulleri* (Merriam), *Aphroidophyllum inceptor* sp. n., *Embolophyllum carbonense* sp. n., *Lyrielasma* sp. B, *Neomphyma karpinskiense* Shurygina, *N.* sp. A (talus specimen only), *Toquimaphyllum vulgare* sp. n., *T. sentum* sp. n., *Ecnomiophyllum merriami* gen. and sp. n., *E. simpsoni* (Merriam), *E. fascis* sp. n., and *E.* sp. A.

IKES CANYON, TOQUIMA RANGE

Beds assigned to the upper Roberts Mountain Formation by Merriam, but which are in the upper McMonnigal Limestone of current usage (Murphy and Anderson, 1991, Fig. 2) in the Mill Canyon plate, exposed in Ikes Canyon Window (USGS Locality M1103), contain *Stylopleura berthiaumi* Merriam, *S. nevadensis* Merriam, and *Tonkinaria simpsoni* Merriam (1974a, p. 52, unfigured). This fauna is dated by its occurrence within the range zone of *Monograptus birchensis*, earlier identified as *M. praehercynicus* (Merriam and McKee, 1976, p. 22; Johnson and Oliver, 1977, p. 1467).

Neaxon sp. A. occurs at the 19A, 28, and 29 levels in middle Lochkovian (*eleanorae* and *trigonicus* Zones) pelagic facies of the McMonnigal Limestone in section TO I, on the south side of Ikes Canyon (Murphy, 1989b, Fig. 4, 6). Merriam (1974a, and *in* Merriam and McKee, 1976) described *Neomphyma nevadense* (Merriam) and *Toquimaphyllum johnsoni* Merriam from the McMonnigal Limestone, 200 m northwest of summit 8474 (Copper Mountain, USGS Locality M1114) near Ikes Canyon. The corals overlie occurrences of *Monograptus birchensis* and are almost certainly middle Lochkovian. *Neomphyma karpinskiense* Shurygina, from the lower part of the Tor Limestone at USGS Locality M1397, 2.4 km southeast of the summit of Whiterock Mountain, may also be part of this fauna.

Impoverished hemipelagic coral faunas, consisting only of *Neaxon* sp. A and a rare unidentified species of *Lythophyllum*, occur in the upper Lochkovian part of the McMonnigal Limestone. The best dated of these faunas occur in the *pandora* beta Zone at levels 24 and 26 in IK IV section, on the south side of Ikes Canyon (Murphy, 1989b, Fig. 4). J. G. Johnson (personal communication, 1990) identified *Cyrtinaella causa* Community brachiopods of Nevada Faunal Interval 4 from IK IV 26.

Merriam (1974a) described three species, named *Cyathophylloides fergusoni*, *Arachnophyllum kayi*, and *Neomphyma crawfordi*, from USGS Locality M1088, on the June Canyon plate in Ikes Canyon. Merriam believed the corals came from the Masket Shale of Kay and Crawford (1964) and that they were from his Lower Silurian, Llandovery Zone A. If Merriam's description of the M1088 locality is correct, we infer that the corals must have come from an *omus* alpha Zone debris-flow in the McMonnigal Limestone or, less likely, from a younger Lochkovian level in the Tor Limestone. *Neomphyma crawfordi* is acceptable as a species of *Neomphyma*, and may well be a Lochkovian coral. The other two species are unrecognizable as described and figured, although Merriam's assignments to genera were certainly incorrect. All three corals are presumed to be Lochkovian but are not considered further in this work.

ROBERTS MOUNTAINS

Two acid-etched, lower Lochkovian silicified coral faunas are available from the Roberts Mountains Formation of BC II section (Berry and Murphy, 1975, Table 10, Fig. 11, 12). The base of the Lochkovian in this section is taken at the 567 ft level, with the entry of *Monograptus uniformis*; higher, at 598 ft, 608 ft and 609 ft, a lower Lochkovian trilobite, *Warburgella rugulosa eureka*, has been described and named by Alberti et al. (1971, 1972). The lower of the two coral faunas occurs with *Icriodus hesperius* and is from the 630 ft level. It consists of *Tryplasma duncanae* Merriam, *Stylopleura berthiaumi* Merriam, and an unidentified species of *Lyrielasma*. The higher fauna comes from 638.5 ft and occurs with both *I. hesperius* and *Criteriognathus eurekaensis*. Many of the available corallites from the higher fauna are unidentifiable, due to coarse and incomplete silicification, but *Tryplasma duncanae* Merriam and *Stylopleura berthiaumi* Merriam are present. *Egosiella?* sp., figured previously as *Cladopora* sp. (Merriam 1974a, Plate 3, Fig. 14; Merriam and McKee, 1976, Plate 1, Fig. 5), and a ptenophyllid, which appears to be transitional between *Lyrielasma* and *Tonkinaria*, are also present. Although the two Birch Creek faunas are at least 19.2 to 21.8 m above the base of the Devonian, they are, nevertheless, low in the lower Lochkovian, which ranges 150 m or more above them in the section.

Merriam (1974a, p. 15 and in Merriam and McKee, 1976, p. 9) described a coral fauna from his unit 3 of the type Roberts Mountains Formation exposure at USGS Locality M1100, on the northwest slope of Roberts Creek Mountain (Berry and Murphy, 1975, Fig. 3). As originally described, the fauna comprised *Tryplasma duncanae* Merriam, *Stylopleura berthiaumi* Merriam, and *Tonkinaria simpsoni* Merriam. New discoveries have enlarged it by addition of poorly preserved specimens of *Stylopleura* sp. A, *Briantia* sp. A, *Lyrielasma* sp. indet., and *Neomphyma* sp. A. It is also known now that the fauna occurs 45 m above the lowest known occurrence of *Icriodus hesperius* in the section and is accompanied by conodonts identified by T. T. Uyeno as *Criteriognathus eurekaensis* and *Belodella* sp.

TRANSITION FROM SILURIAN TO DEVONIAN RUGOSA IN NEVADA

Few of the very latest Silurian coral faunas are known, and almost none has been described from any of the circum Paleopacific regions of the world. Turnover of rugose coral genera at, or close to, the Silurian-Devonian boundary is best known in the Tien Shan fold belt but in most other regions is very difficult to quantify precisely. Analysis of data available for Lochkovian genera found in Nevada reveals that three of the genera (17%), *Mucophyllum*, *Rhizophyllum*, and *Tryplasma*, are genera that were well established before the end of Silurian time and continued through the Lochkovian with little or no reduction in numbers or diversity. Seven others (39%), *Stylopleura*, *Stauromatidium*, *Breviphrentis*, *Carlinastraea*, *Neomphyma*, *Lyrielasma*, and *Toquimaphyllum*, are rare, or uncommon in pre-Lochkovian faunas but assume much greater importance in Devonian assemblages. The remaining eight genera (44%) have no known representation in pre-Lochkovian faunas.

The strong Silurian influence on the earliest Devonian coral faunas of Nevada is neither exceptional nor unexpected. Indeed, most of the well-known Lochkovian faunas, Eastern Americas Realm excepted, are more Silurian in aspect, due mostly to tryplasmatid holdovers, which are confined to one genus in Nevada, and to holmophyllid and entelophyllid holdovers, which are completely unknown in Nevada. Domination of Devonian coral faunas by Devonian genera was achieved globally only in Pragian time, following a late Lochkovian extinction event (Oliver and Pedder, 1994).

BIOGEOGRAPHIC IMPLICATIONS

The most commonly employed index for binary similarity comparisons between Paleozoic biogeographic provinces is the Otsuka Coefficient (OC) (Cheetham and Hazel, 1969). Its application to Devonian coral faunas, and the bearing such coefficients have on the elucidation of Devonian geography, was most recently discussed by Pedder and Oliver (1990). To optimize results, those authors discarded poorly dated faunas, eliminated long-ranging basin-dwelling genera from their computations, made efforts to ensure that the boundaries of regions compared were rational in light of current plate tectonic doctrine, and above all, attempted to apply uniform generic diagnoses to all faunas and regions. We use the same approach here and repeat the notion that Otsuka Coefficients of less than 20 indicate low similarity, whereas those of 30 or more indicate high similarity, in situations where the taxonomic category employed is a genus of Devonian shelf-dwelling Rugosa.

Here we restrict numerical comparisons to regions from which 10 or more Lochkovian rugose genera are known. Nevertheless, small faunas from the probable Lochkovian parts of the Vaughn Gulch Limestone in the Inyo Range and Hidden Valley Dolomite in the Funeral Mountains of southeastern California are clearly from the same biogeographic province as the Nevada Lochkovian corals. Lochkovian corals in clasts from Member 3 of the Gazelle Formation, in the Yreka terrane of northern California, likely also belonged to the same Lochkovian province.

The greatest OC value obtained from comparisons made between Nevada and more distant regions with large Lochkovian Rugosa faunas is 39, and not surprisingly is the OC value between Nevada and Yukon Territory. OC values between Nevada and the eastern Ural Mountains and eastern Australia are both 33. Between Nevada and the western Urals, combined with Dolgiy Island and western Novaya Zemlya, the OC value is 32. Comparisons between Nevada and the Canadian Arctic islands, the Altay-Sayan, and the Tien Shan fold belts all give OC values of 30. Such high figures contrast with OC values of only 11 or 16 (depending on certain genus identity calls) obtained from a comparison of Nevada with the Balkhash terrane of Kazakhstan. All of these regions mentioned are considered part of the Devonian Old World Faunal Realm. Our new data give a very low OC value of only 8 for a comparison of Nevada with the Appalachian fold belt of the Eastern Americas Realm.

Several points emerge from a consideration of the new OC binary similarity comparisons relative to Scotese's Lochkovian reconstruction (Golonka et al., 1994, Fig. 19). The high levels of similarity between Nevada and the regions of Yukon Territory, the Canadian Arctic islands, western Ural Mountains, and eastern Australia are to be expected. In Scotese's reconstruction, all lie between tropical latitudes of approximately 15° S and 18° N in positions that would allow larvae, with the help of oceanic staging posts, and perhaps even mature corals on pumice rafts (Jokiel, 1990) to be transported over large distances by surface currents. However, the eastern Australian list of 19 Lochkovian genera includes five genera from the Melbourne-Mathina terrane and seven others from the Gamilaroi terrane, and it is possible that neither terrane was accreted to the Australian craton in Lochkovian time (Glen et al., 1992; Flood and Aitchison, 1992).

The high degree of similarity between Lochkovian Rugosa of Nevada and those of present-day eastern Ural Mountains, which are placed between 25° and 30° N latitude on the Scotese

Lochkovian reconstruction, would be better explained if the northern part of the present-day eastern Urals were moved a few degrees to a lower Lochkovian latitude. The Lochkovian coral facies now forming part of the Altay-Sayan fold belt are placed on the Siberian craton at latitudes of about 45° N on the Scotese Lochkovian reconstruction. This is very difficult to reconcile with the OC comparison value of 30 between that region and Nevada. The Lochkovian coral facies of the Tien Shan fold belt were in some part of the Turkestan Paleozoic Ocean and are omitted from the Scotese Lochkovian map.

The very low similarity between the Lochkovian faunas of Nevada and those of the Balkhash terrane of Kazakhstan may be due to, or the reasons may have been largely masked by, the terrigenous nature of the Pridolian and Lower Devonian facies of the Balkhash terrane. Problems associated with the enigmatic benthic faunas of the Balhkash terrane have been discussed previously from the perspectives of both coral (Pedder and Oliver, 1990) and brachiopod workers (Talent et al., 1987; Hou and Boucot, 1990) without firm conclusions. The even greater dissimilarity between Nevada Lochkovian Rugosa of the Old World Realm and those of the Appalachian fold belt, in the Eastern Americas Realm, is best explained by the relatively high Lochkovian latitude of the Appalachian corals, shown as about 30° S on Scotese's reconstruction, and by the barrier effect of a transcontinental arch. Scotese's Lochkovian map shows this arch to be breached, whereas most North American Devonian biogeographers believe that it was breached in Pragian, but not Lochkovian, time.

SYSTEMATIC PALEONTOLOGY[1]

All of the new material illustrated from Nevada is deposited and registered in the U. S. National Museum of National History, Smithsonian Institution, Washington, D.C.

Localities used by Merriam are recorded in the Menlo Park locality catalog of the U. S. Geological Survey; their numbers have the prefix M. The abbreviation USGS is used to identify them throughout this work. Merriam's (1972, p. 41; 1974a, p. 57) and Merriam and McKee's (1976, p. 39) works include registers of all the USGS localities cited in this work.

Subclass RUGOSA Milne Edwards and Haime, 1850
Order CYSTIPHYLLIDA Nicholson, 1889
(in Nicholson and Lydekker, 1889)
Suborder CYSTIPHYLLINA Nicholson, 1889
(in Nicholson and Lydekker, 1889)
Family CYSTIPHYLLIDAE Milne Edwards and Haime, 1850
Subfamily CYSTIPHYLLINAE Milne Edwards and Haime, 1850

Genus *Lythophyllum* Wedekind, 1925

Lythophyllum WEDEKIND, 1925, p. 32.
Nardophyllum WEDEKIND, 1925, p. 36.
Plagiophyllum WEDEKIND AND VOLLBRECHT, 1931, legend to Plate 17 (Plate 3 of reprints); WEDEKIND AND VOLLBRECHT, 1932, p. 113.
Wedekindophyllum STUMM, 1949, p. 39.
Patridophyllum ULITINA, 1968, p. 86.

Type species. *Lythophyllum marginatum* Wedekind, 1925, p. 33, Plate 6, Fig. 32, 33. Lower Stenophyllumschichten, middle Middle Devonian; Dachsberg near Gerolstein, Eifel region, Germany. The type stratum is probably in the Lower Givetian Loogh Formation.

Diagnosis. Solitary Cystiphyllidae with centered or excentric septal crusts of fused trabeculae, or nontrabeculate sclerenchyme, based more on the dissepiments than on the wall of the corallite. Isolated septal spines may also be present.

Remarks. A fuller account of the systematics of *Lythophyllum* and a more extended synonymy of the genus have been given by Pedder (1985b, p. 290). Here, only the salient points are briefly reiterated. *Nardophyllum* and *Plagiophyllum* share the same type species, now referred to as *Lythophyllum exzentricum* (Wedekind, 1925, Plate 9, Fig. 59). *L. exzentricum* differs from *L. marginatum* only by having thicker septal crusts with longer fused spines. *Patridophyllum paternum* Ulitina (1968, Plate 18, Fig. 1; Plate 19, Fig. 1), the type species of *Patridophyllum*, has centered, or slightly excentric, septal crusts, with more horizontally and vertically fused spines than the types of the other genera. Given the considerable variation in septal morphology within species referred to *Lythophyllum*, differences between the type species of *Patridophyllum* and *Lythophyllum* are not sufficient to retain *Patridophyllum* as a separate genus. Stumm (1949) proposed the name *Wedekindophyllum* because he believed that *Lythophyllum* Wedekind ought to be spelled *Lithophyllum*, which would make it a homonym of *Lithophyllum* Müller (1859, p. 62). However, the proposed change of spelling was illegal. In any case, *Lithophyllum* Müller (1859, p. 62) is, itself, a misspelling of *Lithophyllium* Müller (1858, p. 154, 155; 1859, p. 52). *Lithophyllium* Müller is a radiolarian genus. The long-ranging and widely distributed genus *Litophyllum* Etheridge (1899, p. 178), which in Nevada occurs in the Lower Denay Limestone of the Roberts Mountains, is a chaetetid sponge and should not be confused with either *Lythophyllum* or *Lithophyllium*.

Lythophyllum sp. A
Figure 3: 6, 7

Remarks. A worn, incomplete specimen (USNM 488380) from the Windmill Limestone depositional breccia on the ridge leading from the entrance of Coal Canyon to summit 6909, on the east side of the canyon, is the only material available of this species. Its corallum is similar in size and shape to those of *Lythophyllum lonense* (Stumm) from Pragian or Emsian strata in Nevada and *L. thorsteinssoni* Pedder from Lochkovian strata on Prince of Wales Island in the Canadian Arctic Archipelago. The Windmill Limestone form is distinguished from these species by

Figure 3 (on facing page). 1, 2, 4, 5: *Rhizophyllum* sp. E; 1, 2, 5, longitudinal and transverse sections, USNM 488381; 4, longitudinal section, USNM 488382. 3, 8: *Tryplasma duncanae* Merriam, exterior views of acid-etched specimen, USNM 488392. 6, 7: *Lythophyllum* sp. A, longitudinal and transverse sections, USNM 488380. 9–18: *Pararhizophyllum parvulum* gen. and sp. nov.; 9, 10, exterior calicular (counter side up) and counter surface views, holotype, USNM 488383; 11, interior view of operculum, USNM 488388; 12, exterior side view (counter side down), USNM 488386; 13, view from cardinal side, USNM 488385; 14, oblique calicular view (counter side down), USNM 488387; 15, view from cardinal side, USNM 488384; 16, interior view of operculum, USNM 488389; 17, 18, exterior views of opercula, USNM 488390, 488389; all acid-etched specimens.

[1]Pedder is solely responsible for the zoological systematics of this chapter.

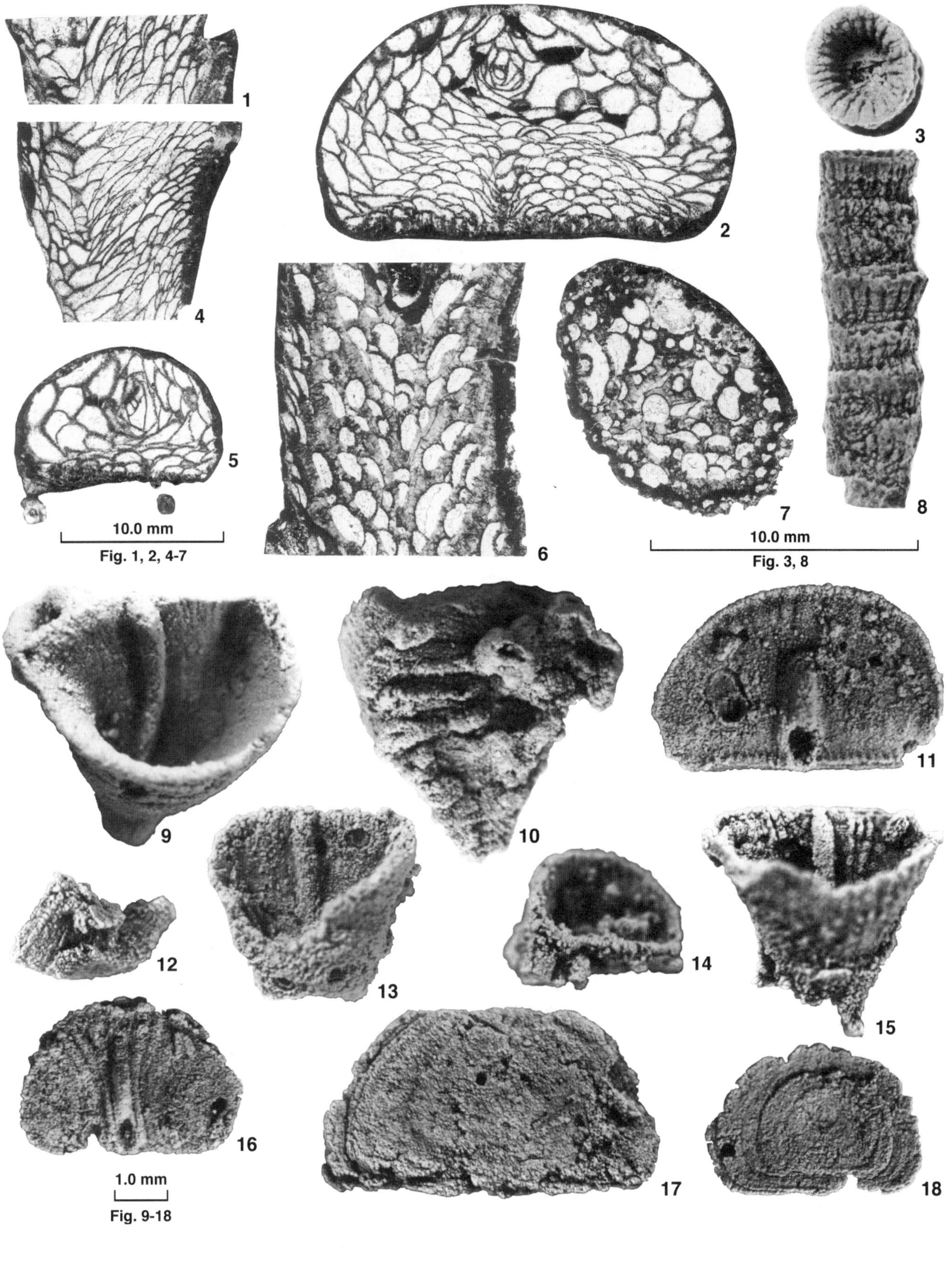
1
2
3
4
5
6
7
8
10.0 mm
Fig. 1, 2, 4-7
10.0 mm
Fig. 3, 8
9
10
11
12
13
14
15
16
17
18
1.0 mm
Fig. 9-18

having more uniformly distributed septal crust, which does not form distinct cones, and by having steeply inclined dissepiments in the periaxial region.

Family CALCEOLIDAE King, 1846

Remarks. Weyer (1996) has demonstrated that the widely used family taxon Goniophyllidae Dybowski (1873, p. 332) is a junior synonym of Calceolidae King (1846, p. 28). Stolarski (1993, p. 175-176) has suggested that the Calceolidae (known to him as Goniophyllidae) may have arisen directly from the currently monospecific Cothoniidae Jell and Jell (1976, p. 181), and that the Rugosa evolved as an Ordovician offshoot of the cothoniid-calceolid lineage. However, the main feature uniting both families—the presence of an operculum—seems outweighed by several other factors. These include the radially septate and fossulate nature of the operculum in *Cothonion* and the apparent absence of intermediate forms between the early Middle Cambrian *Cothonion sympomatum* Jell and Jell and the earliest known calceolids of late Llandovery age. Pending acquisition of new data, the Calceolidae are better retained in the Cystiphyllida than in the Cothoniida Oliver and Coates (1987, table 11.1, p. 177–178) where Stolarski suggested they may belong.

Subfamily CALCEOLINAE King, 1846

Diagnosis. See Johannessen's (1993, p. 125) diagnosis of the Goniphyllinae.

Genus *Rhizophyllum* Lindström, 1866

Rhizophyllum LINDSTRÖM, 1866a, p. 287; LINDSTRÖM, 1866b, p. 411.
Platyphyllum, LINDSTRÖM, 1883a, p. 68; LINDSTRÖM, 1883b, p. 40.
Teratophyllum LANG, SMITH, AND THOMAS, 1940, p. 102, 132.

Type species. *Calceola gotlandica* Roemer, 1856, p. 798. Silurian; east coast of the Isle of Gotland, Sweden. Lindström (1883b, p. 27) implied that the type locality is on the hill known as Laubacke, which suggests that the type horizon is in the Ludlow Eke Beds, or underlying Hemse Group.

Diagnosis. See Hill and Jell (1969, p. 535) and Hill (1981, p. F110).

Remarks. *Platyphyllum* Lindström was proposed for a calceolid species from the Silurian of northeastern Sichuan, China, that was believed to differ from species of *Rhizophyllum* by having better-developed septa. However, the generic name is a junior homonym of *Platyphyllum*, used by Audinet-Serville (1831, p. 145) for a genus of orthopteran insects. The junior homonym was replaced by *Teratophyllum* Lang, Smith, and Thomas. Hill and Jell (1969, Text-fig. 2.1a–2.2b) figured newly prepared thin sections of two of Lindström's specimens of *T. sinense* (Lindström), the type species of both *Platyphyllum* Lindström and *Teratophyllum* Lang, Smith, and Thomas, and concluded that the degree of septal development in the species does not justify the retention of *Teratophyllum*.

Rhizophyllum sp. E

Figure 3: 1, 2, 4, 5

Rhizophyllum sp. cf. *R. enorme* Etheridge. OLIVER, 1964, p. D156, Fig. 153.3a, 153.5a, b; OLIVER *in* Johnson, 1965, p. 371; OLIVER *in* Johnson, 1970, p. 44; MERRIAM, 1974a, p. 16; PEDDER, 1975, p. 285; JOHNSON AND OLIVER, 1977, p. 1464.
Rhizophyllum cf. *R. enorme* (Oliver). MERRIAM *in* Merriam and McKee, 1976, p. 15.

Remarks. Strusz (1970, p. 283, Plate 34, Fig. 1–10; Plate 35, Fig. 1–5; Text-fig. 2, 3) and Pedder (unpublished) have studied considerably more material of *Rhizophyllum enorme* Etheridge from the Garra Formation of New South Wales than was available when Oliver compared a single specimen from the late Lochkovian Windmill Limestone depositional breccia of the Simpson Park Range with Etheridge's species. Also, since Oliver's 1964 work, two additional specimens of *Rhizophyllum* (USNM 488381, 488382) have come to light during preparation of other corals from the upper 2 m of the Windmill Limestone depositional breccia on the east side of Coal Canyon, approximately 150 m south of the canyon entrance. Strusz (1970, p. 288) was correct in believing that the Nevada species is almost certainly distinct from *R. enorme*, but one of the reasons behind this belief—that the septa in the Nevada species are more crowded than those in *R. enorme*—may not be valid. Oliver's original specimen has about 28 septa along 14 mm of the flat counter side. The better of the two new Nevada specimens has about 26 septa along 19.5 mm of the same flat side; this density is not substantially different from the density of septa on the counter side of *R. enorme*.

Differences between the Nevada species and *R. enorme* are as follows: *R. enorme* is larger, with maximum known length and width of 120 mm and 60 mm, compared with about 35 mm and 28 mm in the Nevada species. The counter septum is larger relative to adjacent septa, and dissepiments and tabellae are coarser relative to the size of the corallite in the Nevada species. Two of the three known Nevada specimens have prominent radiciform processes emanating from the proximal region of the corallite (this part of the other specimens was apparently not preserved); no such structure has been observed in large samples of *R. enorme* from New South Wales.

The Nevada species is left in open nomenclature and designated species E to distinguish it from *Rhizophyllum* species A to D of Oliver (1964).

Genus *Pararhizophyllum* novum

Type species. *Pararhizophyllum parvulum* sp. nov.

Diagnosis. Calceolid coral with minute solitary calceoloid corallum. A prominent counter septum extends along the interior midline of the flattened wall of the corallite. Other septa represented only by a few low ridges or aligned trabecular granules. Horizontal skeletal elements and laminated sclerenchyme, such as characterize the genus *Calceola* (Hill and Jell, 1969, p. 544), absent. Operculum a single subsemicircular plate; exterior surface of operculum with essentially isometric growth lines; interior with a prominent, somewhat elongated pit situated close to the middle of the straight margin of the operculum. Away from the straight margin, along the midline radius of the operculum, the pit passes into a short, well-elevated medial opercular septum (term from Johannessen, 1993, p. 123).

Remarks. The closest relative to the new genus is *Rhizophyllum*. *Pararhizophyllum* differs from *Rhizophyllum* by its entirely different order of size, its lack of dissepiments and tabellae, and by its counter septum, which is excessively large relative to other septa. The large number of minute specimens available of the type species argues against the possibility that the type collection represents a near uniformly young population of a species of *Rhizophyllum*, although it is reasonable to regard the new genus as a neotenous derivative of that genus. Besides having laminated sclerenchyme, well illustrated by Birenheide (1990, Plate 1, Fig. 5–8), adult stages of *Calceola* are differentiated from the new genus by their large size, lack of radiciform processes, and the anisometric growth of their opercula. The most obvious difference between *Pararhizophyllum* and juvenile stages of *Calceola* is in the opercula of the two genera. The interior of the juvenile operculum (width 2.0 to 2.3 mm) of *Calceola* has two pairs of major lateral septa that are as prominent and coarse as the medial opercular septum (Stolarski, 1993, Fig. 1.7-1.9).

Rhytidophyllum Lindström resembles *Pararhizophyllum* in size but lacks a counter septum in both the corallite and on the operculum; its operculum also lacks a pit to receive the counter septum of the corallite (Johannessen, 1995, p. 40, Fig. 4P). Like *Rhytidophyllum*, *Pararhizophyllum* is currently monospecific.

Derivation of name. Greek, *para* = close, *rhiza* = root, *phyllon* = leaf, a reference to the closeness of the new genus to *Rhizophyllum*.

Pararhizophyllum parvulum sp. nov.
Figure 3: 9–18

Material. Holotype (corallite), USNM 488383; eight paratypes, corallites, USNM 488384–488387, and opercula, USNM 488388-488391. There are also eight opercula excluded from the type series. All of this material is silicified and is from the middle Lochkovian, *trigonicus* Zone, east of Coal Canyon Fault, northern Simpson Park Range (394 ft level, section SP VII, sample 15; see Berry and Murphy, 1975, Fig. 14; Springer and Murphy, 1994, Fig. 2).

Description. Corallum solitary, calceoloid, variably curved in side view, equilateral or slightly curved in plan view; apical angle 40° to 65°. Lower (counter) surface of corallite strongly wrinkled and giving rise to numerous short, fine radiciform processes; no evidence of the counter septum on the exterior of the corallite. Some of the wrinkles of the flattened counter surface extend around to the curved cardinal surface—but only just; most of the cardinal surface is smooth, except where it is sufficiently worn to expose aligned granular traces of trabeculae. None of the five known corallites is complete; maximum length of corallite estimated to be 5.5 mm; maximum width, obtained from an exceptionally large operculum, 6.3 mm. Septal apparatus dominated by a very prominent medial opercular septum situated along the interior midline of the flattened side of the corallite; other septa suppressed or reduced to a few—generally three or less—lines of trabeculae on the cardinal and counter sides of the corallite. No dissepiments or tabellae. Sclerenchymal thickening also absent, so that the wall of the corallite is thin.

Operculum single, in the form of a semicircular to subsemicircular plate. Exterior smooth, except for isometric growth lines. Interior of operculum provided with a somewhat elongated pit to receive the counter septum of the corallite. This pit is situated near the middle of the straight edge of the operculum; away from the straight edge, the pit passes into a short, prominent medial opercular septum. Length of the pit and septum together equal about half of the radius of the operculum.

Remarks. Lindström (1883b, Plate 9, Fig. 4) illustrated a young specimen of *Rhizophyllum tennesseense* (Roemer), which in some respects resembles *Pararhizophyllum parvulum*. From other illustrations (Lindström, 1883b, Plate 3, Fig. 14; Amsden, 1949, Plate 29, Fig. 1–5; Scrutton, 1985, Plate 3.3.9) and descriptions (Amsden, 1949, p. 111; Sutherland, 1965, p. 14) it is clear that *Rhizophyllum tennesseense*, which comes from the Silurian, Ludlow Series age, Brownsport Formation of Tennessee, is much larger, with a maximum length and width of 23 and 28 mm. Furthermore, it has normal *Rhizophyllum*-like septa, a thick wall, many dissepiments, and some tabellae. The new species is also reminiscent of a Llandovery species from New South Wales described by McLean (1974, p. 33, Plate 6, Fig. 10–13) in open nomenclature as a questionable species of *Rhizophyllum*. The Australian Llandovery species is larger than *Pararhizophyllum parvulum*, with a maximum length and width of 14.0 and 11.5 mm, and its interior is apparently largely filled with sclerenchyme, although some specimens have a few dissepiments in addition to the sclerenchyme.

Derivation of name. Latin, *parvulus, -a, -um*, meaning very small.

Suborder PHOLIDOPHYLLINA Wedekind, 1927

Remarks. Ivanovskiy (1975, p. 47) placed the Mucophyllidae in the Suborder Streptelasmatina of the Order Rugosa. Hill (1981, p. F72, F175) also placed the Mucophyllidae in the Streptelasmatina, but the suborder was assigned to the Order Stauriida in her classification. Despite this measure of agreement at the suborder level, the present author prefers the view, put forward by Wang (1950, p. 228) and later adopted by other Asian workers (e.g., Wang et al., 1987; He and Chen, 1989; Lavrusevich, 1993?), that the Mucophyllidae are related more to the Order Cystiphyllida than to the Streptelasmatina. Apart from the preponderance of solitary coralla and the absence of dissepiments or presepiments in most mucophyllids, there is very little similarity between them and typical Ordovician and Silurian Streptelasmatidae. Nevertheless, differences are pronounced; examples being the absence of rhabdacanthine microstructure from the streptelasmatids, the presence of axial structures formed of lobed inner septal edges and axially elevated tabularial surfaces, and also the presence of septal fossulae in the streptelasmatids.

If, as is done here, the Mucophyllidae are assigned to the Order Cystiphyllida, it is appropriate to recognize two suborders within the order. One of these suborders, the Cystiphyllina, accommodates families whose early forms have a well-developed dissepimentarium. These comprise the Cystiphyllidae, Calceolidae, Digonophyllidae and the Holmophyllidae if they are separated from Cystiphyllidae. The other suborder holds those families that lack horizontal skeletal elements or are in possession of tabulae but not dissepiments or presepiments. These are the Palaeocyclidae, Tryplasmatidae (= Pholidophyllidae), and Mucophyllidae. The name Pholidophyllina, given to this second suborder, is an emendation of Wedekind's (1927, p. 9) Pholidophyllida, which in his classification was a "section" of the Suborder Cystiphyllacea.

In Hill's (1981) classification, corals considered definitely to be Rugosa are divided into two large orders, the Cystiphyllida and Stauriida; the small group of corals comprising the Order Heterocorallia is attached questionably to the Rugosa. It is interesting to note that whereas the Hill classification recognizes 16 suborders in the Stauriida, there is no suborder level division of the Cystiphyllida. Fedorowski's (1991) important work assigns the Heterocorallia and a few genera previously placed in the Rugosa to the Dividocorallia, a new subclass of Anthozoa, although, as Wrzolek (1993) pointed out, some aspects of Fedorowski's classification require further investigation.

Family TRYPLASMATIDAE Etheridge, 1907
Subfamily TRYPLASMATINAE ETHERIDGE, 1907

Genus *Tryplasma* Lonsdale, 1845

Cyathophyllum Subgenus *Tryplasma* LONSDALE, 1845, p. 613 (in part).
Pholidophyllum LINDSTRÖM, 1871a, p. 925; LINDSTRÖM, 1871b, p. 125.
Acanthodes DYBOWSKI, 1873, p. 334, 364.
Spiniferina PENECKE, 1894, p. 592.
Stortophyllum WEDEKIND, 1927, p. 30, 31.

Type species. Cyathophyllum (Tryplasma) aequabile Lonsdale, 1845, p. 613, Plate A, Fig. 7, 7a. "Silurian"; Kakva River 15 or 20 versts (16.0 to 21.3 km) below Kakvinski Station (Murchison et al., 1845, p. 400), eastern slope of the northern Ural Mountains, Russia. Ivanovskiy and Shurygina (1975, p. 16) indicated that the type horizon is probably in the *Favosites regularissimus* Zone, which was considered to be Eifelian by them in 1975, but is now known to be Emsian. The type specimens may be in the British Museum (Natural History) but presently are lost (Rosen and Wise, 1980, p. 149). A neotype, chosen by Ivanovskiy and Shurygina (1975, p. 15), is an approximate topotype and comes from the *Favosites regularissimus* Zone on the left bank of Kakva River, 2.5 km above the Karpinsk Highway. It has been figured by Ivanovskiy and Shurygina (1975, Plate 1, Fig. 1) and Hill (1981, Fig. 40,2a, b).

Diagnosis. See Hill (1981, p. F98), but modify statement concerning tabulae to read: tabulae complete or incomplete forming flat to concave tabularial surfaces.

Remarks. Pholidophyllum was erected for a species having epithecal scales. Subsequently, Lindström (1883b, p. 63–64) acknowledged that the genus was synonymous with *Tryplasma* but argued, contrary to current rules, that the name should be conserved because Lonsdale misinterpreted the septal spines in the weathered holotype of *T. aequabile* to be foramina. Dybowski proposed *Acanthodes* for diaphragmatophorous cyathophyllinid corals with septa (Längsscheidewände) made up of rows of spines. However, the name is preoccupied by *Acanthodes* Agassiz (1833, p. 19) for a Permian fish and was replaced by *Spiniferina* Penecke. As far as the original crude figures permit an opinion, the type species of *Spiniferina* is a typical Gotland Silurian species of *Tryplasma* (Dybowski, 1873, Plate 1, Fig. 11–11c). *Stortophyllum* is based on a coral that differs from the neotype of *Tryplasma aequabile* only in having less complete tabulae and generally more deeply concave tabularial surfaces. Hill (1981, p. F103) and some other recent authors, mostly Russian, have retained *Stortophyllum*, but Ivanovskiy (1969, p. 33; 1976, p. 166) and the present author believes the two genera to be synonymous. Ivanovskiy's (1969) text-figure 4 neatly demonstrates a complete gradation from flat to concave tabularial morphologies in Wedekind's (1927) Gotland species of *Stortophyllum* and *Pholidophyllum*. *Pholidophyllum* is correctly regarded as a synonym of *Tryplasma* by most, if not all, modern authorities. It should be noted that Hill's (1981, Fig. 40,2b) illustration of the longitudinal section of the neotype of *T. aequabile* is upside down, making the concave tabularial surfaces appear convex.

Tryplasma is a rare genus in the Devonian of North America and in Nevada is apparently confined to the Lochkovian part of the system.

Tryplasma duncanae Merriam, 1974
Figure 3: 3, 8

? *Tryplasma*. DUNCAN, 1956, p. 221 (in part), Plate 23, Fig. 3a only.
Tryplasma duncanae MERRIAM, 1974a, p. 37, Plate 1, Fig. 26–28; MERRIAM *in* Merriam and McKee, 1976, Plate 3, Fig. 8, 9; JOHNSON AND OLIVER, 1977, p. 1464.

Remarks. The type occurrence of this species is evidently in the lower Lochkovian coral assemblage in the type Roberts Mountains Formation exposure on the northwest slope of Roberts Creek Mountain (USGS Locality M1100). Additional specimens that would confirm this occurrence have not been collected at the type locality. However, we do have two silicified specimens from the lower Lochkovian of BC II section (Berry and Murphy, 1975, Fig. 12) in the Roberts Mountains. The better of our two specimens (USNM 488392), which is from the 638.5 ft level in the section (the other is from the 630 ft level), is slightly larger than the specimen illustrated in Figure 28 of Merriam's 1974a Plate 1 but narrower than the holotype. It has 21 major septa at approximately 4.5 mm diameter compared with about 22 at approximately 6.5 mm diameter (calculated from illustrations and their stated magnification) in the holotype.

Unfigured fragmentary material, identified as *Tryplasma* sp. cf. *T. duncanae*, was reported by Merriam and McKee (1976, p. 13, 17), from what is believed to be the middle Lochkovian in the southern Tuscarora Mountains (USGS Locality M1313) and northern Simpson Park Range (USGS Locality M1384).

Family MUCOPHYLLIDAE Hill, 1940

Remarks. See Pedder (1985a, p. 588) for an explanation of the authorship of this family.

Genus *Mucophyllum* Etheridge, 1894

Mucophyllum ETHERIDGE, 1894, p. 12.

Type species. Mucophyllum crateroides Etheridge, 1894, p. 18, Plates 3, 4. Upper Silurian, Hume or Bowning Series; Hatton's Corner and Old Limekiln Ridge, Humewood, both near Yass, and Quedong, all New South Wales. The lectotype, chosen by Hill (1940b, p. 400), is the specimen illustrated in Etheridge's 1894 Plate 4, Figure 3. It comes from a probable Ludlow Series bed on Old Limekiln Ridge, Humewood.

Diagnosis. See Hill (1981, p. F175).

Mucophyllum oliveri Merriam, 1974
Figure 4: 7–9

Mucophyllum sp. OLIVER, 1964, p. D151; OLIVER *in* Johnson, 1965, p. 371; OLIVER *in* Johnson, 1970, p. 44.
Mucophyllum oliveri MERRIAM, 1974a, p. 42, Plate 5, Fig. 1–6; PEDDER, 1975, p. 285; MERRIAM *in* Merriam and McKee, 1976, Plate 4, Fig. 1–4; JOHNSON AND OLIVER, 1977, p. 1464; OLIVER AND JOHNSON, 1977, p. 108.

Remarks. The holotype and paratype both came from Windmill Limestone talus on the east side of Coal Canyon, northern Simpson Park Range (USGS Loc. M1108). We have three additional specimens from the east side of Coal Canyon. One is a float specimen (USNM 488395) from close to SP VII section (Berry and Murphy, 1975, Fig. 14), with a matrix that yielded Lochkovian *trigonicus* Zone conodonts; the other two (USNM 488393, 488394) are from the Lochkovian *Quadrithyris* Zone debris flow in the upper Windmill Limestone, approximately 300 m south of the canyon entrance.

The exterior of *Mucophyllum oliveri* is not distinctive, since it resembles several figures of Silurian members of the genus. It especially resembles Pridolian corals from the Maikotta Beds of southern Turkestan Range in the Tien Shan fold belt that Lavrusevich (1971, p. 43, Plate 7) identified as *M. crateroides* Etheridge. It differs from most previously illustrated members of the genus by having thick fibrous layers in the tabularium. In USNM 488394, one of these layers is 7.5 mm thick.

Genus *Stylopleura* Merriam, 1974

Stylopleura MERRIAM, 1974a, p. 34.

Type Species. Stylopleura berthiaumi Merriam, 1974.

Diagnosis. See Pedder (1985a, p. 588).

Stylopleura berthiaumi Merriam, 1974

Stylopleura berthiaumi MERRIAM, 1974a, p. 35, Plate 3, Fig. 6–20; PEDDER, 1975, p. 285; MERRIAM *in* Merriam and McKee, 1976, p. 31, Plate 1, Fig. 1–9; Plate 2, Fig. 8, 9; JOHNSON AND OLIVER, 1977, p. 1464; OLIVER AND JOHNSON, 1977, p. 108; HILL, 1981, p. F178, Fig. 103,1.
Stylopleura sp. b MERRIAM *in* Merriam and McKee, 1976, p. 32, Plate 1, Fig. 10–12; OLIVER AND JOHNSON, 1977, p. 108.
Stylopleura sp. c, cf. *S. berthiaumi* Merriam. MERRIAM *in* Merriam and McKee, 1976, p. 32 (in part?), Plate 1, Fig. 13?, 14?, 15.

Remarks. The holotype came from Merriam's "Silurian" Coral zone D *Tonkinaria simpsoni* assemblage, approximately 573 m (1880 ft) above the base of the type section of the Roberts Mountains Formation on the northwest side of Roberts Creek Mountain (USGS Locality M1100). Johnson and Oliver (1977, p. 1464) deduced that the type stratum is pre-late Lochkovian. T. T. Uyeno has recovered *Criteriognathus eurekaensis* from beds yielding the *Tonkinaria simpsoni* assemblage on Roberts Creek Mountain, which strongly suggests that the holotype of *Stylopleura berthaumi* came from a lower Lochkovian horizon.

Figure 4 (on facing page). 1–4: *Stylopleura* sp. A, transverse and longitudinal sections, USNM 488397. 5, 6: *Stylopleura nevadensis* Merriam, longitudinal and transverse sections, USNM 488396. 7–9: *Mucophyllum oliveri* Merriam; 7, 9, longitudinal and transverse sections, USNM 488393; 8, exterior calicular view of naturally weathered specimen before sectioning, USNM 488394.

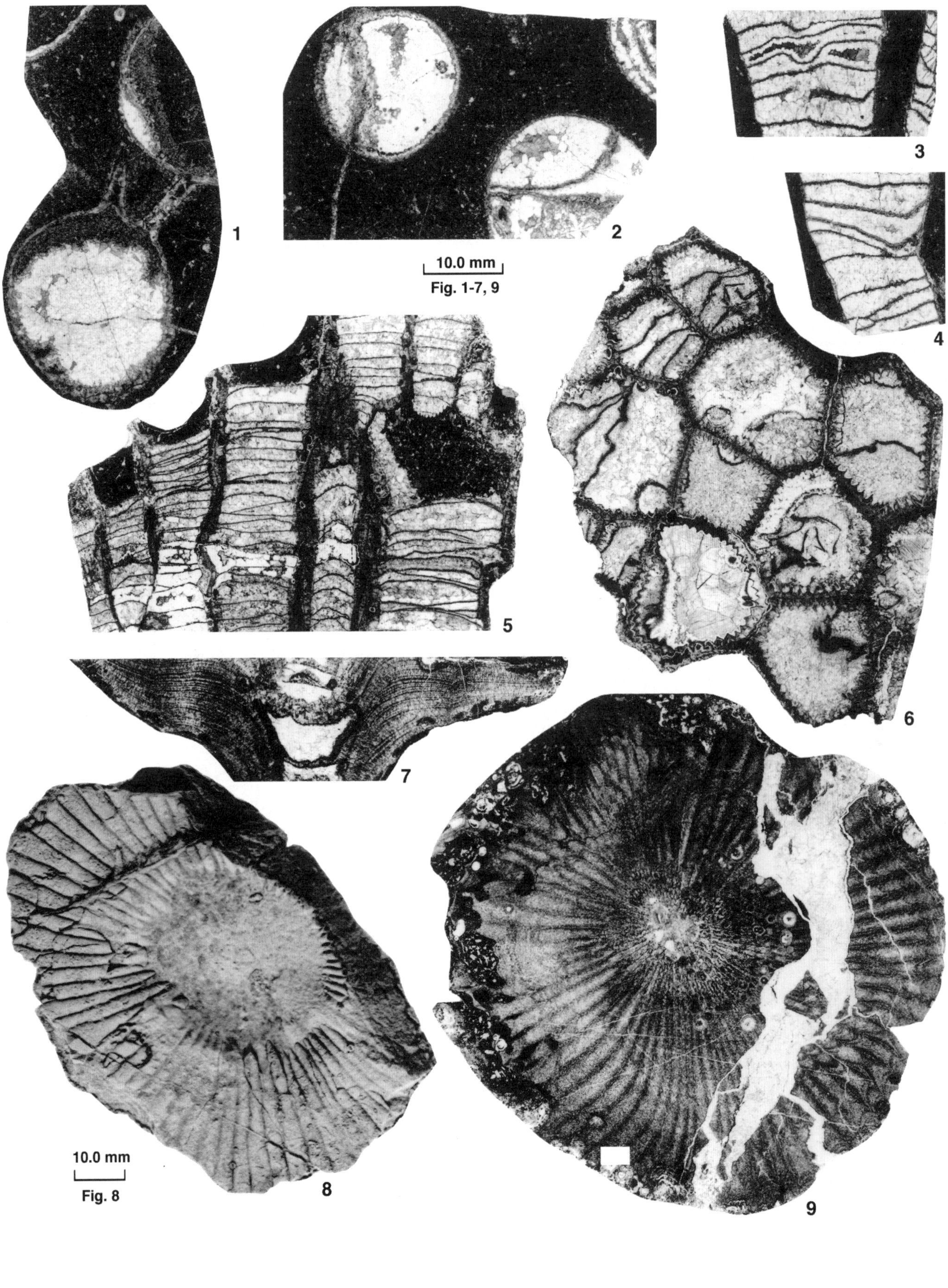
1
2
3
4
10.0 mm
Fig. 1-7, 9
5
6
7
8
9
10.0 mm
Fig. 8

Merriam assigned isolated, acid-etched, fragmentary corallites of this species, from probable lower Lochkovian beds in the southern Tusacarora Mountains (USGS localities M1314 and M1315), to *Stylopleura* sp. b. He was reluctant to refer them to *S. berthiaumi* only because the material did not show budding. In all other respects the corallites are identical to other isolated corallites of *S. berthiaumi*. Other occurrences of *Stylopleura berthiaumi* are at USGS Locality M1103, in the lower Lochkovian of Ikes Canyon, Toquima Range (Merriam, 1974a, Plate 3, Fig. 6–8; Merriam and McKee, 1976, Plate 2, Fig. 8, 9); middle Lochkovian at USGS Locality M1317 on the west side of Coal Canyon (Merriam and McKee, 1976, Plate 1, Fig. 15); and from the upper Lochkovian on the east side of Coal Canyon (USGS Loc. M1106), northern Simpson Park Range (Merriam, 1974a, Plate 3, Fig. 18–20; Merriam and McKee, 1976, Plate 1, Fig. 7–9).

Stylopleura nevadensis Merriam, 1974
Figure 4: 5, 6

Stylopleura nevadensis MERRIAM, 1974a, p. 35, Plate 2, Fig. 11, 12, 15, 16; Plate 15, Fig. 1–4; PEDDER, 1975, p. 285; MERRIAM *in* Merriam and McKee, 1976, p. 32, Plate 2, Fig. 3–6; JOHNSON AND OLIVER, 1977, p. 1464; OLIVER AND JOHNSON, 1977, p. 108.
? Stylopleura sp. cf. S. nevadensis "n. gen., n. sp.," MERRIAM *in* Merriam and McKee, 1976, p. 32, Plate 2, Fig. 7.

Remarks. Stylopleura nevadensis differs from *S. berthiaumi* Merriam by having more closely spaced corallites with less pronounced calicular expansions, so that the corallum is subcerioid or even cerioid locally. Its corallites are larger, typically with diameters below the calicular expansions of 15 to 18 mm, versus 10 to 13 mm in *S. berthiaumi*. Septa in *S. nevadensis* have thicker bases forming a septal stereozone, whereas in *S. berthiaumi* the septa are mostly fine and ridgelike below the calicular expansions. Merriam noted that the relative scarcity of connecting processes in *S. nevadensis* may be more apparent than real and suggested that the species is possibly only a subspecies of *S. berthiaumi*. However, the differences are more than just relative abundances of connecting processes, and there is no need at present to demote the taxonomic status of *S. nevadensis*.

The form figured by Merriam in 1976 as *Stylopleura* sp. cf. *S. nevadensis* n. gen., n. sp. [sic] is, like the type of *S. nevadensis*, from the Windmill Limestone of Coal Canyon (USGS Locality M1381). It seems to be a typical example of *S. nevadensis*, and Merriam gave no reason for separating it from *S. nevadensis*.

The oldest known specimens of *Stylopleura nevadensis* are from the lower Lochkovian of northern Simpson Park and Toquima Ranges; the youngest well-dated occurrences are in the upper Lochkovian part of the Windmill Limestone on the east side of Coal Canyon. The holotype and specimen figured in this work (USNM 488396) are from approximately the same upper Lochkovian part of the Windmill Limestone breccia and locality, close to SP VII section in Coal Canyon (Berry and Murphy, 1975, Fig. 14). Age of the paratype figured from the "Masket Shale" (Merriam 1974a, Plate 2, Fig. 11, 12) of the Mill Canyon plate in the Toquima Range (USGS Locality M1103) has not been established. Merriam and McKee (1976, p. 22) imply that it came from the Roberts Mountains Formation, but it is more likely to have come from some Lochkovian part of the McMonnigal Limestone (see Fig. 2 of Murphy and Anderson, 1991).

Stylopleura sp. A
Figure 4: 1–4

Remarks. Only a part of one poorly preserved, silicified specimen (USNM 488397) of this species is available. It is from the lower Lochkovian, *hesperius* Zone, 234.1- to 236.5-m (768 to 776 ft) level of the section published by Klapper and Murphy (1975, Fig. 9), in the type Roberts Mountains Formation exposure on the northwest side of Roberts Creek Mountain. *Stylopleura* sp. A is characterized by the large diameter, 19 to 23 mm, of its widely spaced corallites and by its feeble septa. In places the septa are no more than corrugations of the corallite wall.

Stylopleura sp. A recalls a coral from the *hesperius* Zone in the Royal Creek headwaters area of Yukon Territory, illustrated as a new species of *Stylopleura* by Pedder (*in* Jackson et al., 1978, Plate 25, Fig. 7, 12–14). The two corals may be conspecific, although the Yukon Territory form is smaller, with a known maximum mean diameter of only 15 mm.

Family STAUROMATIDIIDAE Pedder and Oliver, 1982

Genus *Stauromatidium* Pedder and Oliver, 1982

New genus 6 OLIVER AND PEDDER, 1979, p. 244 (in part).
Stauromatidium PEDDER AND OLIVER, 1982, p. 3 (in part).

Type species. Cyathophyllum marylandicum Swartz, 1913, p. 204, Plate 21, Fig. 1, 2. "Lower Devonian," Helderberg Formation, Keyser Member; Keyser, West Virginia, Hyndman, Pennsylvania, and Cash Valley, Maryland. Pedder and Oliver (1982, p. 5, Plates 3–5) revised the species and chose the specimen from Hyndman, illustrated in Figure 1 of Swartz's Plate 21, to be lectotype. Thin sections of this specimen are illustrated in Figures 1 and 2 of Pedder and Oliver's Plate 3. In 1982 it was undecided whether the lectotype is late Pridolian or early Lochkovian; now it is known to be from the Pridolian part of the Keyser Limestone (W. A. Oliver Jr., personal communication, 1992).

Diagnosis. See Pedder and Oliver (1982, p. 3), but modify first statement to read: Solitary, ceratoid to cylindrical corallum.

Remarks. The genus *Fascidium* has been proposed by Ulitina (1992, p. 13) for two fasciculate species, *Stauromatidium strigosum* Pedder and Oliver (type of *Fascidium*) and *S. trigemma* (Davis), that were included in *Stauromatidium* by Pedder and Oliver in 1982.

Stauromatidium sp. C

Kodonophyllum sp. b MERRIAM *in* Merriam and McKee, 1976, p. 30 (in part), Plate 5, Fig. 23, 24 (not Fig. 22).
Calostylus? sp. MERRIAM *in* Merriam and McKee, 1976, p. 37, Plate 7, Fig. 7, 8.
Stauromatidium sp. cf. S. *marylandicum* (Swartz). PEDDER AND OLIVER, 1982, p. 5 (in part), Plate 6, Fig. 3–8 (not Fig. 1, 2); PEDDER *in* Potter et al., 1990, p. 67.

Remarks. Possibly as many as 10 specimens of this species are available, all from USGS Locality M1314, in the lower Lochkovian on Boulder Creek ("Bootstrap Hill" of Merriam), southern Tuscarora Mountains. Pedder and Oliver discussed the material and pointed out that the coral depicted in Figure 22 of Merriam and McKee's (1976) Plate 5 has lamellar septa. Although a longitudinal section is not available, the coral possibly belongs to a streptelasmatid genus. *Stauromatidium* sp. C also occurs in Lochkovian limestone clasts in member 3 of the Gazelle Formation on the Yreka terrane of northern California (Potter et al., 1990).

This species of *Stauromatidium* is assigned the letter C to distinguish it from *S.* spp. A and B of Pedder and Oliver (1982).

Stauromatidium sp. D

Stauromatidium sp. cf. *S. marylandicum* (Swartz). PEDDER AND OLIVER, 1982, p. 5 (in part), Plate 6, Fig. 1, 2 (not Fig. 3–8).

Remarks. Stauromatidium sp. D is known from a single fragmentary specimen from the lower part of the Windmill Limestone, as defined by Johnson (1965), that is from the upper middle Lochkovian, on the western side of Coal Canyon, northern Simpson Park Range. It is very similar to *S. marylandicum* but is slightly larger (corallite diameter 10.8 mm versus 4.7 to 9.1 mm in *S. marylandicum*) and has 36 major septa, compared with a maximum of only 30 in *S. marylandicum*. *Stauroma-*

tidium sp. C is smaller than *S.* sp. D, with mean corallite diameters of 6.5 to 7.5 mm, and has more isolated monacanths around the axis.

Order STAURIIDA Verrill, 1865
Suborder STREPTELASMATINA Wedekind, 1927
Family STREPTELASMATIDAE Nicholson, 1889
(in Nicholson and Lydekker, 1889)
Subfamily BREVIPHYLLINAE Hill, 1981

Diagnosis. Subfamily of streptelasmatid corals commonly with periodically expanded calices. Early (Lochkovian) forms lack dissepiments and generally have a thick septal stereozone; late forms (Emsian) have a relatively narrower septal stereozone and may have elongate dissepiments. Cardinal fossula absent in early forms, variable in later forms. Septa withdrawn from axis, irregularly amplexoid but lacking vermiform lobes. Tabulae broad in periaxial region, downturned usually with a curved shoulder in outer tabularium. Trabeculae coarse, unaligned or poorly aligned around the periphery of calicinal expansions.

Remarks. The present author considers the Subfamily Nevadaphyllinae Hill (1981, p. F160), erected for two specimens, to be synonymous with the Breviphyllinae Hill (1981, p. F166).

Genus *Breviphrentis* Stumm, 1949

Breviphrentis STUMM, 1949, p. 13.

Type species. Amplexus invaginatus Stumm, 1937, p. 427, Plate 53, Fig. 2; Plate 54, Fig. 2. Basal 152 m (500 ft) of the Nevada Limestone; a little to the southeast of the highest point of Atrypa Peak, 11.3 km (7 mi) southwest of Eureka, Nevada. Two different specimens were labeled holotype in Stumm's (1937) plate captions. Oliver (1993, p. B19) chose the specimen illustrated in Figure 2 of Stumm's Plate 53 as lectotype. Sections cut from this specimen had previously been illustrated by Hill (1981, Fig. 95,1a, b) as holotype and were reillustrated by Oliver (1993, Plate 13, Fig. 1, 2).

Recent work on the ridge extending southeastward from Atrypa Peak in the Fish Creek Range has demonstrated that the type stratum is mostly likely to be 7.5 to 7.8 m above the base of the Coils Creek Member of the McColley Canyon Formation. J. I. Valenzuela-Ríos obtained conodonts from above and below this bed. Forms isolated from the lowest 10 cm of the Coils Creek Member include *Icriodus trojani*, *Polygnathus laticostatus*, *P. inversus*, *P. inversus* transitional to *P. serotinus*, *Pandorinellina exigua* subsp. cf. *exigua*, and *P. steinhornensis* subsp. From 8.4 to 8.5 m above the base of the Coils Creek Member, that is, 0.6 to 0.7 m above the probable type stratum of *Breviphrentis invaginata*, conodonts recovered include *Icriodus trojani*, *Polygnathus inversus*, *P. inversus* transitional to *P. serotinus*, and *Pandorinellina exigua exigua*. Conodonts convincingly date the probable type stratum as upper *inversus* Zone, that is, Emsian.

Remarks. Hill (1981, p. F166) was uncertain whether *Breviphrentis* and *Breviphyllum* Stumm (1949, p. 25) are distinct genera. The two genera are certainly closely related; however, *Breviphyllum* is distinguished by its narrow septal stereozone and by abundant elongate dissepiments in adult stages. It also has fewer, less pronounced calicinal expansions than *Breviphrentis*.

Oliver (1993, p. B19) concluded from a study of material in the U. S. National Museum that three species had been included in *Breviphrentis invaginata* (Stumm) by Merriam (1974b, p. 42). Oliver referred the more coarsely trabeculate forms with calicinal expansions, such as the specimen illustrated by Merriam (1974b, Plate 16, Figs. 12, 13), to one of the unnamed species excluded from *B. invaginata* sensu stricto. However, following study of new material from the probable type locality and stratum of *B. invaginata*, Oliver (in Galle, 1995, p. 62) now regards the forms with calicinal expansions and coarse trabeculae to be typical of *B. invaginata*. This view was supported by Galle and is also the view of the present author. Two typical, and probably topotypic, specimens (USNM 488573, 488574) of *Breviphrentis invaginata* (Stumm), from the ridge extending southeastwards from Atrypa Peak, Fish Creek Range, Bellevue Peak Quadrangle, are illustrated in Figure 5:1, 2, 4–6 herein.

Li (1980, p. 28, Plate 5, Fig. 1) erected a new species named *Kodonophyllum*(?) *curovesense* for two Upper Silurian corals from the left bank of West Chonkuchak River, northeastern Fergana, in the Tien Shan Hercynian fold belt. The fine structure of the septal stereozone in the two specimens is recrystallized. Li wrote that if the fine structure proved not to be that of *Kodonophyllum*, the species would likely be one of *Breviphrentis*. In all observable respects Li's species is similar to *Breviphrentis johnsoni* described in this work and, despite the poor preservation of its types, is clearly better attached to *Breviphrentis* than *Kodonophyllum*. Since it comes from close to the base of the overlying Lochkovian and younger Aravan Suite, it is probably Pridolian and therefore, likely to be the oldest species of *Breviphrentis* yet named.

***Breviphrentis johnsoni* sp. nov.**
Figures 5: 3, 6: 1–11, 7: 1–4

? *Enterolasma?* sp. OLIVER *in* Johnson, 1965, p. 371; OLIVER *in* Johnson, 1970, p. 44.

Material. Holotype, USNM 488398, talus from *Quadrithyris* Zone depositional breccia, east side of Coal Canyon near entrance. Paratype, USNM 488399, 0 to 2.0 m below top of *Quadrithyris* Zone depositional breccia, east side of Coal Canyon, approximately 150 m south of entrance. Paratype, USNM 488400, *Quadrithyris* Zone depositional breccia, ridge leading from entrance of Coal Canyon to summit 6909 ("Pyramid Hill" of Merriam and McKee, 1976, Fig. 2) on east side of Coal Canyon, SE 1/4 sec. 17, T25N, R49E. Paratypes, USNM 488401–488404, 4.0 to 4.3 m above base of *Quadrithyris* Zone depositional breccia, section SP I (Berry and Murphy, 1975, Fig. 14), east side of Coal Canyon Fault, NW 1/4 NE 1/4 sec. 20, T25N, R49E. All of the material is from upper Lochkovian Windmill Limestone, in Horse Creek Valley Quadrangle, northern Simpson Park Range.

Description. Coralla of type series solitary, except for four parricidally produced buds in one paratype (USNM 488403). Corallites subcylindrical between periodic calicinal expansions. Frequency of calicinal expansions extremely variable. Some fragmentary corallites have none, whereas as many as eight are crowded over a vertical distance of 10 mm in parts of paratype USNM 488400, imparting a frilled appearance to the specimen. Typically, one to three expansions occur over a vertical distance of 15 mm. All known specimens fragmentary. Maximum preserved length of corallite about 55 mm. Greatest mean diameter of unexpanded parts of the corallite 21.5 mm; greatest mean diameter of distal rims of expanded calices 25.5 mm. Mean diameters of rare buds <9 mm. Calices about 8.5 mm deep, with convex bases; calicinal walls vertical in unexpanded parts of the corallite. Expanded parts of calices slope inward, mostly angled 25° to 45° to horizontal.

Septa radially arranged in two orders, thickened peripherally to form a septal stereozone 1.0 to 4.5 mm, commonly 2.0 to 3.5 mm thick. Septal counts 32 × 2 to 38 × 2 in most adult corallites (mean diameter >16 mm); the largest paratype (USNM 488400) has 44 × 2 septa at a mean diameter of 21.5 mm. The two orders of septa are indistinguishable in the septal stereozone and on the calicinal expansions. Adaxial parts of septa smooth and locally irregularly amplexoid in the case of the major septa. Most major septa terminate near the axis or up to 3.0 mm from it. Minor septa may not project adaxially from the septal stereozone, especially where the stereozone is thick. Where the septal stereozone is thin, minor septa typically project 0.5 to 1.5 mm inward from it. Rare discrete fragments of major septa occur in the periaxial region of some specimens, including the holotype. Septa trabeculate in the stereozone and on the calicinal expansions. Trabeculae very poorly preserved with unknown microstructure, but possibly rhabdacanthine near rims of calicinal expansions. Trabeculae nondivergent viewed in longitudinal section. Diameter of trabeculae 0.3 to 0.5 mm. Septa apparently nontrabeculate in tabularium.

Although rare inwardly sloping plates occur locally around the periphery of the tabularium, true convex dissepiments or presepiments are absent. Tabulae broad but mostly incomplete; spaced from 14 to 24 over a vertical distance of 10 mm in the periaxial region, less crowded around the periphery. A weak tabularial fossula may or may not be present. Tabularial surfaces generally slightly convex over most of the tabularium, with a downturned and rounded outer shoulder. Sclerenchymal thickening absent from most tabularial surfaces; maximum thickness of sclerenchyme 0.6 mm.

Remarks. The two previously described species most resembling *Breviphrentis johnsoni* sp. nov. are the type species of the genus, *B. invaginata* (Stumm, 1937), and *B. joae* Galle (1995, p. 61, Plates 1, 2) from subsurface beds of likely Eifelian age in the Moravian region of the Czech Republic. *B. invaginata* is a larger species (diameters of the unexpanded parts of adult corallites and rims of expanded calices, as much as 26.0 mm and 34.0 mm respectively) and has more septa (as many as 46 × 2 in large specimens). It differs also by having a weak cardinal fossula, a relatively narrower septal stereozone, less aligned trabeculae on the calicinal expansions, and longer minor septa, which tend to bend toward an adjacent major septum and locally may even be contratingent with a major septum. *B. joae* is less well known to the present author. It is similar in size to *B. johnsoni* sp. nov., but has only 20 × 2 to 30 × 2 septa. In addition, it appears to have a narrower septal stereozone, more amplexoid major septa, and fewer calicinal expansions (called rejuvenescence flanges by Galle, 1995, caption to Fig. 4 of Plate 2). *Breviphrentis*(?) *curovesensis* (Li), discussed under the remarks concerning the genus, differs from *B. johnsoni* by being smaller, with a diameter of 10 to 12 mm, and having shorter major septa.

Oliver (*in* Johnson and Oliver, 1977, p. 1464) reported that a coral from the *Quadrithyris* Zone breccia in Coal Canyon, which he had identified earlier as a questionable species of *Enterolasma*, is a streptelasmatid coral and should not have been listed. The coral in question is possibly *Breviphrentis johnsoni*, since the new species is the only common streptelasmatid coral known from the breccia.

Derivation of name. Patronym for the late J. G. Johnson, to whom this special paper is dedicated.

Family KODONOPHYLLIDAE Wedekind, 1927

Remarks. The presence of a few dissepiments in some specimens of *Briantia* may necessitate its removal from the Kodonophyllidae sensu stricto. However, other features, especially the prominent axial structure and thick septal stereozone, are typical of the family.

Genus Briantia Barrois, 1889

Briantia BARROIS, 1889, p. 44.
Symphyphyllum SPASSKIY *in* Bul'vanker et al., 1968, p. 14.

Type species. Briantia repleta Barrois, 1889, p. 45, Plate 2, Fig. 1. Calcaire d'Erbray, Lower Devonian; Chateau-Briant, just east of Erbray, Lower Loire, France. The type stratum is upper Pragian or lower Emsian. Thin sections of the holotype have been prepared under the supervision of P. Semenoff-Tian-Chansky. Photographs of them, kindly made available by W. A. Oliver, Jr. are reproduced for the first time in Figure 10: 1 and 2 herein.

Description. Corallum solitary, ceratoid to turbinate. Calice dominated by a broad axial boss. Both orders of septa similarly dilated around the periphery, forming a wide septal stereozone. Major septa relatively attenuate in outer tabularium, becoming dilated toward the axis, commonly irregularly coalesced and rotated in axial region. Minor septa confined to septal stereozone or only just projecting from it. Septal arrangement radial, despite shortening of the cardinal septum. Dissepiments moderately long, very little inflated, confined to narrow horizontal openings in the septal stereozone and rare to the point of being absent from most sections. Tabulae incomplete, mostly closely spaced, forming tabularial surfaces with a broad periaxial dome surrounded by a much narrower peripheral trough. Sclerenchyme periodically invests the tabularial surfaces. No tabular fossula. Periaxial tabulae, sclerenchyme, and dilated adaxial septal ends combine to construct a locally dense axial structure, manifest in the calice as a boss.

Remarks. Photographs of the transverse and longitudinal thin sections of the holotype of *Briantia repleta* do not clearly reveal its septal microstructure.

The age of the Erbray Formation is usually considered to be Emsian, but age diagnostic fossils in the 280-m-thick limestone are rare. It has yielded *Polygnathus* sp. cf. *P. dehiscens* of Emsian age, but the lower part evidently passes laterally into the "Angers grauwacke," which is thought to be Pragian (Morzadec et al., 1989, Fig. 2 and p. 411). This is supported by Gourvennec (1989, p. 406), who regards *Eospirifer davousti* (Verneuil) from the Erbray Limestone to be a Pragian brachiopod.

Only one of seven specimens included in the type series of *Symphyphyllum styliferum* Spasskiy, the type species of *Symphyphyllum*, has been illustrated. No dissepiment can be detected in either of the two poor thin sections. Absence of dissepiments may prove to be a basis for separation of *Symphyphyllum* from *Briantia*. At the moment, the two genera seem most likely to be synonymous. The occurrence of *Symphyphyllum styliferum* was originally cited as "*Rensellaria*" horizon of the Coblenzian Baragash Suite on Ganin Stream in the Gorno Altay, Russia. In current stratigraphic terminology, it likely comes from the upper Lochkovian–Pragian Yakushin Beds.

Briantia mulleri (Merriam, 1974)
Figure 8: 1–7

Kodonophyllum mulleri MERRIAM, 1974a, p. 41, Plate 4, Fig. 3–7; PEDDER, 1975, p. 285; MERRIAM *in* Merriam and McKee, 1976, p. 30, Plate 4, Fig. 5, 6.
Briantelasma? mulleri (Merriam). JOHNSON AND OLIVER, 1977, p. 1464; OLIVER AND JOHNSON, 1977, p. 108.

Diagnosis. Large ceratoid to turbinate species of *Briantia* with maximum diameter of about 42 mm. Adult septal stereozone 5.0 to 13.0 mm thick. Number of septa at maturity 47 × 2 to 54 × 2. Gaps in septal stereozone very rare. No dissepiment seen, probably altogether absent.

Remarks. This is the largest of the three named species of *Briantia*. It is more similar to *B. stylifera* (Spasskiy) than to *B. repleta* Barrois. *B. stylifera* differs from *B. mulleri* most clearly by having about the same maximum number of septa (48 × 2 estimated from an incomplete transverse section) at a much smaller maximum diameter of about 17 mm.

B. mulleri occurs in the Windmill Limestone depositional breccia on both sides of Coal Canyon, Simpson Park Range, but has not been found at any other locality where upper Lochkovian rocks are exposed. The holotype (USNM 159386) is from USGS Locality M1107, on the east side of Coal Canyon, approximately 415 m south of the canyon entrance. Both figured specimens are also from the east side of Coal Canyon. One of them (USNM 488405) is from an unrecorded level in the Windmill Limestone breccia on the ridge leading from the entrance

Figure 5 (on facing page). 1, 2, 4–6: *Breviphrentis invaginata* (Stumm); 1, 5, 6, transverse section, part of a longitudinal section, and part of a transverse section, approximate topotype, USNM 488573, Emsian, 7.5 to 7.8 m above base of Coils Creek Member, McColley Canyon Formation, ridge leading to Atrypa Peak, Fish Creek Range, Nevada; 2, 4, longitudinal and transverse sections, approximate topotype, USNM 488574, talus from lower 7 m of Coils Creek Member, McColley Canyon Formation, same locality as Figures 5: 1, 5, 6. 3: *Breviphrentis johnsoni* sp. nov., longitudinal section, holotype, USNM 488398.

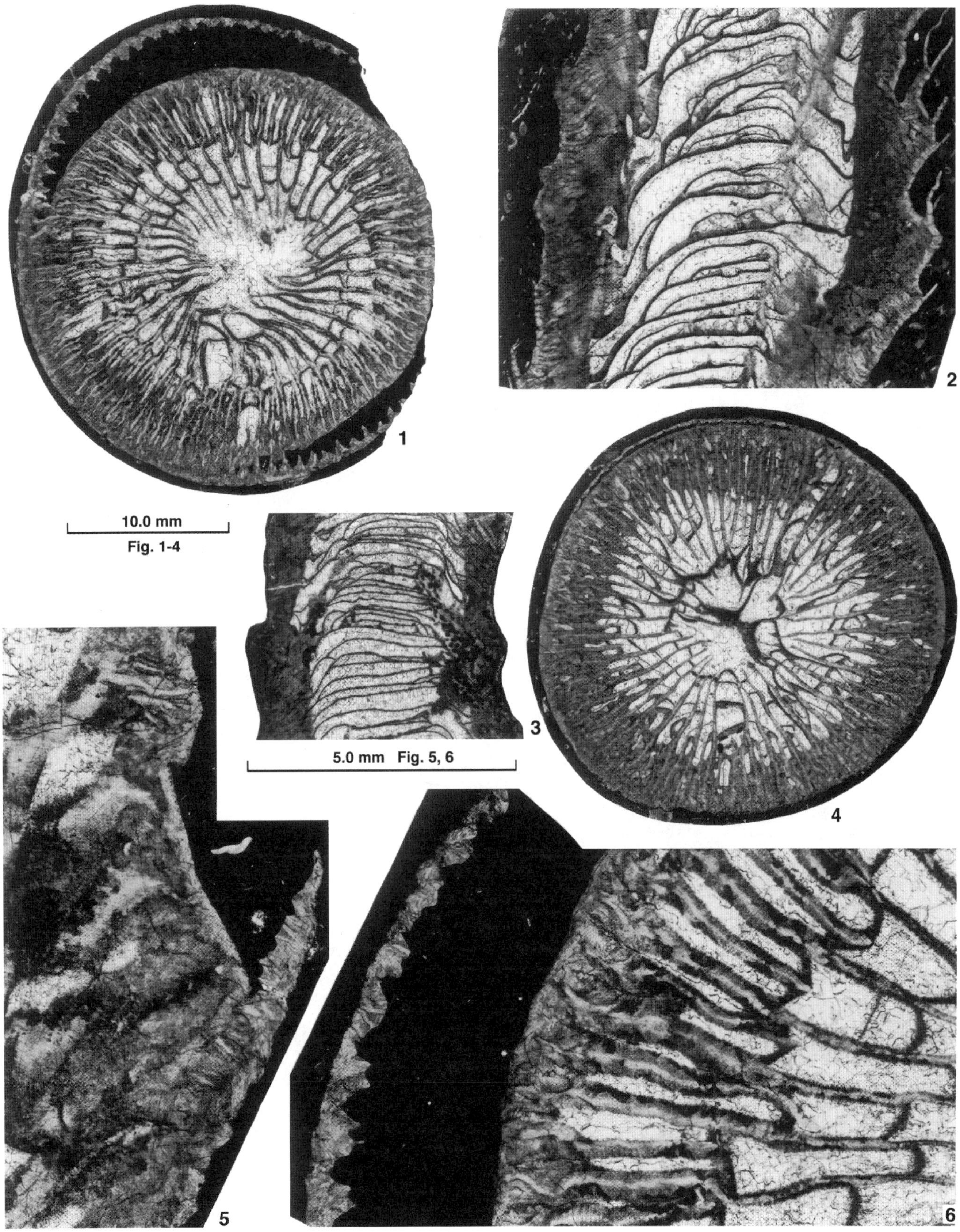
1
2
10.0 mm
Fig. 1-4
3
4
5.0 mm Fig. 5, 6
5
6

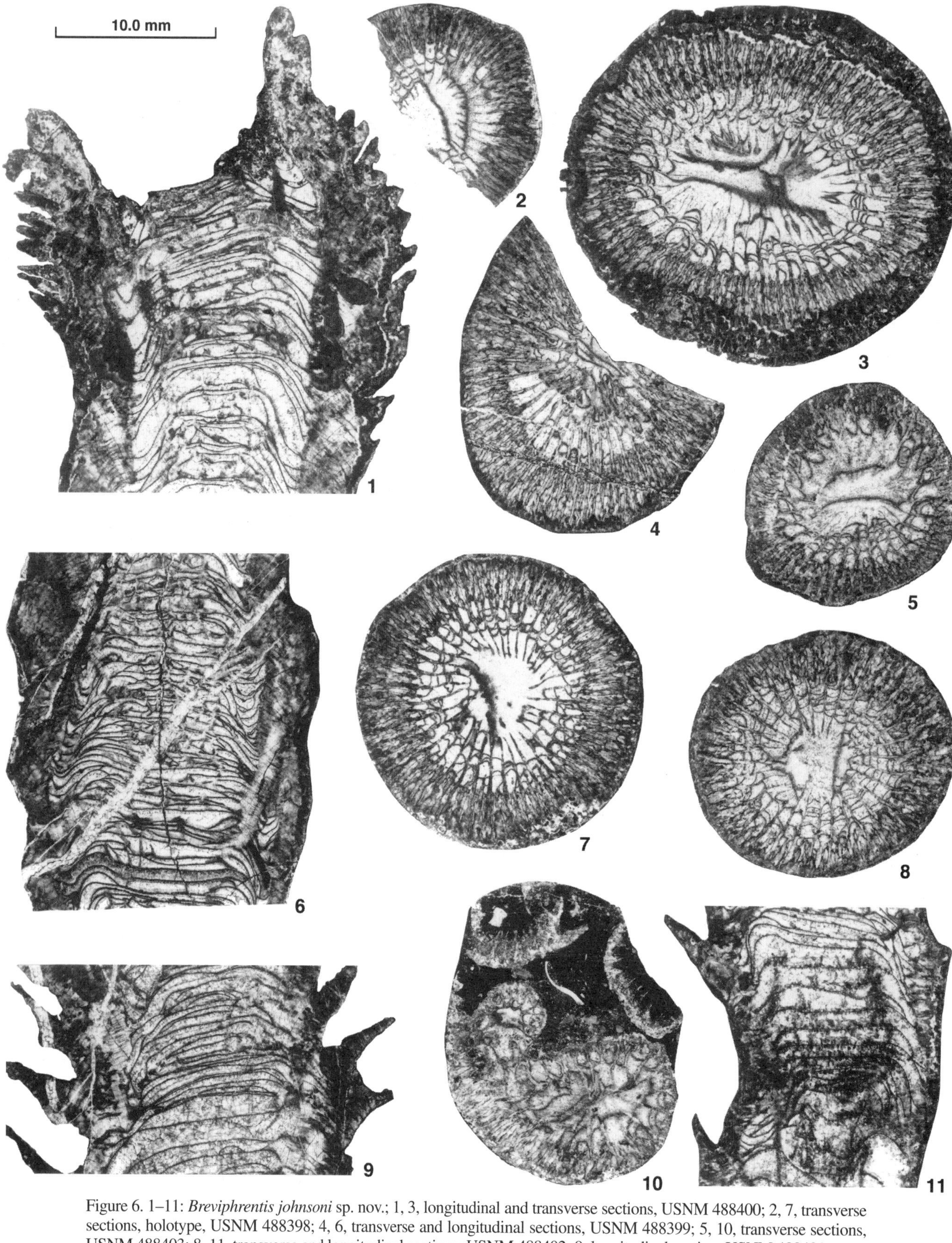

Figure 6. 1–11: *Breviphrentis johnsoni* sp. nov.; 1, 3, longitudinal and transverse sections, USNM 488400; 2, 7, transverse sections, holotype, USNM 488398; 4, 6, transverse and longitudinal sections, USNM 488399; 5, 10, transverse sections, USNM 488403; 8, 11, transverse and longitudinal sections, USNM 488402; 9, longitudinal section, USNM 488401.

of the canyon to summit 6909 ("Pyramid Hill" of Merriam and McKee, 1976, Fig. 1); the other (USNM 488406) is from the upper 2 m of the same unit, approximately 150 m south of the canyon entrance.

Briantia sp. A
Figure 9: 1, 2

Remarks. Owing to silicification and dolomitization, the best of the available material of this form (USNM 488407), which has only been found in the *hesperius* Zone in the lower Lochkovian part of the type Roberts Mountains Formation exposure (Berry and Murphy, 1975, Fig. 3), is too poorly preserved to warrant a name. It is similar to *Briantia mulleri* (Merriam) and is possibly ancestral to it. It differs from *B. mulleri* by being ceratoid rather than turbinate. It is also smaller (maximum diameter about 30 mm) and has fewer septa (maximum number 44 × 2), and its tabulae are more widely spaced than those of *B. mulleri.*

Suborder METRIOPHYLLINA Spasskiy, 1965
Family LACCOPHYLLIDAE Grabau, 1928
Subfamily NEAXONINAE Hill, 1981

Remarks. An unpublished revision of *Petraia decussata* Münster, the type species of *Petraia* Münster (1839, p. 42), announced in Weyer (1984, p. 4, 5) is said to demonstrate a close relationship between *Petraia* and *Neaxon*. As a result, Weyer regards the Laccophyllidae to be synonymous with the Petraiidae de Koninck (1872, p. 113). He also regards the subfamilies Friedbergiinae Ró kowska (1969, p. 78) and Neaxoninae Hill (1981, p. F196) to be synonymous with the nominate subfamily of the Petraiidae. Since these synonymies apparently depend on the choice of a lectotype for *Petraia decussata*, the present author follows Weyer (1989, p. 5) in retaining *Neaxon* in the Neaxoninae and Laccophyllidae until the revision of *P. decussata* is published.

Genus *Neaxon* Kullmann, 1965

Neaxon KULLMANN, 1965, p. 81(49).

Type species. Neaxon regularis Kullmann, 1965, p. 83(51), Text-Fig. 6, Plate 2, Fig. 1–4. Unnamed upper Emsian marls and limestone; a few hundred meters east of Mudá, southeastern Cantabrian Mountains, Spain. The only other species found in the type stratum is a goniatite fragment identified by Kullmann (1965, p. 48[16]) as *Mimagoniatites* sp. A. In 1965, the type stratum was believed to be upper Emsian, although an early Eifelian age was thought to be possible. Now, the type stratum is believed unequivocally to be upper Emsian.

Diagnosis. See Oliver (1992, p. 140).

Remarks. The type species of *Neaxon* has coarse monacanthine trabeculae (Kullmann in Oliver, 1992, p. 140); Weyer (1984, p. 5) emphasized that such trabeculae are an essential diagnostic element of the genus. Oliver (1992, p. 140–142) has summarized the known distribution of *Neaxon* and has pointed out that the holotype and three or four paratypes of *Syringaxon foerstei* Merriam (1973, p. 22, Plate 2, Fig. 1–10; 1974b, p. 39, Plate 1, Fig. 1–16), from the Pragian and possibly Emsian parts of the Rabbit Hill Limestone of Nevada, are referable to it. However, the presence or absence of coarse monacanthine trabeculae in *Neaxon foerstei* cannot be demonstrated owing to the silicified state of the available material.

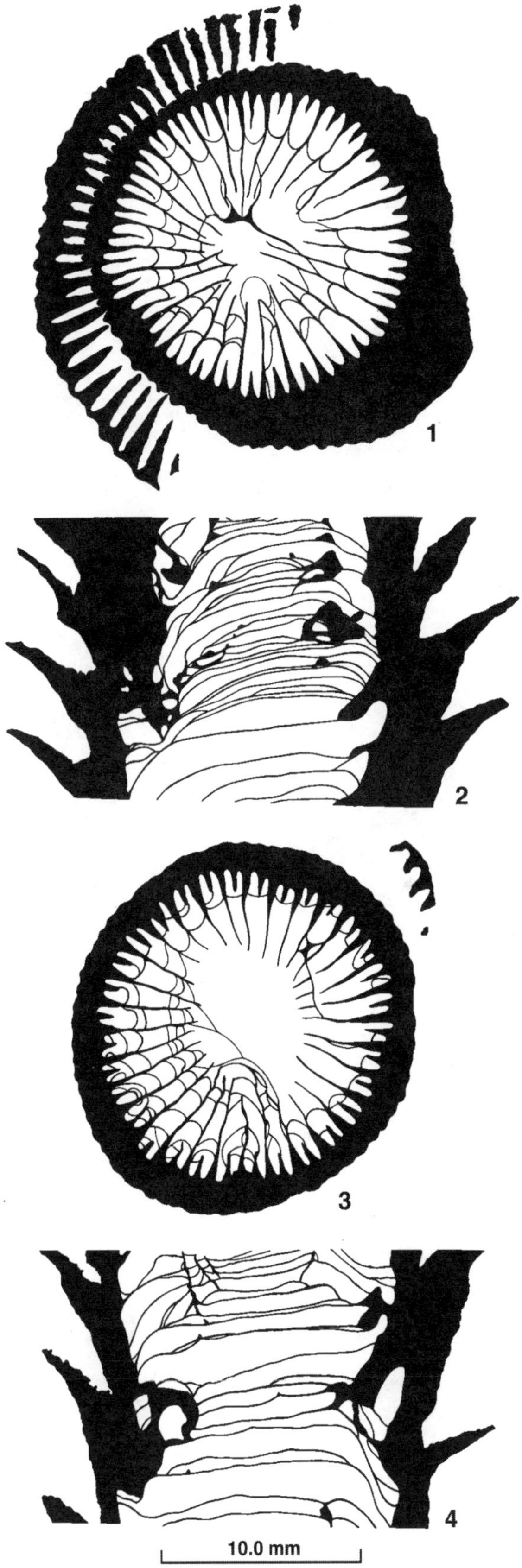

Figure 7. 1–4: *Breviphrentis johnsoni* sp. nov., silhouette figures of two transverse and two longitudinal sections, USNM 488401. Structures offset by the larger calcite veins have been restored to their original positions relative to each other during tracing; also, very limited areas around parts of the peripheries have been restored.

Figure 8. 1–7: *Briantia mulleri* (Merriam); 1, 3, 5, two transverse sections and a longitudinal section, USNM 488405; 2, 4, transverse and longitudinal sections, USNM 488406; 6, 7, exterior side and calicular views of naturally weathered specimen before sectioning, USNM 488405.

Neaxon sp. A
Figure 10: 5–7

Remarks. Small, slim silicified specimens, apparently referable to *Neaxon* sensu lato, have been recovered by acidization from the *eleanorae* to *pandora* beta Zones in the McMonnigal Limestone of the Toquima Range, Wildcat Peak Quadrangle. Diameters of their corallites below the calice are generally less than 5.0 mm, and they typically have no more than 14 septa. By comparison, the holotype of *Neaxon foerstei* (Merriam) has 28 septa and a diameter of 12.0 mm at the base of the calice. The Toquima Range material is too poorly preserved to reveal any trabeculae that might have been present.

Figured specimens are from the *eleanorae* and *trigonicus* Zones in TO 1 section (Murphy, 1989b, Fig. 4, 6) at the 19A (USNM 488408), 28 (USNM 488409), and 29 (USNM 488410) levels. Unfigured specimens come from the *pandora* beta Zone in levels 24 and 26 in the nearby IK IV section (Murphy, 1989b, Fig. 4, Table 1).

Suborder PTENOPHYLLINA Wedekind, 1927
Family SPONGOPHYLLIDAE Dybowski, 1873

Remarks. The very close similarities and morphological gradations between *Spongophyllum* and small species of *Xystriphyllum*, between *Carlinastraea* and *Australophyllum*, and between *Neomphyma* and *Lyrielasma* put in doubt the validity of the Ptenophyllidae Wedekind, 1923, as a family distinct from the Spongophyllidae Dybowski, 1873. A compromise, relegating the Ptenophyllinae as a subfamily of the Spongophyllidae, was adopted by Birenheide (1978, p. 132) and Pedder (1980, p. 606), and is adopted again here, pending a fuller review of the problem. Other subfamilies, including the Actinocystinae Wedekind, 1927, and Redstoneainae Pedder, 1980, are also assigned to the Spongophyllidae but are not represented in the Lochkovian faunas of central Nevada.

Subfamily SPONGOPHYLLINAE Dybowski, 1873

Diagnosis. Spongophyllidae generally having a more slender corallite and narrower marginarium than the *Ptenophyllinae*. Septa smooth, attenuate, and either lacking or having very fine trabeculae in the marginarium. Septal flanges normally absent; if present, very delicate and confined to the tabularium. Naotic septa also absent or very rare, except on the expanded calice of the atypical *Tonkinaria*. Tabulae commonly less closely spaced than in other subfamilies of the Spongophyllidae. Presepiments outnumber dissepiments in most genera and species. No significant sclerenchymal thickening of septa and marginarial surfaces.

Genus *Carlinastraea* Merriam, 1976

Carlinastraea MERRIAM *in* Merriam and McKee, 1976, p. 32.

Type species. Carlinastraea tuscaroraensis Merriam *in* Merriam and McKee, 1976, p. 33, Plate 6, Fig. 5–7; Plate 7, Fig. 1–4. Roberts Mountains Formation, 385 to 400 m above base and 55 to 70 m below top of exposure; "Bootstrap Hill," Boulder Creek, USGS Locality M1313, southern Tuscarora Mountains. Merriam and McKee (1976, p. 17) considered the associated megafossil fauna to be "suggestive" of Merriam's Great Basin Silurian coral zone D or E. However, the type stratum of *C. tuscaroraensis* is about 215 m above graptolites identified by W. B. N. Berry (Merriam and McKee, 1976, p. 18) as *Monograptus uniformis* and *M. angustidens*, which indicates a post-*uniformis* Zone Lochkovian age. *C. tuscaroraensis* is considered to be a junior synonym of *C. halysitoides* (Etheridge).

Diagnosis. Cerioid Spongophyllidae; increase marginarial and nonparricidal. Corallites slender in the type and most other species. Intercorallite wall commonly moniliform. Septa much reduced, locally completely suppressed in the marginarium apart from peripheral septal ridges of the intercorallite wall. Very delicate septal flanges and minute vepreculae present in some species. Relatively large presepiments dominate the marginarium. Tabularium narrow. Tabulae mostly complete, flat to concave in small species, incomplete and variably shaped in larger species.

Figure 9. 1, 2: *Briantia* sp. A, silhouette figure of transverse section, and photograph of part of the same section showing very poor preservation, USNM 488407.

Remarks. Species of *Carlinastraea* were referred to either *Spongophyllum* or *Neomphyma* in pre-1976 literature. *Spongophyllum* differs from *Carlinastraea* by having longer major septa, which in most species only rarely degenerate to septal crests through disruption by presepiments. *Neomphyma* is distinguished by its fasciculate corallum. The type species of *Australophyllum*, discussed herein under *Toquimaphyllum*, is usually considered to be a ptenophyllid coral and yet is very similar to the type species of *Carlinastraea*, which has always been classified as a spongophyllid genus, although Yu and Jell (1990, p. 183) noted that it may have more affinities with the ptenophyllids. For operational purposes of the present work, *Australophyllum* is provisionally distinguished from *Carlinastraea* by its fewer presepiments, resulting in more continuous septa, and by the greater thickness of septa, apparently caused by the presence of distinct, albeit very slender, septal trabeculae throughout the marginarium.

Carlinastraea halysitoides (Etheridge, 1918)

Spongophyllum halysitoides ETHERIDGE, 1918, p. 49, Plate 7, Fig. 1–3; HILL, 1942, p. 161, Plate 2, Fig. 5; SOSHKINA, 1952, p. 93 (in part), Plate 30, Fig. 106 (upper and lower right figures only); LAVRUSEVICH, 1971, p. 44, Plate 8, Fig. 2, 3. Not (?) KHOA AND THOM, 1980, p. 90, Plate 24, Fig. 3. Not BUL′VANKER, 1958, p. 132, Plate 64, Fig. 1, 2.

Spongophyllum halysitoides var. *minor* Hill. SOSHKINA, 1949, p. 123 (in part), Plate 47, Fig. 1 (only). Not HILL, 1940a, p. 162, Plate 3, Fig. 3 [see Pedder, 1985a, p. 598].

Neomphyma rosiforme ZHELTONOGOVA, 1961, p. 81, Plate S-20, Fig. 3; CHEREPNINA, 1971, Plate 25, Fig. 1; ZHAVORONKOVA *in* Tyazheva and Zhavoronkova, 1972, p. 40, Plate 19, Fig. 4, 5; Plate 20, Fig. 1.

Neomphyma pseudofritchi SOSHKINA *in* Soshkina and Dobrolyubova, 1962, Plate 19, Fig. 1 (nomen nudum).

Spongophyllum originale Zhmaev. SHURYGINA, 1968, p. 134, Plate 59, Fig. 3; TSYGANKO, 1981, p. 50, Plate 14, Fig. 1; IVANIYA AND CHEREPNINA, 1986, p. 84, Plate 5, Fig. 11. Not *S. originale* ZHMAEV *in* Kraevskaya, 1955, p. 214, Plate 41, Fig. 3.

Spongophyllum pseudofritchi (Soshkina). GUO, 1976, p. 92, Plate 28, Fig. 3; GUO, 1978, p. 62, Plate 19, Fig. 1; YU AND LIAO, 1978, p. 259, Plate 3, Fig. 6.

Carlinastraea tuscaroraensis MERRIAM *in* Merriam and McKee, 1976, p. 33, Plate 6, Fig. 5–7; Plate 7, Fig. 1–4; OLIVER AND JOHNSON, 1977, p. 108.

Carlinastraea sp. cf. *tuscaroraensis* MERRIAM *in* Merriam and McKee, 1976, Plate 8, Fig. 1, 2.

? *Spongophyllum cyathophylloides* Etheridge. YU AND LIAO, 1978, p. 260, Plate 3, Fig. 4. Not *S. cyathophylloides* Etheridge, 1911 [see Jell and Hill, 1970, p. 102].

Carlinastraea sp. PEDDER *in* Jackson et al., 1978, p. 38, Plate 25, Fig. 8, 10; Plate 28, Fig. 1, 4.

Carlinastraea halysitoides (Etheridge). PEDDER, 1985a, p. 593, Plate 70.1, Fig. 24–33; LIAO AND ZHENG, 1986, p. 626, Plate 1, Fig. 2, 3.

Diagnosis. Species of *Carlinastraea* with mean adult corallite diameters of 2.9 to 6.0 mm, although most fall within the range of 3.8 to 4.2 mm. Septa greatly reduced, other than septal bases commonly absent from marginarium, not clearly differentiated into major and minor orders. Adult septal counts 22 to 32 at the intercorallite wall and zero to 16, but mostly no more than 12, around the tabularium. Presepiments large, in one to four, rarely five, rows. Tabulae mostly complete and concave; spaced 14 to 50, typically 22 to 32, per 10.0 mm of vertical distance.

Remarks. A fuller synonymy, including unillustrated references and additional negative synonymies, has been given by Pedder (1985a).

Broadly interpreted, as it is here, the species has the widest geographic distribution known for any Lochkovian rugose coral. The type occurrence is in one of the apparently lower Lochkovian (*woschmidti* Zone) limestones that comprise the Nemingha Member of the Drik Drik Formation, near Moonbi, on the Gamilaroi terrane of New South Wales. In Nevada, the species was found by Merriam in the Windmill Limestone, in the northern Simpson Park Range (USGS Locality M1384), as well as in the Upper Roberts Mountains Formation at the type locality of *Carlinastraea tuscaroraensis*, in the southern Tuscarora Mountains. Other known occurrences are as follows: limestone clasts in member 3 of the Gazelle Formation, Yreka terrane of northern California; lower Black Stuart Formation, Cariboo Mountains, Quesnel Lake map area, southern British Columbia; debris-flow beds on the slope from carbonate platform to basin shales of the Road River Formation, Royal Creek headwater area, Yukon Territory; Ovinparma Horizon and Siyak Beds on the western slopes of the northern and southern Ural Mountains; Saraynoy Horizon, eastern slope of the northern Ural Mountains; Tomchumysh Horizon and Remnev Beds in the Salair and Gornoy Altai regions of the Altay-Sayan Caledonian fold belt; upper Kunzhak and Shishkat horizons of northern Zeravshan and southern Turkestan Ranges of the Tien Shan Hercynian fold belt; upper Qaganhebu Formation, Bateaobao area, in the Nei Mongolian Hercynian fold belt; Shanjiang Formation, Lijiang County, northwestern Yunnan, between the Maniganggo and Litang sutures, in the Shaluli Shan arc of authors; Erhtaokou Formation, Jilin, North China platform; lower Mullamuddy Formation, Queens Pinch area, in the Lachlan fold belt of New South Wales. Excepting the Nei Mongolian and Jilin occurrences, which are possibly Pridolian (Li et al., 1983; Wang and Li, 1986), all of the certain occurrences of *Carlinastraea halysitoides*, including those in the clasts in the Gazelle Formation, are Lochkovian on evidence of other faunal groups.

We have not recollected the species from Nevada, and the precise ages of the two known occurrences in the state are in doubt. At the Boulder Creek locality, the *Carlinastraea* beds are above the *uniformis* graptolite zone, and on the west side of Coal Canyon, they are apparently low in the *Quadrithyris* Zone depositional breccia and probably in some upper part of the *trigonicus* conodont zone.

Carlinastraea ampla sp. nov.

Figure 11: 1–8

Material. Holotype, USNM 488411, unrecorded level in the Lochkovian part of the McMonnigal Limestone; midway between sections TO I and IK IV (Murphy, 1989b, Fig. 4), about 720 m west-southwest of the mouth of Ikes Canyon, Wildcat Peak Quadrangle, Toquima Range. Paratype, USNM 488412, unrecorded level in the Lochkovian

Figure 10 (on facing page). 1, 2: *Briantia repleta* Barrois, transverse and longitudinal sections, holotype, Faculté Libre des Science d'Angers, Davy Collection 588, upper Pragian or lower Emsian, Calcaire d'Erbray, Chateau-Briant, Lower Loire, France. 3, 4, 9: *Lyrielasma* sp. B, two transverse sections and a longitudinal section, USNM 488427. 5–7: *Neaxon* sp. A, exteriors of acid-etched specimens; 5, USNM 488410; 6, USNM 488409; 7, USNM 488408. 8: *Toquimaphyllum spongophylloides* (Foerste), part of a transverse section, approximate topotype, Ludlow Series, Hume Limestone Member of the Silverdale Formation, Hatton's Corner, near Yass, New South Wales, Australia.

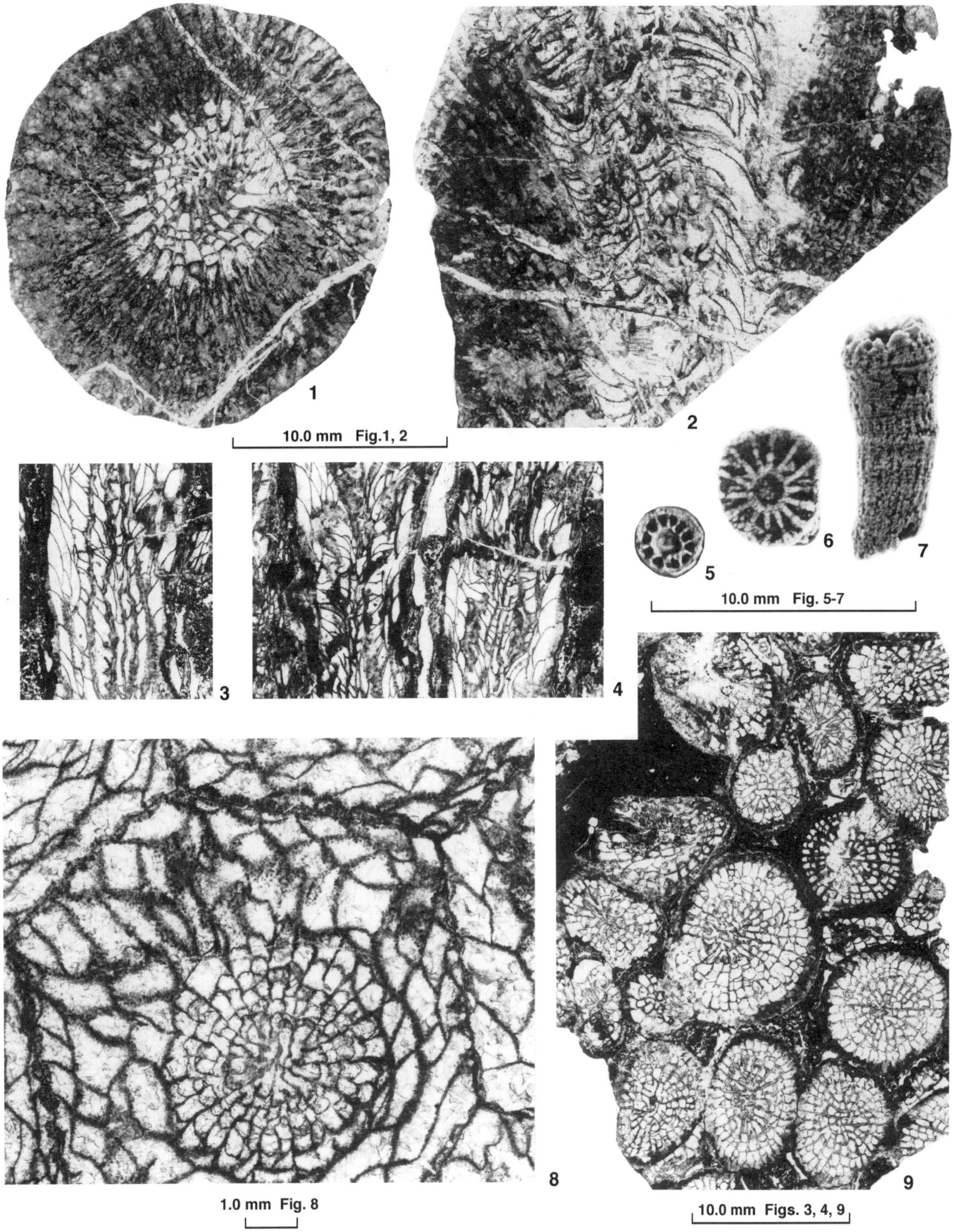
1
2
10.0 mm Fig.1, 2
3
4
5
6
7
10.0 mm Fig. 5-7
8
1.0 mm Fig. 8
9
10.0 mm Figs. 3, 4, 9

part of the McMonnigal Limestone; about halfway up to crest of ridge (lower of two outcrops on ridge) opposite cabin on south side of Ikes Canyon, Wildcat Peak Quadrangle, Toquima Range.

Description. Both types, although incomplete, have typical cerioid spongophyllinid coralla with many prismatic corallites. Increase marginarial and nonparricidal. Offsets develop slowly, but intercorallite walls separate daughter and parent corallites early, when offset diameter is less than 2.0 mm. Mean diameter of adult corallites 9.0 to 11.5 mm in holotype, 8.0 to 10.5 mm in paratype.

Intercorallite walls moniliform, comprised of wedge-shaped septal bases that are delicately denticulate along their adaxial edge; locally, the adaxial edge grades into a very short, attenuate, lamellar septum. A few similarly short and attenuate septa occur as crests on the more marginal presepiments. Septal arrangement radial or more commonly weakly bilateral in the tabularium. Septa irregularly differentiated into two orders in the inner marginarium and tabularium; undifferentiated at the corallite wall. Major septa thin, smooth, straight, or only slightly wavy; almost all occur as crests on presepiments, and most extend to the axial or periaxial region. Minor septa occur exclusively as short crests and are fewer than the major septa. Number of major septa in adult tabularia 14 to 22; number of undifferentiated septal bases in mature corallites 40 to 44. Microstructure very imperfectly preserved. Septal bases appear to be trabeculate, whereas septal crests appear to be nontrabeculate.

Adult marginaria include four, less commonly three, five, or six, rows of large presepiments and a few dissepiments. Boundary between marginarium and tabularium sharp. Most tabulae appear complete in sections cut across, or close to, the short axis of a tabularium that is elliptical in transverse section and incomplete in other longitudinal sections. Tabularial surfaces variably shaped but commonly appear concave in sections transecting the narrower parts of the tabularium. Tabulae spaced 15 to 30, usually 18 to 24, per 10.0 mm of vertical distance.

Remarks. Although the number of rows of presepiments is similar in *Carlinastraea halysitoides*, Etheridge's species is much smaller (typical mean adult corallite diameter 3.8 to 4.2 mm in *C. halysitoides*; 8.0 to 11.5 mm in *C. ampla*) and has fewer septal crests and bases (22 to 32 septal bases in *C. halysitoides*; 40 to 44 in *C. ampla*). A Lochkovian coral from the western slope of the southern Ural Mountains, named *Neomphyma giganteum* (Shurygina) by Zhavoronkova (*in* Tyazheva and Zhavoronkova, 1972, p. 41, Plate 20, Fig. 2) is similar to *Carlinastraea ampla* and may be conspecific with it. However, Shurygina's trivial name *giganteum* was originally published in combination with the genus *Spongophyllum* (Shurygina, 1968, p. 135, Plate 59, Fig. 1, 2), making it a junior primary homonym of *S. giganteum* Etheridge (1899, p. 158, Plate 20, Fig. 1–3; Plate 38, Fig. 3). *S. giganteum* Etheridge is a common Eifelian species in the Moore Creek and correlative limestones of the Gamilaroi terrane of New South Wales (Jones, 1932, p. 54, Text-Fig. 1; Hill, 1942, p. 161, Plate 4, Fig. 15; Pedder et al., 1970a, Plate 23, Fig. 1, 2) and cannot be dismissed as a forgotten name.

No coral resembling *Carlinastraea ampla* has been found in either of the comparatively well-known lower or upper Lochkovian faunas of central Nevada. In the range chart of Figure 2, *C. ampla* is placed questionably in the middle Lochkovian with other poorly dated species from the Toquima Range, although there are no independent data to support this very tentative age assessment.

Genus *Neomphyma* Soshkina, 1937

Neomphyma SOSHKINA, 1937, p. 76, 98.

Type species. Neomphyma originatum Soshkina, 1937, p. 77, 98, Plate 15, Fig. 3, 4. "Upper Ludlow"; road between Petropavlovskoe Works and the village of Pokrovskoe, Severoural′sk region on the eastern slope of the northern Ural Mountains. The type stratum is Lochkovian (Shurygina, 1968, p. 120, 163; Ivanovskiy and Shurygina, 1975, p. 34). In 1968 it was assigned to the Middle Petropavlovsk Suite, but it is now considered to be part of the Saraynoy Horizon (Shurygina, 1977, p. 44). Soshkina's illustrations are silhouette figures. A photograph of the transverse section of the holotype corallite was published by Ivanovskiy and Shurygina (1975, Plate 16, Fig. 4) and Ivanovskiy (1976, Plate 8, Fig. 3) and was reproduced by Hill (1981, Fig. 150,1b).

Diagnosis. Corallum fasciculate with subcylindrical corallites; increase marginarial, known to be both parricidal and nonparricidal. Septa nontrabeculate, attenuate, smooth-sided, except for some crenulations, inside a variable but commonly prominent, trabeculate septal stereozone. Major septa long, variably disrupted by presepiments, and commonly somewhat bilaterally arranged about the axis. Minor septa very short, locally confined to the septal stereozone. Marginarium dominated by large elongate dissepiments and presepiments. Tabularium well differentiated from marginarium, commonly elliptical in plan view, appearing very narrow in some longitudinal sections.

Remarks. Neomphyma is a characteristic genus in numerous shallow-water Lochkovian faunas, especially on the western North American craton and both slopes of the Ural Mountains and in the Altay-Sayan and southern Tien-Shan fold belts.

Ivanovskiy (*in* Ivanovskiy and Shurygina, 1975, p. 34) thought *Neomphyma* to be no more than a subgenus of *Spongophyllum*, but most workers believe that its fasciculate growth form distinguishes it from *Spongophyllum*. The trabeculate stereozone of *Neomphyma* is possibly another distinguishing feature. Hill (1981, p. F240) suggested that *Paleogrypophyllum* Ivaniya et al. (1968, p. 88) may be synonymous with *Neomphyma*. *P. spiraleforme* Ivaniya et al. (1968, p. 88, Plate 7, Fig. 31–34), the type species of *Paleogrypophyllum*, is based on two corals of probable Lochkovian age from the Remnev Suite of Northern Altay, in the Russian part of the Altay-Sayan fold belt. They have never been adequately described or illustrated, and the genus was overlooked by Spasskiy (1977) and Ivaniya and Cherepnina (1986). One of the poor figures seems to support the report in the original description of the presence of contratingent minor septa. If these are, in fact, present, *Paleogrypophyllum* is unlikely to be synonymous with *Neomphyma*.

Neomphyma karpinskiense Shurygina, 1968

Figures 12: 12, 13, 15: 21: 1, 2

Neomphyma karpinskiense SHURYGINA, 1968, p. 137, Plate 60, Fig. 1; Plate 61, Fig. 1; SHURYGINA, 1977, p. 44.
Kyphophyllum sp. t MERRIAM *in* Merriam and McKee, 1976, p. 36, Plate 10, Fig. 14–17.
Neomphyma sp. t (Merriam). OLIVER AND JOHNSON, 1977, p. 108.

Remarks. The holotype and only other specimen figured from the type area come from Lochkovian reef sediments on Gornovaya River, west of Karpinsk, on the eastern slope of the central Ural Mountains. In 1968 the type horizon was assigned to the Middle Petropavlovsk Suite; now it is referred to the Saraynoy Horizon. Merriam's specimen, first identified as *Kyphophyllum* sp. t, is from reef facies in the lower Tor Limestone of the June Canyon plate exposed in the Toquima Range (USGS Locality M1397) and is likely to be upper *delta* or lower *pesavis* Zone age in the old Cordilleran conodont zonation (Murphy and Anderson, 1991, p. 303). Our figured specimen (USNM 488413) is from the *Quadrithyris* Zone depositional breccia in the upper Windmill Limestone on the ridge leading from the east side of the entrance to Coal Can-

Figure 11 (on facing page). 1–8: *Carlinastraea ampla* sp. nov.; 1, 2, 4, 6, 8, parts of two longitudinal sections, two longitudinal sections, and a transverse section, holotype, USNM 488411; 3, 5, 7, two longitudinal sections and a transverse section, USNM 488412.

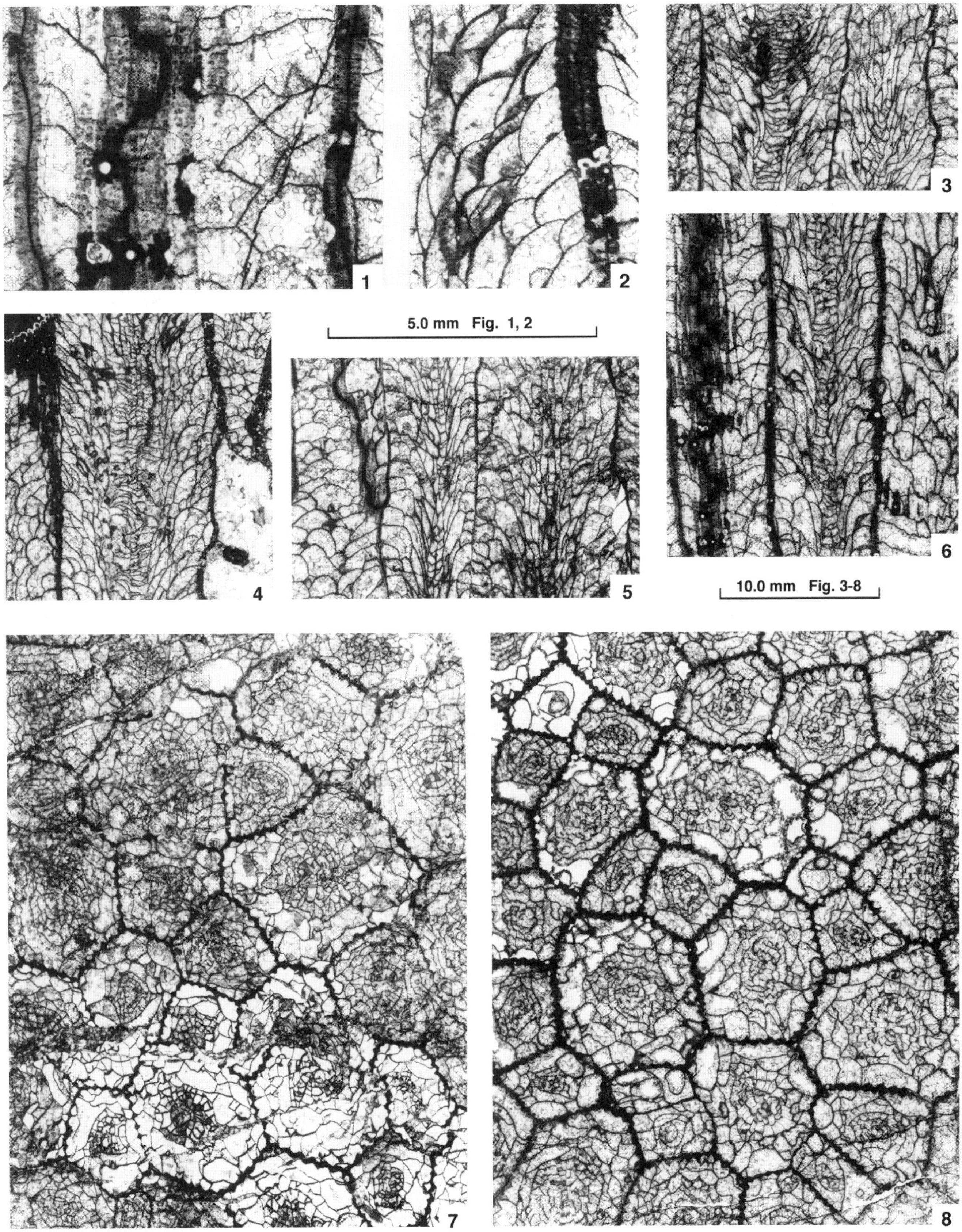
1
2
3
5.0 mm Fig. 1, 2
4
5
6
10.0 mm Fig. 3-8
7
8

yon to summit 6909 ("Pyramid Hill" of Merriam and McKee, 1976, Fig. 2), SE 1/4 sec. 17, T25N, R49E, Horse Creek Valley Quadrangle, northern Simpson Park Range.

Maximum mean corallite diameters in specimens of this very typical species of *Neomphyma* vary from 8.0 to 9.5 mm. Large corallites have 19 × 2 or 20 × 2 septa. The two orders of septa are highly differentiated, and the tabularium is markedly elliptical in transverse section.

Neomphyma nevadense (Merriam, 1974)

Kyphophyllum sp. MERRIAM, 1972, p. 37, Plate 8, Fig. 6.
Kyphophyllum nevadense MERRIAM, 1974a, p. 47, Plate 13, Fig. 1–4; MERRIAM *in* Merriam and McKee, 1976, p. 5, Plate 9, Fig. 1–3.
Neomphyma nevadense (Merriam). PEDDER, 1975, p. 285; JOHNSON AND OLIVER, 1977, p. 1464; OLIVER AND JOHNSON, 1977, p. 110.

Remarks. This species is not typical of *Neomphyma* because of its locally zigzag carinate and probably also flanged and trabeculate septa and is only provisionally retained in the genus. One complete corallum (USNM 159425) and a few fragments only have been collected. All are from the approximately middle Lochkovian part of the McMonnigal Limestone, 200 m northwest of summit 8474 (Copper Mountain, USGS Locality M1114) near Ikes Canyon, Toquima Range.

Neomphyma sp. A
Figure 12: 1, 5, 10, 14

Remarks. One of the two figured specimens of this species is USNM 488414, from the *hesperius* Zone in the Roberts Mountains Formation type section on the northwest side of Roberts Creek Mountain (Berry and Murphy, 1975, Fig. 3). The other is USNM 488415, a loose specimen from the *Quadrithyris* Zone depositional breccia in the Windmill Formation on the east side of Coal Canyon, close to the upper part of SP VII section of Berry and Murphy (1975, Fig. 14), northern Simpson Park Range. Both specimens are badly preserved as a result of silicification.

The species has a narrow septal stereozone; adult corallites have mean diameters of 12.0 to 16.0 mm, 22 to 26 major septa, and three to 10 rows of large elongate dissepiments and presepiments. Apart from its narrow septal stereozone, *Neomphyma* sp. A is typical of the genus, although it does not fall within the morphologic range of any previously described species. It is only slightly larger and has only a few more septa than a similar middle Lochkovian coral from the Royal Creek headwaters area of Yukon Territory, illustrated by Pedder (*in* Jackson et al., 1978, Plate 29, Fig. 1, 4) as *Neomphyma* sp. undet. Should the two forms prove synonymous, the species would be better founded on the comparatively well preserved material from Royal Creek.

Neomphyma sp. B

Kyphophyllum sp. b MERRIAM *in* Merriam and McKee, 1976, p. 36, Plate 5, Fig. 15–21.
Neomphyma sp. b Merriam. OLIVER AND JOHNSON, 1977, p. 108.

Remarks. All of the figured material assigned to this species by Merriam came from lower Lochkovian beds, dated by *Monograptus uniformis* and *M. angustidens*, on Boulder Creek (USGS "Bootstrap Hill" localities M1314, M1315) in the southern Tuscarora Mountains. Unfigured material, identified as *Kyphophyllum?* cf. *K.* sp. b by Merriam (*in* Merriam and McKee, 1976, p. 17), occurs with *Carlinastraea halysitoides* higher in the same section (USGS Locality M1313). The higher collection is likely to be about middle Lochkovian age. *Neomphyma* sp. B is a small species with a narrow septal stereozone, mean adult corallite diameters of about 7.5 mm, and about 20 long major septa disrupted by only a few presepiments. Although the species is small, adult corallites appear to have on the order of six rows of dissepiments.

Genus *Tonkinaria* Merriam, 1974

Tonkinaria MERRIAM, 1974a, p. 51.

Type species. Tonkinaria simpsoni Merriam, 1974. See below.

Diagnosis. Solitary to weakly colonial spongophyllinid coral. Offsets in colonial forms only very rarely attain the size and shape of the protocorallite. Protocorallite with a ceratoid proximal part, passing upward through a turbinate stage to a patellate distal part. Offsets parricidal, budding from the marginal part of the flared calice. Septa in ceratoid part of corallite, clearly differentiated, with both orders of septa at least partly in the form of septal crests; minor septa mostly spinose. Septa undifferentiated and naotic over the middle and outer parts of a fully flared calice. Elongate dissepiments and presepiments present in ceratoid part of corallite; large but little inflated presepiments present on flared calice. Tabulae moderately closely spaced, forming depressed tabularial surfaces.

Remarks. Specimens of *Tonkinaria simpsoni* from the type locality occur in a dolomitized lime grainstone-packstone comprising ooids and subordinate peloids, corals, and pelmatozoan fragments. The corals are silicified and distorted by solution and other fractures. Septa in these poorly preserved specimens of *T. simpsoni* appear to have fine ptenophyllinid lobes, but this is not certain. A few specimens, one figured by Merriam (1974a, Plate 7, Fig. 1, 2) and Merriam and McKee (1976, Plate 5, Fig. 1, 2), have offsets appearing from below the expanded calice. It is not known whether these are adventitious or not.

Tonkinaria simpsoni Merriam, 1974
Figure 12: 2–4, 6–9, 11

? *Entelophyllum.* DUNCAN, 1956, p. 221 (in part), Plate 23, Fig. 5a, b only.
? *Mycophyllum* sp. MERRIAM, 1963, p. 38.
Tonkinaria simpsoni MERRIAM, 1974a, p. 52, Plate 7, Fig. 1–8; MERRIAM *in* Merriam and McKee, 1976, Plate 5, Fig. 1–14; JOHNSON AND OLIVER, 1977, p. 1464; HILL, 1981, p. F243, Fig. 150, 3a-d.
? *Tonkinaria* cf. *T. simpsoni* MERRIAM, 1974a, Plate 7, Fig. 9.

Remarks. Our new specimens (USNM 488416–488419 and unregistered material) are believed to be exact topotypes. Details of their occurrence are the same as those given earlier for the holotype of *Stylopleura berthiaumi*, which has the same type stratum and locality as *T. simpsoni.*

Merriam published illustrations of only four thin sections, and his description of the species was brief. The following comments are offered as a supplement to Merriam's description. No more than one generation of calicinal offsets has been observed in the type population. The maximum number of offsets from a single protocorallite is three. Maximum diameter of the ceratoid part of corallites varies from 10.0 to 14.0 mm, maximum diameter of expanded calice is 33.0 mm, and the maximum height of the protocorallite is estimated to be about 30.0 mm. Rare rejuvenescences of the ceratoid part of the corallite may follow a modest expansion below the ultimate full expansion of the calice. Some specimens have numerous fine radiciform processes. The larger transverse section illustrated by Merriam (1974a, Plate 7, Fig. 5) and Mer-

Figure 12 (on facing page). 1, 5, 10, 14: *Neomphyma* sp. A; 1, 5, transverse and longitudinal sections, USNM 488414; 10, 14, longitudinal and transverse sections of very poorly preserved specimen, USNM 488415. 2–4, 6–9, 11: *Tonkinaria simpsoni* Merriam, silicified material; 2, longitudinal section, USNM 488416; 3, 8, 9, transverse sections, USNM 488417; 4, 6, longitudinal and transverse sections, USNM 488418; 7, 11, longitudinal and transverse sections, USNM 488419. 12, 13, 15: *Neomphyma karpinskiense* Shurygina, two longitudinal sections and a transverse section, USNM 488413. See also Figure 21: 1, 2.

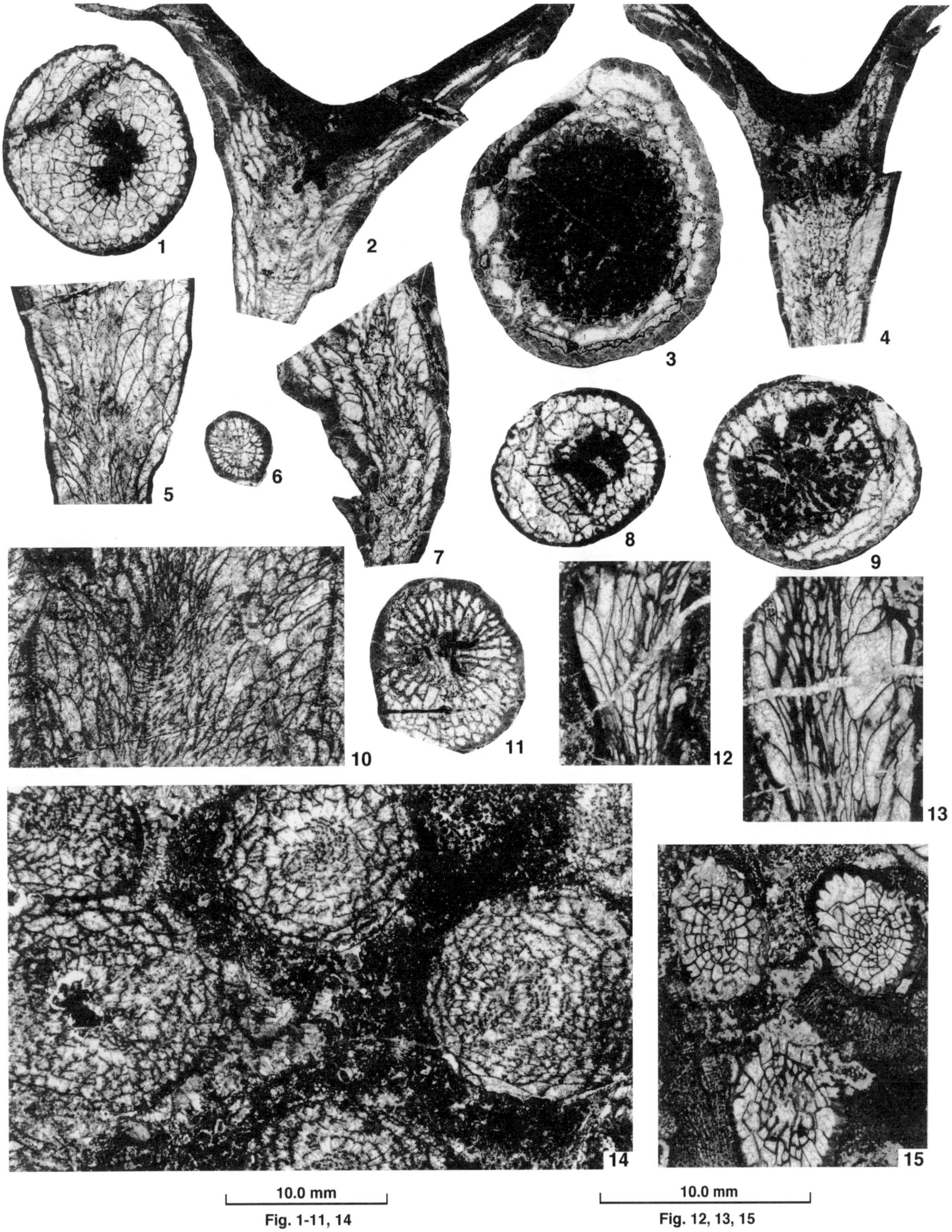
1
2
3
4
5
6
7
8
9
10
11
12
13
14
15
10.0 mm
Fig. 1-11, 14
10.0 mm
Fig. 12, 13, 15

riam and McKee (1976, Plate 5, Fig. 3) and reproduced by Hill (1981, Fig. 150,3c) is atypical—presepiments and septal crests are normally present at this diameter (10.8 mm). The naotic septa of the calice, well shown in several of Merriam's illustrations, are developed over large elongate presepiments, the outermost ones of which apparently extend to the rim of the calice.

Merriam (1974a, p. 51) believed two drawings of corals identified by Duncan (1956) as *Entelophyllum*, "from the Roberts Mountains Formation of Nevada," were possibly based on specimens of *Tonkinaria simpsoni*. This cannot be confirmed from the illustrations themselves. Merriam's 1963 identification of abundant *Mucophyllum* from approximately the same stratigraphic level and locality as the holotype of *T. simpsoni*, almost certainly referred to *T. simpsoni*. *Mucophyllum* has not been seen or mentioned from this locality since. On the other hand, *T. simpsoni* which resembles a small *Mucophyllum* in exterior view, is abundant at the locality.

In addition to the type occurrence in the *hesperius* Zone on Roberts Creek Mountain, Merriam and McKee (1976, Plate 5, Fig. 10–14) illustrated exteriors of acid-etched specimens of *Tonkinaria simpsoni* from probable middle Lochkovian beds on the west side of Coal Canyon (USGS Locality M1383). The species may also occur in the lower part of the Hidden Valley Dolomite in the Funeral Mountains of California (USGS Locality M1097), based on a specimen identified by Merriam (1974a, Plate 7, Fig. 9). We have no independent dating for this specimen.

Subfamily PTENOPHYLLINAE Wedekind, 1923

Diagnosis. Spongophyllidae mostly with medium to very large size corallites and broad marginaria. Calice deep, with base well below the upper edges of the periaxial parts of major septa. Septa trabeculate in marginarium; poorly aligned trabeculae locally form flanges, and vepreculae may also form in the inner dissepimentarium and outer tabularium. Dissepiments outnumber presepiments in most genera and species and normally surround all tabularia. True naotic septa present in several genera. Tabulae typically much disrupted by major septa and their flanges, closely spaced in all genera.

Genus *Aphroidophyllum* Lenz, 1961

Aphroidophyllum LENZ, 1961, p. 505.

Type species. Aphroidophyllum howelli Lenz, 1961, p. 505, Plate 3, Fig. 1, 2. Lenz recorded that the species was known from only two widely separated occurrences of the Hume Formation in the Lower Mackenzie Valley. The original label accompanying the holotype states that the specimen, which has also been figured by Pedder (1971, Plate 9, Fig. 1) and Hill (1981, Fig. 147.1a, b), comes from the upper member of the Hume Formation at Lac à Jacques, co-ordinates 66° 05'N, 127° 28'W. The type stratum is late Eifelian age.

Diagnosis. See Pedder (1971, p. 46) but replace the term aphroid with amural, a term introduced by Scrutton (1988, p. 68).

Remarks. As used here, the genus is unlikely to be monophyletic.

Aphroidophyllum inceptor sp. nov.

Figure 13: 1–4

Aphroidophyllum sp. PEDDER, 1975, p. 285, Fig. 56, 57.

Material. Holotype, USNM 219172, *Quadrithyris* Zone depositional breccia, Windmill Limestone, east side of Coal Canyon, approximately 300 m south of entrance, close to the upper part of section SP V11 (Berry and Murphy, 1975, Fig. 14), SE 1/4 sec. 17, T25N, R49E, Horse Creek valley Quadrangle, northern Simpson Park Range.

Description. Massive amural corallum, shape and dimensions unknown, but before sectioning comprised more than 25 corallites. Axes of adjacent adult corallites mostly 15 to 25 mm apart.

Septal structure in the tabularium different from that in the middle and outer marginarium. In the tabularium and innermost marginarium septa radially arranged and well differentiated into two orders. Major septa commonly interrupted by presepiments in outermost tabularium and innermost marginarium, and uninterrupted throughout most of the tabularium; major septa long, many with ptenophyllinid flanges in periaxial region. Minor septa short, discontinuous in vertical sense, confined to the peripheral region of the tabularium. Septal counts in adult corallites 19 × 2 to 23 × 2. Septal structure in marginarium between adjacent tabularia consists of very short trabeculae, of unknown fine structure, embedded in thin layers of sclerenchyme overlying presepiments. Ill-defined naotic septa present on some marginarial surfaces.

Intercorallite walls absent. Presepiments abundant and, relative to other features of the coral, large. Surfaces of marginarium broadly concave midway between adjacent tabularia and strongly convex immediately outside the tabularia. Tabulae well differentiated from innermost presepiments and rare dissepiments, closely spaced; as many as 45 occur over a vertical distance of 10.0 mm. Tabularial surfaces generally slightly depressed around the outer parts of tabularia, and somewhat elevated and disrupted by septa and septal flanges in the periaxial region.

Remarks. The type and other Emsian and Eifelian species of *Aphroidophyllum* are not similar to *A. inceptor* and were likely derived from species of *Taimyrophyllum* or *Pseudochonophyllum*. No Pragian or early Emsian species of *Aphroidophyllum* is known. The Lochkovian *A. inceptor* probably evolved from a species of *Toquimaphyllum* by loss of intercorallite walls and is not likely to be ancestral to any of the species previously referred to *Aphroidophyllum*.

Derivation of name. Latin, *inceptor*, meaning a beginner.

Genus *Embolophyllum* Pedder, 1967

Embolophyllum PEDDER, 1967, p. 10.
Hezhangophyllum KONG *in* Kong and Huang, 1978, p. 124.

Type species. Acanthophyllum asper Hill, 1940c, p. 252, Plate 9, Fig, 3, ?4. "Lower Middle Devonian"; first karst on Cave Flat road from Wee Jasper, New South Wales. Additional photographs of the holotype have been published by Strusz (1966, Plate 85, Fig. 2), and the species has been revised by Pedder (1967, p. 11, Text-Fig. 5, 6; Plate 2, Fig. 11, 13–15; Plate 3, Fig. 6, 9; Plate 4, Fig. 1, 6). The type occurrence is probably in a middle Emsian stratum, 4.5 to 18.0 m above the base of the Taemas Limestone, between Wee Jasper and section 1 of Pedder et al. (1970b, Text–Fig. 1).

Diagnosis. Solitary to weakly colonial Ptenophyllinae. Corallites ceratoid to trochoid. Offsets budded parricidally or nonparricidally. Septal bases typically expanded to wedge-shape, forming a variable, usually thin, septal stereozone. Septa mostly uninterrupted and consisting of well-aligned trabeculae, throughout the dissepimentarium. Major septa extend to periaxial region and, in some species, are weakly vepreculate near the boundary between dissepimentarium and tabularium. Major septa flanged in most species, commonly strongly so, in the tabularium. Dissepimentarium with numerous, predominantly inwardly sloping dissepiments; presepiments rare, confined to outer marginarium.

Remarks. Hezhangophyllum is based on limited material of a single species, named *H. variseptatum* Kong (*in* Kong and Huang, 1978, p. 125, Plate 40, Fig. 11), from the Emsian Pochiao Formation at Tiekuangshan, on the Yangzi Plate near Hezhang, Guizhou Province, southern China. Kong was aware of the close similarity between *Hezhangophyllum* and *Embolophyllum*, but proposed *Hezhangophyllum* because of the presence of an inner wall and presepiments in *H. variseptatum*. However, presepiments are developed to almost the same extent in *Embolophyllum asper* (Pedder, 1967, Text-Fig. 5b; Plate 4, Fig. 6), and it is most doubtful whether a variable inner wall of sclerenchymal origin, such as the one in *H. variseptatum*, is sufficient to justify a separate genus.

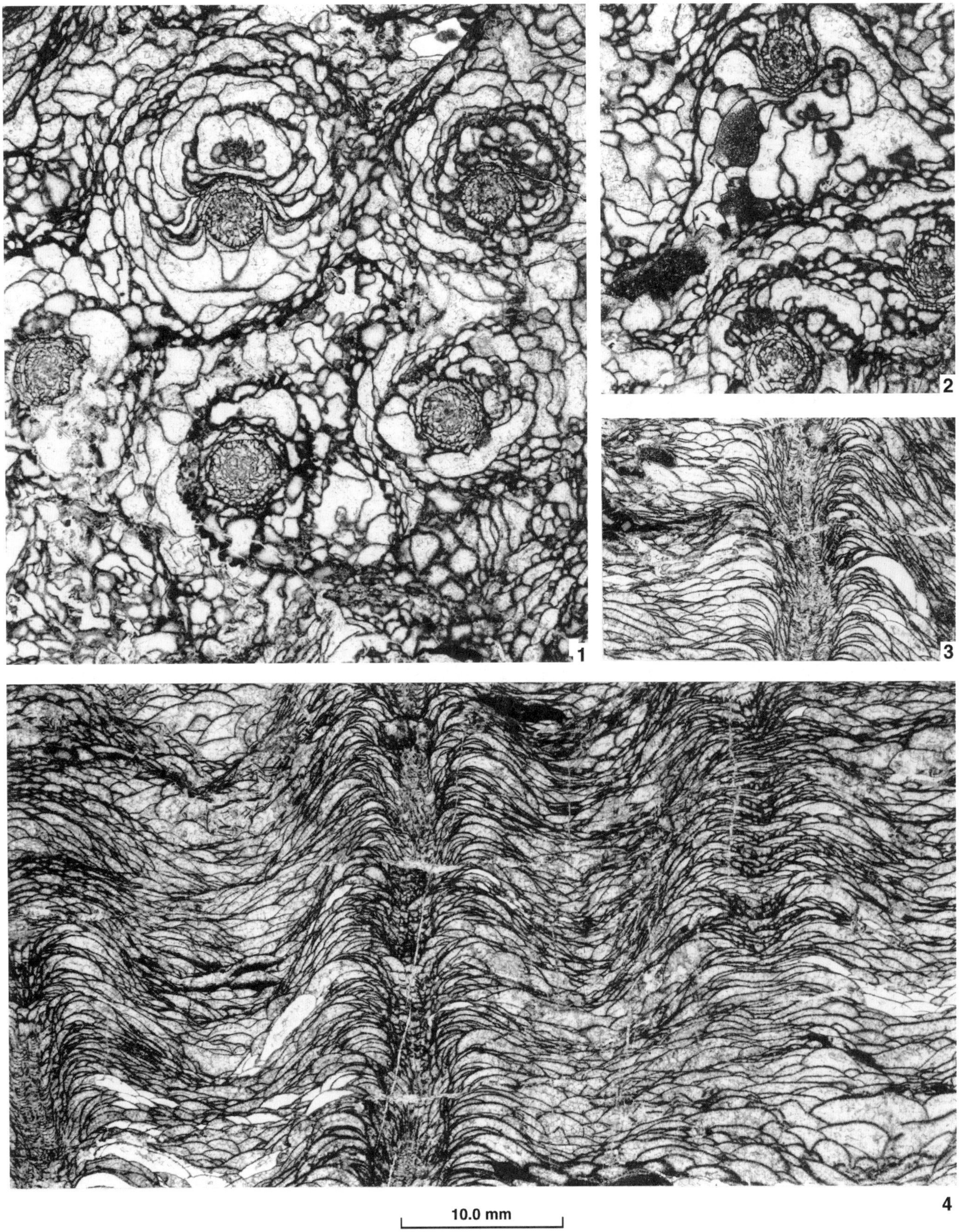

Figure 13. 1–4: *Aphroidophylum inceptor* sp. nov., two transverse and two longitudinal sections, holotype, USNM 219172.

Embolophyllum carbonense sp. nov.
Figure 14: 1–14

Cyathactis? sp. MERRIAM, 1974a, p. 16, 30, Plate 8, Fig. 5, 6.
? *Chonophyllum* sp. MERRIAM, 1974a, Plate 8, Fig. 8, 11, 12 only. Not Plate 8, Fig. 9, 10 (unrecognizable).
Neostringophyllum sp. broad sense. PEDDER, 1975, p. 285; JOHNSON AND OLIVER, 1977, p. 1464; OLIVER AND JOHNSON, 1977, p. 110.

Material. Holotype, USNM 488420, and paratype, USNM 488421, *Quadrithyris* Zone depositional breccia, ridge leading from entrance of Coal Canyon to summit 6909 ("Pyramid Hill" of Merriam and McKee, 1976, Fig. 2) on east side of Coal Canyon. Paratype, USNM 488422, 2.0 to 3.5 m below top of *Quadrithyris* Zone depositional breccia, east side of Coal Canyon, approximately 150 m south of entrance. Paratype, USNM 488423, as above but 0 to 2.0 m below top of *Quadrithyris* Zone depositional breccia. Paratype, specimen figured by Merriam (1974a, Plate 8, Fig. 5), *Quadrithyris* Zone depositional breccia ("Silurian coral-rich limestone breccia"), east side of Coal Canyon, near entrance (USGS Loc. M1106). Paratype, USNM 488424, as above but from a talus block. Paratype, USNM 488425, unrecorded level in *Quadrithyris* Zone depositional breccia, east side of Coal Canyon, approximately 300 m south of entrance. All of the above material is from the upper Lochkovian Windmill Limestone, SE 1/4 sec. 17, T25N, R49E, Horse Creek Valley Quadrangle, northern Simpson Park Range. Paratype, USNM 488426, 4.0 to 4.3 m above base of the *Quadrithyris* Zone depositional breccia in the Lochkovian Windmill Limestone, section SP I (Berry and Murphy, 1975, Fig. 14), west side of Coal Canyon Fault, NW 1/4 NE 1/4 sec. 20, T25N, R49E, Horse Creek Valley Quadrangle, northern Simpson Park Range.

Description. Corallum solitary, ceratoid to trochoid, maximum mean diameter 26.0 mm, but typically less than 20.0 mm below the calice. Rejuvenescences and radiciform processes both common. Calice deep, distance from rim to upper edges of septa in periaxial region of mature coralla about 14 mm; distance from rim to uppermost tabularial surface of mature coralla about 18 or 19 mm. Upper part of calicular wall steeply inclined; lower part vertical, or nearly so.

Septal bases expanded to form a narrow septal stereozone, 0.02 to 1.0 mm wide at thinnest points between adjacent septa. Septa in two orders, but essentially undifferentiated in the dissepimentarium. Septal arrangement radial with slight periaxial rotation or weakly bilateral about an elongate cardinal septum. Septa moderately thick in the dissepimentarium, commonly a little crenulate and only slightly carinate due to irregular alignment of trabeculae; only locally in a very few specimens are they as carinate as depicted in Figure 14: 11. Major septa extend unequally to the periaxial region and are lobed in the normal ptenophyllinid manner in the tabularium. Minor septa project about 1.0 mm into the tabularium and calice. A few presepiments disrupt both orders of septa, especially adjacent to openings leading to radiciform processes. Septal counts in corallites of 9.0 to 15.5 mm diameter 26 × 2 to 28 × 2, and in corallites of 16.0 to 26.0 mm diameter 29 × 2 to 31 × 2.

Dissepiments moderately to mostly steeply inclined; at maturity, typically in five to 10 rows, maximum 16 rows. Presepiments rare, only slightly larger than the dissepiments and normally disrupting only one or two septa. Dissepiments and presepiments overhang openings to radiciform processes where a dissepimentarial surface forms a funnel leading into a radiciform process. Thin sclerenchyme invests a few dissepimentarial surfaces. Very short, unaligned trabeculae appear to be present on dissepimentarial surfaces inside some adult calices but are absent from surfaces of other dissepiments. Boundary between dissepimentarium and tabularium distinct. Tabulae incomplete, much disrupted by major septa, closely spaced, 20 to 40 counted over a vertical distance of 10.0 mm. Tabularial surfaces commonly inwardly sloping around the margin of the tabularium and irregularly elevated periaxially.

Remarks. Embolophyllum carbonense is not likely to be confused with any of the typical and generally younger Australian or Asian species of *Embolophyllum*, because of its solitary form, abundant radiciform processes, and locally unaligned trabeculae in the outer dissepimentarium. The latter feature is taken to indicate a phylogenetic relationship between *E. carbonense* and *Ecnomiophyllum merriami* n. gen. and sp., described elsewhere in this chapter. On purely morphological grounds, the nearest previously described species appears to be *Embolophyllum etheridgei* Pedder (Pedder et. al., 1970b, p. 223, Plate 42, Fig. 1–3, 5–10, 15–17) from upper Emsian strata near Wee Jasper, New South Wales. *E. etheridgei* is distinguished from the new species by its wider septal stereozone (up to 3.5 mm wide), fine vepreculae in the inner dissepimentarium, and consistently smooth septa in the outer dissepimentarium.

Derivation of name. Latin, *carbo, -onis*, meaning coal, and suffix *-ensis*, a reference to the species' occurrence in Coal Canyon.

Genus *Lyrielasma* Hill, 1939

Lyrielasma HILL, 1939, p. 243.
? *Salairophyllum* BESPROZVANNYKH, 1968, p. 111.

Type species. Cyathophyllum subcaespitosum Chapman, 1925, p. 112, Plate 13, Fig. 15–16b. "Silurian (Yeringian)"; Cave Hill, Lilydale, Victoria, Australia. Revisions of the species have been published by Hill (1939, p. 244, Plate 14, Fig. 1–6; Plate 15, Fig. 6, 7) and Pedder (1967, p. 5, text-fig. 2a, b, 3a–4, Plate 1, Fig. 1–14; Plate 2, Fig. 1–9). The holotype came from a Pragian olistolith exposed in Lilydale Quarry. *C. subcaespitosum* Chapman is a junior primary homonym of *C. subcaespitosum* Meek (1873, p. 470), a Carboniferous coral from Idaho, revised by Sando (1965, p. E11, Plate 4) as a species of *Dorlodotia* Salée. Chapman's trivial name *subcaespitosum* has been replaced by *chapmani* Pedder (1967, p. 5).

Diagnosis. See Hill, 1981, p. F238.

Remarks. Although specimens of *Lyrielasma chapmani* collected at the type locality are fragmentary corallites, they likely derived from fasciculate coralla. The holotype and paratype are almost certainly fasciculate, and a specimen figured by Philip (1962, Plate 28, Fig. 6, 7) is definitely so. Assuming *Lyrielasma* to be a colonial genus, *Salairophyllum*, based on *Pilophyllum angustum* Zheltonogova (1961, p. 78, Plate S-17, Fig. 5; Plate S-18, Fig. 1) from the Lochkovian Tomchumysh Beds of the Salair, is distinguished from *Lyrielasma* by its probable, but not certain, solitary form.

Hill (1981, p. F235) regarded *Acmophyllum* Sytova (1968, p. 60) as a possible synonym of *Lyrielasma*. However, besides being exceptionally small and certainly solitary, *A. armatum* Sytova (1968, p. 60, Plate 4, Fig. 7; Plate 5, Fig. 1), the type species from the Lochkovian Borschov Beds of Podolia, has a very narrow stereozone and minute dissepiments. In the author's opinion it is neither *Lyrielasma* nor *Spongophylloides* Meyer, where it was placed by Ivanovskiy (1976, p. 12). Any revision of *Acmophyllum* Sytova should take into account the previous use of the name by Karsch (1890, p. 21, 24) for an orthopteran insect.

Lyrielasma sp. B
Figure 10: 3, 4, 9

Remarks. This very typical species of *Lyrielasma* is represented by a fasciculate corallum (USNM 488427) from the *Quadrithyris* Zone

Figure 14 (on facing page). 1–14: *Embolophyllum carbonense* sp. nov.; 1, 2, side and calicular views of an acid-etched specimen, USNM 488424; 3, 4, 11, transverse section, part of a transverse section, and a longitudinal section, USNM 488422; 5, 9, 10, two transverse sections and a longitudinal section USNM 488426; 6, 7, transverse and longitudinal sections, USNM 488421; 8, 12–14, two transverse sections, part of a longitudinal section, and a longitudinal section, holotype, USNM 488420.

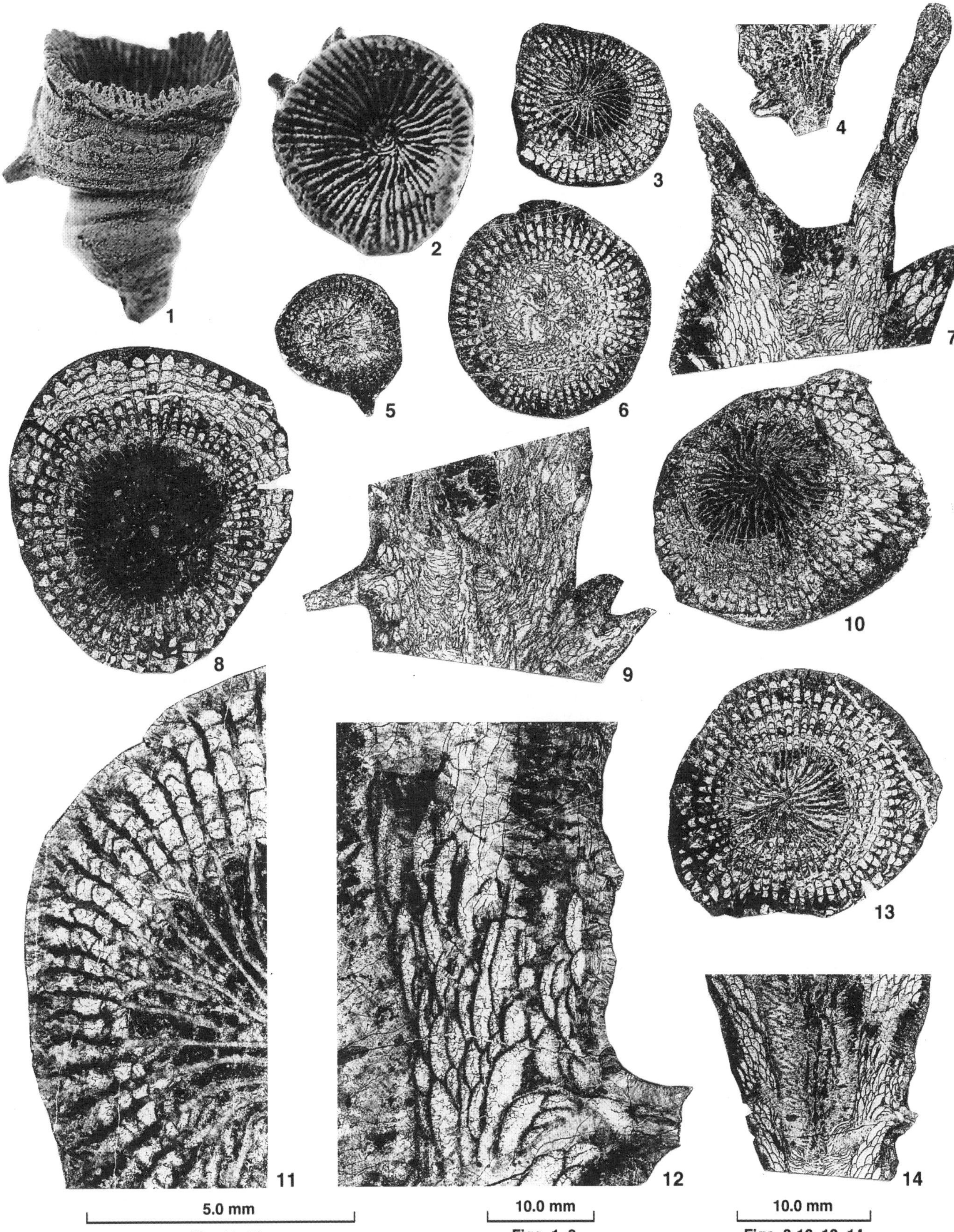
1
2
3
4
5
6
7
8
9
10
11
12
13
14
5.0 mm
Fig. 11, 12
10.0 mm
Figs. 1, 2
10.0 mm
Figs. 3-10, 13, 14

breccia in the Windmill Limestone on the ridge leading from the entrance of Coal Canyon to summit 6909 ("Pyramid Hill" of Merriam and McKee, 1976, Fig. 2). It is very similar to *L. chapmani* Pedder but differs in being slightly smaller, with a maximum mean corallite diameter of 10.5 mm (versus about 13.0 mm in *L. chapmani*) and by having a maximum of 22 × 2 septa (as many as 30 × 2 in *L. chapmani*) and a slightly narrower stereozone, although the width of the stereozone is variable in *L. chapmani.*

For the moment, the species is left in open nomenclature and is given the letter B to distinguish it from *Lyrielasma* sp. A, from the Givetian of western Canada, which is to be illustrated in publication by R. A. McLean.

Genus *Toquimaphyllum* Merriam, 1974

? *Kozlowiaphyllum* RUKHIN, 1938, p. 34.
Australophyllum (Toquimaphyllum) MERRIAM, 1974a, p. 54.

Type species. Australophyllum (Toquimaphyllum) johnsoni Merriam, 1974a, p. 55 (in part), Plate 11, Fig. 1, 2, 7 (only). "Probably in McMonnigal Limestone of Kay and Crawford (1964)," "Upper Silurian coral Zone E"; 201 m ("one-eighth of a mile") northwest of summit 8474 (Copper Mountain) at altitude of 8,300 ft, Ikes Canyon area, Dianas Punch Bowl Quadrangle, Toquima Range, Nevada. The McMonnigal Limestone in the Copper Mountain area, Mill Canyon Plate, ranges from Ludlow, *chimaera* Zone, to Pragian, *sulcatus* Zone (Murphy and Anderson, 1991, p. 301, Fig. 2).

Diagnosis. Cerioid Ptenophyllinae with presepiments of the outer marginarium periodically covered by short unaligned trabeculae, of unknown fine structure, embedded in thin sclerenchyme; presepiments and trabeculae not forming a naotic marginarium.

Remarks. The much needed revision of the many cerioid ptenophyllinid corals with presepiments is not attempted here. If all belong to one genus, as some believe to be the case, the appropriate name for the genus would be *Kozlowiaphyllum. Kozlowiaphyllum* is based on a single corallum, named *K. pentagonum* by Rukhin (1938, p. 34, 99, Plate 3, Fig. 9, 10), from the "upper Upper Silurian" on the right bank of Yasachnaya River below the mouth of Tarynakh River, upper Kolyma Basin, northeastern Russia. Rukhin's illustrations are good but are restricted to two figures of transverse sections of a few corallites. There are only a few small presepiments in each corallite, and no trabecula is visible. The holotype is lost (Ivanovskiy, 1976, p. 94), and the genus has been abandoned or overlooked by current Russian workers. This may be justified, although it is not possible to agree with Ivanovskiy's view that *Kozlowiaphyllum* is a junior synonym of *Spongophyllum.*

Spongophyllum cyathophylloides Etheridge (1911, p. 7, Plate A, Fig. 3; Plate C, Fig. 1, 2) from the Emsian Douglas Creek Limestone of Queensland, which is the type species of *Australophyllum* Stumm (1949, p. 34), has been revised most recently by Jell and Hill (1970, p. 102), and three additional topotypes have been studied for the present work. Longitudinal sections intersecting septal bases are commonly maculate, suggesting the presence of scattered trabeculae in the septal bases, such as were described by Jell and Hill. Unaligned trabeculae do not occur outside, nor do they project beyond, the thin, smooth but slightly wavy septa, and ptenophyllinid flanges are absent. Also, the species is considerably smaller, with mean corallite diameters mostly in the order of 6 or 7 mm, than the Nevada species referred to *Toquimaphyllum.*

A Ludlow-age coral occurring in the Bowspring and Hume limestone members of the Silverdale Formation in the Yass Syncline, New South Wales, was described by Foerste (1888, p. 131, 137, Plate 13, Fig. 16, 17), and given the provisional name—pending description of the calice—of *Endophyllum spongophylloides*. The coral was restudied by Jones (1932, p. 52, Plate 3, Fig. 3, 4) and Hill (1940b, p. 408, Plate 13, Fig. 3–5), in both works as *Spongophyllum spongophylloides* (Foerste). Merriam (1974a, p. 55) referred the species to *Toquimaphyllum*, evidently on account of its many presepiments. No previous author has drawn attention to the abundant, very short trabeculae scattered over many of the presepimentarial surfaces and in the variable septal bases of the species. These trabeculae, here illustrated in Figure 10: 8, together with the ptenophyllinid tabularium and septal flanges, support Merriam's assignment of the species to *Toquimaphyllum.*

Toquimaphyllum johnsoni Merriam, 1974

Australophyllum sp. E MERRIAM, 1972, p. 40, Plate 4, Fig. 1–3.
Australophyllum (Toquimaphyllum) johnsoni MERRIAM, 1974a, p. 55 (in part), Plate 11, Fig. 1, 2, 7 only (not Plate 11, Fig. 5, 6; ? Plate 15, Fig. 17 = *T. vulgare* sp. nov.); MERRIAM *in* Merriam and McKee, 1976, p. 35 (in part), Plate 6, Fig. 8–10.
Kozlowiaphyllum johnsoni (Merriam). OLIVER AND JOHNSON, 1977, p. 108.
Toquimaphyllum johnsoni Merriam. HILL, 1981, p. F243, Fig. 153.1a, b.
? *Australophyllum johnsoni* (Merriam). PEDDER in Potter et al., 1990, p. 67.
Not *Australophyllum (Toquimaphyllum)* cf. *johnsoni* MERRIAM, 1974a, Plate 11, Fig. 3, 4 (= *T. vulgare* sp. nov.), Plate 12, Fig. 1–3 (= ? *T. vulgare* sp. nov.).
Not *Kozlowiaphyllum johnsoni* (Merriam). PEDDER, 1975, p. 285 (= *T. vulgare* sp. nov.); nor ? JOHNSON AND OLIVER, 1977, p. 1464 (= ? *T. vulgare* sp. nov.).

Remarks. We have no additional material of this, the type species of *Toquimaphyllum*. The type occurrence is given under the genus *Toquimaphyllum*, and differences between *T. johnsoni* and *T. vulgare*, which previously has been included in it, are discussed under *T. vulgare*. The middle Lochkovian range attributed to *T. johnsoni* in Figure 2 is uncertain and is based on its occurrence above *Monograptus birchensis* as well as its absence from the large upper Lochkovian fauna at Coal Canyon, northern Simpson Park Range. Clearly, the range could be younger than it is tentatively shown to be in Figure 2.

The doubtful Pedder (in Potter et al., 1990) reference refers to a single fragmentary coral with some phaceloid corallites from a Lochkovian clast in member 3 of the Gazelle Formation, in the China Mountain Quadrangle, northern California.

Toquimaphyllum vulgare sp. nov.

Figures 15: 1–7, 16: 1–4

? *Spongophyllum?* sp. OLIVER *in* Johnson, 1965, p. 371; OLIVER *in* Johnson, 1970, p. 44.
? *Australophyllum* sp. v MERRIAM, 1973, p. 25, Plate 9, Fig. 7.
Australophyllum (Toquimaphyllum) johnsoni MERRIAM, 1974a, p. 55 (in part), Plate 11, Fig. 5, 6; ? Plate 15, Fig. 17 (only). Not Plate 11, Fig. 1, 2, 7 (= *T. johnsoni* Merriam).
Australophyllum (Toquimaphyllum) cf. *johnsoni* MERRIAM, 1974a, Plate 11, Fig. 3, 4; ? Plate 12, Fig. 1–3.
Australophyllum (Australophyllum) sp. v MERRIAM *in* Merriam and McKee, 1976, p. 35 (probably in part), Plate 8, Fig. 8, 9. Probably not Plate 8, Fig. 10, 11.
Kozlowiaphyllum johnsoni (Merriam). PEDDER, 1975, p. 285; ? JOHNSON AND OLIVER, 1977, p. 1464.

Material. Holotype, USNM 488428, and paratype, USNM 488429, *Quadrithyris* Zone depositional breccia, ridge leading from entrance of

Figure 15 (on facing page). 1–7: *Toquimaphyllum vulgare* sp. nov.; 1–3, transverse section, part of a longitudinal section, and a longitudinal section, holotype, USNM 488428; 4, 6, longitudinal and transverse sections, USNM 488429; 5, part of a transverse section, USNM 488432; 7, naturally weathered surface of part of a corallum, USNM 488433.

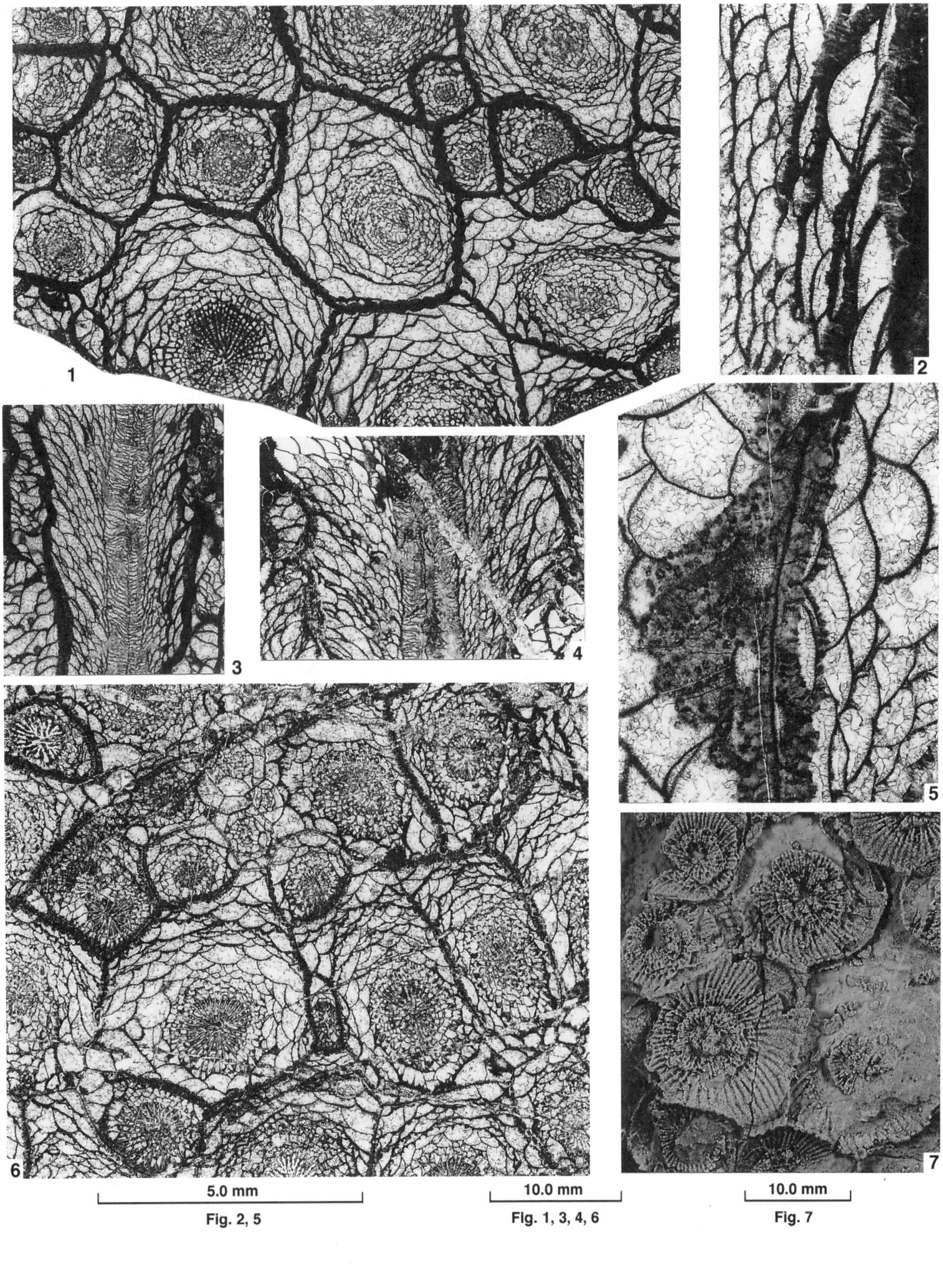

5.0 mm
Fig. 2, 5

10.0 mm
Fig. 1, 3, 4, 6

10.0 mm
Fig. 7

Coal Canyon to summit 6909 ("Pyramid Hill" of Merriam and McKee, 1976, Fig. 2) on east side of Coal Canyon. Paratype, USNM 488430, 2.0 to 3.5 m below top of *Quadrithyris* Zone depositional breccia, east side of Coal Canyon, approximately 150 m south of entrance. Paratype, USNM 159421, *Quadrithyris* Zone depositional breccia, east side of Coal Canyon, near entrance, USGS Locality M1108. Paratype, USNM 159422, *Quadrithyris* Zone depositional breccia, east side of Coal Canyon, near entrance at altitude of about 6,300 ft, USGS Locality M1106. Paratypes, USNM 488431, 488432, *Quadrithyris* Zone depositional breccia, east side of Coal Canyon, approximately 300 m south of entrance, USGS Locality M1450, which is also the upper part of section SP VII (Berry and Murphy, 1975, Fig. 14). Paratype, USNM 488433, talus from *Quadrithyris* Zone depositional breccia, east side of Coal Canyon, near entrance. All of the above material is from the upper Lochkovian Windmill Limestone, SE 1/4 sec. 17, T25N, R49E, Horse Creek Valley Quadrangle, northern Simpson Park Range. Paratype, USNM 488434, 4.0 to 4.3 m above base of *Quadrithyris* Zone depositional breccia, upper Lochkovian Windmill Limestone, section SP I (Berry and Murphy, 1975, Fig. 14), west side of Coal Canyon Fault, NW 1/4 NE 1/4 sec. 20, T25N, R49E, Horse Creek Valley Quadrangle, northern Simpson Park Range.

Description. Corallum cerioid with maximum observed diameter and height of about 350 mm and 250 mm respectively. Increase marginarial and nonparricidal; offsets commonly single, not necessarily originating in the corner of the parent corallite; as many as four contiguous offsets may develop simultaneously from a single parent corallite. Corallites polygonal; walls between angles straight or slightly curved. Mean diameters of adult corallites 11.0 to 18.5 mm, except in paratype USNM 488419, where the greatest mean corallite diameter is 20.0 mm. Inwardly sloping calicular platforms periodically formed of very short, unaligned trabeculae set in a thin layer of sclerenchyme overlying a presepimentarial surface. Ptenophyllinid calicular pit with distal septal edges high above the calicular floor. Corallite walls very variable. In places where the septal bases are not clearly formed, thickness of the combined intercorallite walls of adjacent corallites may be only 0.05 mm, but where the septal bases are well developed, thickness of the combined walls is as much as 1.5 mm.

Septal arrangement radial to very weakly bilateral in tabularium. In the middle and outer dissepimentarium, septa reduced to mostly thin, smooth to slightly wavy crests or may be almost entirely suppressed. Septal crests generally well developed in inner dissepimentarium, so that in transverse section, few discontinuities appear in the major septa in this part of the corallite. Septa in the tabularium continuous and generally slightly thickened with mostly weak ptenophyllinid flanges. Major septa extend to axis or close to it but are not axially rotated; minor septa only just enter the tabularium. Septal counts in adult corallites 19 × 2 to 29 × 2, except in largest corallites of paratype USNM 488419, where they are as high as 31 × 2. Short, unaligned trabeculae of indiscernible fine structure are visible in septal bases and on some presepimentarial surfaces in the least poorly preserved specimens of the type series; similar but longer trabeculae are also visible in the outer parts of some septa.

Horizontal skeletal elements of the outermost and middle regions of the marginarium consist almost entirely of large inwardly inclined presepiments. These give way to smaller, more steeply inclined dissepiments in the innermost region of the marginarium. Typically, the presepiments and dissepiments occur in four to nine rows. Boundary between innermost dissepiments and tabulae well defined. Tabulae incomplete, closely spaced, 26 to 45 counted over a vertical distance of 10.0 mm. Tabularial surfaces variably shaped, commonly deeply depressed periaxially. Diameter of tabularium generally 2.0 to 3.5 mm, maximum 4.0 mm. Ratios of the diameter of tabularium to the diameter of an adult corallite vary from 0.15 to 0.28.

Remarks. The new species most resembles *Toquimaphyllum johnsoni* Merriam and *T. landerense* (Merriam, 1973, p. 24, Plate 9, Fig. 3–6), both of which are imperfectly known. *T. johnsoni* has finer and fewer septa (12 × 2 to 19 × 2 in corallites with mean diameters of 16.0 or 17.0 mm diameter) and flatter outer presepiments. *T. landerense* has a maximum of 21 × 2 septa; both orders of its septa, especially the minor, are apparently more fully developed in the outer marginarium, and the maximum diameter of its tabularium is at least 4.8 mm. *T. vulgare* is one of the most abundant upper Lochkovian corals in the northern Simpson Park Range. If the corals from Mazourka Canyon described as *Australophyllum* (A.) sp. v and *A. (Toquimaphyllum)* cf. *johnsoni* by Merriam belong to the species, then the species also occurs in the probable upper Lochkovian part of the Vaughn Gulch Limestone in the northern Inyo Mountains of California.

Derivation of name. Latin, *vulgaris, -is, -e*, meaning common.

Toquimaphyllum sentum sp. nov.

Figures 17: 1–2, 18: 1–6

? *Endophyllum* sp. OLIVER, 1964, p. D-151; OLIVER *in* Johnson, 1965, p. 371; OLIVER *in* Johnson, 1970, p. 44.

Australophyllum (Australophyllum) sp. c MERRIAM *in* Merriam and McKee, 1976, p. 34, Plate 8, Fig. 3–7.

Kozlowiaphyllum sp. PEDDER, 1975, p. 285; ? JOHNSON AND OLIVER, 1977, p. 1464.

Material. Holotype and eight paratypes. USNM 488435 (holotype), and USNM 488436, *Quadrithyris* Zone depositional breccia, ridge leading from entrance of Coal Canyon to summit 6909 ("Pyramid Hill" of Merriam and McKee, 1976, Fig. 2) on east side of Coal Canyon. USNM 488437, 488438, 2.0 to 3.5 m below top of *Quadrithyris* Zone depositional breccia, east side of Coal Canyon, approximately 150 m south of entrance. USNM 488439, 488440, and 488441, *Quadrithyris* Zone depositional breccia, east side of Coal Canyon, approximately 300 m south of entrance, USGS Locality M1450, which is also the upper part of section SP VII (Berry and Murphy, 1975, Fig. 14). USNM 488442, talus from *Quadrithyris* Zone depositional breccia, east side of Coal Canyon, near entrance. All of the above material is from the upper Lochkovian Windmill Limestone, SE 1/4 sec. 17, T25N, R49E, Horse Creek Valley Quadrangle, northern Simpson Park Range. Paratype, talus specimen figured by Merriam and McKee (1976), almost certainly from the *Quadrithyris* Zone depositional breccia in the upper Lochkovian Windmill Limestone, west of Coal Canyon Fault, about 152 m west of Coal Canyon bottom at altitude of 6,400 ft (USGS Locality M1318), Horse Creek Valley Quadrangle, northern Simpson Park Range.

Description. Corallum cerioid with some marginal fasciculate corallites commonly supported by radiciform processes. Most colonies less than 150 mm in diameter and 100 mm in height. Increase marginarial and nonparricidal. In the prepared part of the type series, only one offset is developed at a time from a corner of any parent corallite. Corallites polygonal with straight or slightly curved walls and mean adult diameters of 17.0 to 30.0 mm. Calicular platforms predominantly inwardly sloping, in places flat or very rarely slightly everted. Surfaces of calicular platforms periodically underlain by short unaligned trabeculae embedded in thin sclerenchyme. Calicular pit with subcylindrical wall; distal septal edges high above the floor of the calice. Intercorallite walls with or without septal bases; combined thickness of adjacent corallite walls with prominent septal bases as much as 1.5 mm; least combined thickness of adjacent walls 0.15 mm, Septal arrangement mostly radial with weak axial whorl. Septa commonly complete in juvenile corallites, rarely so in mature corallites. Nontrabeculate parts of septal crests in the outer marginarium thin, smooth, or crenulate. Septal trabeculae in outer marginarium commonly poorly aligned, making the septa highly carinate in places. Septa continuous, dilated, especially in

Figure 16 (on facing page). 1–4: *Toquimaphyllum vulgare* sp. nov.; 1, 2, transverse and longitudinal sections, USNM 488433; 3, 4, longitudinal and transverse sections, USNM 488431.

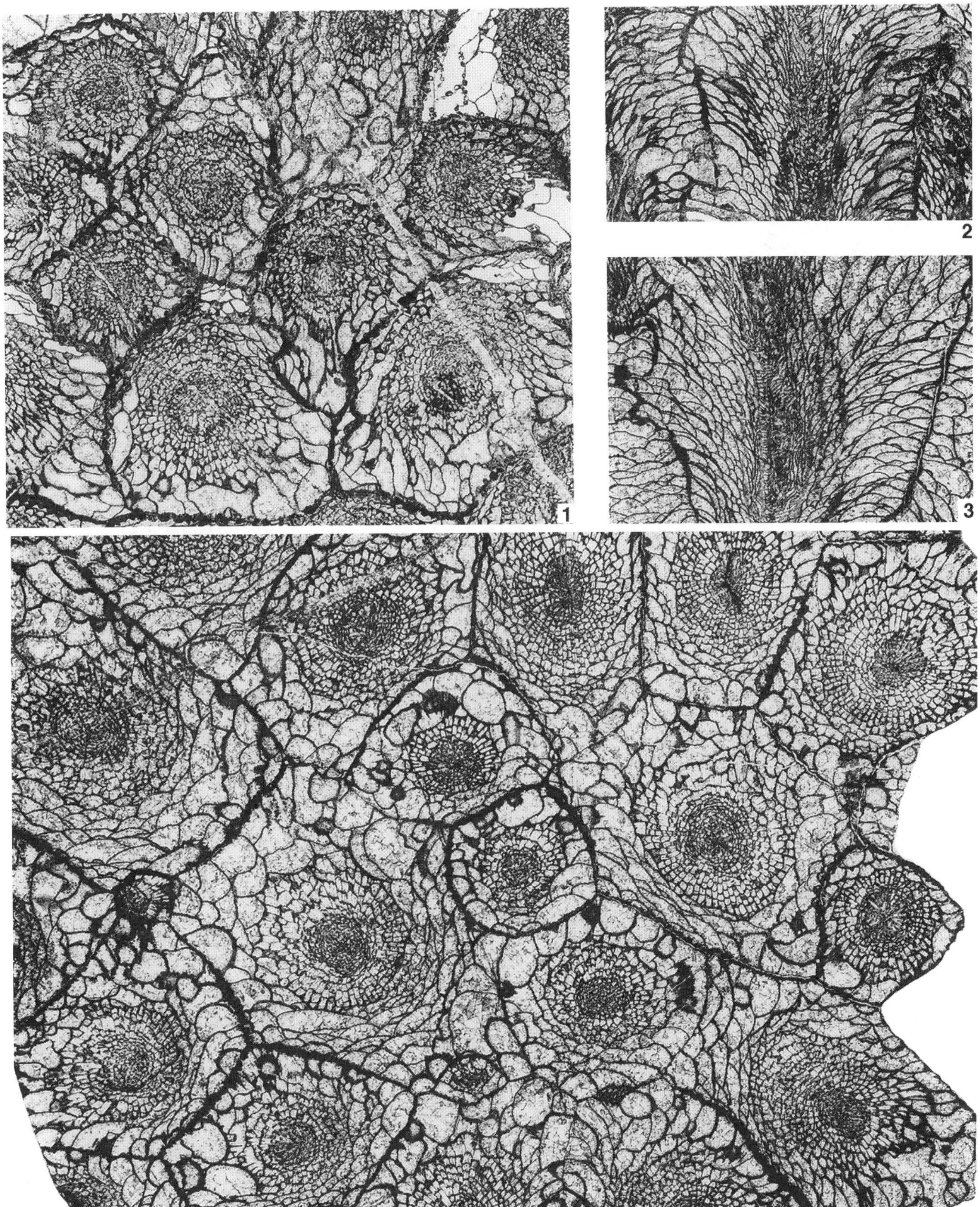
1
2
3
4
10.0 mm

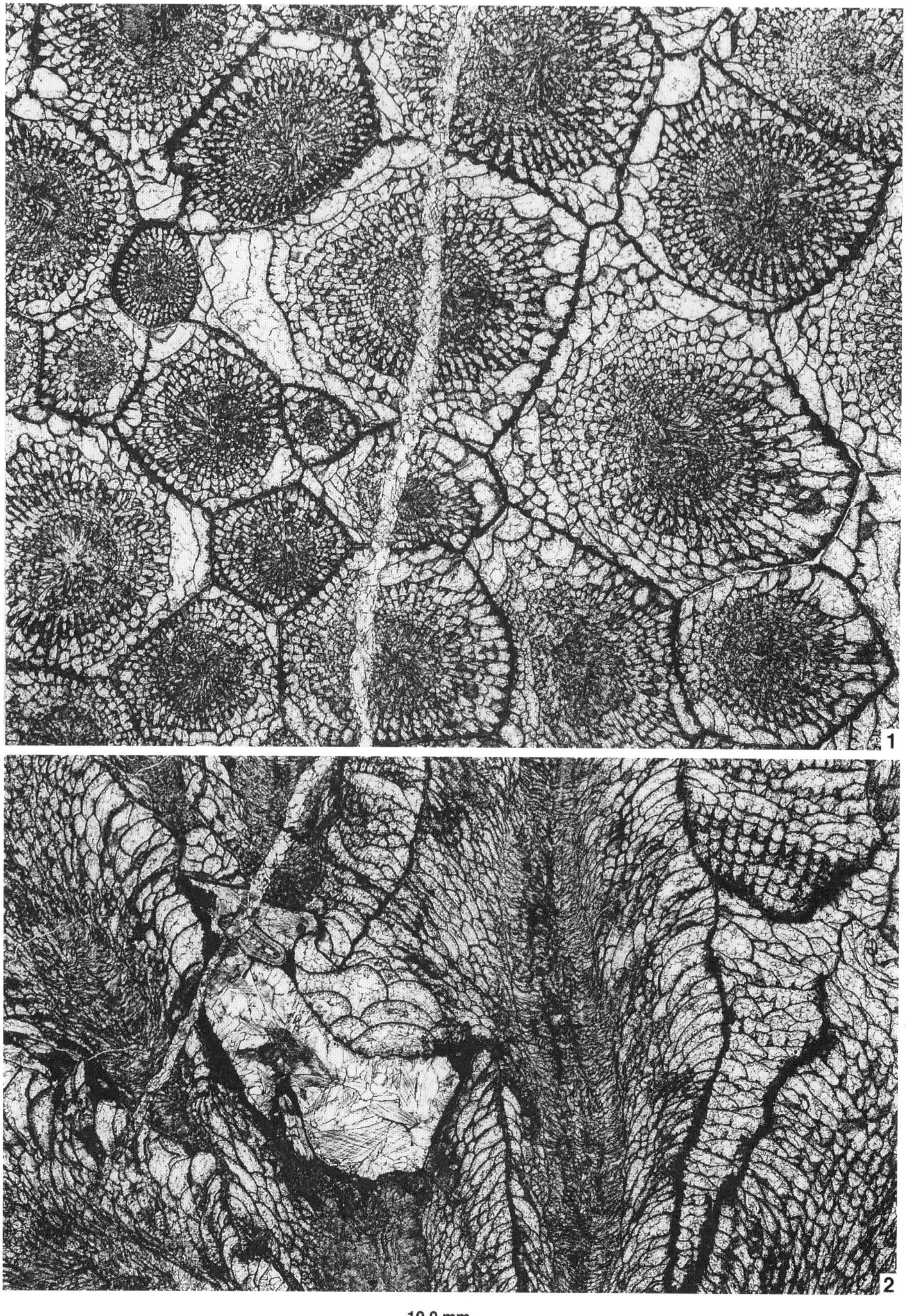

1

2

10.0 mm

Figure 17. 1, 2: *Toquimaphyllum sentum* sp. nov., transverse and longitudinal sections, holotype, USNM 488435.

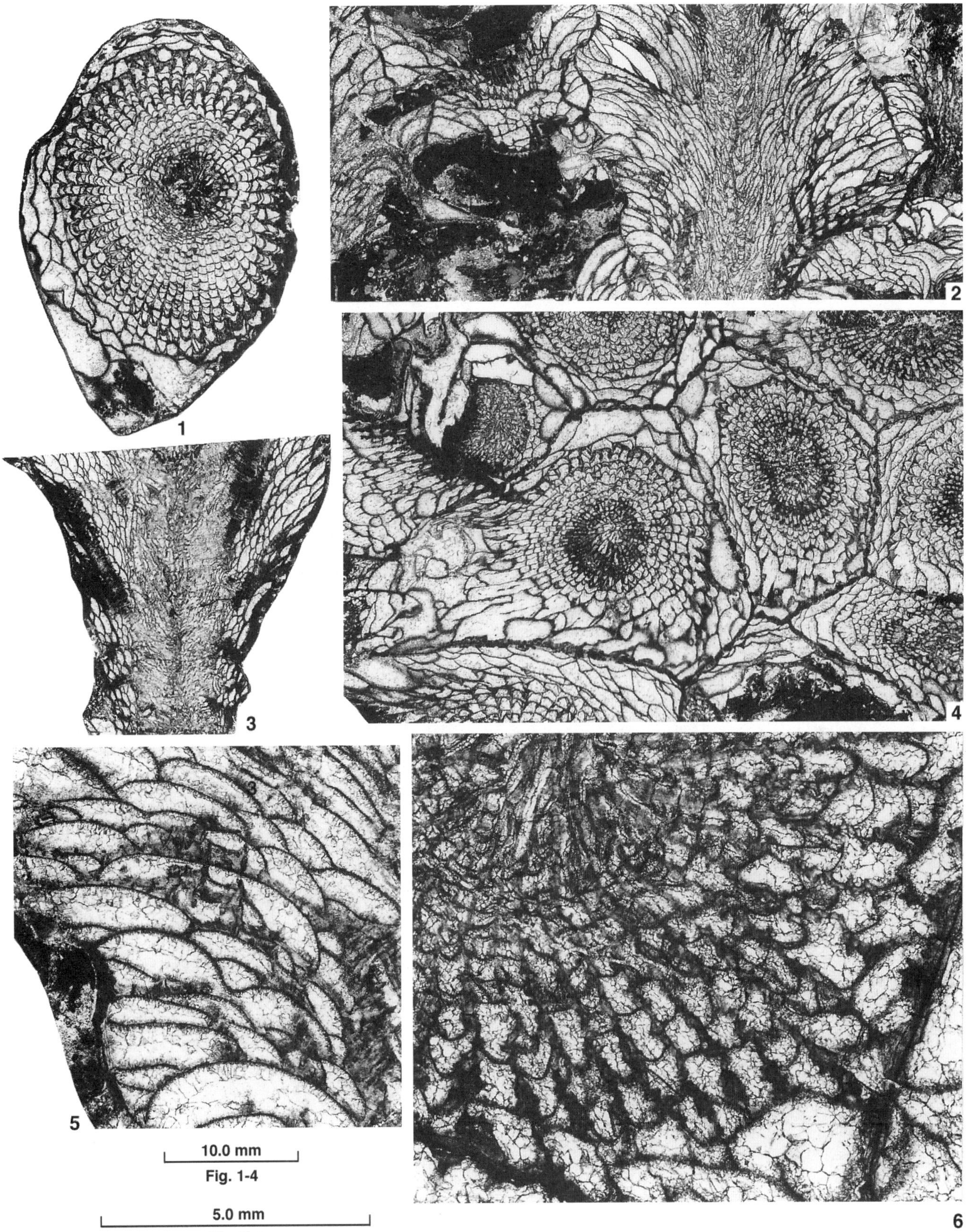

Figure 18. 1–6: *Toquimaphyllum sentum* sp. nov.; 1, 3, transverse and longitudinal sections of protocorallite, USNM 488440; 2, 4, 5, longitudinal section, transverse section, and part of a longitudinal section, USNM 488436; 6, part of a transverse section, holotype, USNM 488435.

some juvenile corallites, and strongly flanged in the inner dissepimentarium and tabularium. Major septa reach the periaxial region to form an axial whorl; minor septa do not extend inward from the outer tabularium. Septal counts in adult corallites 23 × 2 to 31 × 2.

Horizontal skeletal elements of the outermost marginarium dominated by large, flat-lying to inwardly sloping presepiments; smaller, more inwardly inclined presepiments and some dissepiments dominate the middle marginarium. The innermost marginarium includes steeply inclined to vertical dissepiments but no presepiments. Boundary between dissepimentarium and tabularium sharp. Tabulae much disrupted by septa and their flanges, closely spaced, 25 to 35 counted over a vertical distance of 10.0 mm. Tabularial surfaces generally shaped as an irregular inverted cone. Diameter of tabularium 4.0 to 8.0 mm. Ratios of the diameter of the tabularium to that of a mature corallite 0.20 to 0.34.

Remarks. Although the corallum of *Toquimaphyllum sentum* is normally smaller than that of *T. vulgare*, corallite dimensions, especially those of the tabularium, which is one of the most stable features in ptenophyllinid corals, are greater (dt 2.0 to 3.5 mm in *T. vulgare*, 4.0 to 8.0 mm in *T. sentum*; adult dt:dc 0.15 to 0.28 in *T. vulgare*, 0.20 to 0.34 in *T. sentum*). Large adult corallites of *T. vulgare* just overlap the size of small adult corallites of *T. sentum*. However, dilation and strong carination of the septal skeleton in parts of the marginarium in *T. sentum* are also important distinctions between the two species and are decisive in separating large corallites of *T. vulgare* from small corallites of *T. sentum*.

Derivation of name. Latin, *sentus, -a, -um*, meaning thorny or rough.

Genus *Ecnomiophyllum* novum

New genus 23 Oliver and Pedder, 1979, p. 242.

Type species. Ecnomiophyllum merriami sp. nov.

Diagnosis. Solitary, ceratoid to trochoid ptenophyllinid corallites that either remain solitary or become protocorallites. Increase both parricidal and nonparricidal, generally producing no more than two generations of offsets. Offsets commonly fail to grow to maturity. Septal trabeculae not regularly palisaded to form uniformly smooth or weakly carinate septa but set at various angles in the marginarium to form flanges and even vepreculae. Typically wide marginarium includes dissepiments and presepiments. Presepiments commonly bear short trabeculae embedded in thin crusts of sclerenchyme, as well as septal crests. Fully naotic septa not developed.

Remarks. The genera most resembling *Ecnomiophyllum* are *Embolophyllum* and *Toquimaphyllum*. *Embolophyllum* is distinguished from the new genus by its well-palisaded septal trabeculae in the marginarium, illustrated in numerous publications (e.g., Pedder, 1967, Plate 6, Fig. 6; Strusz, 1968, Plate 2, Fig. 1b, 2a, 3a, 3c), and by its much rarer and smooth-surfaced presepiments. The cerioid form of the mature corallum distinguishes *Toquimaphyllum* from *Ecnomiophyllum*.

Although the diagnosis for *Ecnomiophyllum* was framed only on the Nevada species referred to the genus, in most respects it fits the description and illustrations of a coral from the Garra Formation of New South Wales, named *Australophyllum?* sp. cf. *thomasae* (Hill and Jones, 1940) by Strusz (1966, p. 586, Plate 94, Fig. 2–5; Plate 95, Fig. 1–3). The finely trabeculate septal crusts are well developed in Strusz's coral but flanges are rare in the marginarium, which is dominated by presepiments and has very few continuously lamellar septa. From Sorentino's (1989) conodont studies in a stratigraphic section of Garra Formation 3.5 km south of the outcrop that yielded Strusz's coral, and from what is known of the diachronous (younging northward) age trend of the Garra Formation between Molong and Guerrie, Strusz's possible species of *Ecnomiophyllum* is likely to be a lower Pragian coral.

Two or three other Devonian species, previously identified as *Spongophylloides* sp. and currently known only from very limited material, may prove to belong to *Ecnomiophyllum*. One of these is a peripherally worn coral, prepared as a single transverse section (Yu and Liao, 1978, Plate 3, Fig. 3), from the Lochkovian part (Li, 1988) of the Shanjiang Formation in the Shaluli Shan arc of northwestern Yunnan. Others are two fragmentary corals discussed and illustrated by Oliver and Galle (1971, p. 56, Plate 9, Fig. 4–6) from the Pragian Koneprusy Limestone of the Barrandian terrane, in the Czech Republic.

A late Lochkovian coral from the Lower Pribalkhash Horizon on the Balkhash terrane of Kazakhstan, named *Australophyllum crassiseptatum* by Kaplan (Sytova and Kaplan, 1975, p. 74, Plate 14, Fig. 2), needs to be reconsidered as a possible species of *Ecnomiophyllum*. The species is apparently based on a single specimen and, as presently described and illustrated, cannot be assigned to any genus with certainty. The illustrations show coarse trabeculae much like those of *Ecnomiophyllum*. The colony is described as massive; however, the figures show only two corallites, which, although attached to each other, appear to be subcylindrical.

Derivation of name. Greek, *eknomios* = unusual, and *phyllon* = leaf, in Latin form.

Ecnomiophyllum merriami sp. nov.

Figure 19: 1–12

Salairophyllum? sp. MERRIAM, 1974a, p. 56, Plate 12, Fig. 6–8; MERRIAM *in* Merriam and McKee, 1976, p. 5.
"*Salairophyllum*" sp. PEDDER, 1975, p. 285.
New genus 23 OLIVER AND PEDDER, 1979, p. 242.

Material. Holotype, USNM 488443, 0 to 2.0 m below top of *Quadrithyris* Zone depositional breccia, east side of Coal Canyon, approximately 150 m south of entrance. Paratype, USNM 488444, 2.0 to 3.5 m below top of *Quadrithyris* Zone depositional breccia, same locality as holotype. Paratype, USNM 488445, unrecorded level in *Quadrithyris* Zone depositional breccia, same locality as holotype. Paratype, USNM 488446, unrecorded level in *Quadrithyris* Zone depositional breccia, east side of Coal Canyon, approximately 300 m south of entrance. Paratype, USNM 488447, talus from *Quadrithyris* Zone depositional breccia, east side of Coal Canyon, near entrance. All of the above material is from the upper Lochkovian Windmill Limestone, SE 1/4 sec. 17, T25N, R49E, Horse Creek Valley Quadrangle, northern Simpson Park Range. Paratype, USNM 159424, talus probably from *Quadrithyris* depositional breccia ("probably from upper Silurian coral zone E"), east side of coal Canyon, about 965 m ("six-tenths of a mile") south of entrance, USGS Locality M1117. Paratype, USNM 488448, lower 11.0 m of *Quadrithyris* Zone depositional breccia, section SP I (Berry and Murphy, 1975, Fig. 14), west side of Coal Canyon Fault. Paratype, USNM 488449, 4.0 to 4.3 m above base of *Quadrithyris* Zone depositional breccia, same section SP I. With the unlikely, although possible, exception of USNM 159424, the last three paratypes are from the upper Lochkovian Windmill Limestone in NW 1/4 NE 1/4 sec. 20, T25N, R49E, Horse Creek Valley Quadrangle, Northern Simpson Park Range.

Description. Mature corallum a ceratoid to trochoid protocorallite, with maximum mean diameter 25.0 mm and a small fasciculate cluster of one or two generations of narrower offsets. Maximum mean diameter of offsets 16.0 mm. Increase marginarial, parricidal in one paratype, nonparricidal in another; as many as four offsets produced simultaneously from a parent corallite. Radiciform processes abundant, commonly uniting offsets of the same generation. Calice deep, lower part cylindrical, with upper edges of major septa more than 8.0

→

Figure 19. 1–12: *Ecnomiophyllum merriami* sp. nov.; 1, 5, transverse and longitudinal sections, USNM 488445; 2, 6, 8, 10, 11, two transverse sections, a longitudinal section, and parts of a longitudinal section and a transverse section, holotype, USNM 488443; 3, transverse section, USNM 488446; 4, transverse section, USNM 488449; 7, transverse section, USNM 488447; 9, 12, longitudinal and transverse sections, USNM 488448.

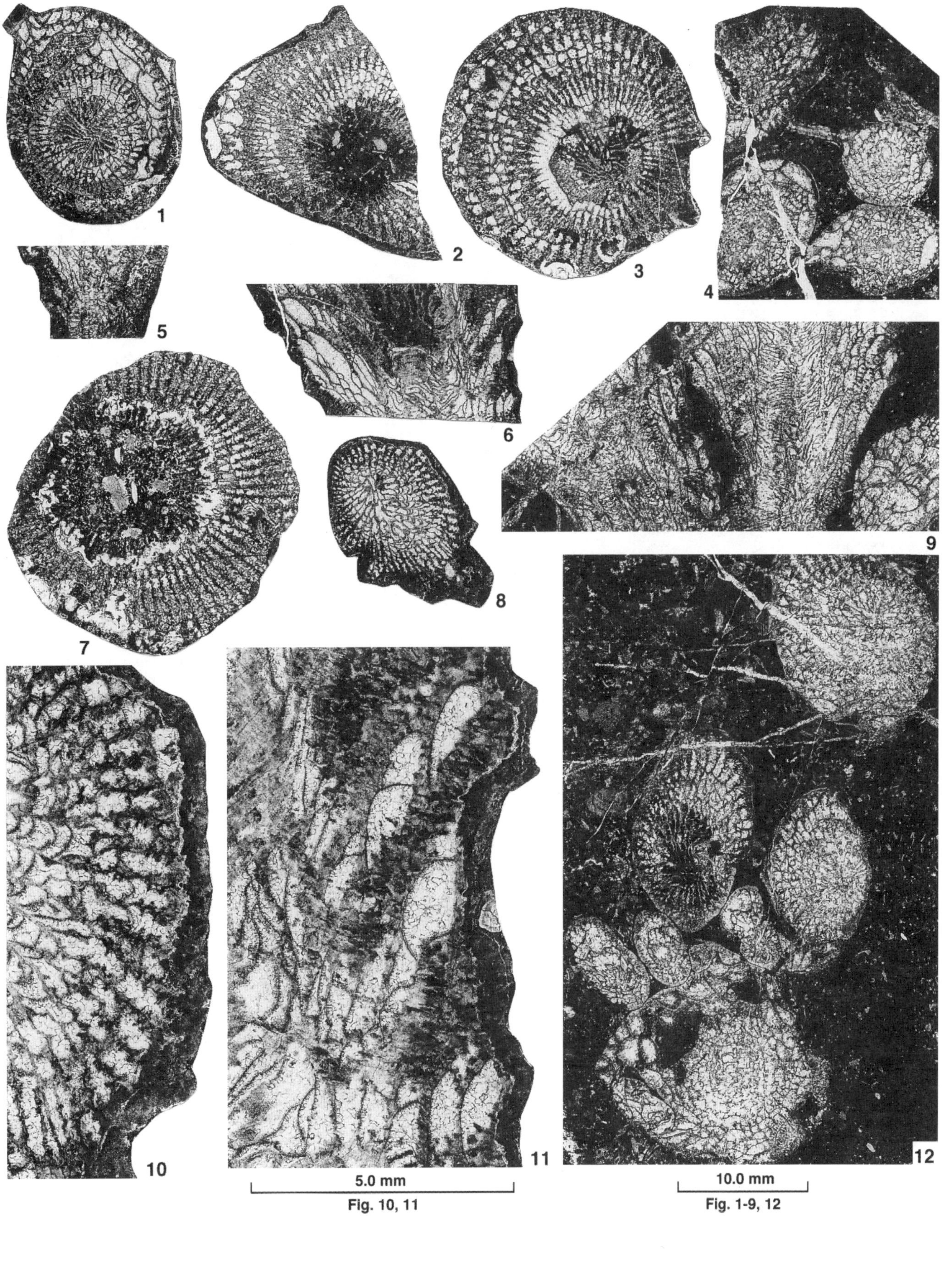
1
2
3
4
5
6
7
8
9
10
11
12
5.0 mm
Fig. 10, 11
10.0 mm
Fig. 1-9, 12

mm above the base of the calice; upper part with inwardly sloping platform.

Septa in two orders, poorly differentiated in dissepimentarium, radially arranged, typically with a weak axial whorl. Septa strongly carinate to vepreculate in dissepimentarium, especially in protocorallite, owing to divergence of short, scattered trabeculae. Septa and sclerenchyme locally form a peripheral septal stereozone, as much as 5.0 mm thick, in the protocorallite. Where presepiments disrupt the septa, the septal stereozone is commonly rejuvenated inside the outer presepiments. Major septa with ptenophyllinid flanges in tabularium, typically weakly rotated in periaxial region. Minor septa more disrupted than major septa, some confined to dissepimentarium, although most just penetrate the tabularium. Septal counts in protocorallites with mean diameters of 10.0 to 25.0 mm 26 × 2 to 33 × 2. Maximum septal count in offsets 25 × 2.

Marginarium includes two to nine rows of dissepiments and presepiments, both more or less uniformly inclined adaxially, except the small innermost dissepiments, which are mostly vertical or nearly so. Boundary between dissepimentarium and tabularium sharp. Tabulae incomplete, disrupted by septal flanges, closely spaced, 19 to 26 counted over a vertical distance of 10.0 mm. Outer tabularial surfaces mostly inwardly sloping; periaxial tabularial surfaces concave or convex. Diameter of adult tabularium in protocorallite 7.0 to 8.0 mm. Diameter of offset tabularia less than 5.5 mm.

Remarks. The distinctive trabeculae and the stereozone disrupted by presepiments remove *Ecnomiophyllum merriami* from any previously described ptenophyllinid coral. Merriam's figured specimen, here included in the type series, was questionably identified as an undetermined species of *Salairophyllum* by Merriam. The type species of *Salairophyllum* is *S. angustum* (Zheltonogova, 1961, p. 78, Plate S-17, Fig. 5; Plate S-18, Fig. 1), originally described from the Lochkovian Tomchumysh Beds of the Kuznetsk Basin, Russia. According to the original monograph and subsequent work by Besprozvannykh (1968, p. 111, Plate 49, Fig. 4) and Shurygina (1968, p. 128, Plate 55, Fig. 1–3), *S. angustum*, which is also known from the Lochkovian of the eastern Urals, is a solitary cylindrical coral, lacking both the presepiments and peculiar trabeculae of *E. merriami*.

Derivation of name. Patronym for the late C. W. Merriam.

Ecnomiophyllum simpsoni (Merriam, 1974)

Figure 21: 3–8

Papiliophyllum elegantulum Stumm. OLIVER, 1964, p. D151.
? chonophylloid corals. OLIVER *in* Johnson, 1965, p. 371.
Chonophyllum simpsoni MERRIAM, 1974a, p. 53, Plate 8, Fig. 1–4; MERRIAM *in* Merriam and McKee, 1976, p. 5, 15; Plate 7, Fig. 5, 6; Plate 9, Fig. 9, 10.
Dubrovia simpsoni (Merriam). PEDDER, 1975, p. 285.
"Chonophyllum" simpsoni Merriam. JOHNSON AND OLIVER, 1977, p. 1464; OLIVER AND JOHNSON, 1977, p. 108.
Not *Chonophyllum* sp., cf. *C. simpsoni* MERRIAM, 1974a, Plate 8, Fig. 7 (= ? *Ecnomiophyllum* sp. A).

Diagnosis. Large solitary, trochoid species of *Ecnomiophyllum* with maximum mean diameter of 50 mm, an estimated maximum height of more than 90 mm, and a maximum of at least 46 × 2 septa. Marginarium with as many as 20 rows of presepiments and dissepiments, inclined to form a locally flat or even everted calicular platform.

Remarks. The only two specimens definitely attributed to the species by Merriam were two fragmentary specimens of adult individuals, collected from talus derived from the *Quadrithyris* Zone depositional breccia of the Windmill Limestone on the east side of Coal Canyon near its mouth (USGS Locality M1108). We have a large specimen (USNM 488450) that closely resembles the holotype and paratype, from the *Quadrithyris* Zone debris-flow on the east side of Coal Canyon, approximately 300 m south of the Canyon entrance, and another, smaller coral (USNM 488451), tentatively interpreted as a young example of the same species. The latter is from 2.0 to 3.5 m below the top of the same debris-flow, on the east side of Coal Canyon, approximately 150 m south of its entrance. All known specimens are from SE 1/4 sec. 17, T25N, R49E, northern Simpson Park Range.

Although no offset has been seen in this species, traces of fine trabeculae on the surfaces of presepiments, which are one of the principal diagnostic features of *Ecnomiophyllum*, can be seen in Merriam's (1974a, Plate 8, Fig. 1) illustration of the holotype. Similar trabeculae show well in the transverse section of USNM 488451, illustrated in Figure 21: 8.

Oliver's 1964 identification of *Papiliophyllum elegantulum* in the *Quadrithyris* Zone breccia at Coal Canyon was later withdrawn (Oliver *in* Johnson, 1965, p. 371) and eventually updated as *"Chonophyllum" simpsoni* (Oliver *in* Johnson and Oliver, 1977).

Ecnomiophyllum fascis sp. nov.

Figure 20: 6–9

Kyphophyllum sp. c MERRIAM *in* Merriam and McKee, 1976, p. 5, 15, Plate 9, Fig. 4–8.
Neomphyma sp. c. OLIVER *in* Oliver and Johnson, 1977, p. 108.

Material. Holotype, USNM 488452, *Quadrithyris* Zone depositional breccia, ridge leading from the entrance of Coal Canyon to summit 6909 ("Pyramid Hill" of Merriam and McKee, 1976, Fig. 2) on east side of Coal Canyon, SE 1/4 sec. 17, T25N, R49E. Paratype, specimen illustrated by Merriam and McKee (1976), "upper beds of Roberts Mountains Formation," which, in this context, are the same *Quadrithyris* Zone depositional breccia, Coal Canyon, 850 m south of canyon entrance on east side, NW 1/4 NE 1/4 sec. 20, T25N, R49E, USGS Locality M1380, shown on Figure 2 of Merriam and McKee (1976). Both types are from the Windmill Limestone, Horse Creek Valley Quadrangle, northern Simpson Park Range.

Description. Corallum colonial, consisting of a cluster of nine ceratoid to cylindrical corallites in the holotype, probably fewer in the paratype. Mode of increase unknown. Maximum mean diameter of corallites approximately 28 mm in the holotype and 31 mm in the paratype. Some corallites attached to each other or united with others by means of radiciform processes.

Outer wall thin relative to maximum diameter of corallite, 0.2 to 1.0 mm wide. Septa in two, radially to very weakly bilaterally arranged orders; mostly disrupted by presepiments around the margins of large corallites. Greatest number of septa 28 × 2 in holotype, 32 × 2 in paratype. Major septa extend to the periaxial region but are generally not axially rotated. Minor septa confined to the marginarium; in fully developed corallites most occur only as septal crests. Septa thin and delicately crenulate in some corallites, especially those of the paratype. Other septa have strong ptenophyllinid flanges in the marginarium as well as the tabularium. Some of these flanges, shown in Figure 20: 7 and 8, approach the morphology of vepreculae.

Horizontal skeletal elements of adult marginaria typically comprise six to nine rows of elongate outer presepiments and similarly shaped inner dissepiments. Mean diameter of adult tabularium 5.0 to 8.0 mm. Tabulae incomplete, closely spaced, about 20 occurring over a vertical distance of 10.0 mm. Atypically for a ptenophyllinid coral, the outermost parts of some tabulae locally grade into inner dissepiments.

Remarks. The species most resembling *Ecnomiophyllum fascis* is

Figure 20. 1–5: *Ecnomiophyllum* sp. A, two transverse sections, a longitudinal section, and parts of longitudinal and transverse sections, USNM 488453. 6–9: *Ecnomiophyllum fascis* sp. nov., longitudinal section, part of a transverse section, part of a longitudinal section through the marginarium, and a transverse section, holotype, USNM 488452.

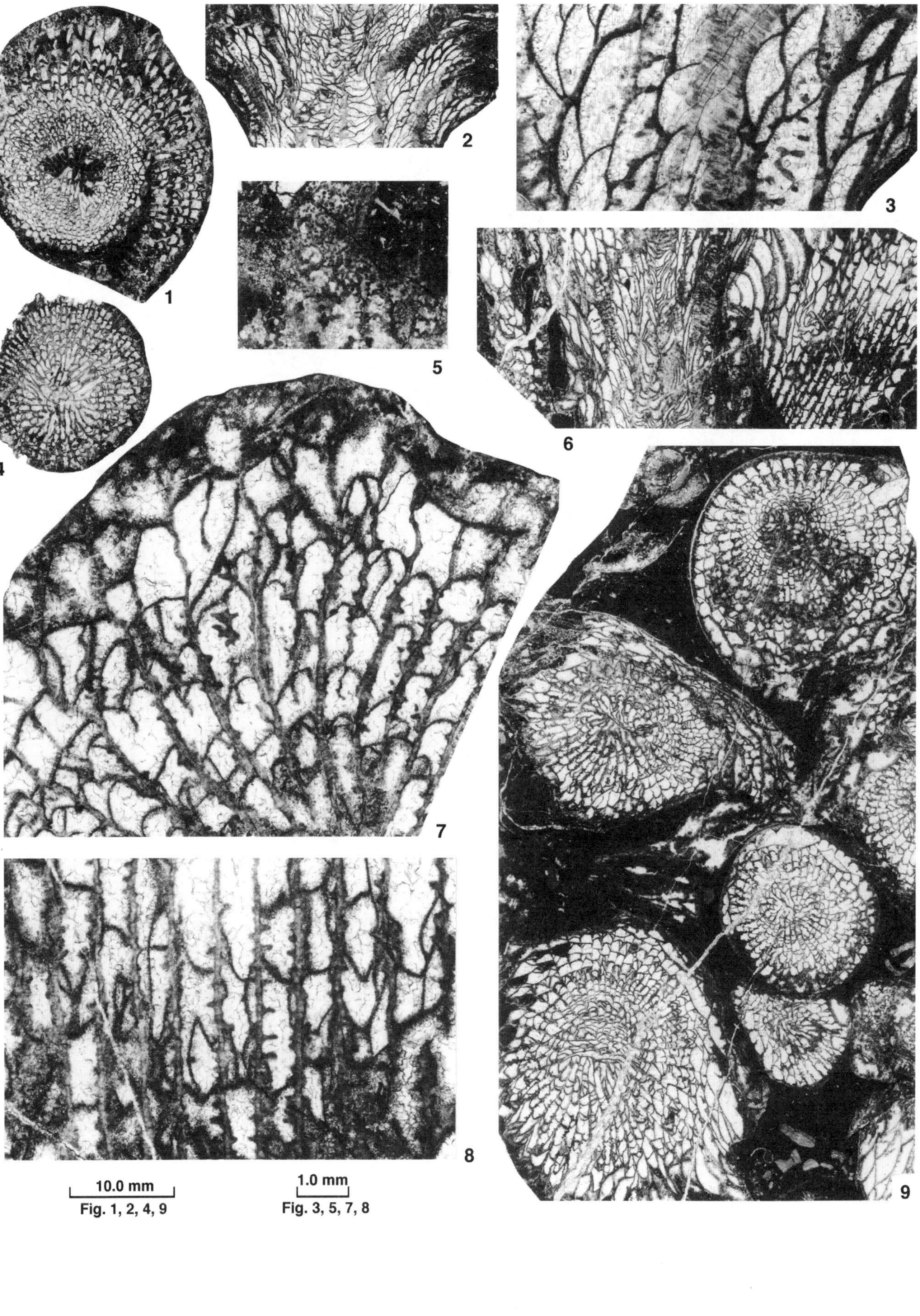
1
2
3
4
5
6
7
8
9
10.0 mm
Fig. 1, 2, 4, 9
1.0 mm
Fig. 3, 5, 7, 8

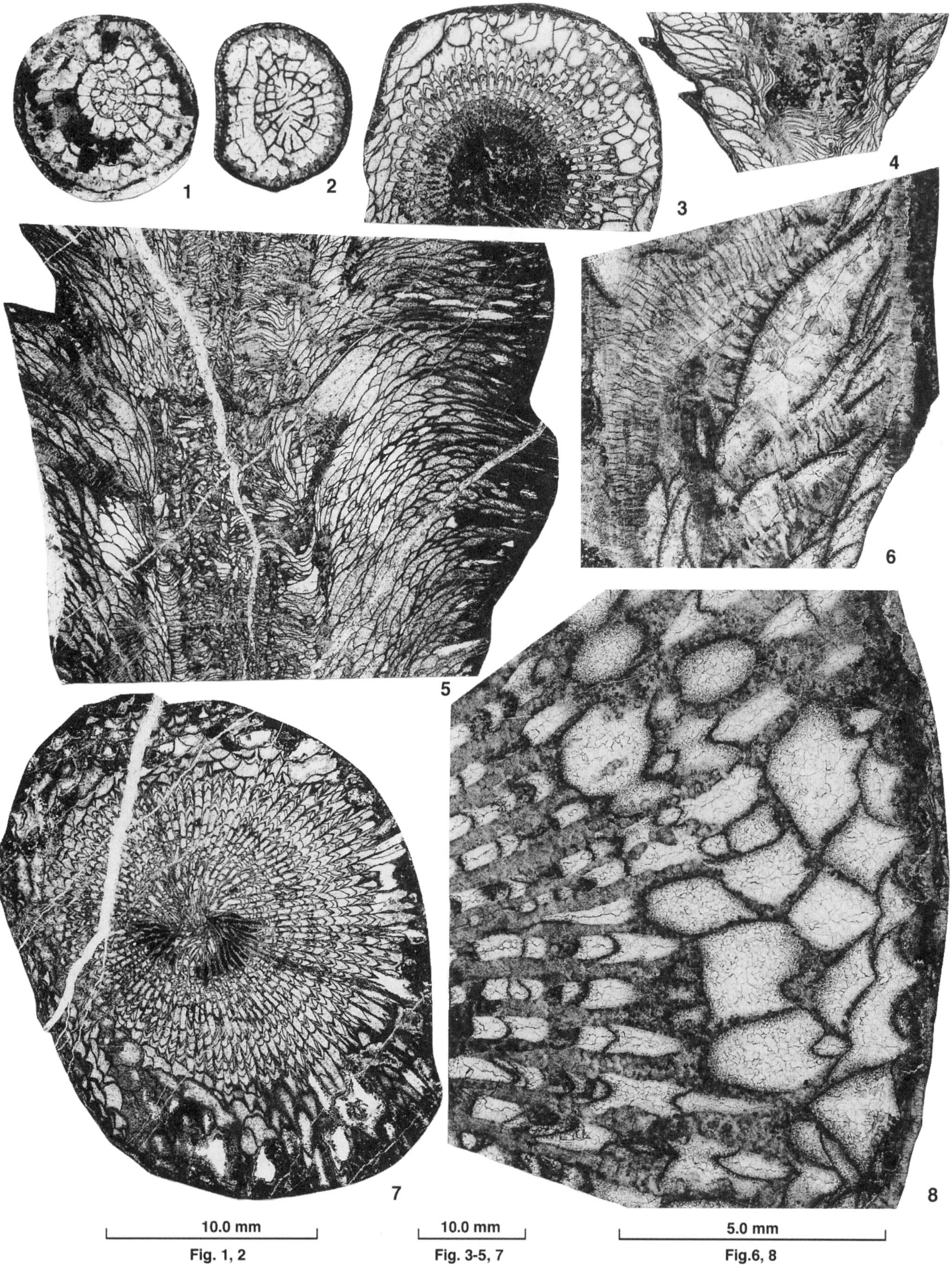

Figure 21. 1, 2: *Neomphyma karpinskiense* Shurygina, two transverse sections from same corallum, USNM 488413; see also Figure 12: 12, 13, 15. 3–8: *Ecnomiophyllum simpsoni* (Merriam); 3, 4, 6, 8, transverse section, longitudinal section, part of a longitudinal section, and part of a transverse section, USNM 488451; 5, 7, longitudinal and transverse sections, USNM 488450.

E. merriami sp. nov. However, *E. merriami* is smaller, it has a septal stereozone, and its septa are coarser and more heavily carinate.

Derivation of name. Latin, *fascis*, meaning a cluster.

Ecnomiophyllum sp. A
Figure 20: 1–5

? *Chonophyllum* sp., cf. *C. simpsoni* MERRIAM, 1974a, Plate 8, Fig. 7.

Remarks. This informal taxon is based primarily on a single, much worn corallite (USNM 488453), from 2.0 to 3.5 m below the top of the *Quadrithyris* Zone depositional breccia, on the east side of Coal Canyon, approximately 150 m south of its entrance. The specimen illustrated by Merriam (1974a, Plate 8, Fig. 7) was collected by A. J. Boucot in 1964, apparently from the same depositional breccia, on the east side of Coal Canyon near its entrance (USGS Locality M1110). Both specimens are from the upper Lochkovian Windmill Limestone, SE 1/4 sec. 17, T25N, R49E, Horse Creek Valley Quadrangle, northern Simpson Park Range.

USNM 488453 has 34 × 2 septa at an estimated mean corallite diameter of about 32 mm, and is therefore slightly larger than the largest corallites of *Ecnomiophyllum fascis* but significantly smaller than adult specimens of *E. simpsoni*. More importantly, it differs from *E. fascis* by having flatter outer presepiments, which are coated with short trabeculae and thin sclerenchyme, visible in Figure 20: 5. In addition to the size difference, it is distinguished from *E. simpsoni* by having more prominent ptenophyllinid septal lobes in the marginarium, seen in Figure 20: 3.

No longitudinal section is available of the specimen collected by Boucot; strictly, the specimen is unrecognizable and is only very tentatively assigned to *Ecnomiophyllum* sp. A. Merriam made it appear more similar to *E. simpsoni* than it is, by magnifying the illustration ×4 and placing it next to an illustration of *E. simpsoni*, enlarged ×2.

ACKNOWLEDGMENTS

W. A. Oliver Jr., U.S. Geological Survey, Washington D.C., allowed us to reproduce two of his previously unpublished photographs of the type species of *Briantia*. T. T. Uyeno, Geological Survey of Canada, Calgary, identified conodonts from a Roberts Mountains Formation sample. J. I. Valenzuela-Ríos, previously a Fulbright Fellow, identified conodonts from Atrypa Peak and assisted in the field in 1993. We are most grateful for all of this help.

Critical readers R. A. McLean, Amoco Canada Petroleum Company Ltd., Calgary, and W. A. Oliver Jr., suggested several improvements to the original manuscript.

We are also grateful to the following staff of the Geological Survey of Canada, Calgary, for cooperation in the production of this work: G. N. Edwards is responsible for photomechanical work and digital half-toning of the illustrations; D. R. Then constructed the range chart of Figure 2; and B. C. Rutley photographed all of the corals illustrated from Nevada.

REFERENCES CITED

Agassiz, J. L. R., 1833, Recherches sur les poissons fossiles,. . ., volume 1: Neuchâtel. [Not seen]

Alberti, G. K. B., Haas, W., and Ormiston, A., 1971, Discovery of the trilobite *Warburgella rugulosa* (Alth, 1874) in Gedinnian strata of central Nevada: Neues Jahrbuch für Geologie und Paläontologie Monatshefte, Jahrgang 1971, no. 4, p. 193–194.

Alberti, G. K. B., Haas, W., and Ormiston, A. R., 1972, *Warburgella rugulosaeureka* n. ssp. (Trilobita, Proetidae) aus dem Unter-Gedinnium von Zentral-Nevada (U.S.A.): Senckenbergiana lethaea, v. 53, p. 123–133.

Amsden, T. W., 1949, Stratigraphy and paleontology of the Brownsport Formation (Silurian) of western Tennessee: Yale University, Peabody Museum of Natural History, Bulletin 5, viii + 138 p.

Audinet-Serville, J. G., 1831, Revue méthodique des insectes de l'ordre des Orthoptères: Annales des Sciences Naturelles, Paris, v. 22, p. 28–65, 134–168, 262–292. [Not seen]

Barrois, C., 1889, Faune du Calcaire d'Erbray (Loire Inférieure): Société Géologique du Nord Mémoires, v. 3, pt. 1, p. 1–348.

Berry, W. B. N., and Murphy, M. A., 1975, Silurian and Devonian graptolites of central Nevada: University of California Publications in Geological Sciences, v. 110, 109 p.

Besprozvannykh, N. I., 1968, Rugozy Tom'chumyshskikh sloev Salaira, *in* Ivanovskiy, A. B., ed., Korally pogranichnykh sloev silura i devona Altae-Sayanskoy gornoy oblasti i Urala: Moskva, Izdatel'stvo Nauka, p. 110–116, 160, 161.

Birenheide, R., 1978, Rugose Korallen des Devon, *in* Krömmelbein, K., ed., Leitfossilien begründet von Georg Gürich, No. 2: Berlin, Gebrüder Borntraeger, vi + 265 p.

Birenheide, R., 1990, Zur palökologischen Bedeutung des Stereoplasmas in cystimorphen Deckelkorallen: Natur und Museum, v. 120, no. 5, p. 129–138.

Bul'vanker, E. Z., 1958, Devonskie chetysrekhluchevye korally okrani Kuznetskogo basseyna: Leningrad, Vsesoyuznyy Nauchno-issledovatel'skiy Geologischeskiy Institut, text 212 p. and atlas.

Bul'vanker, E. Z., Goryanov, V. B., Ivanovskiy, A. B., Spasskiy, N. Ya., and Shchukina, V. Ya., 1968, Novye predstaviteli chetyrekhluchevykh korallovykh polipov SSSR, *in* Markovskiy, B. P., ed., Novye vidy drevnikh rasteniy i bespozvonochnykh SSSR: Moskva, Vsesoyuznyy Nauchno-Issledovatel'skiy Geologicheskiy Institut (VSEGEI), issue 2, part 2, p. 14–45, 304–343.

Chapman, F., 1925, New or little-known fossils in the National Museum. Part XXVIII. Some Silurian rugose corals: Royal Society of Victoria Proceedings, new series, v. 37, p. 104–118.

Cheetham, A. H., and Hazel, J. E., 1969, Binary (presence-absence) similarity coefficients: Journal of Paleontology, v. 43, p. 1130–1136.

Cherepnina, S. K., 1971, Rugozy zhedinskikh otlozheniy Gornogo Altaya (remnevskie sloi), *in* Ivanovskiy, A. B., ed., Rugozy i stromatoporoidei paleozoya SSSR, Trudy II Vsesoyuznogo sympoziuma po izucheniyu iskopaemykh korallov SSSR, issue 2: Moskva, Izdatel'stvo Nauka, p. 89–91, 143.

Duncan, H., 1956, Ordovician and Silurian coral faunas of western United States: U.S. Geological Survey Bulletin 1021-F, p. i–iii, 209–236.

Dybowski, W. N., 1873–1874, Monographie der Zoantharia sclerodermata rugosa aus der Silurformation Estlands, Nord-Livlands und der Insel Gotland, . . .: Archiv für die Naturkunde Liv-, Ehst- und Kurlands, Ser. 1, v. 5, p. 257–414 (1873), 415–532 (1874).

Etheridge, R., Jr., 1894, Description of a proposed new genus of rugose coral (*Mucophyllum*): New South Wales Geological Survey Records, v. 4, p. 11–18.

Etheridge, R., Jr., 1899, On the corals of the Tamworth district, chiefly from the Moore Creek and Woolomol limestones: Geological Survey of New South Wales Records, v. 6, p. 151–182.

Etheridge, R., Jr., 1907, A monograph of the Silurian and Devonian corals of New South Wales: With illustrations from other parts of Australia. Part II: The genus *Tryplasma*: Geological Survey of New South Wales Memoirs, Palaeontology, no. 13, p. i–ix, 41–102.

Etheridge, R., Jr., 1911, The Lower Palaeozoic corals of Chillagoe and Clermont. Part 1: Queensland Geological Survey Publication 231, 8 p.

Etheridge, R., Jr., 1918, Two remarkable corals from the Devonian of New South Wales: Australian Museum Records, v. 12, p. 49–51.

Fedorowski, J., 1991, Dividocorallia, a new subclass of Palaeozoic Anthozoa: L'Institut Royal des Sciences Naturelles de Belgique Bulletin, Sciences de la Terre, v. 61, p. 21–105.

Flood, P. G., and Aitchison, J. C., 1992, Late Devonian accretion of the Gamilaroi Terrane to eastern Gondwana: Provenance linkage suggested by the first appearance of Lachlan Fold Belt-derived quartzarenite: Australian Journal of Earth Sciences, v. 39, p. 539–544.

Foerste, A. F., 1888, Notes on Paleozoic fossils: Scientific Laboratories of Denison University Bulletin, v. 3, p. 117–137.

Galle, A., 1995, The Breviphrentis-dominated coral faunule from the Middle Devonian of Moravia, Czech Republic: Cesky Geologicky ústav Vestník, v. 70, no. 2, p. 59–70.

Glen, R. A., Scheibner, E., and VandenBerg, A. H. M., 1992, Paleozoic intraplate escape tectonics in Gondwanaland and major stike-slip duplication in the Lachlan orogen of southeastern Australia: Geology, v. 20, p. 795–798.

Golonka, J., Ross, M. I., and Scotese, C. R., 1994, Phanerozoic paleogeographic and paleoclimatic modeling maps, *in* Embry, A. F., Beauchamp, B., and Glass, D. J., eds., Pangea: Global environments and resources: Canadian Society of Petroleum Geologists Memoir 17, p. 1–47.

Gourvennec, R., 1989, Radial microornament in spiriferid brachiopods and paleogeographical implications: Lethaia, v. 22, p. 405–411.

Grabau, A. W., 1928, Palaeozoic corals of China. Part I. Tetraseptata. II. Second contribution to our knowledge of the streptelasmoid corals of China and adjacent territories: Palaeontologia Sinica, ser. B, v. 2, fasc. 2, 175 p.

Guo Shengzhe, 1976, Tetracoralla, *in* The Inner Mongolian Autonomous Region Geological Office and the Northeast China Institute of Geological Sciences, eds., Atlas of palaeontology of the North China region, Inner Mongolia volume, pt. 1, Palaeozoic Era: Beijing, Geological Publishing House, p. 63–101 [Chinese only].

Guo Shengzhe, 1978, Late Silurian tetracorals from northern Bailingmiao of the Autonomous Region of Inner Mongolia: Professional Papers of Stratigraphy and Palaeontology, no. 6, p. 50–68 [Chinese only].

He Yuanxiang, and Chen Jianqiang, 1989, Skeletal structures and classification of the Order Cystiphyllida of China, Chapter 3, *in* Wang Hongzhen et al., Classification, evolution and biogeography of the Palaeozoic corals of China: Beijing, Science Press, p. 61–71 [Chinese with English summary].

Hill, D., 1939, The Devonian rugose corals of Lilydale and Loyola, Victoria: Royal Society of Victoria Proceedings, new series, v. 51, p. 219–256.

Hill, D., 1940a, The Middle Devonian rugose corals of Queensland, II. The Silverwood–Lucky Valley area: Royal Society of Queeensland Proceedings, v. 51, p. 150–168.

Hill, D., 1940b, The Silurian Rugosa of the Yass-Bowning district, N.S.W.: Linnean Society of New South Wales Proceedings, v. 65, p. 388–420.

Hill, D., 1940c, The Lower Middle Devonian rugose corals of the Murrumbidgee and Goodradigbee rivers, N.S.W.: Royal Society of New South Wales Journal and Proceedings, v. 74, p. 247–276.

Hill, D., 1942, The Devonian rugose corals of the Tamworth district, N.S.W.: Royal Society of New South Wales Journal and Proceedings, v. 76, p. 142–164.

Hill, D., 1981, Coelenterata, Supplement 1, Rugosa and Tabulata, *in* Robison, R. A., ed., Treatise on invertebrate paleontology, Part F: Boulder, Colorado, Geological Society of America (and University of Kansas Press) xl + 762 p. (2 v.).

Hill, D., and Jell, J. S., 1969, On the rugose coral genera *Rhizophyllum* Lindström, *Platyphyllum* Lindström and *Calceola* Lamarck: Neues Jahrbuch für Geologie und Paläontologie Monatshefte, Jahrgang 1969, no. 9, p. 534–551.

Hou Hongfei, and Boucot, A. J., 1990, The Balkhash-Mongolia-Okhotsk region of the Old World Realm (Devonian), *in* McKerrow, W. S., and Scotese, C. R., eds., Palaeozoic palaeogeography and biogeography: Geological Society of London Memoir 12, p. 297–303.

Ivaniya, V. A., and Cherepnina, S. K., 1986, Devonskie korally-rugozy Zapadnoy Sibiri (atlas-opredelitel′): Tomsk, Izdatel′stvo Tomskogo Universiteta, 136 p.

Ivaniya, V. A., Kosareva, E. G., and Fedorovich, A. I., 1968, Novye materialy po faune devona Gornogo Altaya (rugozy), *in* Ivaniya, V. A., ed., Novye materialy po stratigrafii i paleontologii nizhenego i srednego paleozoya Zapadnoy Sibiri: Tomskiy Gosudarstvennyy Universitet imeni V. V. Kuybysheva Trudy, ser. geol., v. 202, p. 83–100.

Ivanovskiy, A. B., 1969, Korally semeystv Tryplasmatidae i Cyathophylloididae (rugozy): Moskva, Izdatel′stvo Nauka, 112 p.

Ivanovskiy, A. B., 1975, Rugozy: Akademiya Nauk SSSR, Sibirskoe Otdelenie, Institut Geologii i Geofiziki Trudy, issue 242, 124 p.

Ivanovskiy, A. B., 1976, Ukazatel′ rodov rugoz: Moskva, Izdatel′stvo Nauka, 256 p.

Ivanovskiy, A. B., and Shurygina, M. V., 1975, Reviziya rugoz Urala: Akademiya Nauk SSSR, Sibirskoe Otdelenie, Institut Geologii i Geofiziki Trudy, issue 218, 66 p.

Jackson, D. E., Lenz, A. C., and Pedder, A. E. H., 1978, Late Silurian and Early Devonian graptolite, brachiopod and coral faunas from northwestern and arctic Canada: Geological Association of Canada Special Paper 17, 159 p.

Jaeger, H., 1989, Devonian Graptoloidea, *in* McMillan, N. J., Embry, A. F., and Glass, D. J., eds., Devonian of the world: Calgary, Canadian Society of Petroleum Geologists Memoir 14, v. 3, p. 431–438. [imprint 1988]

Jell, J. S., and Hill, D., 1970, Revision of the coral fauna from the Devonian Douglas Creek Limestone, Clermont, central Queensland: Royal Society of Queensland Proceedings, v. 81, p. 93–119.

Jell, P. A., and Jell, J. S., 1976, Early Middle Cambrian corals from western New South Wales: Alcheringa, v. 1, p. 181–195.

Johannessen, W. H., 1993, Species of the Silurian operculate coral genus *Goniophyllum*: Geologiska Föreningens i Stockholm Förhandlingar, v. 115, p. 119–143.

Johannessen, W. H., 1995, Species of the Silurian operculate rugose coral genera *Araeopoma* and *Rhytidophyllum*: Geologiska Föreningens i Stockholm Fördhandlingar, v. 117, p. 31–41.

Johnson, J. G., 1965, Lower Devonian stratigraphy and correlation, northern Simpson Park Range, Nevada: Bulletin of Canadian Petroleum Geology, v. 13, p. 365–381.

Johnson, J. G., 1970, Great Basin Lower Devonian Brachiopoda: Geological Society of America Memoir 121, xi + 421 p.

Johnson, J. G., 1977, Lower and Middle Devonian faunal intervals in central Nevada based on brachiopods, *in* Murphy, M. A., Berry, W. B. N., and Sandberg, C. A., eds., Western North America: Devonian: University of California, Riverside Campus Museum Contribution 4, p. 16–32.

Johnson, J. G., 1990, Lower and Middle Devonian brachiopod-dominated communities of Nevada, and their position in a biofacies-province-realm model: Journal of Paleontology, v. 64, p. 902–941.

Johnson, J. G., and Murphy, M. A., 1969, Age and position of Lower Devonian graptolite zones relative to the Appalachian standard succession: Geological Society of America Bulletin, v. 80, p. 1275–1282.

Johnson, J. G., and Murphy, M. A., 1984, Time-rock model for Siluro-Devonian continental shelf, western United States: Geological Society of America Bulletin, v. 95, p. 1349–1359.

Johnson, J. G., and Oliver, W. A., Jr., 1977, Silurian and Devonian coral zones in the Great Basin, Nevada and California: Geological Society of America Bulletin, v. 88, p. 1462–1468.

Johnson, J. G., Sandberg, C. A., and Poole, F. G., 1991, Devonian lithofacies of western United States, *in* Cooper, J. D., and Stevens, C. H., eds., Paleozoic paleogeography of the western United States—II: Society of Economic Paleontologists and Mineralogists, Pacific Section, v. 67, p. 83–105.

Jokiel, P. L., 1990, Transport of reef corals into the Great Barrier Reef: Nature, v. 347, p. 665–667.

Jones, O. A., 1932, A revision of the Australian species of the coral genera *Spongophyllum* E. & H. and *Endophyllum* E. & H. with a note on *Aphrophyllum* Smith: Royal Society of Queensland Proceedings, v. 44, p. 50–63.

Karsch, F. A. F., 1890, Ueber die von Herrn Dr. R. Büttner in Westafrica gesammelten Tettigiden-Arten: Entomologische Nachrichten, Jahrgang 16, p. 17–27.

Kay, M., and Crawford, J. P., 1964, Paleozoic facies from the miogeosynclinal to the eugeosynclinal belt in thrust slices, central Nevada: Geological Society of America Bulletin, v. 75, p. 425–454.

Khoa, N. D., and Thom, N., 1980, Phu lop San ho bon tia (Rugosa), *in* Dánh,

T., and 33 others, Hoa thach dac trung o mien bac Viêt Nam (Characteristic fossils in the north of Vietnam): Hanoi, Science and Technics Publishing House, p. 86–94, 327–330, 499–507.

King, W., 1846, Remarks on certain genera belonging to the Class Palliobranchiata: Annals and Magazine of Natural History, including Zoology, Botany and Geology, ser. 1, v. 18, p. 26–42, 83–94.

Klapper, G., 1977, Lower and Middle Devonian conodont sequence in central Nevada, with contributions by D. B. Johnson, *in* Murphy, M. A., Berry, W. B. N., and Sandberg, C. A., eds., Western North America: Devonian: University of California, Riverside Campus Museum Contribution 4, p. 33–54.

Klapper, G., and Murphy, M. A., 1975, Silurian–Lower Devonian conodont sequence in the Roberts Mountains Formation of central Nevada: University of California Publications in Geological Sciences, v. 111, vi + 62 p. [imprint 1974]

Kong Lei, and Huang Yunming, 1978, Tetracoralla, *in* Stratigraphy and Palaeontology Work Team, eds., Fossil atlas of southwestern China, Guizhou section, Volume 1: Beijing, Geological Publishing House, p. 35–161, 739–759 [Chinese only].

Koninck, L. G. de, 1872, Nouvelles recherches sur les animaux fossiles du terrain carbonifère de la Belgique, comprenant la classification méthodique et la synonymie de toutes les espèces connues,. . ., pt. 1: Bruxelles, F. Hayez, imprimeur, iv + 178 p. Also: L'Académie Royale des Sciences, des Lettres et des Beaux-Arts de Belgique Mémoires, v. 39, p. 1–178. [not seen]

Kraevskaya, L. N., 1955, Tetrakorally nizhnego i srednego devona, *in* Khalfin, L. L., ed., Atlas rukovodyashchikh form iskopaemykh fauny i flory Zapadnoy Sibiri, Volume 1: Moskva, Gosudarstvennoe Nauchnotekhnicheskoe Izdatel'stvo, p. 206–218, 396–413.

Kullmann, J., 1965, Rugose Korallen der Cephalopodenfazies und ihre Verbreitung im Devon des südöstlichen Kantabrischen Gebirges (Nordspanien): Akademie der Wissenschaften und der Literatur in Mainz, Abhandlungen der Mathematisch-Naturwissenschaftlichen Klasse, Jahrgang 1965, no. 2, p. 33–168 [also paginated 1–136].

Lang, W. D., Smith, S., and Thomas, H. D., 1940, Index of Palaeozoic coral genera: London, British Museum (Natural History), vii + 231 p.

Lavrusevich, A. I., 1971, Nekotorye rugozy iz pozdnesiluriyskikh i rannedevonskikh otlozheniy Tsentralnogo Tadzhikistana, *in* Lavrusevich, A. I., ed., Upravlenie Geologiya Soveta Ministrov Tadzhikskoy SSR, issue 4, Paleontologiya i stratigraphiya: Moskva, Izdatel'stvo Nedra, p. 33–52, 123–135.

Lavrusevich, A. I., 1993?, Devonskaya sistema. Otryad Rugosa (Rugozy), *in* Dzhalilov, M. R., ed., Atlas iskopaemoy fauny i flory Tadzhikistana. Ordovik. Silur. Devon.: Dushanbe, Akademiya Nauk Tadzhikskoy SSR, Institut Geologii, p. 171–177, 388–389, 402–411, 418–423, 426–429. [imprint 1991]

Lenz, A. C., 1961, Devonian rugose corals of the Lower Mackenzie Valley, Northwest Territories, *in* Raasch, G. O., ed., Geology of the Arctic, Proceedings of the First International Symposium on Arctic Geology, Calgary, Alberta, 1960, Volume 1: Toronto, University of Toronto Press, p. 500–514.

Li Vu Din, 1980, Novye vidy rugoz iz verkhnesiluriyskikh i devonskikh otlozheniy severo-vostochnoy Fergany: Paleontologicheskiy Zhurnal, 1980, no. 3, p. 28–33.

Li Wenguo, Rong Jiayu, Dong Deyuan, Yang Daorong, Su Yangzheng, and Wang Yifu, 1983, Silurian and Devonian rocks of the Bateaobao area in Darhan Mumingan Joint Banner, Inner Mongolia, *in* Li Wenguo, Rong Jiayu, and Dong Deyuan, eds., Silurian and Devonian rocks and faunas of the Bateaobao area in Darhan Mumingan Joint Banner Inner Mongolia: People's Publishing House of Inner Mongolia, p. 1–25 [Chinese with English abstract].

Li Yun, 1988, No Upper Silurian strata occurring in the Alengchu section in Mingyin of Lijiang: Dicengxue Zazhi (Journal of Stratigraphy), v. 12, p. 239–241 [Chinese only].

Liao Weihua, and Zheng Chunzi, 1986, Early Devonian *Carlinastraea* fauna from Erhtaokou Formation of Jilin: Acta Palaeontologica Sinica, v. 25, p. 622–635 [Chinese with English summary].

Lindström, G., 1866a, Några iakttagelser öfver Zoantharia rugosa: Öfversigt af Kongliga Vetenskaps-Akademiens Förhandlingar, v. 22 (1865), p. 271–294.

Lindström, G., 1866b, Some observations on the Zoantharia Rugosa: Geological Magazine, decade 1, v. 3, p. 356–362, 406–414.

Lindström G., 1871a, Om operkularbildningen hos några nutida och siluriska koraller: Öfversigt af Kongliga Vetenskaps-Akademiens Förhandlingar, Årgång 27 (1870), p. 921–926.

Lindström G., 1871b, On some operculated corals, Silurian and recent: Geological Magazine, decade 1, v. 8, p. 122–126.

Lindström, G., 1883a, Obersilurische Korallen von Tshau-tiën im nordöstlichen Theil der Provinz Sz'-tshwan, *in* Richthofen, F. von, China. Ergebnisse eigener Reisen und darauf gegründeter Studien, Volume 4: Berlin, Verlag von Dietrich Reimer, p. 50–74.

Lindström, G., 1883b, Om de Palaeozoiska formationernas operkelbärande koraller: Bihang till Kongliga Svenska Vetenskaps-Akademiens Handlingar, v. 7, no. 4, p. 1–112. [imprint 1882]

Lonsdale, W., 1845, Description of some characteristic Palaeozoic Corals of Russia, Appendix A, *in* Murchison, R .I., Verneuil, E. de, and Keyserling, A. von, The geology of Russia in Europe and the Ural Mountains Volume 1: London, John Murray, p. 591–634.

McKee, E. H., Merriam, C. W., and Berry, W. B. N., 1972, Biostratigraphy and correlation of McMonnigal and Tor Limestones, Toquima Range, Nevada: American Association of Petroleum Geologists Bulletin, v. 56, p. 1563–1570.

McLean, R. A., 1974, Cystiphyllidae and Goniophyllidae (Rugosa) from the Lower Silurian of New South Wales: Palaeontographica, Abteilung A, v. 147, p. 1–38.

Meek, F. B., 1873, Preliminary paleontological report, consisting of lists and descriptions of fossils, with remarks on the ages of the rocks in which they were found, etc., etc.: U.S. Geological Survey of the Territories Annual Report 6 (1872), p. 429–518.

Merriam, C. W., 1963, Paleozoic rocks of Antelope Valley, Eureka and Nye counties, Nevada: U.S. Geological Survey Professional Paper 423, iii + 67 p.

Merriam, C. W., 1972, Silurian rugose corals of the Klamath Mountains region, California: U.S. Geological Survey Professional Paper 738, iv + 50 p.

Merriam, C. W., 1973, Paleontology and stratigraphy of the Rabbit Hill Limestone and Lone Mountain Dolomite of central Nevada: U.S. Geological Survey Professional Paper 808, v + 50 p.

Merriam, C. W., 1974a, Silurian rugose corals of the central and southwest Great Basin: U.S. Geological Survey Professional Paper 777, iv + 66 p. [imprint 1973].

Merriam, C. W., 1974b, Lower and Lower Middle Devonian rugose corals of the central Great Basin: U.S. Geological Survey Professional Paper 805, v + 83 p.

Merriam, C. W., and McKee, E. H., with a section on conodonts by J. W. Huddle, 1976, The Roberts Mountains Formation, a regional stratigraphic study with emphasis on rugose coral distribution: U.S. Geological Survey Professional Paper 973, iv + 51 p.

Milne Edwards, H., and Haime, J., 1850, A monograph of the British fossil corals, Introduction and chapters 1–7: London, Palaeontographical Society, lxxxv + 71 p.

Morzadec, P., Paris, F., Plusquellec, Y., Racheboeuf, P., and Weyant, M., 1989, Devonian stratigraphy and paleogeography of the Armorican Massif (western France), *in* McMillan, N. J., Embry, A. F., and Glass, D. J., eds., Devonian of the world: Calgary, Canadian Society of Petroleum Geologists Memoir 14, v. 1, p. 401–420. [imprint 1988]

Müller, J. P., 1858, Einige neue bei St. Tropez am Mittelmeer beobachtete Polycystinen und Acanthometren aus den Abbildungen: Königliche Preussische Akademie der Wissenschaften zu Berlin Monatsberichte, Gesammtsitzung vom 11. Februar 1858, p. 154–155.

Müller, J. P., 1859, Über die Thalassicollen, Polycystinen und Acanthometren des Mittelmeeres: Königliche Preussische Akademie der Wissenschaften zu Berlin Abhandlungen, Physikalisch-Mathematische Klasse, 1858, p. 1–62.

Münster, G., 1839, Beiträge zur Petrefacten-kunde . . ., no. 1: Bayreuth, Buchner, vii + 124 p. [Not seen]

Murchison, R. I., Verneuil, E. de, and Keyserling, A. von, 1845, The geology of Russia in Europe and the Ural Mountains, Volume 1: London, John Murray, xxiv + 700 p.

Murphy, M. A., 1989a, Central Nevada, *in* Holland, C. H., and Bassett, M. G., eds., A global standard for the Silurian System: National Museum of Wales, Geological Series, no. 9, p. 171–177.

Murphy, M. A., 1989b, Lower Pragian boundary (Lower Devonian) and its application in Nevada, *in* Ziegler, W., ed., 1st International Senckenberg Conference and 5th European Conodont Symposium (ECOS V) Contributions III, Frankfurt/Main, 1988: Courier Forschungsinstitut Senckenberg, v. 117, p. 61–70.

Murphy, M. A., and Anderson, M. E., 1991, Devonian formations in the Toquima Range, central Nevada, *in* Cooper, J. D., and Stevens, C. H., eds., Paleozoic paleogeography of the western United States—II: Society of Economic Paleontologists and Mineralogists, Pacific Section, v. 67, p. 299–310.

Murphy, M. A., and Berry, W. B. N., 1983, Early Devonian conodont-graptolite collation and correlations with brachiopod and coral zones, central Nevada: American Association of Petroleum Geologists Bulletin, v. 67, p. 371–379.

Nicholson, H. A., and Lydekker, R., 1889, A manual of palaeontology for the use of students with a general introduction on the principles of palaeontology, Volume 1 (third edition): Edinburgh, William Blackwood and Sons, xviii + 885 p.

Oliver, W. A., Jr., 1964, New occurrences of the rugose coral *Rhizophyllum* in North America, *in* Geological Survey Research 1963: U.S. Geological Survey Professional Paper 475-D, p. 149–158.

Oliver, W. A., Jr., 1992, Corals from the Turkey Creek Limestone (Lower Devonian), southern Oklahoma, *in* Chaplin, J. R., and Barrick, J. E., eds., Special papers in paleontology and stratigraphy: A tribute to Thomas W. Amsden: Oklahoma Geological Survey Bulletin 145, p. 137–159.

Oliver, W. A., Jr., 1993, The Siphonophrentidae (rugose corals, Devonian) of eastern North America, *in* Sando, W. J., ed., Shorter contributions to paleontology and stratigraphy, 1992: U.S. Geological Survey Bulletin 2024, p. i–iv, B1–B32.

Oliver, W. A., Jr., and Coates, A. G., 1987, Phylum Cnidaria, *in* Boardman, R. S., Cheetham, A. H., and Rowell, A. J., eds., Fossil invertebrates: Palo Alto, Blackwell Scientific Publications, p. 140–193.

Oliver, W. A., Jr., and Galle, A., 1971, Rugose corals from the Upper Kon prusy Limestone (Lower Devonian) in Bohemia: Sborník Geologich ch V d. Paleontologie, v. P, no. 14, p. 35–106.

Oliver, W. A., Jr., and Johnson, J. G., 1977, Notes on Upper Silurian and Devonian coral zones in the Great Basin, *in* Murphy, M. A., Berry, W. B. N., and Sandberg, C. A., eds., Western North America: Devonian: University of California, Riverside Campus Museum Contribution 4, p. 107–111.

Oliver, W. A., Jr., and Pedder, A. E. H., 1979, Rugose corals in Devonian stratigraphical correlation, *in* House, M. R., Scrutton, C. T., and Bassett, M. G., eds., The Devonian System: Special Papers in Palaeontology no. 23, p. 233–248.

Oliver, W. A., Jr., and Pedder, A. E. H., 1994, Crises in the Devonian history of the rugose corals: Paleobiology, v. 20, p. 178–190.

Pedder, A. E. H., 1967, *Lyrielasma* and a new related genus of Devonian tetracorals: Royal Society of Victoria Proceedings, new ser., v. 80, p. 1–30.

Pedder, A. E. H., 1971, Two new aphroid corals from the Middle Devonian Hume Formation of western Canada, *in* Norford, B. S., and Irish, E. J. W., eds., Contributions to Canadian paleontology: Geological Survey of Canada Bulletin 192, p. 45–61.

Pedder, A. E. H., 1975, Sequence and relationships of three Lower Devonian coral faunas from Yukon Territory, *in* Report of Activities, Part B: Geological Survey of Canada Paper 75-1B, p. 285–295.

Pedder, A. E. H., 1980, Devonian corals of late Eifelian age from the Ogilvie Formation of Yukon Territory: Canadian Journal of Earth Sciences, v. 17, p. 594–616.

Pedder, A. E. H., 1985a, Lower Devonian rugose corals of Lochkovian age from Yukon Territory, *in* Current Research, Part A: Geological Survey of Canada Paper 85–1A, p. 587–602.

Pedder, A. E. H., 1985b, Lochkovian (Early Devonian) rugose corals from Prince of Wales and Baillie Hamilton islands, Canadian Arctic Archipelago, *in* Current Research, Part B: Geological Survey of Canada Paper 85-1B, p. 285–301.

Pedder, A. E. H., and Oliver, W. A., Jr., 1982, *Stauromatidium* and Stauromatidiidae, new genus and family of Upper Silurian and Lower Devonian rugose corals: Geological Survey of Canada Bulletin 352, v + 43 p.

Pedder, A. E. H., and Oliver, W. A., Jr., 1990, Rugose coral distribution as a test of Devonian palaeogeographic models, *in* McKerrow, W. S., and Scotese, C. R., eds., Palaeozoic palaeogeography and biogeography: Geological Society of London Memoir 12, p. 267–275.

Pedder, A. E. H., Jackson, J. H., and Ellenor, D. W., 1970a, An interim account of the Middle Devonian Timor Limestone of north-eastern New South Wales: Linnean Society of New South Wales Proceedings, v. 94, p. 242–272.

Pedder, A. E. H., Jackson, J. H., and Philip, G. M., 1970b, Lower Devonian biostratigraphy in the Wee Jasper region of New South Wales: Journal of Paleontology, v. 44, p. 206–251.

Penecke, K. A., 1894, Das Grazer Devon: Kaiserlich-Königlich Geologische Reichsanstalt (Bundesanstalt) Wien Jahrbuch, v. 43 (1893), Heft 4, p. 567–616.

Philip, G. M., 1962, The palaeontology and stratigraphy of the Siluro-Devonian sediments of the Tyers area, Gippsland, Victoria: Royal Society of Victoria Proceedings, new series, v. 75, p. 123–246.

Potter, A. W., and 10 others, 1990, Early Paleozoic stratigraphic, paleogeographic, and biogeographic relations of the eastern Klamath belt, northern California, *in* Harwood, D. S., and Miller, M. M., eds., Paleozoic and early Mesozoic paleogeographic relations; Sierra Nevada, Klamath Mountains, and related terranes: Geological Society of America Special Paper 255, p. 57–74.

Roemer, F. A., 1856, Bericht von einer geologisch-paläontologischen Reise nach Schweden: Leonhard und Bronn's Neues Jahrbuch für Mineralogie, Geognosie und Geologie, Jahrgang 1856, p. 794–815. [Not seen]

Rosen, B. R., and Wise, R. F., 1980, Revision of the rugose coral *Diphyphyllum concinnum* Lonsdale, 1845 and historical remarks on Murchison's Russian coral collection: British Museum (Natural History) Bulletin, Geology Series, v. 33, no. 2, p. 147–155.

Ró kowska, M., 1969, Famennian tetracoralloid and heterocoralloid fauna from the Holy Cross Mountains (Poland): Acta Palaeontologica Polonica, v. 14, p. 5–187.

Rukhin, L. B., 1938, Nizhnepaleozoiskie korally i stromatoporoidei verkhney chasti basseyna r. Kolymy: Materialy po izucheniyu Kolymsko-Indigirskogo kraya, ser. 2, Geologiya i geomorfologiya, issue 10: Leningrad, Gostrest Dal'stroya, 119 p.

Sando, W. J., 1965, Revision of some Paleozoic coral species from the western United States, *in* Contributions to paleontology: U.S. Geological Survey Professional Paper 503-E, p. i-iii, 1–38.

Scrutton, C. T., 1985, Subclass Rugosa, *in* Murray, J. W., ed., Atlas of invertebrate macrofossils: Harlow, England, Longman in collaboration with the Palaeontological Association, p. 13–30.

Scrutton, C. T., 1988, Patterns of extinction and survival in Palaeozoic corals, *in* Larwood, G. P., ed., Extinction and survival in the fossil record: Oxford, Clarendon Press, Systematics Association Special Volume 34, p. 65–88.

Shurygina, M. V., 1968, Pozdnesiluriyskie i rannedevonskie rugozy vostochnogo sklona Severnogo i Srednego Urala, *in* Ivanovskiy, A. B., ed., Korally pogranichnykh sloev silura i devona Altae-Sayanskoy gornoy oblasti i Urala: Moskva, Izdatel'stvo Nauka, p. 117–150, 161–164.

Shurygina, M. V., 1977, Rugozy, *in* Sycheva, E. S., and Shimanova, L. N., eds., Biostratigrafiya i fauna rannego devona vostochnogo sklona Urala: Moskva, Izdatel'stvo Nedra, p. 43–51, 161–163.

Sorentino, L., 1989, Conodont assemblages spanning the Lochkovian-Pragian (Early Devonian) boundary at Eurimbla, central New South Wales, *in* Ziegler, W., ed., 1st International Senckenberg Conference and 5th European Conodont Symposium (ECOS V) Contributions III, Frankfurt/Main, 1988: Courier Forschungsinstitut Senckenberg, v. 117, p. 81–115.

Soshkina, E. D., 1937, Korally verkhnego silura i nizhnego devona vostochnogo i zapadnogo sklonov Urala: Akademiya Nauk SSSR, Paleozoologocheskiy Institut Trudy, v. 6, issue 4, 155 p.

Soshkina, E. D., 1949, Devonskie korally Rugosa Urala: Akademiya Nauk SSSR, Paleontologicheskiy Institut Trudy, v. 15, issue 4, 160 p.

Soshkina, E. D., 1952, Opredelitel′ devonskikh chetyrekhluchevykh korallov: Akademiya Nauk SSSR, Paleontologicheskiy Institut Trudy, v. 39, 127p.

Soshkina, E. D., and Dobrolyubova, T.A., 1962, Otryad Evenkiellida, *in* Sokolov, B. S., ed., Osnovy paleontologii, Volume 2, Gubki, arkheotsiaty, kishechnopolostnye, chervi: Moskva, Izdatel'stvo Akadmii Nauk SSSR, p. 333–339.

Spasskiy, N. Ya., 1965, Osnovy sistematiki devonskikh chetyrekhluchevykh korallov, *in* Sokolov, B. S., and Ivanovskiy, A. B., eds., Rugozy paleozoya SSSR, Trudy I Vsesoyuznogo simpoziuma po izucheniyu iskopaemykh korallov SSSR, issue 3: Moskva, Izdatel'stvo Nauka, p. 80–90.

Spasskiy, N. Ya., 1977, Devonskie rugozy SSSR (sistematika, stratigraficheskoe i geograficheskoe znachenie): Leningrad, Izdatel'stvo Leningradskogo Universiteta, 344 p.

Springer, K. B., and Murphy, M. A., 1994, Punctuated stasis and collateral evolution in the Devonian lineage of *Monograptus hercynicus*: Lethaia, v. 27, p. 119–128.

Stolarski, J., 1993, Ontogenetic development and functional morphology in the early growth-stages of *Calceola sandalina* (Linnaeus, 1771), *in* Oekentorp-Küster, P., ed., Proceedings, International Symposium on Fossil Cnidaria and Porifera, 6th, Münster, Germany, 9–14. September 1991: Courier Forschungsinstitut Senckenberg 164, p. 169–177.

Strusz, D. L., 1966, Spongophyllidae from the Devonian Garra formation, New South Wales: Palaeontology, v. 9, p. 544–598.

Strusz, D. L., 1968, On *Cyathophyllum mansfieldense* Dun 1898: Lower Devonian, Loyola, Victoria: Royal Society of Victoria Proceedings, new series, v. 81, p. 11–18.

Strusz, D. L., 1970, *Rhizophyllum* and *Calceola* from the Devonian of New South Wales, *in* Palaeontological Papers, 1967: Commonwealth of Australia, Bureau of Mineral Resources, Geology and Geophysics Bulletin 108, p. 281–303.

Stumm, E. C., 1937, The Lower Middle Devonian tetracorals of the Nevada Limestone: Journal of Paleontology, v. 11, p. 423–443.

Stumm, E. D., 1949, Revision of the families and genera of the Devonian tetracorals: Geological Society of America Memoir 40, viii + 92 p.

Sutherland, P. K., 1965, Henryhouse rugose corals: Oklahoma Geological Survey Bulletin 109, 92 p.

Swartz, C. K., 1913, Coelenterata, *in* Maryland Geological Survey, Lower Devonian (text) and Devonian Plates (atlas): Baltimore, The Johns Hopkins Press, p. 195–227.

Sytova, V. A., 1968, Tetrakorally skal′skogo i borshchovskogo gorizontov Podolii, *in* Balashov, Z. G., ed., Siluriysko-devonskaya fauna Podolii: Leningrad, Izdatel′stvo Leningradskogo Universiteta, p. 51–71.

Sytova, V. A., and Kaplan, A. A., 1975, Tetrakorally, *in* Menner, V. V., ed., Kharakteristika fauny pogranichnykh sloev silura i devona Tsentral′nogo Kazakhstana: Moskva, Nedra, Materialy po geologii Tsentral′nogo Kazakhstana, v. 12, p. 61–76, 194–195, 220–226.

Talent, J. A., Gratsianova, R. T., Shishkina, G. R., and Yolkin, E. A., 1987, Devonian faunas in relation to crustal blocks: Kazakhstan, Mongolia, Northern China, *in* Werner, R., and Ziegler, W., eds., Neue Daten zur Biostratigraphie, Paläogeographie und Paläontologie des Devons in Eurasien und Australien: Courier Forschungsinstitut Senckenberg 92, p. 225–233.

Tsyganko, V. S., 1981, Devonskie rugozy severa Urala: Leningrad, Lenigradskoe Otdelenie Izdatel′stva Nauka, 220 p.

Tyazheva, A. P., and Zhavoronkova, R. A., 1972, Korally i brakhiopody pogranichnykh otlozheniy silura i nizhnego devona zapadnogo sklona Yuzhnogo Urala: Moskva, Izdatel′stvo Nauka, 183 p.

Ulitina, L.M., 1968, Devonskie korally tsistifilliny Zakavkaz′ya: Akademiya Nauk SSSR, Paleontologicheskiy Institut Trudy, v. 113, 100 p.

Ulitina, L. M., 1992, Novye devonskie stauromatidiidy (rugozy) iz Mongolii, *in* Grunt, T. A., ed., Novye taksony iskopaemykh bespozvonochnykh Mongolii: Sovmestnaya Rossiysko-Mongol'skaya Paleontologicheskaya Ekspeditsiya Trudy, issue 4, p. 12–14, 131.

Verrill, A. E., 1865, Classification of polyps: (extract condensed from a synopsis of the polypi of the North Pacific Exploring Expedition, under captains Ringgold and Rodgers, U.S.N.): Essex Institute Proceedings, v. 4, p. 145–152.

Wang Chengyuan, and Li Dongjin, 1986, Conodonts from the Ertaogou Formation in central Jilin Province: Acta Micropalaeontologica Sinica, v. 3, p. 421–428 [Chinese with English abstract].

Wang, H. C. (Hongzhen), 1950, A revision of the Zoantharia Rugosa in the light of their minute skeletal structures: Royal Society of London Philosophical Transactions, ser. B, v. 234, p. 175–246.

Wang Hongzhen, He Yuanxiang, and Chen Jianqiang, 1987, Skeletal structures and classification of the Order Cystiphyllida (Rugosa): Geoscience, v. 1, p. 3–14 [Chinese with English abstract and captions].

Wedekind, R., 1923, Die Gliederung des Mitteldevons auf Grund von Korallen: Gesellschaft zur Beförderung der gesamten Naturwissenschaften zu Marburg Sitzungsberichte, Jahrgang 1922, p. 24–35.

Wedekind, R., 1925, Das Mitteldevon der Eifel, Eine biostratigraphische Studie, Teil 2, Materialien zur Kenntnis des mittleren Mitteldevon: Gesellschaft zur Beförderung der gesamten Naturwissenschaften zu Marburg Schriften, v. 14, no. 4, vii + 85 p.

Wedekind, R., 1927, Die Zoantharia Rugosa von Gotland (Bes. Nordgotland). Nebst Bemerkungen zur Biostratigraphie des Gotlandium: Sveriges Geologiska Undersökning, Ser. Ca, no. 19, 95 p.

Wedekind, R., and Vollbrecht, E., 1931, Die Lytophyllidae des mittleren Mitteldevon der Eifel. Part 1: Das Tatsachenmaterial: Palaeontographica, v. 75, p. 84–110.

Wedekind, R., and Vollbrecht, E., 1932, Die Lytophyllidae des mittleren Mitteldevon der Eifel. Part 2: Die systematische Erfassung des Tatsachenmateriales: Palaeontographica, v. 76, p. 95–115.

Weyer, D., 1984, *Neaxon cheilos* n. sp. aus dem Unterfamenne von Schleiz im Thüringer Schiefergebirge (Anthozoa, Rugosa; Oberdevon): Magdeburg Kulturhistorisches Museum, Abhandlungen und Berichte für Naturkunde und Vorgeschichte, v. 12, no. 5, p. 3–16. [imprint 1982]

Weyer, D., 1989, *Neaxon muensteri*, eine neue Koralle aus dem europäischen Oberfamenne: Magdeburg Kulturhistorisches Museum, Abhandlungen und Berichte für Naturkunde und Vorgeschichte, v. 14, p. 3–16.

Weyer, D., 1996, Calceolidae versus Goniophyllidae (Anthozoa, Rugosa; Silur-Devon): Museum für Naturkunde Magdeburg, Abhandlungen und Berichte für Naturkunde, v. 19, p. 69–71.

Wrzolek, T., 1993, Affinities of the Heterocorallia: Acta Palaeontologica Polonica, v. 38, p. 119–120.

Yu Changmin, and Jell, J. S., 1990, Early Devonian rugose coral fauna from the Shield Creek Formation, Broken River Embayment, north Queensland, *in* Jell, P. A., ed., Devonian and Carboniferous coral studies: Association of Australasian Palaeontologists Memoir 10, p. 169–209.

Yu Changming, and Liao Weihua, 1978, Lower Devonian rugose corals from Alengchu of Lijiang, northwestern Yunnan: Acta Palaeontologica Sinica, v. 17, p. 245–266 [Chinese with English abstract].

Zheltonogova, V. A., 1961, Siluriyskaya sistema. Podklass Tetracoralla (Rugosa). Tetrakorally, *in* Khalfin, L. L., ed., Biostratigrafiya paleozoya Sayano-Altayskoy gornoy oblasti, Volume 2, Sredniy paleozoy: Sibirskiy Nauchno-Issledovatel′skiy Institut Geologii, Geofiziki i Mineral′nogo Trudy, issue 20, p. 33–36, 74–88, 152–169 [imprint 1960].

Published with Permission of the Director General, Geological Survey of Canada.

Manuscript Accepted by the Society February 27, 1997

Printed in U.S.A.

Index

[Italic page numbers indicate major references]

C

Q

R

S

T

Y

Z

Printed in U.S.A.